Developments in Environmental Modelling, 7

Mathematical Models in Biological Waste Water Treatment

Developments in Environmental Modelling

Series Editor: S.E. Jørgensen
Langkaer Vaenge 9,
3500 Vaerløse,
Copenhagen,
Denmark

1.   ENERGY AND ECOLOGICAL MODELLING
edited by W.J. Mitsch, R.W. Bosserman and J.M. Klopatek
1981 839 pp.

2.   WATER MANAGEMENT MODELS IN PRACTICE:
A CASE STUDY OF THE ASWAN HIGH DAM
by D. Whittington and G. Guariso
1983 xxii + 246 pp.

3.   NUMERICAL ECOLOGY
by L. Legendre and P. Legendre
1983 xvi + 419 pp.

4A.   APPLICATION OF ECOLOGICAL MODELLING IN
ENVIRONMENTAL MANAGEMENT, PART A
edited by S.E. Jørgensen
1983 viii + 735 pp.

4B.   APPLICATION OF ECOLOGICAL MODELLING IN
ENVIRONMENTAL MANAGEMENT, PART B
edited by S.E. Jørgensen and W.J. Mitsch
1983 viii + 438 pp.

5.   ANALYSIS OF ECOLOGICAL SYSTEMS: STATE-OF-THE-ART
IN ECOLOGICAL MODELLING
edited by W.K. Lauenroth, G.V. Skogerboe and M. Flug
1983 922 pp.

6.   MODELLING THE FATE AND EFFECTS OF
TOXIC SUBSTANCES IN THE ENVIRONMENT
edited by S.E. Jørgensen
1984 vii + 342 pp.

Developments in Environmental Modelling, 7

Mathematical Models in Biological Waste Water Treatment

Edited by

Professor S.E. JØRGENSEN
Langkaer Vaenge 9, 3500 Vaerløse, Copenhagen, Denmark

Dr. M.J. GROMIEC
Instytut Meteorologii, i Gospodarki Wodnej, 01-673 Warszawa u. Podlesna 61, Poland

ELSEVIER
Amsterdam — Oxford — New York — Tokyo   1985

ELSEVIER SCIENCE PUBLISHERS B.V.
Sara Burgerhartstraat 25
P.O. Box 211, 1000 AE Amsterdam, The Netherlands

ELSEVIER SCIENCE PUBLISHING COMPANY INC.
52 Vanderbilt Avenue
New York, NY 10017, U.S.A.

ISBN 0-444-42535-7 (Vol. 7)
ISBN 0-444-41948-9 (Series)

CONTENTS

INTRODUCTION

S.E. JØRGENSEN AND M.J. GROMIEC

The pollutants present in waste water can be removed either physically or converted by means of chemical or biological processes. Unit processes applied for waste water treatment can be classified as primary, secondary and tertiary. The first term refers mainly to physical unit operations, the second one includes chemical and biological unit processes, and the last one might apply to combinations of all previously mentioned processes.

This book deals with mathematical modelling of biological waste treatment processes. In general, biological processes have proved to be very effective for the removal of dissolved biodegradable organics from waste water. The organics are converted into gases and biological solids to be subsequently removed during a clarification process.

Waste water treatment processes, which occur in reactors, can operate on batch or continuous bases. Almost all biological waste water reactions operate as continuous-flow units, which can be divided into complete-mixed reactors, plug-flow reactors, and back-mixed reactors.

The biological processes for waste water treatment can be classified as aerobic, or anaerobic. The unit biological processes are further divided into suspended-growth, attached-growth, or a combination of the two. In suspended-growth processes microorganisms are maintained in suspension in the waste water. Attached--growth (fixed-film) processes are carried out by microbial slimes, attached to an inert medium.

Biological waste water treatment processes are mainly used for the removal of the carbonaceous organic materials in waste water, nitrification, denitrification and stabilization.

The aerobic biological treatment processes currently used in practice include various modifications of the activated sludge process, waste stabilization pond systems, aerated lagoons, the trickling filtration process, rotating biological contactors, fixed-bed nitrification reactors, the aerobic digestion process, etc.

The principal nitrogen conversion and removal occurs by means of nitrification and denitrification processes. The aerobic

biological processes are used to oxidize ammonia nitrogen to the
nitrate form, and the anaerobic processes are applied to reduce
nitrate nitrogen to molecular nitrogen. Nitrification is a chemo
autotrophic process, and can be carried out in such treatment
units as activated sludge, trickling filters and rotating contactors. Denitrification is carried out by heterotrophic organisms
and a carbon material is needed to serve as a carbon and energy
source. The treatment units used for denitrification can be ponds,
suspended culture systems or packed-bed reactors.

The activated sludge process is the main representative of
suspended-growth aerobic processes. This process is a biological
waste water treatment method in which a mixture of waste water
and microorganisms are mixed and aerated or oxygenated. The sludge
is separated from the treated waste water in a secondary clarifier
and it is returned to an aeration tank as needed. The following
modifications of the conventional activated sludge process have
been applied in waste water treatment: the tappered aeration
process, the step aeration process, the contact stabilization
process, the extended aeration process, the Kraus process, etc.
Also, pure oxygen has been used as a substitute for air in the
activated sludge process.

A variation of the activated sludge process is the oxidation
ditch, in which the waste water circulates and a surface-type
aerator provides aeration and circulation. A technical modification
of the oxidation ditch concept (identical to extended aeration
process) is the Carrousel process.

Waste stabilization pond systems are applied as an alternative
to high-rate biological treatment processes. An anaerobic waste
stabilization pond is an unmixed basin in which anaerobic micro-
organisms remove organic materials. A facultative waste stabilization pond is a basin in which dissolved oxygen is available
in the upper zone through algae and natural re-aeration, while
the bottom zone remains in an anaerobic condition. An aerobic
waste stabilization pond is an unmixed basin in which aerobic
microorganisms remove organics in the waste water and algae
provide oxygen to maintain an aerobic condition. This type of
pond is frequently used as a polishing unit. A pond applied for
the reduction of indicator or disease caused on microorganisms is
called a maturation pond.

An aerated lagoon (mechanically aerated pond) is a basin which
is mechanically mixed and aerated by means of mechanical aerators.

Two types of aerated lagoons can be considered, the aerobic lagoons and aerobic-anaerobic aerated lagoons. This type of pond may function as an aerobic or facultative waste water treatment system. Ponds and lagoons may operate separately, in series or in parallel.

Attached-growth aerobic biological waste water treatment processes include trickling filters, rotating biological contactors, fixed-bed nitrification reactors, etc.

The trickling filtration process is composed of a trickling filter and a clarifier. The trickling filter consists of a packed bed covered with biological slime. The materials used for the medium, range from rocks and redwood slats to corrugated plastic sheets and plastic rings. Trickling filters are generally classified according to the applied hydraulic and organic loadings as well as the method of operation. The following types are in common use: low-rate filters (sometimes referred as standard-rate filters), high-rate filters, and super rate-filters. The super rate-filters are normally called "roughing" filters and are used for high-strength wastes. The trickling filter plants may use either single-stage or multi-stage filtration.

In recent years, there has been a number of modified attached growth processes proposed for the waste water treatment. Fixed-bed aerobic reactors are used to achieve nitrification. A system using plastic panels submerged in an aeration basin is called the fixed activated sludge process and this intriguing idea has been known as Hay's process.

A rotating biological contactor is a continuous flow unit composed of a tank with a series of disks (normally made of plastic). The disks, covered with biological slime, are partially submerged in waste water and rotate slowly through the liquid.

Anaerobic treatment processes prove to be useful for handling biodegradable organic wastes. Anaerobic treatment is used for the degradation and breakdown of organic solids or for the breakdown of soluble organics to gasous end products. Anaerobic treatment processes have also been used for the treatment of high strength organic industrial wastes.

The most common anaerobic attached-growth biological processes are the anaerobic filter process and anaerobic waste stabilization ponds.

Anaerobic treatment by means of suspended-growth processes has proven to be useful for handling biodegradable organic wastes. The basic types of anaerobic digestion are: the conventional anaerobic digestion process and the high-rate aerobic digestion process. Two-stage digestion is essentially a combination of a high--rate reactor and a conventional reactor operated in series. An anaerobic contact process is used for the treatment of organic wastes with low solids content, and the most significant feature of this process is sludge solids recycling.

Aerobic digestion is an alternative method of treating the organic wastes. Two types of the aerobic digestion process are in common use: the conventional-air aerobic digestion and pure-oxygen aerobic digestion. Thermophilic aerobic digestion is another modification of the aerobic process.

This book covers most of the above-mentioned processes of the treatment of biological waste water in respect to mathematical models for the design and/or practical control. The applicability of the various models and limitations are also presented in this book.

In addition, since the final decision is based on a cost estimate, examples of coupling process models with economic models are included.

However, biological processes for waste water treatment have been used for at least a hundred years. Yet, there is a continuous need for better understanding of these processes. In this respect, mathematical modelling of biological waste water treatment offers a number of potential benefits. An important potential of modelling is related to the design. Steady-state models are normally used for the design of various biological processes, and they can also be used for prediction of their performance.

Due to the dynamic character of biological processes, time--variant models are necessary. Dynamis models are used to describe the operation of biological waste water treatment processes and to establish the most effective real-time control strategy. These models still require further development, since most biological processes are not yet fully understood with respect to the inter- action of process variables. The optimum performance of any biological waste water treatment process is to produce a specific effluent with a minimum consumption of energy, materials, etc. Therefore, the advantage of mathematical modelling of biological waste water treatment processes stem from the improvements in process performance and the potential savings in construction and operating costs (minimalization of materials and energy consumption).

However, in spite of much research progress in the mathematical modelling of biological waste water treatment processes, there are still many problems that must be solved. In particular, the coupling of models for the individual processes into an overall model for an entire waste water treatment plant is an important task.

The development of waste water treatment has for many years mainly been based on empirical ideas, while theory has often lagged behind practice. Theoretical developments have been slow, because of the large degree of variation in biological treatment processes. The action of random biological populations on wastes of a highly variable composition gives variable results from which it is difficult to discern any general principles.

Mathematical modelling has, however, enabled considerable progress to be made in the theory of waste treatment processes. Numerical methods and computations have been widely used for the last 10-20 years to find answers to the sets of questions which would be insoluble by means of analytical methods.

The use of modelling in ecology and chemical engineering has developed rapidly during the last 10-15 years. It is, therefore, not surprising that in the same period modelling has found widespread application within the field of waste treatment (which also has the characteristic features of ecosystems and chemical reactants).

Biological treatment plants are based upon the same biological processes that are found in ecosystems. Therefore, a number of papers on models of biological treatment processes have recently been published in various scientific journals, including "Ecological Modelling".

The ecological modeller can learn from these papers, when he has to model the same processes in the (even more variable) ecosystem and the modeller of biological treatment plants can of course learn from the experience which the ecologist or environmental manager gains from the more complex ecosystems.

The development of biological treatment models follows approximately the same procedure as used in ecological modelling, see for instance Jørgensen, 1983.

A biological treatment is obviously much less complex than an ecosystem. Therefore, it is possible to give a reasonably accurate description of the processes, that take place in a biological waste treatment plant, and the number of state variables needed

in the models is limited. To a certain extent, a biological treatment plant can be considered as a microcosm model of microbiological processes in ecosystems.

Models of waste treatment processes are widely used to optimize
or find the most cost effective solution. This implies that an
accurate model is needed, which is possible with sufficient data
material. This type of model use is more developed for water
treatment than for ecosystems, although ecological-economic models
are under rapid development.

Biological treatment models should be developed from:

1) an intimate knowledge of the system,
2) the composition of the waste water,
3) the environment in which the plant will operate,
4) the questions to be answered, and
5) the available data.

As a result, a large variety of these models exist, as in
ecology. We have, therefore, found it necessary to present not
only models of different processes - activated sludge, trickling
filter, rotating biological contactor, waste stabilization pond,
nitrification, denitrification, anaerobic digestion, etc. - but
also to present different approaches used to find solutions used
in different types of waste water. In the same way that a general
model of a lake does not exist, we do not have a model for an
activated sludge plant.

We feel that we have managed to show the state-of-the-art in
the field by the presentation of a wide spectrum of models, but
it has only been possible due to the large number of experts, who
have contributed to the book. The book was written by an international group of experts from Europe, The United States, Japan
and Australia and it represents an excellent collection of papers
dealing with the subject matter.

REFERENCES

Jørgensen, S.E., 1983. The modelling procedure. In: S.E. Jørgensen
    (Editor), Application of Ecological Modelling in Environmental
    Modelling, Part A. Elsevier, Amsterdam, pp. 5-15.

PRINCIPLES OF MATHEMATICAL MODELING OF BIOLOGICAL WASTEWATER
TREATMENT PROCESSES

V.A. VAVILIN

2.1. INTRODUCTION: NON-LINEAR MODELS OF PROCESSES WHICH DETERMINE THE QUALITY OF WATER

An indicative hydrobiological response to anthropogenic pollution is the change in the species content of an aquatic population
which, together with its environment, presents a complex dynamic
system. In order to describe the system formally, which always
involves idealized models of the system, internal variables and
external parameters should be defined /Fig. 2.1/.

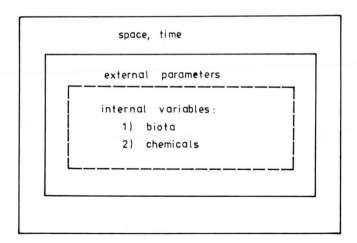

Fig. 2.1. Elements of a mathematical model.

2.1.1. Simulation models

The development of any model designed to describe a particular
object can only be effective if the problem which should be solved
through the model is well defined, so that the model's parameters
and variables can be clearly specified. The solution obtained with
the help of the model should give the answer to the problem.

The variables and parameters depend upon the processes which take place in the aquatic ecosystem and whose description is incorporated into the model. The particular processes to be included depends largely upon the characteristic "normal" time scale that corresponds to the model; the normal time scale reflects normal processes, and allows very fast and very slow processes, which occur in the real ecosystem, to be ignored.

Most models of aquatic ecosystems are described with differential equations. Non-linear differential equations can be used to simulate complex interactions between the ecosystem's variables.

The development of simulation models involves numerous simplifications concerning the processes that take place in the system. Thus, in ecological studies it is customary to discern different trophic levels while simulation models are usually based on such aggregated variables as total biomass which involves many micro-organism species and total concentration of organic matter.

The development of simulation modelling techniques using modern computers has created markedly better possibilities for quantitative description of natural phenomena. However, in modelling real aquatic systems, this approach cannot help compensate for the scientist's inadequate knowledge of the laws that govern the processes under study or for the absence of quantitative data on the system.

2.1.2. Linear and non-linear models

There are two groups of specialists dealing with water quality management problems i.e. the engineers who prepare the designs and take part in the realization of water management measures, and the scientists whose task includes the evaluation of possible effects of such measures and the development of methods to increase their efficiency.

The traditional approach of engineers is based on relatively simple "linear" concepts of the natural processes. The engineer tries to derive simple input - output relations stressing one or several variables. The internal mechanism of the aquatic system which responds to an external disturbance is of little interest to him. The input-output relations are usually obtained through uncomplicated linear models. Such "engineering" models include,

in effect, the Streeter-Phelps model which describes the process
of self-purification in rivers, and the Vollenweider model which
describes the eutrophication of water bodies.

At the same time, scientists /primarily biologists/ know and
understand the complexity of laws that govern the behaviour of
aquatic ecosystems which contain numerous species of micro-
organisms. As the spectrum of characteristic time scales is wide
and some of them have the order of the human lifespan , these
laws are little known. Simulation models which contain scores of
variables can provide a tool with which to study at least the
most important features of an object. Yet, such models involve
several unknown coefficients, and their estimation can be a formi-
dable problem. Note that further complication of a simulation
model does not necassarily lead to a better description of the
underlying "mechanics" of the processes that take place in the
real system.

Non-linear models of biological processes are a step beyond
linear engineering models /Vavilin, 1983b/. Being relatively
uncomplicated, these models allow one to understand in which
situations the linear engineering estimates can be applied and
how the range of their validity can be extended.

2.1.3. Coupled variables

The quality of water is characterized by biotic and chemical
components /see Fig. 2.1/. Simple models contain only a single
variable; thus, the decay of organic matter, L, is often described by a differential equation of the first order with respect
to the pollutant concentration

$$dL/dt = - k L \qquad\qquad /2.1/$$

where k = decay rate constant.

The exponential growth of the biomass, B, traditionally,
described by the equation

$$dB/dt = \mu B \qquad\qquad /2.2/$$

where μ = specific growth rate.

Cleary, the coefficients k and μ in these simple relationships present complex functions of both external parameters and internal variables of the actual ecosystem.

With long time scales, the non-linear logistic equation of Verhulst and Pearl can be used instead of the linear Eq 2.2

$$dB/dt = \hat{\mu} B \left(B_\infty - B\right)/ B_\infty \qquad\qquad /2.3/$$

where B_∞ = limit biomass, and $\hat{\mu}$ = rate constant.

The logistic equation formally shows that the biomass growth has an upper limit.

Equations 2.1 to 2.3 describe independent variations of biological and chemical variables. In actual situations, however, these variations are usually interdependent. A classic example of such coupled variables is given by the Monod model

$$dB/dt = \mu \left(L\right) B$$
$$dL/dt = -\mu \left(L\right) B/Y \qquad\qquad /2.4/$$

where Y = yield coefficient of biomass.

Its specific growth rate attains a saturation state with respect to the substrate concentration

$$\mu \left(L\right) = \frac{\mu_m L}{k_L + L} \qquad\qquad /2.5/$$

where μ_m = maximum specific rate, and

$\qquad k_L$ = half-saturation constant.

According to Eq. 2.4

$$L = L_0 + \left(B_0 - B\right) /Y \qquad\qquad /2.6/$$

where L_0 and B_0 = initial conditions.

Substitution of Eq. 2.6 into the first of Eq. 2.4 under the condition $k_L \gg L_0$ yields the logistic Eq. 2.3 for the biomass where $\hat{\mu}/B_\infty = \mu_m/k_L Y$ and $B_\infty = YL_0 + B_0$. Thus from the Monod model a logistic equation is obtained with the population growth being determined by the storage of the limiting substrate. The logistic equation solution is:

$$B = B_\infty / \left[1 + \left(B_\infty / B_o - 1 \right) e^{-\mu t} \right] \qquad /2.7/$$

If $B_o \gg L_o$ $\left(\Delta B \ll B_o \right)$, the variable B in Eq. 2.4 can be regarded as a parameter and, taking into account Monod's relation /Eq. 2.5/ the following equation with respect to L is obtained:

$$\frac{dL}{dt} = - \frac{\mu_m B_o}{Y} \cdot \frac{L}{k_L + L} \qquad /2.8/$$

Clearly, when $k_L \gg L_o$, Eq. 2.8 is transformed into the traditional Eq. 2.1 of the first order with respect to substrate and $k = \mu_m B_o / k_L Y$.

An extension of the time interval makes it necessary to take into account the biomass auto-oxidation processes /Fig. 2.2/.

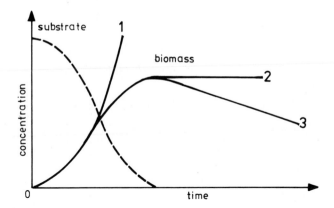

Fig. 2.2. Choosing a model corresponding to the time interval:
1,2,3 - models represented by Eq. 2.2, 2.3 and 2.9, respectively.

Then Eq. 2.4 are rewritten as:

$$\frac{dB}{dt} = \mu \left(L \right) B - k_d B$$

$$\frac{dL}{dt} = -\mu \left(L \right) B/Y \qquad /2.9/$$

Numerous models of the Monod type with coupled variables, B and L, currently exist. Initially, they were only used to describe the growth of pure micro-organism cultures on simple substrates, but later, models of this type were used to try to describe multi-species systems with aggregated variables such as biological treatment systems with multi-component pollutants. The aggregated character of variables is, in fact, ignored in such situations.

The Monod hyperbolic relation /Eq. 2.5/ is also used as an elementary function in complex simulation models which describe interactions between different trophic levels. Evidently, such "canonization" of the Monod relationship can harldy be justified because multi-component/multi-species systems can behave in a much more complicated manner.

On the whole the selection of necessary variables and parameters and the choice of particular functions which link them presents a rather loose problem which is very difficult to formalize. Implicitly, one can give preference to relatively simple macro-dynamic models which have up to three variables and just a few coefficients. The adequacy of the selected model to the actual system depends on the range of variations of the variables and parameters both in time and space.

Further discussion will involve different biological treatment models in which the process of self-purification runs its course intensively in an artificial system under high concentrations of micro-organisms and dissolved oxygen. It will be shown that formal non-linear models which present extentions of the Monod model, can rather closely describe the biological treatment process with its mechanism determined by the multi-component/multi-species ecosystem of micro-organisms.

2.2. DYNAMICS OF BIOMASS GROWTH AND SUBSTRATE CONSUMPTION

Formal description and development of adequate mathematical models of real objects is now regarded as a necessary element of any scientific research. To a large extent, this approach was promoted by the possibility to conduct highly specialized experiments. The generalization of the results obtained in modelling specific phenomena has, in fact, led to the creation of quite strict and orderly physical theories.

As for ecological systems, such formalization of knowledge is still in its initial stage and the most impressive results have been obtained in those fields where experimental data are reproducible. This refers, in particular, to population microbiology; dynamics of microbiological populations forms the subject of many experimental studies.

Traditional studies in chemical kinetics have been instrumental in creating concepts of microbiological technology.

2.2.1. Equations of formal chemical kinetics

In formal chemical kinetics, mono- or bi-molecular reactions are taken as elementary stages. Then, taking into account the law of acting masses, the system of differential equations which describe the rates of variation of the i-th component's concentration can be written as

$$dL_i/dt = \beta_i^k L_k + \gamma_i^{kl} L_k L_e \qquad \qquad /2.10/$$

with, a linear integral $\sum \mu_i L_i$ = constant

where $L_i \geqslant 0$ and the coefficients are subject to the conditions

$$\beta_i^k = \begin{cases} 0 \\ |\beta_k^k| \end{cases} ; \quad \gamma_i^{kl} = \begin{cases} 0 \\ |\gamma_k^{kl}| \end{cases} ; \quad \gamma_k^{kl} = \gamma_e^{lk} ; \quad \beta_k^k \leqslant 0; \quad \gamma_k^{kl} \leqslant 0.$$

In particular, the last of these conditions means a "non-chemical" behaviour /that is, disobedience to the law of acting masses/ of functions which formally describe autokatalisis,

$$\beta_k^k > 0, \quad \gamma_k^{kl} = \gamma_l^{lk} > 0.$$

In studying kinetics of complex chemical systems described by Eq. 2.10, the quasi-stationary concentrations technique is frequently applied. A large number of variables, namely, the fast variables can then be excluded by using passage to the limit. This leads to a simple system of differential equations, with rather cumbersome non-linear right sides, which is compared with experimental data. When only one variable remains Eq. 2.10 can be written as

$$dL/dt = \rho \left(L, k_1, \ldots, k_n \right) \qquad\qquad /2.11/$$

where $k_1, \ldots, k_n$ = coefficients which depend on β_i^k, γ_i^{kl} and μ_i.

The mechanism of a chemical reaction is regarded as known when elementary stages for Eq. 2.11 /described by Eq. 2.10/ are found. This approach helps find the mechanism of autokatalitic reactions, zero-order decay reactions, etc.

Korzukhin /1967/ proved a theorem providing an algorithm that allows one to construct a chemical system /Eq. 2.10/ with any given dynamics. However, this is difficult to realize in practice and, moreover, the solution is far from being unique. Strenuous efforts now being undertaken to describe the mechanism of auto-oscillations in chemical reactions /Zhabotinsky, 1974/ can serve as an example. In the case of relaxational oscillations, it is effective to choose two coupled variables, a fast one, L_2, and a slow one, L_1,

$$\varepsilon \, dL_2/dt = f \left(L_1, L_2 \right)$$
$$dL_1/dt = \varphi \left(L_1, L_2 \right)$$

where ε = a small parameter.

Eq. 2.11 describes the process that takes place in a batch reactor when all necessary reagents are fed into the reactor at the initial stage and then sustained for a proper time. There are two types of reactors for continuous flow of liquid; an ideal complete-mixing reactor with intense mixing of the contents, and an ideal plug-flow reactor with no mixing /diffusion/ along the axis of the flow so that the liquid passes the apparatus as a compact mass. The respective stationary equations are:

$$L_o - L_e - T \rho \left(L_e \right) = 0 \quad /\text{complete-mixing reactor}/ \qquad /2.12/$$

$$dL/dz = - T \rho \left(L \right) \quad /\text{plug-flow reactor}/ \qquad\qquad /2.13/$$

where the dimensionless variable z takes values between 0 and 1 along the axis of the flow and

$$L = \begin{cases} L_o & \text{when } z = 0 \\ L_e & \text{when } z = 1 \end{cases}$$

L_o, L_e being the reagent's input and output concentrations respectively. The function $\rho\left(L\right)$ is often approximated with a power
law of n-th order reaction with respect to the reagent concentration

$$\rho\left(L\right) = kL^n \qquad\qquad\qquad /2.14/$$

2.2.2. Models of pure micro-organisms culture growth

In population microbiology it became necessary to create
a model which would relate two variables, namely, the limiting
substrate and micro-organisms' concentrations. Such a model was
first proposed by Monod /1942/. The Monod relationship /Eq. 2.5/
reflects the fact that the ability of an individual cell to process a substrate , L , is limited when a high substrate concentration "saturates" the cell. At low L values, there is a linear
relation between the rate of treatment and the substrate concentration.

This property is characteristic of another well-known formula
/Teissier, 1942/:

$$\mu = \mu_m \left(1 - e^{-L/k}\right) \qquad\qquad /2.15/$$

where k = a constant.

Eqs. 2.5 and 2.15 can be substituted by a piecewise-linear
approximation which is directly related to the so-called two-phase
model:

$$\mu = \begin{cases} \mu_m, & L \gg L_{cr}, \quad\text{first phase} \\ \mu_m \ L/L_{cr} & L < L_{cr} \ \text{second phase} \end{cases} \qquad /2.16/$$

Here, L_{cr} is a critical substrate concentration such that $L_{cr} \approx 2k_L$.
At the first phase, an excess of nutrients is postulated, while
during the second phase the specific growth rate is linearly
dependent on the substrate concentration.

The effect of "saturation" is described with several other
relationships such as

$$\mu = \mu_m \ L^r \ /(k_L^r + L^r) \qquad /\text{see Moser, 1958}/ \qquad /2.17/$$

and

$$\mu = \mu_m L / (\hat{k}_L B + L) \quad /\text{see Contois, 1959}/ \qquad /2.18/$$

where, r = an empirical constant.

Eq. 2.5 and Eqs. 2.15-2.18 satisfy the following relationships:

$$f(L) = \mu / \mu_m \leqslant 1 \quad \text{and } \lim f(L) = 1 \text{ as } L \to \infty.$$

High substrate concentrations are known to inhibit the growth of micro-organisms. The Haldane relationship /see Haldane, 1930/ taken from fermentative kinetics is often used to describe this effect,

$$\mu = \mu_m L / (k_L + L + L^2/k_i) \qquad /2.19/$$

where k_i = an inhibition constant.

Scores of different formulae are currently used to describe major kinetic laws govering the growth of micro-organisms. On the whole, models of the Monod type account for the most general phenomena, such as the growth and decay of micro-organisms, and saturation of growth rate with respect to substrate.

The Monod relationship /Eq. 2.5/ is simular to that of Michaelis-Menten which pertains to the rate of fermentative reactions. In accordance with the "weak link" concept, the slowest stage in the chain of the substrate ferementative transformations puts a limit on the rate of the overall process. Then, the hyperbolic relationship given by Eq. 2.5 can be regarded as a result of inter-cellular mono- and bi-molecular biochemical interactions. This interpretation of the Monod formula along with the simplicity of estimation of its coefficients explain why microbiologists have "canonized" the formula.

When other types of microbiological reactors are designed, the equations similar to Eqs. 2.12 and 2.13 should each contain two variables, i.e. substrate and micro-organisms´ concentrations. Thus, in the case of the continuous-flow complete-mixing reactor and the Monod model, the system is

$$- B_e + T \frac{\mu_m B_e L_e}{k_L + L_e} = 0$$

Table 2.1. Some models applied for designing continuous-flow reactors

Model	Oxidation rate	Equations for output substrate concentration	
		Complete-mixing reactor	plug-flow reactor
1	2	3	4
1. Zero-order	$k_o B$	$L_e = L_o - a$	$L_e = L_o - a$
2. First-order	$k_1 BL$	$L_o = L_o/(1+a)$	$L_e = L_o e^{-a}$
3. N-th order	$k_n BL^n$	$(L_o - L_e)/L_e^{\,n} - a = 0$	$L_e = \left\{ L_o^{1-n} - a(1-n) \right\}^{1/(n-1)}$
4. Monod	$k_v BL/(k_1 + L)$	$L_e = P + \sqrt{P^2 + k_L L_o}$ $\quad P = (L_o - k_L - a)/2$	$L_e - L_o - k_L \ln L_o/L_e - a = 0$
5. Moser	$\dfrac{k_v B (L/L_k)^n}{1 + (L/L_k)^n}$	$\dfrac{(L_o - L_e)\left[1 + (L_o/L_k)^n\right]}{(L_e/L_k)^n} - a = 0$	$\dfrac{L_k}{1-n}\left\{ \left(\dfrac{L_o}{L_k}\right)^{1-n} - \left(\dfrac{L_e}{L_k}\right)^{1-n} \right\} + L_o - L_e - a = 0$
6. Modified Moser	$\dfrac{k_v B (L/L_k)^n}{1 + (L/L_k)^m}$	$\dfrac{(L_o - L_e)\left[1 + L_o/L_k^{\;m}\right]}{(L_e/L_k)^n} - a = 0$	$\dfrac{1-n}{1-n+m}\left\{ \left(\dfrac{L_o}{L_k}\right)^{1-n+m} - \left(\dfrac{L_e}{L_k}\right)^{1-n+m} \right\} + \left(\dfrac{L_o}{L_k}\right)^{1-n} + \left(\dfrac{L_e}{L_k}\right)^{1-n} - \dfrac{1(1-n)}{L_k} = 0$

Table 2.1. /continued/

1	2	3	4
7. Haldane	$\dfrac{k_v BL}{k_L + L + L^2/k_i}$	$\dfrac{(L_o - L_e)(k_L + L_e^2/k_i)}{L_e} - a = 0$	$(L_o - L_e)\left(1 + \dfrac{L_o - L_e}{2k_i}\right) + k_L \ln \dfrac{L_o}{L_e} - a = 0$
8. Yerusa-limsky	$\dfrac{k_v BL}{(k_L + L)(k_B + B)}$	$L_e = P + \sqrt{P^2 + k_L L_o}$ $\quad$ $P = \left(L_o - k_L - \dfrac{a}{k_B + B}\right)/2$	$L_o - L_e + k_L \ln L_o/L_e - a/(k_B + B) = 0$
9. Teissier	$k_v B\left(1 - e^{-L/L_k}\right)$	$(L_o - L_e)\left(1 - e^{-L/L_k}\right) - a = 0$	$L_e = L_k \ln\left\{1 + \left(e^{L_o/L_k} - 1\right) e^{-a/L_k}\right\}$
10. Contois	$k_v BL/(\hat{k}_L B + L)$	$L_e = P + \sqrt{P^2 + \hat{k}_L BL_o}$ $\quad$ $P = \left(L_o - \hat{k}_L B - a\right)/2$	$L_o - L_e + \hat{k}_L B \ln L_o/L_e - a = 0$

Note: L_o = input pollutant concentration; B = micro-organisms concentration

T = detention time of wastewater in the reactor; a = $k_B T$

$$L_o - L_e - T \frac{\mu_m B_e L_e}{Y \left(k_L + L_e \right)} = 0 \qquad\qquad /2.20/$$

The system has two solutions, a trivial one $/L_e = L_o, B_e = 0/$ and

$$L_e = k_L / \left(\mu_m T - 1 \right), \qquad B_e = Y \left(L_o - L_e \right) \qquad /2.21/$$

which is meaningless unless $0 \leqslant B_e \leqslant YL_o$ and $0 \leqslant L_e \leqslant L_o$.
The parameter T must satisfy the condition

$$\frac{1}{\mu_m} \left(1 + k_L / L_o \right) \leqslant T \leqslant \infty . \quad \text{If} \quad T < \frac{1}{\mu_m} \left(1 + k_L / L_o \right), \qquad /2.22/$$

the micro-organisms can be washed out of the reactor; that is, the growth of the biomass cannot compensate for its decrease.

If the increment of micro-organisms' concentration in the reactor is small /that is, $\triangle B \ll B_o/$, it is sufficient to consider only the variable substrate concentration while the concentration of micro-organisms can be regarded as a parameter of the system. A summary of several models used to compute continuous-flow reactors is given in Table 2.1. Note that the efficiency of the process defined as

$$E = \left(L_o - L_e \right) / L_o \qquad\qquad /2.23/$$

can be quite different for complete-mixing and plug-flow reactors.

Pure cultures of micro-organisms were assumed to be homogeneous in the above-mentioned models. Yet, the micro-organisms' physiology is known to vary considerably along different parts of the growth curve. Several models that take into account physiological variations of populations are known /Tsuchia et al., 1966/. Still, the classic Monod model can be viewed as a basis for the formal description of microbiological systems while the hyperbolic Monod relationship $\rho \left(B, L \right) = \rho_m BL/k_L + L$

has actually replaced the bi-linear interaction formula $\rho \left(B, L \right) = kBL$

which presents ad elementary stage of biochemical reactions. Note that linear and bi-linear interactions lie in the basis of the well-known Lotka-Volterra ecological model that describes system of the predator – prey type /Lotka, 1925/.

2.2.3. Models of growth of mixed micro-organisms´ cultures

Processes of growth of pure micro-organisms´ cultures on mixed substrates were studied in many experiments. Monod /1942/ found that when the substrate was formed by a mixture of glucose and sorbitol, the substrates were removed one by one: first glucose, then sorbitol. Monod named this phenomena "diauxia". His observation was later confirmed by other authors. When the media consists of more than two substrates, the effect of polyauxia can be observed. The reason for this effect seems to lie in the fact that metabolic products generated during the process of decay of easily-oxidized substances accumulate in the cell and inhibit the synthesis of the enzymes responsible for the consumption of other substrates. Some authors, however reported that the characteristic diauxic effect was not observed in their experiments.

Since the substrate which is consumed by micro-organisms in biological treatment systems is sewage water, the system is usually a multi-component/multi-species one. Each species of micro-organism possesses a specific type of metabolism of organic combinations. Apart from the species which consume a wide spectrum of substrates, there are species which can consume only specific substrates in systems of biological treatment.

The diauxia phenomena was repeatedly observed in biological treatment systems /Gaudy and Gaudy, 1972/. However, many authors observed simultaneous removal of different substrates /see for example, Tischler and Eckenfelder, 1969/.

The dynamics of the total biomass ,B, and consumption of complex pollutant ,L, shown in Fig. 2.3 is described by the system:

$$dB/dt = kB \left(B_\infty - B\right)$$
$$dL/dt = -KBL/Y \qquad\qquad /2.24/$$

where, B satisfies the logistic Eq. 2.3.

The limit biomass value depends on the initial conditions $B_\infty = f/B_o, L_o/$. However, the traditional approach is based on the Monod-type models applied to aggregated variables B and L. Note that the model´s coefficients, such as maximum growth rate and auto-oxidation rate constant, that correspond to an "average" species of micro-organism are fictitious in this case /Vavilin,

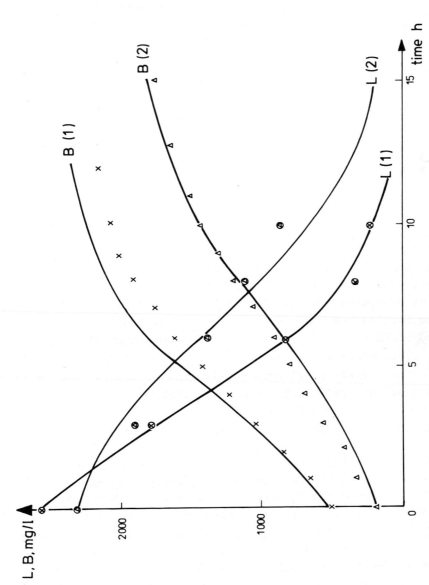

Fig. 2.3. Dynamics of total biomass growth B and substrate consumption L /synthetic domestic wastewater/ according to model 2.24 for k = 1.7·10⁻⁴ l/mg.hr, Y = 0.75, using experimental data from Keshavan et al. /1964/

1980/ and can hardly be regarded as constants.

Assume that micro-organisms in a heterogeneous system are highly specialized, that is, each pollutant is oxidized by only one bacteria species while the oxidation of individual components follows the classic Monod model. This leads to the following system of equations:

$$dB_i/dt = \frac{\mu_{mi} \, B_i \, L_i}{k_{Li} + L_i} - k_{di} \, B_i$$

/2.25/

$$dL_i/dt = - \frac{1}{Y_i} \, \frac{\mu_{mi} \, B_i \, L_i}{k_{Li} + L_i} \quad , \quad i = 1,2, \ldots, N,$$

where N = total number of components.

Here, the succession of micro-organisms is observed, that is, the contributions of different micro-organisms to the overall process of oxidation are changing along the kinetic curve.

If the initial concentrations, B_{io}, of different micro-organisms differ substantially, a seemingly consecutive consumption of substrates, i.e., diauxia, can occur.

With continuous-flow systems, slowly growing species can be washed out of the system /see section 2.2.2/.

2.3. FORMAL MODELS OF BIOLOGICAL TREATMENT

Biological treatment has long been used to remove organic substances from wastewater. In activated sludge reactors /aeration tanks/, micro-organisms exist in suspended form and move along with the liquid, while in fixed-film reactors micro-organisms attach themselves to the surface of the media forming a biofilm.

In real situations, municipal wastewater is always a multi-component substrate - pollutant while activated sludge and biofilm involve different types of micro-organisms. The measured variables usually include the total concentration of the pollutant measured in BOD, COD or TOC units and total biomass. Municipal wastewater which flows into a biological treatment plant normally already contains a share of micro-organisms which consume

some components of sewage. The micro-organisms actively multiply
in the system. In the long run, the efficiency of the treatment
process with respect to specific polluting components and,
finally, to the generalized pollutant, depends on what particular species will multiply enough to dominate the ecosystem.

Physico-chemical conditions in biological treatment systems
can differ widely. Thus, psychrophiles appear at low temperatures
and thermophiles at high temperatures. Oxygen or nutrient shortage and high or low pH values lead to the predominance of those
species which are better adapted to the conditions in the system.

Species composition and relative importance of individual
species in the activated sludge or biofilm microflora are determined by the character of the processed substance and by the
physico-chemical conditions and technological regime of the
treatment system. All this explains the great diversity of species of micro-organisms, found by microbiologists, in biological
treatment systems.

2.3.1. Formulation of the problem

In order to solve many problems in the desig of aerobic biological treatment systems it is necessary to know the dependence
of the pollutant concentration in treated wastewater upon such
parameters as input pollutant concentration, L_o, micro-organisms
concentration, B, detention time of liquid waste in the reactor,
$T = V/q$ /where V is reactor's volume, and q discharge of wastewater per unit time/ and temperature, t^o,

$$L_e = f \left(L_o, \ B, \ T, \ t^o, \ldots, \ \alpha_1 \ldots, \ \alpha_n \right) \qquad \qquad /2.26/$$

As the biomass is growing during the treatment process and
is being removed by a mechanical action an expression for the
biomass increment is also required, i.e.

$$\Delta B = \varphi \left(L_o, \ B, \ T, \ t^o, \ldots, \ \beta_1, \ldots, \ \beta_k \right) \qquad \qquad /2.27/$$

An expression for the amount of oxygen consumed during the
treatment process,

$$\Delta O_2 = \psi \left(L_o, \ B, \ T, \ t^o, \ldots, \ \gamma_1, \ldots, \ \gamma_l \right) \qquad \qquad /2.28/$$

is often required in addition to Eqs. 2.26 and 2.27.
Coefficients $\alpha_1, \ldots, \alpha_n$, $\beta_1, \ldots, \beta_k$, $\gamma_1, \ldots, \gamma_l$ in Eqs.
2.26-2.28 should be estimated in advance on the basis of respective experimental data.

When the values L_o, B, q, t^o, etc. are known, the necessary
volume estimate of the reactor can be found from Eq. 2.26 by
fixing the value of the pollutant concentration, L_e. Equation
2.27 enables the estimation of the capacities required for further processing of the biomass, while the system of aeration of
the liquid waste can be computed on the basis of Eq. 2.28.

The emirical functions given by Egs. 2.26-2.28 can be fitted
in accordance with respective sets of experimental data. Actually,
Eqs. 2.26-2.28 present the input-output relationships of a "black
box" representing a particular treatment system. The efficiency
of emirical functions Eq, 2.26-2.28 in designing biological
treatment systems depends initially on the quality and scope of
the experimental data which were used to estimate the coefficients
$\alpha_1, \ldots, \alpha_n$, $\beta_1, \ldots, \beta_k$, $\gamma_1, \ldots, \gamma_l$.

The more traditional approach, however, lies in solving respective equations for the treatment process rate, rather than in
the direct estimation of the input-output relationships /Eqs.
2.26-2.28/.

A natural criterion which reflects the model's adequacy to the
experimental data is chosen in the form

$$\sigma = \sqrt{\frac{1}{N-k} \sum_{i=1}^{N} \left(L_e^m - L_e^c\right)^2} \qquad \qquad /2.29/$$

where L_e^m, L_e^c = the measured and computed concentrations respectively,

   N = number of experiments and

   k = number of coefficients in the model.

If the range of variations of L_e is large, the model's
adequacy is better defined by the value

$$\hat{\sigma} = \sqrt{\frac{1}{N-k} \sum_{i=1}^{N} \left(\frac{L_e^m - L_e^c}{L_e^m}\right)^2} \qquad \qquad /2.30/$$

which is similar to the relative error of the concentration estimate. The general technique used to find the model's coefficients consists in the minimization of σ or $\hat{\sigma}$ values.

Note that models are rarely chosen on the basis of the criterion given by Eqs. 2.29 or 2.30. In other words, it is implicitly assumed that the model used in each particular study is the only possible one. In some studies /see for example Hanumanulu, 1970/ different models are compared but the range of variation of their parameters, L_o, B, T, t^o, is small thus preventing the selection of a correct model. Under such conditions, divergent models can give equal values of σ or $\hat{\sigma}$.

2.3.2. Traditional models

Descriptions of aerobic biological treatment traditionally start from concepts developed in chemical technology. Respective equations /see Section 2.2/ are written in accordance with the type of the reactor /batch, complete-mixing or plug-flow/. Thus, if the first order Eckenfelder model /Eckenfelder, 1967/ is taken as the basis, the equations for the specific rate of treatment take the forms

$$v = \frac{1}{B}\frac{dL}{dt} = - \rho \left(L \right) = - kL \qquad\qquad /2.31/$$

for batch reactor,

$$v = \frac{1}{B}\frac{dL}{dz} = - T\rho \left(L \right) = - TkL \qquad\qquad /2.32/$$

for plug-flow reactor, and

$$v = \frac{L_o - L_e}{BL} = \rho \left(L_e \right) = kL_e \qquad\qquad /2.33/$$

for complete-mixing reactor, where z = dimensionless distance along the reactor's axis

When, as usually happens, the concentration of micro-organisms is high enough /that is $\Delta B \ll B_o$/, the solutions of the above equations lead to the expressions

$$L_e = L_o \exp \left(- kBT \right) \qquad\qquad /2.34/$$

for batch or plug-flow reactors, and

$$L_e = L_o/(1 + kBT)$$ /2.35/

for complete-mixing reactor.

In reactors with a biofilm, it can safely be assumed that the concentration of micro-organisms is proportional to the media's specific surface area, α.

An analysis /Vavilin, 1981a, 1982a/ showed that in designing biological treatment reactors it was first necessary to find a correct expression for the transfer function /Eq. 2.26/ while the type of the equation used to obtain it, was of no importance at all. Thus, there is no need, at least as a first approximation, to discern between types of reactors in accordance with hydrodynamic conditions of the liquid flow /Figs. 2.4-2.8/. This fact is explained, initially by the aggregated nature of the measured variables, L and B.

The transfer functions given by Eqs. 2.34 and 2.35 can be further improved by introducing the following modifications /Vavilin, 1982a/

$$L_e = L_o \exp (-kBT)^n$$ /2.36/

and

$$L_e = L_o/ \left[1 + (kBT)^n \right]$$

where n = an additional coefficient.

In the case of models given by Eqs. 2.34-2.36, $\eta = L_o/L_o = f(BT)$.

If variations of L_o are large, the traditional first order relationships /Eqs. 2.34 and 2.35/ become less adequate to fit the experimental data; these are then better described by the following formulae derived from the Grau-Oleszkiewicz-Eckenfelder models /Grau et al., 1975; Oleszkiewicz, 1976; Adams et al., 1975/:

$$L_e = L_o \exp (- kBT/L_o)$$ /2.37/

$$L_e = L_o/ (1 + kBT/L_o)$$ /2.38/

Using the Monod hyperbolic relationship the biological treatment rate can be expressed as

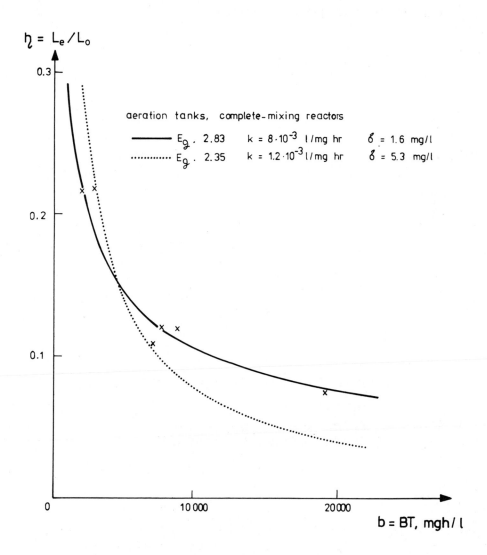

Fig. 2.4. Dependence of $\eta = L_e/L_0$ upon parameter b = BT averaged
for six municipal treatment plants /complete-mixing
reactors/ according to Vavilin /1982a/

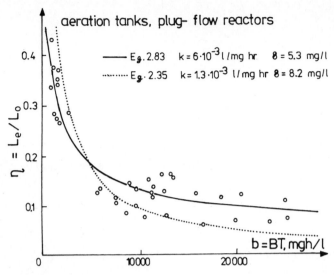

Fig. 2.5. Dependence of $\eta = L_e/L_0$ upon parameter b = BT averaged
for 36 municipal treatment plants /plug-flow reactors/
according to Haseltine /1955/

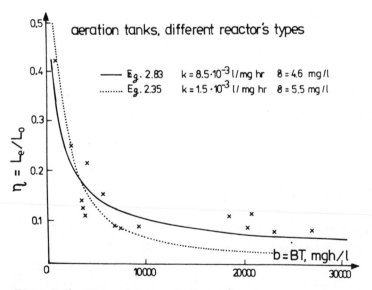

Fig. 2.6. Dependence of $\eta = L_e/L_0$ upon parameter b = BT averaged
for 15 cities of the Russian Soviet Federated Republic
for both complete-mixing and plug-flow reactors, according to Vavilin /1983b/

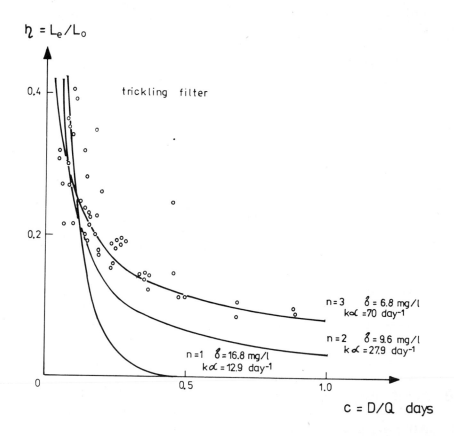

$\eta = L_e/L_o$

trickling filter

0.4

0.2

n=3 δ = 6.8 mg/l
kα =70 day^{-1}

n=2 δ = 9.6 mg/l
kα =27.9 day^{-1}

n=1 δ =16.8 mg/l
kα =12.9 day^{-1}

0   0.5   1.0

$c = D/Q$ days

Fig. 2.7. Dependence of $\eta = L_e/L_o$ upon parameter $c = D/Q$ for a
fixed-film reactor with gravel media treating municipal
sewage water according to model 2.78 /Case n = 1 coincides with model 2,34/ using experimental data from
Keefer and Meisell /1952/

$$v = \frac{L_o - L_e}{BT} = \rho\left(L_e\right) = \frac{\rho_m L_e}{k_L + L_e} \qquad /2.39/$$

so that

$$L_e = P + \sqrt{P^2 + k_L L_o} \qquad /2.40/$$

where,

$$P = \left(L_o - k_L - \rho_m BT\right)/2$$

In selecting a model it is natural to prefer one which contains
fewer coefficients and these coefficients should be such that

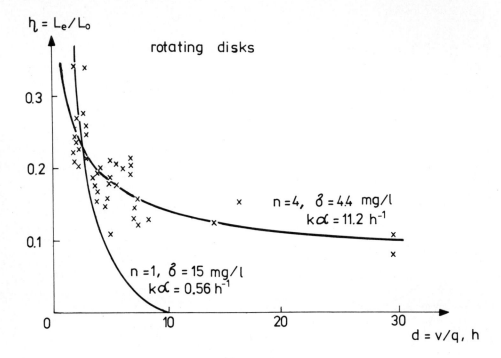

Fig. 2.8. Dependence of $\eta = L_e/L_0$ upon parameter $d = V/q$ for five
rotating disks /Case n = 1 coincides with Eq. 2.34/
using experimental data from Benjes /1979/

they can easily be found either analytically or graphically. Thus,
in the cases of the two first order models given above

$$k = \frac{1}{BT} \ln L_o/L_e \qquad\qquad /2.41/$$

or

$$k = \frac{L_o - L_e}{BTL_e} \qquad\qquad /2.42/$$

respectively.

The coefficients of the Monod model /Eq. 2.39/ are found from
the linear forms

$$1/v = k_L/\rho_m \cdot 1/L_e + 1/\rho_m \qquad\qquad /2.43/$$

$$L_e/v = k_L/\rho_m + L_e/\rho_m \qquad\qquad /2.44/$$

The Grau model

$$v = \frac{L_o - L_e}{BT} = \rho\left(L_e\right) = \rho_m \frac{L_e}{L_o} \qquad /2.45/$$

can be improved by introducing a correction to account for
hardly-oxidized substances, $L\infty$, so that the model takes the
form /Adams et al., 1975/

$$v = \frac{L_o - L_e}{BT} = \rho\left(L_e\right) = \rho_m \frac{L_e - L\infty}{L_o} \qquad /2.46/$$

$$vL_o = \rho_m L_e - \rho_m L\infty \qquad /2.47/$$

where, $\quad L_e = \dfrac{L_o + L\infty \, \rho_m \, BT/L_o}{1 + \rho_m \, BT/L_o} \qquad /2.48/$

Some models of biological treatment are given, with comments,
in Table 2.2. The choice of a "correct" model depends on the
treatments´ technological regime; that is, on the set of the
system´s parameters, L_o, B, T, etc. Traditional studies have not
taken these considerations into account.

The rate of biological treatment can be described, besides the
above models, by power functions of parameters L_o, B, T, e.g.

$$v = \frac{L_o - L_e}{BT} = kL_o^{l_1} \, B^{l_2} \, T^{l_3} \qquad /2.49/$$

where, l_1, l_2 and l_3 = empirical coefficients /see for example
Särner, 1981/.

According to the model given by Eq. 2.49, the rate of the
process is independent of the treatment level. However, the analysis of some data shows that values l_1, l_2, and l_3 depend upon
the regime of treatment process; that is on the characteristic
values of parameters L_o, B, T. Thus, for any chosen set of coefficients, the model /Eq. 2.49/ etc."works" within a relatively
narrow range of variation of its parameters.

Comparison of the results obtained for traditional formal
models with experimental data frequently reveals substantial
discrepancies. Some authors were surprised to find a weak correlation between the rate of treatment and the concentration of the

Table 2.2. Some models of biological treatment processes

Form of function $L_e = L_e\ /L_0, B, T/$	Expression for treatment rate $\varrho\ /L_e/$ or $\varrho\ /L/$	General description
1	2	3
1. $L_e = L_0 e^{-kBT}$	$\varrho\ (L) = kL$	The Eckenfelder model proposed for designing plug-flow aeraton tanks and trickling filters. Better adapted to describe rough treatment.
2. $L_e = L_0 e^{-(kBT)^n}$	-	A modification of Model 1 proposed by Eckenfelder for designing trickling filters. Applicable to aeration tanks as well.
3. $L_e = L_0 /\ (1 + kBT)$	$\varrho(L_e) = kL_e\ \varrho(L) = kL^2/L_0$	The Eckenfelder model proposed for designing complete-mixing reactors. Gives value v biased negatively at high η an positively at low η. Can be used to design trickling filters and rotating disks.
4. $L_e = L_0 /\ \left[1 + (kBT)^n\right]$	-	A modification of Model 3 proposed by Eckenfelder for designing trickling filters. Applicable to aeration tanks as well.
5. $L_e = L_0 /(1 + \hat{k}T)$	$\varrho\ (L_e) = kL_e/B$	The McKinney model proposed for designing complete-mixing reactors. Often gives better results than the Eckenfelder Model 3. Here $\hat{k} = KB = $ const; therefore, this model actually accounts for a lower "activity" of micro-organisms at high values of B. "Works" poorly when variations of T are large.

Table 2.2. /continued/

1	2	3
6. $L_e = L_0/(1 + \varrho_m BT/L)$	$\varrho(L_e) = \varrho_m L_e/L_0$ $\varrho(L) = \varrho_m (L_e/L_0)^2$	The Grau model proposed for designing complete-mixing reactors.
7. $L_e = L_0 e^{-\varrho_m BT/L_0}$	$\varrho(L) = \varrho_m L/L_0$	The Grau model proposed to describe batch experiments. Known as Oleszkiewicz model for designing trickling filters. Poor correspondence to experimental data in case of high-level treatment.
8. $L_e = \dfrac{L_0 + L_\infty \varrho_m BT/L_0}{1 + \varrho_m BT/L_0}$	$\varrho(L_e) = \varrho_m \dfrac{L_0 - L_\infty}{L_0}$	The Adams-Eckenfelder model proposed for designing complete-mixing reactors. Can be used to design trickling filters and rotating disks as well
9. $L_e = L_0 \times \dfrac{\sqrt{1 + 4kBT} - 1}{2kBT}$	$L_e = kL_e^2/L_0$	Author's model proposed for designing aeration tanks, trickling filters and rotating disks at high-level treatment.
10. $L_e = P + \sqrt{P^2 = k_L L_0}$ $P = (L_0 - k_L - \varrho_m BT)/2$	$\varrho(L_e) = \varrho_m \dfrac{L_e}{k_L + L_e}$	The classical Monod model. In this form, is applied to design complete-mixing reactors. Can also be used effectively to design plug-flow reactors, trickling filters and rotating disks in case of rough treatment.
11. $L_e = \hat{P} + \sqrt{P^2 + k_L BL_0}$ $\hat{P} = (L_0 - \hat{k}_L B - \varrho_m BT)/2$	$\varrho(L_e) = \varrho_m \dfrac{L_e}{\hat{k}_L B + L_e}$	The Contois model – a modification of Monod model.

Table 2.2. /continued/

1	2	3
12. $$\sqrt{L_e} = -P + \sqrt{P^2 + L_0}$$	$\rho(L_e) = k\sqrt{L_e}$	A model of 1/2-order. Satisfactory agreement to experimental data in case of rough treatment.
13. $$L_e = L_0 / \sqrt[n-1]{1 + (n-1)\,kBT}$$	$\rho(L) = kL\left(L/L_0\right)^{n-1}$	The Fair model proposed for designing plug-flow reactors. Good for describing high-level treatment.
14. $$L_e = L_0 / \sqrt[n-1]{1 + (n-1)\,kBTL_0^{n-1-p}}$$	$\rho(L) = kL^n/L_0^p$	A modification of Model 13.

Notes: 1. In case of trickling filters or rotating disks, B is proportional to the media´s or disk´s specific surface.

2. Contact time T D/Q^m should be substituted for T = D/Q in case of trickling filter.

generalized pollutant in the reactor or the total concentration
of micro-organisms. Apparently, the main reason is the aggregated
nature of generalized variables, L and B.

If the sludge or biofilm does not contain species that consume
the main components of the municipal wastewater, the treatment
process does not occur even at high values of B. Thus, the capability of sludge or biofilm to process a particular type of wastewater is determined by the amount of active micro-organisms it
contains. This amount depends upon the process of accumulation
of inert biomass.

2.3.3. Formalized scheme and some theoretical concepts of aerobic biological treatment

The process of aerobic biological treatment occurring in such
reactors as aeration tanks or fixed-film reactors, can be roughly
depicted as shown in Fig. 2.9.

The municipal wastewater which enters the reactor has concentrations L_o at the input and L_e at the output. The retention time
of sewage water in the reactor is $T = V/q$ where, V = volume of
the reactor and q = discharge of municipal wastewater.

A definite concentration of micro-organisms which exist in the
reactor either as suspended particles of activated sludge or as
a biofilm attached to the media's surface is maintained in the
reactor. Apparently, in the case of a trickling filter their
concentration, B, is proportional to the specific surface of the
media /surface per unit volume of the reactor/. The process occurs
at a temperature, t^o. The municipal wastewater in the reactor is
saturated with oxygen either by a mechanical action or, more
ofter naturally in aeration tanks or trickling filters, respectively.

In the aeration tank, activated sludge is highly concentrated
due to its re-circulation. Re-circulation of treated wastewater
in the trickling filter can help improve the efficiency of the
treatment process.

It should be noted that in a trickling filter, the retention
time, T, depends on the length of the time interval, T_c, during
which the wastewater is in direct contact with the biofilm. This
interval depends upon the "glueing" of wastewater to the media's
surface. The time, T_c, was experimentally found to be directly

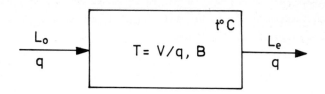

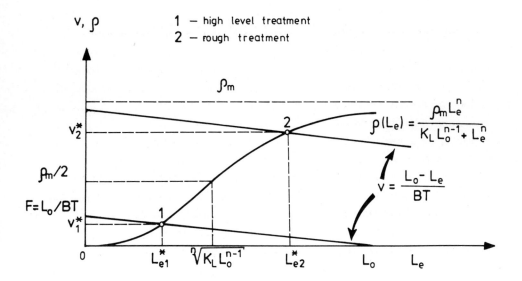

Fig. 2.9. Formalized scheme and generalized model /Eq. 2.50/ of aerobic biological treatment

proportional to the ratio D/Q^m where D = filter depth, Q = q/A specific discharge of sewage liquid and A = surface of the filter cross-section.

Several theoretical concepts can be formulated concerning idealized reactors for aerobic biological treatment /Vavilin, 1982b, c; 1983/:

1. Reactors which realize the process of aerobic biological treatment "re-process" a definite spectrum of substrate-pollutants each of which represents are component of a generalized multi-component pollutant. The rate of treatment depends on the oxidation rates of indivudual components and their contributions

to the overall rate of oxidation. As retention time, T, and micro-organisms concentration, B, grow, the share of "re-processed" hardly-oxidized components of the generalized pollutant also grows. The spectrum of "re-processed" reagents shifts to easily-oxidized components as the pollutant concentration, L_o, at the reactor's input increases.

2. The microbiological content of sludge particles in an aeration tank or in biofilm of a fixed-film reactor is formed in accordance with the spectrum of "re-processed" reagents. Consequently, the species content depends not only upon the type of municipal wastewater and physico-chemical conditions, but also on the operational regime of the reactor. The stable distribution of species can be characterized with the organic loading on the system, $F = L_o/BT$. If the loading is high, the prevailing species in the microflora of activated sludge or biofilm are better adapted to consume easily-oxidized reagents.

3. Inert biomass is accumulated in the process of biological treatment. Its relative content grows with diminishing organic loading on the system.

4. When there is a transfer to a new type of multi-component pollutant, to new physico-chemical conditions or to any other technological regime of the treatment process a definite time /adaptation time/ will be required to form a new stable distribution of species in the microflora of activated sludge or biofilm.

5. The concentration of generalized pollutant is normally measured in units of biochemical /BOD/ or chemical /COD/ consumption of oxygen. The ratio COD/BOD grows during the process of biological treatment because the spectrum of remaining reagents shifts to hardly-oxidized components. Therefore, the rate constants with respect to COD are smaller than those with respect to BOD.

6. The type of reactor /plug-flow or complete-mixing/ is not important, as a first approximation, for treatment of multi-component wastewater because the major role in the treatment process belongs to zero order reactions of oxidation of individual components. In batch reactors, the type of kinetic curve depends upon physico-chemical conditions and technological regime which determine the initial multi-species system of micro-organisms. Only cycled barch experiments can create respective stable distri-

butions of species of micro-organisms.

7. In contrast to aeration tanks, fixed-film reactors "scan" the species distribution of the biofilm along the flow of liquid sewage in accordance with the spectrum of processed substrates.

8. The law which governs the treatment process in determined by the treatment level. Under rough treatment, micro-organisms of the sludge or biofilm are "saturated" with the components of the substrate-pollutant which prevail in the reactor so that the rate of treatment attains a maximum; this corresponds to the zero order reaction with respect to generalized pollutant, L. The biomass increment is also highest under these conditions. Under high-level treatment, the rate of treatment obeys an equation which corresponds to the n-th order reaction with respect to the generalized pollutant, L.

9. Under high-level treatment, the spectrum of "re-processed" reagents shifts to hardly-oxidized components. In this case, there is a weak dependence of treatment efficiency upon the micro-organisms concentration, B, and temperature, t^o. Under rough treatment, this correlation is more important; in addition the rate of treatment can be limited by the concentration of dissolved oxygen.

10. If the efficiency of treatment varies substantially, or the number of treatment plants under consideration is high, or the extent of attenuation of BOD or COD concentration is being studied, experimental data can be adequately described by generalized models of special types. Traditional models can give biased results. When efficiency varies within a narrow range, different models "work" in a similar manner.

11. Municipal wastewater is treated thoroughly /high-level treatment/ and therefore the process of treatment is described with the n-th order model with respect to the generalized pollutant, L.

12. It seems that the type of biological treatment plant with most prospects is one with reactors which combine individual properties of aeration tanks and fixed-film reactors. The fixation of microflora in plug-flow aeration tanks, using special types of media possessing high specific surface, substantially increases the biomass of micro-organisms and helps attain their sequential distribution along the flow of liquid waste so that components of the pollutant are removed one by one. In reactors of this type

the retention time can be fairly long.

2.3.4. Generalized model of biological treatment

Analysis of different models and processing of respective data as led to the formulation of generalized models of biological treatment /Vavilin, 1981b, 1982b, c/. In particular,

$$v = \frac{L_o - L_e}{BT} = \rho\left(L_e\right) = \frac{\rho_m L_e^n}{k_L L_o^{n-1} + L_e^n} = \frac{\rho_m L_e\left(L_e/L_o\right)^{n-1}}{k_L + L_e\left(L_e/L_o\right)^{n-1}} \qquad /2.50/$$

where ρ_m = maximum specific rate of treatment,

k_L = half-saturation constant, and

$n \geqslant 1$ = additional coefficient.

For the sake of simplicity, the traditional equation was used:

$$\triangle B = Y\left(L_o - L_e\right) - k_d BT \qquad /2.51/$$

Y = yield coefficient of generalized substrate into micro-organisms´ biomass and

k_d = coefficient of decay and auto-oxidation of net biomass of micro-organisms[*].

As a rule, the increment of micro-organisms´ biomass during time, T, is much smaller than the concentration of micro-organisms in the reactor, i.e. $\triangle B = B_e - B_o \ll B_o$, where B_e and B_o are concentrations of micro-organisms at the reactor´s output and input respectively.

In the differential form, Eq. 2.50 is written as

$$v = -\frac{1}{B}\frac{dL}{dt} = \rho/L/ = \frac{\rho_m L^n}{k_L L_o^{n-1} + L^n} = \frac{\rho_m L\left(L/L_o\right)^{n-1}}{k_L + L\left(L/L_o\right)^{n-1}} \qquad /2.52/$$

and

$$L = \begin{cases} L_o \text{ when } t = 0 \\ L_e \text{ when } t = T \end{cases}$$

---

[*] It will be shown later that Eq. 2.51 can be incorrect under low organic loading, $F = L_o/BT$

In models represented by Eqs. 2.50 and 2.52, the rate of treatment depends not only upon the absolute value of outlet pollutant concentration, L_e, but also upon the level of treatment determined by the quantity $\eta = L_e/L_o$.

The treatment process described by the generalized model /Eq. 2.50/ is shown graphically in Fig. 2.9. The function ρ /L_e/, of the type given by Eq. 2.50 can be looked upon as a static characteristic which can be used to compute reactors possesing a wide set of operational regimes. High or low organic loadings, F, correspond mainly to easily- or hardly-decomposed re-processed reagents respectively.

Apparently, $v = \rho_m/2$ when $L_e = \sqrt[n]{k_L L_o^{n-1}}$; that is, the pollutant concentration in the reactor which corresponds to the half-saturation condition, is an increasing function of the pollutant concentration in municipal wastewater at the reactor's input. In this regard, it should be noted that the generalized model /Eq. 2.50/ can be approximated with the Grau model /Eq. 2.45/ much better than with the traditional first order Eckenfelder model /Eq. 2.33/, as shown in Fig. 2.10.

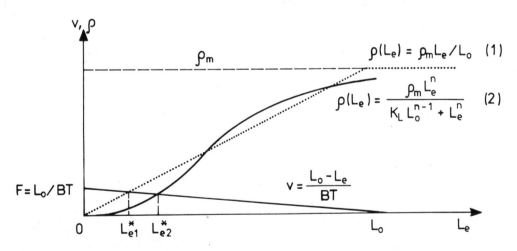

Fig. 2.10. Approximating generalized model /Eq. 2.50/ with Eq. 2.45

The values of postive coefficients ρ_m, k_L and n depend, first of all, upon the type of wastewater /Vavilin, 1982d/. Specifically, in the case of a single substrate, n = 1 thus leading to the classic Monod relationship, while good approximations for hardly-oxidized multi-component pollutants can be obtained at n = 2 for the model represented by Eq. 2.50 and n = 3 for the model represented by Eq. 2.52.

An example of a graphical solution of Eq. 2.50 at n = 2 in the case of paper industry wastewater is shown in Fig. 2.11.

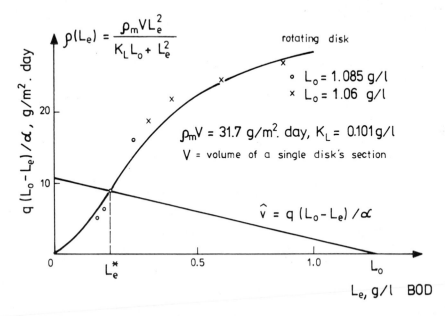

Fig. 2.11. Graphic solution of Eq. 2.50 using experimental data from Paolini et al. /1979/

Under a varying concentration, L_o, models /Eqs. 2.50 and 2.52/ should be regarded with respect to a criterion which coincides with the product of pollutant concentration L_o and a factor whose value depends upon the ratio of the output and input concentrations $\eta = L_e/L_o$, i.e.

$$\xi = L_e \, \eta^{\,n-1} \qquad\qquad /2.53/$$

In the course of treatment, the value η, which depends upon the treatment level, determines the amount of re-processed components in a complex pollutant. The smaller the value η, the more hardly-oxidized components re-processed. Eq. 2.50 can be re-written as

$$v = \frac{L_o - L_e}{BT} = \rho\,(\,\xi\,) = \frac{\rho_m \, \xi}{k_L + \xi} \qquad\qquad /2.54/$$

Examples of this relationship for an aeration tank /batch experiments/ and a rotating disk at n = 2 are shown in Fig. 2.12 and 2.13. Eq. 2.54 describes, with a good degree of approximation, the experiments under large variations of the criterion ξ ; that is, under large variations of treatment level. Fig. 2.14 shows average data for 36 aeration tanks oxidizing municipal wastewater. It can be seen that Eq. 2.54 also corresponds rather well to a wide set of reactors under different organic loadings.

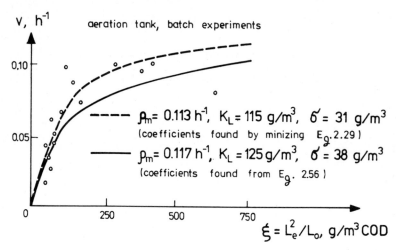

Fig. 2.12. Application of Eq. 2.54 for an aeration tank /batch experiment/, n = 2, pollutant is measured in COD units using experimental data from Chudoba /1969/

The coefficients ρ_m and k_L in the function $\rho / \xi /$ given by Eq. 2.54 can be found from the linear forms /Fig. 2.15/

$$1/v = k_L / \rho_m \cdot 1/\xi + 1/\rho_m \qquad /2.55/$$

or

$$\xi/v = k_L / \rho_m + \xi/\rho_m \qquad /2.56/$$

Note that under high L_e the linear form of Eq. 2.56 gives smaller values σ than those corresponding to Eq. 2.55.

The wastewater retention time, T, in the reactor can be found from Eq. 2.50 as

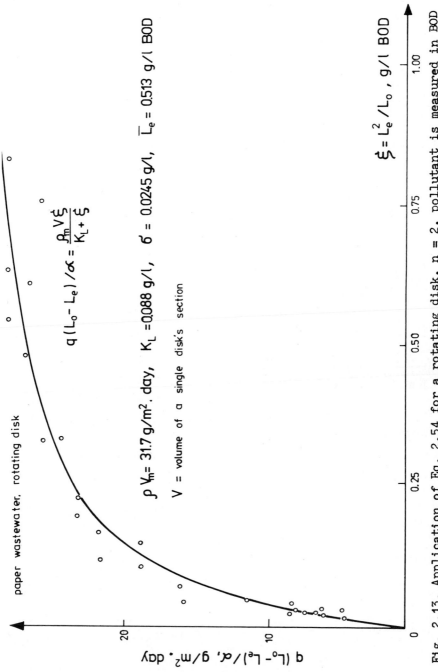

paper wastewater, rotating disk

$$q(L_o - L_e)/\alpha = \frac{\rho_m V \dot{\xi}}{K_L + \dot{\xi}}$$

$\rho V_m = 31.7\ g/m^2 \cdot day$, $K_L = 0.088\ g/l$, $\sigma = 0.0245\ g/l$, $\bar{L}_e = 0.513\ g/l$ BOD

V = volume of a single disk's section

$\dot{\xi} = L_e^2/L_o$, g/l BOD

q $(L_o - L_e)/\alpha$, g/m² day

Fig. 2.13. Application of Eq. 2.54 for a rotating disk, n = 2, pollutant is measured in BOD
units, using experimental data from Paolini et al. /1979/

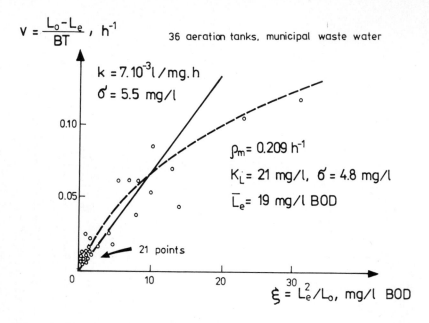

Fig. 2.14. Application of Eqs. 2.54, 2.81 for 36 aeration tanks,
n = 2, pollutant is measured in BOD units, using
experimental data from Haseltine /1955/.

$$T = \frac{L_o - L_e}{\rho_m B \left\{ L_e^n / (k_L L_o^{n-1} + L_e^n) \right\}} \qquad\qquad /2.57/$$

At the same time, the concentration $L_e = L_e^{*}$ is determined by
the formula

$$L_e^{n+1} - L_e^n \left(L_o - \rho_m BT \right) + L_e k_L L_o^{n-1} - k_L L_o^n = 0 \qquad /2.58/$$

If n = 2, Eq. 2.58 can be solved analytically in Cardano form
/Vavilin, 1982b/:

$$L_e = \gamma + \sqrt[3]{\alpha + \sqrt{Q}} + \sqrt[3]{\alpha - \sqrt{Q}} \qquad\qquad /2.59/$$

where

$$\gamma = \frac{L_e - \rho_m BT}{3}, \qquad \alpha = \gamma^3 - \frac{k_L L_o}{2} \left(\gamma - L_o \right), \qquad \beta = -\gamma^2 + k_L L_o/3,$$

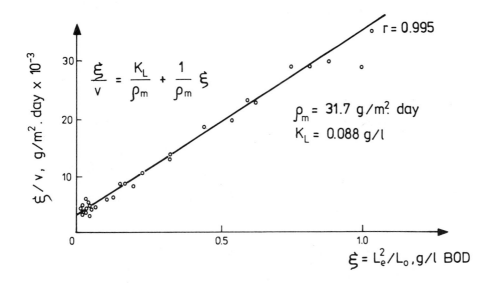

Fig. 2.15. Graphical estimation of coefficients in Eq. 2.54
according to linear form of Eq. 2.56; experimental
data correspond to Fig. 2.13.

and $Q = \alpha^2 + \beta^3 \geqslant 0$. If $\beta < 0$ and $Q \leqslant 0$ then

$$L_e = \gamma + y_{1,2,3} \qquad\qquad /2.60/$$

Of the three roots of L_e, the one chosen as the solution should
be positive. Here, $y_1 = 2\sqrt{-\beta} \cdot \cos \varphi/3$;
$y_{2,3} = -2\sqrt{-\beta} \cos\left(\varphi/3 \pm \frac{2\pi}{3}\right)$

where, $\cos \varphi = \alpha\sqrt{(-\beta)^3}$ and the square root has the some
sign as α.

Fig. 2.16 shows the "origin" of different orders, n, in multi-
component pollutant oxidation when individual components are oxidized according to the zero order reaction. It is easy to understand that the same order with respect to the generalized pollutant will be followed at the initial part of the kinetic curve
under sufficiently high concentration, L_o. The order of the
reaction is increased along with the treatment time as progressively hardly-oxidized reagents are processed.

46

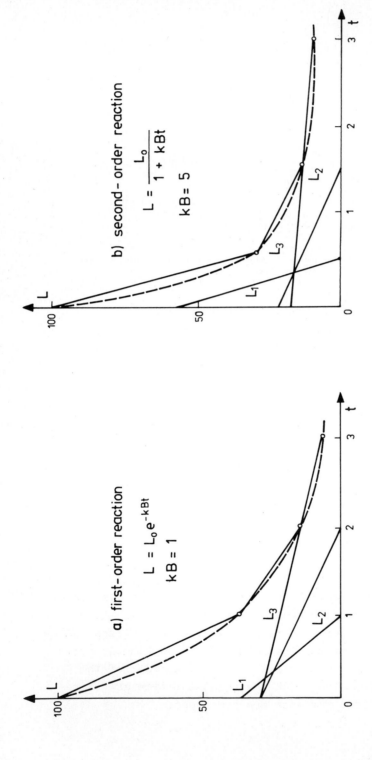

Fig. 2.16. Kinetic curves with respect to generalized pollutant corresponding to /a/ first order and /b/ second order reaction and a three-component pollutant /individual components are oxidized according to zero order reaction/according to Vavilin and Vasiliev /1981/

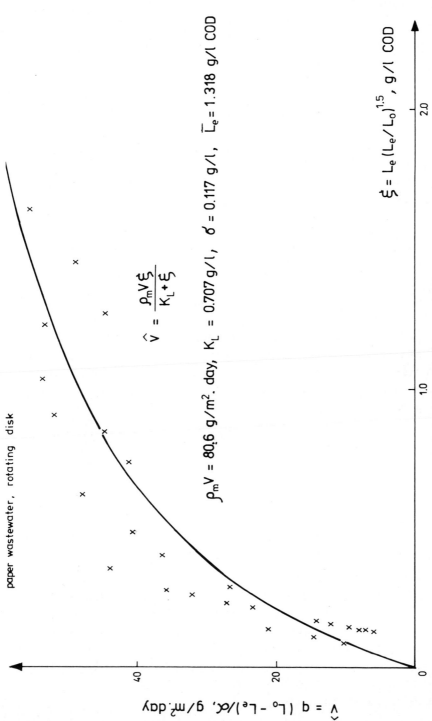

Fig. 2.17. Application of Eq. 2.54, n = 2.5, pollutant is measured in COD units, using experimental data from Paolini et al. /1979/

The values of coefficients ρ_m, k_L and n also depend upon
the method of measuring the complex pollutant concentration.
Eq. 2.54 is shown in Fig. 2.17 for paper industry wastewater
measured in COD units. Comparison of Figs. 2.13 and 2.17 shows
that COD units require higher n values and lower ratios ρ_m/k_L.

As a rule, pollutants exist in municipal wastewater in both
dissolved and suspended forms. The analysis of some data, e.g.,
those given in Toerber et al. /1974/, showed that respective rate
constants of dissolved BOD or COD were higher than those of total
BOD or COD: the order, n, can also be different.

2.3.5. Diffusion model of a single substrate-pollutant biological treatment

It has already been shown that the process of oxidation of
simple substrates can be described rather well by a specific form
of generalized models /Eqs. 2.50 and 2.52/ with n = 1 and the
classic Monod relationship. Respective examples for activated
sludge and biofilm are given in Fig. 2.18.

A diffusion model describing biological treatment of a single
substrate-pollutant was suggested in Vavilin and Vasiliev /1977/
and Vavilin et al. /1980/. The substrate's rate of consumption
by spherically-symmetric sludge particles was shown to follow
the expression

$$\hat{v}\,(L) = \hat{v}_m \left[1 - \left(\frac{r_{cr}}{R}\right)^3 + \frac{3r_{cr}}{\varkappa^2 R^3} \left(\varkappa r_{cr}\,\mathrm{cth}\,\varkappa r_{cr} - 1\right)\right] \qquad /2.61/$$

where,

$$\hat{v}_m = \frac{4}{3}\,\pi R^3\,\rho_d\,\frac{\mu_m}{Y}\ n = \frac{\mu_m B}{Y}, \qquad /2.62/$$

$\varkappa = \sqrt{\rho_d \mu_m\,/YD_i L_{cr}}$,

ρ_d = dry biomass density of respective micro-organisms in activated sludge particles,

L_{cr} = a critical substrate concentration which requires a change
of the reaction's order from zero to one with respect to
individual species of micro-organisms and in accordance
with a piecewise-linear relationship,

D_e, D_i = substrate's diffusion coefficients in water and activated
sludge particles respectively,

R = sludge particle radius,

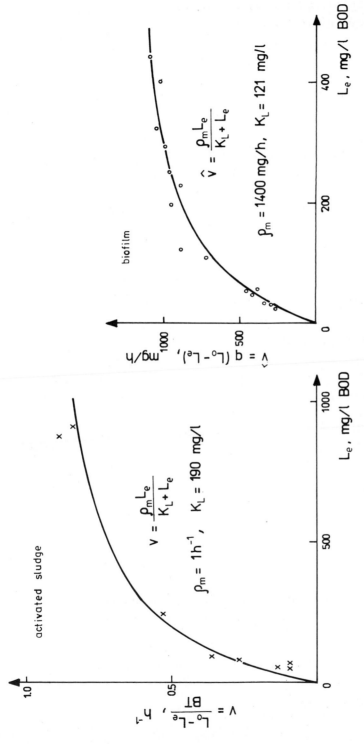

Fig. 2.18. Monod relation /Eq. 2.39/ for activated sludge and biofilm treating glucose; experimental data for activated sludge from Krishnan and Gaudy /1969/, and figure for biofilm from Kornegay and Andrews /1969/

Y = yield coefficient of micro-organisms,

μ_m = their maximum growth rate,

n = number of particles per unit volume of the reactor, and

B = activated sludge concentration.

The value r_{cr} is found from the expression for the substrate concentration in the solution L_s, i.e.,

$$L_s = L_{cr}\left\{\left[\frac{1}{6} + \frac{r_{cr}^3}{3R^3} - \frac{r_{cr}^2}{2R^2} + \frac{1}{3}\frac{\delta}{R}\frac{D_i}{D_e}\left(1 - \frac{r_{cr}^3}{R^3}\right)\right]\left(\text{æ}R\right)^2 + \right.$$

$$+ \left[\frac{r_{cr}}{R}\left(1 - \frac{r_{cr}}{R}\right)\text{cth}\left(\text{æ}r_{cr}\right) + \frac{\delta}{R}\frac{D_i}{D_e}\frac{r_{cr}^2}{R^2}\text{cth}\left(\text{æ}R\right)\right]\text{æ}R +$$

$$\left. + \frac{r_{cr}}{R}\left(1 - \frac{\delta}{R}\frac{D_i}{D_e}\right)\right\} \qquad /2.63/$$

where, δ = diffusion layer thickness.

In accordance with Eqs. 2.61 and 2.63, the diffusion model describes the kinetics of a variable order reaction /from zero to first/. The average order of the kinetic curve is determined by the parameters of the model. Thus, the zero order /i.e., $r_{cr} \to 0$/ will occur when the value $\hat{v} = \hat{v}_m$ which is found from Eq. 2.62 and then

$$L_s \geqslant L_s^{(0)} = L_{cr}\left[1 + \left(\text{æ}R\right)^2\left(\frac{1}{6} + \frac{1}{3}\frac{\delta}{R}\frac{D_i}{D_e}\right)\right] \qquad /2.64/$$

The reaction's order will be equal to one when

$$L_s \leqslant L_s^{(1)} = L_{cr}\left[1 + \frac{\delta}{R}\frac{D_i}{D_e}\left(\text{æ}R\text{cth}\text{æ}R - 1\right)\right] \qquad /2.65/$$

while

$$\hat{v}\left(L_s\right) = \hat{v}_m\frac{3\left(\text{æ}R\text{cth}\text{æ}R - 1\right)}{\text{æ}^2R^2\left[1 + \frac{\delta}{R}\frac{D_i}{D_e}\left(\text{æ}R\text{cth}\text{æ}R - 1\right)\right]} \cdot \frac{L_s}{L_{cr}} \qquad /2.66/$$

The kinetics of a single substrate oxidation can be approximated by a relationship which leads to the zero order reaction at high concentrations, L, and the first order reaction at low concentrations. The estimation of half-saturation constant, k_L, carried out by Vavilin and Vasiliev /1981/ yielded the following formula

$$k_L \approx L_{cr}\left[\left(\frac{1}{3} - \frac{1}{2}5/3 + \frac{1}{6}\frac{\delta}{R}\frac{D_i}{D_e}\right)\varkappa^2 R^2 + \frac{1}{2}\right] \qquad /2.67/$$

It should be noted that the function $\hat{v}$ (L_s) computed for the diffusion model, is well approximated by the Monod hyperbolic curve.

The actual shape of sludge particles is far from being spherical. Yet, the radius of a sphere which has the same volume - to - surface area ratio, V/S, as sludge particles can be chosen as an equivalent. Measurements of the ration, R $\sim$ V/S, for different types of sludge were show by Mueller et al., 1966, to have on equivalent radius of between 20 and 50 mμ while the average radius of individual bacteria was about 1 mμ. Thus, it is clear that activated sludge particles should have higher half-saturation constants, k_L /Fig. 2.19/.

The transition from zero to first order reactions for **large** particles occurs in a wide range of substrate concentration values while for small particles this range is quite narrow and the ratio $\hat{v}$ $(L)/\hat{v}$ $_m$ practically coincides with the form postulated for individual micro-organisms which form sludge particles. A more complicated relationship has been used for individual micro-organisms in Vavilin and Vasiliev /1982b/.

In general, mass-transfer processes do not affect sludge consumption by small particles when the radius of particles satisfies the condition $R \ll \frac{1}{\varkappa} = \left(YD_i L_{cr}/Q_d \mu_m\right)^{0.5}$.

On the contrary, when the parameter $\varkappa R \geqslant 1$, the function $\hat{v}$ (L) will be determined initially by diffusion processes while the half-saturation constant in the formal Monod model has the meaning of diffusion resistance of the substrate that penetrates sludge particles.

The zero order consumption of individual substrates by activated sludge has been stressed in several studies /Tischler and Eckenfelder, 1969; Chudova, 1969/. Observe that according to Eqs. 2.62 and 2.67, both maximum rate, $\hat{v}$ $_m$, and half-saturation

52

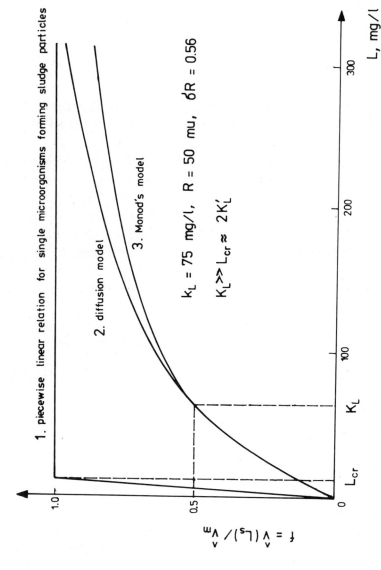

Fig. 2.19. Dependence of relative oxidation rate upon substrate concentration L for diffusion model /Eq. 2.2/ and Monod model /Eq. 2.3/, according to Vavilin and Vasiliev /1983a/

constant, k_L, are proportional to the dry biomass density, ρ_d, of respective micro-organisms in sludge particles. In the case of multi-component municipal wastewater, when the sludge contains numerous species of micro-organisms, the value ρ_d can be small for some species thus leading to the zero order kinetics. If simple substrates disappear according to the zero order reaction, the transfer function, $L_e = f(L_o, B, T)$, will be the same for both complete-mixing and plug-flow reactors. Approximately equal efficiency of multi-component municipal wastewater treatment in both types of reactors has been stressed by several authors /see, for example Toerber et al., 1974/.

The diffusional mechanism of simple substrate oxidation with biofilm is descussed in many studies /Atkinson and Fowler, 1974; La-Motta, 1975; Williamson and McCarty, 1976/. An estimate of half-saturation constant for a biofilm which consumes a single substrate-pollutant is given in Vavilin et al. /1980/. Instead of the Monod model, a good approximation to the diffusion model solution is given by the relationship which corresponds to the one-half order reaction with respect to substrate /Harremoës, 1977/.

2.3.6. Models of rough treatment of multi-component pollutants

Eq. 2.54 yields two types of behaviour of the function $\rho\left(\xi\right)$, namely, rough treatment $\left(\xi \gg k_L\right)$ and high-level treatment $\left(\xi \ll k_L\right)$. At high pollutant concentrations in the reactor, micro-organisms in the sludge or biofilm become "saturated" with predominant components of the substrate-pollutant, the treatment rate reaches a maximum and corresponds to the zero order reaction with respect to the general pollutant so that

$$v = -\frac{1}{B}\frac{dL}{dt} = \rho_m = \text{constant} \qquad /2.68/$$

The biomass increment is maximum in this case. In accordance with Eq. 2.68

$$L_e = L_o - \rho_m BT \qquad /2.69/$$

As the concentration of the pollutant in the reactor decreases, a good approximation to the generalized model /Eq. 2.52/ is given by a model of the order 1/n, i.e.

$$v = -\frac{1}{B}\frac{dL}{dt} = kL^{1/n} \qquad\qquad /2.70/$$

where, $L = L_o$ when $t = 0$, and
$\quad\quad\; L = L_e$ when $t = T$.

The solution of Eq. 2.70 is

$$L = \left(L_o^{\frac{n-1}{n}} - kB\,\frac{n-1}{n}\,t\right)^{\frac{n}{n-1}} \qquad\qquad /2.70a/$$

Substitution into Eq. 2.70 yields

$$\frac{dL}{dt} = -\frac{kB}{L_o^{1-1/n} - kB\left(1 - 1/n\right)t}L \qquad\qquad /2.71/$$

This gives the following formula for the rate constant of the quasi-first order reaction, with rough treatment

$$k_1 = \frac{kB}{L_o^{1-1/n} - kB\left(1 - 1/n\right)t} \qquad\qquad /2.72/$$

It is seen from Eq. 2.72 that Eq. 2.70 can only be used when

$$t < L_o^{1-1/n}/kB\left(1 - 1/n\right)$$

Instead of the $/n^{-1}/$ order models, the Monod model

$$v = -\frac{1}{B}\frac{dL}{dt} = \frac{\rho_m\,L}{k_L + L} \qquad\qquad /2.73/$$

can also be used in the case of rough treatment.

Alternatively, one may put $1/n = 1/2$ so that

$$v = -\frac{1}{B}\frac{dL}{dt} = kL^{1/2} \qquad\qquad /2.74/$$

In the case of rough treatment, good approximations to the generalized model /Eq. 2.50/ are given by the Monod model /Eq. 2.39/, or by the relationship

$$v = \frac{L_o - L_e}{BT} = kL_e^{1/2} \qquad\qquad /2.75/$$

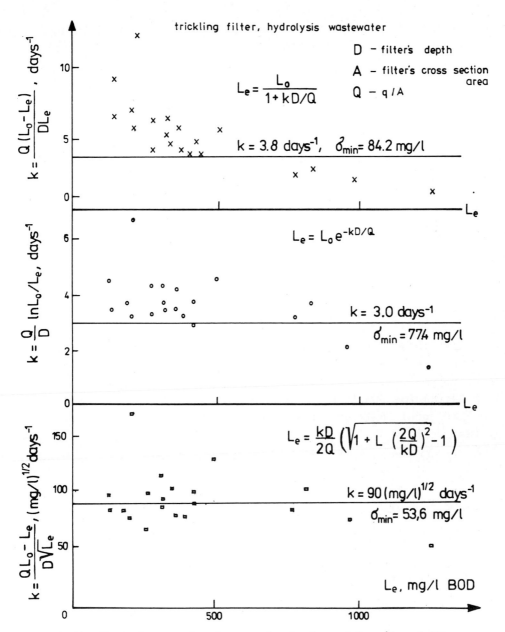

Fig. 2.20. Dependence of rate constants in models /Eq. 2.33;
2.31; 2.75/ upon L_e /rough treatment/ using, experimental data from Voronov et al. /1981/

Fig. 2.20 shows an example of data analysis for rough trickling filter treating hydrolysis industry wastewater. The traditional first order model does not fit so well as the model 2.75 having a one-half order. The rate constant of the quasi-first order is seen to be under-estimated for low L_e or L_e/L_o values and over-estimated for high L_e or L_e/L_o values.

2.3.7. Models of high-level treatment of multi-component pollutants

With high-level treatment when $\xi \ll k_L$ which involves transition to hardly-oxidized components, the generalized model /Eq. 2.52/ is written as

$$v = - \frac{1}{B} \frac{dL}{dt} = \rho(L) = \frac{\rho_m}{k_L} \left(\frac{L}{L_o}\right)^{n-1} L \qquad /2.76/$$

so that the rate constant of the quasi-first order

$$k_1 = \frac{\rho_m}{k_L} \left(\frac{L}{L_o}\right)^{n-1} \qquad /2.77/$$

is a decreasing function of treatment level. Eq. 2.76 is called the Fair model /Fair and Gayer, 1954/. Integration of Eq. 2.76 yields

$$L = L_o \sqrt[n-1]{1 + k(n-1) BT} \qquad /2.78/$$

where, $k = \rho_m/k_L$.

Substitution of Eq. 2.78 into Eq. 2.76 gives

$$\frac{dL}{dt} = - \frac{k B}{1 + k(n-1) Bt} L \qquad /2.79/$$

so that the rate constant of the quasi-first order with high-level treatment is

$$k_1 = \frac{k B}{1 + k(n-1) Bt} \qquad /2.80/$$

If $k(n-1) bT \gg 1$ the rate constant, k_1, does not depend upon $\varepsilon = kB$.

Instead of the generalized model /Eq. 2.50/, in the case of high-level treatment

$$v = \frac{L_o - L_e}{BT} = (L_e) = k \left(L_e/L_o \right)^{n-1} L_e \qquad\qquad /2.81/$$

An example is given in Fig. 2.14.

Rough estimates of the coefficients of Eq. 2.81 can be obtained from the linear form

$$\ln \left\{ \frac{L_o - L_e}{BTL_e} \right\} = \ln k + \left(n - 1 \right) \ln L_o/L_e.$$

Observe that Eq. 2.76 at $n = 2$ and Eq. 2.81 at $n = 1$ have the same transfer function $L_e = L_o\left(1 + kBT \right)$.

With hardly-oxidized reagents of multi-component municipal wastewater, specific forms of the generalized model /Eq. 2.52/ at $n = 3$ or the generalized model /Eq. 2.53/ at $n = 2$ are known to behave well. They give the transfer functions

$$L_e = L_o/ \sqrt{1 + 2 kBT} \qquad\qquad /2.82/$$

or

$$L_e = L_o \left\{ \frac{\sqrt{1 + 4k\ BT} - 1}{2\ k\ BT} \right\} \qquad\qquad /2.83/$$

respectively. Both forms yield approximately equal absolute errors, σ.

Examples of data analysis in cases of high-level treatment of multi-component municipal wastewater in an aeration tank, trickling filter and rotating disk are shown in Fig. 2.21-2.23. Traditional first order models /Eqs. 2.31 or 2.33/ do not fit so well as higher order models, for example, those given by Eqs. 2.82 or 2.83. It is seen from these figures that the value of the constant rate of the quasi-first order, k_1, is correlated with $\eta = L_e/L_o$. Thus, the application of the first order model leads to biased results: the reaction rate, v, is over-estimated when η is small and under-estimated when η is large.

When coefficients are graphically estimated from linear forms it is also necessary to determine δ or $\hat{\delta}$ since the correlation

58

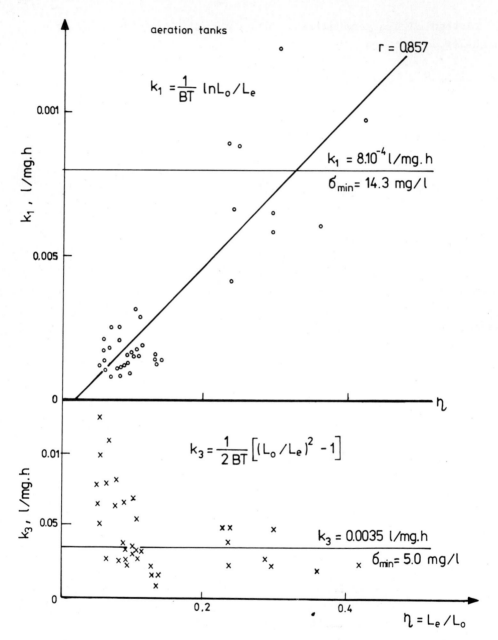

Fig. 2.21. Dependence of rate constant of the first $/k_1/$ and third $/k_3/$ order reaction upon $\eta = L_0/L_0$ for 36 plug-flow reactors treating municipal wastewater /fixed values k_1 and k_3 correspond to minimal variance/, using experimental data from Haseltine /1955/

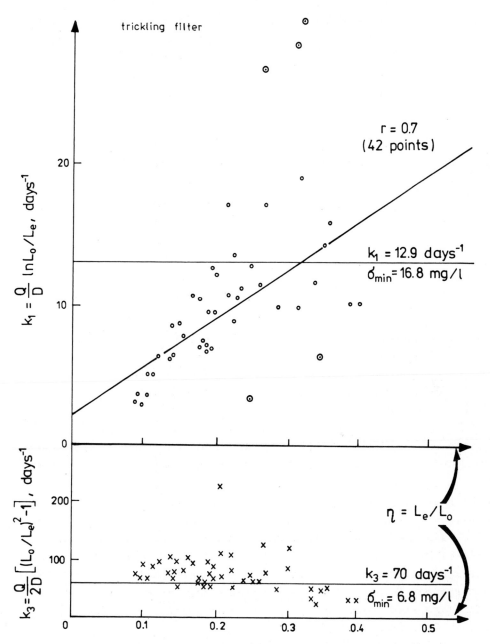

Fig. 2.22. Dependence of rate constant of the first /k_1/ and third /k_3/ order reaction upon $\eta = L_e/L_0$ for a trickling filter treating municipal wastewater /fixed values k_1 and k_3 correspond to minimal variance/, using experimental data from Keefer and Meisel /1952/

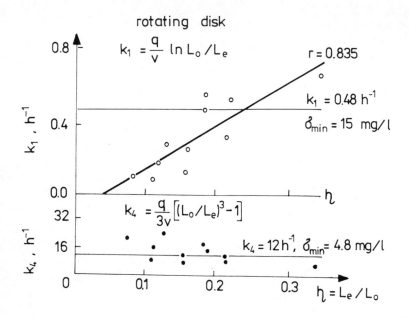

Fig. 2.23. Dependence of rate constant of the first /k_1/ and
fourth /k_4/ order reaction upon $\eta = L_e/L_o$ for a rotating disk /fixed values k_1 and k_4 correspond to minimal variance/ using experimental data from Benjes
/1979/

coefficient, r, which is close to unity, does not necessarily prove
the model's adequacy if the linear form has been chosen erroneou-
sly. An example is provided by Suschka /1980/.

High-level treatment of some types of domestic wastewater is
sometimes described on the basis of the Monod model /Eqs. 2.39 or
2.73/. Its coefficients are usually found graphically and the
value of half-saturation constant, k_L, is found to be rather high;
this means a transition to first order models /Eqs. 2.33 or 2.31/
with respect to substrate. Computations of σ show that these
models are also inferior to higher order models, e.g. those given
by Eqs. 2.82 or 2.83.

The value $\xi = L_e/L_o /L_o/^{n-1}$ serves as a criterion which enables
the differ entiation between rough and high-level treatments.
Some characteristic values of ξ used in computations of reactors
of different types are shown in Table 2.3.

Table 2.3 Values of criterion $\mathcal{M}$ and respective models for designing treatment plants /from Vavilin /1983a/

Reactor, sewage or wastewater	Experimental data, g/m³				Model		
	Mean value $\mathcal{M}= L_e^2/L_0$	$\mathcal{M}$ min	$\mathcal{M}$ max	Mean value L_e	Equation	σ min g/m³	Coefficients
1	2	3	4	5	6	7	8
Aeration tanks							
1. Laboratory aeration tank, pepton /Adams et al., 1975/	1.9	0.96	2.5	25	83	1.8 /5.7/	$k=9.4 \cdot 10^{-7} m^3/g \cdot s^{-1}$
2. Laboratory aeration tank, industrial wastewater /Paolini & Varioli, 1982/	16	0.4	68	123	40	18 /24/	$\rho_m = 6.8 \cdot 10^{-5} s^{-1}$ $k_L = 700 \; g/m^3$
Trickling filters							
3. Municipal treatment plant /Keefer & Meisel, 1952/	6.3	0.84	20	26	83	6.6 /9.6/	$k_\alpha = 1.28 \cdot 10^{-3} s^{-1}$
4. Pilot plant, hydrolisis wastewater /Voronov et al., 1981	250	17.4	1.046	395	40	52 /84/	$\rho_m = 8.9 \cdot 10^{-4} g/m^2 \cdot s^{-1}$ $k_L = 510 \; g/m^3$

Table 2.3 /continued/

1	2	3	4	5	6	7	8
Rotating disks							
5. Five different rotating disks, municipal sewage water /Benjes, 1979/	5.2	0.75	15.1	25	84	4.5 /7.0/	$k\alpha = 1.56 \cdot 10^{-3}$ s^{-1}
6. Laboratory rotating disk, industrial wastewater /Paolini et al., 1979/	173	0.3	848	384	40	50 /195/	$\rho_m \alpha = 3.15 \cdot 10^{-2}$ $g/m^2 \cdot s^{-1}$ $k_L = 54$ g/m^3

Note: All waste or sewage water except pepton is measured in BOD units /pepton in TOC units/. Minimal values σ'_{min} for first-order Eq /35/ are given in parenthesis.

2.4. DEPENDENCE OF BIOLOGICAL TREATMENT RATE ON TECHNOLOGICAL CHARACTERISTICS OF TREATMENT PLANTS

The design of biological treatment plants should take into account a wide set of possible technological regimes of operation; a range of parameter values should be computed /Vavilin, 1982d/. This point is further discussed below.

2.4.1. Mean retention time of wastewater in the reactor

Kinetics of oxidation of generalized organic matter, measured in BOD, COD or TOC units, can be formed as a sum of kinetic curves of simple substrates´ oxidation. Eckenfelder and his co-authors /Tischler and Eckenfelder, 1969; Siber and Eckenfelder, 1980/ showed that simple substrates are oxidized simultaneously and that kinetics of their oxidation frequently follows a zero order reaction with respect to pollutant concentration. Consequently, for the i-th simple substrate:

$$L_i = L_{io} - k_i \ BT \quad \text{if} \quad 0 \leqslant t \leqslant L_{io}/k_i \ B$$

$$L_o = 0 \ \text{if} \ t > L_{io}/k_i \ B$$

where, L_{io} = initial concentration of the i-th pollutant.

B = micro-organisms concentration, and

k_i = rate constant of zero order oxidation reaction corresponding to the i-th component.

Let m simple substrates be oxidized. Arrange them according to the order of their disappearance, i = 1, 2,...., 1,..., m - 1, m. The net kinetic curve is formed of m linear sections. To any 1-th section such that $1 \geqslant 1$ and $1 \leqslant m$ corresponds the value

$$L_e^\Sigma = \sum_{i=1}^{m} \left(L_{oi} - \frac{k_i \ L_{o,1-1}}{k_{1-1}} \right) - \sum_{i=1}^{m} k_i \ Bt$$

$$= \sum_{i=1}^{m} \left\{ L_{oi} - k_i \left[L_{q1-1}/k_{1-1} + Bt \right] \right\} \ .$$

The relationship holds if $L_{o,1-1}/k_{1-1}\,B \leqslant t_1 \leqslant L_{o1}/k_1\,B$

where, $L_{oo} = 0$ and $L_{oo}/k_o = 0$.

The net kinetic curve is determined with m coefficients, k_i, and m initial conditions, L_{oi}. This dependence ban be expressed by an equation which clearly presents a formal approximation to the kinetic curve /Vavilin, 1980/, $dL/dt = - \rho\,/B,L/$.

The following equation taken from formal chemical kinetics is often used: $\rho\,/B,L/ = k\,BL^n$.

If t is small, that is $t < L_{01}/k_1\,B$, the reaction's order with respect to the generalized pollutant $L^{\Sigma} = \sum\limits_{i=1}^{m} L_i$ equals zero /n = 0/, namely:

$$L^{\Sigma}\,(t) = \sum_{i=1}^{m} \left(L_{oi} - k_i\,Bt\right) = L_o - \rho_m\,Bt$$

where, $\rho_m = \sum\limits_{i=1}^{m} k_i$, and $L_o = \sum\limits_{i=1}^{m} L_{oi}$.

As the length of the time interval, t, grows the share of "re-processed" hardly-oxidized components also grows. This calls for higher reaction orders, n = 1, 2,...

A growing actual order of reaction resulting from lower substrate concentration /longer retention times/ is "automatically" taken into account in the generalized model /Eq. 2.52/. Fig. 2.24 shows kinetic curves of municipal sewage oxidation corresponding to the model represented by Eq. 2.52.

In continuous flow reactors, the value $a_i = k_i B$ cannot be assumed constant for different retention times, T, because a new retention time means a new stationary distribution of species of micro-organisms. This problem will be discussed in the next chapter which deals with the multi-component/multi-species model.

When designing reactors which oxidize a definite type of municipal wastewater there is usually a given range of values for the retention time T: $T_{min} \leqslant T \leqslant T_{max}$.

If $\Delta T = T_{max} - T_{min}$ is small then, other conditions being equal, the different models "work" more or less equally effectively /that is, they give almost equal values of σ /.

Rough treatment which corresponds to a relatively short retention time, T, can be described by a model of 1/n-th order given

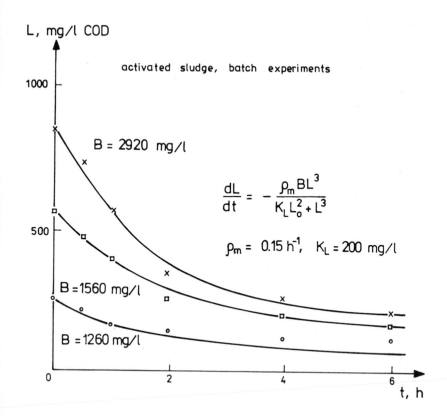

Fig. 2.24. Kinetic curves of municipal wastewater oxidation
according to model /Eq. 2.52/, using experimental
data from Chudoba /1969/

by Eq. 2.70. In this case, a longer interval, Δ T, and longer
times, T_{min} and T_{max}, result in a higher rate constant of the
quasi-first order: this is in accordance with Eq. 2.72. If the
1/n-th order model /Eq. 2.70/ is changed to the first order model
/Eq. 2.31/, the results become biased; the value L_e is under-estimated when $T \approx T_{min}$ and over-estimated when $T \approx T_{max}$ /Fig. 2.25/.

High-level treatment which corresponds to relatively long
retention time, T, can be described by a model of n-th order
given by Eq. 2.76. In this case, a longer interval Δ T, and longer
times, T_{min} and T_{max}, lead to a smaller rate constant of the quasi-
first order reaction: this is in accordance with Eq. 2.80. If the
n-th order model /Eq. 2.76/ is changed to the first order model

66

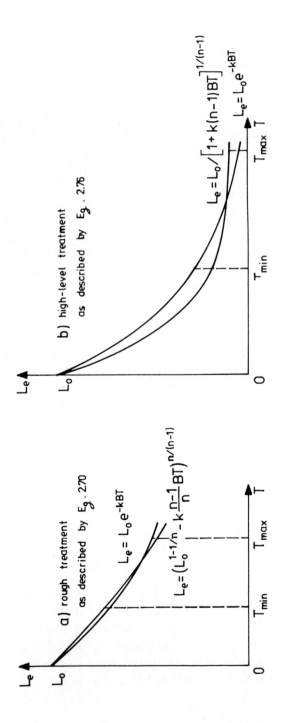

Fig. 2.25. Approximating rough and high-level treatment with first order model /Eq. 2.31/ and the "source" of bias.

/Eq. 2.31/ the results become biased: the value L_e is over-estimated when $T \approx T_{min}$ and under-estimated when $T \approx T_{max}$ /Fig. 2.25/.

2.4.2. Concentration of micro-organisms in the reactor

When designing reactors, there is usually given range of values
B_1: $B_{min} \leqslant B \leqslant B_{max}$.

If $\triangle B = B_{max} - B_{min}$ is small then, other conditions being
equal, the different models "work" more or less equally effectively /that is, they give almost equal values σ /.

In the case of high-level treatment which corresponds to relatively high values of B, the rate constant of the quasi-first
order given by Eq. 2.80 does not depend upon B if $k \left(n-1\right) BT \gg 1$:
this is in accordance with Eq. 2.79.

When designing aeration tanks for use the Eckenfelder model
/Eq. 2.33/ is sometimes changed to the McKinney /1962/ model which
satisfies the equation $k^{Eck} B = k^{Mck}$ = constant.

Thus, a higher value of B means that relatively hardly-oxidized
reagents are "captured" and this in turn results in a smaller rate
constant. This explains why the McKinney model may give practically the same value of σ as the Eckenfelder's. However, with
rough treatment, both these models are markedly inferior to higher
order models /Vavilin, 1981a/.

Similarly, when designing fixed-film reactors the first order
model with specific media surface does not necessarily lead to
much smaller values of σ.

The first order models /Eq. 2.31 and 2.33/ lead to biased
results of the same character as those discussed for retention
time, T.

2.4.3. Input pollutant concentration

In accordance with generalized models /Eqs. 2.50 and 2.52/, the
rate of treatment grows with higher values of L_o. However, the
effect on the rate of treatment becomes stronger for high-level
treatment /see Särner, 1981/. For the zero order model /Eq. 2.69/,
the treatment rate does not depend upon L_o. In the case of rough
treatment described by Eq. 2.70, the rate constant of the guasi-
first order k_1 is a decreasing function of L_o in accordance with
Eq. 2.72. This effect has been taken into in formal models 2.37

and 2.38 of Grau, Oleszkiewicz and Eckenfelder. These models are
in better agreement with experimental data than traditional first
order models 2.34 and 2.35.

In the case of high-level treatment, the rate constant of the
quasi-first order k_1 does not depend upon L_o, in accordance with
Eq. 2.80.

2.4.4. <u>Temperature and concentration of dissolved oxygen</u>

In generalized models /Eqs. 2.50 and 2.52/, the maximum rate
should be regarded as a function of temperature, t^o, and concentration, C, of dissolved oxygen, i.e.: $\rho_m = \rho_m \left(t^o, C\right)$.

Specifically, the well-known Arrhenius relationship, leads to
the equation

$$\rho_m = \rho_m \; \mathcal{æ}t^o - \hat{t} \tag{/2.84/}$$

where, $\mathcal{æ}$ = temperature coefficient, and
$\hat{t}$ = fixed temperature usually 20^oC.

The rate of biological treatment increases along with temperature but the effect of temperature is relatively less marked with
high-level treatment. Since $k = \rho_m/k_L$, the dependence of the rate
constant of the quasi-first order upon temperature is much stronger, in the case of rough treatment than that of high-level,
treatment; this is in accordance with Eqs. 2.72, 2.80.

Fig. 2.26 shows the results of experiments with a laboratory
rotating biological contactor at different temperatures. It is
easily seen that at high values $\eta = L_e/L_o$, which correspond to
small T = V/q and, consequently, high organic loading $F = L_o/\alpha T$,
the rate constant of the quasi-first order is strongly dependent
upon temperature /higher values of $\mathcal{æ}$/. The dependence is weaker
at higher temperatures. In general, it can be seen that the quasi-
first order model leads to biased results: the rate is under-estimated for high η and over-estimated for low η . With the first
order model /Fig. 2.26a/ this corection for temperature results
in slightly smaller values of σ .

More adequate models are not biased and, if the model is good,
correction for temperature can lead to a much smaller σ /see
Fig. 2.26b/.

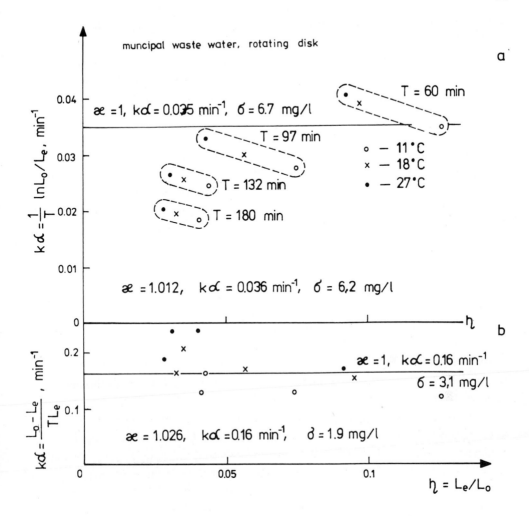

Fig. 2.26. Dependence of rate constants upon $\eta = L_o/L_0$ at
   different temperatures according to model /Eq. 2.76/:
   a - n = 1, b - n = 2, using experimental data from
   Ellis and Banaga /1979/

In traditional first order models, the value æ of correction
for temperature is usually fixed in accordance with a given type
of treatment plant. Thus, for aeration tanks it is often taken
bo be 1.047 and for trickling filters, 1.035. At the same time,
it has been found in several studies that this correction factor
depends upon the operational regime. Specifically, for activated
sludge systems, æ increases with loading /Eckenfelder, 1967;

Wuhrmann, 1968; Novack, 1974/. This fact is easily derived from Eq. 2.80.

Since the maximum specific rate of treatment, ρ_m, increases with the concentration of dissolved oxygen, C, it should be noted that the effect of C on the rate of treatment is smaller for high-level treatment.

Thus, traditional models of biological treatment which are applied nowadays, may give biased results when the ranges of parameter values are wide. The latter include retention time of wastewater in the reactor, T, concentration of micro-organisms, B, input pollutant concentration, L_o, temperature, t^o, and dissolved oxygen concentration, C. The error can be avioded by using generalized models /Vavilin, 1982d/.

2.5. DEPENDENCE OF BIOLOGICAL TREATMENT RATE ON MICRO-ORGANISMS' SPECIES COMPOSITION OF ACTIVATED SLUDGE OR BIOFILM

It has already been stated that the type of treated water, physico-chemical conditions and technological regime of the treatment plant determine both the species composition and relative importance of individual species in the microflora of suspended growth or biofilm of fixed-film reactors.

It has been proved in numerous studies that the species composition changes when the operational regime of treatment plant is changed. Thus, Lighthart and Loew /1972/ and Günter /1976/ found that the species composition depends on the loading and temperature, t^o. The wastewater before it comes to the aeration tank, normally contains all the bacteria which will multiply in the activated sludge during biological treatment.

As shown in Pike and Carrington /1972/, the population formed in an activated sludge system is simpler than that formed in fixed-film reactors. The spectrum of treated components in a fixed-film reactor is changing downstream of the flow of liquid waste in the reactor thus changing the species composition in the biofilm /Rogovskaya, 1967; Torpey et al., 1971/.

2.5.1. Aggregated variables and integral parameters in formal models of biological treatment

Biological treatment systems form mixed populations of micro-organisms. Yet, their traditional "engineering" descriptions are

based on such aggregated variables as the total mass of micro-
organisms and net pollutant concentration expressed in BOD, COD or
TOC units. The total mass of micro-organisms is regarded here as
a parameter since it is assumed to remain constant at a sufficiently high level.

The introduction of integral parameters allows one to construct
models with fewer coefficients. Consider, as an example, the Grau
model /Eq. 2.45/. Introducing an integral parameter

$$F = L_o/BT \qquad\qquad /2.85/$$

which is called organic loading, and taking into account Eq. 2.38,
Eq. 2.45 can be re-written as:

$$v = \frac{L_o - L_e}{BT} = \frac{\rho_m F}{\rho_m + F} \qquad\qquad /2.86/$$

If $F \ll \rho_m$, i.e. $L_e \ll L_o$, this equation is transformed to the
following identity:

$$v = \frac{L_o - L_e}{BT} \approx F = \frac{L_o}{BT} \qquad\qquad /2.87/$$

while for $F \gg \rho_m$:

$$v \approx \rho_m \qquad\qquad /2.88/$$

Thus, the input-output relationship derived from the Grau model is
based on the integral parameter given by Eq. 2.85. According to
Eq. 2.86, equal organic loadings on the plant mean equal treatment rates.

The Grau model /Eq. 2.45/ contains only one coefficient while
the empirical model /Eq. 2.49/ has four coefficients.

The generalized models /Eqs. 2.50 and 2.52/ involve the integral parameter given by Eq. 2.53. The treatment rate remains constant when this parameter does not change. It has already been
shown above that a less complicated model can be adopted to compute reactors in accordance with the range of variation of ξ
/Table 2.3/.

An example of the application of dimensionless parameters to
compute biological treatment reactors is given by Tuček et al.

72

/1971/. Note that by designating $\alpha = v/\rho_m$, $\beta = \rho_m/F$, $\gamma = k_L/L_e$, and $\delta = k_L/\xi$, one can arrive at equations which give relationships between the dimensionless parameters contained in Eqs. 2.86, 2.39, and 2.50.

An important integral characteristic is the biomass age, or mean retention time of biomass in the reactor

$$\theta = \frac{B}{\Delta B/T}$$
/2.89/

where, ΔB = the increment of micro-organisms concentration during the time interval, T.

The age of sludge is used to derive its sedimentation characteristics which, in turn, affect the concentration of sludge in the aeration tank /Vasilev and Vavilin, 1982/.

In order to estimate the biomass increment, Eq. 2.51 is usually used. It can be re-written as

$$\mu = \frac{\Delta B}{BT} = \frac{1}{\theta} = Y\frac{\Delta L}{BT} - k_d = Y_v - k_d$$
/2.90/

where, μ = specific growth rate of biomass.

Applying this equation to Eq. 2.45 one obtains, according to Vandevenne and Eckenfelder /1981/:

$$L_e = \frac{L_o(1+k_d\theta)}{Y\rho_m\theta}$$
/2.91/

while, in the case of Eq. 2.50

$$L_e = \left\{ k_L L_o^{n-1} \frac{1+k_d\theta}{(Y\rho_m - k_d)\theta - 1} \right\}^{1/n}$$
/2.92/

Assuming the rate constant, k, in the traditional Eckenfelder model /Eq. 2.33/ to be a decreasing function of age, θ, one can improve this model substantially, that is, obtain much smaller values of δ /Vavilin, 1981a/. A lower activity of aged sludge is usually explained as the result of the accumulation of inert residuals. However, another reason is the different "spectra" of treated components which correspond to different ages of sludge. Thus, easily- or hardly-oxidized "re-processed" components correspond to the "young" or "old" sludge /i.e., to high or low organic loadings/ respectibely.

The mean retention time of biomass in the reactor can also be used to compute fixed-film reactors /Koug, 1979/.

2.5.2. Some consequences of a change in the spectrum of treated substrates and species composition of micro-organisms

1. With the transition to a new type of treated organic pollutants, new chemical physical conditions or a new technological regime of treatment plant, the normal operational regime of the plant will be restored only after some time interval /the adaptation time/. During this interval, there is also transition to a new stable distribution of species in the microflora of activated sludge or biofilm.

2. It is well known that the rate of biological treatment depends on the sludge viability, McKinney gives values of ν between 30 and 50 per cent for domestic wastewater and 10 per cent under the conditions of extended aeration. By modyfying the Sinclair and Topiwala model, Grady and Roper /1974/ found the sludge viability and the general decay rate constant to be decreasing functions of age, i.e.

$$\nu = \frac{1/\theta + b_n}{1/\theta + b_n + d} \qquad /2.93/$$

$$k_d = \left(\frac{1}{\theta} + b_n\right) \cdot \frac{1/\theta + b_v + d}{1/\theta + b_n + d} - \frac{1}{\theta} \qquad /2.94/$$

where, b_n and b_v = auto-oxidation constants of live and dead, respectively, cells, and

d = decay rate constant.

In studying the dependence of ν upon the specific rate of sludge growth $\mu = 1/\theta$, Weddle and Jenkins /1971/ showed that an abrupt increase in μ occurred at some values of ν. Thus, at practically the same sludge viability ν, there is a sudden increace in the specific rate of biomass growth and, consequently, the treatment rate $v = \Delta L/BT = (L_o - L_e)/BT$.

It is natural to explain this phenomenon by the transition to treatment of easily-oxidized components, which is followed by a respective change in the species composition in activates sludge.

A similar dependence has been observed experimentally for such biochemical indicators as the content of ATP in sludge, and dehydrogenase activity /Weddle and Jenkins, 1971; Nelson, and Lawrance, 1980/.

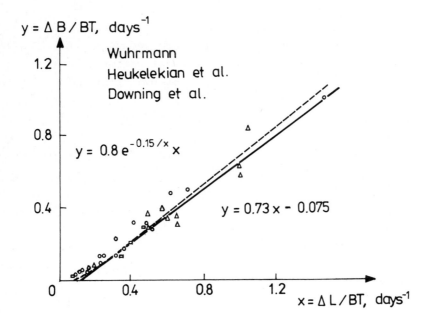

$y = \Delta B / BT$, days^{-1}

1.2

Wuhrmann
Heukelekian et al.
Downing et al.

0.8

$y = 0.8\,e^{-0.15/x}\,x$

0.4

$y = 0.73\,x - 0.075$

0   0.4   0.8   1.2

$x = \Delta L / BT$, days^{-1}

Fig. 2.27. Function $\mu = \mu\,(v)$ for domestic sewage according to /Eqs. 2.90, 2.95/ derived by Silveston /1967/ and the author respectively.

3. The increment of sludge biomass or the decrement of generalized pollutant is usually related to Eq. 2.51 or, in another form, to Eq. 2.90. At the same time, Gaudy and his colleagues /see for example Ramanathan and Gaudy, 1972/ noticed that such coefficients as Y should not be assumed constant for heterogenous cultures. The dependence $\mu = \Delta B/BT = f\,(v)$ is shown in Fig. 2.27 for municipal wastewater. It should be noted that for small μ and v corresponding to long age θ /small loading/, the linear dependence /Eq. 2.90/ does not hold. For the observed biomass yield coefficient, Y_{obs} when $\mu = Y_{obs}v$, Sherrard and Schroeder /1970/ suggested the relationship $Y_{obs} = Y^{*}\exp\,(-k\theta)$,

where, k = constant. Then $Y_{obs} \rightarrow 0$ as $\theta \rightarrow \infty$, and $Y_{obs} \rightarrow Y^{\divideontimes}$ as $\theta \rightarrow 0$.

The relationship

$$\mu = Y^{\divideontimes} e^{-k/v} v,$$ /2.95/

which describes experimental data better for small values of v /see Fig. 2.27/, can be used instead of linear dependence /Eq. 2.90/. Its coefficients are found from the linear form

$$\ln \mu /v = \ln Y^{\divideontimes} - k/v.$$

If the biomass yield coefficient, Y, can be assumed approximately constant, then the coefficient k_d proves to be a decreasing function of age θ; this has been noticed in many studies /for example Goodman and Englande, 1974/. Note that the Grady and Roper model mentioned above, which takes into account the accumulation of inert biomass with growing age, is only capable of describing a slowly decreasing function $k_d = k_d (\theta)$. This fact is seen from Eq. 2.94. A steeper dependence can be obtained by also taking into account the fact that a growing age, θ, leads to a transition to a new species distribution in sludge with smaller characteristic values of k_d.

4. The oxygen consumption and decrease in the concentration of generalized pollutant are usually related through the equation

$$R = \frac{\triangle O_2}{BT} = a + a'$$ /2.96/

where, a = coefficient of oxygen utilization during the process of biomass production, and

a' = oxygen consumption rate necessary to maintain the biomass activity.

The dependence given by Eq. 2.96 is shown in Fig. 2.28a,b and for both total and viable biomass of sludge. The coefficient a' is seen to be much higher for viable biomass while the coefficient a is much higher for total biomass. This is easily understandable: in the case of viable biomass, the growing function $R = R (\mu)$ can be explained by the transition to treatment of relatively easily-

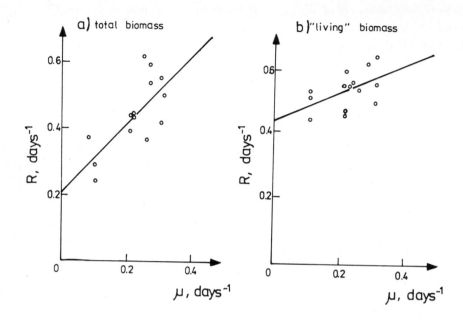

Fig. 2.28. Dependence /Eq. 2.96/ for /a/ total biomass and /b/
viable biomass, according to Benefield et al. /1979/

oxidized substrates with growing organic loading /i.e. shorter
sludge age θ/ and respective changes in the species composition;
in the case of total biomass, this effect is accompanied by
a change in the concentration of live micro-organisms thus leading
to higher values of a.

5. Under difinite conditions, an activated sludge treatment
system can fail. A typical example is given by the sludge bulking
phenomenon when sludge becomes difficult to settle,so that its
concentration in the aeration tank drops abruptly.

Sludge bulking occurs when the organic loading on the sludge
is either large /oxygen shortage/ or small /Palm et al., 1980/.
Hardly-oxidized components are mainly consumed in the latter case.
Sludge bulking is often related to disproportionately high multi-
plication of filamentous forms of micro-organisms, though some
authors state that specific species of micro-organisms are responsible for this phenomenon.

6. Two major techniques used to intensify the biological
treatment process are:

a. increasing the biomass of active micro-organisms, and

b. specialization of microbic cenoses to consume individual groups of organic components contained in the wastewater.

This can be achieved through a spatial succesion of micro-organisms fixed along the flow of liquid waste when a plug-flow reactor is filled with a special type of media.

2.5.3. Model of activated sludge or biofilm adaptation to a multi-component pollutant

In Vavilin and Vasiliev /1982a, 1983/, models were proposed which describe the adaptation of activated sludge microflora to a multi-component pollutant. The models were based on the following principles:

1. The micro-organisms contained in the activated sludge are highly specialized, that is, each component of the pollutant is oxidized by only one species of bacteria;

2. The oxidation of individual components follows the Monod or zero order reaction, and

3. The decay of active microflora leads to the accumulation of inert biomass which is oxidized by an auto-oxidation process.

In describing cycling batch experiments as well as the process of continuous biological treatment in plug-flow reactors, the following system of differential equations was considered:

$$dB_i^{(n)} /dt = \left(\mu_i - k_{di} \right) B_i^{(n)}$$

$$dL_i^{(n)} /dt = - \mu_i B_i^{(n)} /Y_i$$

$$d\tilde{B}_i^{(n)} /dt = \alpha_i k_{di} B_i^{(n)}. \hspace{3cm} /2.97/$$

where,

$L_i^{(n)}$ = concentration of the i-th component of the pollutant,

$B_i^{(n)}$ = concentration of the i-th species of the microflora which oxidizes the i-th component of the
pollutant during the n-th period of sludge retention in the reactor /the n-th cycle/,

$\tilde{B}_i^{(n)}$ = inert biomass concentration attained after the
i-th species has decayed,

78

μ_i = specific growth rate of micro-organisms belonging
to the i-th species,

k_{di} = decay and auto-oxidation rate constant,

α_i = coefficient of transformation of active biomass into
inert biomass, and

Y_i = biomass yield coefficient.

If there are N-1 components, the total concentration of activated sludge and pollutant are:

$$L^{(n)} = \sum_{i=1}^{N-1} L_i^{(n)} \qquad\qquad /2.98/$$

$$B^{(n)} = \sum_{i=1}^{N} B_i^{(n)} \qquad\qquad /2.99/$$

where, B_N = inert biomass concentration.

The specific growth rate of micro-organisms belonging to the i-th species can be described by the hyperbolic Monod function

$$\mu_i = \mu_{mi} L_i / (k_{Li} + L_i) \qquad\qquad /2.100/$$

where, μ_{mi} = maximum specific growth rate corresponding to the i-th species, and

k_{Li} = half-saturation constant.

If the half-saturation constant is small so that $k_{Li} \ll L_{0i}$, the reaction's order will be close to zero for a wide range of substrate concentrations: in this case $\mu_i \approx \mu_{mi}$.

The relative content of the i-th species of micro-organisms reached by the end of the n-th cycle is

$$P_i^{(n)} = \frac{B_{10}^{(n)} + \Delta B_i^{(n)}}{B_0 + \sum_{i=1}^{n} \Delta B_i} \qquad\qquad /2.101/$$

As the species are impossible to discern during the process of increasing biomass removal, the concentration of the i-th species by the beginning of the $(n+1)$-th cycle is

$$B_{10}^{(n+1)} = P_i^{(n)} B_0 \qquad\qquad /2.102/$$

Knowing the values μ_i, k_{di}, Y_i, α_i and the content of waste-water and assuming an initial vector $P_i^{(0)}$ of species distribution in the sludge microflora, it is possible to determine the composition of species in the sludge after n cycles. The stable distribution can be estimated if n is sufficiently large, say $n > 100$. It is also possible to trace the change in the kinetic curve $L(t)$ occurring during the process of adaptation of activated sludge; this characteristic is often studied in experiments.

Fig. 2.29 shows the kinetic curve of pollutant oxidation in unadapted sludge when the relative concentrations coincide for all species of micro-organisms, and the curves achieved after n cycles of adaptation. Respectively, the relative distribution of species is different for the latter case. In the formal description of kinetic curves $L(t)$, the reaction's order with respect to the generalized substrate, L, is growing **during** the process of adaptation.

An important quantity which determines the treatment process is the loading on sludge, $F = L_0/BT$. The model of adaptation discussed above shows that the share of species that treat hardly-oxidized components increases as F decreases /see Fig. 2.30/. At the same time, the inert biomass concentration also increases and the kinetic curves for the generalized pollutant are changed accordingly.

For a constant loading, the distribution of species changes only slightly even under different values of L_0, B and T, and therefore, this parameter is indicative of the net activity of the sludge.

Variations of some characteristics of activated sludge are shown in Figs. 2.31, 2.32, and 2.33 in accordance with model 2.97. Their comparison with respective experimental curves mentioned in the previous section shows that the model /Eq. 2.97/ which takes into account both variations in the species composition and accumulation of inert biomass in activated sludge can be used to give a qualitative description of experimentally observed phenomena.

A multi-component/multi-species model for biofilm is discussed in Vavilin and Vasiliev /1983/.

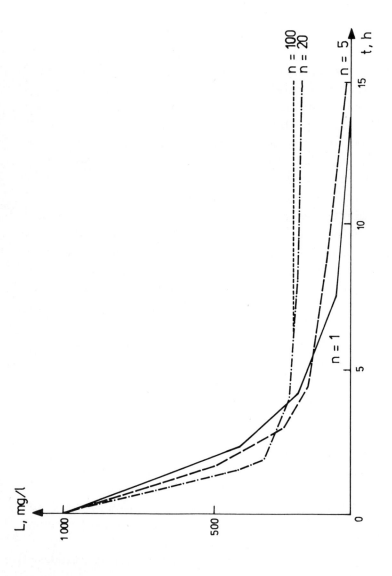

Fig. 2.29. Dynamics of multi-component substrate oxidation for different degrees of sludge adaptation /for inadapted sludge n = 1/ according to Vavilin and Vasiliev /1983b/

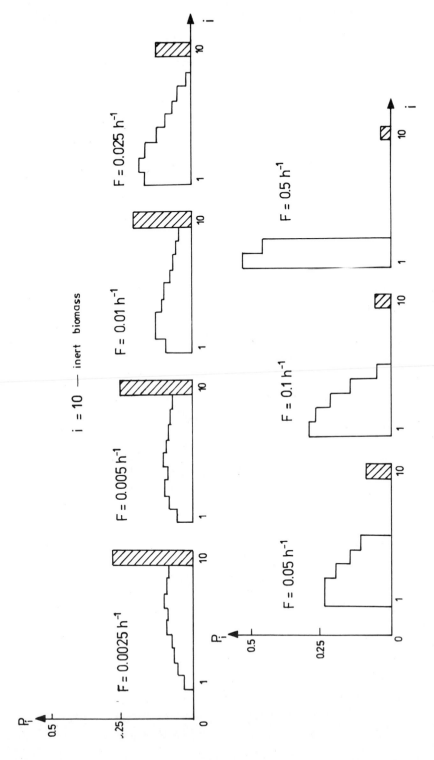

Fig. 2.30. Stationary distribution of relative species content for different organic loadings, according to Vavilin and Vasiliev /1983b/

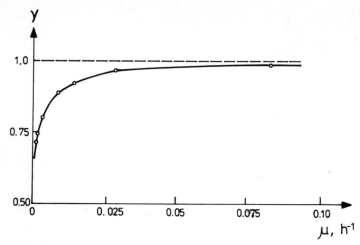

Fig. 2.31. Dependence of relative content of viable biomass
upon biomass specific growth rate according to multi-
component/multi-species model, according to Vavilin
/1983a/

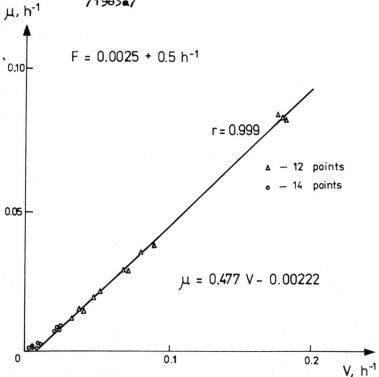

Fig. 2.32. Dependence /2.90/ according to multi-species/
multi-component model for a wide range of organic
loadings, according to Vavilin /1983a/

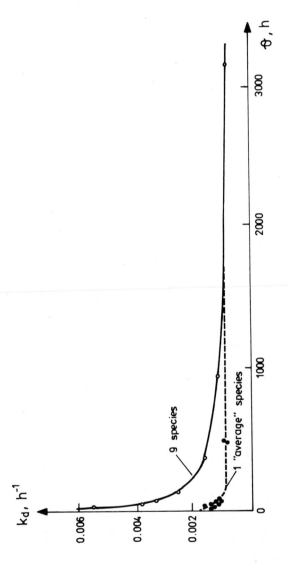

Fig. 2.33. Behavior of total biomass decay coefficient according to multi-component/multi-species model, according to Vavilin /1983a/

84

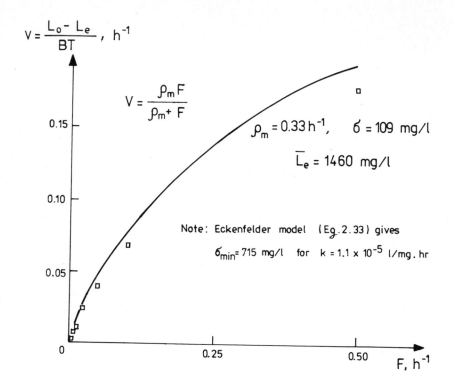

$$V = \frac{L_o - L_e}{BT}, \ h^{-1}$$

$$V = \frac{\rho_m F}{\rho_m + F}$$

$\rho_m = 0.33\,h^{-1}, \quad \delta = 109\ mg/l$

$\overline{L}_e = 1460\ mg/l$

Note: Eckenfelder model (Eg. 2.33) gives

$\delta_{min} = 715\ mg/l \quad for \quad k = 1.1 \times 10^{-5}\ l/mg.hr$

Fig. 2.34. Dependence /2.36/ for data obtained through the multi-component/multi-species model, according to Vavilin /1983a/

2.5.4. Multi-component/multi-species model and formal models of biological treatment

Fig. 2.34 shows the results of analysis of data obtained in accordance with the formal Grau model /Eq. 2.45/ for a multi-component/multi-species model. The simple Grau model a fairly good approximation to the multi-component/multi-species model. In analyzing the results obtained with the latter model, it has been stressed that the distribution of micro-organisms' species can be characterized by the value of loading, $F = L_0/BT$. Consequently, the rate of total pollutant oxidation is also determined with loading F. The formal Grau model presents a successful description of function $v = v/F/$ obtained for the multi-component/multi-species model.

Both the Grau model /Eq. 2.45/ and the Eckenfelder model /Eq. 2.33/ are first order models with respect to the total con-

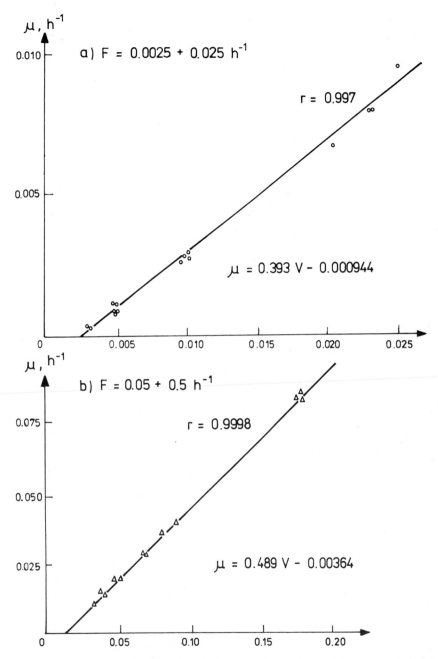

Fig. 2.35. Dependence /2.90/ according to multi-component/
multi-species model for /a/ low and /b/ high
organic loadings, according to Vavilin /1983a/

centration of the pollutant. The analysis of kinetic curves shown
in Fig. 2.29 reveals a better agreement between the experiment
and high order formal models such as, those given by Eqs. 2.50
and 2.52.

The increment of total biomass as a function of total pollutant loss is shown for different ranges of organic loading, F, in
Fig. 2.35. The linear dependence given by Eq. 2.90 is adopted. It
is seen that at large F the biomass yield coefficient is slightly
higher while the biomass decay coefficient is much higher. At the
same time, when the data corresponding to low loadings, F, are
added to those which correspond to high loadings, the value of Y,
remains practically constant but the value of k_d is changed substantially /Fig. 2.35b and Fig. 2.32/.

Thus, simple formal models of biological treatment may present
both successful and poor approximations to complex multi-compo-
nent/multi-species models /Vavilin, 1983a/.

2.6. CONCLUDING REMARKS: FROM MODELS TO THEORY OF BIOLOGICAL
   TREATMENT

Mathematical modelling of natural processes usually involves
several stages:
1. abstract modelling when the model developed on the basis
   of general considerations and common sense is little compared with experimental data;
2. concrete modelling when the models constructed on the basis
   of experimental data are compared with each other according
   to a chosen criterion and the best models are chosen; and
3. theoretical modelling which enables one to understand the
   mechanisms of the phenomena under consideration.

This last stage is the most important purpose of mathematical
modelling.

Any real natural system is multi-dimensional, but in formali-
zing existing knowledge, this multi-dimensional space is inevita-
bly, though not necessarily conciously reduced, so that just a few
variables and coefficients are kept. The universal success in far
from being warranted.

When designing biological treatment plants it is common to use
formal models of biological treatment. The penalty for their simplicity, that is, for the small number of coefficients they con-

tain, is the essential inadequacy of formal models. Conversion
to models of a new type which are capable of taking into account
the heterogeneity of the pollutant and biomass, the system´s inner
structure, allows one to solve finer problems and leads finally
to an understanding of the mechanism of biological treatment.

Attempts to "explain" non-linear formal models revealed the
necessity to create a multi-component/multi-species model of biological treatment. Clearly, even such a complicated model presents only a rough approximation to the phenomena under study.

Thus, it has been assumed in the model that the characteristics
of individual species remain unaffected by changes in external
parameters; this is obviously far from reality.

The traditional formalism derived from chemical kinetics does
not necessarily work for systems possessing an inner structure.
Even for pure cultures of micro-organisms, batch experiments
cannot be equated with continous experiments. This is demonstrated most clearly in the case of multi-species system. Under the
stationary conditions of a continuous experiment, a specific
stable distribution of species exists which is determined by the
system´s external parameters. In a batch experiment, the initial
stable distribution of species is determined by the conditions
under which the system existed previously, and this stable distribution is being changed during the batch experiment; only after
many cycles of batch experiment can the system attain a new stable
distribution.

In order to describe the interaction between different trophic
levels inside the framework of aggregated variables, it seems more
justifiable to apply, instead of the classic Monod relationship,
its generalization of the type suggested here for describing biological treatment.

If the yield coefficient of the total biomass in a multi-species system can roughly be regarded as a constant, the total decay
coefficient of the biomass is strongly dependent upon external
parameters.

In general, it can be concluded that three hierarchic levels
can be distinquished /Fig. 2.36/:
1. cultures of micro-organisms;
2. aggregates of micro-organisms; and
3. associations of micro-organisms.
Each of these levels can be described by a rather complex model

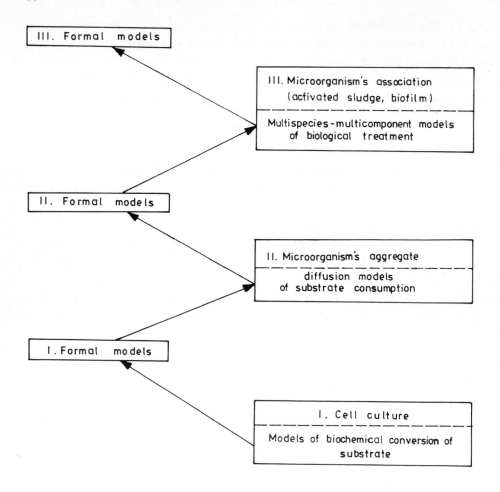

Fig. 2.36. System of models describing biological treatment
process.

which would reflect the "mechanism" of respective processes. At
the same time, relatively simple formal models, which present the
system as a black box, are also admissible. In developing complex
non-formal high-level models, simple input/output relationships,
or formal models corresponding to lower levels should be used as
elementary units.

The appropriate level of study is determined to a large extent
by the specific problem under consideration. The development of
an idealized scheme of the level's mechanism presents an important step in the development of any mathematical model while the
understanding of the object under study is a necessary condition

for successful idealization.

In this discussion, rather long-term processes have been analyzed which should be taken into account in the design of biological treatment plants. In order to control the process of treatment on the real time scale one should also consider faster processes such as the pollutant´s "sorption" by activated sludge or
biofilm. This latter process presents a complicated physico-
chemical and biological phenomenon.

References

Adams, C.E., Eckenfelder, W.W., and Hovious, J.C., 1975. A Kinetic
    model for design of completely-mixed activated sludge
    treating variable strength industrial wastewater. Water Res.,
    9, 37–42.
Atkinson, B., and Fowler, H.W., 1974. Advances in Biochemical
    Engineering, vol. 3, Springer-Verlag, New York, N.Y.
Benefield, L., Lawrence, D., and Randall, C., 1979. The effect of
    sludge viability of biokinetic coefficient evaluation. J. WPCF,
    51, 187–194.
Benjes, H.H, 1979. Small community wastewater treatment facili-
    ties – biological treatment system. In: Design seminar Handout.
    Small wastewater treatment facilities. Cincinatti, Ohio, U.S.
    EPA, 1–91.
Borghi, M.D., Palazzi, E., and Ferraiolo, G., 1977. The role of
    kinetic phenomena on the design of rotating biological sur-
    face. Chimica Ind. /Milan/, 59; 77–81.
Chudova, I., 1969. Discussion. In: Forth Ind. Waste Conf. on Water
    Poll. Res. /Prague/, Pergamon Press, London, 375–378.
Contois, D.E., 1959. Kinetics of bacterial growth: relationship
    between population density and specific growth rate of con-
    tinuous culture, 1–2, 40.
Eckenfelder, W.W., 1967. Comparative biological waste treatment
    design. J. Sanit. Eng. Div. ASCE, 93, 157–170.
Ellis, K.V., and Banaga, S.E.I., 1979. A study of rotating disk
    treatment units operating at different temperatures. Water
    Pol. Control, 75, 74–81.
Fair, G.M., and Geyer, J.C., 1954. Water supply and wastewater
    disposal, John Wiley and Sons, N.Y.

Gaudy, A.F., and Gaudy, E.T., 1972. Biological concepts for design and operation of the activated sludge process. U.S.EPA.

Goodman, B.L., and Englande, A.J. 1974. A unified model of the activated sludge process. J. WPCF, 46, 312-332.

Grady, C.P.L., and Roper, R.E., 1974. A model for the bio-oxidation process which incorporates the viability concept. Water Res., 8, 471-482.

Grau, P., Dohanios, M., and Chudova, J. 1975. Kinetics of multi-component substrate removal by activated sludge. Water Res., 9, 637-642.

Günter, L.I., 1976. Doctoral thesis. K.D.Pamfilov Academy of Municipal Economy, Moscow /in Russian/.

Haldane, J.B.S., 1930. Enzymes. Longmans, London.

Hanumanulu, V., 1970. Performance of deep trickling filters by five methods, J. WPCF, 42, 1446-1457.

Harremoës, P., 1977. Half-order reactions in biofilm and filter kinetics. Vatten, 2, 122-143.

Haseltine, T.R., 1955. A rational approach to the design of activated sludge plants. Water Sew. Works, 102, 487-495.

Keefer, C.E., and Meisel, J., 1952. Remodelled trickling filters. Water Sew. Works, 99, 277-279.

Keshavan, K., Behn, V.C., and Ames, W.F., 1964. Kinetics of aerobic removal or organic wastes. J. Sanit. Eng. Div. ASCE, 90, 99-126.

Kornegay, B.H., and Andrews, J.F., 1968. Kinetics of fixed-film biological reactor. J. WPCF, 40, 460-468.

Korzukhin, M.D., 1967. Mathematical modelling of chemical system kinetics I-III, In: Oscillating processes in biological and chemical systems. Ed. G.M. Frank, Nauka Publishers, Moscow, 231-251 /in Russian/.

Koug, M.F., 1979. Practical design equations for trickling filter process. Biotechn. Bioeng., 21, 417-431.

Krishnan, P., and Gaudy, A.F., 1966. Mechanism and kinetics of substrate utilization at high biological solids concentration. In: Proc. 21st Purdue Ind. Waste Conf., Purdue Univ., 495-510.

La Motta, E.J., 1976. External mass transfer in biological film reactor. Biotechnol. Bioeng., 18, 1359-1370.

Lighthart, B., and Loew, G.A., 1972. Identification key for bacteria cultures from an activated sludge plant., J. WPCF, 44, 2078-2085.

Lotka, A.J., 1925. Elements of physical biology. Williams and Wilkins, Baltimore.

Monod, J., 1942. Recherchés sur la croissance des cultures bacteriennes. Herman, Paris.

Moser, H., 1958. The dynamics of bacterial population maintained in the chemostat. Carnegie Inst. Washington, Pub. No. 614, Washington, D.C.

McKinney, R.E., 1962. Mathematics of complete-mixing activated sludge. J. Sanit. Eng. Div. ASCE, 88, 87-95.

Mueller, J.A., Voelkel, K.G., and Boyle, W.C., 1966. Nominal floc diameter related to oxygen transfer. J. Sanit. Eng. Div. ASCE, 92, 9-20.

Nelson, P.O., and Lawrance, A.Wm., 1980. Microbial viability measurements and activated sludge kinetics. Water Res., 14, 217-226.

Novak, J.T., 1974. Temperature substrate interactions in biological treatment. J. WPCF, 46, 1984-1994.

Oleszkiewicz, J., 1976. Rational design of high-rate trickling filters, based on experimental data. Env. Protection Engineering, 2, 85-105.

Palm, J.C., Jenkins, D., and Parker, D.S., 1980. Relationship between organic loading, dissolved oxygen concentration and sludge settleability in completely-mixed activated sludge process. J. WPCF, 52, 2484-2506.

Paolini, A.E., Sebastiani, E., and Varioli, G., 1979. Development of mathematical models for the treatment of industrial wastewater by means of biological rotating disk reactors. Water Res., 13, 751-761.

Paolini, A.E., and Varioli, G., 1982. Kinetic consideration on the performance of activated sludge reactors and rotating biological contactors. Water Res., 16, 155.

Pike, E.B., and Carrington, E.G., 1972. Recent developments in the study of bacteria in the activated sludge process. Water Pol. Control. 71, 583-605.

Ramanathan, M., and Gaudy, A.F., 1972. Sludge yields in aerobic systems. J. WPCF, 44, 441-450.

Roe, P.C., and Bhagat, S.K., 1982. Adenosine triphosphate as a control parameter for activated sludge processes. J. WPCF, 54, 244-254.

Rogovskaya, Tz., 1967. Biochemical method of industrial water treatment. Stroiizdat Publishers, Moscow /in Russian/.

92

Sárner, E., 1981. Removal of dissolved and particulate organic matter in high-rate trickling filters. Water Res., 15, 671-678.

Sherrard, J.H., and Schroeder, E.D., 1972. Relationship between the observed cell yield coefficient and mean cell residence time in the completely-mixed activated sludge process. Water Res., 6, 1039-1049.

Siber, S., and Eckenfelder, W.W., 1980. Effluent quality variation from multi-component substrate in activated sludge process. Water Res., 14, 471-476.

Silveston, B., 1967. Biological waste treatment: Notes for a course at the University of Waterloo.

Suschka, J., 1980. Bio-oxidation in continuous activated sludge process. Water Res., 14, 197-206.

Teissier, G., 1942. Croissence des populations bacteriennes et quantite d'aliment disponsible. Rev. Sci., 209.

Tischler, L.T., and Eckenfelder, W.W., 1969. Linear substrate removal in activated sludge process. In: Fourth Ind. Waste Conf. on Water Poll. Res /Prague/. Pergamon Press, London, Section II, 1-14.

Toerber, E.D., Paulson, V.L,- and Smith, H.S., 1974. Comparison of completely-mixed and plug-flow biological systems. J. WPCF, 46, 1995-2014.

Torpey, W.N., Heulekian, H., Kaplovsky, A.J., and Epstein, R., 1971. Rotating disks with biological growths prepare waste-water for disposal or reuse. J. WPCF, 43, 2182-2188.

Tsuchia, H.M., Frederickson, A.G., and Aris, R., 1966. Dynamics of microbial cell populations. Adv. Chem. Eng., 6, 125-206.

Tuček, F., Chudova, J., and Madera, V., 1971. Unified basis for design of biological aerobic treatment processes. Water Res., 6, 647-680.

Vandevenne, L., and Eckenfelder, W.W., 1980. A comparison of models for completely-mixed activated sludge treatment design and operation. Water Res., 14, 561-566.

Vasiliev, V.B., and Vavilin, V.A., 1982. Activated sludge system design under wide variation of organic loading. Biotechnol. Bioeng., 24, 2337-2355.

Vavilin, V.A., 1980. Phenomenological modelling of biodegradation of multi-component substrates. Env. Protection Engineering, 6, 319-343.

Vavilin, V.A., 1981a. Theory of aerobic biological treatment processes and unified model for aeration tank and trickling filter design. Rep. Acad. Sci. USSR, 256, 759-762 /in Russian/.

Vavilin, V.A., 1981b. Generalized model and mechanism of aerobic biological treatment. Doklady AN SSSR, 258, 1269-1273 /in Russian/.

Vavilin, V.A., 1982a. Models and design of aerobic biological treatment processes. Acta hydrochem. hydrobiol., 19, 211-242.

Vavilin, V.A., 1982b. The theory and design of aerobic biological treatment. Biotechnol. Bioeng., 24, 1721-1747.

Vavilin, V.A., 1982c. Generalized model of aerobic biological treatment. Ecol. Modelling, 17, 157-173.

Vavilin, V.A., 1982d. The effect of temperature, of inlet pollutant concentration and of micro-organisms' concentration on the rate of biological treatment. Biotechnol. Bioeng., 24, 2609-2625.

Vavilin, V.A., 1983a. Dependence of biological treatment rate on species composition in activated sludge or biofilm. Part 2. From models to theory. Biotechnol. Bioeng., 25.

Vavilin, V.A., 1983b. Non-linear models of biological treatment and self-purification processes in rivers /in press/. Nauka Publishers, Moscow /in Russian/.

Vavilin, V.A., and Vasiliev, V.B., 1977. Application of Monod's formula to biological waste treatment of flocs of activated sludge. IFAC - Symp. Control Mechanisms in Bio- and Ecosystem. Bio communication, Leipzig, 5, 194-200.

Vavilin, V.A., and Vasiliev, V.B., 1981. Evaluation of mathematical models for aeration tank design. Vodnye Resursy, 2, 132-145 /in Russian/.

Vavilin, V.A., and Vasiliev, V.B., 1982a. The model of activated sludge ecosystem adaptation to multi-component pollutant. Doklady AN SSSR, 267, 1012-1016 /in Russian/.

Vavilin, V.A., and Vasiliev, V,B., 1983a. On mechanism of simple substrate utilization by activated sludge particles. Ecol. Modelling, 18, 27-34.

Vavilin, V.A., and Vasiliev, V.B., 1983b. Dependence of biological treatment rate on species composition in activated sludge or biofilm. Part 1. Biological treatment model with ecosystem adaptation. Biotechnol. Bioeng., 25.

Vavilin, V.A., Vasiliev, V.B., and Kuzmin, S.S., 1980. Modelling the biochemical oxidation processes in artificial conditions of waste treatment plants. Ecol. Modelling, 10, 105-137.

Voronov, Yu., V., Ivchatov, A.L., Vavilin, V.A., and Kuzmin, S.S., 1981. Design of first-step trickling filters. Hidroliznoje proizvodstvo, 6, 19-21 /in Russian/.

Weddle, C.L., and Jenkins, D., 1971. The viability and activity of activated sludge. Water Res., 5, 621-640.

Williamson, K.J., and McCarthy, P.L., 1976. A model of substrate utilization by bacterial films. J. WPCF, 48, 9-23.

Wuhrmann, K., 1968. Research developments in regard to concept and base values of activated sludge system. In: Water Resources Symp., Univ. of Texas Press, Austin, p. 143.

Zhabotinsky, A.M., 1974. Periodic auto-oscillation in concentration variables. Nauka Publishers, Moscow /in Russian/.

COMPREHENSIVE ACTIVATED SLUDGE PROCESS DESIGN

W. Wesley Eckenfelder, Mervyn C. Goronszy and Andrew T. Watkin

1. INTRODUCTION

This chapter develops a design procedure for the activated sludge process based on material balances and the concept of active mass. The relationships that govern the activated sludge process are:

$$\text{organics} + a'O_2 + N + P \xrightarrow[k]{\text{cells}} a \text{ new cells} + CO_2 + H_2O + \text{non-degradable soluble residue}$$

BOD, COD, TOC

$$\text{cells} + b'O_2 \longrightarrow CO_2 + H_2O + N + P + \text{Non degradable cellular residue}$$

The parameters necessary to generate a process design are the fraction of organics removed oxidized for energy, denoted by the coefficient a', the fraction of organics removed and synthesized into biomass, denoted by the coefficient a, a reaction rate coefficient k, and the rate of endogenous oxidation b. These coefficients are obtained from the literature, from experience elsewhere, or from laboratory or pilot plant studies on specific wastewaters. The design parameters employed in this chapter are shown in Figure 1.

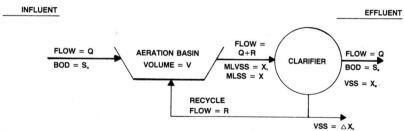

Figure 1. Design Parameters

2. PREDICTION OF EXCESS BIOLOGICAL SOLIDS

Material balances have been recently developed by Quirk and

Eckenfelder (1984), which enable the degradable fraction of solids in an activated sludge system to be readily determined. When these solids are recirculated, endogenation reduces the degradable fraction present in the reactor. When material balance relationships are written for such a closed system the degradable fraction at any operating condition is readily predictable. Incorporation of the degradable fraction concept into the material balance is critical in obtaining reliable predictions for excess biological solids production.

2.1   Determination and Definition of Degradable Fraction

The degradable fraction of an activated sludge mixed liquor is determined by a batch endogenous respiration study. The endogenation of mixed liquor volatile suspended solids follows first order kinetics and is illustrated in Figure 2.

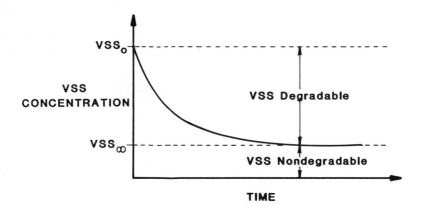

Figure 2.   Batch Endogenation of Mixed Liquor
Degradable and nondegradable fractions are calculated based on the results of the endogenous respiration study as from:

$$X_d = \frac{VSS_o - VSS_\infty}{VSS_o} \text{ and } X_n = \frac{VSS_\infty}{VSS_o}$$

Studies have shown that for practical purposes the degradable fraction, as it relates to the time frame of the activated sludge process, may be determined experimentally in approximately 30 days. The degradable fraction has been found to vary from 0.8 to 0.40 for various wastes and sludge ages ranging from 2 to 40 days.

The importance of degradable fraction in excess sludge generation has been shown and incorporated in a design model by Eckenfelder (1980) and is illustrated in this chapter. Degradable fraction as an activity parameter is also illustrated in a kinetic expression for the activated sludge process.

2.2 Model Development for Excess Biological Growth

A relatively simple relationship between degradable fraction and sludge age can be developed by material balance formulation. Equations (1) and (2) are obtained from degradable and nondegradable mass balances around a completely mixed biological reactor for a soluble substrate.

$$X_d \Delta X_v = aX_d'S_r - bX_d X_v t \tag{1}$$

$$X_n \Delta X_n = aX_n'S_r \tag{2}$$

X_d' is the degradable portion of newly generated biomass and is theoretically defined as the degradable fraction of the biomass at a sludge age of zero days. The determination and typical range of values of X_d' is provided in Table 1.

By equating S_r in Equations (1) and (2) the following expression relating the degradable fraction to sludge age is obtained.

$$X_d = \frac{X_d'}{1 + bX_n'\Theta_c} \tag{3}$$

By inverting Equation (3) a linear form between $1/X_d$ and Θ_c may be obtained which facilitates the determination of X_d'.

$$\frac{1}{X_d} = \frac{bX_n'}{X_d'} \Theta_c + \frac{1}{X_d'} \tag{4}$$

Figure 3 shows the correlation between degradable fraction and sludge age using data generated by Eckhoff and Jenkins (1967).

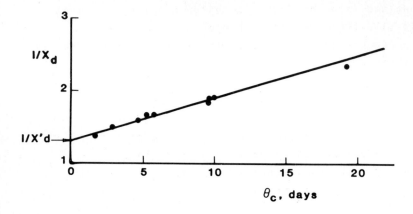

Figure 3.   Linear Relationship between Degradable Fraction
and Sludge Age

Values of X_d' have generally been found to range between 0.75 and
0.81 as indicated in Table 1.   These relatively consistant values
of X_d' permit X_d to be readily calculated, once the endogenous rate
constant is known, from Equation (3).

Table 1.   Experimentally Derived Values of X_d' and X_n'

Substrate	X_d'	X_n'	Reference
Skim Milk and Phenol	0.778	0.222	Rhodes (1984)
Glucose, phenol, and sulfanilic acid	0.750	0.250	Siber and Eckenfelder (1980)
Egg Albumen and Beef Extract	0.781	0.219	Eckhoff and Jenkins (1967)
Skim Milk	0.810	0.190	Upadhyaya and Eckenfelder (1975)

Excess biological solids production can then be determined by the
use of Equation (3) along with the appropriate kinetic
formulation.

Excess biological sludge can be related to sludge age as:

$$\Theta_c = \frac{X_v t}{\Delta X_v} \tag{5}$$

ΔX_v can also be expressed

$$\Delta X_v = aS_r - bX_d X_v t \tag{6}$$

Combining Equations (5) and (6) yields the expression

$$\Theta_c = \frac{X_v t}{aS_r - bX_d X_v t} \tag{6a}$$

Combining Equations (1) and (3) yields the expression

$$\Theta_c = \frac{X_v t}{a\dfrac{X_d'}{X_d} S_r - bX_v t} \tag{6b}$$

In cases where both Θ_c and X_d are unknown, Θ_c can be computed by combining Equations (3) and (6a) or Equations (3) and (6b) for soluble substrates to yield

$$\Theta_c = \frac{-(aS_r - bX_v t) + \sqrt{(aS_r - bX_v t)^2 + 4(abX_n'S_r)(X_v t)}}{2abX_n'S_r} \tag{6c}$$

When Θ_c is computed from Equation (6c), X_d can be computed from Equation (3). For the case where influent solids are involved such as in domestic wastewater treatment, Equation (6c) needs to be modified to account for the fraction of volatile suspended solids in the influent which is not degraded. This fraction contributes to the overall mass of solids accumulated in the aeration tank, the wasting program and hence the sludge age. Modifying Equation (6) yields

$$\Delta X_v = aS_r + X_{iND} - bX_d X_v ft \tag{6d}$$

and

$$\Theta_c = \frac{X_v t}{aS_r + X_{iND} - bX_d X_v ft} \tag{6e}$$

Proceduraly, the solution of this equation for Θ_c is found by trial and error. Initial values of X_{iND} and f need to be calculated from a first estimate of Θ_c given by Θ_n. The calculation is repeated until the values of Θ_c given by the above

equation and the mass balance coincide. An initial estimate of
the reduction in volatile solids equivalent to the initial
estimate of Θ_c (ie Θ_n) is obtained from
%VSS reduction = $2.84 + 35.07 \log_{10} \Theta_c$.
These data are an estimate only as they refer to aerobic digestion
of mixed primary solids and activated sludge and not for digestion
of the smaller particle size fraction which passes through the
primary settling unit. Equation (6c) becomes

$$\Theta_c = \frac{-(aS_r + X_{iND}) - X_v tb(X_n' + X_d'f)}{2bX_n'(aS_r + X_{iND})}$$ (6f)

$$+ \frac{\sqrt{[(aS_r + X_{iND}) - X_v tb(X_n + X_d f)]^2 + 4X_v tbX_n(aS_r + X_{iND})}}{2bX_n'(aS_r + X_{iND})}$$

3. OXYGEN REQUIREMENTS

The oxygen requirements for an activated sludge reactor may
be determined by a mass balance on oxygen utilization. If
nitrification is assumed to be negligible and substrate
utilization is based on BOD, then the following expression can be
developed for soluble substrates.

$$r_r = a'S_r + b'X_d X_v t$$ (7)

4. NUTRIENT REQUIREMENTS

Aerobic organisms require nutrients in the form of nitrogen
and phosphorous and other trace elements for optimal activity.
The trace elements required are usually present in sufficient
quantity in the carrier water. In some cases, however, N and P
have been shown to be deficient. Nitrogen and phosphorus are
frequently deficient in industrial wastewaters and must be added
to insure efficient treatment of the wastewater. Domestic
wastewater has an excess of these nutrients. The deficiency of
either of these nutrients may cause dispersed or filamentous
growth. A ratio of BOD:N:P of 100:5:1 will insure an excess of
nutrients. The requirements may be more exactly determined
through the relationships:

$$N = 0.123 \frac{X_d}{X_d'} \Delta X_v + 0.07 \frac{X_d' - X_d}{X_d'} \Delta X_v$$ (8)

$$P = 0.026 \, \frac{X_d}{X_d'} \, \Delta X_v + 0.01 \, \frac{X_d'-X_d}{X_d'} \, \Delta X_v \qquad (9)$$

Equations (8) and (9) apply for soluble wastewaters in which the excess sludge represents biomass. If influent VSS are present in the wastewater Equations (8) and (9) must be appropriately adjusted.

5. KINETICS OF SUBSTRATE REMOVAL - INDUSTRIAL WASTEWATERS

Several authors have suggested that active mass should be incorporated in kinetic models for substrate removal; however, to date this has been only a theoretical argument. The material balance does not of itself determine the amount of substrate remaining in solution. This is determined by the kinetics of substrate metabolism. Once this kinetic relationship is known the substrate remaining will respond in a predictable fashion to the active mass present in the reactor. It is the material balance which dictates what that active mass will be and establishes the framework for the performance relationships which can be derived for an activated sludge plant.

The degradable fraction has been proposed as an indicator of the viable or active proportion of a systems' biomass for soluble substrates by Eckenfelder and Upadhyaya (1975) who demonstrated good correlation between the degradable fraction, ATP concentration, oxygen uptake rates and dehydrogenase enzyme concentration versus sludge ages ranging from 10 to 40 days in steady state continuous reactors. If it is accepted that degradable fraction is an indicator of the active or viable biomass in a system, then by inference the active fraction of a system's biomass will vary with sludge age. This effect is incorporated in the kinetic design equations for biological systems.

The degradable fraction is of importance when determining excess sludge quantities generated by the activated sludge process. However, for substrate removal kinetics the active fraction will be used for model correlations. The active fraction is the fraction of total viable cell mass in system and can be defined as follows:

$$X_a = \frac{X_d}{X_d'}$$

Established kinetic models for the activated sludge process have historically failed to incorporate substrate utilization rates which are dependent on the active mass. Eckenfelders' (1980) first order kinetic model for completely mixed reactors will serve as an illustration of this point and is presented as Equation (10).

$$\frac{S_r}{X_v t} = K \frac{Se}{So} \tag{10}$$

Here the specific substrate utilization rate, $S_r/X_v t$, is dependent on the mixed liquor volatile suspended solids concentration. Equation (1) which was derived from a mass balance of degradable solids is rewritten below as Equation (11) to illustrate the effect of active mass on growth kinetics.

$$\frac{1}{\Theta_c} = a \frac{S_r}{X_a X_v t} - b \tag{11}$$

It is apparent that the specific utilization rate must incorporate active mass instead of simply volatile mass. Equation (10) can be rewritten based on active mass as follows:

$$\frac{S_r}{X_a X_v t} = K' \frac{Se}{So} \tag{12}$$

It follows that Equation (12) more accurately describes substrate utilization kinetics because removal is based on active mass and not just volatile mass. This concept is intuitively obvious upon studying Figure 3, where it is seen that the biological viability of a system is significantly decreased at higher sludge ages. The older kinetic formulation described by Equation (10) can be shown to over estimate MLVSS concentration at low sludge ages and under estimate MLVSS concentration at high sludge ages. Also since it has been demonstrated that active mass is a nonlinear function of sludge age the older kinetic formulation, Equation (10), cannot by inference be a true linear relationship.

5.1 Biosorption

Studies by several investigators have shown that many soluble organics exhibit a rapid sorption when contacted with activated sludge. This sorption occurs in contact times of 2-10 minutes. The sorption phenomona can be related to floc load, which can be defined as mgCOD, BOD, TOC/gMLSS, MLVSS or mass of biodegradable solids. Some results are shown in Figure 4.

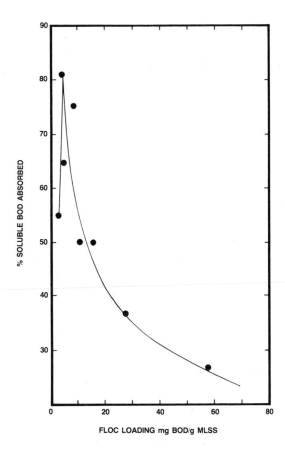

Figure 4. Biosorption versus Floc Loading

Oxygen is consumed at a rapid rate following sorption. This effect is shown in Figure 5 where measured oxygen consumption vs substrate removal is shown for either batch or continuous conditions. Initial substrate removal velocities are high indicating accumulation of organics within the biomass. As the aeration time increases oxygen consumption and substrate removal eventually come into balance yielding the constants a' and b' of Equation (7). The point at which this balance occurs is the point at which metabolism of absorbed substrate has been completed.

104

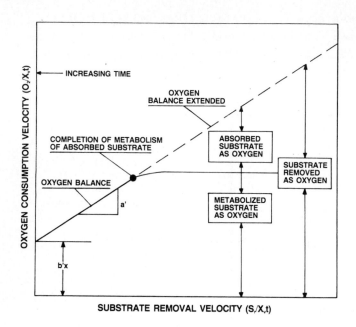

Figure 5. Oxygen Consumption Velocity versus Substrate
Removal Velocity

6. KINETICS OF SUBSTRATE REMOVAL - MUNICIPAL WASTEWATERS

Depending on its origin, wastewater may contain a significant
fraction of its nondegradable organic compounds in particulate
form. For domestic wastewater, it has been shown (7) that the
principal organic components (whether as COD or BOD) can be
identified (the respective percentages are shown in parentheses)
as a soluble fraction (40%), a colloidal fraction (10%) and a
settleable fraction (50%). Domestiic wastewater, even after
primary settling typically contains a significant fraction of its
organics in colloidal and suspended form. On a percentage basis
the fractions after settling are of the order of, soluble fraction
- 56%, non-settleable particulates - 30% and the colloidal
fraction 14%.

In an activated sludge plant a number of reaction mechanisms
may be postulated for these three basic components. Initially the
particulate fraction is biocoagulated and or physically enmeshed
by the activated sludge matrix. A minor fraction of the
particulate material may be absorbed onto active biomass. The
enmeshed and biocoaglated solids undergo extracellular enzymatic

breakdown whereby the large complex organic molecules of the
particulate substrate are converted to a readily degradable
soluble form. Breakdown of particulate substrates is very slow by
comparison with synthesis of soluble substrates, the limiting rate
being determined by the rate of extracellular breakdown. The
fraction of the particulate substrate which is solubilized is
prirmarily a function of the solids residence time. This also
governs the active fraction of the biomass and hence the
population of extracellular enzymes that are available for the
solubilization reactions. The size distribution of the
particulates also affect the rate of breakdown. There is also a
non-biodegradable fraction of the particulate substrate.

For the soluble substrate component there is an initial
absorption of a part of the soluble fraction and biological
degradation of the remaining soluble fraction in keeping with the
kinetics of the system. The absorption and synthesis mechanisms
operate in conjunction. Absorption of soluble substrate is a
function of the active fraction of the biomass and the value of
the maximum fraction of soluble substrate that can be absorbed by
the active fraction of biomass. Stated simply, if the active
fraction of biomass has a residual of absorbed soluble substrate,
its ability to attain the maximum absorption potential is reduced
(constrained) by the fraction of substrate that is already
absorbed. This is of particular importance in plug-flow
configured systems.

Within the range of recycle ratios normally used in practice
activated sludge usually has the ability to completely
biocoagulate all particulate substrate entering the biological
reactor over the diurnal cycle. In a plug-flow configuration this
is accomplished within the initial 20% of the reactor assuming
sufficient mixing etc. In domestic waste treatment, without
primary settling, breakdown of the biocoagulated degradable
particulate fraction is essentially completely accomplished in
about 7-10 days. This can be verified from batch aeration studies
on activated sludge containing settled primary solids. For
wastewaters having received primary settling, the same mechanisms
occur but with a reduced biocoagulated fraction. As with the
particulate substrate component there is a fraction of the soluble
substrate component which is non-biodegradable.

By way of example, consider a settled sewage having a COD of
around 300 mg/l. Using the fractional percentages referred to

previously, some 130 mg/l of COD is removeable as a result of
initial biocoagulation and enmeshment leaving some 170 mg/l to be
accounted for by absorption and biodegradation mechanisms. The
past history of the biomass is important as this factor determines
the absorption potential of the biosolids and ultimately affects
the concentration of soluble substrate which remains to be
biodegraded. In this context, the extent to which absorption of
soluble substrate takes place is a function of the biodegradable
fraction of the biomass solids coming into contact with the waste,
the relative magnitude of the floc loading that is experienced and
the extent to which that biomass has used up its absorption
potential.

This has very practical ramifications particularly with
plug-flow configured plants. Should the hydraulic retention time
be reduced under increased flow conditions, say during diurnal
hydraulic load variation, particulate and soluble substrate
removal can easily be inadequate. Additionally this circumstance
can be exacerbated by a situation brought about by a reduction in
absorption potential of the return biomass due to the fact that
the biomass is still experiencing a relatively high state of
absorption. These effects can be gauged by the ratio of the
magnitude of the specific oxygen utilization rate of the return
biomass to the maximum value attained in the initial section of
the aeration chamber. It is important that the biomass remain
under aeration for an adequate time period in order to reestablish
its absorption potential, i.e. use up existing cellular storage
and accumulation products. Repeated operation with only partial
biomass absorption capacity available will not only generally lead
to a degradation of effluent quality, but will also provide
conditions which can ultimately lead to the generation of a
non-flocculating biomass.

A simple behavioral model which can be used to describe the
removal of carbonaceous material in a plug-flow configured
activated sludge process may be postulated as:

$$S_o - S_e = S_{BC} + S_{AB} + S_{BD} \tag{13}$$

The model is shown schematically in Figure 6. (This model
neglects to account for non-degradable fractions for both
particulate and soluble substrate).

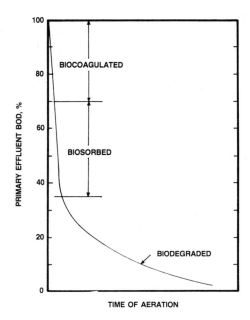

Figure 6. Schematic Plug Flow Model For Domestic Wastes

The first order equation with respect to soluble substrate concentration defines the S_{BD} term as:

$$\frac{S_e}{S_i} = e^{-kt} \tag{14}$$

In some cases a second order reaction equation may more appropriately describe the biodegradation term, particularly when one or more of the degradable components has a very low removal rate. The effect of initial substrate concentration in batch oxidation, (which is in effect a simulated plug-flow system), has been studied by Tischler (1973) and Grau (1975) who proposed a modified equation:

$$\frac{S_e}{S_i} = e^{-K_1 X_v t / S_i} \tag{14a}$$

Introduction of the biodegradable fraction of the biomass yields equation (14b).

$$\frac{S_e}{S_i} = e^{-K_p X_a X_v t / S_i}$$

(14b)

The use of the model is somewhat emperical with respect to the S_{BC} and S_{AB} terms. Orders of magnitude have been established for these two terms from batch absorption studies on activated sludge using sewage that has received primary settlement (Goronszy et al., 1980).

Figure 4 summarizes these results as percent soluble BOD absorbed versus floc loading (as mg BOD/g MLSS). Use of Figure 4 is predicated by the fact that the specific oxygen utilization rate (SOUR) of the return activated sludge is not elevated much above 7 mg O_2/g MLSS/hr as a sludge with a higher initial SOUR would not necessarily give the same absorptive response.

This model can be used for system design of both cyclically aerated and plug-flow configured domestic wastewater treatment plants. The total oxygen requirement can be calculated from the BOD to be satisfied using overall mass balance equations. The design procedure assumes that the model described by Equation (13) can be applied and that the aeration intensity can be estimated accordingly so that the aeration system can be designed to provide the required intensity at expected loading rates which will vary diurnally and perhaps seasonally. In order to effectively match aeration intensity with the expected level of treatment it is necessary to provide a design which will achieve a dispersion number appropriate to the plug-flow hydraulic conditions that are desired.

The dispersion number, as described by Levenspiel (1962), can be used to express the degree of longitudinal mixing in an activated sludge plant. For the ideal cases of 'plug-flow' and 'complete-mixing' the dimensionless dispersion number has values of 0 and ∞ respectively. All real systems have values between these two extremes as depicted in Figure 7. Good settling sludges have been identified with plug-flow hydraulic regime activated sludge plants (Thomlinson and Chambers, 1979 a,b; Chudoba et al., 1973).

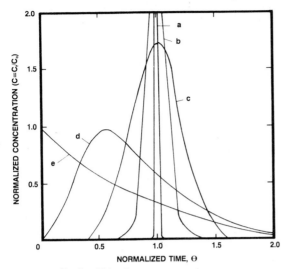

a, Plug flow, D/uL = 0
b, Small amount of dispersion, D/uL = 0.002
c, Intermediate amout of dispersion, D/uL = 0.02
d, Large amount of dispersion, D/uL = 0.2
e, Completely mixed, D/uL = ∞

Figure 7.   Dispersion Model Prediction for Various Degrees
of Longitudinal Mixing (after Levenspiel, 1962).

While the mean velocity of flow in an aeration tank varies
with the diurnal fluctuations in sewage flowrate and is not
constant as strictly required by the experimental method used for
determining the dispersion number, it is possible to derive an
expression relating the dispersion number to the physical
dimensions of the aeration tank and the prevailing operating
conditions.   The dispersion number is a dimensionless number which
can be considered analagous to diffusion superimposed on
convective flow where:

$$N_D = \frac{D}{UL} \tag{15}$$

and
$$U = \frac{Q(1 + r)}{8.6 \times 10^4 A} \tag{15a}$$

and $A = WH$

Hence

$$N_D = \frac{8.6 \times 10^4 DWH}{LQ(1 + r)} \qquad (15b)$$

Data obtained from tracer experiments in 24 activated sludge plants have indicated a value of D = 0.068 m^2/S can be used to estimate the dispersion number with an accuracy of ± 15% (Boon et al., 1983). Values of N_D less than about 0.1 should ensure good plug-flow hydraulics in practice. In such cases it can be assumed that an average 60% of the organic load is satisfied in the first 200% of the aeration rank volume where the length to width ratio of the tank >20. In the remaining volume of the aeration tank about 15% of the load is satisfied in the next 20% of the aeration tank volume followed by 10,10,5% for each following 20% of the tank volume.

Diurnal variations in rate of flow of wastewater and in concentrations of BOD to be satisfied and ammoniacal nitrogen to be oxidized result in significant variations in oxygen demand. Appropriate peaking factors need to be used when calculating the oxygen demand diurnal variations. To avoid lack of dissolved oxygen limiting the rate of treatment, the period for which the total demand should be calculated is equal to the retention time of the mixed liquor (recycled sludge plus sewage) in the aeration tank.

7.   SLUDGE QUALITY CONTROL

At one time sludge bulking was thought to be caused primarily by Sphaerotilus natans and was attributed to a wide variety of causes. In recent years more than 28 indicative filament types have been identified (Eikelboom, 1975; Richard et al., 1982) caused by specific design or operating variables. These are low dissolved oxygen, low F/M, nitrogen and/or phosphorus deficiency, low pH or in some cases septic wastewater. The dominant filament types caused by these conditions are shown in Table 2.

7.1   <u>Low Dissolved Oxygen</u>

Palm et al. (1980) related dissolved oxygen to filamentous bulking. Consider Figure 8. Under low organic loading with a low specific oxygen uptake rate, oxygen is able to diffuse through the floc even at concentrations of 1 mg/L or less, thereby maintaining the floc fully aerobic. This condition favors growth of the floc

Table 2. Dominant Filament Types Indicative of Activated
Sludge Operation Problems (Eikelboom, 1975).

Suggested Causative Condition	Indicative Filament Types
Low DO	S. natans, H. hydrossis, type 1701
Low F/M	M. parvicella, H. hydrossis, types 021N, 0041, 0675, 0092, 0581, 0961, and 0803
Nitrogen and/or phosphorus	S. natans, Thiothrix spp., type 021N; possibly H. hydrossis and types 0041 and 0675
Low pH	fungi
Septic wastes	Thiothrix spp., Beggiatoa spp, type 021N

forming organisms. When the F/M is raised to 0.4, thereby
increasing the oxygen uptake rate, oxygen can no longer penetrate
the floc resulting in anoxic or anaerobic conditions. This

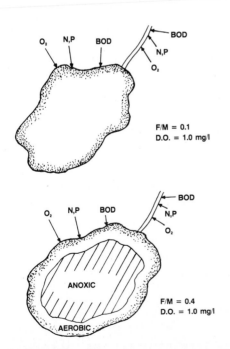

Figure 8. Schematic of Oxygen Penetration into Flocs

condition will favor growth of the filaments. The results of Palm et al. (1980) have been replotted as related to F/M in Figure 9. It becomes apparent that process economics requires an optimization between operating dissolved oxygen and organic loading.

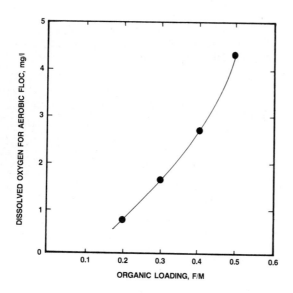

Figure 9. Relationship between F/M and Dissolved Oxygen to Maintain Aerobic Flocs.

7.2 Low F/M

Low F/M bulking implies a food limiting situation for the floc forming organisms in which the biological environment favors the growth of specific filaments. In the presence of readily degradable organics at low concentration some filaments have a higher growth rate than the floc forming organisms. In addition, diffusion may become limiting in some cases as was shown for oxygen in Figure 8. Complete mix reactors do not work in these cases and will usually result in extreme sludge bulking. It has also been shown that the use of a plug-flow reactor or a cyclically aerated reactor with an initial high F/M will favor growth of floc forming organisms. For many organic wastewaters an initial short term contact unit or selector will result in a rapid absorption of the organics by the floc forming organisms, thereby limiting the food available to the filaments. Contact times of 5 to 15 minutes have been successfully employed for sludge bulking control.

7.3 Nutrient Deficiency

Many industrial wastewaters require nutrient addition. Failure to feed adequate nutrients has resulted in filamentous sludge bulking. The mechanism is probably similar to that shown for oxygen in Figure 8. It was shown in a pulp and paper activated sludge plant that filaments developed when the effluent ammonia nitrogen fell below 1.5 mg/L.

7.4 Sludge Bulking Control

Sludge bulking control can be achieved by chlorination of the sludge to selectively kill the filaments. Palm et al (1980) have indicated chlorine dosages of 5-15 mg Cl_2/g MLSS/day. In nitrifying plants the chlorine dosage should not exceed 4.5 mg Cl_2/g MLSS/day since the aerobic nitrifiers tend to concentrate in the outer layers of the floc. For conventional plants the chlorine should be applied to the return activated sludge but industrial plants with long aeration periods require direct chlorination of the mixed liquor to be effective. In some cases the application of an anoxic or anaerobic stage will suppress the filaments. In one industrial plant a 3 hour anoxic stage completely eliminated the filaments but resulted in a higher SVI than an oxygen plant with floc containing some binding filaments.

8. NITRIFICATION

Nitrification is the biological oxidation of ammonia to nitrate with nitrite formation as an intermediate. The microorganisms involved are the autotrophic species Nitrosomonas and Nitrobacter which carry out the synthesis-oxidation reaction in two steps:

$$55\,NH_4^+ + 5\,CO_2 + 76\,O_2 \xrightarrow{\text{Nitrosomonas}} C_5H_7NO_2 + 52\,H_2O + 109\,H^+ 54\,NO_2^-$$
$$+54\,NO_2^-$$

$$400\,NO_2^- + 5\,CO_2 + NH_4^+ + 195\,O_2 + 2\,H_2O$$
$$\xrightarrow{\text{Nitrobacter}} C_5H_7NO_2 + 400\,NO_3^- + H^+$$

The cell yield for Nitrosomonas has been reported as 0.05-0.29 mgVSS/mg NH_3-N and for Nitrobacter 0.02-0.08 mgVSS/mg NH_3-N (Wong Chong and Loehr, 1975). A value of 0.15 mgVSS/mg NH_3-N is usually used for design purposes (U.S. EPA, 1975). The maximum specific growth rate of Nitrosomonas has been reported as 0.01-0.05 mgVSS/mg VSS.hr, and that of Nitrobacter averages 0.04 mgVSS/mgVSS.hr. It is generally accepted that the specific growth rate of Nitrobacter is higher than the growth rate of Nitrosomonas and hence there is no accumulation of nitrite in the process. Under these conditions the growth rate of Nitrosomonas will control the overall reaction.

Oxygen consumption ratios are 3.22 mg O_2/mg NH_3-N oxidized and 1.11 mg O_2/mg NO_2-N oxidized.

For effective nitrification to occur the sludge age must be greater than the reciprocal of the growth rate of Nitrosomonas. Shorter sludge ages will result in a washout of these organisms. Several investigators have reported nitrification to occur over a temperature range of 5-45°C with the optimum range being 25-35°C. The critical sludge age can be computed from the relationship (Downing et al., 1964):

$$\Theta_c = 2.13e^{0.098(15-T)} \tag{16}$$

For single sludge systems design for nitrifying conditions is usually based on Monod kinetics even though in most practical circumstances a zero order kinetic model can be used to fit the nitrification reaction.

8.1 Nitrification Kinetics

For single sludge systems the mass of nitrifiers per litre of reactor volume ($\dot{N}$ mg/L) is not easily determined. The nitrifiers (n) are inseparably contained with the heterotrophic matrix, the fractional content of which varies. $\dot{N}$ cannot be controlled in the same way as the mass of heterotrophs (X mg/L). In a practical sense wasting of X imposes a specific daily wasting rate on the mass of nitrifiers causing them to vary in mass in response to load fluctuations and to assume a specific rate of growth which approaches the imposed specific daily wasting rate. In practice the reactor is overdesigned with respect to average treatment requirements in order to keep effluent fluctuations within acceptable limits.

As an element of biomass passes through a plug-flow reactor, the ammonia nitrogen concentration in the associated liquid phase changes from a high of Ni(initial NH_3-N concentration after dilution of No by the recycle flow) at the inlet end of the reactor to a low of N_1 at the effluent end.

An average nitrifier growth rate can be stipulated (μna), assuming that all NH_3-N defined by (Ni - N_1) is oxidized and that there is no other addition to (Ni - N_1):

$$\mu na = \frac{a_N(Ni - N_1)}{\dot{N}\, t_n} \tag{17}$$

For a plug-flow reactor, the average nitrifier growth rate is given by:

$$\mu na = \frac{\mu max\ (No - N1)}{No - N1 + (1 + r)\ K_N \ln\left(\dfrac{No + r\ N1}{(1 + r)\ N1}\right)} \tag{18}$$

The following can be used for Equation (18):

$$K_N = 10^{(0.051T - 1.158)}, \text{ mg/L } NH_3 N \tag{19}$$

$$\mu max = 0.47 e^{[0.0998(T-15)]}\left(\frac{DO}{1.3 + DO}\right)^{[1-0.8333(7.2-pH)]}, \text{ day}^{-1} \tag{20}$$

In using Equation (17) the value of r will normally be in excess of unity if the maleffects of denitrification on effluent quality in the secondary settling unit are to be minimized. Current practice employs recycle ratios up to 2.5 in cases where the system does not employ biological denitrification and where conditions can be expected to enhance the effects of dissimilatory nitrogen removal in the solids-liquid separation unit.

8.2 Nitrification Rates

The nitrification rate can be calculated from the ammonia oxidation rate by recognizing that the nitrifiers are only a fraction of the total mass of biological solids in the system. Typical rates of nitrification are shown in Figure 10.

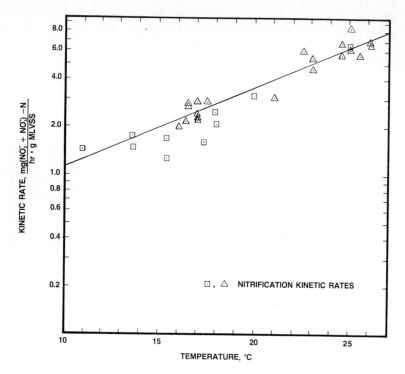

Figure 10.   Nitrification Kinetics;   (Eckenfelder, 1984),
⊡(Goronszy and Barnes, 1982).

The fraction of nitrifiers can be estimated from

$$f_N = \frac{a_N \, N_r}{a_N N_r + a S_r} \tag{21}$$

For a combined carbon oxidation nitrification system a
reasonable estimate for a_N of 0.15 g/gNH$_3$-N removed has been
suggested (U.S. EPA, 1975) and for a, 0.55 g/g BOD removed. It
should be emphasized that the values obtained for the nitrifier
fraction are estimates only and are not supported by measurements.
As an example for 160 mg/L of BOD$_5$ removed and 23 mg/L of
TKN oxidized, f_N can be calculated to be:

$$f_N = \frac{0.15 \times 23}{0.15 \times 23 + 0.55 \times 160}$$

$$= 0.038$$

9. SAFETY FACTOR SELECTION

Duirnal peaking in load requires the selection of a safety factor in order to achieve the desired extent of nitrification, initially calculated for a steady state condition, under transient loading conditions. TKN load variations have a significant impact on nitrification kinetics, and ammonia breakthrough can occur under peak load situations. The prevention of breakthrough requires the selection of a safety factor in excess of the ratio of the peak to average ammonia load. The application of a safety factor amounts to a buffering of the reactor against sudden increases in load. The safety factor used for plug-flow systems can be shown by kinetic theory to be less than that which needs to be used in a complete mix system for the same conditions. In the absence of data and in order to be conservative a minimum safety factor of 2 is recommended to account for diurnal peak loads. For complete mix configurations a safety factor of up to 2.5 has been recommended (U.S. EPA, 1975).

Example I - Industrial Wastewater

An industrial wastewater with a flow of $1890m^3$/day and a BOD_5 of 450 mg/L is to be treated in a completely mixed activated sludge reactor to achieve an effluent soluble BOD of 30 mg/L. The average wastewater temperatures are 20^oC and 10^oC for summer and winter respectively. Calculate:

1) The aeration basin volume
2) net sludge production
3) oxygen, nitrogen and phosphorus requirements.

The following parameters were defined for this wastewater:

K', days^{-1}	7.9
a	0.6
b, days^{-1}	0.12
a'	0.50
b', days^{-1}	0.17
X_d	0.80
Θk	1.065
Θb	1.04

1) Aeration Basin
Summer Conditions

From Equation (12)

$$\frac{S_r}{X_a X_v t} = K'\frac{Se}{So} = 7.9\left(\frac{30}{450}\right) = 0.527$$

Calculate Θ_c using Equation (11)

$$\frac{1}{\Theta_c} = a\frac{S_r}{X_a X_v t} - b = 0.6(.527) - 0.12 = 0.196$$

$\Theta_c = 5.1$ days

Calculate X_d from Equation (3) and

$$X_d = \frac{X_d'}{1 + bX_n \Theta_c} = \frac{0.8}{1 + 0.12(0.2)(5.1)} = 0.71$$

$$X_a = \frac{X_d}{X_d'} = \frac{0.71}{0.80} = 0.891$$

Calculate $X_v t$ using Equation (12)

$$X_v t = \frac{S_r}{X_a \left(K'\dfrac{Se}{So}\right)} = \frac{450 - 30}{0.891(0.527)} = 894$$

Check F/M for low D.O. filamentous bulking control, ie. for D.O. = 2.5 mg/l the F/M $\leq$ 0.4 (Figure 9).

$$\frac{F}{M} = \frac{S_o}{X_v t} = \frac{450}{894} = 0.50 \text{ (too high)}$$

Use F/M = 0.4

$$X_v t = \frac{S_o}{F/M} = \frac{450}{0.4} = 1,125 \frac{mg - day}{L}$$

Assume X_v = 2000 mg/L

$$t = \frac{1,125}{2,000} = 0.56 \text{ days}$$

The basin volume is

$V = Qt = 1890(0.56) = 1,060 m^3$

Recalculate Θ_c for the new value of $X_v t$ by Equation (6c):

$$\Theta_c = \frac{-117 + \sqrt{(117)^2 + 4(6.048)(1125)}}{2(6.048)} = 7.1 \text{ days}$$

X_d and X_a become 0.68 and 0.855 respectively

Winter Conditions - to check X_v

First correcting K' and b for temperature

$K_{10°C} = 7.9(1.065)^{10-20} = 4.2 \text{ days}^{-1}$

$b_{10°C} = 0.12(1.04)^{10-20} = 0.081 \text{ day}^{-1}$

Again using Equation (12)

$$\frac{S_r}{X_a X_v t} = 4.2(\frac{30}{450}) = 0.28$$

Θ_c for the winter condition is:

$$\frac{1}{\Theta_c} = 0.6(0.28) - 0.081 = 0.087$$

$\Theta_c = 11.5 \text{ days}$

$X_d = 0.67$ and $X_a = 0.84$

From Equation (12)

$$X_v t = \frac{420}{0.84(0.28)} = 1780 \quad \frac{\text{mg-day}}{\text{L}}$$

$$X_v = \frac{1780}{0.56} = 3200 \text{ mg/L}$$

2) Excess Biological Sludge Production
    Summer Conditions
    Using Equation (1)

$$\Delta X_v = \frac{aX_d'S_r}{X_d} - bX_v t = \frac{0.6(0.8)(420)}{0.68} - 0.12(1125)$$

$\Delta X_v = 161 \text{ mg/L}$

$= (161)(1.89) = 305 \text{ kg VSS/day}$

Winter Conditions

$$\Delta X_v = \frac{0.6(0.8)(420)}{0.67} - 0.081(1780)$$

$$= 157 \text{ mg/L}$$

$$= 157(1.89) = 297 \text{ kg VSS/day}$$

3) Oxygen Requirements

Using Equation (7) the summer oxygen requirement is:

$$r_r = a'S_r + b'X_dX_vt = 0.5(420) + 0.17(0.68)(1125)$$

$$= 340 \text{ mg/L}$$

$$= 340(1.89) = 643 \text{ kg } O_2/\text{day}.$$

the winter requirements adjusting b' for temperature is:

$$b'_{10°C} = 0.17(1.04)^{10-20} = 0.115$$

$$r_r = 0.5(420) + 0.115(0.67)(1780) = 347 \text{ mg/L}$$

$$= 347(1.89) = 657 \text{ kg } O_2/\text{day}$$

4) Nutrient Requirements

summer nutrient requirements using Equations (8) and (9) are;

$$N = .123(0.855)(305) + 0.07(\frac{0.80 - 0.68}{0.80})(305)$$

$$= 35 \text{ kg N/day}$$

$$P = 0.026(0.855)(305) + 0.01 (\frac{0.80 - 0.68}{0.80})(305)$$

$$= 7.2 \text{ kg P/day}$$

winter nutrient requirements are:

$$N = 0.123(0.84)(297) + 0.07 (\frac{0.80 - 0.67}{0.80})(297)$$

$$= 34 \text{ kg N/day}$$

$$P = 0.026(0.84)(297) + 0.01(\frac{0.80 - 0.67}{0.80})(297)$$

$$= 7.0 \text{ kg P/day}$$

Example 2 - Domestic Wastewater

BOD_5 influent $= 170$ mg/L
Suspended solids influent $= 170$ mg/L
Soluble/Colloidal BOD influent $= 120$ mg/L (S_o)
BOD_5 effluent $= 20$ mg/L (10 mg/L soluble)
Influent N $= 25$ mg/L (N_o as biodegradable TKN)
Effluent N $= 2$ mg/L (N_1 as ammonia N)
Temperature: T $= 10°C$ winter

pH   = 7.2
DO   = 2 mg/L
MLVSS   = 3000 mg/L (initial assumption for first
                                      estimate)
Sludge recycle ratio (r) = 0.5 - 1.0
Safety factor (SF)   = 2.0
Flow   = 15120 m^3/day
a   = 0.55
$b20^{o}C$ = 0.10 $days^{-1}$
$bT^{o}C$ = $b20 (1.004)^{T-20}$
a'   = 0.6
$b'_{10^{o}C}$ = 0.10

Stirred specific volume index (SSVI) = 100 - 119 ml/g
Hindered settling velocity given by (Johnstone et al., 1979)

$$u = 5.3e^{-0.4067X}, \ m/hr$$

$$K_N = 10^{[0.051(10)-1.158]} = 0.225 \ mg/L$$

$$\mu max = 0.47e^{[0.098(10-15)]} \left(\frac{2}{1.3 + 2} \right) = 0.175 \ day^{-1}$$

$$\mu na = \frac{0.175 \ (\ 17 \ )}{(17) + (1+2) \ x \ 0.225 \ \frac{\ln[17 + 2 \ x \ 2]}{(1 + 2) \ 2}} = 0.167 \ day^{-1}$$

Minimum solids retention time for nitrification at pH, temperature
and DO

$$= \frac{1}{0.167}$$

$$= 6.1 \ days$$

Design solids retention time for nitrification,

$$tn = 6.1 \ x \ SF$$
$$= 12.2 \ days$$

Mean hydraulic retention time for the limiting recycle condition
can be calculated from Equation (6a).

$$tc = \frac{\Theta_N \ a \ (So - Se)}{X_v (1 + bX_d \Theta_N)}$$

$$= \frac{12.2 \times 0.55 \times 160}{3000(1 + 0.068 \times 0.69 \times 12.2)}$$

$$= 5.4 \text{ hrs } (0.22 \text{ days})$$

Other values of tc for values of r up to 2 and the tabled values of X_v are summarized in Table 3b.

TABLE 3a. Design summary

$r^{(1)}$	N_i mg/L	N_R mg/L	$X_v^{(1)}$ mg/L	$tn^{(2)}$ hrs	S_{ot} mg/L	$S_o^{(3)}$
0.25	15.6	13.6	1700	7.3	138	98
0.5	13.3	11.3	2250	4.6	117	83
1.0	10.5	8.5	3375	2.3	90	65
2.0	7.7	5.7	4125	1.3	63	47

(1) Data (Johnstone et al., 1979) for 2xADWF
(2) Excluding safety factor
(3) BOD soluble/colloidal

TABLE 3b. Design summary

r	tc hrs	Floc Load$^{(4)}$ mgBOd/g MLSS	S(Soluble) Remaining mg/L	t_b hrs	$V^{(5)}$ m^3
0.25	7.7	43	67	6.2	11500
0.50	4.9	28	53	3.3	8694
1.0	2.4	14	33	1.0	5796
2.0	1.3	8.5	19	0.3	4914

(4) MLVSS/MLSS = 0.75
(5) Based on t_n x SF

Sludge yield from organic removal may be estimated from Equation (6),

$$\Delta X_v = aS_r - bX_d X_v t_c$$

$$= 0.55 \times 160 - 0.1 \times 0.65 \times 3000 \times 0.2$$

= 49 mg/L (assuming influent BOD is primarily soluble/
colloidal).

NH_3-N synthesized = 12%.

= 0.12 x 49
= 6 mg/l

Limiting nitrogen conditions therefore require that (25-6-2) mg/L of NH_3-N be oxidized i.e. 17 mg/L. The rate of nitrification at 10^oC can be taken from Figure 10 (1.1 mg NH_3-N/g MLVSS/hr).

The initial concentration of NH_3-N to be oxidized depends on the recycle rate. For example for r = 1.0,

$$Ni = \frac{1x19 + 1x2}{2}$$
= 10.5 mg/L

For this case the MLVSS = 3375 mg/L and

$$tn = \frac{(10.5 - 2)}{1.1x3.375}$$
= 2.3 hrs

Values for other recycle ratios are shown in Table 3a.

Alternatively the hydraulic retention time can be calculated using Equation (17).

$$\mu na = \frac{a_N (Ni - N1)}{\dot{N}\ tn}$$

and,

$$Hn = tn(1 + r)$$

$$f_N = \frac{a_N\ N_r}{a_N N_r + a\ S_r}$$

$$\dot{N} = f_N\ X_v$$

For this example the value of a_N is 0.15,

$$f_N = \frac{0.15\ x\ 8.5}{0.15\ x\ 8.5 + 0.55\ x\ 80}$$

= 0.0282
and $\dot{N}$ = 0.0282 x 3375
= 95.2 mg/L

124

It follows that

$$t_n = \frac{0.15 \ (8.5)}{95.2 \ x \ 0.167}$$
$$= 1.9 \ hrs$$

This procedure also affords a reasonable estimate of the critical time parameter for nitrification.

In order to meet maximum diurnal loading conditions and still comply with the effluent target value it is necessary to apply the peaking or safety factor, the magnitude of which approximates to the peak to average load that is experienced. A doubling of the influent ammonia concentration means that the retention time also needs to be doubled in order to complete the zero order oxidation reaction. A minimum operating condition, based on influent flow rate, therefore requires an hydraulic retention of 9.2 hours.

BOD REMOVAL

Initial BOD removal occurs by biosorption and biocoagulation mechanisms at the inlet end of the aeration tank. The extent of the absorption is very much dependent on the floc loading (assuming that the return sludge is near to its endogenous respiration rate).

Soluble/colloidal BOD   =   120 mg/L
Particulate   BOD   =   50 mg/L

Floc loading conditions at the inlet of the aeration tank (Table 3b) are calculated as follows (as an example assume r = 1),

Consider   r = 1

Soluble BOD in RAS   =   10 mg/L

Soluble/colloidal BOD = $\frac{1 \ x \ 120 \ + \ 2 \ x \ 10}{2}$
at inlet
                         = 65 mg/L
Floc load   =   65
                            3.375
                         = 14 mgBOD/g MLVSS

From Figure 4, percent soluble BOD biosorbed is 32 mg/L (assuming MLVSS/MLSS = 0.75). The BOD reductions which can be attributed to the biosorption mechanism are summarized in Table 3b.

Using Equation (14)

$$\frac{Se}{Si} = e^{-K_1 X_v t / S_i}$$

An appropriate value of K_{o1} taken from full scale plant data (Chambers et al., 1983) at $11^{o}C$, is 0.288 days^{-1}. Using this value gives an hydraulic retention for aeration of 1 hour. Values for other recycle ratios are shown in Table 3b.

Variation of the recycle ratio effects the concentration and hydraulic retention of the biomass in the aeration tank. Principle controlling factors on the biomass concentration are its settleability and the sizing of the secondary settling unit. This aspect is addressed using mass flux theory in order to provide a maximum solids handling capacity for each set of hydraulic conditions which are incorporated in the total design. Given that the biomass is in a nitrifying condition and that good sludge settlement characteristics can be maintained variation to biomass concentration via sludge recycle directly governs the time necessary to achieve a desired level of nitrification. Solids data used in Table 3a relate to a sludge having an SSVI at 3500 mg/l in the range 100-119 ml/g (Johnstone et al., 1979). Using the $10^{o}C$ nitrification rate of 1.10 mgNH_3-N/g MLVSS/hr the effect of recycle rate variation for specific operating conditions is illustrated (assuming MLVSS/MLSS = 0.75 and average influent NH_3-N of 25 mg/L) in Tables 3a and 3b.

Selecting an MLVSS of 3375 mg/L with r = 1.0 the required tank volume is 5796 m^3, which gives tn = 4.6 hrs (which includes the SF of 2).

Check on sludge Age.

It has been shown that for primary effluent the influent volatile suspended solids are all essentially degraded in a sludge age in excess of 10 days. The operational sludge age can therefore be calculated from Equation (6c) and yields

Θ_c = 33 days
and
X_d = 0.55

[The value of t used in Equation (6c) is based on the ADWF retention time (9.2 hrs).]

The long sludge age, as compared to the 12.2 days originally determined by applying a safety factor to the minimum growth conditions, is obtained as a result of the hydraulic peaking factor which was applied in order to meet peak load diurnal conditions. It is apparent then, that the sludge age will be determined by the hydraulic retention time that is necessary to meet the nitrification requirements.

$$\Delta X_v = aS_r - bX_d X_v t$$
$$= 0.55 \ (170-10) \times 15120 \times 10^{-3} - 0.068 \times 0.55 \times 3375.5796 \times 10^{-3}$$
$$= 0.554(170-10) - 0.068 \times 0.55 \times 3375 \times \frac{9.2}{24}$$

$$= 88 - 48$$
$$= 40 \ mg/L$$
$$\Delta X_v \ mass = 40 \times 15120 \times 10^{-3}$$
$$= 604 \ kg/day$$

Oxygen Computation

Carbonaceous
$$O_2 = a'S_r + b'X_d X_v t$$
$$O_2 = 0.6(170-10) + 0.1 \times 0.55 \times 3375 \times \frac{9.2}{24}$$

$$= 96 + 71$$
$$= 167 \ mg/L$$
$$kg \ O_2 = 167 \times 15120 \times 10^{-3}$$
$$= 2525 \ kg/day$$
$$= 105.2 \ kg/hr$$

Nitrogenous
$$O_2 = 17 \times 15120 \times 10^{-3} \times 4.33$$
$$= 1113 \ kg/day$$
$$= 46.4 \ kg/hr$$

Total oxygen = 3638 kg/day
$$= 152 \ kg/hr$$

An approximate oxygen demand distribution for sizing aeration intensity is shown in Table 4 which accords with experimental data taken from a number of plug flow systems (Boon et al., 1983). Nitrification is assumed to be constant throughout the aeration tank. Maximum nitrification rates will be achieved if the dissolved oxygen concentration throughout the aeration tank is not less than 2 mg/l.

TABLE 4. Variation of proportion of oxygen demand along
          the length of a plug-flow aeration tank
          (L:W = 20:1)

Proportion of aeration tank volume %	Proportion of oxygen demand for carbonaceous oxidation %	
	Average	Range Diurnally
20	60	40-85
20	15	5-20
20	10	5-15
20	10	5-15
20	5	<1-10

Process oxygen requirements for each 20 percent section of tank
volume are summarized in Table 5 and show the aeration system
design is required to have a turn down capacity of 10:1 for
maximum to minimum conditions. The value of α used in the design
of the aeration system will not remain constant. The efficiency
of transfer of oxygen into aerated liquor varies with the degree
of purification of the wastewater. Experiments with a fine-bubble
aeration system, operating in a plug-flow unit, have shown
variations in α from 0.3 at the start of the aeration tank (where
the waste water initially contacts the recycled sludge) to 0.8 at
the completion of treatment (high quality nitrified
effluent)(Chambers et al., 1983).

In order to minimize the effects of α variations, current
design practice uses L:W ratio of not less than 12:1 in order to
achieve 0.6 to 0.8   variation in α .

For the purposes of this design assume α in the first two 20
percentile sections is 0.6 and in the remaining three is 0.8.

TABLE 5. Aeration intensity, design summary.

Proportion of Tank Volume %	Carbonacous Oxygen Demand in Each Section kg/hr			Carbonacous + Nitrogenous Oxygen Demand in Each Section kg/hr		
	Average	Range Min	Max	Average	Range Min	Max
1st 20	64	42	90	73	52	100
2nd 20	16	5.5	21	25	15	31
3rd 20	11	5.5	16	20	15	25
4th 20	11	5.5	16	19	15	25
5th 20	5	1	10	15	10	20

Tank Dimensions

Total Aeration tank volume = (9.2) hrs at ADWF

$$= 5796 \ m^3$$

As a minimum design for 2 aeration basins each of $2900m^3$

$$N_D = \frac{0.068 \ W. \ H}{L. \ Q \ (1 + r)}$$

Influent/basin = 87.5 l/sec

$$= .0875 \ m^3/sec.$$

Select W = H = 5 m.

L = 116 m (L:W = 23:1)

$$N_D = \frac{0.068 \times 5 \times 5}{116 \times 0.0875 \ (1 + r)}$$

$$= \frac{0.148}{(1 + r)}$$

when r = 1
$$N_D = 0.084$$

which should yield adequate plug-flow conditions

REFERENCES

(1)   Boon, A.G., Chambers, B. and Collinson, B. "Energy
      Saving in the Activated Sludge Process." Report
      from Water Research Centre. Stevenage, UK. 1983.

(2)   Chambers, B., Robertson, P. and Thomas, V.K. "Energy
      Saving - Optimization of Fine-Bubble Aeration."
      Report from Water Research Centre. Stevenage,
      UK., February 1983.

(3)   Chudoba, J., Grau, P., and Ottova, V. "Control of
      Activated Sludge Filamentous Bulking - II."
      Selection of Micro-organisms by Means of a
      Selector." Water Research, 7, 1389-1406, 1973.

(4)   Downing, A.L., Painter, H.A., and Knowles, G.,
      "Nitrification in the Activated Sludge Process" J.
      Inst. Sew. Purif., 130, 1964.

(5)   Eckenfelder, W.W., Principles of Water Quality Management,
      CBI Publishing Co., Boston, 1980.

(6)   Eckenfelder, W.W., Goronszy, M.C., Quirk, T.P., "The
      Activated Sludge Process: State-of-the-Art"
      CRC Press, in Publication, 1984.

(7) Eckhoff D.W. and D. Jenkins, "Activated Sludge Systems
     Kinetics of the Steady and Transient States," Sanitary
     Engineering Research Laboratory, Report 67-12, University
     of California, 1967.

(8) Eikelboom, D.H. "Filamentous Organisms Observed
     in Activated Sludge." Water Research, 1975,
     9, 365-388.

(9) Goronszy, M.C., Barnes, D., and Irvine, R.L.,
     "Intermittent Biological Waste Treatment Systems-
     Process Considerations." Water, A.I. Chem. Eng.
     Symp. Ser. 77, 129-136, 1980.

(10) Goronszy, M.C. and Barnes, D., "Nitrogen Removal in
     Intermittently Decanted and Aerated Activated
     Sludge Systems." Process Biochemistry, 17, 35-41,
     1982.

(11) Grau, P. et al. "Kinetics of Multi-component subtrate
     Removal by Activated Sludge." Water Research,
     9, 637, 1975.

(12) Gujer, W., "The Effect of Particulate Organic Material
     on Activated Sludge Yield and Oxygen Requirement."
     Prog. Wat. Tech., 12, 79-95, 1980.

(13) Johnstone, D.W.M., Rachwal, A.J., and Hanbury, M.J.,
     "Settlement Characteristics and Settlement Tank
     Performance in the Carrousel System." Water
     Pollution Control, 78, 337-356, 1979.

(14) Levenspiel, O. Chemical Reaction Engineering, New York
     and London, John Wiley and Sons, 1962.

(15) Palm, J.H. Jenkins, D. and Parker, D.S. "The
     Relationship between Organic Loading, Dissolved
     Oxygen Concentration and Sludge Settleability in
     the Completely Mixed Activated Sludge Process."
     Jour. Wat. Pollut. Control Fed., 1980, 52,
     2484-2506.

(16) Quirk,T.P. and W.W. Eckenfelder, "Prediction and
     Utilization of the Active Mass in Activated
     Sludge RI Analysis and Design," 57th Annual WPCF
     Conference, New Orleans, 1984.

(17) Rhodes, M., "Effect of Organic Loading on Degradable
     Fraction," MS Thesis, Vanderbilt University, 1984.

130

(18) Richard, M.G., Jenkins, D. Hao, O., and Shimizu,
     G. "The Isolation and Characterization of Filamentous
     Microorganisms from Activated Sludge Bulking Report
     No. 81-2." Sanitrary Engineering and Environmental
     Health Research Laboratory, Univ. of California,
     Berkley, 1982.

(19) Rickert, D.A., and Hunter, J.V., "Colloidal Matter in
     Wastewaters and Secondary Effluents." Journal
     Water. Pollut. Control Fed., 44, 1, 134-139,
     1972.

(20) Siber S. and Eckenfelder, W., "Effluent Quality Variation
     from Multicomponant Substrates in the Activated Sludge
     Process," Water Research, 14, 471, May 1980.

(21) Tischler, "A Mathematical Study of the Kinetics of
     Biological Oxidation" MS Thesis, University of
     Texas, 1973.

(22) Tomlinson, E.J., and Chambers, B. "The Use of Anoxic
     Mixing Zones to Control the Settleability of
     Activated Sludge." Technical Report TR116, Water
     Research Centre. Stevenage, UK. 1979a.

(23) Tomlinson, E.J., and Chambers, B. "The Effect of
     Longitudinal Mixing on the Settleability of
     Activated Sludge." Technical Report TR122,
     Water Research Centre. Stevenage, UK. 1979b.
     1975.

(24) Upadhyaya, A.K. and W.W. Eckenfelder, "Biodegradable
     Fraction as an Activity Parameter of Activated Sludge"
     Water Research, 9:8:691, August, 1975.

(25) U.S. Environmental Protection Agency, Process Design
     Manual for Nitrogen Control, USEPA, October, 1975.

(26) Wong Chong, G.M., and Loehr, R.C., "The Kinetics of
     Microbial Nitrification". Water Research, 9, 2,
     1099, 1975.

Nomenclature

A   - sectional area, m^2

a   - cell yield coefficient mgVSS/mg BOD,COD,TOC

a'   - oxygen utilization coefficient, mgO_2/mg BOD,COD,TOC

a_N - yield factor, nitrification (mg nitrifier biomass
         produced/mg NH_3-N oxidized)

b   - endogenous rate constant, days^{-1}

b'   - oxygen utilization coefficient for endogenous respiration, days^{-1}

D   - coefficient of axial dispersion, m^2/sec

F/M   - food to microorganism ratio, days^{-1}

f   - fraction biological volatile solids

fn   - fraction of nitrifiers

H   - aeration tank side water depth, m

H_n - hydraulic retention time in nitrification reactor
         $H_n = V_n/Q$, days.

K   - kinetic rate coefficient based on X_v, days^{-1}

K'   - kinetic rate coefficient based on X_a, days^{-1}

K_1 - specific first order rate coefficient based on volatile solids, days^{-1}

K_N - Michaelis-Menton constant in nitrification, mg/L

k   - overall first order rate coefficient, days^{-1}

k_p - specific first order rate coefficient based on active mass, days^{-1}

L   - total length of aeration tank, m

MLSS - mixed liquor suspended solids, mg/L

MLVSS - mixed liquor volatile suspended solids, mg/L

N   - concentration of nitrifiers, mg/L
         $= fnX_v$

N_1 - NH$_3$-N concentration at exit of plug flow reactor, mg/L

N_i - NH$_3$-N concentration at inlet of plug flow rector, mg/L

N_D - dispersion number, dimensionless

N_r - nitrogen reduced, mg/L

Q   - wastewater flow, m^3/day

R   - sludge recycle flow, m^3/day

r   - R/Q

r_r - oxygen utilization rate, mgO$_2$/L

R_N - nitrification rate, mgNH$_3$-N oxidized/mg nitrifying VSS.day)

S_o - influent organic concentration, mg/L

S_e - effluent organnic concentration, mg/L

S_r - substrate removal across reactor, mg/L

S_i - soluble BOD,COD,TOC remaining after initial absorption mg/L

S_{BD} - organics reduction via biocoagulation, mg/L

S_{AB} - organics reduction via initial absorption, mg/L

S_{BD} - organics reduction via biodegradation, mg/L

SVI   - sludge volume index, ml/g

SSVI - stirred specific sludge volume index, ml/g

SF   - safety factor

t - hydraulic retention time, days

t_b - hydraulic retention time based on carbonaceous demand, hrs

t'_n - hydraulic retention time for nitrification calculated on growth rate, hrs

t_n - flow through time in plug flow nitrification reactor

$$t_n = H_n/(1 + r), \text{ days}$$

T - temperature, $^{\circ}C$

U - mean velocity of flow, m/s

u - hindered settling velocity, m/hr

V - aeration volume, m^3

W - aeration tank width, m

X - mixed liquor suspended solids, mg/L

X_a - active fraction, dimensionless

X_d - degradable fraction, dimensionless

X'_d - degradable fraction upon generation, dimensionless

X_n - non-degradable fraction, dimensionless

X'_n - non-degradable fraction upon generation, dimensionless

X_v - mixed liquor volatile suspended solids, mg/L

X_r - recycle suspended solids, mg/L

X_v - excess biological volatile solids, mg/L

Θ_c - sludge age or mean cell residence time, days

Θ_n - mean cell residence time for nitrification, days

μna   - average specific nitrifier growth rate for one reaction cycle of a plug flow reactor, days^{-1}

μmax - maximum specific nitrifier growth rate at tank, DO and pH condition, days^{-1}

α - oxygen transfer coefficient of the waste

MODELLING OF EFFLUENT QUALITY CONTROL FOR
ACTIVATED SLUDGE PLANTS

POUL ERIK SØRENSEN, HEAD OF WASTEWATER -,
SOLID- AND TOXIC WASTE DIVISION.

NATIONAL AGENCY OF ENVIRONMENTAL PROTECTION
MINISTRY OF THE ENVIRONMENT
29, STRANDGADE, COPENHAGEN K.
DENMARK.

INTRODUCTION.

In two earlier papers Westberg (1967,1969) from Sweden has
introduced the concept of the totally controlled activated sludge
process. He defined it as a process where a constant desired effluent quality could be maintained. He concluded from a mathematical model that this goal could only be achieved if an external
sludge storage was available from which sludge could be extracted
or delivered during high or low loading periods respectively.

For practical reasons it is difficult to operate an activated sludge process with an external sludge storage, as it
would require an extra aeration basin.

However, as it will be shown the step-feed configuration of
the activated sludge process offers the availability for sludge
storage without needing an additional tank.

THE STEP-FEED PROCESS.

Many operational benefits are possible for the activated
sludge process by control of the wastewater feed distribution
to the aeration tank. As a first benefit, this permits control
of the hydraulic and sludge residence times. This enables the
operating engineer to exert control of the effluent quality,
the sludge age, and nitrification as well as to optimize energy
utilization and sludge production. As a second benefit, step feed
can be used beneficially in preventing gross process failure by
hydraulic overloading or sludge bulking.

Figure 1 illustrates this form of control in a case with

three separate tanks. The resultant sludge solids concentration are shown for four different feed patterns. It is seen that in cases 2, 3 and 4, sludge tends to accumulate in the first tanks. If the distribution of the wastewater is controlled, it will be possible during periods of low organic loading to use a feed system such as that in Case 4, where only part of the sludge is directly used while the rest is being stabilized. With increased

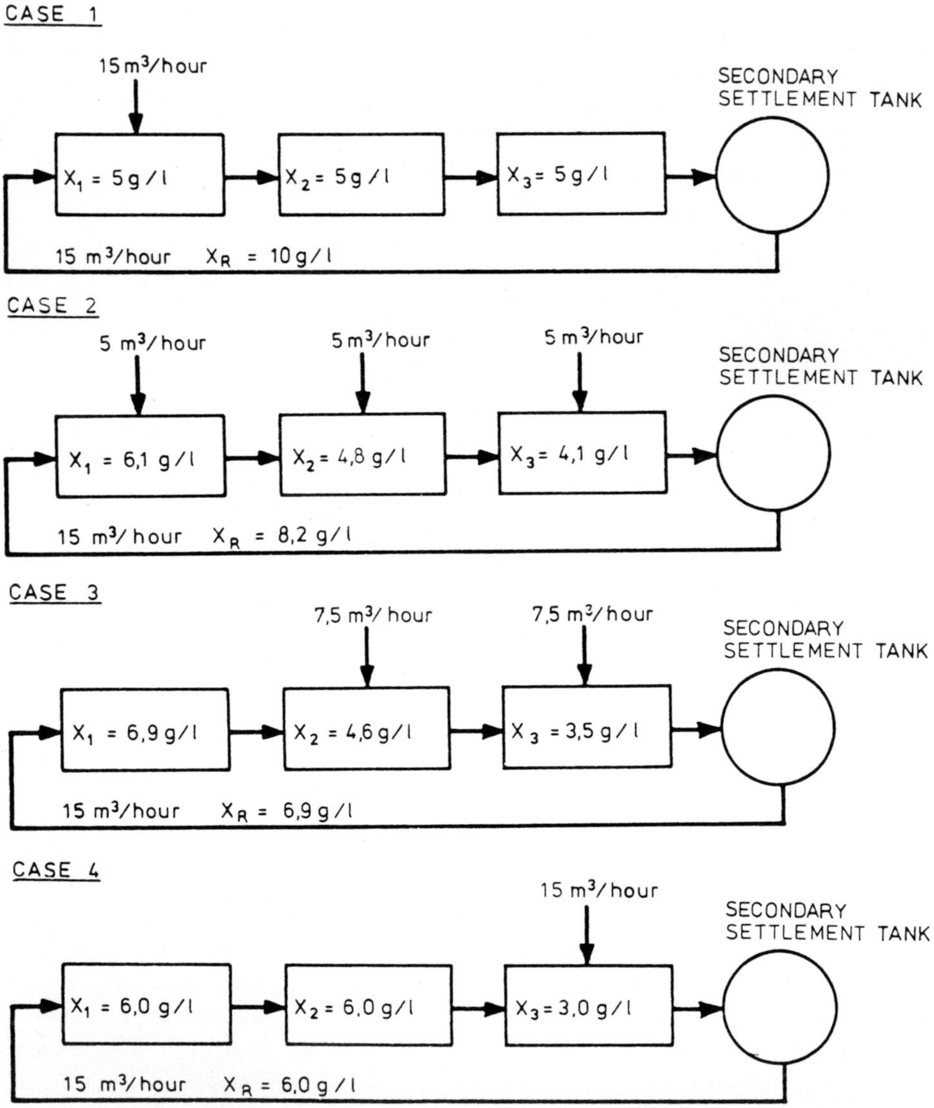

CASE 1

CASE 2

CASE 3

CASE 4

Figure 1: Examples of how sludge can be stored by changed distribution of the wastewater inflow to the aeration tanks.

organic loading, the system can, for instance, first be changed over to Case 3, where the sludge mass in two tanks is now used. If the loading rises further, Case 2 and finally Case 1, can be used. The wastewater on its way through the plant now comes into contact with larger and larger proportions of the sludge mass, so that the resultant purification efficiency rises. The ability to store sludge in the first tanks, and the ability to control residence time and sludge flux to the settler makes the above-mentioned benefits possible.

Torpey (1948) was the first to report on successful applications of a step feed strategy. By shifting the wastewater feed from the middle to the back end of an aeration tank, he was able to cure cases of bulking sludge.

Andrews et al. (1972, 1974, 1976) have described how different step feed strategies affect the sludge flux to the settler, the sludge concentrations in the tanks, and the sludge age.

To obtain experimental results of such operation, the control strategy was evaluated as part of a two-year pilot plant study investigating control schemes for the activated sludge process. The study was conducted during 1976-78 by the Danish Water Quality Institute for the National Agency of Enviromental Protection, Denmark, and the City of Copenhagen. The results have been published by Sørensen (1977, 1979, 1980)

In 1982 a full scale evaluation was performed. This study has not yet been published.

KINETIC EQUATIONS.

The cost of carrying out a large number of field experiments was high. Consequently, for financial reasons, it was not possible to perform all the experiments that would have been desirable. In order to obtain a greater amount of knowledge, a number of simulations were therefore carried out.

The data obtained from the experiments were used to calibrate a dynamic mathematical model of the process. The kinetic equations used in the model were, with some alterations, essentially the equations developed by Busby and Andrews (1975) and Stenstrom (1975, 1979).

The model was a structured model consisting of the following basic equations:

BOD removal $\qquad : r_S = R_{XS} \cdot \dfrac{S}{S + K_S} \cdot XT$

Formation of active mass $: r_{XA} = R_{XA} \cdot \dfrac{XS/XT}{K_{XS} + XS/XT} \cdot XA$

Formation of stored mass $: r_{XS} = r_S - \dfrac{r_{XA}}{Y1}$

Formation of inert mass $\quad : r_{XI} = Y2 \cdot R_{XI} \cdot XA$

Oxygen uptake rate $\qquad : r_{OX} = \left(\dfrac{1 - Y1}{Y1}\right) \cdot r_{XA} + \left(\dfrac{1 - Y2}{Y2}\right) \cdot r_{X1}$

The only major difference between this model and the models of
Busby and Andrews, and Stenstrom is the BOD removal term. This
was found by mathematical optimization to be nearly first order
and not dependent on the amount of stored mass, which was found
to be small.

In Table 1 is given a list of symbols and values of the
reaction rate and stoichimetric constants determined by mathematical optimization.

The DO in the plant was controlled at 2.0 mg/l in the last
tank for all simulations. The return sludge was pumped at a fixed
rate proportional to the inflow. It was thus assumed that sludge
was returned from the settler at the same rate in which it entered. This means that the settler did not limit the influence of
the step feed strategies. ·

Andrews and Lee (1972) have demonstrated the effect of a step
feed change on a settler with a certain mass of sludge stored.
In such a case the sludge stored in the settler will disappear,
when the feed is shifted from the front end to the back end of
the settler. This is due to the decrease in sludge flux to the
settler. This will increase the temporary impact of the shift.

With the model it was possible to simulate different step
feed patterns and thus solve the difficult problem of how to
establish the optimum pattern for obtaining a desired effluent
quality. The problem is not easy to solve. One reason is that
variation in the BOD removal are only partly followed by changes
in the sludge respiration. This was observed in all the intensive
day and night measurements.

Symbol		Values
r_S	BOD removal rate	mg/l/h
r_{XS}	Rate of formation of stored biomass	mg/l/h
r_{XA}	Rate of formation of active biomass	mg/l/h
r_{XI}	Rate of formation of inert biomass	mg/l/h
r_{OX}	Rate of oxygen uptake	mg/l/h
R_{XS}	Rate constant for BOD removal	$0,15 \text{ hour}^{-1}$
R_{XA}	Rate constant for formation of active biomass	$0,45 \text{ hour}^{-1}$
R_{XI}	Rate constant for formation of inert biomass	$0,10 \text{ hour}^{-1}$
K_S	Michaelis constant	150 mg/l
K_{fS}	Saturation constant	0,2 mg stored biomass/mg total solids
Y1	Yield constant	0,7 mg active biomass/mg stored biomass
Y2	Yield constant	0,3 mg inert biomass/mg active biomass
S	Substrate concentration	mg BOD/l
XS	Fraction of sludge solids which is stored	mg MLSS/l
XA	Fraction of sludge solids which is active	mg MLSS/l
X1	Fraction of sludge solids which is inert	mg MLSS/l
XT	Total sludge solids	mg MLSS/l

Table 1 Symbols

As can be seen from figure 2 which is real data, there is a much bigger variation in the BOD removal than in the oxygen uptake rate. On the average, the BOD removal varied by a factor

138

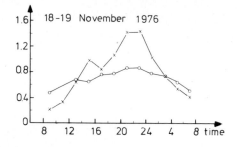

x = kg BOD removed / hour

O = kg O_2 transferred / hour

Figure 2: Curves showing BOD Removal Rate and Oxygen Uptake
Rate as Function of Time.

of six between maximum and minimum during 24 hours, whereas the
oxygen uptake rate only varied by a factor of two. This means
that the activated sludge process is itself able to even out the
dynamic differences occurring in the BOD loading. The reason for
this phenomenon is probably that some storage of matter takes
place during high loaded periods and that the activity of the
sludge increases during these periods. This leads to an increased
endogenous respiration which only slowly decreases when the high
load is removed.

Another problem is the changes in sludge concentrations in
the tanks, which is brought about by changes in the feed distribution.

These optimization problems can be solved by using a feed
forward model.

MATHEMATICAL MODEL.

A flow chart of the pilot-plant is shown in figure 3 to
facilitate comprehension.

A dynamic model of the processes can be expressed in the
following equations.

139

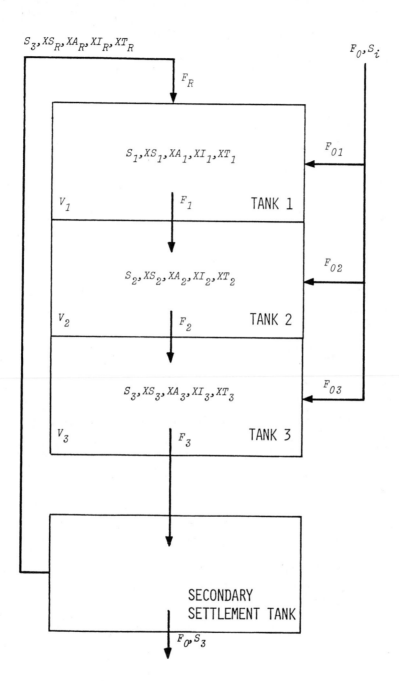

Figure 3: Flow chart for pilot plant.

TOTALS

XT1 = XS1 + XA1 + XI1

XT2 = XS2 + XA2 + XI2

XT3 = XS3 + XA3 + XI3

Xtotal = V_1 . XT1 + V_2 . XT2 + V_3 . XT3

BOD BALANCES

$$\frac{dS_1}{dt} = \frac{F_{01} \cdot S_i + F_R \cdot S_3 - F_1 \cdot S_1}{V_1} - r_{XS1}$$

$$\frac{dS_2}{dt} = \frac{F_{02} \cdot S_i + F_1 \cdot S_1 - F_2 \cdot S_2}{V_2} - r_{XS2}$$

$$\frac{dS_3}{dt} = \frac{F_{03} \cdot S_i + F_2 \cdot S_2 - F_3 \cdot S_3}{V_3} - r_{XS3}$$

The following balances are performed similar to the BOD balances. Only the equations for tank 1 are shown for these balances.

BALANCE OF STORED MATTER

$$\frac{dXS_1}{dt} = \frac{F_R \cdot XS_R - F_1 \cdot XS_1}{V_1} + r_{XS1} - \frac{r_{XA1}}{Y_1}$$

BALANCE OF ACTIVE BIOMASS

$$\frac{dXA_1}{dt} = \frac{F_R \cdot XA_R - F_1 \cdot XA_1}{V_1} + r_{XA1} - \frac{r_{XI1}}{Y_2}$$

BALANCE OF INERT BIOMASS

$$\frac{dXI_1}{dt} = \frac{F_R \cdot XI_R - F_1 \cdot XI_1}{V_1} + r_{XI1}$$

DISSOLVED OXYGEN BALANCE

$$\frac{dC_1}{dt} = \frac{F_R \cdot C_R - F_1 \cdot C_1}{V_1} - r_{OX1} + COX1$$

COX1 is the oxygenation capacity of the aeration equipment in tank 1.

AERATION CONTROL.

The supply of oxygen to the tanks is controlled by means of a proportional integral regulator (PI-reg.), which controls the air blowers speed on the basis of measurement of the oxygen concentration in one of the three tanks.

The dynamics of the treatment plant are characterised by large time constants (of the order of magnitude of hours), and the regulator setting is therefore not critical. Suitable values of the proportional amplification (PRC) and the integration time (TAUI) can, for instance, be found by systematic optimization methods or by trial and error.

The regulating part of the simulation program is described below:

For the oxygen probe, say, tank 1 an electrical signal, SEL, is obtained. In the present control system it is not necessary to take into account 1st or 2n- order (asymmetrical) transfer function in the probe, as the time constants in this will more often than not be less than 0,5 min.

The error signal E is obtained by subtraction of the relevant value of C1, the oxygen concentration in tank 1, and CSETP, the oxygen setpoint concentration.

In the regulator part itself, the manipulated error signal EM is produced from the error signal E.

$$EM = PRC \cdot EBEG/TAUI + PRC \cdot E + (PRC/TAUI) \cdot \int_{o}^{t} Edt$$

or in FORTRAN

$$EM = PRC \cdot E + (PRC/TAUI) \cdot INTGRL \ (EBEG,E)$$

The meaning of the first part of the equation, the proportional part, can be varied by changing the proportional amplification, PRC. The second part, the integral section, depends on the PRC and the integration time constant, TAUI, and its meaning in the whole expression can be changed by adjustment of TAUI. If TAUI is made very large, the integration effect disappears and the system is proportionally regulated.

The blower speed, BL, then depends on the regulator output.

MODELLING OF DISTRIBUTION OF WASTEWATER TO AERATION TANKS.

This is done by forming three matrices, each of which controls the feeding of wastewater to a tank depending on the blower speed.

If, for instance, the following wastewater distribution is desired:

Blower speed:	Wastewater to:
200 - 350	Tank 3
350 - 500	Tank 2 + tank 3
500 - 650	Tank 1 + tank 2 + tank 3
650 - 800	Tank 1

the following three functions are formed:

$$A1 = \left\{ \begin{array}{cccccccc} 200 & 350 & 350.1 & 500 & 500.1 & 650 & 650.1 & 800 \\ 0 & 0 & 0 & 0 & 0.33 & 0.33 & 1 & 1 \end{array} \right\}$$

$$A2 = \left\{ \begin{array}{cccccccc} 200 & 350 & 350.1 & 500 & 500.1 & 650 & 650.1 & 800 \\ 0 & 0 & 0.5 & 0.5 & 0.33 & 0.33 & 0 & 0 \end{array} \right\}$$

$$A3 = \left\{ \begin{array}{cccccccc} 200 & 350 & 350.1 & 500 & 500.1 & 650 & 650.1 & 800 \\ 1 & 1 & 0.5 & 0.5 & 0.33 & 0.33 & 0 & 0 \end{array} \right\}$$

CALIBRATION OF THE MODEL AND SIMULATIONS.

Calibration of the model was done with the aid of an optimization programme developed for CSMP. The optimization technique was "the complex method of Box". This optimization programme was programmed in Fortran and is written by Michael Stenstrom and published in his Ph. D.-theses (1975).

The calibrations are done by reading in tables of measured values in the programme and stating which constants it is desired to have optimized. With the aid of a series of iterations the programme then optimizes the constants so that the best possible correlations is achieved between measured and simulated values.

The results of the optimizations are shown in table 1.

In figure 4 is shown a comparison between measured and simulated values. It can be seen that the structured model is able to predict the levelling out of the oxygen uptake rate in proportion to the BOD-removal.

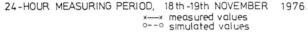

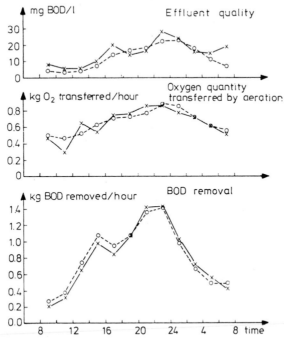

Figure 4: Comparison between measured and simulated values.

Figure 5 shows results of simulations of five different distribution patterns of the incoming wastewater.

a) All wastewater to Tank 3
   (Corresponds to a contact stabilization plant)

b) Step areation
   (The wastewater distributed equally to the three tanks)

c) Plug flow
   (All wastewater fed to tank 1)

d) Controlled wastewater distribution. Imaginary BOD feedback control is imposed to control the effluent BOD to 20/mg/l. Flow is shifted between tank 1 and 3. This case will give the best possible control.

e) Controlled wastewater distribution. Flow to tank 1 controlled proportional to the oxygen consumption rate.

The highest possible degree of purification is obtained with plug flow, while the lowest degree is obtained with feeding to tank 3 throughout the 24-hour period.

144

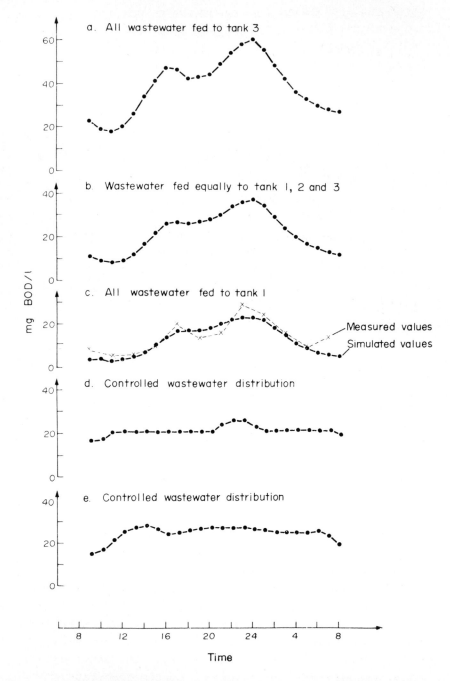

Figure 5: Model calculations. BOD in effluent with 5 different
patterns of wastewater feed to aeration tanks.
Measured data are plotted in case c.

Other feed patterns give results lying between these two. It is seen that even in case d it is not possible to get a perfectly constant effluent quality. In periods with either very high or very low loadings, step feed does not provide sufficient control.

CONTROL OF NITRIFICATION IN A PURE OXYGEN ACTIVATED SLUDGE PROCESS.

A simular strategy to the above mentioned was applied in a study for the city of Houston carried out in 1979. The study is published in a paper by Andrews, Sørensen and Garrett (1980).

In this study the step feed strategy was used to control the ammonia concentration in the effluent of a two step pure oxygen process. The purpose was to obtain an optimal level of ammonia to minimize the cost of chlorination of the treated wastewater. The control strategy was tested for a period of two weeks in a pilot plant. The strategy was successful and the calculated savings in chlorine costs are approximately 250.000 $ per year.

CONCLUSIONS.

The described dynamic model and the optimization method has proved very successful in effluent quality control experiments in Copenhagen, Denmark and Houston, Tx, USA.

But the model and the technique can be used in all forms of kinetic studies of activated sludge plants.

REFERENCES.

1. Niels Westberg, "A Study of the Activated Sludge Process as a Bacterial Growth Process". Water Research 1, 795-803 (1967).
2. Niels Westberg, "An Introductory Study of Regulations in the Activated Sludge Process". Water Research 3, 613-621 (1969).
3. John F. Andrews and Chin R. Lee, "Dynamics and Control of a Multi-Stage Biological Process". Proc.IV IFS: Ferment.Technol. Today 55-63 (1972).
4. John F. Andrews, Review Paper: "Dynamic Models and Control Strategies for Wastewater Treatment Processes". Water Research 8, 261-289 (1974).
5. John F. Andrews, Michael K. Stenstrom and Heinrich O. Buhr, "Control Systems for the Reduction of Effluent Variability from the Activated Sludge Process". Progress in Water Technology 8, No. 2, 41-68 (1976).

146

6. W.B. Torpey, "Practical Results of Step Aeration". Sewage Works Journal 20, 781 (1948).

7. Poul Erik Sørensen, "Evaluation of Control Schemes for the Activated Sludge Process Using Large Pilot Scale Experiments". Progress in Water Technology 9, 419-425 (1977).

8. Poul Erik Sørensen, "Pilot-Scale Evaluation of Control Schemes for the Activated Sludge Process". Contributions from the Water Quality Institute, No. 1 (1979).

9. Poul Erik Sørensen, "Evaluation of Operationel Benefits to the Activated Sludge Process Using Feed Control Strategies". Progress in Water Technology, 12 pp 109-125. (1980).

10. J.B. Busby and J.F. Andrews, "Dynamic Modelling and Control Strategies for the Activated Sludge Process". Journal Water Pollution Control Federation, 47, 1055 (1975).

11. M.K. Stenstrom and J.F. Andrews, "Real-Time Control of Activated Sludge Process". Journal of the Environmental Engineering Division, American Society of Civil Engineers, Vol. 105, 245-260 (1979).

12. M.K. Stenstrom, "A Dynamic Model and Computer Compatible Control Strategies for Wastewater Treatment Plants". Ph.D. Dissertation, Clemson University, Clemson, S.C. (1975).

13. John F. Andrews, Poul Erik Sørensen and M.T. Garrett, "Control of Nitrification in the Oxygen Activated Sludge Process". Progress in Water Technology, 12 (1980).

MATHEMATICAL MODEL FOR DUAL-POWER LEVEL, MULTICELLULAR (DPMC) AERATED LAGOON
SYSTEMS

LINVIL G. RICH

1 INTRODUCTION

Present wastewater treatment practice relies heavily on high-maintenance,
mechanically-complex systems. Such systems may be optimal for the treatment of
wastewaters with medium to high discharge rates, or those with special treatment requirements. However, for wastewaters with relatively low discharges and
low-to-moderate concentrations of biodegradable organics, high-maintenance,
mechanically-complex systems may be less than optimal. The cost of these systems, both capital and operational, and the requirement for high level operator
skills can limit their feasibility for such applications. Consequently, a need
exists for reliable, low-maintenance, mechanically-simple wastewater treatment
systems capable of producing effluents of acceptable quality. Aerated lagoons
properly designed meet this need.

Dual-power level, multicellular (DPMC) aerated lagoon systems discussed here,
although mechanically-simple, have complex configurations and are operated in
such a way as to make their design difficult and not amenable to the precise
application of theoretical principles. However, sufficient information is now
available in the literature to permit a rational development of design methodology based on approximations and reasonable assumptions and which is relatively
easy to use. It is to this end that the chapter has been written.

2 PERFORMANCE HISTORY

The performance of aerated lagoon systems is generally evaluated in terms of
the total five-day, 20°C biochemical oxygen demand (BOD_5), and the total suspended solids (TSS) discharged in the effluent. The total BOD_5 is a function
of the soluble five-day, 20°C biochemical oxygen demand and the total suspended
solids concentration in the effluent.

Table 1 summarizes the performance of six aerated lagoon systems located in
the Piedmont region of South Carolina (Rich, 1978). Each system consisted of
an aeration cell followed by an unaerated polishing pond, an arrangement popular
in systems constructed during the period from 1965 to 1975. Furthermore, each
system was characterized by long retention times. Either through intention or
miscalculation both the aeration cells and the polishing ponds operated with
retention times ranging from one to as many as four weeks each. Note should be
made of the relatively high level of suspended solids found in the effluents

TABLE 1.

Aerated lagoon performance. (Effluent characteristics expressed in milligrams per liter).

System	50%[a] BOD$_5$	TSS	90%[b] BOD$_5$	TSS	Number of samples
1	29	49	49	79	34
2	30	44	52	86	34
3	24	49	37	73	34
4	29	52	47	102	25
5	22	39	38	65	26
6	32	64	56	108	26
Avg	28	50	47	86	

[a]Values exceeded in 50% of effluent samples.
[b]Values exceeded in 10% of effluent samples.

from these systems. Fifty percent of the time, the TSS ranged from 39 to 64 mg/ℓ, and in one out of ten samples from each of two systems the TSS exceeded 100 mg/ℓ.

High suspended solids concentrations contribute also to higher biochemical oxygen demands. From the same study (Rich, 1978), the average correlation between effluent BOD$_5$ and effluent TSS was found to be

$$BOD_5 = 13.0 + 0.31 \text{ TSS} \tag{1}$$

Equation 1 suggests that, on an average, the BOD$_5$ contributed by the suspended solids at least equals that in a soluble form.

High effluent suspended solids measured during the study was the result of excessive algal growth both in the aeration cells and in the polishing ponds. During a period of dry weather over which no apparent changes in algal species occurred, the correlation between the total suspended solids and chlorophyll a concentrations in the effluents of four cells aerated at power levels of about 1 W/m^3 was found to be (White & Rich, 1976)

$$TSS = 21.3 + 143 \text{ chlorophyll a} \tag{2}$$

with a correlation coefficient of 0.89. Equation 2 suggests that if algae had not been present, the TSS on an average, would not have exceeded 22 mg/ℓ.

In summary, 1) algal growth can, and does, increase suspended solids in effluents from aerated lagoon systems, 2) increases in effluent suspended solids will increase the total effluent BOD$_5$, and 3) in the absence of algal cells, an average effluent suspended solids concentration significantly below 30 mg/ℓ can be achieved. It is apparent, therefore, that the objectives in aerated lagoon design should be 1) the reduction of soluble BOD_5 to a relatively low level, and 2) the suppression of algal growth.

3 SOLUBLE BOD_5 REMOVAL
3.1 Steady-state model

For present purposes, two types of aerated lagoons are defined. Lagoons in which the settleable solids are maintained in suspension will be referred to here as being completely suspended. Those in which few, if any, of the settle-able suspended solids are maintained in suspension will be referred to as being partially suspended.

It has been shown that the combination of a completely-suspended cell fol-lowed by one or more partially-suspended cells theoretically requires less lagoon volume to achieve a given soluble BOD_5 concentration in the system's effluent than does a single cell of either type (Kormanik, 1972; Tikhe, 1975; Rich & White, 1977). Such a system, called a dual-power level, multicellular (DPMC) aerated lagoon system, is shown as a flow diagram in Fig. 1.

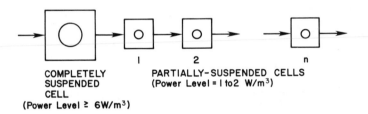

COMPLETELY
SUSPENDED
CELL
(Power Level ≥ 6W/m³)

PARTIALLY-SUSPENDED CELLS
(Power Level = I to2 W/m³)

Fig. 1. Dual-power level, multicellular aerated lagoon system.

A relationship between the hydraulic retention times in the first (completely-suspended) cell and in the following (partially-suspended) cells to yield a given soluble BOD_5 in the system effluent has been developed from steady-state mass balances of BOD_5 across each cell in the system (Rich, 1982c).

$$(V/Q)_j = \frac{\left[\dfrac{1 + k_d(V/Q)_1}{S_e(\hat{\mu}/K_S)(V/Q)_1}\right]^{1/n} - 1}{(\hat{\mu}/YK_S)X_a} \qquad (3)$$

where $(V/Q)_j$ = hydraulic retention time in each of the partially-suspended cells (d); $(V/Q)_1$ = hydraulic retention time in the completely-suspended cell (d); n = number of equal-sized, partially-suspended cells in series; μ = maximum specific growth rate (d^{-1}); K_S = saturation constant (mg/ℓ); Y = growth yield; k_d = specific decay rate (d^{-1}); S_e = soluble BOD_5 in system effluent (mg/ℓ); X_a = average biomass concentration remaining in suspension in partially-suspended cells (mg/ℓ).

For comparison purposes, Eq. 3 has been solved for four different systems using a typical set of values for the equation parameters and variables. The comparison is illustrated in Fig. 2 where the total hydraulic retention time in

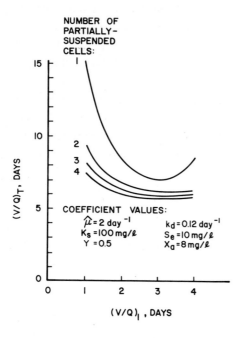

Fig. 2. Total retention time in dual-power level, multicellular lagoon system as a function of retention time in the first (completely-mixed) cell.

each system is plotted as a function of the retention time in the first (completely-suspended) cell. The systems differ in the number of partially-suspended cells. It is to be noted that the total hydraulic retention time required to achieve the objective of 10 mg/ℓ of soluble BOD_5 in the effluent from the last cell is less for those systems with the larger number of partially-suspended cells. Furthermore, by using a larger number of partially-suspended cells, the total retention time required in the system is less sensitive to the retention time provided for in the first cell, especially for values of the latter exceeding one day.

The advantages cited above are enhanced even more by further increase in cell number. However, greatest enhancement is realized as the number of partially-suspended cells is increased from one to four.

3.2 Parameter estimation

In system design, two factors should be considered relative to the hydraulic retention time in the completely-suspended cell. Power requirements in this cell may be many times (on a unit volume basis) those in the partially-suspended cells. Consequently, longer retention times in the former result in higher power costs. On the other hand, if the hydraulic retention time in the completely-suspended cell is too short, washout of bacterial biomass may occur, resulting in process failure. As a design limit, the minimum retention time can be calculated with an expression which avoids washout by a safety factor (Rich, 1982c).

$$(V/Q)_1 = f \frac{K_S + S_0}{\hat{\mu} S_0} \tag{4}$$

where S_0 = average total BOD_5 of the influent wastewater (mg/ℓ).

The value of the washout safety factor, f, that should be used depends upon the level of risk that the designer wishes to assume. From information available, it appears that f should be at least 3 (Lawrence & McCarty, 1970).

The use of Eqs. 3 and 4 requires a knowledge of the values of the kinetic coefficients - Y, $\hat{\mu}$, and K_S. Procedures are available for their determination (Sundstrom & Klei, 1979; Benefield & Randall, 1980). Values for four different types of wastewater are listed in Table 2. The value of the specific decay rate, k_d, appears to be relatively independent of the wastewater being treated (Rich, 1982a). Furthermore, the value of $(V/Q)_j$ calculated with Eq. 3 is relatively insensitive to the value of k_d used in the calculation. For these reasons, it is suggested that a value of k_d equal to 0.25 d^{-1} at 20°C can be used for determining the value of the variable in Eq. 3.

TABLE 2
Biological parameters for several types of wastewater.

	$\hat{\mu}_{20}$, d^{-1}	K_S, mg/ℓ	Y, mg/mg	Reference
Domestic sewage	13.2	120	0.5	(Jorden et al., 1971)
Pork processing*	17.3	107	0.56	(Fleming, personal communication, 1980)
Shrimp processing	18.5	85.5	0.50	(Horn & Pohland, 1973)
Soybean	12.0	355	0.74	(Jorden et al., 1971)

*After preliminary anaerobic treatment

Temperature will influence the values of $\hat{\mu}$ and k_d. Such influence can be expressed quantitatively by (Bartsch & Randall, 1971)

$$\hat{\mu} = \hat{\mu}_{20} \ (1.10)^{T-20} \tag{5}$$

and (Randall et al., 1975)

$$k_d = k_{d20} \ (1.05)^{T-20} \tag{6}$$

where T = temperature, °C.

At temperatures greater than 20°C, values of k_d do not appear to differ significantly from that at 20°C (Randall et al., 1975).

The average biomass concentration remaining in suspension in the partially-suspended cells, X_a, is the most difficult variable in Eq. 3 to evaluate. For most temperatures, configurations, and sizes that would normally be encountered in the design of an lagoon system to treat a domestic wastewater, it has been estimated that the value of X_a will vary from 7 to 9 mg/ℓ (Rich, 1982c). However, in spite of the key role that X_a plays in Eq. 3, the total BOD_5 in the effluent will be relatively insensitive to the particular concentration selected. A group of researchers (Eckenfelder et al., 1972) report the relationship

$$(BOD_5)_t = (BOD_5)_s + 0.84 \ X_a \tag{7}$$

where $(BOD_5)_t$ and $(BOD_5)_s$ are the effluent total and soluble BOD_5, respectively. A plot of Eq. 7 is presented in Fig. 3 for a specific cell-size combination and a specific set of typical parameter values. Although one may question the validity of Eq. 7, one cannot deny the trade-off occurring between the soluble BOD_5 and the biomass concentration in the determination of the effluent total BOD_5.

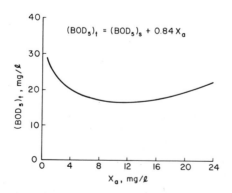

Fig. 3. Sensitivity of total effluent BOD_5 to change in biomass concentration.

3.3 Estimating lagoon temperatures

The use of Eq. 3 depends upon an estimate of the water temperature for the calculation of $\hat{\mu}$ and k_d. Relationships are available with which aerated lagoon temperatures during the coldest period of the year can be estimated from a knowledge of the appropriate values of the surface heat exchange coefficient and the equilibrium temperature (Rich, 1976).

In a multicellular lagoon system with a series configuration, water temperature will vary from cell to cell. Precise determination of temperature levels in each cell requires an iterative approach, which is time consuming, and in most cases, not justifiable considering the error introduced by assuming generalized values of the surface heat exchange coefficient and the equilibrium temperature. A shorter but less precise approach is to consider the entire system as consisting of a single cell, and to assume the water temperature to be 5°C above the ambient air temperature during the coldest week of the year, or 2°C, whichever is the maximum. Most of the temperature drop will occur in the first cell where both the difference between the ambient air and the waste-water temperature, and the turbulence are the greatest. Since the retention time in this cell will be relatively short (less than 2 or 3 days) and since most of the BOD_5 removal will take place there, it appears reasonable to assume an overall, effective wastewater temperature of a few degrees above the ambient.

3.4 Model sensitivity

Using a typical set of parameter values, Eq. 3 was solved for a soluble effluent BOD_5 of 10 mg/ℓ. Subsequently, using the same retention times, $(V/Q)_1$ and $(V/Q)_j$, a series of solutions were made, in each of which the value of a single variable was changed from the value used in the initial solution. In this way, the consequence of using incorrect parameter values on the soluble effluent BOD_5 predicted by the model could be estimated.

Figs. 4 and 5 illustrate the results of the analysis. The families of curves found therein relate the effluent soluble BOD_5 to different values of the parameters normalized to the values used in the initial solution. Effluent soluble BOD_5 is shown to be relatively insensitive to changes in the values of the biological parameters, Y and k_d, but quite sensitive to changes in $\hat{\mu}$ and K_s. From Fig. 5 it is seen that changes in the values of X_a and T result in only moderate changes in the effluent soluble BOD_5.

4 FACTORS INFLUENCING ALGAL GROWTH

4.1 Contribution to effluent suspended solids

Because the sedimentation rates of many species of algae are low, algae are frequently a major component of the suspended solids in an aerated lagoon efflu-

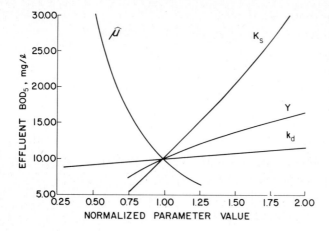

Fig. 4. Influence of biologic parameters on effluent soluble BOD_5. $[(V/Q)_1=1.5$ d; $(V/Q)_j=1.1$ d; n=3; T=5°C; $X_a=8$ mg/ℓ; $\hat{\mu}_{20}=13.0$ d^{-1}; $k_{d20}=0.24$ d^{-1}; Y=0.5; $K_s=120$ mg/ℓ].

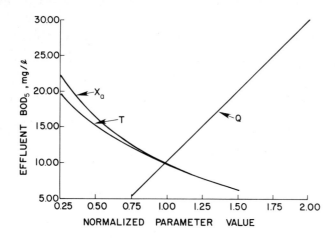

Fig. 5. Influence of design parameters on effluent soluble BOD_5. $[(V/Q)_1=1.5$ d; $(V/Q)j=1.1$ d; n=3; T=°C; $X_a=8$ mg/ℓ; $\hat{\mu}_{20}=13.0$ d^{-1}; $k_{d20}=0.24$ d^{-1}; Y=0.5; $K_s=120$ mg/ℓ].

ent. Consequently, the fraction of suspended solids that is nonsettleable can be controlled by creating conditions that will minimize the growth of algae. All features that can be applied to a system design to limit the growth of algae, contribute significantly to the reduction of both suspended solids and their biochemical oxygen demand.

4.2 <u>Cell depth</u>

Light is a major factor in the growth of algae. If significant color or turbidity is present in the lagoon system, growth will be limited. Depth is also
important. Light is absorbed exponentially through the water column. For a
given hydraulic retention time, an increase in lagoon depth will result in a
decrease in surface area, thereby significantly reducing the light imput per
unit volume of lagoon cell.

4.3 <u>Aerator-power intensity</u>

Algal concentrations are often correlated with chlorophyll a measurements.
Figure 6 illustrates the influence that aeration power intensity has on the mean
chlorophyll a concentration and, hence, the algal biomass concentration. Data
points on the figure are averages of measurements made in six operating aerated
lagoon cells treating domestic wastewaters (Rich, 1978). From these points, it
is seen that at aerator power intensities of 5-6 W/m^3 the mixing intensity and
thus, the suspended material, in a lagoon is sufficient to suppress the growth
of algae.

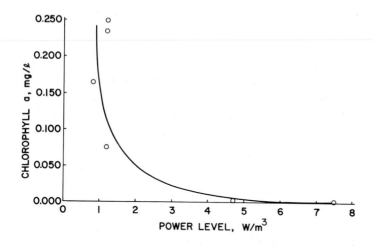

Fig. 6. Mean values of chlorophyll a concentrations in effluents from aeration
cells as a function of aerator-power input.

4.4 <u>Hydraulic retention time</u>

Figure 7 illustrates the influence of hydraulic retention time on the effluent
suspended solids from two polishing pond systems, each receiving secondary effluent from an activated sludge plant treating a domestic wastewater (Water Pollution Research Laboratory, 1973). One system consisted of a single cell and the
other, four cells in series. At total retention times less than two and a half

days, the annual mean concentrations of suspended solids in the effluents from both systems decreased as the result of flocculation and sedimentation within the ponds. Beyond two and a half days, the concentrations in the effluents of both systems increase, such increase being the result of algal growth.

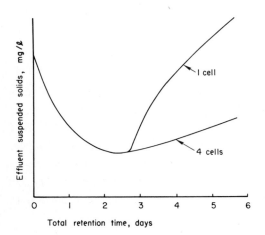

Fig. 7. Influence of multicellular configuration on effluent suspended solids of polishing ponds.

4.5 Multicellular configurations

Figure 7 also illustrates the influence of multicellular configurations on algal suppression. Beyond two and a half days the suspended solids concentrations in the effluents of both systems increase. However, it is observed that for the system consisting of four cells in series, no significant growth occurs for total retention times as long as four to five days.

Although other factors may also be involved, the difference between the performances of the two systems can be explained in terms of the hydraulics of an ideal system. Fig. 8 compares the residence time distribution of a three-cell system, completely mixed and with cells of equal size, to that of a single cell. Only 6.2 percent of the effluent from the three-cell system has a retention time greater than twice the average retention time, as compared with 13.5 percent for the single-cell system. For a four-cell system, the percent drops to 4.2. For the range of growth rates to be expected in a natural system, the differences in these percentages are significant in suppressing the growth of algae (Rich, 1983b).

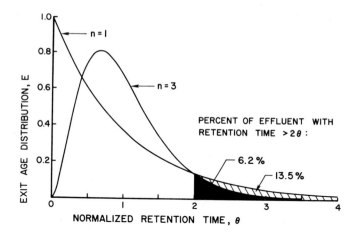

Fig. 8. Residence time distribution curves for completely-mixed tanks in series.

5 SOLIDS STABILIZATION

The partially-suspended cells should have a bottom area sufficient to accommodate the benthal stabilization of the biodegradable fraction of the solids that settle in the cells. In small cells, sludge accumulations for periods longer than one year will occupy relatively large fractions of the total cell volume and, hence, will influence significantly the hydraulic retention time. Consequently, solids removal on an annual basis is desirable. To ensure that few of the biodegradable solids are carried over from one annual cycle to the next, the loading rate of such solids should not exceed the rate at which the solids can be expected to be stabilized over the annual cycle. Based on this consideration, the bottom area required for each cell must be at least equal to (Rich, 1982c)

$$A_j = \frac{F_1 Q Y S_0}{n L_c} \qquad (8)$$

where A_j = bottom area of each partially-suspended cell (m^2); Q = flow rate of wastewater through the system (m^3/d); L_c = limiting biodegradable solids loading $(g/m^2 d)$; F_1 = solids decay factor.

The solids decay factor, F_1 is a function of the hydraulic retention time in the first (completely-suspended) cell and the annual average temperature of the wastewater in the lagoon system. The latter can be assumed to be equal to the average annual air temperature. Values of the decay factor can be obtained from Table 3 (Rich, 1980).

Little information is available on the rates at which biodegradable solids are decomposed under conditions in which the solids are continually being added

TABLE 3

Solids decay factors

(V/Q)$_1$,days	5	T, °C 10	15	20
1.0	0.812	0.772	0.727	0.676
1.5	0.774	0.728	0.677	0.621
2.0	0.743	0.693	0.639	0.581
2.5	0.717	0.666	0.609	0.549
3.0	0.695	0.641	0.533	0.523

to the deposit. From that which is available it appears that in temperate regions, even for loadings of biodegradable solids as high as 80 g/m^{2}d, one can expect complete destruction of such solids over an annual cycle (Rich et al., 1980). For design purposes, one can set L$_c$ in Eq. 8 to this value.

6 SLUDGE ACCUMULATION

If no biodegradable solids are to be carried over from one annual cycle to the next, sludge accumulation will be a function primarily of the nonbiodegradable fraction. The volume of the annual accumulation of sludge can be estimated with (Rich, 1980).

$$V = \frac{365 \ Q \ X_i}{x \ \rho} \tag{9}$$

where V = volume (m^3); X$_i$ = inert component of the wastewater suspended solids concentration (mg/ℓ); x = weight fraction of solids in sludge; ρ = water density = 10^6 g/m^3.

The concentration of inert suspended solids in the influent wastewater, X$_i$, will consist of the inorganic and the nonbiodegradable volatile suspended solids. This concentration can be estimated by aerating the wastewater in a batch mode for a period of 30 days in the dark (to prevent algal growth) and at room temperature. The term X$_i$ is assumed to be equal to the concentration of solids remaining.

The weight fraction of solids, x, in a sludge formed at the bottom of a lagoon cell can be expected to be about 4 percent (Balasha & Sperber, 1975; Rich et al., 1980).

7 AERATOR-POWER LEVELS

7.1 Complete solids suspension

Power levels required to maintain a given turbulence for mixing, the suspension of settleable solids, or both, are a function of several factors - concentration of suspension, lagoon size and geometry, and the type of aeration system used. Because of the high degree of equipment specificity involved, equipment manufacturers' recommendations should be sought. In practice, however, engineers use generalized relationships, based on experience, to make preliminary estimates of power requirements early in the design activity. One such relationship for low-speed, mechanical surface aerators to maintain all settleable solids in suspension is (Associated Water and Air Resources, personal communication, 1976)

$$p = 0.004X + 5 \tag{10}$$

when $X < 2000$ mg/ℓ.

The suspended solids concentration in an aerated lagoon in which all solids are maintained in suspension will equal the influent suspended solids concentration plus the concentration of the suspended solids produced in the system minus the concentration of the influent suspended solids that have been degraded. For domestic wastewaters, the suspended solids concentration will occasionally be as high as 200 mg/ℓ (Rich, 1978). Solving Eq. 10 for $X = 200$, yields a power level of 5.8 W/m^3, a value close to that (6 W/m^3) which is reported in the literature for low-speed, mechanical aerators in an experimental but full-scale lagoon cell (Fleckseder & Malina, 1970).

7.2 Threshold for solids suspension

Figure 9 is a plot of suspended solids concentration as a function of aerator-power level. The data were obtained from a full-scale, lagoon cell fitted with a variable speed, surface aerator and operated at a retention time long enough for algal growth to occur (Fleckseder & Malina, 1970). The data points indicate that above a certain minimum suspended solids concentration (55 mg/ℓ) established by the nonsettleable fraction, the suspended solids concentration that can be maintained in suspension increases linearly with aerator-power level. If algal growth had been suppressed and the concentration of the nonsettleable solids had been 22 mg/ℓ (as predicted by Eq. 2), then a straight line drawn through the points that establish the linearity extrapolates to a value of about 2 W/m^3. Therefore, it appears that in the absence of algae the threshold power level for settleable solids suspension is about 2 W/m^3. This value is supported by results from a field study (National Council of the Paper Industry for Air and Stream Improvement, 1971).

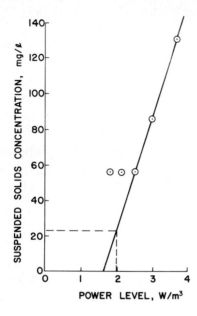

Fig. 9. Suspended solids concentration as a function of aerator-power level.

7.3 Oxygen requirements

For design, the maximum oxygen demand in the first (completely-suspended) cell can be estimated with (Rich, 1982c)

$$R_{O_2} = 6.24 \times 10^{-5} \, Q \, S_0 \tag{11}$$

where R_{O_2} = maximum oxygen demand (kg/h); Q = wastewater flow rate (m^3/d); S_0 = average influent BOD_5 (mg/ℓ).

Equation 11 includes a safety factor of 1.5 as a safeguard during those periods when influent conditions vary from the average.

The maximum oxygen demand exerted in the water column of a partially-suspended cell as the result of residual BOD_5 in the effluent of the preceding cell would be expected to occur in the winter, whereas that exerted by the bottom deposit would occur during the summer. For typical conditions, the maximum demand exerted by the bottom deposit is significantly larger than the water column demand. A convenient approach to specifying aeration equipment for the partially-suspended cells is to base it on the maximum deposit demand occurring during summer conditions, a time when the other demand is minimal. Such demand can be estimated with (Rich, 1982c)

$$R_{O_2} = 4.16 \times 10^{-5} \, A_j \, B \tag{12}$$

where A_j = bottom area of each partially-suspended cell (m^2); B = maximum
bottom deposit oxygen demand (g/m^2d).

In temperate regions, values of the maximum bottom deposit demand, B, can be
expected to reach 150 g/m^2d (Rich, et al., 1980).

Those biodegradable solids that do not decompose in such a way as to exert
an oxygen demand can be expected to be converted to methane through anaerobic
decomposition. Temperature permitting, active methane production will occur in
solids deposits even though the water above the deposit contains as much as 5-6
mg/ℓ of oxygen (Rich, 1981).

The aerator capacity required for the provision of oxygen can be computed
with

$$p = 10^3 \frac{R_{O_2}}{N \, V} \tag{13}$$

where p = power level (W/m^3); V = volume (m^3); N = expected aerator performance
(kgO_2/kWh).

For preliminary calculations, the expected aerator performance, N, is often
assumed to be 1.25 kgO_2/kWh.

If the value of the power level, p, in the partially-suspended cells computes
to be greater than 2, the lagoon depth should be increased to a value where p =
2 W/m^3. Such increase, of course, will result in increased retention time in
the lagoon cell, thereby increasing the potential for algal growth.

7.4 Diffused-air aeration

Equation 10 applies to power requirements to be met with mechanical surface
aerators. When diffused-air aeration is used, the power requirement for solids
suspension is about two-thirds greater. Conversion of power level in watts per
cubic meter to cubic meters of air (at standard conditions) per minute is made
with

$$Q_a = 9.74 \times 10^{-4} \, p \tag{14}$$

where Q_a = air flow rate (m^3/min).

For diffused-air aeration, the air requirement to provide oxygen can be calculated with

$$Q_a = 0.061 \frac{R_{O_2}}{E} \tag{15}$$

Transfer efficiencies, E, typically vary between 0.05 and 0.15.

Air diffusers placed in the partially-suspended cells should be perched
above the maximum anticipated sludge level. Failure to do so may interfere
with settling and the consolidation of deposits.

8 CRITICAL DESIGN DETAILS

The wastewater should be screened prior to being introduced to the first cell. Cell depths should be as great as possible, consistent with the limitations placed on retention time and bottom surface area. The influent pipe to the first cell should discharge at a shallow depth below the aerators. Effluent from the first cell should be withdrawn from behind a surface baffle. Effluents from all other cells should be withdrawn at the surface without surface baffling so as to minimize retention of any algae that might be generated in the system. Riprap or a concrete apron should be placed at the water line to prevent erosion.

9 LIMITATIONS

The constraints on retention time limit the use of the DPMC systems to wastewaters with a BOD_5 less than about 300 mg/ℓ. Wastewaters with greater BOD_5 concentrations will generate greater quantities of biodegradable solids. The latter require larger bottom areas in the partially-suspended cells, thereby violating retention time limitations. Furthermore, there is evidence that few settleable solids will be generated from wastewater with a BOD_5 of less than 100 mg/ℓ (National Council of the Paper Industry for Air and Stream Improvement, 1971). Like all other biological treatment processes, the lagoon process is inhibited by toxic materials. Where such materials are likely to be present, pretreatment for their removal must be incorporated in the system.

10 DESIGN EXAMPLE

A dual-power level, four cell aerated lagoon system is to be designed to treat a domestic wastewater, the average flow of which is 3785 m^3/d. The BOD_5 and inert solids concentration of the wastewater is 200 and 133 mg/ℓ, respectively. The system is to be located where the ambient air temperature during the coldest week of the year is 0°C and the average air temperature for the year is 15°C. (Such temperature conditions are representative of those for the southeastern region of the United States.)

Solution procedure:
1. Determine retention time in first cell. From Table 2, $\hat{\mu}_{20}$ = 13.2 d^{-1}, K_S = 120 mg/ℓ, and Y = 0.50. Since 5°C > 2°C, Eqs. 4 and 5 are solved for T = 5°C.

$$\hat{\mu} = 13.2(1.10)^{5-20} = 3.2 \ d^{-1}$$

$$(V/Q)_1 = 3 \ \frac{120+200}{3.2(200)} = 1.50 \ d$$

2. Determine retention time in second, third, and fourth cells. Assuming that $k_{d20} = 0.25$ d^{-1}, Eq. 6 is solved for T = 5°C.

$$k_d = 0.25(1.05)^{5-20} = 0.12 \ d^{-1}$$

Assuming X_a = 8 mg/ℓ, Eq. 3 is solved for S_e = 10 mg/ℓ.

$$(V/Q)_j = \frac{\left[\frac{1+0.12(1.50)}{10(3.2/120)(1.50)}\right]^{1/3}-1}{(3.2/0.5\cdot120)8} = 1.02 \ d$$

3. Determine bottom area in second, third, and fourth cells. From Table 3, for T = 15°C and $(V/Q)_1$ = 1.50 d, F_1 = 0.677. Then from Eq. 8

$$A_j = \frac{0.677(3785)(0.5)(200)}{3(80)} = 1068 \ m^2$$

4. Determine minimum water depth in second, third, and fourth cells.

$$D_w = \frac{(V/Q)_j Q}{A_j} = \frac{1.02(3785)}{1068} = 3.61 \ m$$

5. Determine maximum sludge depth in second, third, and fourth cells. From Eq. 9

$$V = \frac{365(3785)(133)}{0.04(10^6)} = 4595 \ m^3$$

$$D_s = \frac{V}{nA_j} = \frac{4594}{3(1068)} = 1.43 \ m$$

6. Establish depth of second, third, and fourth cells. Assuming that cold weather conditions occur when sludge depth is only one third the maximum depth,

$$D = D_w + (D_s/3) = 3.61 + (1.43/3) = 4.09 \ m$$

A depth of 4.09 m will provide the required retention time for soluble BOD_5 removal during the coldest period of the year.

7. Determine power level in first cell. From Eqs. 11 and 13, the power level required for oxygen will be

$$R_{O_2} = 6.24\times10^{-5} (3785)(200) = 47.24 \ kg/h$$

$$p = 10^3 \frac{47.24}{1.25(1.5)(3785)} = 6.65 \text{ W/m}^3$$

Since the power requirement for oxygen exceeds that required for suspension (6 W/m^3), use 6.65 W/m^3.

8. Determine the power levels in the second, third, and fourth cells. From Eqs. 12 and 13

$$R_{O_2} = 4.16 \times 10^{-5} (1068)(150) = 6.66 \text{ kg/h}$$

$$p = 10^3 \frac{6.66}{1.25(1.02)(3785)} = 1.38 \text{ W/m}^3$$

Since 1.38 < 2.00, power levels of 1.5 W/m^3 will be installed. However, during operation, O_2 in these cells will be monitored and when oxygen demand permits, only 1 W/m^3 will be used.

11 PERFORMANCE OF DPMC SYSTEMS

The general concept of the dual-power level, multicellular aerated lagoon system has been followed in the upgrading of existing facultative lagoon systems. Although the upgraded systems often fail to conform strictly to those criteria that have been established for optimal design, they do have multicellular configurations, dual-aeration levels, and most importantly, operating retention times much shorter than those provided for in the older aerated lagoon systems.

Preliminary performance data for the upgraded systems are impressive. Effluent data for three systems are listed in Table 4. The data presented therein are from samples collected over at least one annual cycle. Of particular interest are the low suspended solids concentrations. Such data clearly indicate a superiority of performance over that of the older systems listed in Table 1. On an operating basis, the performance of the upgraded systems compare favorably with the performance of activated sludge package plants. See Table 5. There, the average performance of the three DPMC systems is compared to that of 11 package plants.

TABLE 4

Performance of DPMC aerated lagoon systems[a]

(Effluent characteristics expressed in milligrams per liter)

System	50%[b] BOD_5	TSS	90%[c] BOD_5	TSS	Number of samples
1	12	15	28	27	74
2	19	9	37	32	16
3	22	13	37	19	27
Avg	18	13	34	29	

[a]Data provided by Environmental Analytics, Lexington, SC.
[b]Values exceeded in 50% of effluent samples.
[c]Values exceeded in 10% of effluent samples.

TABLE 5

Comparison of DPMC lagoon performance with activated sludge package plant performance. (Effluent characteristics expressed in milligrams per liter).

	50%[c] BOD_5	TSS	95%[d] BOD_5	TSS	Number
Package Plants[a]	15	20	50	60	11
DPMC Lagoons[b]	18	13	39	36	3

[a]Guo, P.H.M. et al. "Evaluation of Extended Aeration Activated Sludge Package Plants." Jour. Water Poll. Control Fed., 53, 1(1981) 33-42.

[b]Data provided by Environmental Analytics, Lexington, SC.
[c]Values exceeded in 50% of effluent samples.
[d]Values exceeded in 5% of effluent samples.

12 CONCLUSIONS

The DPMC aerated lagoon system offers a low-cost alternative to high-maintenance, mechanically complex secondary treatment systems, especially for the treatment of small domestic wastewater discharges. Properly designed, these systems will discharge effluents low in both suspended solids and biochemical oxygen demand. Removal of the soluble component of the biochemical oxygen demand is enhanced by the multicellular configuration of the system and the high concentration of bacterial biomass kept in suspension in the first cell. Low effluent suspended solids are achieved by the multicellular configuration and by limiting the hydraulic retention time. For domestic wastewaters treated in the southeastern region of the United States, the retention time in the first cell should be no more than 1.5 days and that in each of the other cells should be no more than about 1.0 days. Cell depth should not be less than three meters. Because of the relatively small size of the system, nonbiodegradable solids that accumulate at the bottom of the second, third, and fourth cells should be removed annually. However, these solids will be highly stable and ready for land disposal. The systems should not be used for treating wastewaters with BOD_5 of greater than 300 mg/ℓ or less than 100 mg/ℓ.

REFERENCES

Balasha, E. and Sperber, H.J., 1975. Treatment of domestic wastes in an aerated lagoon and polishing pond. Water Research, 9:43-49.

Bartsch, E.H. and Randall, C.W., 1971. Aerated lagoons - a report on the state of the art. Journal of the Water Pollution Control Federation, 43:699-708.

Benefield, L.D. and Randall, C.W., 1980. Biological Process Design for Wastewater Treatment. Prentice-Hall, Inc., Englewood Cliffs, N.J., 526 pp.

Eckenfelder, W.W., Jr., Magee, C.D., and Adams, C.D., 1972. A rational design procedure for aerated lagoons treating municipal and industrial wastewaters. In: Proceedings, 6th International Water Pollution Research, 18-23 June 1972, at Jerusalem, Israel.

Fleckseder, H.R. and Malina, J.F., Jr., 1970. Performance of the aerated lagoon process. Center for Research in Water Resources, University of Texas, Austin, Texas, Technical Report, EHE-70-22, CRWR-71, 110 pp.

Horn, C.R. and Pohland, F.G., 1973. Characterization and treatability of selected shellfish processing wastes. In: Proceedings of the 28th Annual Industrial Waste Conference, May 1973, at Lafayette, Indiana: 819-831.

Jorden, W.L. Pohland, F.G., and Kornegay, B.H., 1971. Evaluating treatability of selected industrial wastes. In: Proceedings, 26th Purdue Industrial Waste Conference, May 1971, at Lafayette, Indiana: 514-529.

Kormanik, R.A., 1972. Design of two-stage aerated lagoons. Journal of the Water Pollution Control Federation, 44: 451-458.

Lawrence, A.W. and McCarty, P.L., 1970. Unified basis for biological treatment, design and operation, Journal of the Sanitary Engineering Division of the American Society of Civil Engineers, 96: 757-778.

National Council of the Paper Industry for Air and Stream Improvement, 1971. A study of mixing characteristics of an aerated stabilization basin. New York, N.Y., Stream Improvement Technical Bulletin No. 245, 64 pp.

Randall, C.W., Richard, J.B., and King, P.H., 1975. Temperature effects on aerobic digestion kinetics. Journal of the Environmental Engineering Division of the American Society of Civil Engineers, 101:795-811.

Rich, L.G., 1976. Improved waste-treatment systems design based on the natural thermal environment. Water Resources Research Institute, Clemson University, Clemson, S.C., Report No. 64, 90 pp.

Rich, L.G., 1978. Solids control in effluents from aerated lagoon systems. Water Resources Research Institute, Clemson University, Clemson, S.C., Report No. 73, 105 pp.

Rich, L.G., 1980. Low-Maintenance, Mechanically-Simple Wastewater Treatment Systems, McGraw-Hill, New York, N.Y., 211 pp.

Rich, L.G., 1981. Stabilization characteristics of deposits of biologic
     solids generated in aerated lagoons. Water Resources Research Institute,
     Clemson University, Clemson, S.C., Report No. 92, 30 pp.
Rich, L.G., 1982a. A cost-effective system for the aerobic stabilization and
     disposal of waste activated sludge solids. Water Research, 16: 535-542.
Rich, L.G., 1982b. Influence of multicellular configurations on algal growth
     in aerated lagoons. Water Research, 16: 929-931.
Rich, L.G., 1982c. Design approach to dual-power aerated lagoons. Journal of
     the Environmental Engineering Division of the American Society of Civil
     Engineers, 108: 532-548.
Rich, L.G. and White, S.C., 1977. BOD_5 removal from aerated lagoon systems.
     Water and Sewage Works, 124: R21-R23.
Rich, L.G., Tarnowski, D.S., and Bryant, C.W., Jr., 1980. Benthal
     stabilization of aerated lagoon solids. Paper delivered at 53rd Annual
     Conference, Water Pollution Control Federation, 1 October 1980, at Las
     Vegas, Nevada, 22 pp.
Sundstrom, D.W. and Klei, H.E., 1979. Wastewater Treatment. Prentice-Hall,
     Inc., Englewood Cliffs, N.J., 444 pp.
Tikhe, M.L., 1975. Aerofac aerated lagoons. Journal of the Water Pollution
     Control Federation, 47: 626-629.
Water Pollution Research Laboratory, 1973. Treatment of secondary sewage
     effluents in lagoons. Notes on Water Pollution, No. 63, 4 pp.
White, S.C. and Rich, L.G., 1976. How to design aerated lagoon systems to
     meet 1977 effluent standards - experimental studies. Water and Sewage Works,
     123: 85-87.

NOTATION

A_j = bottom area of each partially-suspended cell, m^2

B = rate of benthal oxygen demand, g O_2/m^2d

BOD_5 = 5-day, 20°C biochemical oxygen demand, mg/ℓ

$(BOD_5)_s$ = soluble BOD_5, mg/ℓ

$(BOD_5)_t$ = total BOD_5, mg/ℓ

K_S = saturation constant, mg/ℓ

k_d = specific decay rate, d^{-1}

F_1 = solids decay factor

L_C = limiting biodegradable solids loading on
        partially-suspended cells, g/m^2d

N = expected aerator performance, kg O_2/kW h

n = number of equal-sized, partially-suspended cells in
     series

p = power level, W/m^3

Q = average wastewater flow rate, m^3/d

R_{O_2} = oxygen requirement rate, kg/h

S_0 = total BOD_5 of influent wastewater, mg/ℓ

S_e = soluble BOD_5 in effluent of last cell, mg/ℓ

T = lagoon water temperature, °C

TSS = total suspended solids, mg/ℓ

V = volume, m^3

$(V/Q)_1$ = hydraulic retention time in completely-suspended
       cell, d

$(V/Q)_j$ = hydraulic retention time in each of the
       partially-suspended cells, d

X = suspended solids concentration, mg/ℓ

X_a = biomass concentration in system effluent, mg/ℓ

X_i = inert fraction of the suspended solids concentration, mg/ℓ

x = weight fraction of solids in sludge

Y = growth yield

$\hat{\mu}$ = maximum specific growth rate, d^{-1}

ρ = water density, g/m^3

MATHEMATICAL MODELS FOR WASTE STABILIZATION PONDS

Jack J. Fritz, National Research Council, Washington, D.C.

TABLE OF CONTENTS

1.0 <u>INTRODUCTION</u>

The oxidation or stabilization pond remains the simplest and most economical method of wastewater treatment suited for rural areas and small communities. It is the first method of treatment other than simply dumping wastes into a nearby body of water and, therefore, has special relevance for developing countries where more costly techniques are prohibitive. Sophisticated or expensive equipment is not required and significant public health benefits result when the pond is properly maintained. However, this method of treatment is deceptively simple and has fallen into disfavor due to land requirement, effluent organics concentration and dependence on variable environmental factors. However, most problems result from a basic lack of understanding of the biochemical mechanisms and improper operation. These facts encouraged the pursuit of research in this area.

Currently there are some 5,000 ponds in existence in the United States alone (Reynolds, <u>et al</u>., 1977) and in at least 39 countries (Gloyna, 1971). There are several types of ponds usually categorized by the method of treatment, aerobic or anaerobic, and the rate of organic loading. Ponds may be arranged in series, or in parallel and often with recirculation schemes. This work deals with facultative ponds, which are by far the most common and resemble natural ecosystems. They can be characterized simply as having an aerobic liquid layer underlaid by an anaerobic benthic layer. Organic wastewater is stabilized in the aerobic layer through bacterial action with oxygen provided by algal photosynthesis in the sunlit upper reaches.

Treatment in facultative ponds through the synthesis of aerobic bacteria from organic substrate produces nutrients and inorganic forms of carbon used by algal cells for growth and photosynthesis, thereby producing the required oxygen. This symbiotic relationship between algae and bacteria is well known but has not been quantitatively described in the literature.

As algal cells age and sink into the benthic layer, they ferment, creating the products of anaerobic digestion, hence the name facultative ponds. Nutrients are released into the upper liquid layer for incorporation into suspended biomass. In poorly operating ponds, slow recycling of nutrients and inorganic carbon can lead to the eventual buildup of the benthic layer where periodic cleaning may be required. The clarified effluent remains near the top and is released continuously at the opposite end from the influent.

Until recently, pond design procedures were based on observed data from selected site specific locations. At best, they were approximate and generally dealt with BOD performance only. Mathematical relationships were simply based on area organic loading and residence time. Comprehensive models which were able to predict performance based on the variety of physical and biochemical factors did not exist. Non-steady state simulation of biomass and biochemical species subject to environmental mechanisms has remained absent from the literature until the early 1970s. In addition, a model incorporating the temporal changes in climate and hydrologic constraints was required before a design procedure could be fully developed.

The goal of recent research was, therefore, the development of comprehensive, non-steady state process models of facultative stabilization pond behavior using the principle of conservation of mass (Fritz, et al., 1979; Ferrara, et al., 1980). These simulations were intended to model existing ponds and serve as possible design tools.

2.0 BIOCHEMICAL PROCESSES

Facultative stabilization ponds are used to treat organic wastewaters by natural purification through both anaerobic and aerobic processes. Sensitive to climate, principally temperature, sunshine and wind mixing, they are ideal for tropical and semi-tropical areas. Aerobic stabilization is effected via facultative or aerobic heterotrophic bacteria with oxygen supplied through algal photosynthesis. Anaerobic biodegradation of settled solids occurs in the benthic sludge layer through the action of heterotrophic acid and methane forming bacteria. This combined aerobic-anaerobic lagoon process is the principal method of treatment for organic wastewaters in many parts of the world.

Ponds are often arranged in series or in parallel, with or without recirculation, to achieve a desired effluent organics concentration. Usually, long hydraulic residence time and substantial land areas are required. Where land is available, lagooning is an economical treatment method for rural and small communities. Reduction in BOD varies depending on location, organic loading and other factors to be described. Reductions of 30 to 90 percent are not uncommon (Gloyna, 1976). Design rationales for the construction of stabilization ponds are primarily based on empirical techniques developed by examining data from various sites.

Stabilization of waste organics proceeds as a result of the symbiotic relationship between algae supplying oxygen for photosynthesis and bacteria oxidizing influent organics to carbon dioxide which is used for algal growth. The principal source of energy is solar radiation providing adequate temperatures for bacterial growth through interfacial heat transfer and supplying short wave solar radiation needed for algal photosynthesis. Fig. 1 is a schematic diagram of the processes occurring in a stabilization pond ecosystem. The fundamental aerobic and anaerobic processes can be represented as:

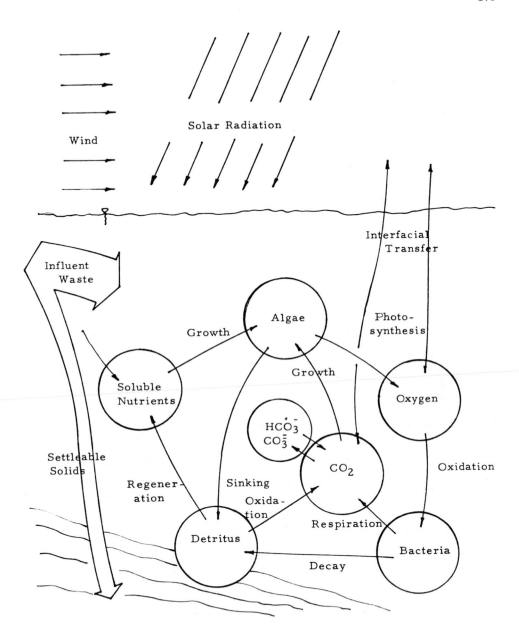

FIGURE 1 Stabilization Pond Ecosystem

stabilization of organics

    aerobic bacteria + influent organic material + $O_2 \rightarrow$ more

    aerobic bacteria + nutrients + H_2O + CO_2 (1)

algal growth

    algae + CO_2 + H_2O + solar radiation $\rightarrow$ more algae + O_2 (2)

anaerobic benthic processes

    anaerobic bacteria + influent organic matter + settled algal

    and bacterial biomass $\rightarrow$ more anaerobic bacteria + nutrients

    + CO_2 + H_2O + CH_4 (3)

    Algal growth proceeds due to the presence of inorganic carbon in the form
of CO_2 and HCO_3^- from bacterial oxidation of organic matter and influent
alkalinity. Other nutrients necessary for growth are NH_3, NO_3^- and
PO_4^{-3} originating in the influent and from benthic processes.

    Algal species most encountered are Euglena, Chlorella and Chlamydomonas
(McKinney, et al., 1970). There are several empirical studies which attempted
to define the stoichiometric relationship between nutrients, algal growth and
oxygen generation. Table 1 gives the empirical composition formulas which
resulted.

    It has also been suggested that the same N to P ratio (Cordeiro, 1974) as
found in bacterial studies be used for algal species. Hoover and Porges
(1952) initially reported the formula $C_5H_7O_2N$ to be representative for
bacterial cultures. Carbon, nitrogen and phosphorus composition ratios by
weight of 41:7:1, respectively, were used by Fritz (1979), corresponding to
the determination of Stumm and Morgan (1970).

    Stoichiometric oxygen production for algae is 1.3 moles of O_2 per mole of
CO_2 utilized based on the Stumm relationship. Actual field studies found
Chlorella to produce 1.22 moles of O_2 per mole of CO_2 utilized
(Stoltenberg, 1962). Reaeration due to wind action supplies a much smaller
proportion of the system's oxygen requirements. Because of intense
photosynthetic activity during the daylight hours, oxygen super-saturation may
reach as high as 40 mg/l (McKinney, 1970) causing some loss of oxygen to the
atmosphere. Yet, during the hours of darkness, the pond may become anaerobic
throughout as a result of continual demand for oxygen by bacteria.

TABLE 1

Empirical Formulas for Algal Species

Empirical Formula	Species	Source
$C_{5.7}H_{9.8}O_{2.3}N$	Euglena	Fogg (1953)
$C_{7.62}H_{8.08}O_{2.53}N$	---	Oswald (1953)
$C_{5.9}H_{9.4}O_{2.7}N$	Chlorella	Richardson (1969)
$C_{106}H_{263}O_{110}N_{16}P$	---	Stumm & Morgan (1970)
$C_6H_{11.1}O_{2.7}N$	Chlorella	Ward & King (1976)
$C_{105}H_{147}O_{42}N_{21}P$	---	Sandoval et al. (1976)

Variations in pH are also common during the daily cycles because of utilization of CO_2 by algae, thereby causing dissociation of HCO_3 to produce a hydroxyl ion. Variations in pH from 7 to 10 can occur during the latter part of the daylight hours. As pH rises above 9, bacterial activity begins to diminish, causing a reduction in CO_2 production and thereby possibly limiting subsequent algal growth. In addition, oxygen varies diurnally as a result of photosynthetic activity as well. Fig. 2 and 3 show typical dirunal variations in both pH and oxygen.

A fraction of the influent BOD may settle to the bottom as it enters the pond depending upon the influent structure and is joined by settling bacterial and algal biomass. Anaerobic digestion proceeds, especially at temperatures above 15°C, through the action of acid and methane formers. Soluble products enter the aerobic layer above to be further oxidized in bacterial and algal synthesis. A properly operating facultative pond should always be aerobic in the liquid layer and anaerobic in the benthic sludge layer.

3.0 HYDRODYNAMIC PROCESSES

The hydraulic regimes of waste stabilization ponds have not been well defined in favor of modeling biological performance. Researchers have selected either the complete mix or plug flow model. Clearly, most ponds fit neither definition satisfactorily which led to the emergence of more complex models in the 1970s.

As early as 1961, Marais used a complete mix model and first order kinetics to characterize the relationship in BOD_5 reduction. Subsequent research focused primarily on empirically defining the rate constants for ponds throughout the world. Thirumurthi (1974) was among the first to suggest the use of a dispersion model based on the Whener and Wilhelm (1956) work with chemical reactors. However, this approach required pond performance data to determine the kinetic constants making it of limited use as a predictive

176

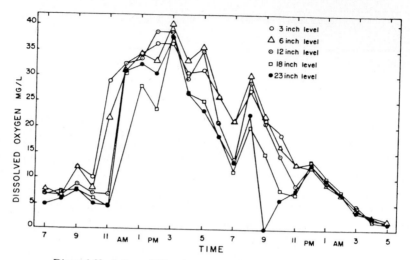

Diurnal Variation of Dissolved Oxygen at Five Levels of the Bear Creek
Lagoon, Columbia, Missouri, on August 18-19, 1968.

FIGURE 2   Oxygen Variation in Stabilization Ponds (from King, 1976)

tool. A similar model was developed by Watters (1973) at Utah State
University simulating the flow regime as reactors in series. Both complete
mix and plug flow reactors were used in various configurations to fit data
from experimental ponds. As a predictive tool, its use was limited because
dye tracer studies were required before a representative configuration could
be determined.

In recent work carried out by Ferrara and Harleman (1981), a return flow
model was proposed which included a forward active zone between inlet and
outlet and return zones along the periphery. However, in order to define
these transport processes, the relative sizes of these zones and the
dispersion coefficient had to be known. Solution of the conservation of mass
equation in one dimension was required:

$$\frac{\partial C}{\partial t} + v \frac{\partial C}{\partial x} = \frac{\partial}{\partial x} \left[D \frac{\partial C}{\partial x} \right] - kC \qquad (4)$$

in which C is the concentration of a substance; v is the velocity through the
section; D is the longitudinal dispersion coefficient; k is the first order
decay coefficient of substance C; x is the longitudinal coordinate and t is
time.

This equation was written for both active and return zones and solved
simultaneously using numerical methods. Utah State University pond data

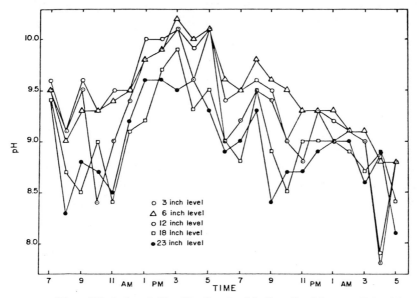

Diurnal Variation of pH at Five Levels of the Bear Creek Lagoon, Columbia, Missouri, on August 18-19, 1968.

FIGURE 3   pH Variation in Stabilization Ponds (from King, 1976)

suggests that neither ideal plug flow nor complete mix performance was achieved. However, the researchers do suggest that the complete mix model compares well with dye tracer studies under steady state conditions and appears to be a reasonable approximation of pond performance without the aid of calibration procedures.

Future progress in this area appears to be limited by the lack of research funds and the emphasis on the use of capital intensive secondary treatment plants. However, as construction funds also diminish in the 1980s, less expensive ways of treating wastewater will again be sought and additional research may be required.

4.0   TRADITIONAL EMPIRICAL DESIGN METHODS

4.1   Design Rationale:

Facultative pond design criteria have been largely based upon techniques developed through analysis of empirical data gathered at sites located in the warmer regions of the United States. These criteria vary widely with location becoming primarily rule of thumb. Despite extensive literature dealing with ponds, few comprehensive, non-steady state, mass balance process models are being applied to predict performance based upon the combination of wastewater characteristics and environmental factors. Until recently, few complete data sets that are necessary to build a comprehensive model have existed. Adequate understanding of the fundamental biochemical processes and their quantitative description have also been lacking prior to recent research into the eutrophication process (Middlebrooks, et al., 1974).

Application opportunities for a non-steady state rational model are numerous; for example, it could be the starting point in the development of an inclusive model simulating production of energy through anaerobic digestion of algae. The overall efficiency of solar energy conversion by this process is considered to be between 2 and 3 percent (Odum, 1971; Benemann, et al., 1977).

On the assumption that approximately 30 tons of dry algae can be grown per acre per year (Oswald, et al., 1957), with an average annual light intensity of 300 ly/day and methane production at 0.38 m^3/kg of COD destroyed, while algal cells represent a COD of 1.244 mg/mg of cell, absolute energy potential from one square meter of pond is 1.3 MJ/year. The resulting efficiency of solar energy conversion is 2.5 percent. Although this figure is somewhat low, it must be remembered that less capital outlay is required with the exception of a digester, land and algal harvesting equipment, in contrast with other energy sources. In addition, the cost of waste treatment can be used as a credit since that capital expenditure will not be required. More precise analyses have been carried out in the United States under the Department of Energy sponsorship; specific details can be found in the International Bioenergy Directory, 1981.

4.2   Effluent Requirements:

Effluent discharge standards for secondary treatment from wastewater treatment plants also apply to facultative stabilization ponds. In the United States, as a result of PL 92-500, the Environmental Protection Agency has set standards at the following levels:

BOD_5

1. Mean of a 30 consecutive day sample cannot exceed 30 mg/l.
2. Mean of a 7 consecutive day sample cannot exceed 45 mg/l.
3. Mean of a 30 consecutive day sample cannot exceed 15 percent of the mean of the influent for the same period.

The identical regulations apply to suspended solids. Limits have also been established for fecal coliforms. Specifically, the mean for a 30 consecutive day sample cannot exceed 200 fecal coliforms per 100 ml or 400 for a 7 consecutive day sample.

4.3   Current Approaches to Pond Design:

Surface organic loading rates, kg BOD/hectare, and retention time remain the traditional guidelines for pond design. Population loading rates and minimal depth criteria are also to be found among various regional standards. Table 2 cites empirical loading rates reported at various U.S. locations. Retention time varying from 20 to 180 days is not uncommon (Canter, et al., 1969). Although early attempts (Oswald, 1957) at establishing design methods considered the connection between sunshine, temperature, algal growth and resulting oxygen production, a comprehensive model did not emerge. Simple loading rates continue to be used as the principal guidelines.

More recent contributions to pond design are from Oswald (1976), Gloyna (1976), Marais (1970), Mara (1975), Thirumurthi (1974) and Larsen (1974).

4.3.1 Marais (1970)

Initial theoretical design concepts (Marais, 1961) were based on completely mixed reactor theory, the existence of wind shear and first order kinetics. The resulting formulation yielded reasonable results when tested using data from South African ponds. Fig. 4 is a schematic diagram of the completely mixed reactor simulating the pond. A mass balance for organic substrate, was written as follows:

$$\frac{dS}{dt} = \frac{S_i Q_i}{V} - \frac{S Q_o}{V} - k_1 S \tag{5}$$

in which S_i = inflowing BOD concentration, mg/l; S = BOD concentration in the pond, mg/l; k_1 = first order BOD decay coefficient, day^{-1}; Q_i = influent flow rate, l/day; Q_o = effluent flow rate, l/day and V = pond volume, l.

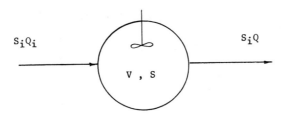

$S_i Q_i$ $S_i Q$

V , S

FIGURE 4   Completely Mixed Reactor System

TABLE 2
Some Examples of Empirical Aereal Organic Loading Rates

Location	Organic Loading Rate kg BOD/hectare-day	Reference
Texas	13 - 250	Towne et al. (1957)
Oklahoma	22 - 67	McKinney et al. (1970)
Missouri	50 - 90	McKinney et al. (1970)
California	10 - 179	McKinney et al. (1970)
Ohio	112	McKinney et al. (1970)
Utah	15 - 60	Middlebrooks et al. (1974)
New Hampshire	20 - 40	Bowen (1977)
South Dakota	20	McKinney et al. (1970)

Solving Eq. 5 for steady-state conditions, that is dS/dt = 0, and $Q_i = Q_o = Q$, assuming no seepage, evaporation or precipitation, one obtains:

$$\frac{S}{S_i} = \frac{1}{1 + k_1 (V/Q)} \tag{6}$$

Substituting hydraulic residence time into Eq. 6, $V/Q = \theta$, the following expression emerges:

$$\frac{S}{S_i} = \frac{1}{1 + k_1 \theta} \tag{7}$$

Equation 7 is the kinetic model most often applied in the design of ponds. In 1970, Marais refined his model incorporating the effects of a benthic sludge layer by settling a fraction of the influent BOD and providing for gas generation. He posed two mass balance equations, reduction of BOD in the sludge and gas evolution from that layer.

First the rate of change of sludge BOD mass was written as:

$$\frac{dS_t}{dt} = -K_s S_t + i_s S_i (Q_i \times 10^{-6}) \tag{8}$$

in which, K_s = first order BOD decay coefficient for sludge mass, day^{-1}; S_t = total mass sludge BOD, kg and i_s = fraction of the influent BOD which settles to the benthos.

The rate of gas evolution from the sludge was written as:

$$\frac{dV_g}{dt} = C_v S_q K_s S_t \tag{9}$$

in which, V_g = volume of gas produced, m^3; C_v = volume of gas liberated

per unit mass of BOD destroyed, m^3/kg BOD and s_g = fraction of sludge BOD
which decomposes into rising gases.

Secondly, Marais wrote the substrate mass balance for the entire pond as
consisting of the influent BOD dispersed throughout the pond, the fraction
lost to anaerobic processes, outflow and the portion lost in first order
decay. That balance is written as follows:

$$\frac{dS}{dt} = i_p \frac{S_i Q_i}{V} + s_p \frac{K_s S_t}{V} 10^6 - \frac{Q_o}{V} - k_1 S \qquad (10)$$

in which i_p = fraction of influent BOD dispersed throughout the pond; and
s_p = fraction of BOD lot from the sludge due to anaerobic processes whose
products enter the liquid layer.

Again, assuming equal inflow and outflow, $Q_o = Q_1 = Q$, constant
influent BOD and temperature and recalling that at equilibrium dS/dt = 0 and
$dS_t/dt = 0$, the steady state solution for S_t and S can be written as:

$$S_t = \frac{i_s S_i Q}{10^6 K_s} \qquad (11)$$

and

$$S = \frac{S_i}{k_1 \theta + 1} (i_p + s_p i_p) \qquad (12)$$

Marais further suggested the following values for the fractional
coefficients:

$$i_s = 0.4 \text{ to } 0.6; \; s_p = 0.4; \; s_g = 0.6; \; C_v = 0.62 \; m^3/kg \text{ BOD}$$

The accuracy of Eqs. 7 through 12 is also dependent on empirical
coefficients k_1 and K_s. Their values will vary with depth, distance from
the influent, temperature, organic loading, oxygen availability and other
wastewater characteristics. For K_s, Marais posed the following:

$$K_s = 0.002(1.35)^{-(20-T)} \qquad (13)$$

in which T = sludge temperature, °C.

Table 3 lists empirical values for k_1. They vary considerably for
ponds. Temperature, thought to be the most critical component in determining
k_1, is usually modeled by an Arrhenius type expression:

$$k_1 = k_{20} \beta^{T-20} \qquad (14)$$

in which, $1.05 \le \beta \le 1.07$ (Mara, 1976) and k_{20} = first order BOD decay
coefficient at 20°C, day^{-1}.

TABLE 3

First Order BOD Decay Coefficients for Stabilization Ponds

Temperature °C	$k_1(\text{day}^{-1})$	Reference
--	0.17 - 0.60	Marais & Shaw (1961)
20.0	0.09 - 0.29	Abbot (1946)
20.0	0.10 - 0.27	Ballinger et al. (1962)
22.6	0.0831	Click (1970)
23.6	0.0462	Click (1970)
20.0	0.10 - 0.30	Com.San.Eng. (1946)
20.0	0.172 - 0.241	Gotaas (1948)
20.0	0.110 - 0.125	Greenfield (1926)
20.0	0.077	Herman (1957)
26.0	0.057	Meenaghan et al. (1963)
20.0	0.165 - 0.259	Moore et al. (1950)
20.0	0.39	Suwannakarn (1963)
20 - 25	0.02 - 0.09	Thirumurthi
23	0.114 - 0.148	Thomas (1950)
7.8 - 24.4	0.029 - 0.075	Thirumurthi (1974)

Marais, however, suggested that the following temperature expression be used in conjunction with his previously stated mass balance model:

$$k_1 = 1.2(1.085)^{-(35-T)} \qquad (15)$$

4.3.2 Thirumurthi (1974)

Thirumurthi suggested that the accurate determination of k_1, was the key to the design process. He proposed further modification to include factors for the effects of toxic wastes and magnitude of the organic loading. Dimensionless correction factors were developed using pond data from Nova Scotia and Missouri. He initially defined a set of standard conditions as:

1. Pond temperature at 20°C.
2. Organic loading of 67.2 kg BOD/hectare-day.
3. No toxic chemicals.
4. Minimum solar energy of 100 ly/day.
5. No benthic load.

When actual conditions deviated from the above standard, he suggested using correction factors. His expression for k_1 is written as:

$$k_1 = k_s C_T C_o C_{tox} \qquad (16)$$

in which, k_s = the first order BOD removal coefficient at standard conditions, day^{-1}; C_T = temperature correction factor, 1.036^{T-20}; C_o = organic load factor and C_{tox} = toxicity factor.

The organic load factor C_o, depends on k_s and organic load L. It was derived by Thirumurthi based upon data by Neel, et al., (1961) for Missouri ponds.

$$C_o = 1 - \frac{.083}{K_s} [\log_{10} (\frac{67.2}{L})] \tag{17}$$

in which, L is kg BOD/hectare-day.

As the organic loading decreases, the decay coefficient usually decreases, requiring additional time for stabilization. With no industrial wastes, C_{tox} is equal to unity, but in the presence of certain toxic chemicals, algal cells cannot grow or do so only at reduced rates. Table 4 lists suggested values for C_{tox} for selected organic chemicals. The overall effect of the coefficients is to modify the value of k_1 if conditions deviate from standard.

Thirumurthi also sought to modify the well known completely mixed model developed by Marais to one exhibiting some plug flow behavior. Because ponds fit neither flow regimes, Eq. 7 may be extended for reactors in series in order to simulate a more representative flow phenomena. Eq. 7 can be modified as follows (Mara, 1976):

$$\frac{S}{S_i} = \frac{1}{(1 + k_1\theta)^n} \tag{18}$$

in which n = number of ponds.

Taken in the limit, that is as $n \to \infty$, the following expression results:

$$\frac{S}{S_1} = e^{-k_1 \theta} \tag{19}$$

which is the equation for ideal plug flow conditions.

A more exact representation of the flow regime, using a longitudinal dispersion coefficient, will yield increased accuracy in determining performance. Drawing upon the work of Whener and Wilhelm (1956), Thirumurthi (1974) posed Eq. 20 as an approximation for dispersed flow conditions, the regime between the extremes of plug flow and complete mix.

$$\frac{S}{S_1} = \frac{4ae^{1/2d}}{(1+a)^2 e^{a/2d} - (1-a)^2 e^{-a/2d}} \tag{20}$$

TABLE 4

Toxicity Correction Factors (from Thirumurthi, 1974)

Organic Chemical	Concentration of Organic Chemical in the Pond Influent (mg/l)	Suggested Values of C_{tox}
Methanoic acid	180 360	2.0 16.0
Ethanoic acid	270	1.6
Propanoic acid	180	2.65
Hexanoic acid	200 300	1.3 5.0
1-Butanol	4,000	2.0
Octanol	150 200	2.0 4.0
Malthane	70 140	2.5 8.0
25 percent DDT in 67 percent xylene solvent and 8 percent emulsifier	100 125	2.5 a
"Ortho" 29 percent Malathion 10 percent BHC 43 percent solvents 18 percent inert ingredients	100 340	2.5 a

[a]Complete destruction of algal life.

in which, $a = (1 + 4k_1 \theta d)^{1/2}$; d = dimensionless dispersion number, D/ul; u = mean velocity, m/h; l = length of travel, m and D = longitudinal dispersion coefficient, m^2/h.

In the case of ideal plug flow, d = 0, with no longitudinal dispersion, Eq. 20 reduces to Eq. 19. When mixing is complete, d = ∞, thereby reducing Eq. 20 to Eq. 7. Fig. 5 is a representation of Eq. 20. With the dimensionless product θk_1, the percentage of BOD remaining in the effluent can be determined once the dispersion number is known. As is evident from the graph, for any value of k_1 and a desired level of effluent BOD, retention time in a plug flow reactor is minimum, yielding a small reactor, while in a completely mixed system, retention time is maximum, requiring a large reactor.

4.3.3 Gloyna (1957)

The objective of the foregoing kinetic studies was to develop a technique for pond sizing to meet specified criteria, usually the avoidance of anaerobic conditions and an acceptable effluent BOD concentration. Herman and Gloyna (1958) were among the first to size ponds based upon operating experience from Texas. A minimum depth of one meter was specified and their original expression was developed using an organics concentration of 200 mg BOD/1.

$$V = NQ \frac{S_1}{200} \theta_o \ \exp \ [T_o - T]C \tag{21}$$

in which, V = pond volume, m^3; Q = influent flow rate, m^3/d/capita; θ_o = optimal residence time, days; N = population served; T_o = optimal pond temperature, °C and C = Arrhenius constant = 0.0693.

Both optimal temperature and residence time were used in the development of the model; therefore, the author cautions the engineer in the use of this expression. Derived for three ponds in series, its accuracy is diminished when influent BOD is not near 200 mg/1. Due to such limitations, in subsequent work with Huang (1967), Gloyna recast Eq. 21 in the following form:

$$V = 1.75 \times 10^{-2} \ Q \ L' \ [1.085^{35-T}]ff' \tag{22}$$

in which, V = pond volume, m^3; Q = flow rate, m^3/d; L' = ultimate BOD, mg/1; f = algal toxicity factor, usually = 10; f' = sulfide or other chemical oxygen demand factor, usually = 1.0, for $SO_4^=$ ion concentration less than 500 mg/1.

Fig. 6 illustrates the hydraulic retention time necessary to provide a 90 percent BOD reduction for various liquid temperatures and influent organics concentration using Eq. 21. A 7-day residence time was selected for computation of these curves thought to represent a hydraulic regime between the ideal extremes.

4.3.4 Oswald (1957)

Early attempts at linking performance to environmental factors were made by Oswald (1957) in correlating oxygen production with solar energy and algal

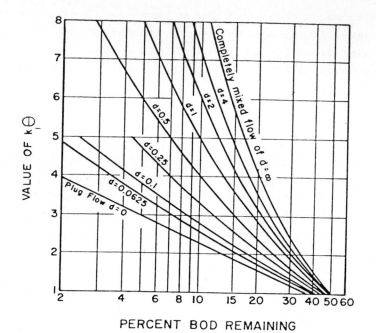

PERCENT BOD REMAINING

FIGURE 5   Non-Ideal Reactor Design Formula Chart (from Thirumurthi, 1974)

growth. It was shown that for <u>Chlorella</u>, light energy conversion required to produce one milligram of oxygen was 3.68 calories during photosynthesis. For a given efficiency of energy conversion, the quantity of oxygen produced was written as:

$$W = FI/3.68 \qquad (23)$$

in which, W = oxygen production, mg/l - day; F = oxygenation factor or weight of oxygen produced per unit BOD satisfied; and I = solar radiation, cal/cm^2 - day.

A value of 1.6 to 1.8 for F was considered appropriate for maximum BOD removal. Higher values accompany increased pH which retards microorganisms activity, while at lower values, insufficient oxygen is available for bacterial needs.

In 1970, Oswald, <u>et al</u>., proposed a design relationship based on anaerobic benthic processses. Gas generation as a function of temperature was stated as:

$$G = 5.16(T - 15) \qquad (24)$$

in which, G = rate of gas production, m^3/ha-da and T = pond liquid temperature.

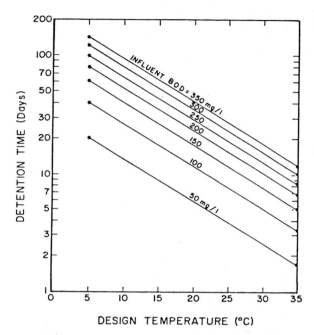

FIGURE 6 Hydraulic Retention Time as a Function of Temperature and Influent
BOD Concentration (from Gloyna, 1976)

Oswald postulated that light was limiting algal growth, reasoning that in
most pond effluents there remain nutrients which have not been absorbed. He
suggested that ponds be as shallow as possible, with maximum depth determined
by:

$$h = \frac{\ln I_o - I_h}{X_a \alpha} \qquad (25)$$

in which, h = penetration depth of sunlight into the algal culture, cm;
I_o = light intensity at the surface, cal/cm^2 - min; I_n = light intensity
at depth h, cal/cm^2 - min; X_a = concentration of algae, mg/l and α = light
absorption coefficient, 1/mg - cm.

4.3.5 Larsen (1974)

A more comprehensive attempt at incorporating environmental constraints
into pond design was made by Larsen (1974). His model consists of five
dimensionless numbers developed by regressing data at a small New Mexico
pond. Based upon BOD, flow rate, radiation, wind speed, and water
temperature, the system of dimensionless numbers was outlined as follows:

$$MOT = \frac{\text{Surface Area (Solar Radiation)}}{\text{Influent Flow (Influent BOD)}}$$

$$RED = \frac{\text{Influent BOD - Effluent BOD}}{\text{Influent BOD}}$$

$$TTC = \frac{\text{Wind Speed (Influent BOD)}}{\text{Solar Radiation}}$$

$$TEMPR = \frac{\text{Pond Liquid Temperature}}{\text{Air Temperature}}$$

$$DRY = \text{Relative Humidity}$$

$$MOT = (2.47^{RED} + 2.47^{TTC} + 24.9/TEMPR + 150.0/DRY) \qquad (26)$$

Larsen's model, as well as those of others, are based on annual trends and averages, yielding limited information for short-term pond performance prediction. Little of the earlier work found in the literature dealt with biochemical variation on a diurnal basis of the carbonate system dynamics. Fluctuations of pH and dissolved oxygen have not been modeled for ponds or eutrophic lakes. The development of non-steady state process models composed of principal biochemical components solved in discrete time steps throughout the daily cycle is a recent innovation. Such models are necessary for proper design and could be used in the performance analysis of existing pond systems.

5.0 RECENT MODELING EFFORTS

The modeling of water quality in lakes, streams and estuaries as an applied science dates back approximately 15 years, with early work carried out by Chen (1972), DiToro, O'Connor, and Thomann (1970). Elaborate models have been developed to predict the impact of various pollution control strategies (Middlebrooks, et al., 1974). The very first models (Streeter and Phelps, 1925) were developed as a tool to determine oxygen levels as indicators of the condition of water bodies subject to waste inputs. Recent, more elaborate models (Thomann, 1974), examine quantitatively the cumulative and synergistic effects of combined waste loads on various biochemical species. However, application and reliability of large scale processes is still somewhat limited, due to biochemical, hydrodynamic and environmental factors, as well as inherent mathematical complexity. Attempts at applying dynamic process modeling to wastewater ponds is not found in the early literature. Traditional treatment process models simulate only substrate and microorganism kinetics with environmental constraints incorporated in gross rate constants.

The basic water quality modeling technique is the application of conservation of mass in time differential form of biochemical species such as BOD, O_2, algae, and other components within a finite complete mix element. In large complex models, many such elements are linked together by mass fluxes with the hydrodynamic diffusion equation solved within each element.

Efforts at defining the biochemical species found in waste stabilization ponds on a non-steady state basis began in earnest during the late 1970s (Fritz, et al., 1979; Ferrara and Harleman, 1980). This work borrowed much conceptual theory from water quality modeling carried out a decade earlier. The basic technique was to develop several simultaneous conservation of mass equations for the various biochemical species solved via numerical methods on a digital computer. This approach was relatively efficient, allowing for much "tinkering" of biokinetic coefficients during calibration.

The model to be developed in the next section is distinct from water quality models in that (1) it is based on the stoichiometric relationships existing as a result of the carbon, oxygen, nitrogen and phosphorous content of algae and bacteria; (2) all biochemical concentrations are high in contrast with dilute systems; (3) the model simulates both short-range, i.e., daily, and long-range, i.e., annual pond characteristics; (4) the model includes a comprehensive heat balance and carbonate equilibrium determination; (5) benthic activity is modeled; and (6) case studies are used in model calibration.

5.1 Biochemical Model Development

The following development of a non-steady state model is representative of the new class of models which are currently being developed (Ferrara and Avci, 1982; Pano and Middlebrooks, 1982).

Conservation of mass equations are developed for interactive chemical and biomass species reacting to environmental factors that affect process mechanisms. Figure 7 is a schematic diagram for a completely mixed pond with a detritus system. Influent and effluent concentrations consist of soluble COD, dissolved oxygen, bacterial mass, algal mass, inorganic carbon, organic and inorganic phosphorus, organic nitrogen, ammonia, nitrate and alkalinity as outlined in Table 5. The detritus or anaerobic sludge layer is coupled to the completely mixed module through a continuous interchange of mass flows. This layer functions as a "black box" in which anaerobic processes digest both bacterial and algal biomass that has settled. Inorganic nitrogen and phosphorus as well as carbon dioxide and methane are returned to the liquid layer as nutrients for biomass growth.

190

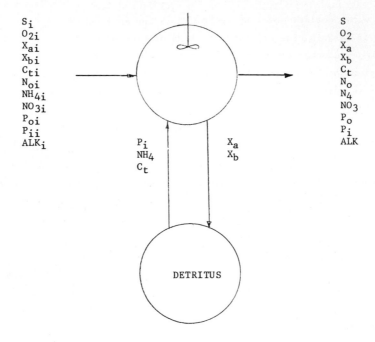

FIGURE 7   Schematic Diagram of Completely Mixed Pond with Detritus System

This analysis simulates pond characteristics as a function of time subject
to site specific environmental physical factors for the selected time period,
the day or month. Environmental factors determine reaction rates and
therefore species concentration. Energy variables are solar radiation, wind
shear and air temperature. The hydrologic components are flow, precipitation,
evaporation and seepage. Balances are made for both energy and hydrologic
variables during each computational step. Energy from wind shear is assumed
to govern the mixing regime and interfacial transfer of oxygen and carbon
dioxide. Thermal energy inputs are short-wave atmospheric radiation,
long-wave atmospheric radiation, and the influent wastewater temperature.
Thermal energy outputs are due to conduction, long-wave radiation, reflection
and evaporation. Energy for algal photosynthesis is provided by short-wave
solar radiation penetrating the air-water interface.

Deterministic, time dependent functions are used for each of the hydrologic
components. Exogenous variables, influent flow rate, evaporation,
precipitation, seepage and wind velocity are considered constant for any
particular day. Air temperatures are represented by daily and annual

TABLE 5

Stabilization Pond Process Model Variables

Influent Concentration, mg/l		Reactor and Effluent Concentration, mg/l	
S_i	= soluble organic substrate (COD)	S	= soluble organic substrate (COD)
O_{2i}	= dissolved oxygen	O_2	= dissolved oxygen
X_{bi}	= bacterial biomass	X_b	= bacterial biomass
C_{ti}	= total in organic carbon	C_t	= total inorganic carbon
P_{oi}	= soluble organic phosphorus	P_o	= soluble organic phosphorus
P_{ii}	= soluble inorganic phosphorus	P_i	= soluble inorganic phosphorus
N_{oi}	= soluble organic nitrogen	N_o	= soluble organic nitrogen
NH_{4i}	= ammonia nitrogen	NH_4	= ammonia nitrogen
NO_{3i}	= nitrate nitrogen	NO_3	= nitrate nitrogen
ALK_i	= alkalinity as $CaCO_3$	ALK	= alkalinity as $CaCO_3$
X_{ai}	= algal biomass	X_a	= algal biomass
		D	= detrital mass, mg/cm^2

sinusoidal functions. Endogenous variables, solar radiation and pond liquid temperature are computed for each time step based upon local time, latitude, longitude, altitude, cloud cover and air temperature.

5.1.1 General Mass Balance

In a continuous flow biological reactor, a complete materials balance includes terms for substance produced or consumed in biochemical reactions, inflow, outflow and accumulation or depletion. Such a general mass balance for either a biotic or abiotic substance can be written as follows for a completely mixed reactor element:

$$\sum (\text{reactions}) + (\text{inflow}) = (\text{outflow}) + (\text{accumulation}) \qquad (27)$$

If C is substance concentration, and reactor volume V is constant, with outputs and inputs solely due to liquid flows, Equation 27 can be written in time differential form as:

$$V \sum (r_c) + Q_i C_i = Q_o C + V \frac{dC}{dt} \qquad (28)$$

where r_c = volumetric reaction rate of substance, mg/l-day; C_i = influent concentration of substance, mg/l; C = effluent concentration of substance, mg/l; $V(dC/dt)$ = volumetric rate of change of substance in the reactor, mg/day; and n = number of reactions that involve the substance.

If the substance is a reactant or biochemical formulation of other products, then r_c is positive. If no reactions occur that involve the substance, then r_c equals zero. If the influent flow rate is equal to the effluent flow rate and the volume is constant, the hydraulic residence time, $0 = V/Q$, can be substituted into Equation 28 to obtain:

$$\frac{dC}{dt} = \frac{C_i}{\theta} - \frac{C}{\theta} + \Sigma (r_c) \tag{29}$$

Coupled mass balances of the form of Equation 29 for continuous flow completely mixed reactors are written for each biochemical component modeled. These equations form the basis of kinetic interaction among the chemical and biomass species being considered.

The biochemical reaction rates are summarized in Table 6. Several of the reaction rates are discussed below.

5.1.2 Bacterial Growth

The mass balance equation for substrate concentration for a continuous completely mixed reactor is:

$$\frac{dS}{dt} = \frac{S_{is}}{\theta} - \frac{S}{\theta} + r_s \tag{30}$$

where S = substrate concentration as COD, mg/l; S_{is} = soluble influent COD concentration, mg/l; and r_s = volumetric reaction rate of substrate due to microbial uptake, mg/l-day and is given in Table 6. The soluble influent COD that goes into the mass balance is computed as:

$$S_{is} = S_i (1 - F_r + F_n) \tag{31}$$

where S_i = influent total or raw COD, mg/l; F_r = fraction representing refractory influent COD; and

$$F_n = \begin{cases} -e^{0.16T} & \text{if } T \leq 15°C \\ \\ (T - 15)^2/100 & \text{if } T > 15°C \end{cases} \tag{32}$$

TABLE 6

Mass Balance Reaction Rates

Reaction	Volumetric Reaction Rate

Bacterial Growth:

Degradable COD
Utilitization

$$r_s = -kX_b \left[\frac{S}{K_s+S}\right]\left[\frac{O_2}{K_{02}+O_2}\right]\left[\frac{N_i^*}{K_{bn}+N_i}\right]\left[\frac{P_i}{K_{bp}+P_i}\right]$$

Bacterial Growth $\qquad r_{xb1} = -Yr_s - k_bX_b$

Sedimentation $\qquad r_{xb2} = -Yr_s - k_bX_b$

Algal Growth:

Algal Growth

$$r_{xal} = X_a \left\{\mu_a f(L)f(T)\left[\frac{CO_2}{K_{CO2}+CO_2}\right]\left[\frac{N_i}{K_{an}+N_i}\right]\left[\frac{P_i}{K_{ap}+P_i}\right]-k_a\right\}$$

Sedimentation $\qquad r_{xa2} = -s_aX_a$

Oxygen Generation and Utilization:

Interfacial Transfer $\qquad r_{1do} = \frac{A}{V}K_{102}(O_s-O_2)$

Photosynthesis $\qquad r_{2do} = 1.244\ r_{xal}$

Bacterial Decay $\qquad r_{3do} = -\left[(1-1.42Y)(-r_s) + 1.42\ k_bX_b\right]$

Nitrification $\qquad r_{4do} = -4.57\ r_{3am}$

Organic Nitrogen:

Bacterial & Algal Decay $\qquad r_{1no} = 0.124\ k_bX_b + 0.063\ k_aX_b$

Organic N $\longrightarrow$ NH$_3$ $\qquad r_{2no} = -\alpha_N N_o$

Ammonia:

Organic N $\longrightarrow$ NH$_3$ $\qquad r_{1am} = -r_{2no}$

Algal & Bacterial Growth $\qquad r_{2am} = 0.063(r_{xal}+k_aX_a) - 0.124(-Yr_s)$

Nitrification

$$r_{3am} = -\frac{\hat{\mu}_N}{Y_N}\left[\frac{NH_4}{K_N+NH_4}\right]\left[\frac{O_2}{K_{NO2}+O_2}\right]\left[C_{pH}\right]\left[C_T\right]$$

Benthic Regeneration $\qquad r_{4am} = R_N/d$

Nitrate:

Nitrification $\qquad r_{1na} = r_{3am}$

Algal & Bacterial Growth $\qquad r_{2na} = r_{2am}$

Organic Phosphorus:

Bacterial & Algal Decay $\qquad r_{1op} = 0.009\ k_aX_a + 0.024\ k_bX_b$

Organic P $\longrightarrow$ Inorganic P $\qquad r_{2op} = -\alpha_p P_o$

TABLE 6 (Continued)

Mass Balance Reaction Rates

Reaction	Volumetric Reaction Rate

Inorganic Phosphorus:

Algal & Bacterial Growth $\quad r_{1ip} = 0.009 \ (r_{xal} + k_a X_a) - 0.024(-Yr_s)$

Organic P $\quad$ Inorganic P $\quad r_{2ip} = r_{2op}$

Benthic Regeneration $\quad r_{3ip} = R_p/d$

Total Inorganic Carbon Generation and Utilization:

Interfacial Transfer $\qquad r_{1ct} = \dfrac{12}{44} \dfrac{A}{V} K_{1CO2}(CO_{2s}-CO_2)$

Bacterial Decay $\qquad r_{2ct} = \dfrac{12}{32} r_{3do}$

Photosynthesis $\qquad r_{3ct} = -1.314 \ r_{xal}$

Benthic Decay $\qquad r_{4ct} = R_c \ C_m/d$

Alkalinity:

Nitrification $\qquad r_{1a} = -7.14 \ r_{3am}$

Detritus:

Sedimentation $\qquad r_{1d} = (1/d) \ (s_a X_b + s_b X_b)$

Upwelling $\qquad r_{2d} = U_r D$

N_1 = total nitrogen concentration (ammonia and nitrate)

when $T \leq 15°C$, F_n represents the fraction of influent total COD that goes into temporary storage by sedimentaion plus that which flows through the pond without being oxidized; however, when $T > 15°C$, F_n represents the COD that is released from temporary storage as a fraction of influent total COD. Because COD solubility is temperature dependent, some COD will settle and some will reach the effluent before it is oxidized if $T \leq 15°C$. However, if $T > 15°C$ some benthic sludge will rise to the surface and increase COD and suspended solids, especially in the spring.

F_r was assumed to be 10 percent of influent total COD.

For heterotrophic bacteria the balance is:

$$\frac{dX_b}{dt} = \frac{X_{bi}}{\theta} - \frac{X_b}{\theta} + r_{xb1} + r_{xb2} \qquad (33)$$

where X_b = concentration of bacterial biomass, mg/l; X_{bi} = influent bacterial concentration, mg/l; and r_{xb1} and r_{xb2} are, respectively, the volumetric settling rate of microorganisms, mg/l-day.

Growth kinetics of bacterial and algal batch cultures using the concept of a limiting nutrient, a mixed order reaction, are based on Monod kinetics. This formulation is similar to the well-known Michaelis-Menten expression used in enzyme kinetics. The model for heterotrophic bacterial growth rate, r_{xb1}, is based on the availability of substrate S, as COD, and nutrients necessary for cell synthesis. Growth reduction due to low nitrogen and phosphorus concentrations is expressed algebraically as the product of the Michaelis-Menten formulations for those nutrients. Because aerobic bacteria require oxygen to metabolize wastes, O_2 concentration is also considered rate limiting as shown in Table 6 for r_s.

Following Eckenfelder (1970), the substrate utilization kinetic coefficient k is modeled as:

$$k = k_{20} \; \beta^{T - 20} \tag{34}$$

where T = temperature, °C; β = exponential temperature coefficient and k_{20} = specific rate of substrate utilization at 20°C, day^{-1}.

When bacteria respire, they do so at higher rates with elevated temperatures. Therefore, the bacterial decay coefficient k_b is similarly written as:

$$k_b = k_{20} \; \beta^{T - 20} \tag{35}$$

where k_{b20} = decay coefficient at 20°C, day^{-1}.

5.1.3 Algal Growth

Algal growth is limited by carbon dioxide, total inorganic nitrogen and total inorganic phosphorus availability as shown in Table 2.2 for r_{xal}. The balance for algal biomass is:

$$\frac{dX_a}{dt} = \frac{X_{ai}}{\theta} - \frac{X_a}{\theta} + r_{xal} + r_{xa2} \tag{36}$$

where X_a = concentration of algal biomass, mg/l; X_{ai} = influent algal biomass concentration, mg/l; r_{xal} = volumetric algal growth rate, mg/l-day; and r_{xa2} = volumetric algal settling rate, mg/l-day, as shown in Table 2.2.

The algal respiration coefficient k_a, in day^{-1}, is computed as suggested by DiToro et al. (1970) as:

$$k_a = C_r T \tag{37}$$

where, C_r = linear multiplier, $°C^{-1}-day^{-1}$.

5.1.4 Oxygen Generation and Utilization

The availability of oxygen for bacterial decay of soluble organics is of prime importance if the system is to be maintained without lapsing into anoxic conditions. Wide variations in concentration, from supersaturation to depletion, can occur during the diurnal cycle. Oxygen is added to the liquid layer by inflow, interfacial transfer and algal photosynthesis. It is removed by bacterial and algal respiration, outflow and interfacial transfer when supersaturation occurs. Therefore, the balance for oxygen is:

$$\frac{dO_2}{dt} = \frac{O_{2i}}{\theta} - \frac{O_2}{\theta} + r_{1do} + r_{2do} + r_{3do} + r_{4do} \qquad (38)$$

where O_2 = concentration of oxygen, mg/l; O_{2i} = influent concentration of oxygen, mg/l; r_{1do} = volumetric rate of oxygen interfacial transfer, mg/l-day; r_{2do} = volumetric rate of photosynthetic oxygen production, mg/l-day; r_{3do} = volumetric rate of bacterial oxygen utilization, mg/l-day; and r_{4do} = volumetric rate of oxygen used in nitrification, mg/l-day. Expressions for r_{1do}, r_{2do}, r_{3do} and r_{4do} are given in Table 6.

The interfacial transfer coefficient K_{102} is defined for shallow lakes and lagoons by Banks and Herrera (1977) and computed through:

$$K_{102} = \frac{1}{d} (0.384 \ W^{.5} - 0.088W + 0.0029 \ W^2) \qquad (39)$$

where d = pond depth, m, and W = wind velocity, km/hr. Saturation concentrations for dissolved oxygen and carbon dioxide are computed from nonlinear functions of temperature and altitude.

Photosynthetic oxygen production is modeled using Equation 40. This is the stoichiometric relationship posed by Stumm and Morgan (1970). There are 1.244 milligrams of oxygen produced for each milligram of algal cells synthesized.

$$106 \ CO_2 + 16 \ NO_3^- + HPO_4^= + 122 \ H_2O + 18H^+ \longrightarrow C_{106}H_{263}O_{110}N_{16}P + 138 \ O_2 \qquad (40)$$

Equation 41 is used to determine oxygen utilization for respiration by bacterial cells. This assumes a cellular composition of $C_5H_7NO_2$ for bacteria as given by Hoover and Porges (1952).

$$C_5H \ NO_2 + 5 \ O_2 \longrightarrow 5 \ CO_2 + 2 \ H_2O + NH_3 \qquad (41)$$

Equations 40 and 41 are the basis for the stoichiometric relationships applied throughout the model.

5.1.5 Nitrogen Balances

The mass balance for soluble organic nitrogen is:

$$\frac{dN_o}{dt} \equiv \frac{N_{oi}}{\theta} - \frac{N_o}{\theta} + r_{1no} + r_{2no} \tag{42}$$

where N_o = organic nitrogen concentration, mg/1; N_{oi} = influent organic
nitrogen concentration, mg/1; r_{1no} = volumetric rate of organic nitrogen
production from bacterial and algal decay, mg/1-day and r_{2no} = volumetric
reaction rate of organic nitrogen transformation to NH_3, mg/1-day. The
balance for ammonia and nitrate is, respectively:

$$\frac{dNH_4}{dt} = \frac{NH_{4i}}{\theta} - \frac{NH_4}{\theta} + r_{1am} + P\,r_{2am} + r_{3m} + r_{4am} \tag{43}$$

and,

$$\frac{dNO_3}{dt} = \frac{NO_{3i}}{\theta} = \frac{NO_3}{\theta} + r_{1na} + P_2 r_{2na} \tag{44}$$

where NH_4 = ammonia concentration, mg/1; NH_{4i} = influent ammonia
concentration, mg/1; r_{1am} = volumetric reaction rate of organic nitrogen to
NH_3, mg/1-day; r_{2am} = volumetric reaction rate for algal and bacterial
uptake of ammonia, mg/1; r_{3am} = volumetric reaction rate for nitrification
of ammonia, mg/1; r_{4am} = volumetric reaction rate for benthic regeneration
of ammonia, mg/1; NO_{3i} = nitrate concentration, mg/1; r_{1na} = volumetric
reaction rate for nitrification of nitrate, mg/1-day; r_{2na} = volumetric
reaction rate for algal and bacterial use of nitrates, mg/1-day,
P_1 = preference factor for ammonia, and P_2 = preference factor for
nitrate. If NH_4^+ = 0, then P_1 = 0 and P_2 = 1. However, if $NH_4^+ \neq 0$,
then P_1 = 1 and P_2 = 0.

Nitrogen enters the stabilization pond in wastewaters containing ammonia,
organic nitrogen, and nitrate. Nitrogen fixation from the atmosphere can be
achieved by some algal species; however, this process is not modeled. Organic
nitrogen from biomass or wastewater, in the form of proteins, is broken down
by hydrolysis into amino acids, and that, in turn, results in the formation of
ammonia through decomposition by bacteria.

Initially part of the soluble ammonia combines with H^+ to form ammonium
ions as follows:

$$NH_3 + H^+ \longrightarrow NH_4^+ \tag{45}$$

This tends to raise the pH. Oxidation proceeds through the action of autotrophic Nitrosomonas and Nitrobacter bacteria to sequentially produce nitrite and nitrate.

$$NH_4^+ + 2O_2 \rightarrow H_2O + NO_3^- + 2H^+ \qquad (46)$$

These reactions require 4.57 mg of oxygen for each mg of ammonia nitrified as N. Formation of nitrate is more rapid than formation of nitrite because Nitrobacter require about 3 times as much substrate as Nitrosomonas to obtain the same amount of energy. Therefore, nitrite concentration is always low and nitrification is usually rate limited by the activity of Nitrosomonas. Thus, nitrite is not modeled.

Denitrification is also not modeled. If the facultative pond is operating properly denitrification should only occur near the benthic region. Currently there is no information on denitrification processes as they occur in facultative ponds.

The reaction rate coefficient for organic nitrogen conversion to ammonia, a_N in day^{-1}, is computed following Di Toro, et al. (1975) as:

$$a_N = 0.002 \; T \qquad (47)$$

Nitrification is described as a single step process with functional links to temperature, dissolved oxygen and pH. The sensitivity of autotrophic nitrifying bacteria, specifically Nitrosomonas, to these constraints has been investigated by Downing (1966). Using the Monod model, the Nitrosomonas growth rate coefficient, r_{3am} is computed as shown in Table 6. The half saturation constant for Nitrosomonas, K_N, is computed from Downing's model as:

$$K_N = 10^{0.051T} - 1.58 \qquad (48)$$

High or low values of pH inhibit Nitrosomonas growth, particularly at values less than 7 and greater than 9. Following Anthonisen et al. (1976) it is postulated that the existence of free ammonia and nitrous acid inhibit nitrifying organisms by causing differences between intercellular and extracellular pH. Following Downing (1966) this relationship is modeled as:

$$C_{pH} = [1 - 0.833(7.2 - pH)] \text{ for pH} < 7.2$$

$$\qquad (49)$$

$$C_{pH} = 1.0 \text{ for pH} \geq 7.2$$

where C_{pH} = Nitrosomonas growth inhibiting factor for pH.

The temperature dependence factor, C_T, is also modeled following Downing (1966) as:

$$C_T = \exp [0.098(T - 15)] \tag{50}$$

Algae as well as autotrophic bacteria takes up both ammonia and nitrate but they prefer ammonia (Golterman 1975), therefore, preference factors, p_1 and p_2, are introduced. Ammonia must be depleted before nitrate begins to be utilized for cell synthesis.

5.1.6 Phosphorus Balances

Phosphorus generally occurs in wastewaters as orthophosphates, polyphosphates, and organic phosphorus that readily becomes orthophosphate. For simulation purposes, phosphorus is divided into organic and inorganic categories. Sources for organic phosphorus are inflow, algal and bacterial respiration, whereas sinks are outflow and transformation into organic phosphorus. The principal sink for inorganic phosphorus is assimilation by biomass for cell synthesis.

The balance for organic phosphorus is:

$$\frac{dP_o}{dt} = \frac{P_{oi}}{\theta} - \frac{P_o}{\theta} + r_{1op} + r_{2op} \tag{51}$$

where P_o = organic phosphorus concentration, mg/l; P_{oi} = influent organic phosphorus concentration, mg/l; r_{1op} = volumetric reaction rate for organic phosphorus production from bacterial and algal decay, mg/l-day; and r_{2op} = volumetric reaction rate for inorganic to organic phosphorus transfer, mg/l-day.

The model by DiToro et al. (1975) is used to compute the coefficient for organic to inorganic phosphorus transfer, a_p, day^{-1}, as:

$$a_p = 0.02 \, T \tag{52}$$

The inorganic phosphorus balance is computed from:

$$\frac{dP_i}{dt} = \frac{P_{ii}}{\theta} - \frac{P_i}{\theta} + r_{1ip} + r_{2ip} + r_{3ip} \tag{53}$$

where P_i = inorganic phosphorus concentration, mg/l; P_{ii} = influent inorganic phosphorus concentration, mg/l; r_{1ip} = volumetric reaction rate for assimilation of inorganic phosphorus by biomass for cell synthesis, mg/l-day; r_{2ip} = volumetric reaction rate for organic to inorganic phosphorus transfer, mg/l-day; and r_{3ip} = volumetric reaction rate for inorganic phosphorus production from benthic regeneration, mg/l-day.

5.1.7 Carbon Balance

The behavior of the carbonate system in aquatic environments is well
documented (Stumm and Morgan, 1970; Loewenthal and Marais, 1976) and has been
simulated in water quality modeling applications (DiToro et al., 1970; Yeasted
and Shane, 1976). Middlebrooks et al. (1974) have conducted research to
determine the limiting nutrients controlling algal growth related to severe
eutrophication problems in fresh water lakes and streams. King (1970)
indicated that in highly eutrophic environments inorganic carbon was thought
to be rate limiting due to slow diffusion of carbon dioxide across the
air-water interface. However, because of bio-oxidation there is more carbon
dioxide available in ponds. Goldman et al. (1974) noted that this may
increase the algal growth rate to the point where nitrogen and phosphorus
become rate determining.

As photosynthesis proceeds during the daylight hours, algal cells
assimilate inorganic carbon and the H^+ ion concentration decreases due to
the shift in equilibrium of H_2CO_3, HCO_3, and CO_3 ions. Characteristically, pH rises as sunlight becomes more intense and more algae cells are
synthesized. If the pH rises above 9, biological processes will begin to be
inhibited.

The balance for total inorganic carbon is written as:

$$\frac{dC_t}{dt} = \frac{C_{ti}}{\theta} - \frac{C_t}{\theta} + r_{1ct} + r_{2ct} + r_{3ct} + r_{4ct} \tag{54}$$

where C_t = total inorganic carbon concentration, mg/l; C_{ti} = influent
total inorganic carbon concentration, mg/l; r_{1ct} = volumetric reaction rate
for carbon dioxide, mg/l-day; r_{2ct} = volumetric reaction rate for carbon
dioxide production from bacterial decay, mg/l-day; r_{3ct} = volumetric
reaction rate for carbon dioxide utilization in photosynthesis, mg/l-day; and
r_{4ct} = volumetric reaction rate for carbon dioxide production from benthic
decay, mg/l-day. The concentration of total inorganic carbon is written as:

$$[C_t] = [CO_2] + [HCO_3^-] + [CO_3^=] \tag{55}$$

Interfacial transfer of carbon dioxide from the atmosphere supplies only a
small fraction of photosynthetic requirements. The carbon dioxide interfacial
transfer coefficient is obtained by multiplying the oxygen transfer
coefficient by the ratio of carbon dioxide to oxygen molecular weights.
Oxidation of soluble organics and bacterial respiration in the liquid layer
produces carbon dioxide in proportion to oxygen consumed and is therefore
related to r_{3do}.

Pond waters may become very alkaline as carbon dioxide is removed as a result of algal growth. If no precipitates are forming, alkalinity is considered conservative because reversible ionization reactions will not affect the mass balance. During nitrification, the release of hydrogen ions from ammonia produces acidity. There are 2 milliequivalents of alkalinity destroyed per mole of NH_4 oxidized. This is equivalent to 7.14 mg of alkalinity per mg NH_3 nitrified. The balance for alkalinity is:

$$\frac{d\ ALK}{dt} = \frac{ALK_i}{\theta} - \frac{ALK}{\theta} + r_{1a} \tag{56}$$

where ALK = alkalinity concentration, mg/l; ALK_i = influent alkalinity concentration, mg/l; and r_{1a} = volumetric reaction rate for depletion of alkalinity by nitrification, mg/l-day. Alkalinity is defined as:

$$[ALK] = [HCO_3^-] + 2[CO_3^=] + [OH^-] - [H^+] \tag{57}$$

5.1.8 Detritus Balance

Algae, bacteria and other suspended matter settle to form a benthic sludge layer in facultative ponds. Anaerobic processes recycle nutrients and release carbon dioxide and methane into the overlaying aqueous layer. The detrital mass balance is simply:

$$\frac{dD_m}{dt} = r_{1d} + r_{2d} \tag{58}$$

where D_m = active detrital mass, mg/cm^2; r_{1d} = volumetric reaction rate for algal and bacterial sedimentation, mg/l-day; and r_{2d} = volumetric reaction rate for regeneration form benthic layer, mg/l-day. A regeneration rate, U_r, is computed as:

$$U_r = U_{r20}\ \beta^{T-20} \tag{59}$$

where U_{r20} = regeneration rate at 20°, day^{-1}, β = Arrhenius temperature constant; and T = temperature, °C.

An initial value for D_m is not necessary because incoming and regenerated mass are independent. Throughout the annual cycle, dD_m/dt is positive because the sludge layer is known to accumulate in ponds. However, during specific periods, spring for example, the time rate of change of D_m may be negative due to a high instantaneous reaction rate as stored organics begin to be metabolized.

Because the benthos acts as a source of nutrients, expressions describing the release of nitrogen, phosphorus, and carbon from the benthos are written as:

$$R_N = U_r \quad D_m \quad (0.063 S_a + 0.124 S_b) \tag{60}$$

$$R_P = U_r \quad D_m \quad (0.009 S_a + 0.024 S_b) \tag{61}$$

$$R_C = U_r \quad D_m \quad (0.358 S_a + 0.531 S_b) \tag{62}$$

where R_N = regenerated nitrogen from the benthos, mg/cm^2-day; S_a = algal cells as a fraction of settling biomass; S_b = bacterial cells as a fraction of settling biomass; R_P = regenerated phosphorus from the benthos, mg/cm^2-day; and R_C = inorganic carbon regenerated from the benthos, mg/cm^2-day. Also:

$$S_a = \frac{s_a X_a}{s_a X_a + s_b X_b} \tag{63}$$

$$S_b = \frac{s_b X_b}{s_a X_a + s_b X_b} \tag{64}$$

Carbon release from the detrital mass is in the form of carbon dioxide and methane. A representation of the anaerobic digestion of bacterial biomass is:

$$6C_5H_7O_2N + 18 \; H_2O \longrightarrow 15 \; CH_4 + 15 \; CO_2 + 6NH_3 \tag{65}$$

Because only carbon dioxide is used for algal synthesis, r_{4ct} contains a coefficient C_m to delineate the fraction of carbon released as CO_2.

The linkages between the various biochemical and biomass components are illustrated in Figure 8.

6.0 PHYSICAL MODEL DEVELOPMENT

Photosynthetic productivity is governed by light and temperature as well as nutrient concentration. Accurate determination of available light and bulk liquid temperature is necessary for model operation. Light and temperature can vary considerably throughout the seasons causing wide fluctuations in biological activity depending upon the location of the site.

6.1 Response to Light

The penetration and intensity of solar radiation reaching algal cells is one of the principal factors affecting growth and the resulting photosynthetic oxygen production. Because of algae and other suspended matter, incoming

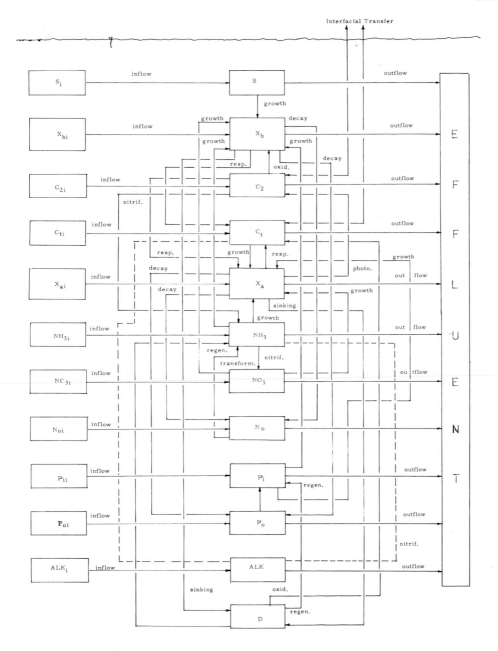

FIGURE 8   Conceptual Model of Biochemical and Biomass Interactions for
Stabilization Pond

light is absorbed, scattered, and, therefore, attenuated as a function of depth and turbidity. Based on a vertical path, light intensity will decay exponentially with depth. From the Beer-Lambert Law this is expressed as:

$$I_h = I_o e^{-\epsilon h} \tag{66}$$

where I_h = light intensity reaching depth h, cal/cm^2-day; I_o = light intensity penetrating the surface, cal/cm^2-day; h = depth, cm; and ϵ = light attenuation coefficient, cm^{-1}. The expression for ϵ proposed by DiToro et al. (1970) for suspended algal cells was modified to include bacterial biomass. The value for   is computed by:

$$\epsilon = \epsilon_w + 0.17(X_a + X_b) \tag{67}$$

where ϵ_w = attenuation coefficient through clear water, cm^{-1}. Because most facultative ponds are shallow, an average light intensity I_{av} was computed for the assumption of a homogeneous suspension as:

$$I_{av} = \frac{I_o}{d} (1 - e^{-\epsilon d}) \tag{68}$$

where d = pond depth, cm.

The development of an expression for computing I_o is given in the next section.

Steele (1965) presented a photosynthetic rate coefficient as a function of light. This coefficient incorporates the feature of inhibition of biological activity at higher than optimal intensities and represented by:

$$f(L) = \frac{I_{av}}{I_m} \exp(1 - \frac{I_{av}}{I_m}) \tag{69}$$

where I_m = optimal light intensity, cal/cm^2-day.

6.2   Temperature Sensitivity

The response of aquatic biological growth systems to temperature changes has been modeled by several workers (Lassiter and Kearns, 1974; DiToro et al., 1970; Lehman et al., 1975). At temperatures less than optimal, the growth rate increases exponentially with temperature. As the temperature approaches the optimal point, the rate of change in the growth rate approaches zero and becomes negative as it moves above optimal. The growth rate declines almost exponentially for temperatures exceeding the optimal. For this study the Beta curve, a special right-skewed Gamma distribution, is used to model growth response to temperature changes. Oswald (1977) stated that, for most mesophilic species of algae found in stabilization ponds, maximum growth occurs between 20° and 30°C and that it decreases to zero above 35°C. Written in coefficient form, the Beta distribution is:

$$f(T) = \frac{T^5(1 - T/T_1)^2}{T_m^5 (1 - T/T_m)^2} \tag{70}$$

where T = pond liquid temperature °C; T_m = 25°C; and T_1 = 35°C.

The values of f(L) and f(T) are then used to compute r_{xal} as shown in Table 6. A heat balance is made for each computational step in order to determine pond liquid temperature as explained in Section 6.3. Computed temperature is considered homogeneous throughout the pond and any change as a result of heat transfer is considered instantaneous.

6.3   Pond Heat Balance

A technique for computing the bulk liquid temperature in shallow and slow moving bodies of water is necessary in order to compute biochemical and biomass species concentration. Existing methods for computing temperatures have been developed for power industry studies of cooling ponds and involve complex techniques that incorporate transport phenomena and require measurement of solar radiation (Edinger et al., 1968, 1974; Brady et al., 1969).

Surface heat exchange processes are slow and involve large areas. Because there is usually limited mixing in stabilization ponds, dispersion and advection occur slowly. This causes temperature gradients to be small and somewhat insensitive to rapid changes in climatology. However, as seasons change, bulk water temperature follows the variation in solar radiation and ambient air temperature. Depending on the mixing regime, stratification can occur if there is little wind shear and the pond is more than a few feet in depth.

The temperature change of a shallow body of water is a function of short-wave solar radiation, evaporation, influent flow and temperature, conduction and water surface temperature. The individual components affecting the transfer of heat into and out of water through its surface are listed below (Edinger et al., 1974):

Incoming I_s = Short-wave solar radiation, cal/m^2-day

H_a = Long-wave atmospheric radiation, cal/m^2-day

Outgoing I_{sr} = Reflected short-wave solar radiation, cal/m^2-day

H_{ar} = Reflected long-wave atmospheric radiation, cal/m^2-day

H_w = Long-wave water surface radiation, cal/m^2-day

H_e = Evaporation losses, cal/m^2-day

H_c = Conduction losses or gains, cal/m^2-day

<u>Incoming</u> I_s = Short-wave solar radiation, cal/m^2-day

H_a = Long-wave atmospheric radiation, cal/m^2-day

<u>Outgoing</u> I_{sr} = Reflected short-wave solar radiation, cal/m^2-day

H_{ar} = Reflected long-wave atmospheric radiation, cal/m^2-day

H_w = Long-wave water surface radiation, cal/m^2-day

H_e = Evaporation losses, cal/m^2-day

H_c = Conduction losses or gains, cal/m^2-day

The net heat entering or leaving a body of water through the surface, H_{net}, can be computed as follows:

$$H_{net} = I_s + H_a - (I_{sr} + H_{ar} + H_w + H_e + H_c) \qquad (71)$$

A positive value for H_{net} indicates that heat has been added whereas a negative value for H_{net} indicates that heat has been lost from the body of water. An expanded treatment of this subject is given in Fritz <u>et al</u>. (1980). Figure 9 illustrates the heat flux components and their interaction.

A heat balance for the entire pond consists of terms for inflow, outflow, and interfacial heat transfer. The heat added by the pond inflow and the heat removed by the pond outflow are computed from Eq. 72 and 73, respectively.

$$H_i = \rho \, Q_i c_p (T_i - T) \qquad (72)$$

$$H_i = \rho \, Q_o c_p (T_i - T_o) \qquad (73)$$

where, H_i = influent heat, kcal/day; H_o = effluent heat, kcal/day; ρ = density of water, kg/m^3; T_i = influent temperature, deg. C; T_e = effluent temperature, deg. C; T = pond liquid temperature, deg. C; c_p = specific heat of water, kcal/kg-deg. C; Q_i = influent flow rate, m^3/day and Q_o = effluent flow rate, m^3/day.

6.4 Heat Balance Equation

The rate of change in heat per unit volume can now be computed as:

$$\rho \, C_p \, \frac{dT}{dt} = \frac{H_i}{V} - \frac{H_o}{V} + \frac{H_{net}}{d} \qquad (74)$$

where, V = pond volume, m^3 and d = depth, m.

Substituting Eqs. 72 and 73 into Eq. 74, rearranging and simplifying yields:

$$\frac{dT}{dt} = \frac{H_{net}}{\rho \, C_p d} + \frac{1}{V} \, [Q_i(T_i - T) - Q_o(T - T_e)] \qquad (75)$$

Because T_e = T for a completely mixed reactor, one can finally write:

$$\frac{dT}{dt} = \frac{H_{net}}{\rho\, C_p d} + \frac{Q_i}{V}\ (T_i - T) \qquad (76)$$

where, at time $t = t_o$, $T = T_o$.

Equation 76 is solved for each time step interval to compute the time profile for a mean pond temperature.

6.5   Hydrologic Balance

The pond outflow, Q_o, is computed as:

$$Q_o = Q_i + Q_p - Q_e - Q_s \qquad (77)$$

where Q_p = inflow as precipitation, m^3/day; Q_e = outflow as evaporation, m^3/day; and Q_s = outflow as seepage, m^3/day. This assumes that the volume remains constant, i.e., $dV/dt = 0$. The values of Q_i, Q_p, Q_e, and Q_s are required for each time period.

Because the outflow rate is equal to inflow rate only when the last three terms in Eq. 77 sum to zero, an approximate hydraulic residence time θ_a is defined as:

$$\theta = \frac{2V}{Q_i + Q_o} \qquad (78)$$

for use in the mass balance computations.

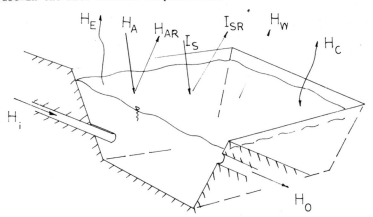

FIGURE 9   Heat Balance Components

7.0   MATHEMATICAL SOLUTION TECHNIQUES

7.1   Runge-Kutta Method

The mass balance equations for substrate, bacterial and algal biomass, oxygen, organic nitrogen, ammonia, nitrate, organic phosphorous, inorganic phosphorus, inorganic carbon, alkalinity and detritus, constitute 12 ordinary, first-order, nonlinear differential equations that must be solved simultaneously. They are considered stable in that the solution is convergent with initial conditions imposed that are of the order of empirical values known to exist. Because a closed form, analytical solution is difficult or impossible to derive, the fourth-order Runge-Kutta or Kutta-Simpson technique is used. This method was chosen because of its simplicity in programming, low truncation error, fast convergence and requirement for only initial conditions. The reader is referred to Ketter and Prawel (1969) for details as to the procedure. The error associated with the fourth-order Kutta-Simpson techniques is of the order $(\Delta t)^5$. Because there is no direct method of computing the truncation error, sample calculations were carried out and deviations from a known result noted. Usually, the value of Δt is halved until the computed result is constant or to within the desired degree of accuracy. Thereafter, deviation from the base result can be obtained by doubling the time step interval with each set of computations. However, as step size is reduced, round-off errors accumulate thus decreasing accuracy. There is no simple procedure for determining the optimal time step increment that will yield both accuracy and minimize round-off error. However, to determine a suitable step increment, particularly for full-year simulation runs, computations were carried out using various intervals. A 4-hour time step was the largest period yielding consistent results compared to smaller intervals and was, therefore, selected as being appropriate.

7.2   Carbonate Equilibria Computation

Carbon dioxide is cycled through the mass balance equations by the processes of photosynthesis, interfacial mass transfer and respiration. With each time step interval, pH and all carbonate species must be computed. The carbonate reactions involved are as follows:

$$CO_2 \; + \; H_2O \; \longrightarrow \; H^+ \; + \; HCO_3^- \qquad\qquad (79)$$

$$HCO_3^- \; \longrightarrow \; H^+ \; + \; CO_3^= \qquad\qquad (80)$$

$$H_2O \; \longrightarrow \; H^+ \; + \; OH^- \qquad\qquad (81)$$

Alkalinity, or the buffer capacity of water, is defined as its molar capacity to neutralize an acid and is due to the presence of bicarbonates, carbonates, and hydroxides found in water supplies and was defined in Eq. 57. With C_t, total inorganic carbon, as defined from Eq. 55, the concentration of all carbonate species can be computed.

Within the carbon equilibria subroutine, pH and carbonate species are determined by solving a fourth-order polynomial via the Newton-Raphson method. The definition of CO_2 - ACY is:

$$[CO_2 - ACY] = [C_t] - [ALK] \qquad (82)$$

Since ALK is conservative, CO_2 - ACY is a function of additions or depletions from C_t. The principal component of CO_2 - ACY is carbon dioxide, and within a small range of pH, could be approximated by CO_2 only. DiToro (1976) gives an analysis of the error associated with this simplification, at pH levels below 6 or above 7 significant errors result. Because pH values encountered in ponds are generally above 7, CO_2 - ACY is defined by Eq. 82.

With the equilibrium constants K_1, K_2, K_w, and the terms in Eq. 82, species concentration and pH can be determined by solving a fourth-order polynomial as follows:

$$f([H^+]) = A_4[H^+]^4 + A_3[H^+]^3 + A_2[H^+]^2 + A_1[H^+] + A_o \qquad (83)$$

where,

$$A_1 = -K_1K_2K_w \qquad (84)$$

$$A_2 = K_1K_2 - K_1[CO_2 - ACY] - K_w \qquad (85)$$

$$A_3 = [ALK] + K_1 \qquad A_4 = 1.0 \qquad (86)$$

The solution is determined via a Newton-Raphson iteration technique which converges rapidly. Yeasted and Shane (1976) used the same procedure related to acid mine drainage problems. Caution must be exercised in determining a starting value for H^+. If pH is less than the initial guess, the procedure will not converge due to the existence of other solutions. In stabilization ponds, since pH rarely goes below 7, an initial value of 6 is chosen to begin the iteration process.

The differential form of Eq. 83 is written as:

$$f'([H^+]) = 4A_4[H^+]^3 + 3A_3[H^+]^2 + 2A_2[H^+] + A_1 \qquad (87)$$

and the accompanying Newton-Raphson recursive expression for H^+ ion concentration for the n + 1 iterate is:

$$[H^+]_{n+1} = [H^+]_n - \frac{f([H^+])}{f'([H^+])} \qquad (88)$$

When the hydrogen ion concentration is known to be within a required
tolerance, all carbonate species are determined by computing the distribution
coefficients as a function of first and second dissociation constants K_1 and
K_2 according to the following expressions:

$$[CO_2] = C_5 a_0 \tag{89}$$

$$[HCO_3] = C_t a_1 \tag{90}$$

$$[CO_3] = C_t a_2 \tag{91}$$

where,

$$a_0 = (1 + K_1/[H^+] + K_1 K_2/[H^+]^2)^{-1} \tag{92}$$

$$a_1 = ([H^+]/K_1 + 1 + K_2/[H^+])^{-1} \tag{93}$$

$$a_2 = ([H^+]/K_1 K_2 + [H^+]/K_2 + 1)^{-1} \tag{94}$$

8.0 MODEL CALIBRATION

8.1 Evaluation Procedure

Reported data for two existing stabilization ponds were used to calibrate
and evaluate the model. One pond is located in New Mexico and the other is in
Utah. Only the New Mexico example is described here. Local environmental
conditions, influent water quality characteristics, and kinetic coefficients
from the literature were used for calibration. The simulated effluent water
quality characteristics were then compared with the reported findings.

8.2 Simulation of New Mexico Pond

Larsen (1974) monitored environmental and water quality characteristics
daily from August 1, 1972 through July 31, 1973 at a small facultative pond
that treats a combination of municipal and high strength animal wastes. The
pond, located near Albuquerque, New Mexico, is rectangular, 78.1 m by 84.2 m,
0.9 m deep, and located at an elevation of 1718 m. The general pond arrangement is shown in Fig. 10. Mean influent total or raw COD of 913 mg/l for the
sampling period was reduced an average of 53 percent. Because the pond is
small, 6576 m^2, compared to most municipal systems, it was represented by a
single completely mixed reactor. The environmental and influent water quality
conditions reported were assumed to be constant over the 24-hour period.
Tables 7 and 8 list values for the biochemical and physical constants used in
the simulation.

The simulation of solar radiation and pond water temperature are shown in Figs. 11 and 12, respectively. The agreement between computed and observed values were judged to be satisfactory.

The effluent total or raw COD values reported by Larsen included the COD computed from the mass balance equation, the COD in bacterial and algal biomass, the refractory COD, and any COD that flows through the pond without being oxidized. Therefore, the effluent total COD is computed as:

$$S_t = S + 1.42X_b + 1.244X_a + S_i(F_r + F_{ft}) \tag{95}$$

where S_t = effluent total COD, mg/l; S = soluble substrate COD concentration calculated from mass balance, mg/l; F_r = refractory COD; and F_{ft} = the COD

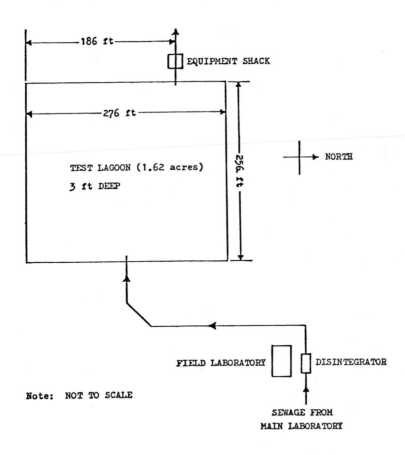

FIGURE 10 Physical Arrangement of the New Mexico Test Pond

TABLE 7

Values for Biochemical Constants for New Mexico Pond

Constraint	Constraint Name	Value	Units	Source
Y	yield coefficient	0.5	mg/mg	Metcalf & Eddy
k_{20}	max. substr. util. rate	5.0	day^{-1}	Metcalf & Eddy
K_s	substr. half sat. const.	50.0	mg/l	Metcalf & Eddy
k_{b20}	bact. decay coeff.*	0.07	day^{-1}	Metcalf & Eddy (1972)
K_{O2}	O_2 half sat. const.	1.0	mg/l	Thibodeaux (1976)
K_{bn}	N_2 bact. half sat. const.	0.01	mg/l	assumed
K_{bp}	P-bact. half sat. const.	0.01	mg/l	assumed
β	Arrhenius const.	1.07	--	Mara (1976)
s_b	bact. settl. rate	0.05	day^{-1}	Canale (1976)
μ_a	algal growth coeff.*	2.0	day^{-1}	Canale (1976)
K_{CO2}	CO_2 half sat. const. 1.0	mg/l	Goldman (1974)	
K_{an}	N_2 algal half sat. const.	0.10	mg/l	Canale (1976)
K_{ap}	P-algal half sat. const.	0.02	mg/l	Canale (1976)
k_a	algal resp. coeff.*	0.08	day^{-1}	DiToro (1970)
α_N	$N_o \rightarrow NH_4$ rate coeff. 0.05	day^{-1}	DiToro (1970)	
$\hat{\mu}_N$	<u>Nitrosom.</u> growth rate	0.008	day^{-1}	assumed
Y_n	yield coeff. <u>Nitrosom.</u>	0.15	mg/mg	assumed
α_P	$P_o - P_i$ rate coeff.* 0.4	day^{-1}	DiToro (1975)	
$C_m,$	CO_2 - C ratio	0.5	--	assumed
s_a	algal settl. rate	0.05	day^{-1}	Canale (1976)
U_{r20}	regeneration rate*	0.09	day^{-1}	Foree (1970)

that flows through the pond without being oxidized, expressed as a fraction of influent total COD. F_{ft} fits the following relationship:

$$F_{ft} = \begin{cases} e^{-0.16T} - e^{-0.37T} & \text{if } T \leq 15°C \\ 0 & \text{if } T > 15°C \end{cases} \qquad (96)$$

The effluent filtered COD, as defined in Larsen's study, is computed as:

$$S_f = S + S_i(F_r + F_{ft}) \qquad (97)$$

where S_f = effluent filtered COD, mg/l, and represents the untreated COD.

Because effluent suspended solids are unusually high, it is assumed that no significant sedimentation of biomass occurs prior to discharge.

TABLE 8

Values for Physical Constants for New Mexico Pond

Variable	Constraint Name	Value	Units	Source
—	longitude	107.0	deg.	Albuquerque, New Mexico
ϕ	latitude	35.0	deg.	Albuquerque, New Mexico
z	altitude	1718.0	m	Larsen (1974)
ϵ_w	H$_2$O light atten. coeff.	1.0	m^{-1}	Stefan (1976)
I_m	algal optimal light int.	250.0	ly/day	DiToro (1975)
I_{sc}	solar constant	2.0	ly min.	Ryan & Stolz (1972)
C	cloud factor	0.5		Ryan & Stolz (1972)
T_{av}	annual av. air temp.	12.8	°C	Larsen (1974)
T_{am}	annual temp. variation	12.2	°C	Larsen (1974)
T_i	influent temp.	20.0	°C	assumed
Q_s	seepage	0.0	m^3/day	assumed
A	pond area	6568.0	m^2	Larsen (1974)

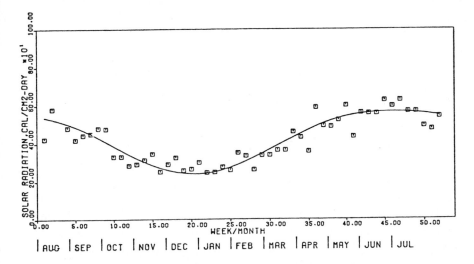

FIGURE 11   Computed and Observed Solar Radiation for New Mexico Pond

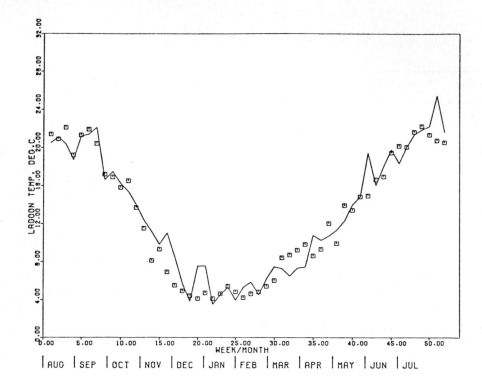

FIGURE 12   Computed and Observed Water Temperature for New Mexico Pond

Computed and observed weekly effluent total and filtered COD are shown in Fig. 13.   The observed and computed monthly mean effluent total COD are compared in Fig. 14.   The COD reduction was judged to be modeled satisfactorily.

Figs. 15 and 16, respectively, present dissolved oxygen and pH profiles throughout a typical diurnal cycle. As photosynthesis proceeds during the daylight hours, particularly at midday, oxygen production is maximized and often reaches supersaturation.   In the early evening, as photosynthesis diminishes, dissolved oxygen decreases due to the continuing demand by bacteria for bioxidation processes.   The period midnight to 6:00 a.m. is most critical. This is when many ponds become anaerobic. When this occurs, hydrogen sulfide is produced which can cause odor problems. When light returns in the morning, oxygen production resumes.

There are also appreciable swings in pH throughout the diurnal cycle due to photosynthetic activity. As carbon dioxide is consumed there is a decrease in H^+ ions and, hence, an increase in the pH. King (1976) and others (Oswald, et al., 1957; Lowenthal and Marais, 1976) have shown that this phenomena occurs daily in stabilization ponds.   There is a much larger diurnal variation

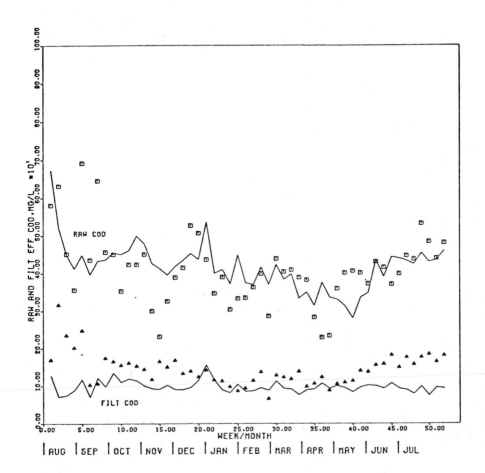

FIGURE 13   Computed and Observed Weekly Effluent COD for New Mexico Pond

in pH at higher temperatures, because during colder days there is less
biological activity. The model, therefore, simulated daily variation of DO
and pH in a manner characteristic of observed results found at many
stabilization ponds.

Observed and computed effluent ammonia and alkalinity are shown in
Fig. 17. The nitrification rate was modified by changing the _Nitrosomonas_
growth rate, μ_N, from 0.008 to 0.002 day^{-1} to achieve agreement with
observed data. The agreement is satisfactory except for the period from

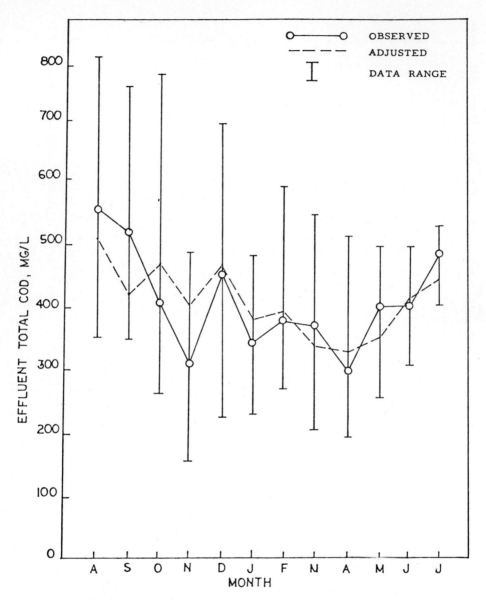

FIGURE 14   Computed and Observed Monthly Mean Effluent Total COD for
New Mexico Pond

mid-March to mid-June.   Fig. 18 compares observed and computed weekly nitrate
concentrations.   From these figures, it appears that there may be a phenomena
that occurs during the spring that was not modeled, such as upwelling or
turnover.   However, the alkalinity and nitrate concentration are not critical
to COD reduction and, hence, pond performance.

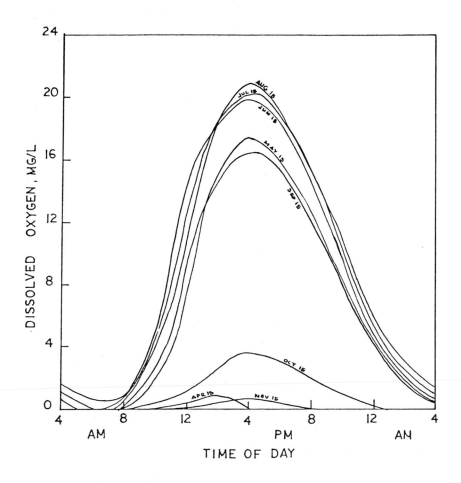

FIGURE 15   Computed Diurnal Variation in Dissolved Oxygen for
New Mexico Pond

Computed values for soluble organic phosphorus and inorganic phosphorus
diverges from Larsen's empirical results as shown in Fig. 19.   Alteration of
the linear coefficient in rate expression Eq. 52 from 0.02 to 0.002 day^{-1}
resulted in a decrease of phosphorus conversion from the organic into the
inorganic form.   However, continued divergence from empirical results may be
due to the assumption that no chemical precipitation or adsorption is assumed
to occur.   Insufficient data was presented by Larsen to evaluate such a
process.

Fig. 20 illustrates computed algal and bacterial biomass.   Both mass
quantities follow the expected pattern.

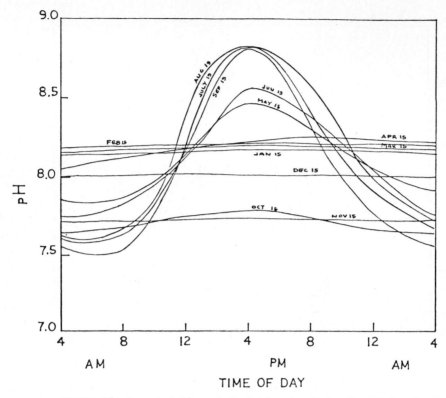

FIGURE 16   Computed Diurnal Variation in pH for New Mexico Pond

9.0   MODEL APPLICATION FOR DESIGN

9.1   Model Uses

The dynamic model presented has several uses. It could form a submodel in the development of an all-inclusive simulation of algal growth from a variety of waste streams. It could also be used to model integrated aquaculture systems. Perhaps its most relevant use remains in the design and simulation of wastewater stabilization ponds or highly loaded eutrophic lakes.

To use the model in pond design, the characteristics of the influent wastewater, physical constraints, environmental conditions, and the biochemical constants must be determined. A pond shape, or simply the number of complete mix reactors connected in series, and volume need to be assumed. The model can then be used to simulate the biochemical dynamics of the pond. The simulation results are then evaluated to determine if effluent requirements are satisfied. If the effluent requirements are not satisfied, then another pond shape or volume must be chosen and the computation repeated. This iterative process continues until a pond shape and volume are chosen such that the effluent requirements are satisfied.

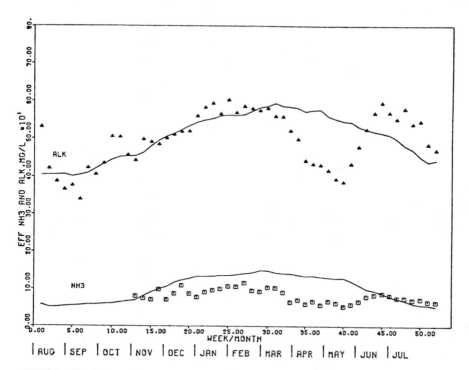

FIGURE 17   Computed and Observed Weekly Ammonia and Alkalinity
            Concentrations for New Mexico Pond

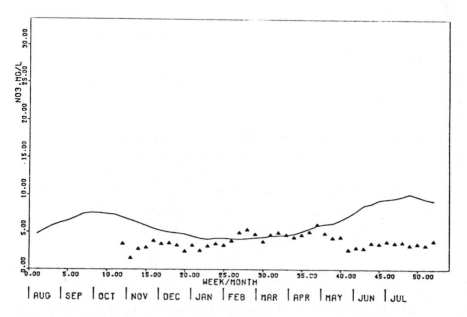

FIGURE 18   Computed and Observed Weekly Nitrate Concentrations for
            New Mexico Pond

220

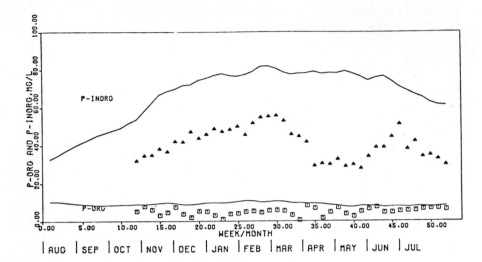

FIGURE 19 Observed and Computed Weekly Organic and Inorganic
Phosphorus Concentrations for New Mexico Pond

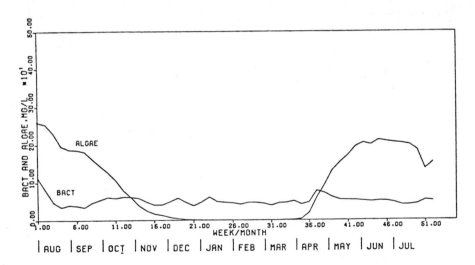

FIGURE 20 Computed Weekly Algal and Bacterial Concentrations for
New Mexico Pond

9.2 Hypothetical Design Example

As an example of the model's application in a semi-tropical environment, a
hypothetical site in northeast Brazil was selected for analysis. The example
has been greatly simplified to demonstrate the methodology of analysis. The
specific location chosen is Campina Grande, Paraiba do Norte, site of Federal
University of Paraiba's experimental ponds that are used for tropical waste

treatment research. Located below the equator, the air temperature
distribution was assumed to be:

$$T_a = 20.0 + 5.0 \ [\cos \ (D_t/365) \ 2\pi] \tag{98}$$

The specific problem is to design a wastewater stabilization pond for an
influent flow rate of 200 m^3/day with the water quality characteristics
listed in Table 9. These characteristics are representative of a medium
strength waste. The only requirement is that effluent total COD be less than
30 mg/l.

Precipitation, evaporation, and seepage were assumed to be zero. This
assumption is valid if the pond has an impermeable liner and the amount of
precipitation is equal to the amount of evaporation for each time interval.
In all probability, the evaporation exceeds the precipitation in this area;
however, the data were not readily available. The physical constants specific
to this site were chosen to be those listed in Table 10. The other physical
and biochemical constants are assumed to be the same as presented earlier.

A pond with an area of 1 hectare and a depth of 1 meter was chosen for the
first trial. The computed effluent total or raw COD is shown in Fig. 21. The
peak effluent total COD is approximately 35 mg/l in December. This is above
the effluent limitation; therefore, a pond with an area of 2 hectare and a
depth of 1 meter is chosen for the second trial.

The effluent total COD for the 2-hectare pond is also shown in Fig. 21.
The peak effluent total COD is approximately 26 mg/l in December. This is
less than the effluent requirement and the design is considered acceptable.

It would be possible to decrease the pond size to obtain a computed
effluent total COD of 30 mg/l. However, a margin of safety is necessary
requiring several recomputations to arrive at the optimal design.

The procedure would be similar if there were effluent constraints on other
water quality characteristics. Figs. 22 through 27 illustrate the time
profile for computed water temperature, algal biomass, bacterial biomass,
ammonia, alkalinity, nitrate, organic phosphorus, and inorganic phosphorus.

The effluent filtered COD remains constant. The principal component of the
effluent filtered COD is refractory COD. This indicates that the soluble COD
is completely oxidized. Because total COD is composed of algal and bacterial
biomass, it tends to follow similar profiles as shown in Figs. 21 and 23. Due
to the tropical nature of the site, growth continues throughout the year, but
reaches particularly high levels during the summer season.

TABLE 9

Assumed Influent Wastewater Characteristics and Initial
Values of Variables for Hypothetical Example, Northeast Brazil Pond

Variable	Value (mg/l)
S_i	500.0
X_{bi}	1.0
O_{2i}	2.0
C_{ti}	60.0
X_{ai}	1.0
NH_{4i}	30.0
N_{oi}	10.0
P_{oi}	5.0
P_{ii}	10.0
NO_{3i}	1.0
ALK_i	200.0

Fig. 27 shows the computed effluent total and filtered COD if the algal settling rate is 0.24 day^{-1} instead of 0.05 day^{-1}. This resulted in significantly lower values for total COD. In fact, the 1-hectare pond would be satisfactory for this value of s_a. This figure is included to demonstrate the sensitivity of pond performance to 1 parameter. More work needs to be done to determine the appropriate values of the physical and biochemical parameters for different field conditions because a change in one of the parameters can significantly affect the performance of the pond. It also indicates that until more confidence can be placed in the biochemical parameters, the model can only be used for preliminary design and evaluation of ponds in conjunction with the empirical methods cited earlier.

10. CONCLUSIONS AND RECOMMENDATIONS

    10.1 Conclusions

As a result of model evaluation using data from two dissimilar facultative pond systems located in New Mexico and Utah in the original study, the following conclusions are presented.

TABLE 10

Values for Physical Constants for Hypothetical Example,
Northeast Brazil Pond

Constraint	Constraint Name	Value	Units	Reference
--	Longitude	36	deg.	N.E. Brazil
ϕ	Latitude	-7	deg.	N.E. Brazil
z	Altitude	100.0	m.	N.E. Brazil
T_{av}	Annual Av. Air Temp.	20.0	deg.C	assumed
T_{am}	Annual Temp. Var.	5.0	deg.C	assumed
T_i	Influent Temp.	20.0	deg.C	assumed
C	Cloud factor	0.5	-	assumed
Q_s	Seepage flow	0.0	m^3/day	assumed
Q_p	Precipitation	0.0	m^3/day	assumed
Q_e	Evaporation	0.0	m^3/day	assumed
A	Pond Area	1.0	hectare	assumed
NR	No. of Reactors	1.0	-	assumed

Solar Radiation and Water Temperature:

   It is possible to acceptably simulate solar radiation and bulk pond water
temperature through the application of equations for net insolation and the
heat balance as outlined. The specific data requirements are longitude,
latitude, altitude, cloud cover fraction, influent temperature, air
temperature distribution, precipitation, evaporation, seepage, and relative
humidity. The model will provide annual or daily profiles of water
temperature and solar insolation for kinetic formulations.

COD Profiles:

   High effluent COD found in many ponds are the result of suspended biomass
grown in the pond as opposed to influent COD flowing through. This underlines
the ability of ponds to treat nutrient-rich wastewaters by incorporation into
algae, but also emphasizes the need for filtration or adequate settling in
order to keep biomass out of the effluent.
   Computation of raw and filtered effluent COD required determination of the
extent of sedimentation of suspended material. With the New Mexico system, it

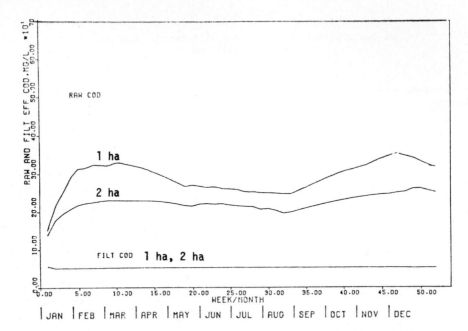

FIGURE 21   Computed Raw and Filtered Effluent COD for Hypothetical
Example, Northeast Brazil Pond, with Pond Areas of 1 and
2 hectares

was reported that all biomass synthesized remained essentially suspended and found its way into the effluent keeping raw COD at high levels. In contrast, the Utah data (not given here) suggested highly efficient settling, keeping raw effluent COD at low levels, the exception occurring during spring overturn. In addition, the Utah ponds treated a low strength waste resulting in less biomass growth in general.

Because of the expected differences in effluent suspended matter, it became necessary to modify the sinking coefficient s_a and the equation for raw effluent COD, S_t, by adding factors for storage and upwelling in the Utah case. These adjustments required a priori knowledge of expected physical performance of a particular system. Such information may be available based on other systems with similar hydraulic characteristics and loadings subject to comparable climatic conditions.

It can be assumed that for single ponds of less than a hectare, significant biomass will be found in the effluent. For larger systems, particularly those with multiple cells, sedimentation of biomass may remove most suspended solids from the effluent stream. This area of analysis will require further work to correlate the extent of suspended matter as a function of particular hydraulic characteristics.

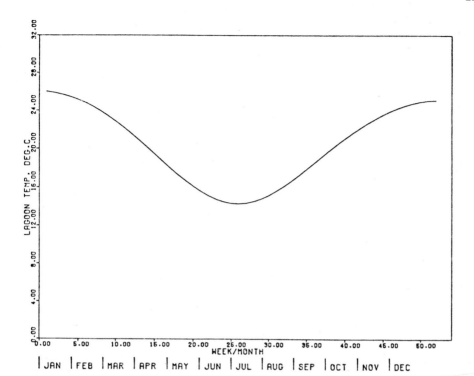

FIGURE 22 Computed Water Temperature for Hypothetical Example,
Northeast Brazil Pond

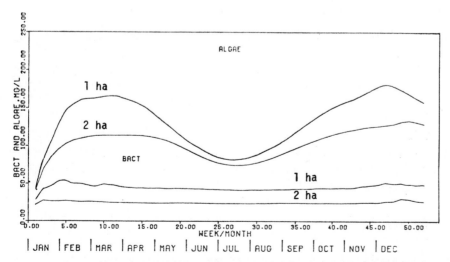

FIGURE 23 Computed Algal and Bacterial Biomass Concentration for
Hypothetical Example, Northeast Brazil Pond, with Pond
Areas of 1 and 2 hectares

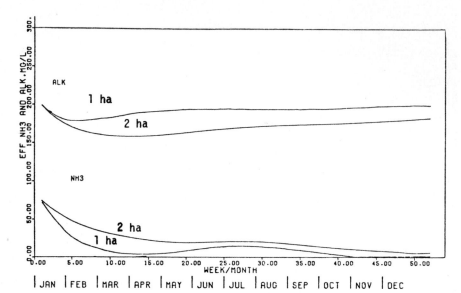

FIGURE 24   Computed Ammonia Concentration and Alkalinity for
Hypothetical Example, Northeast Brazil Pond, with
Pond Areas of 1 and 2 hectares

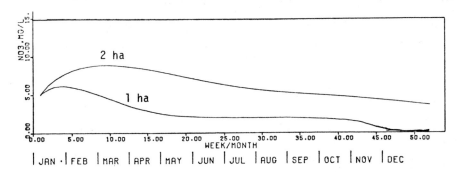

FIGURE 25   Computed Nitrate Concentration for Hypothetical Example,
Northeast Brazil Pond, with Pond Areas of 1 and 2 hectares

Another conclusion drawn was that the refractory portion of influent COD is
higher than anticipated.   Generally, a value of 10 percent of raw influent COD
was considered refractory.   For wastewater having a COD less than 300 mg/l
this assumption may not be valid.

Rate Expressions:

Rate constants commonly used to describe biological processes in
conventional wastewater treatment systems are not directly applicable for use
with stabilization ponds.   Specifically, these include transformation rates of

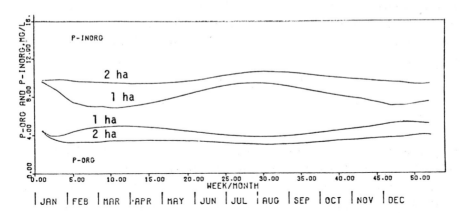

FIGURE 26   Computed Organic and Inorganic Phosphorus Concentrations
             for Hypothetical Example, Northeast Brazil Pond, with Pond
             Areas of 1 and 2 hectares

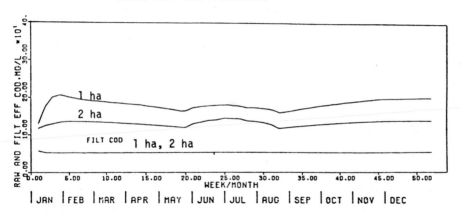

FIGURE 27   Computed Effluent Raw and Filtered COD for Hypothetical
             Example, Northeast Brazil Pond, with Pond Areas of 1 and 2
             hectares, with s_a increased to 0.24 day^{-1}

organic forms of nutrients to the inorganic forms and nitrification rates.
The rate of nitrification found suitable was approximately one-quarter of what
was reported for activated sludge processes (Knowles et al., 1965). Reasons
for the moderation in the rate are the periodic lack of oxygen, temperature
and pH variation, and inadequate mixing. As a result, inorganic nitrogen is
stored during the cold season and utilized when algae begin to synthesize in
the warmer spring (DiGiano and Yi-Siung, 1977).

The transformation of organic phosphorus to inorganic phosphorus found to
be suitable was one-tenth of that used in modeling Lake Erie nutrient concentrations (DiToro et al., 1975). On the other hand, coefficients such as the

rate of substrate utilization k, yield coefficient Y, decay coefficients k_a and k_b, as well as algal growth rate μ_a, were applied as found in the literature (Metcalf and Eddy, 1972) leading to acceptable results. It, therefore, remains to determine how particular rate expressions can be tailored for application to stabilization ponds.

Algal Growth:

Algal growth in facultative stabilization ponds appears to be limited by CO_2 availability as opposed to total inorganic carbon (King, 1974). Use of this assumption was made in the development of this model. This controversy will undoubtedly continue for some time. The studies mentioned, Goldman (1974) and King (1970, 1974), have attempted to deal with the problem, whereas the present work only tests one conclusion applied to a stabilization pond setting. Then nutrient limiting algal growth may not be the same for a eutrophying lake as for a wastewater fed oxidation pond. In some cases it may be phosphorus, nitrogen, CO_2 or C_t with each situation requiring an in situ analysis. For this reason, both phosphorus and nitrogen were included as rate limiting in this model.

Oxygen and pH Profiles:

The model herein developed simulates oxygen concentration and pH throughout the diurnal cycle. Both profiles follow characteristic curves found to be observed at many pond sites. This feature inhibits bacterial growth during low oxygen availability and adverse pH conditions. It also indicates the model's applicability for short-term prediction. Incorporation of carbonate system computations made possible the determination of all carbonate species as well as pH.

Biochemical and Biomass Concentrations:

Simulation of biochemical and biomass species concentration was achieved for both test cases with acceptable accuracy after modification of rate expressions for nitrification and transformation of organic to inorganic forms. Precise description would require more accurate data and the inclusion of physical and biochemical mechanisms beyond the scope of this work.

Solution Technique:

The application of a fourth-order Runge-Kutta solution technique proved to be appropriate for this research. It allowed the user flexibility in determining computational step size while giving an indication of the accuracy of the result. Model equations were of such a form as to be stable, leading to a convergent solution.

Model Limitations:

The model as developed will not simulate periodic occurrences such as spring and fall overturn. It will, however, model the slow diffusion of COD and nutrients from the benthic region into the liquid layer as a result of winter storage. To model sudden biochemical occurrences such as a turnover, functions must be developed on a site specific basis. In order to simulate such periodic behavior as part of a general model, further analysis is necessary which will link that occurrence to known phenomena such as sudden temperature changes. Because additional complete long-term data sets were not available, further work along these lines was not possible.

10.2 Summary

A non-steady-state rational process model has been presented as a first attempt in simulating the behavior of biochemical and biomass components in a wastewater stabilization pond. The model consists of twelve ordinary, first-order, nonlinear differential equations that must be solved simultaneously. The equations represent mass balances for organic substrate, bacterial mass, algal mass, oxygen, organic nitrogen, ammonia, nitrate, organic phosphorus, inorganic phosphorus, inorganic carbon, alkalinity and detritus. In addition, light intensity, a heat balance and a hydrologic balance are computed as well. The model is intended for use as a tool for prediction and evaluation of facultative ponds with varying wastewater characteristics, environmental conditions, and configuration.

10.3 Recommendations

In the course of pursuing this work, it became apparent that significant subject areas pertaining to pond processes have barely been explored. Some are outlined below.

Benthic Processes:

Scant literature is available on benthic processes occurring in ponds with the exception of some laboratory studies of decaying algae (Foree et al., 1973). These processes are complex and require extensive study in order to arrive at a quantifiable theory. Work in the determination of the rate of sludge buildup in order to predict the life of a pond should also follow. Models could be constructed of the benthic processes leading to operating policies which would minimize sludge buildup and odor problems. Investigating the fate of heavy metals and other trace pollutants in the benthic environment has only recently been started.

Nitrification and Denitrification:

Nitrification and denitrification in stabilization ponds are not well understood. There were no studies found in the literature. With the wide variation in oxygen availability as a function of depth, time and temperature as well as the variation in pH, nitrogen system species are especially difficult to trace. In the interest of achieving ammonia reduction by dentrification in order to avoid discharging nutrients into receiving bodies of water, it becomes imperative to determine the mechanisms controlling nitrification and dentrification in ponds.

Hydraulic Regimes:

Hydraulic regimes in ponds and their effect on biological processes have only recently been explored. Attempts at approaching near plug flow conditions with laboratory ponds should be pursued because systems with baffles have been shown to give improved performance (Reynolds et al., 1977). Optimal geometry of pond systems should be developed with emphasis on COD reduction and pathogen removal.

Model Use:

If ponds are to have a continued popularity in rural or small communities, they must meet acceptable effluent standards. Only through proper design can these criteria be met. The purpose of this work has been to develop a model of characteristics and environmental factors. It is hoped the model will be used in the formulation of design techniques. Whether for wastewater treatment or for algal growth as a food or energy source, a fundamental model of pond processes has heretofore been absent from the literature.

LIST OF PRINCIPAL MODEL SYMBOLS

Symbol	Description	Units
A	pond area	m^3
A_t	atmospheric transmission coeff.	--
ALK	pond alkalinity as Ca CO_3	mg/l
ALK_i	influent alkalinity as Ca CO_3	mg/l
C	cloud factor	--
C_m	carbon dioxide/carbon ratio	--
c_p	specific heat of water	kcal/kg-deg.C.
C_{ph}	nitrification pH factor	--
C_r	algal respiration multiplier	$°C^{-1} day^{-1}$
C_t	pond total inorganic carbon con.	mg/l
C_{ti}	influent total inorganic carbon conc.	mg/l
C_T	nitrification temperature factor	--
CO_2	pond carbon dioxide conc.	mg/l
CO_{2s}	saturation carbon dioxide conc.	mg/l
d	pond depth	m
D_m	active detrital mass	mg/cm^2
D_t	Julian day of the year, 1-365	--
e_a	vapor pressure at wind height	mm Hg
e_s	saturated vapor pressure	mm Hg
f(L)	algal light coeff.	--
f(T)	algal temp. coeff.	--
F_{ft}	influent flow through COD fraction	--
F_n	(T ≤ 15°C) influent COD fraction not oxidized / (T ≥ 15°C) COD fraction released from storage	-- / --
F_r	refractory influent COD fraction	--

h_a	hour angle	rad.
H_a	long wave atm. rad.	cal/cm^2-day
H_{ar}	long wave atm. refl. rad.	cal/cm^2-day
H_{an}	net long wave atm. rad.	cal/cm^2-day
H_c	conduction losses or gains	cal/cm^2-day
H_e	evaporative losses	cal/cm^2-day
H_i	influent heat flux	cal/day
H_{net}	net interfacial heat transfer	cal/cm^2-day
H_o	effluent heat flux	cal/day
H_w	long wave water surf. rad.	cal/cm^2-day
HOUR	time of day	--
I	light intensity reaching algal cells	cal/cm^2-day
I_{av}	mean light intensity for pond	cal/cm^2-day
I_m	optimal light intensity	cal/cm^2-day
I_o	light penetrating water surface	cal/cm^2-day
I_s	net insolation reaching surface	cal/cm^2-day
I_{sc}	solar constant	cal/cm^2-day
I_{sr}	reflected short-wave solar radiation	cal/m^2-day
k	max. substrate util. rate	day^{-1}
k_{20}	max. substrate util. rate @ 20^oC	day^{-1}
k_a	algal decay rate	day^{-1}
k_b	bacterial decay rate	day^{-1}
k_{b20}	bacterial decay rate @ 20^oC.	day^{-1}
K_{an}	algal N half sat. coeff.	mg/1
K_{ap}	algal P half sat. coeff.	mg/1
K_{bn}	bacterial N half sat. coeff.	mg/1
K_{bp}	bacterial P half sat. coeff.	mg/1
K_{NO2}	_Nitrosomonas_ O_2 half sat. coeff.	mg/1
K_N	_Nitrosomonas_ N half sat. coeff.	mg/1

K_{CO2}	algal CO_2 half sat. coeff.	mg/l
K_{O2}	bacterial O_2 half sat. coeff.	mg/l
K_s	bacterial COD half sat. coeff.	mg/l
$K_{1\ O2}$	oxygen transfer rate	m/day
K_{1CO2}	carbon dioxide transfer rate	m/day
m	optical air mass coeff.	--
N_i	total inorganic nitrogen conc.	mg/l
N_o	pond organic nitrogen conc.	mg/l
N_{oi}	influent organic nitrogen conc.	mg/l
NH_4	pond ammonia conc.	mg/l
NH_{4i}	influent ammonia conc.	mg/l
NO_3	pond nitrate conc.	mg/l
NO_{3i}	influent nitrate conc.	mg/l
O_2	pond oxygen conc.	mg/l
O_{2i}	influent oxygen conc.	mg/l
O_s	saturated oxygen conc.	mg/l
P_1	ammonia preference fact.	--
P_2	nitrate preference fact.	--
P_i	pond inorganic phos. conc.	mg/l
P_{ii}	influent inorganic phos. conc.	mg/l
P_o	pond organic phos. conc.	mg/l
P_{oi}	influent organic phos. conc.	mg/l
Q_e	evaporation losses	m^3/day
Q_i	influent flow rate	m^3/day
Q_o	effluent flow rate	m^3/day
Q_p	precipitation	m^3/day
Q_s	seepage losses	m^3/day

r_{1a}	rate of depletion of alkalinity by nitrification	mg/1-day
r_{1am}	rate of organic nitrogen to transformation to NH_3	mg/1-day
r_{2am}	rate of biomass uptake of ammonia	mg/1-day
r_{3am}	rate of nitrification of ammonia	mg/1-day
r_{4am}	rate of benthic regeneration of ammonia	mg/1-day
r_{1d}	rate of algal and bacterial sedimentation	mg/1-day
r_{2d}	rate of regeneration from benthic layer	mg/1-day
r_{1do}	rate of interfacial O_2 transfer	mg/1-day
r_{2do}	rate of photosynthetic O_2 production	mg/1-day
r_{3do}	rate of bacterial O_2 utilization	mg/1-day
r_{4do}	rate of nitrification O_2 utilization	mg/1-day
r_{1ct}	rate of interfacial CO_2 transfer	mg/1-day
r_{2ct}	rate of bacterial CO_2 generation	mg/1-day
r_{3ct}	rate of algal CO_2 generation	mg/1-day
r_{4ct}	rate of carbon dioxide production from benthic decay	mg/1-day
r_{1ip}	rate of assimilation of inorganic phosphorus	mg/1-day
r_{2ip}	rate of organic to inorganic phosphorus transfer	mg/1-day
r_{3ip}	rate of inorganic phosphorus from benthic	mg/1-day
r_{1na}	rate of nitrification of nitrate	mg/1-day
r_{2na}	rate of biomass use of nitrate	mg/1-day
r_{1no}	rate of organic nitrogen production from decay	mg/1-day
r_{2no}	rate of organic nitrogen to transformation to NH_3	mg/1-day
r_{1op}	rate of organic phosphorus production from decay	mg/1-day
r_{2op}	rate of organic to inorganic phosphorus transfer	mg/1-day
r_s	COD reaction rate	mg/1-day
r_{xal}	algal growth rate	mg/1-day
r_{xa2}	algal settling rate	mg/1-day

r_{xb1}	bacterial growth rate	mg/l-day
r_{xb2}	bacterial settling rate	mg/l-day
R_C	benthic CO_2 release	$mg/cm^2/day$
R_h	relative humidity	
R_N	benthic ammonia regeneration	mg/cm^2-day
R_P	benthic inorganic phosphorus regeneration	mg/cm^2-day
s_a	algal sinking rate	day^{-1}
s_b	bacterial sinking rate	day^{-1}
S	pond soluble COD conc.	mg/l
S_a	algal fraction of benthic mass	--
S_b	bacterial fraction of benthic mass	--
S_f	effluent filtered COD conc.	mg/l
S_i	influent total COD conc.	mg/l
S_{is}	soluble influent COD conc.	mg/l
S_t	pond total COD conc.	mg/l
S_{up}	upwelled non-soluble COD fract.	--
t	time	day
t_t	true solar time	hour
T	pond liquid temperature	^{o}C
T_a	air temperature	^{o}C
T_{ad}	ampl. daily temp. var., air	^{o}C
T_{am}	ampl. annual temp. var., air	^{o}C
T_{av}	mean annual temperature, air	^{o}C
T_i	influent temperature	^{o}C
T_l	algal upper limiting temp.	^{o}C
T_m	algal optimal temp.	^{o}C
T_s	water surface temp.	^{o}C

U_r	benthic regeneration rate	day^{-1}
U_{r2o}	benthic regeneration rate @ $20^{\circ}C$	day^{-1}
V	pond volume	m^3
W	wind velocity	$km/hour$
X_a	pond algal biomass	$mg/1$
X_{ai}	influent algal biomass	$mg/1$
X_b	pond bacterial biomass	$mg/1$
X_{bi}	influent bacterial biomass	$mg/1$
Y	yield coeff. carbonacous	mg/mg
Y_N	_Nitrosomonas_ yield coeff.	mg/mg
z	site altitude	m
α	solar altitude	deg.
α_N	organic nitrogen trans. rate	day^{-1}
α_P	organic phos. trans. rate	day^{-1}
β	Arrhenius coefficient	--
δ	declination of the sun	rad.
ϵ	light attenuation coeff.	cm^{-1}
ϵ_ω	light atten. coeff. for water	cm^{-1}
θ	hydraulic residence time	day
μ_a	max. algal growth rate	day^{-1}
μ_N	_Nitrosomonas_ growth rate	day^{-1}
$\hat{\mu}_N$	max. _Nitrosomonas_ growth rate	day^{-1}
ϕ	site latitude	deg.

---

*All reaction rates r_i are volumetric reaction rates.

REFERENCES

Anthonisen, A.C., Loehr, R.C., Prakasam, T.B.S., and Srinath, E.G., 1976. Inhibition of Nitrification by Ammonia and Nitrous Acid. Journ. Water Poll. Control Fed., <u>48</u>, 5.

Abbot, W.E., 1946, Twenty Years Experience With Dissolved Oxygen Adsorption Test in China. Journal of the Society of Chemical Industry, <u>65</u>, 87.

Ballinger, D.G., and Lishka, R.J., 1962, Reliability and Precision of BOD and COD Determinations. Journ. Water Poll. Control Fed., <u>34</u>, 470.

Banks, R.B., and Herrera, F.F., Effect of Wind and Rain on Surface Reaeration. Journ. Env. Eng. Div., ASCE, EE3, 489, (1977).

Benemann, J.R., Weissman, J.C., Koopman, B.L., and Oswald, W.J., Energy Production by Microbial Photosynthesis. Nature, <u>268</u>, (1977).

Bente, P.F., The International Bio-Energy Directory, The Bio-Energy Council, Washington, D.C. (1981).

Bowen, S.P., Performance Evaluation of Existing Lagoons, Peterborough, New Hampshire. Tech. Rep., EPA-600/2-77-085, (1977).

Brady, D.K., Graves, W.L., and Geyer, J.C., Surface Heat Exchange at Power Plant Cooling Lakes, Cooling Water Studies for Edison Electric Institute, Report No. 5, Johns Hopkins Univ. (1969).

Canale, R.P., Modeling Biochemical Processes in Aquatic Ecosystems. Ann Arbor Science, Ann Arbor, Mich., (1976).

Canter, L.W., Englande, A.J., and Mauldin, A.F., Loading Rates on Waste Stabilization Ponds. Journ. San. Eng. Div., ASCE, SA6, 1117 (1969).

Chen, C.W., and Orlob, G.T., Ecologic Simulation for Aquatic Environments. Tech. Rep. OWRR C-2044, Water Resources Engineers, Inc., Walnut Creek, Ca., (1972).

Click, C.N., An Evaluation of Some Critical Factors in a Pilot-Sized Deep Facultative Lagoon. Ph.D. Thesis, Clemson University, Clemson, S.C. (1970).

Cordeiro, C.F., Echelberger, W.F., and Verhoff, F.H., "Rates of Carbon, Oxygen, Nitrogen and Phosphorus Cycling Through Microbial Populations in Stratified Lakes. In Modeling the Eutrophication Process, Ed. Middlebrooks, et al., Ann Arbor Science, Ann Arbor, Mich. (1974).

Davis, E.A., Dedrick, C.S., French, C.S., Milner, H.W., Myers, J., Smith, J.H., and Spoehr, H.A., Laboratory Experiments on Chlorella Culture at the Carnegie Institute of Washington Department of Plant Biology. Ed. J.S. Burlew, Algae Culture From Laboratory to Pilot Plant. Carnegie Inst. of Washington Publ. 600, Washington, D.C. (1964).

DiGiano, F.A., and Yi-Siung Su, Nitrogen Transformation in Land Treatment. Journ. Env. Eng. Div., ASCE EE6, 1075, (1977).

DiToro, D.M., Combining Chemical Equilibrium and Phytoplankton Models--A General Methodology. In Modeling and Eutrophication Process in Aquatic Ecosystems, ed. Canale, R.P., Ann Arbor Science, Ann Arbor, Mich. (1976).

DiToro, D.M., O'Connor, D.J., and Thomann, R.V., A Dynamic Model of Phytoplankton Populations in Natural Waters. Manhattan College, Bronx, New York (1970).

DiToro, D.M., O'Connor, D.J., Thomann, R.V., and Mancini, J.L., Phytoplankton-Zooplankton-Nutrient Interaction Model of Western Lake Erie. Systems Analysis and Simulation in Ecology, Vol. 3, Academic Press, New York (1975).

Downing, A.L., Population Dynamics in Biological Systems. Proc. 3rd Interntl. Conf. Water Poll. Research, WPCF, Munich, Germany (1966).

Eckenfelder, W.W., Water Quality Engineering for Practicing Engineers, Barnes & Noble, New York (1970).

Edinger, J.E., Duttweiler, D.W., and Geyer, J.C., The Response of Water Temperature to Meterological Conditions, Water Resources Research, 4, 5, 1137 (1968).

Edinger, J.E., Brady, D.J., and Geyer, J.C., Heat Exchange and Transport in the Environment. Johns Hopkins University, Baltimore, Maryland (1974).

Ferrara, R.A., Avci, C.B., Nitrogen Dynamics in Waste Stabilization Ponds, Journ. Water Poll. Control Fed., 54, 4 (1982).

Ferrara, R.A., Harleman, D.R.F., Dynamic Nutrient Cycle Model for Waste Stabilization Ponds, Journ. Envi. Eng. Div., ASCE, 106, EEI (1980).

Ferrara, R.A., Harleman, D.R.F., Hydraulic Modeling for Waste Stabilization Ponds, ASCE, 107, EE4 (1981).

Fogg, G.E., The Metabolism of Algae, John Wiley & Sons, New York (1953).

Foree, E.G., and Jewell, W.J., The Extent of Nitrogen and Phosphorus Regeneration from Decomposing Algae. In Advances in Water Pollution Research, Proc. 5th Interntl. Conf. Water Poll. Res., Pergamon Press Ltd., London, Eng., (1970).

Foree, E.G. and Scroggin, C.R., Carbon and Nitrogen as Regulators of Algal Growth. Journ. Env. Eng. Div., ASCE, EE5, 639 (1973).

Fritz, J.J., Middleton, A.C., Meredith, D.C., Dynamic Process Modeling of Wastewater Stabilization Ponds, Journ. Water Poll. Control Fed., 55, 11 (1979).

Fritz, J.J., Meredith, D.D., Middleton, A.C., Non-Steady State Bulk Temperature Determination for Stabilization Ponds, Water Research, 14, 5 (1980).

Gloyna, E.F., Waste Stabilization Ponds, World Health Organization, Geneva, Switzerland (1970).

Gloyna, E.F., Facultative Waste Stabilization Pond Design. In Ponds as a Wastewater Treatment Alternative, Ed. by Gloyna, et al., Water Resources Symposium No. 9, College of Engineering, Univ. of Texas, Austin, Texas (1976).

Goldman, J.C., Oswald, W.J., and Jenkins, D., The Kinetics of Inorganic Carbon Limited Algal Growth. Journ. Water Poll. Fed., 46, 554 (1974).

Golterman, H.L., ed., Physiological Limnology: Developments in Water Science 2, Elsevier Scientific Publishing Co., Amsterdam (1975).

Gotaas, H.B., Effect of Temperature on Biochemical Oxidation of Sewage. Sewage and Industrial Wastes, 20, 441 (1948).

Greenfield, R.E., and Elder, A.L., The Effect of Temperature on the Rate of Deoxygenation of Diluted Sewage. Ind. Eng. Chemistry, 18, 281 (1926).

Herman, E.R., Developments of Design Criteria for Waste Stabilization Ponds. Ph.D. Thesis, University of Texas, Austin, Texas (1957).

Herman, E.R. and Gloyna, E.F., Waste Stabilization Ponds: Formulation of design Equations. Sewage and Industrial Wastes, 30, 736(1958).

Hoover, S.R. and Porges, N., Assimilation of Dairy Wastes by Activated Sludge II. The Equation of Synthesis and Rate of Oxygen Utilization. Sewage and Industrial Wastes, 24, 306 (1952).

Huang, J.C. and Gloyna, E.F., Effects of Toxic Organics on Photosynthetic Reoxygenation. Env. Health Eng. Labs., Center for Research in Water Resources, Tech. Rep. EHE-07-6701, CRWR-20, University of Texas at Austin, Austin, Texas (1967).

Ketter, R.L., and Prawel, S.P., Modern Methods of Engineering Computation, McGraw-Hill, New York (1969).

King, D.L., The Role of Carbon in Eutrophication. Journ. Water Poll. Control Fed., 42, 2035 (1970).

King, D.L., Changes in Water Chemistry Induced by Algae. In Ponds as a Waste-water Treatment Alternative, ed. Gloyna, et al., Water Res. Symposium No. 9, University of Texas at Austin, Austin, Texas (1976).

Knowles, G., Downing, A.L., and Barrett, M.J., Determination of Kinetic Constants for Nitrifying Bacteria in Mixed Culture, With the Aid of an Electronic Computer. Journ. General Microbiology, 38, 263 (1965).

Larsen, T.B., A Dimensionless Design Equation for Sewage Lagoons. Ph.D. Thesis, University of New Mexico, Albuquerque, N.M. (1974).

Lassiter, R.R. and Kearns, D.K., Phytoplankton Population Changes and Nutrient Fluctuations in a Simple Aquatic Ecosystem Model. In Modeling the Eutrophication Process. Ed. Middlebrooks, et al., Ann Arbor Science Pub., Ann Arbor, Mich. (1974).

Lehman, J.T., Botkin, D.B. and Likens, G.E., The Assumption and Rationales of a Computer Model of Phytoplankton Population Dynamics. Limnology and Oceanography, 20, 343 (1975).

Lowenthal, R.E., and Marais, G.v.R., Carbonate Chemistry of Aquatic Ecosystems: Vol. I, Ann Arbor Science, Ann Arbor, Mich. (1976).

Mara, D.C., Proposed Design for Oxidation Ponds in Hot Climates. Journ. Env. Eng. Div., ASCE, EE2, 296 (1975).

Mara, D.C., Sewage Treatment in Hot Climates, John Wiley & Sons, London (1976).

Marais, G.v.R., Dynamic Behavior of Oxidation Ponds. Proc. of the 2nd Intntl. Symposium on Waste treatment Lagoons, Kansas City, Kansas (1970).

Marais, G.v.R., and Shaw, V.A., A Rational Theory for the Design of Sewage Stabilization Ponds in Central and South Africa. Transactions, South African Institution of Civil Engineers, 13, 11 (1961).

240

McKinney, R.E., Dornbush, J.N. and Vennes, J.W., "Waste Treatment Lagoons-- State

Meenaghan, G.F. and Alley, F.C., Photosynthesis: A Major or a Minor Role in Waste Stabilization Process. Presented at 14th Oklahoma Industrial Waste Conf., Stillwater, Okla. (1963).

Metcalf & Eddy, Inc., Wastewater Engineering, McGraw-Hill, New York (1972).

Middlebrooks, E.J., Falkenborg, D.H. and Maloney, T.E., Modeling the Eutrophication Process, Ann Arbor Science, Ann Arbor, Mich. (1974).

Moore, E.W., Thomas, H.A., and Snow, W.B., Simplified Method for Analysis of BOD Data. Sewage and Ind. Wastes, 22, 1343 (1950).

National Research Council, Committee on Sanitary Engineering, Sewage Treatment in Military Installations. Journ. Water Poll. Control Fed., 18, 791 (1946).

Neel, J.K., et al., Experimental Lagooning of Raw Sewage, Fayette, Missouri, Experimental Stabilization Ponds, 1957-58. Journ. Water Poll. Control Fed., 39, 1657 (1961).

Odum, H.T., Environment, Power and Society, Wiley-Interscience, New York (1971).

Oswald, W.J., Experiences With New Pond Designs in California. In Ponds as a Wastewater Treatment Alternative, ed. Gloyna, et al., University of Texas at Austin, Austin, Tex. (1976).

Oswald, W.J., The Engineering Aspects of Microalgae. For publication in Handbook of Microbiology, CRC Press, Cleveland, Ohio (1977).

Oswald, W.J., Gotaas, H.B., Ludwig, H.F., and Lynch, V., Algae Symbiosis in Oxidation Ponds III. Photosynthetic Oxygenation. Sewage and Ind. Wastes, 25, 692 (1953).

Oswald, W.J., Gotaas, H.B., Goluecke, C.G., and Kellen, W.R., Algae in Waste Treatment. Sewage and Industrial Wastes, 29, 437 (1957).

Oswald, W.J., Meron, A., and Zabat, M.D., Designing Ponds to Meet Water Quality Criteria. Proc. of the 2nd Intntl. Symposium on Waste Treatment Lagoons, Kansas City, Kansas (1970).

Pano, A., Middlebrooks, E.J., Ammonia Nitrogen Removal in Facultative Wastewater Stabilization Ponds, Journ. Water Poll. Control Fed., 54, 4 (1982).

Reynolds, J.H., Swiss, R.E., Macko, C.A., and Middlebrooks, J.E., Performance Evaluation of an Existing Seven Cell Lagoon System. U.S. EPA Tech. Rep. EPA-600/2-77-086, EPA, Cincinnati, Ohio (1977).

Richardson, B., Orcutt, D.M., Schwerner, H.A., Martinez, C.L. and Wickline, H.E., Effects of Nitrogen Limitations on the Growth and Composition of Unicellular Algae in Continuous Culture. Applied Microbiology, 18, 245 (1969).

Ryan, P.J. and Stolzenbach, K.D., Engineering Aspects of Heat Disposal From Power Generation. Mass. Inst. of Tech., Cambridge, Mass. (1972).

Sandoval, M., Verhoff, F.H. and Cahill, T.H., Mathematical Modeling of Nutrient Cycling in Rivers. In Modeling Biochemical Processes in Aquatic Ecosystems, ed. Canale, R.P., Ann Arbor Science, Ann Arbor, Mich. (1976).

Steele, J.H., Notes on Some Theoretical Problems in Production Ecology. In Primary Production in Aquatic Environments, ed. Goldman, C.R., Mem. Inst. Idrobiol., Univ. of Cal. Press, 383, Berkeley, Cal. (1965).

Stefan, H., Skoglund, T. and Megard, R.O., Wind Control of Algae Growth in Eutrophic Lakes. Journ. Env. Eng. Div., EE6, 1201 (1976).

Stoltenberg, D.H., Algal Metabolism as Related to the Theory of Oxidation Ponds, M.S. Thesis, University of Kansas, Manhattan, Kansas (1962).

Streeter, H.W. and Phelps, E.B., A Study of the Pollution of the Ohio River, Part III, Factors Concerned in the Phenomena of Oxidation and Reaeration. U.S.P.H.S., Public Health Bull. 146, (1925).

Stumm, W. and Morgan, J.J., Aquatic Chemistry, Wiley-Interscience, New York (1970).

Suwannakarn, V., Temperature Effects on Waste Stabilization Ponds. Ph.D. Thesis, University of Texas, Austin, Texas (1963).

Thibodeaux, L.J., An Aqueous Environmental Simulation Model for Mid-South Lakes and Reservoirs. Water Resources Research Center, Rep. No. 41, Univ. of Ark., Fayetteville, Ark. (1976).

Thirumurthi, D., Design Criteria for Stabilization Ponds. Journ. Water Poll. Control Fed., 46, 2094 (1974).

Thirumurthi, D. and Nashashibi, O.I., A New Approach for Designing Waste Stabilization Ponds. Water and Sewage Works, 114, R-208 (1967).

Thomann, R.V., DiToro, D.H., and O'Connor, D.J., A Preliminary Model of Phytoplankton Dynamics in the Upper Potomac Estuary. Journ. Env. Eng. Div., ASCE, EE2, 100 (1974).

Thomas, H.A., Graphical Determination of BOD Curve Constants. Water and Sewage Works, 97, 123 (1950).

Towne, W.W., Bartsch, A.F. and Davis, W.H., Raw Sewage Stabilization Ponds in the Dakotas. Sewage and Ind. Wastes, 29, 377 (1957).

Watters, G.Z., et al., The Hydraulics of Waste Stabilization Ponds, Utah Water Research laboratory, Utah State University, (1973).

Ward, C.H. and King, J.M., Fate of Algae in Laboratory Cultures. In Ponds As A Wastewater Treatment Alternative, ed. Gloyna, et al., The University of Texas at Austin, Austin, Texas (1976).

Wehner, J.F., and Wilhelm, R.H., Boundary Conditions of Flow Reactors. Chem. Eng. Science, 6, 89 (1956).

Yeasted, J.G. and Shane, R., pH Profiles in a River System With Multiple Acid Loads. Journ. Water Poll. Control Fed., 48, 91 (1976).

MATHEMATICAL MODELS FOR THE TRICKLING FILTER PROCESS

JOHN ROBERTS

University of Newcastle,
N.S.W., 2308, Australia.

1. GENERAL

The trickling filter has become one of the most widely used secondary treatment processes for organic liquid wastes. It is one of the oldest engineering biological treatment systems. Low and high rate trickling filters have been extensively investigated over the past forty years and have provided a fertile area for applied and fundamental research into their mode of operation.

The references noted in the following sections are neither extensive or exclusive. They generally have been chosen for either their historical significance or for the detailed information they contain on a particular topic. Important aspects drawn from these references which deal with operating and design parameters affecting performance of trickling filter systems are combined in several mathematical models in the final Sections.

1.1   Fixed film systems

The trickling filter and the more recently developed rotating biological reactor are the two principal fixed film systems in use today. Simply stated, the concept of these systems is to provide a surface on which a microbial film can grow. Waste water containing dissolved organic material flows over the exposed active film and is biologically oxidised to form stable materials and carbon dioxide. From the development of stacked rock beds in the late 1800's until the 1940's, fixed sprays or rotary distributors spread influent across the surface of the packing. With increased loadings, either hydraulic or organic, plastic modular or random packings have been manufactured which provided a much greater void space in the media and correspondingly, less likelihood of flooding or clogging. These fixed media units operate as plug-flow reactors possibly with some dispersion either under aerobic or anaerobic conditions.

Commercial installations of rotating reactors incorporating a fixed microbial film adhering to a cylindrical shaped, horizontally oriented packing have been introduced since the early 1970's. The media is typically made of plastic on which the microbial film alternatively contacts the waste water and air as it rotates through the trough and through the atmosphere. These units are principally used as "package" systems and they operate as completely mixed reactors in series, somewhat like activated sludge units acted under aerobic conditions.

1.2   Description of the process - physical characteristics

The essential features of a trickling filter system are (a) a primary settling tank (b) the fixed film unit and (c) a secondary settling tank shown in Figure 1.1 (Liptak,1974;ASCE,1977;Water Research Centre,1980;Benefield and Randall,1980; Winkler,1981).

Primary sedimentation prior to the trickling filter unit is a function of the concentration of suspended solids in the influent stream, as well as the effective void size of the media. Stone and small sized random synthetic media generally require effective primary settling to minimise clogging-flooding problems. While primary sedimentation may not be necessary for plastic modular media which has large void size, it is advantageous to allow for flow/concentration equilisation and/or pH-nutrient adjustment.

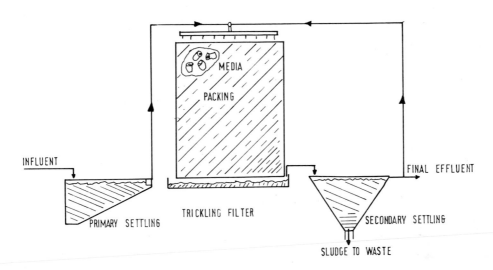

Fig. 1.1   Trickling Filter System

North American practice dictates a maximum surface loading of 1.5 m^3/m^2h and a minimum detention period of 2.0 hours be employed in a primary settling tank. Depending then on the suspended solids concentration in the influent, the removal efficiency will be between 40 and 80 percent. Longer holding times improve this efficiency range upwards to about 60 - 90 percent.

British practice recommends a maximum surface loading of 1.3 - 1.9 m^3/m^2h and a minimum detention period of 1.5 - 2.0 hours.

With reasonably stable operation of a primary settling tank, soluble biodegradable organic material will also be removed. Two hours detention time will eliminate about 30 percent; much longer times will remove about 40 percent.

The *fixed film* unit is made up of a retaining structure, filter media, an overhead influence distributor system and an underdrainage network. Retaining walls can be a very open structure for large graded stone media or concrete walls in cold climates. Synthetic media require only lightweight cladding. Filter shape is related to the type of distributor used. Circular or octagonal filters use rotary distributors while square, rectangular or hexagonal filters may have fixed nozzle distributors.

The filter media can be graded stone or slag which necessitates a strong containing structure and substantial floor supports and underdrains. Newer media are generally plastic modules or random elements (see Section 3.1). Because of the combined weight of stone media together with its structure, the depth of packing is usually between 1 - 3 metres. Plastic media, on the other hand, are usually 3 to 12 metres in depth, with 5 metres being typical.

Settled primary effluent is distributed to the top of the filter and then trickles downward over the media. Organic constituents are removed by adsorption and assimilation by the active microbial film adhering to the packing surfaces. Oxygen necessary for aerobic metabolism is supplied both by dissolved oxygen in the liquid and by air movement through the void space of the media.

Under anaerobic operating conditions, the void space of synthetic packings provides little resistance for the movement of the gases generated.

In filters that have low organic loadings and correspondingly low hydraulic loadings, the microbial film builds up slowly. As this microbial film ages and dies on the surfaces of the packing media, elements of the film peel away or "slough" and is washed from the trickling filter. With high organic loadings and high hydraulic loadings, the microbial film growth is more rapid, ages more quickly and coupled with the greater hydraulic shearing action causes the film to slough from the media surfaces almost continuously.

In all types of trickling filters, some excess biological solids will be produced, washed away, and provision must be made to remove these solids from the effluent.

Secondary sedimentation is necessary because the sloughed organic solids are highly putrescible, have a relatively high oxygen demand, are light and flocculent. These secondary settling tanks are designed on similar principles to those for primary settling, but with detention times between 1.5 to 2.5 hours. Longer residence times allow the accumulating sludge to become anaerobic and tend to float to the surface and overflow with the final effluent. Depending on the concentration of suspended solids entering the settling tank, no more than between 45 and 85 percent can be expected to be removed.

The operation of secondary settling tanks is especially important in the case of high rate trickling filter units. Recirculated effluent should be maintained in an aerobic condition, inferring that the liquid in the settling tank should be at all times aerobic.

Secondary settling tanks then act as activated sludge units with between 1.5 and 2.5 hours residence time, removing between 13 and 16 percent of the remaining biodegradable organic material contained in the incoming effluent. The biodegradable fraction removed is very sensitive to liquid temperature, with negligible removal below about 10°C (see also Section 2.5).

1.3   Process Applications

1.3.1 Organic Removal

The trickling filter system is a secondary method of treatment following primary sedimentation. Most non-toxic organic materials can be satisfactorily biologically decomposed either by aerobic or anaerobic bacteria, given sufficient acclimatisation, when passed through such a unit. Typical conversion efficiencies in excess of 90 percent are to be expected from single or sometimes multi-stage combinations (Dennis and Jennett,1974;ASCE,1977; Mosey,1978; Witt et al.,1980; Blersch,1980; Genung,1980; Kennedy et al.,1982; Sachs et al.,1982; Vanden Berg,1982).

As a consequence, trickling filters are commonly used for the partial or full treatment of municipal and industrial liquid wastes. With partial treatment, filters accept high concentration wastes which would be impossible to handle in activated sludge units. Conversion efficiencies between 50 – 75 percent are common, with the effluent then going to sewer or onwards to treatment by activated sludge.

1.3.2 Nutrient Removal

Biological nitrification may be termed a tertiary two stage process, where ammonium ions are converted to nitrites, nitrates. In aerobic trickling filters, carbonaceous oxidation and nitrification take place simultaneously, with conversion efficiency in excess of 90 percent for ammonium ion. Denitrification is an anaerobic process wherein the nitrate ion is reduced to nitrogen gas and carbon compounds. A supplementary carbon source, such as sewage, sugar or methanol is necessary to provide the energy of reaction for reduction of nitrate ion. Depending on the specific surface area provided, denitrification can be expected to be tending towards 100 percent.(Dosh et al.,1974; Water Research Centre,1974; Roberts,1975; ASCE,1977; Ellis,1980; Winkler,1981).

Phosphorus is only removed to a minor extent in waste water passage through a trickling filter system. Only about one percent of an organic waste content is required for metabolic phosphorus utilisation. For example, a soluble organic biodegradable waste, conc.200 mg/ℓ, would require only about 2 mg/ℓ P for microbial cell utilisation. The actual phosphorus concentration might be 10 mg/ℓ, so that the amount of phosphorus removed would be 20 percent (ASCE, 1977; Water Research Centre,1978).

2.1   Composition and Concentration

2.1.1   Influent

The trickling filter process, as do other biological treatment systems, removes only the biodegradable fraction of any organic influent together with the necessary equivalent proportions of nitrogen and phosphorus available for metabolic activity.

Organic material concentrations can be measured by any of four techniques; one a bacterial oxidation method, two of which are wet oxidation methods and one a total destruction method.

These wet oxidation tests involve the digestion of an influent sample under strong oxidising acid conditions with either potassium permanganate or potassium dichromate. Permanganate oxidation does not entirely destroy an organic component while a dichromate digestion destroys all but aromatic ring compounds. These two wet oxidation tests are known as PV (Permanganate Value) and COD (Chemical Oxygen Demand) and their units are both mg/ℓ of equivalent organic material.

The total destruction technique is an instrumental method of completely oxidising a sample to carbon dioxide and water vapour and the measure is termed TOC (Total Organic Carbon) with a unit expressed as mg/ℓ of equivalent organic material.

These three methods of analysis have the same failing in that the biodegradable fraction of organic compounds are not determined. In addition, these methods do not provide data on the rate of biochemical oxidation.

The most widely used and accepted measure of biodegradable organic material of an influent is the Biochemical Oxygen Demand, under standard conditions of five days biological reaction at $20°C$; the unit of which is mg/ℓ BOD_5.

For a particular waste of consistent composition, that is, not admixed with dissimilar organic wastes, any of the results of PV, COD or TOC can be correlated with BOD_5, as can be seen as illustrated in Figure 2.1 (Stones, 1980).

The intercept on the Y axis is interpreted to be the nondegradable or "refractory" element of a particular organic material. Thus, provided the waste is of consistent composition and not a varying mixture of quite different compounds or influent sources, then far more rapid COD or TOC measurements can be used for comparing process performance.

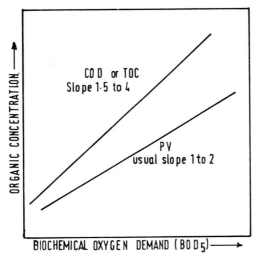

Fig. 2.1 Correlation of BOD with PV, COD, TOC

Typical concentrations for organic wastes range from 50-180 mg/l BOD_5 (North American) domestic sewage to 700-3100mg/l BOD_5, vegetable protein, (ASCE, 1977 and Water Research Centre, 1978).

2.1.2   Effluent

On passing through a trickling filter, an amount of organic material is consumed and converted by the biological film adhering to the packing. For units where concentrations are measured across the filter, that is, feed influent (including any recycle) pumped to the top of the bed and effluent sampled from the base of the filter, then a simple linear correlation is nearly always obtained as shown in Figure 2.2, (Roberts, 1973, 1975).

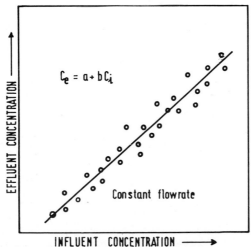

Fig. 2.2 Correlation of effluent-influent concentrations

If the intercept on the Y axis is negative, then strong adsorption
of the particular organic material is occurring on the active biological film.
However, if the intercept is positive, then desorption from the film is
occurring, with excretions of metabolic waste products at low feed
concentrations. A zero intercept can be inferred to mean that the trickling
filter is in a state of dynamic equilibrium. Depending on the size of the
intercept, the shape of the fractional efficiency curve changes, as in
Figure 2.3.

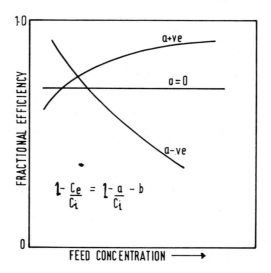

Fig. 2.3 Correlation of fractional efficiency with
feed concentration

2.2   Hydraulic Loading

A major factor which influences performance of a trickling filter is the
surface application rate or hydraulic loading expressed as (total, including
recycled liquid) influent flow rate per cross sectional bed area i.e. m^3/m^2d
units.

Prior to the 1930's, trickling filters had hydraulic loadings usually
ranging between 1 to $5m^3/m^2d$. Since that era, continuing investigations into
type of packing - stone, graded, plastic modular or random, have led to the
following hydraulic classification (ASCE, 1977; Benefield, 1980; Liptak, 1974;
Winkler, 1981):

Low Rate		-	1 to	5 m^3/m^2d
Intermediate		-	5 to	12
High Rate (stone)		-	12 to	45
High Rate (plastic)		-	15 to	105
Roughing		-	65 to	210

These roughing filters are usually high rate, packed with plastic media, which can handle raw macerated unsettled influent as the first stage of a multistage biological treatment facility.

A typical curve of trickling filter performance is shown in Figure 2.4 for two depths of packing. The percent reduction in BOD is for the filter and secondary settling tank together.

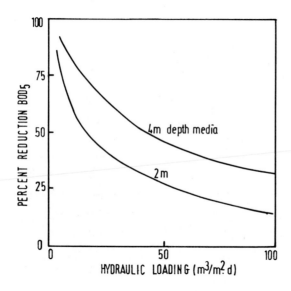

Fig. 2.4   Filter efficiency versus hydraulic loading

The lower limit of hydraulic loading is usually termed the "minimum wetting rate"; below which liquid applied to the packing is insufficient to keep the packing entirely wetted.

The situation is further complicated by the intermittent irrigation of low rate operations, (Water Research Centre, 1978; Winkler, 1981).

Eckenfelder (1971) cites an empirical expression for estimating minimum wetting rate (MWR) of stone packings:

$$MWR = 0.125 \ A_v \ m^3/m^2.day \qquad \text{where } A_v = \text{packing specific surface area}$$

A more general relationship has been developed by the author to accommodate all random and modular packings which takes into account void fraction (ε) and shape factor of the media (ψ), (see also Sec. 3.1,3.2):

$$MWR = 0.043 \, A_v \psi^{2/3}/(1-\varepsilon)^{5/4} \quad m^3/m^2.day$$

The upper limit of hydraulic loading is not the onset of flooding for high void fraction packings. This upper limit is governed both by the requirement of maintaining thin film flow on the one hand, and possible scouring of the active biological film on the other hand.

For stone media and other low voidage packings, however, the upper irrigation rate is that due to flooding and therefore complete loss of operation and filter bed ventilation (Winkler, 1981). A general expression for upper irrigation rate (UIR), developed by the author is as follows:

$$UIR = 242/ \, A_v^{2/3}(1-\varepsilon)^{2.74} \quad m^3/m^2.day \qquad for \varepsilon < 0.5$$

and

$$UIR = 1130/A_v^{2/3}(1-\varepsilon)^{0.51} \quad m^3/m^2.day \qquad for \varepsilon > 0.5$$

For typical commercial packings, estimates of MWR and UIR are noted in the following Table 2.1:

TABLE 2.1   Estimates of Minimum Wetting Rates and Upper Irrigation rates of commercial packings

MEDIA PACKING	MINIMUM WETTING RATE $m^3/m^2.day$	UPPER IRRIGATION RATE $m^3/m^2.day$
Stone 50-100 mm	3.5-5.4	51-104
25-38	7.8-12.0	30-61
Dow-Surfpac	24	258
Crinkleclose	35	138
ICI- Flocor EC	37	290
M	33	143
R	72	142
Mass Transfer Filterpak I	30	185
II	36	136
Plastic Pall Rings 50mm	32	188
25	55	102
Plastic Tellerites 25mm	34	72
Cloisonyl I	55	168
II	51	134

The situation can arise when the irrigation rate is below the desired minimum irrigation rate leading to some drying out of the biological slime, especially in high rate trickling filters.

The effective wetted specific surface area of media (A_e) is then reduced which leads to a drop off in performance of the unit. An estimate of this effective wetted specific surface area can be made from a correlation by Onda et al. (1967), modified here to a simpler form:-

$$A_e/A_v = 1-\exp(- C_p Q^{0.4})$$

where C_p = media packing coefficient evaluated from the assumption that $A_e/A_v = 0.99$ at Q = MWR for any particular packing.

An estimate of fractional effective wetted area can then be made at flow rates below MWR. For example at a fractional MWR =1.0, A_e/A_v=0.99, while at a fractional MWR of 0.125, A_e/A_v = 0.865.

2.3 Recycle

Although recycle of effluent was primarily devised for high rate trickling filters, it has also become an integral design arrangement for low rate units. Reduced night time flow to low rate trickling filters cause the biological film to partially dry out. This results in decreased microbial activity during the morning load period. By recycling effluent for those occurrences of low influent flow, it is possible to maintain the biological film sufficiently wetted and active to immediately accommodate increased loadings, (ASCE, 1977; Benefield, 1980; Grady, 1980).

Recycle of effluent around a high rate trickling filter increases the organic load, reduces the organic concentration, reduces the contact time in the biological film and reduces the removal efficiency of the filter, but possibly increases the efficiency relative to the whole system, (see also Sec. 5), (Water Research Centre, 1978).

Depending on the type of raw influent "food", the incoming concentration to the filter is generally less than 0.5 - 1 kg/m^3. If a high strength influent is to be treated, say 1 - 20 kg/m^3 concentration range, the organic loading needed to achieve the desired removal efficiency may dictate a hydraulic loading below the minimum wetting rate for the media. By using sufficient return flow of effluent, the hydraulic load can be increased to sufficiently dilute the high strength waste, but only slightly increase the organic load.

Recycled effluent can be drawn from before or after the secondary settling tank and returned to the primary settling tank or mixed directly with the influent going to the head of the trickling filter.

If unsettled effluent is recirculated, the filter media must have a large void fraction, to allow any suspended solids to be washed through the bed, thereby preventing any flooding, due to increased liquid holdup or clogging of the packing.

A comprehensive discussion on recycle arrangements is given in
(ASCE, 1977) with the recycle ratio defined as the ratio of recirculated flow
to the average raw influent flow, typically in the range 0.5 to 4.0. For low
volume, high strength waste such as a distillery or penicillin effluent, the
recycle ratio might be as high as 25 : 1, to dampen out fluctuations in
concentration and/or influent flow, (Water Research Centre, 1978; Winkler,
1981).

Several benefits arise from recirculation:

(a) Increased removal of contaminating materials, Recycle ratios
     usually above 10 : 1 are of little value in increasing removal
     but may sometimes be required to reach the desired dilution
     of the toxic chemical for it not to cause inhibition or
     poisoning of the biomass.

(b) Dilution of strong wastes. High recycle ratios are important
     when treating wastes which have high concentration or for
     which the active biological film has a low assimilation
     concentration.

(c) Inoculation of incoming wastes, where otherwise a delay in
     initial biochemical activity may occur.

2.4   Organic Loading

The mass application rate of organic biodegradable material per unit
volume of filter media is termed the organic loading, commonly expressed as
$kg/m^3 d$.

When effluent is recycled, the calculation of both hydraulic and organic
loadings become involved. Some investigators have disregarded the loadings
associated with this recirculated effluent stream when evaluating trickling
filter performance.

While organic loading is an important consideration, other factors
also effect performance so that the specification of organic loading by
itself is not a sufficient criterion for subsequent design. Figure 2.5
shows performance data for a variety of wastes which highlight the
different treatability characteristics of the influent "food" sources.

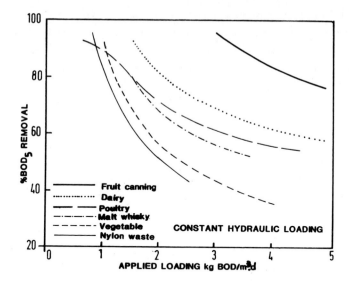

Fig. 2.5 Filter performance for a variety of wastes

2.5   Temperature

The performance of trickling filters is affected by changes in
temperature of the filter biological slime and the liquid flowing over this
film. A decrease in temperature results in a decrease in biological
respiration rate, a decrease in oxygen transfer rate and an increase in
oxygen saturation. The combined effect of these factors results in a
biological slime operating at a lower level of activity leading to a poorer
efficiency. A lower temperature also increases effluent viscosity, thereby
reducing settling velocity of humus in the secondary settling tank. The
overall performance of the trickling filter unit is therefore seen to be
deleteriously affected by reduced temperatures.

2.6   Ventilation and Dissolved Oxygen Concentration

Ventilation of the filter bed is essential for good performance. Air
movement allows carbon dioxide to be carried away and oxygen to be absorbed
by gas phase mass transfer mechanism. If adequate passageways are provided,
the difference in temperature between the ambient air and wastewater-
biological film is normally sufficient to promote air circulation and maintain
aerobic conditions necessary for effective trickling filter operation. As in
a cooling tower, there is heat and mass transfer simultaneously occurring,
resulting in a density gradient of the air contained within the media void
space (ASCE, 1977; Imhoff, 1971; Truesdale, 1963).

Provided that the air passageways are well designed and the trickling filter is not overloaded in its organic influent, then the dissolved oxygen concentration will not fall below 0.2 of the saturated value on exit from the filter unit. Natural ventilation in high rate units provides an adequate supply of oxygen for the biological oxidation process. Tower effluents consistently contain a high concentration of dissolved oxygen, usually in the range 0.5 - 0.7 of saturation.

2.7 Toxic Substances

Some components of industrial liquid wastes above certain low concentrations are toxic to the biological life forms necessary for treatment processes. The fixed film or tricling filter system is considered to be very stable in operation and resistant to shock toxic loadings in comparison to the activated sludge systems.

Some of the more common substances which may be considered toxic are the heavy metal ions of copper, chromium, nickel, zinc, lead, cyanides, arsenicals; sulfonanides, nitro and benzoic derivatives and phenols, (Barch, 1964; Baumann, 1975; Beg, 1982; Jackson, 1970).

In addition to the biofilm's capacity to acclimatise and then oxidise certain toxic substances, the biomass can become acclimatised to low concentrations (between 1 - 5 mg/ℓ) of heavy metal ions which are absorbed onto the film surface. For each mg/ℓ of heavy metal ion or cyanide, the trickling filter performance drops off by about 0.5 percent.

Nitrification is the biological process most sensitive to toxic materials due to autotrophic bacteria being more readily inhibited than the aerobic heterotrophic bacteria. The nitrifying bacteria cannot acclimatise to toxic concentrations above 1 mg/ℓ of heavy metal ions. Organic insecticides and fungicides completely inhibit nitrification.

If the shock load is of relatively short duration, in which time the surface layers of active biofilm become poisoned, then the surface region is killed and subsequently slough off due to the hydraulic irrigation. Fresh biofilm is exposed which has not been poisoned or inhibited, leading to rapid recovery of the trickling filter unit, (Cook and Herning, 1978).

However, a toxic load of long duration would cause catastrophic failure of the biomass to assimilate "food" material, leading to complete maloperation of the system. Recovery would then take weeks instead of hours or a few days.

3. DESIGN PARAMETERS AFFECTING PERFORMANCE

3.1   Random or Modular Packings

The two major properties of a packing material are its specific surface area and void fraction and to a lesser extent, its shape or shape factor. A larger surface area per unit depth allows more biological film per unit volume of filter (Sec. 2.4). A high void fraction allows for considerably higher hydraulic loadings (Sec. 2.2) and better air flow through the column of packing, leading to improved aeration of the wastewater film (Sec. 2.6).

Random packings are those which, as single elements, are simply dumped into the empty tower during installation. The original material was graded river stones, for which the British and American Standard Specifications were written. The newer random packings are manufactured from chemically inert, structurally strong material such as fireclay, polypropylene or rustproof steel. The latter have far thinner walls than fireclay which then provides increased void space.

Modular or regular packings offer the advantages of less resistance to both air flow and liquid flow, due to a high void fraction (0.8 - 0.97) and virtually no occurrences of blockage of the packed tower due to sloughed biofilm.

Packings of FLOCOR, KOROSEAL, SURFPAC are made of several synthetic sheets of wave shaped surfaces, formed into cubes about 0.6 - 1m wide. Other forms, such as CLOISONYLE are long (4-8m) tubes of synthetic material, stacked vertically together to produce a honeycomb combination, (Water Research Centre, 1978).

The physical characteristics of a variety of packings are given in Table 3.1.:

Channeling or maldistribution of liquid can be a problem in deep random packed towers operating at low irrigation rates, because of liquid dispersion away from the top distribution regions on the exposed packing. Modular packings contain and maintain elements of the initial liquid distribution with little to no dispersion and are adequately wetted at relatively low liquid rates.

258

TABLE 3.1 Packing Media Characteristics

MEDIA	SPECIFIC SURFACE m^2/m^3	VOID FRACTION (-)	SHAPE FACTOR (-)
Stone 50-100mm	56.2	0.34-0.49	0.81-0.94
25-38	125	0.34-0.49	0.81-0.94
Dow-Surfpac	82	0.94	0.093
Crinkleclose	187	0.93	0.063
ICI-Flocor EC	95	0.96	0.065
M	177	0.93	0.063
R	200	0.94	0.125
Mass Transfer Filterpak I	120	0.93	0.097
II	190	0.93	0.062
Plastic Pall Rings 50mm	105	0.92	0.164
25	220	0.90	0.189
Plastic Tellerites 25mm	249	0.83	0.203
Cloisonyl I	180	0.95	0.069
II	220	0.94	0.064

3.2   Void Fraction and Shape Factor

Depending on the three dimensional geometry of a packing element, there
is a unique relationship between void fraction (ε) and shape factor (ψ).
This can best be seen from Figure 3.1, (Brown, 1950).

As the void fraction increases, the shape factor decreases. Thick-walled
material, such as stone, ceramic or timber for the same basic element shape,
leads to a higher shape factor and lower void fraction. Conversely, thin-walled
elements of plastic or metal lead to high void fraction and low shape factor.

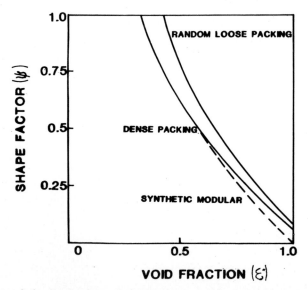

Fig. 3.1 Variation of Shape Factor with Void Fraction

The intercorrelation of void fraction and shape factor are important variables which control minimum wetting rate and upper irrigation rate (see Sec. 2.2).

3.3 Specific Surface Area of Packing

The smaller the elements of a particular media, the larger the specific surface area, as can be seen in Table 3.1.

There is also a considerable increase in performance of a trickling filter with increase in specific surface area, for otherwise constant process conditions, illustrated in Figure 3.2, (Bruce, 1970).

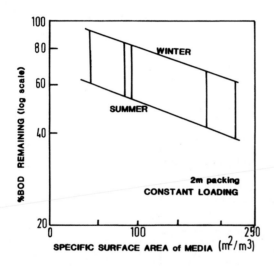

Fig. 3.2 Effect of Specific Surface Area on Filter Efficiency

3.4 Depth of Packing

The older conventional packing, of graded stone, or blast furnace slag, have a high bulk density per unit area of floor, as do the newer random fire clay elements. This then dictates the depth of packing to about 1.5 - 2.5m because of the high bearing pressure exerted on the ground beneath the trickling filter.

Specially fabricated synthetic packings, made of thin-walled plastic or steel, have considerably lower bulk density per unit area of floor. The main factor here, as to maximum depth, is the crushing or collapse pressure when in operation, involving liquid hold up and biofilm weights bearing down on the bottom sections of packing.

260

For most high void fraction media, the minimum depth is about 3 - 4m,
while the maximum depth, without intermediate structural support, is about
6 - 9m, depending on the type of packing. Figure 2.4 showed the improved
performance of trickling filters with increased depth of packing.

3.5   Primary and Secondary Sedimentation

As discussed in Sec. 1.2, the trickling filter system usually includes
primary and secondary settling tanks. Because of past clogging problems in
stone media, there is often a statutory requirement for primary treatment
before any trickling filter. The efficiency of the primary settling tank
has a direct effect on the design and performance of the trickling filter
unit, (ASCE, 1977).

A primary sedimentation tank in many situations also doubles for an
equalisation tank which may have up to six to eight hours' residence time.
Besides removing a considerable percentage of suspended solids, the
primary tank acts as an anaerobic fermentation unit. Figure 3.3 shows the
performance of a well designed unit, (Clarke, 1971).

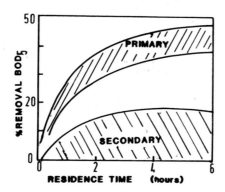

Fig. 3.3 Removal of BOD in settling tanks

With the usual residence time exceeding three hours in primary tanks,
at least one third of the BOD of the raw influent is removed prior to being
pumped or diluted by recycled effluent and passing to the trickling filter
unit.

A secondary settling tank acts as a simple aerobic activated sludge
unit with a residence time between two to four hours. In the northern
hemisphere, the effluent is cooled on passage through the trickling filter
to such an extent that little biological activity takes place. Warmer climates
however, allow sufficient biological activity to cause up to about twenty

percent of the remaining BOD in the effluent to be removed, as also shown
in Figure 3.3.

As one simple example, suppose a raw influent concentration is 375 mg/ℓ
BOD_5. After passing through the primary settling tank the liquid pumped to
the high rate stone media trickling filter contains 25o mg/ℓ BOD_5. Passage
through the packing results in 90 percent removal of organic material, that is
a trickle filter effluent of 25 mg/ℓ BOD_5. After flowing through the secondary
tank, the final effluent is 20 mg/ℓ BOD_5, which gives a 94.7 percent performance
for the entire unit.

4. MATHEMATICAL MODELS DESCRIBING PERFORMANCE

Many investigations have been made in trying to elucidate the fundamentals
involved in the trickling filter process. Factors affecting performance have
been noted in Sec. 2 and 3. Design criteria for the volume of packing
necessary have usually been derived from pilot plant and full-scale operating
data. The analysis of particular data sets which include organic and
hydraulic loadings and type of waste being processed has led to the
development of a number of empirical formulae.

4.1.1 Design Curves - U.S.A. and European Practice

Historically the National Research Council (1946) (ASCE, 1977) and the
Great Lakes Upper Mississippi River Board (1971), (Clarke, 1971), in America
had developed operational data curves for stone media trickling filters of
typical loadings and depths. Both design curves included secondary
sedimentation as an integral part of the overall process efficiency.

The NRC formula was developed from a typical military base effluent
obtained during 1943-44 at loading up to $1kg/m^3$ day and the correlation forced
to 100 percent efficiency at zero loading. The recycling of trickling filter
effluent was also assumed to be beneficial. Because of the wide spread of
data, a relaxation of these two constraints lead to a similar correlation
coefficient to the original, but indicating that recycled effluent decreases
process efficiency, (see Schroader, 1975). Tested on actual domestic sewage
trickling filter performance, the NRC formula under-estimated efficiency by an
average of 22 percent.

The Great Lakes or Ten States correlation was developed as a design
guideline for trickling filters in the colder northern regions of America at
loadings up to 4 kg/m^3 day.

No allowance was made for recirculation, variation in effluent
treatability, hydraulic loading, size of media, or temperature. This Ten State
correlation, tested on actual plant data, under-estimated efficiency by an
average of 13 percent.

European practice followed a similar development where overall removal efficiency, including secondary sedimentation, was correlated with organic loading up to about 1.2 kg/m^3/day. Keefer and Meisel (1952) and Rincke (1967) highlighted the scatter of data, as illustrated in Figure 4.1. Again, no allowance was made for size of stone media, depth of packing, temperature or type of waste.

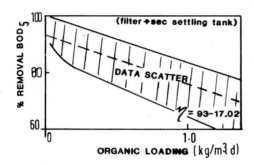

Fig. 4.1 European Design Curve

Current design procedures of the 1970's and 1980's in Europe and North America involve on site pilot plant evaluation of a particular waste using a proprietry media (FLOCOR, SURFPAC etc.) packed to an appropriate depth. Figure 2.5 illustrated typical design curves based on British research. Here too, the overall efficiency is that of the trickling filter and secondary settling tank.

Fullscale performance of a trickling filter unit designed using the above procedure closely matches that of the design curve.

The definition of organic loading used in the design curves has to be carefully noted. In Europe, the organic loading is that which passes *after* primary settlement (but including any recycle flow), to the head of the trickling filter for distribution. Some Corporations in U.S.A., however, use the whole organic loading passing to the primary settling tank. This is not the nett loading which passes through the trickling filter unit.

4.1.2 Empirical Formulae

Based on a multiple regression analysis of existing stone packed trickling filter units, Galler and Gotass (1964, 65) developed a correlation for single stage systems, from which two limits can be determined:

$$\text{for } R=0 \quad C_e/C_p \propto C_p^{0.19} . Q^{0.13} \quad \text{or} \quad C_e/C_p \propto \text{Load}^{0.13} . C_p^{0.06} \tag{4.1a}$$

and $R \gg 1$ $C_e/C_p \propto C_p^{0.19} . Q^{0.13} . R^{0.41}$ or $C_e/C_p \propto Load^{0.13} . C_p^{0.06} . R^{0.41}$ (4.1b)

where C_p = effluent from the secondary settling tank, mg/l BOD_5

$\quad C_p$ = influent before mixing with recycled effluent, mg/l BOD_5

$\quad Q$ = influent flowrate, m^3/d

$\quad R=0$, recycle ratio

$\quad Load = C_p . Q$, kg/d

In terms of recycle ratio then, fractional effluent is independent of R at low values but increases as R becomes relatively large.

A more recent multiple regression by Hammam (1969) for stone packed trickling filters again can be re-arranged to obtain two limits:

for $R=U$ $C_e/C_p \propto C_p^{0.62} . Q^{0.5}$ or $C_e/C_p \propto Load^{0.5} . C_p^{0.12}$ (4.2a)

and $R \gg 1$ $C_e/C_p \propto C_p^{0.62} . Q^{0.5}/R^{0.5}$ or $C_e/C_p \propto Load^{0.5} . C_p^{0.12}/R^{0.5}$ (4.2b)

For this Hammam correlation, Eq. 4.2, fraction effluent is again independent of R at low values, but decreases as R increases which is at variance with the Galler and Gotaas formula.

4.1.3 Empirical Models

Velz (1948) concluded that "the rate of extraction of organic matter per interval of depth of a biological bed is proportional to the remaining concentration of organic matter, measured in terms of its removability". In this statement, Velz was implying a plug flow reactor with first order kinetics, which was expressed as the simple differential equation:

$$\frac{dc}{dz} = - k_B C$$

on integration $C_f/C_i = \exp(-k_B Z)$ (4.3)

$\quad C_f$ = effluent concentration on exit from filter

$\quad k_B$ = biochemical rate coefficient, m^{-1}

$\quad Z$ = trickling filter packed bed depth, m

Velz also observed that k_B decreased as the flow rate increased.

A number of empirical models incorporated residence time of wastewater or flow rate, especially Howland (1958, 60) and Schulze (1960) led to:

$$C_f/C_i = \exp(-k_1 Z/Q^{2/3})$$ (4.4)

$\quad Q$ = hydraulic loading rate to head of trickling filter $m^3/m^2 d$

$\quad k_1$ = first order rate coefficient

In comparing Eq. 4.3 with Eq. 4.4, it is evident that

$$k_B = \frac{k_1}{Q^{2/3}}$$

Eckenfelder (1961) and many subsequent workers have further extended these developments to allow for type of packing and variation of biomass with depth.

$$C_f/C_i = \exp\ (-k_1 z^m /Q^n) \tag{4.5}$$

For high rate trickling filters, assuming uniform biofilm distribution, the exponent m tends to unity, whereas for non-uniform distribution on low rate filters, m tends to 0.66.

The first order rate coefficient in Eq. 4.5 is a function of type of packing, the thickness and distribution of active film, the biochemical characteristics and concentration of the influent, as well as temperature.

4.2   Reactor Models

The trickling filter process as detailed in Sec. 1.2 is very complex and theoretical models have proved impractical for designs because of the large number of unknown coefficients inherent in any particular model (see ASCE, 1977; Benefield & Randall (1980); Liptek, 1974; Water Research Centre (1978); Winkler, 1981).

A sufficiently detailed model, however, once fitted to actual operating data, can explain subtle variations in process conditions or used to interpret performance. The developments which follow are essentially those of Grady and Lim (1980).

4.2.1   Plug Flow with External Resistance to Mass Transfer

The substrate or "food" concentration profile for wastewater flow over an element of packed media is shown idealised in Figure 4.2.

Assuming steady state conditions so that there is no accumulation of substrate and Monod kinetics prevailing, then the rate of substrate utilisation becomes

$$r_s \big| = \frac{-\ r_{max}\ C_s}{K_s + C_s} \tag{4.2.1}$$

where K_s = Monod saturation concentration, kg/m^3

r_{max} = maximum biochemical removal rate of substrate, $kg/m^2 h$

r_s = rate of substrate utilisation, $kg/m^2 h$

C_s = bios film surface "food" concentration, kg/m^3

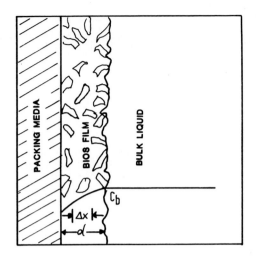

Fig. 4.2   Idealised Concentration Profile to the Bios

From the definition of mass transfer coefficient for the "film theory"

$$k_L (C_b - C_s) = r_s \qquad (4.2.2)$$

where C_b = bulk "food" concentration, kg/m^3

$\qquad k_L$ = mass transfer coefficient, m/h

Since C_s in Eq. 4.2.1 is ill-defined, then Eq. 4.2.1, 2.2, can be solved

for C_s and Eq. 4.2.1 rewritten as

$$r_s = \frac{r_{max} C_b}{K_s + C_b} \eta_e \qquad (4.2.3)$$

where η_e = external effectiveness factor which ranges between

$$0 < \eta_e < 1$$

and $\eta_e = (1+\overline{K})(1-\overline{C})/N_{Da} \qquad (4.2.4)$

where $\overline{K} \quad = K_s/C_b$

$\qquad N_{Da} \quad = \dfrac{r_{max}}{k_L C_b} \quad$ the Damköhler Number

$\qquad \overline{C} \quad = C_s/C_b$

This Damköhler Number is interpreted to be the ratio of maximum biochemical
rate to maximum mass transfer rate.

When N_{Da} >1, the biochemical process is said to be mass transfer limiting and conversely, when N_{Da} <1, the process is said to be biochemical reaction limiting.

For first-order kinetics, $K_s \gg C_s$ and $C_b > C_s$

so that $\overline{K} \gtrsim 1$

and $0 < \overline{C} < 1$

From Eq. 4.2.4, and these limits, noted here:

for $\overline{C} \rightarrow 0$, $\eta_e \rightarrow \dfrac{1 + \overline{K}}{N_{Da}}$

$\overline{C} \rightarrow 1$, $\eta_e \rightarrow 0$

$$(4.2.5)$$

For zero-order kinetics, $K_s \ll C_s$ and $C_b > C_s$

then $\overline{K} \rightarrow 0$

and $0 < \overline{C} < 1$

for $\overline{C} \rightarrow 0$, $\eta_e \rightarrow \dfrac{1}{N_{Da}}$

$\overline{C} \rightarrow 1$, $\eta_e \rightarrow 0$

$$(4.2.6)$$

If we now consider an element of reactor depth, ΔZ, and describe a substrate "food" mass balance:

Change in "food" mass = change in bios mass + loss in wastewater

$$F.\Delta C = -(r_s A_s .\Delta Z + r_C A\Delta Z) \qquad (4.2.7)$$

where A = void space liquid cross sectional area, m^2

A_s = total surface area of bios, m^2

F = flow rate, $m^3 h$

r_C = substrate "food" consumption rate, $kg/m^3 h$

Z = total length of trickling filter reactor, m

Allowing bacterial cell movement into the liquid film due to continual sloughing:

$$X = X_i + Y \Delta C \qquad (4.2.8)$$

where X = bios cell concentration in the liquid, kg/m^3

X_i = cell concentration at the inlet to the reactor, kg/m^3

Y = cell yield from "food", kg bios/kg "food"

and $\mu = \mu_{max} \dfrac{C}{K_s + C}$ the Monod relationship $\qquad (4.2.9)$

μ = substrate utilisation rate, kg/kg.h

μ_{max} = maximum substrate utilisation rate, kg/kg.h

and $r_C = \mu X/Y \qquad (4.2.10)$

substituting Eq. 4.2-8, 9, 10 into Eq. 4.2.7 and taking the limit

$$\frac{dC}{dZ} = - \left[\frac{r_{max} \cdot A_s}{F} \cdot \eta_e + \frac{\mu_{max}}{F} \cdot A \left(\frac{Xi}{Y} + (C_i - C)\right) \right] \cdot \frac{C}{K_s + C} \qquad (4.2.11)$$

with initial conditions $C = C_i$ at $Z = 0$

$$C = C_e \text{ at } Z = Z$$

Eq. 4.2.11 cannot be solved analytically, since the effectiveness factor η_e is a function of "food" concentration (see Eq. 4.2.4.).

If we make the simplification that η_e is constant, and no substrate is removed by bios in suspension, then Eq. 4.2.11 can be re-arranged.

$$\frac{dC}{dZ} = - r_{max} \frac{A_s \cdot \eta_e}{F} \cdot \frac{C}{K_s + C} \qquad (4.2.12)$$

which on integration:

$$K_s \ell n \frac{C_e}{C_i} + (C_e - C_i) = r_{max} \frac{A_s \eta_e}{F} \qquad (4.2.12(a))$$

$C = C_i$ at $Z = 0$

$C = C_e$ at $Z = Z$

This Eq. 4.2.12 is identical in form to that of Kornegay and Andrews (1968).

If the further simplification is made for that of first-order kinetics $K_s >> C$, then

$$\frac{dC}{dZ} = - r_{max} \frac{A_s}{F} \frac{\eta_e}{K_s} \frac{C}{} \qquad (4.2.13)$$

which on integration:

$$\ell n \frac{C_e}{C_i} = - r_{max} \frac{A_s}{K_s F} \eta_e \qquad (4.2.13(a))$$

or $C_e/C_i = \exp(-r_{max} \cdot A_s \cdot \eta_e / K_s F)$ $\qquad (4.2.13(b))$

which is similar in form to that of Eckenfelder, see Eq. 4.2.2, or Atkinson (1974).

If zero order kinetics occur, $K_s << C$, then

$$\frac{dC}{dZ} = r_{max} A_s \eta_e/F \qquad (4.2.14(a))$$

which on integration $C_i - C_e = r_{max} A_s \eta_e/F$ $\qquad (4.2.14(b))$

4.2.2.   Plug Flow with Internal Resistance to Mass Transfer

Let us now consider the mass transport of substrate into the biofilm without any limitation of external mass transfer to the bios from the bulk liquid film, idealised in Figure 4.3.

268

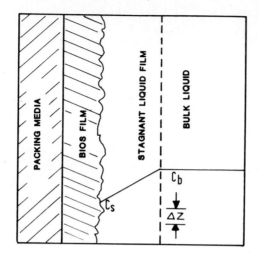

Fig. 4.3   Idealised Concentration Profile into the Bios

Taking an element of biofilm and describing a mass balance on substrate "food":

Diffusional mass transport = change in bios mass

$$D_e \frac{dC}{dx} = r_s A_v A_s \Delta X \tag{4.2.2.1}$$

D_e = effective diffusion coefficient into bios, m^2/h

A_v = specific surface area of bios, m^2/m^3

re-writing as a differential equation

$$D_e \frac{d^2 C}{dx^2} - r_s A_v = 0 \tag{4.2.2.2}$$

substitute for r_s from Eq. 4.2.1

$$D_e \frac{d^2 C}{dx^2} - r_s A_v \frac{C}{K_s + C} = 0 \tag{4.2.2.3}$$

subject to the boundary conditions $C = C_b$ at $x = d$

and $\frac{dc}{dx} = 0$ at $x = 0$

This is a second order non linear differential equation which has no analytic solution.

If we consider a first order kinetic approximation to Eq. 4.2.2.3

that is $D_e \frac{d^2 C}{dx^2} - r_{max} A_v \frac{C}{K_s} = 0$

or $\frac{d^2 C}{dx^2} - r_{max} \frac{A_v}{D_e K_s} . C = 0 \tag{4.2.2.4}$

subject to the same boundary conditions above.

The analytic solution is

$$\frac{C_x}{C_b} = \frac{\cosh(\Phi x)}{\cosh(\Phi)}$$

(4.2.2.5)

where $\Phi^2 = (A_v \ As \ d \ r_{max} \ \frac{C_b}{K_s})/(As \ D_e \ \frac{C_b}{d})$

that is, the ratio of maximum kinetic rate to maximum diffusional rate of mass transport and $\Phi \equiv$ Thiele modulus.

When $\Phi > 1$, the biological process is considered to be mass transport limiting and conversely, when $\Phi < 1$, the process is considered to be biochemically reaction-limiting.

Eq. 4.2.2.4 is an ideal situation which can be modified to allow for some mass transfer limitation within the bios by incorporating an internal effectiveness factor, η_i, where mathematically, $\eta_i = \tanh \Phi/\Phi$

For $\Phi \to 0$, $\eta_i \to 1$ with a gradient $d\eta_i/d\Phi = -\Phi$
which implies an extremely thin biofilm, a slow reaction rate or rapid mass transfer.

For $\Phi >> 1$, $\eta_i \to \frac{1}{\Phi}$ with a gradient $d\eta_i/d\Phi = -1$

which implies a thick biofilm, a fast reaction rate or very slow mass transfer.

We now incorporate the internal effectiveness factor into Eq. 4.2.2.4,

$$\frac{d^2C}{dx^2} - \frac{r_{max} \ A_v \ C_i \ \eta_i}{D_e K_s} = 0$$

(4.2.2.6)

subject to the same boundary conditions.

The analytic solution is

$$C_x/C_b = \frac{\cosh(\sqrt{\Phi \eta_i} x)}{\cosh(\sqrt{\Phi \eta_i})}$$

(4.2.2.7)

so that for $\Phi \to 0$, $\eta_i \to 1$ at $x = 0$, at the solid support surface
$C_x/C_b \to 1$ that is negligible concentration gradient into the bios irrespective of distance into the bio film.

However, for $\Phi >> 1$, $\eta_i \to \frac{1}{\Phi}$

$$C_x/C_b \to \frac{2}{3}(1 + \frac{\bar{x}}{2}) \quad \text{for } 0 < \bar{x} < 1$$

$$\text{where } \bar{x} = \frac{x}{d}$$

that is $\frac{2}{3} < C_x/C_b < 1$

which is an appreciable concentration gradient across the biofilm, and it is in this region when the slow rate of mass transfer affects the overall biochemical reaction rate.

For zero order kinetic approximation, Eq. 4.2.2.3 reduces to

$$\frac{d^2C}{dx^2} - \frac{r_{max} A_v}{De} = 0 \qquad\qquad (4.2.2.8)$$

subject to the same boundary conditions.

The analytic solution is

$$C_b - C_x = \frac{1}{2} \frac{r_{max} A_v}{D_e} (d^2 - x^2)$$

or

$$\frac{C_x}{C_b} = 1 - \frac{1}{2} \frac{(r_{max} d)}{D_e \cdot C_b} A_v (d - \frac{x^2}{d})$$

$$= 1 - \frac{1}{2} N_{Da_k} A_v (d - \frac{x^2}{d}) \qquad\qquad (4.2.2.9)$$

where $N_{Da_k} = \dfrac{r_{max} d}{D_e C_b}$ a kinetic type Damköhler Number

so that for $C_x/C_b \to 1$, that is negligible concentration gradient into the bios when there is a thin biofilm, a slow reaction rate or a high bulk "food" concentration and large diffusion coefficient.

However, for $C_x/C_b \to 0$, implying a large concentration gradient, there must exist a thick bios, a high biochemical reaction rate or low bulk "food" concentration and small diffusion coefficient.

Jennings (1976), Williamson (1976, 81), Schroeder (1976), Mutharasan (1978) and McCarthy (1981) each developed biofilm models based on elements of the general Eq. 4.2.2.3, some including and some excluding the internal effectiveness factor, (see also LaMotta, 1976). Other models which followed similar developments included those of Elmaleh (1978), Frechter (1982), Kong and Yang (1979), Mitchell (1978), Moodie and Greenfield (1979) and Webster (1981).

4.3   Combination of External and Internal Resistance to Mass Transfer

The general case can now be considered which incorporates aspects of the developments in Sec. 4.2.1, 2. If we combine interpretations contained in Figure 4.2, then the overall rate of mass transfer of "food" from bulk liquid is postulated to be a complex mechanism consisting of at least the following steps, (see Roberts, 1973):

1. movement of the "food" from bulk liquid region through the external "stagnant" film to the surface of the bios (mass transfer)

2. movement of the "food" from the surface into the region of active sites (mass transfer)

3. activation of "food" at these active sites (kinetics)

4. biochemical conversion of "food" in the active bios (kinetics)

5. exchange of expiration products at these sites (kinetics)

6. movement of expiration products to the bios surface (mass transfer)

7. movement of these products through the external "stagnant" film out into the bulk liquid (mass transfer)

Steps 1 and 7 are involved with external resistance to mass transport and η_e, 2 and 6 are involved with internal resistance to mass transport and η_i, 3, 4 and 5 are involved with the overall kinetics of the biochemical process.

Any one, or a number, of these steps may be a rate limiting factor. If the activation and material exchange at the active sites is very fast compared to the mass transfer steps, then mass transport considerations are of dominant importance, such as liquid velocity, surface area, film depth and concentration difference between bulk liquid and bios surface. However, if the converse is true, then kinetic factors are dominant – such as the character of bios surface active sites, their number and size, temperature, and fundamental kinetic rate constant for a particular soluble "food".

4.3.1   Overall Effectiveness Factor in a General Equation

If we go back to Sec. 4.2 in the development of a reactor model, the mass transport flux through the "stagnant" liquid film is Eq. 4.2.2, 2.3, 2.6

$$k_L \, (C_b - C_s) = \frac{r_{max} \, C_b}{K_s + C_b} \, \eta_e = \frac{r_{max} \, C_s}{(K_s + C_s)} \, \eta_i \qquad (4.3.1)$$

substituting for C_s in Eq. 4.3.1 and assuming first order kinetics, leads to

$$r_s = K_o, \, C_b$$

with K_o = an overall, kinetic coefficient, m/h

where $$\frac{1}{K_o} = \frac{1}{k_L} + \frac{1}{\left(\dfrac{r_{max}}{K_s}\right)\eta_i} \qquad (4.3.2)$$

or $$\frac{1}{\eta_o} = \left(\frac{r_{max}}{K_s}\right) \frac{1}{k_L} + \frac{1}{\eta_i} \qquad (4.3.3)$$

where η_o = overall effectiveness factor

in $$r_s = \frac{r_{max}}{K_s} \, C_b \, \eta_o \qquad (4.3.4)$$

From Eq. 4.3.3, 3.4, the overall biochemical reaction rate increases with overall effectiveness factor which is itself enhanced by increase in mass transfer coefficient (or liquid velocity), or increase in internal effectiveness factor (thin biofilm).

For the general case involving 4.3.1 and Monod kinetics, Fink et al (1973) developed a numerical solution incorporating dimensionless parameters as shown in Figure 4.3

$$\psi = \frac{k_L d}{D_e} \quad \text{a Sherwood number} \qquad \phi_F = \frac{\phi}{\sqrt{(1+K)}} \qquad \text{a modified Thiele Modulus}$$

$$\text{and} \quad K = \frac{C_b}{K_s}$$

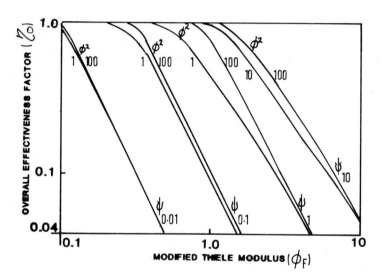

Fig. 4.4 Numerical Solution for Overall Effectiveness Factor

On examination of the trends in Figure 4.3 it is quite apparent that,
$\eta_o \propto 1/\phi_F^2$, ψ and slightly on ϕ

which implies that the overall effectiveness factor increases with

(a) $(1+K)$ or increase in "food" concentration i.e. driving force

(b) $1/\phi_F^2$ or thin biofilm, rapid mass transfer (high liquid velocity)

If we take the limit, $\eta_o \to 1$ then $\phi_F < 0.1$ [no resistance to external mass transfer]

or if $\phi_F \simeq 1$, $\psi > 10$

which implies a very thin biofilm, very rapid mass transfer, slow reaction rate or small rapidly diffusing "food" molecules.

Conversely, if $\eta_o \to 0$, then $\phi_F > 10$ [resistance to external mass transfer] or if $\phi_F \simeq 1$, $\psi < 0.01$ which implies a thick biofilm, slow mass transfer, fast reaction rate or large slowly diffusing "food" molecules.

Using the background information developed in this Section 4.3, we can immediately modify Eq. 4.2.11 and substitute η_o in place of η_e:

$$\text{then } \frac{dc}{dz} = - \left\{ \frac{r_{max} A_s}{F} \eta_o + \frac{\mu_{max} A}{F} \left[\frac{X_i}{Y} + (C_i - C) \right] \right\} \frac{C}{K_s + C} \qquad (4.3.5)$$

with initial conditions $C = C_i$ at $z = 0$

$$C = C_e \qquad z = Z$$

As in Sec. 4.2 special simplifications allow straightforward analytic solution of Eq. 4.3.5:

(i) Kornegay Type (1968) $K_s \ln \frac{C_e}{C_i} + (C_e - C_i) = -r_{max} A_s \frac{\eta_o}{F}$ (4.3.6)

(ii) Eckenfelder Type (1961)
or Atkinson (1974) $\frac{C_e}{C_i} = \exp \left(- \frac{r_{max} A_s}{K_s F} \eta_o \right)$ (4.3.7)

(iii) Zero Order Kinetics $C_i - C_e = \frac{r_{max} A_s}{F} \eta_o$ (4.3.8)

The major difficulty involved in the application of the Grady-Lim (1978) model and solution of Eq. 4.3.5 is the unknown values of the many parameters which have to be incorporated. As Grady and Lim point out, no parameters have been evaluated for any particular wastewater and future research will be necessary in obtaining appropriate coefficients and for validation of this model.

In Sec. 5.1, steady state solutions to Eq. 4.3.5 will be presented which cover a range in parameter values. Notwithstanding these difficulties, the Grady-Lim model does offer interpretive possibilities in a general sense.

4.4   Combined Reactor - Film Model

The development of this model, due to Ames et al (1962), was based on theoretical considerations of film flow reactor operation. This hypothetical reactor consists of a cylinder filled with an oriented or random packed media covered with a biological slime. Influent soluble organic "food" having a Biochemical Oxygen Demand flows downward through the column, with liquid and bios exchanging material. BOD is transferred from the liquid to the slime layer by diffusional mass transport and adsorption. A continuous biochemical reaction occurs on the bios surface and an equilibrium exists between BOD in the solid phase (or bios) and BOD in the solid-liquid interface. Differential

material balances of the "food" component in the liquid and bios result in
a set of simultaneous partial differential equations. Ames' original work is
summarised below.

4.4.1   Ames Model

In the bulk liquid phase:

$$\overline{u}\rho \, \frac{dc}{dz} + K_L A_v (C-C^*) = - h_L \frac{dc}{dt} \qquad (4.4.1)$$

where h_L = liquid holdup per unit volume, kg/m^3

$\quad K_L$ = liquid phase mass transfer coefficient, $kg/m^2 d$

$\quad \overline{u}$ = liquid velocity, m/d

$\quad C^*$ = "food" concentration at liquid-bios interface, kg/m^3

$\quad \rho$ = liquid density, kg/m^3

In the active bios solid phase:

$$K C_s X - K_L A_v (C - C^*) = - h_s \cdot \frac{dC_s}{dt} \qquad (4.4.2)$$

where C_s = "food" concentration in the bios, kg/m^3

$\quad h_s$ = liquid holdup in the bios, kg/m^3

$\quad K$ = quasi first order biochemical rate coefficient, d^{-1}

$\quad X$ = dissolved oxygen concentration, kg/m^3

and to inter-relate surface and bulk concentrations:

$$C^* = \alpha \, C_s + C_r \qquad (4.4.3)$$

where C_r = "food" concentration external to interface, kg/m^3

$\quad \alpha$ = specific adsorption coefficient

Equations (4.4.1-3) can be re-arranged into

$$\frac{dc}{df} = \frac{\overline{u}\rho}{h_L} \frac{dc}{dz} - \frac{K_L A_v}{h_L} (C - C^*) \qquad (4.4.4)$$

and

$$\frac{dC^*}{dt} = \alpha \frac{K_L A_v}{h_s} (C - C^*) - \frac{KX}{h_s} (C^* - C_r) \qquad (4.4.5)$$

Using Laplace transforms on Eq. (4.4.4-5), a unique solution can be obtained
which leads to:

$$C_e = C_r + (C_i - C_r) \exp (- K_m Z/Q) \qquad (4.4.6)$$

where

$$\frac{1}{K_m} = \frac{1}{K_L A_v} + \frac{\alpha}{KX} \qquad (4.4.7)$$

that is [overall] $\equiv$ [mass transfer] + [kinetic terms]

$\quad K_m$ = overall mass transfer coefficient, kg/m^3

$\quad Q$ = $\overline{u}\rho$ = surface irrigation rate, $kg/m^2 d$

$\quad z$ = packed bed depth, m

Amado (1964) revised the analytic solution of Eq. (4.4.4-5) to end up with
Eq. (4.4.6-7) but no published data had appeared until 1973 on the use and
interpretation of the Ames' equation. The author, in a number of publications
(1973, 76) demonstrated the application and versatility of the Ames' type model.

Eq. 4.4.6 can be obtained more directly by a simple derivation.

Let the general case be described by a Velz type differential equation
(1948):

$$\frac{dc}{dz} = -K_m C/Q \qquad\qquad (4.4.4(a))$$

$$\text{where } \frac{1}{K_M} = \frac{1}{K_L} + \frac{1}{K_B} \qquad\qquad (4.4.7(a))$$

C = "food" concentration in bulk phase of liquid

K_m = overall mass transfer coefficient - liquid to bios

K_B = biochemical reaction term - within bios

K_L = liquid phase mass transfer term - liquid to bios

Q = liquid flow rate

and $K_B = a/C^m$ which might vary with flow rate and packed depth

where a = coefficient describing the particular acclimatised bios

m = an exponent characterising the dependence on substrate concentration

re-arranging Eq. 4.4.4(a) by substituting 4.4.7(a)

$$\frac{dc}{C} = -\frac{K_L K_B}{(K_L+K_B)} \frac{dz}{Q} \qquad\qquad (4.4.4(b))$$

$$\text{or, } (\frac{K_L - K_B}{K_L \cdot K_B}) \frac{dc}{C} = \frac{-dz}{Q} \qquad\qquad (4.4.4(c))$$

substituting for K_B

$$(\frac{K_L}{a} C^{m-1} + \frac{1}{K_L C}) \, dc = -\frac{dz}{Q} \qquad\qquad (4.4.4(d))$$

with initial conditions $z = 0 \quad C = C_i - C_r$

$$z = Z \quad C = C_e - C_r$$

Integration of Eq. 4.4.4(d) for the cases when m = 0, 1:

$$\text{for } m = 0, \; \frac{C_e - C_r}{C_i - C_r} = \exp(-\frac{K_L \cdot a}{K_L + a} \cdot Z/Q) = \exp(-K_m Z/Q) \qquad (4.4.4(e))$$

which is identical to the Ames-Amado Eq. (4.4.6),

$$\text{for } m = 1, \; \frac{K_L}{a} (C_e - C_i) + \ln \frac{C_e - C_i}{C_i - C_r} = -K_L Z/Q \qquad\qquad (4.4.4(f))$$

which is identical in form to the Kornegay type Eq. (4.3.6), which is itself
a special simplification of the Grady-Lim model.

Eq. (4.4.4(e)) is itself a general form of the Eckenfelder type empirical Eq. (4.1.12) and Schulz Eq. (4.1.11),

$$\frac{C_e}{C_i} = \exp\left(-k_1 Z/Q^n\right) \tag{4.1.12}$$

As will be shown in Sec. 4.4.3, K_L and subsequently K_m are functions of flowrate, so that for laminar conditions, that is media irrigated at low rates,

$$K_m \, \alpha \, Q^{1/3}$$

and for turbulent conditions, media irrigated at high rates,

$$K_m \, \alpha \, Q^{1/2}$$

By comparing Eq. (4.4.4(e)) and (4.1.12)

$$\frac{k_1}{Q^n} = \frac{K_m}{Q}$$

Theoretically then, $n = 2/3$ for low irrigation rates

$n = 1/2$ for high irrigation rates

4.4.2   Interpretation of Ames-Roberts Model

Now $C_e = C_r + (C_i - C_r) \exp\left(-K_m Z/Q\right)$ (4.4.6)

$\qquad = C_r [1-\exp(\quad)] + \exp(\quad) \cdot C_i$ (4.4.7)

$\qquad = a + b, C_i$ (4.4.8)

where $a = C_r[1-\exp(\quad)]$

$\qquad b = \exp(\quad)$

Simple limits can be obtained:

1.   for $Q \to \infty$, $b \to 1$, then $a \to 0$ and $C_e = C_i$

that is, effluent tends to influent concentration;

2.   for $Z \to \infty$, $b \to 0$, then $a \to C_r$ and $C_e = C_r$

that is, effluent tends to zero, since

as z increases, (b) decreases and either (a) increases or C_r decreases.

Conversely, as Q increases, (b) increases and either (a) decreases or C_r increases.

When the value of C_r is negative, active adsorption of the "food" material is occurring into the bios and this will generally be the situation for a maturing trickling filter or one before sloughing of the bios occurs.

For a positive value of C_r, desorption of excretion products from the bios is occurring or sloughing of bios is occurring which also releases soluble products into the effluent stream.

The situation when C_r is zero occurs with a mature filter in a state of dynamic equilibrium, operating at constant conditions of flow rate, food source and concentration and temperature.

Since $C_e = a + b \, C_i$

the fractional efficiency: $1 - \dfrac{C_e}{C_i} = 1 - \dfrac{a}{C_i} - b = \eta$

Again, some simple limits can be obtained:

1. for $Q \rightarrow \infty$, $b \rightarrow 1$, then $a \rightarrow 0$ and $\eta \rightarrow 0$;

2. for $Z \rightarrow \infty$, $b \rightarrow 0$, then $C_r \rightarrow 0$ and $\eta \rightarrow 1.0$;

3. if a or $C_r \rightarrow 0$, $\eta = 1-b$; that is, a constant efficiency;

4. a or C_r is positive, an increasing efficiency with influent concentration results;

5. a or C_r is negative, efficiency decreases with influent concentration,

each situation being sketched in Figure 2.3.

Another limit can be taken, for the situation when the trickling filter is newly packed and start up is initiated. In this early time period

from Eq. (4.4.7) $K_m \simeq \dfrac{KX}{\alpha}$ since $C_r \rightarrow 0$ (initially no bios)

for $\dfrac{1}{K_L A_v} \ll \dfrac{\alpha}{KX}$

From the definition of α in Eq. 4.4.3

$\alpha \rightarrow 1$ (at initial start up)

Typical values of K, X are (see Roberts, 1973)

Quasi biochemical coefficient $K \simeq 1 \times 10^{-2}$, h^{-1}

Dissolved oxygen concentration, $X \simeq 5 \times 10^{-3}$, kg/m^3

then $K_m \simeq \dfrac{1 \times 10^{-2} \cdot 5 \times 10^{-3}}{1} \simeq 5 \times 10^{-5}$, $kg/m^3 h$

Since $\dfrac{C_e}{C_i} \simeq \exp(-K_m Z/Q)$ since $a \rightarrow 0$

then $\dfrac{C_e}{C_i} \rightarrow 1$ or $C_e \rightarrow C_i$ $\left(\begin{array}{l} Z = 2 - 8m \\ Q = 200 - 10000 \ kg/m^2 h \end{array}\right)$

On initial start up, the effluent concentration approaches the influent concentration, as would be expected in practice.

4.4.3 Determination of Mass Transfer Coefficient, K_L

Many attempts have been made to develop theoretical relationships to describe the functional behaviour of liquid phase mass transfer coefficients with media characteristics and fluid properties. The transfer of material between liquid and film surface in a packed bed is one of the fundamental problems in chemical engineering unit operations.

Empirical correlations of data, based on theoretical principles for simple situations have led to particular forms of relationships, (see Hirase and Mori, 1976).

By definition,

$$N_{Re_X} = \frac{d_x \, u}{\nu} \qquad\qquad \text{Reynolds Number}$$

$$N_{Sc} = \frac{\nu}{D} \qquad\qquad \text{Schmidt Number}$$

$$N_{Sh_X} = \frac{K_L \, d_x}{D} \qquad\qquad \text{Sherwood Number}$$

$$j_m = \frac{a}{N_{Re_X}^{\,n}} \qquad\qquad \text{Mass transfer j factor}$$

where d_x = a linear characteristic dimension, d_ℓ

d_h = an hydraulic diameter, $d_h = \frac{2\,\epsilon}{3(1-\epsilon)} \cdot d_\ell$

u = the superficial velocity of liquid through the packing

ν = the kinematic liquid viscosity

D = the effective diffusivity of "food" through the liquid

K_L = the liquid phase mass transfer coefficient

a, n = empirically fitted coefficients

Usually by definition $j_m = \dfrac{K_L \, N_{Sc}^{2/3}}{u}$

More recent references have shown that:

Miyanchi (1971,72) $a = 1.0$ $n = 2/3$ for $N_{Re_X} < 10$

$\qquad\qquad\qquad a = 0.75$ $n = 1/2$ $\quad N_{Re_X} > 10$

based on $N_{Re_X} = \dfrac{d_h \, u}{\nu}$ and $j_m = \dfrac{K_L \, N_{Sc}^{2/3}}{u}$

Kataoka (1972) $a = 1.85$ $n = 2/3$ for $N_{Re_X} < 10$

$\qquad\qquad\qquad a = 1.15$ $n = 1/2$ $\quad N_{Re_X} > 100$

based on $N_{Re_X} = \dfrac{d_\ell u}{\nu(1-\epsilon)}$

and $j_m = (\dfrac{1-\epsilon}{\epsilon})^{1/3} \cdot \dfrac{K_L \, N_{Sc}^{2/3}}{u/\epsilon}$

Upadhyay (1975) $a = 3.82$ $n = 0.73$ for $N_{Re_X} < 20$

$\qquad\qquad\qquad a = 1.62$ $n = 0.44$ $\quad N_{Re_X} > 20$

based on $N_{Re_X} = \dfrac{d_\ell u}{\nu(1-\epsilon)}$; $j_m = \dfrac{K_L \cdot N_{Sc}^{2/3}}{u}$

For the media involved in trickling filters, the usual characteristic dimension is the specific surface area, A_v which is the inverse of d_ℓ. The Reynolds Number definition then becomes, for example $N_{Re_X} = Q/\nu A_v(1-\varepsilon)$.

Experimental results from multiple full scale pilot units operated by I.C.I. from 1964-1978 in the English village of Buckfastleigh have provided more than 150 unit years of data. These units had been used to evaluate various FLOCOR packings and compared with a variety of stone and synthetic media. The whole plant consisted of up to 20 units operating in parallel on the same domestic-industrial influent with packed depths between 1.21 and 5.48m and irrigation rates between 1.1 and $290m^3/m^2d$, (see Roberts, 1973).

Regression analyses on feed to and effluent from each unit enabled this Ames-Roberts model, incorporating the above determination of mass transfer coefficient, to be evaluated.

The most appropriate mass transfer estimate was that due to *Kataoka* above, and it was also found that $K_m \alpha Q^n$
where $n = 0.31 \pm 0.020$ for $N_{Re_X} < 30$

$\quad$ $n = 0.46 \pm 0.032$ $\quad$ $N_{Re_X} > 30$

which is similar to the theoretical finding of $n = 1/3, 1/2$ in the Kataoka relationship above.

4.4.4 $\quad$ Magnitude of Biochemical Term

For a variety of wastes published by Quirk (1972), the author (1973) obtained a consistent set of parameters. These are compared in Table 4.1, which highlights the differences and trends with decrease in biochemical term values.

TABLE 4.4 $\quad$ Comparison of Individual Parameters for a Variety of Organic Waste

Synthetic media $A_v = 89m^2/m^3$; $Q = 70.5m^3/m^2d$; $X = 5.0 \times 10^{-3}$ kg/m^3

WASTE	OVERALL K_m	MASS TRANSFER $K_L A_V$	BIOCHEMICAL $[K_X/\alpha]$	K_L kg/m^2h	$K_{(h^{-1})}$	$\alpha_{(-)}$
Ragmill	800	1166	2550	13.1	13.2×10^{-3}	2.6×10^{-8}
Slaughterhouse	425	668	1170	7.5	9.6	4.1
Kraft papermill	298	490	560	5.5	8.1	5.3
Whey-sewage	290	481	730	5.4	7.9	5.4
Boxboard mill	260	436	645	4.9	7.5	5.8
Canning	201	347	480	3.9	6.6	6.9

An independent estimate of the biochemical $[\frac{KX}{\alpha}]$ can be made for a particular waste (see Sec. 5.1.2.7), given the media and flowrate.

4.4.4   A Comparison of the Grady-Lim and Ames-Roberts Models

In comparing the general Grady-Lim model with that of Ames-Roberts and in particular Eq. 4.3.2 and 4.4.7, 7(a)

$$\frac{1}{K_o} = \frac{1}{K_L} + \frac{1}{\frac{r_{max}}{K_s} \eta_i} \tag{4.3.2}$$

$$\frac{1}{K_m} = \frac{1}{K_L} + \frac{1}{K_B} \tag{4.4.7(a)}$$

an Overall term = a Mass transfer term + a Biochemical term in each model.

The Ames-Roberts model uses first order kinetics, implying that $K_s >> C$. If this is the case in practice, then the overall effectiveness profile relative to depth is quite flat, see Figure 5.1.

If we compare Eq. 4.3.7, which is the special simplification of the general model, with Eq. 4.4.4:

$$\frac{C_e}{C_i} = \exp\left(-\frac{r_{max} A_s}{K_s F} \eta_o\right) \tag{4.3.7}$$

and

$$\frac{C_e}{C_i} = \exp\left(-\frac{K_m Z}{Q}\right) \quad \text{for } C_r = 0 \tag{4.4.4}$$

then

$$\frac{r_{max} A_s}{K_s F} \eta_o = \frac{K_m Z}{Q}$$

since $\dfrac{A_s}{F} = \dfrac{A_v Z}{Q}$, $\quad \dfrac{r_{max} A_v Z \eta_o}{K_s Q} = \dfrac{K_m Z}{Q}$ provided the dimensions are consistent.

or $[\dfrac{r_{max} A_v}{K_s}]\eta_o = K_m$ \hfill (4.4.9)

The bracketed term is a constant coefficient, independent of flowrate and packed depth, leaving the overall effectiveness factor to be proportional to overall mass transfer coefficient. Some of the interpretations from the Grady-Lim model involving modified Thiele modulus and Damköhler Number can then be transferred to that of Ames-Roberts and vice versa. In particular, in the development leading to Eq. 4.4.4, K_m is independent of depth but varies with flowrate. This implies that the overall effectiveness factor is independent of depth (nearly so in practice) and varies with flowrate, see Figure 5.1.2.

If the simpler Eckenfelder Eq. 4.1.12 is compared with the Grady-Lim Eq. 4.3.7

$$\frac{C_e}{C_i} = \exp\left(- k_1 Z/Q^n\right)$$
(4.1.12)

then

$$\frac{r_{max} A_s}{K_s F} \eta_o = \frac{k_1}{Q^n}$$

Since $\dfrac{A_s}{F} = \dfrac{A_v Z}{Q}$; $\dfrac{r_{max} A_v Z}{K_s Q} \eta_o = \dfrac{k_1 Z}{Q^n}$

or $\left[\dfrac{r_{max} A_v}{K_s}\right]\eta_o = k_1 Q^{1-n}$
(4.4.10)

Again, the bracketed term in Eq. 4.4.10 is a constant coefficient, so that

$$\eta_o \; \alpha \; Q^{1-n}$$

since k_1 is independent of flowrate. This coefficient, k_1, cannot be interpreted as broadly as the Ames-Roberts model parameter, K_m, because k_1 is understood to be only a kinetic rate term.

Recent research by Castaldi and Malina (1982) has discounted diffusional mass transfer into the microbian film as being of significant importance. They conclude that the observed substrate uptake rate is clearly a function of velocity intensity and substrate concentration in the bulk liquid. An abrupt increase in uptake rate occurs as turbulent flow is approached which implies that solids retention time and concentration of bios are not of primary concern as design parameters. The data also indicate that below the limiting substrate concentration, the observed uptake rate is principally a surface reaction, regardless of magnitude of velocity. The conclusion must be drawn that the internal effectiveness factor in the Grady-Lim model tends to unity and is virtually independent of bios thickness. Sec. 4.2.1 - Plug flow with external resistance of mass transfer, idealised in Figure 4.2 would then seem to be the better Grady-Lim model which is also the basis of the Ames-Roberts approach.

5. SIMULATION OF THE TRICKLING FILTER PROCESS

Any model which purports to emulate a process should be capable of :
  (i) accurately describing operation under steady state or dynamic conditions, and
  (ii) useful interpretation of process performance.
As pointed out in Sec. 4.3, the Grady-Lim type model *may* be capable of (ii) above, but because of the unknown range of values to be placed on the many

parameters to be incorporated, (i) may not be fulfilled.

The Ames-Roberts type model, outlined in Sec. 4.4 appears to have the dual capability of emulating steady state and dynamic performance *and* of simple interpretation.

5.1.1   Interpretive Application of the Grady-Lim Model

The general arrangement for the simulation of a trickling filter system is similar to that of Fig. 1. Recycled effluent is returned to the head of the trickling filter and mixed with influent coming from the primary settling tank. Eq. 4.3.5 is modified to allow for this recycle stream:

$$\frac{dc}{dz} = - \{\frac{r_{max} A_s \eta}{F(1+\alpha)}o + \frac{\mu_{max}A}{F(1+\alpha)} [\frac{X_i}{Y} + (C_i - C)]\} \cdot \frac{C}{K_s + C} \qquad (5.1.1)$$

where $C = (C_p + \alpha C_e)/(1+\alpha)$ at $z = 0$

Eq. 5.1.1 has to be solved by a numerical method between the limits $z = 0$ to $z = Z$. A Runge-Kutta-Merson variable step length procedure was found most suitable with computations done on a VAX 11-780 computer system.

In this Grady-Lim model at least six process parameters have to be estimated, namely:

1. a_v $\equiv$ surface area of micro organisms per unit volume of biomass (m^2/m^3) involved in Thiele modulus, ϕ
2. D_e $\equiv$ effective diffusion coefficient, m^2/h involved in Thiele modulus, ϕ, and Sherwood Number, ψ
3. K_s $\equiv$ Monod saturation coefficient, kg/m^3 also involved in Thiele modulus and the modified term.
4. d $\equiv$ biomass effective film thickness, m involved in the Thiele modulus, ϕ, and Sherwood Number, ψ
5. μ_{max} $\equiv$ maximum specific growth rate, h^{-1}
6. r_{max} $\equiv$ maximum specific "food" removal rate, $kg/m^2 h$

These parameters are then added to the known design variables of

A_v $\equiv$ specific surface area for the particular media, m^2/m^3

ε $\equiv$ void fraction for the media

A_c $\equiv$ empty tower cross sectional area, m^2

A $\equiv$ flow cross sectional area = $\varepsilon.A_c$, m^2

A_s $\equiv$ total area of biofilm, m^2

Z $\equiv$ packed bed depth, m

and the operational variables:

$\qquad$ F $\qquad \equiv$ flowrate, m^3/h

$\qquad \alpha \qquad \equiv$ recycle ratio

$\qquad C_p \qquad \equiv$ influent "food" concentration, kg/m^3

It is assumed that there are no cells being circulated in the recycle stream and no active cells in suspension move along with the liquid film as it trickles over the packing, hence $X = 0$.

A close estimate of mass transfer coefficient, k_L, is made from an equation due to Kataoka et al (1972), see Sec. 4.4.3. To evaluate the overall effectiveness factor η_0, the original technique of Fink (1973) was utilised.

Direct numerical integration of Eq. 5.1.1 is possible when $\alpha = 0$. With recycle of effluent, a boundary value problem exists in that the effluent concentration is not immediately determined. A "shooting" method if iterative calculations was employed involving the Newton-Raphson technique to determine the concentration, C_e.

Parameters used by Grady-Lim have been incorporated here in Table 5.1 so that the reader can make a direct comparison of this presentation and that of the originators in the simulations to follow.

TABLE 5.1 Parameters used in Eq. 5.1.1

a_v	$= 100 \ m^2/m^3$	A_v	$= 100 \ m^2/m^3$
D_e	$= 1.392 \times 10^{-5} \ m^2/h$	ε	$= 0.35$
K_s	$= 0.030 \ kg/m^3$	Z	$= 2.00 \ m$
d	$= 1.0 \times 10^{-5} \ m$	Tower cross sectional area	$= 0.02 \ m^2$
μ_{max}	$= 0.1296 \ h^{-1}$	Q	$= 24 \ m^3/m^2 d$
r_{max}	$= 4.18 \times 10^{-3} \ kg/m^2 h$	F	$= 0.02 \ m^3/h$
C_f	$= 0.075, \ 0.30, \ 0.75 \ kg/m^3$	A_c	$= 7.0 \times 10^{-3} \ m^2$
		A_s	$= 200 \ m^2$
j_m	$= \dfrac{a}{N_{Re}^{\ n}}$ for $N_{Re} < 30$	a	$= 1.85, \ n = 2/3$
	$N_{Re} > 30$	a	$= 1.15, \ n = 1/2$
k_L	$=$ Factor $j_m. \ \dfrac{Q}{\varepsilon} \cdot (\dfrac{\varepsilon}{1-})^{1/3} /N_{Sc}^{2/3}$		

where Factor $= 0.346$ to allow a similar value to that of Grady-Lim model in evaluating k_L for these given conditions noted here.

5.1.1.1 Profiles through the Tower for Various Concentrations

Using the parameters of Table 5.1, numerical solutions of Eq. 5.1.1 were obtained and presented in graphical form in Figure 5.1.1.

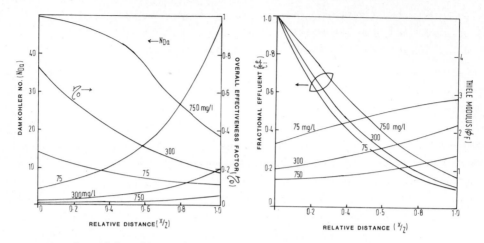

Fig. 5.1.1(a)   Effect of Influence Concentration on Profiles

The fractional concentrations appear to be dropping away exponentially, which if plotted as log (C_e/C_i) vs (x/z) form smooth curves, slightly convex upwards. These curvatures become more pronounced with change in overall effectiveness factor. The higher the initial concentration, the greater the overall effectiveness factor which then decreases with depth. Conversely, the higher the initial concentration, the smaller the Damköhler Number. Above 0.3 kg/m^3 influent concentration, N_{Da} < 1.0, which implies that the biochemical reaction mechanism is rate determining. As "food" concentration decreases on passing through the tower, Damköhler Number rapidly increases, showing there is considerable increase in resistance to mass transfer.

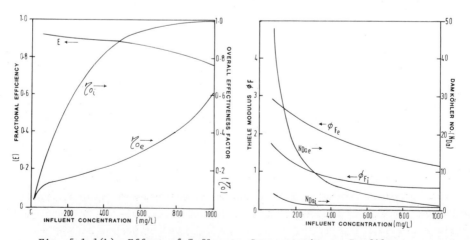

Fig. 5.1.1(b)   Effect of Influence Concentration on Profiles

With constant flowrate but increase in influent concentration, this can be interpreted to be an increase in organic loading ($Q \times C_i$). From Figure 5.1.1(b) we note a drop-off in efficiency with increase in organic loading. The downward trend is similar but not identical in form to that shown in Fig. 2.1.5.

5.1.1.2   Variation in Profiles with Flowrate

As the flowrate is increased, most of the important tower characteristics change considerably, as shown in Fig. 5.1.2. An exponential decrease in removal efficiency occurs which, when correlated as $\log (C_e/C_i)$ vs $(1/Q)$ yields almost a straight line relationship. Because of constant feed concentration, the inlet modified Thiele modulus does not vary; whereas a near exponential decrease in effluent Thiele modulus occurs. This implies a relative magnitude shift from negligible kinetic resistance to considerable kinetic resistance. This effect is coupled with an increase in mass transfer rate due to flowrate increase which significantly decreases the effluent Damköhler Number towards unity. Since $N_{Da} > 1$ infers mass transfer limitations in terms of the overall effectiveness factor, it can be seen that increase in flowrate asymptotically increases effectiveness factor, at the same time as reducing the mass transfer resistance through the Damköhler Number.

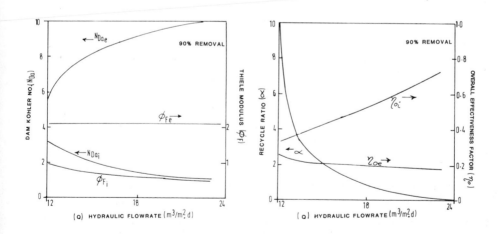

Fig. 5.1.2   Effect of Flowrate on Profiles

With flowrate varying and influent concentration constant, this again can be interpreted as an increase in organic loading. The typical drop-off in tower efficiency with increase in loading almost matches the downward trends shown in Fig. 2.1.6.

5.1.1.3   Variation of Profiles with Recycle Ratio

Figure 5.1.3(a) shows that an increase in recycle ratio at constant influent rate changes the shape of the concentration profile from exponential decay to near linear decrease. At the same time, the profiles for Modified Thiele modulus, Damköhler Number and overall effectiveness factor become less steep. The principal reasons are the same as with previous Sec. 5.1.1.2; due to enhanced mass transfer rate because of a considerable increase in flowrate through the packing.

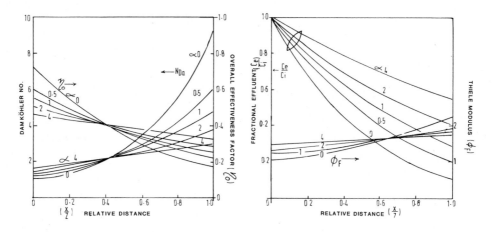

Fig. 5.1.3(a)   Variation of Profiles with Recycle Ratio

The inlet and outlet values for each of the major parameters as functions of recycle ratio are shown in Figure 5.1.3(b). *Actual* fractional effluent across the tricling filter increases with increase in recycle ratio (where $(C_p + \alpha C_e)/(1+\alpha)$). *Overall* fractional effluent increases up to a recycle ratio of 4 : 1 then starts to decrease (where C_p is the feed concentration after primary sedimentation, without effluent mixing). See also Sec. 4.1.2 for empirical interpretations. Recycled effluent dilutes the influent going to the filter causing the inlet overall effectiveness factor to sharply decrease. The twin causes for this are coupled interactions of rapidly increasing modified Thiele modulus due to a slower reaction rate and a slight increase in Damköhler Number inferring continued mass transfer limitations.

Outlet profiles show some interesting trends. The modified Thiele Modulus passes through a minimum which indicates decreasing mass transfer resistance with recycle ratio up to 4 : 1. Overall effectiveness factor and Damköhler Number both rapidly change, again showing a decrease in mass transfer resistance or increase in biochemical rate process limitations.

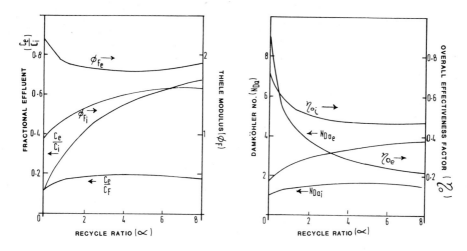

Fig. 5.1.3(b)   Variations of Profiles with Recycle Ratio

5.1.1.4   Effect of Varying Flowrate and Recycle to Maintain 90 Percent
            Efficiency

If the influent flowrate and recycle ratio are varied together for a
fixed packing depth, unique combinations of (Q,α) can maintain a constant
removal efficiency, here chosen to be 90 percent, as illustrated in Fig.
5.1.4.   It shoud be pointed out that the actual flowrates passing to the
trickling filter range from about 23 m^3/m^2d at about $\alpha = 0$ to about 110 m^3/m^2d
at $\alpha = 8$, so that Fig. 5.1.2 can be noted in conjunction with Fig. 5.1.4..

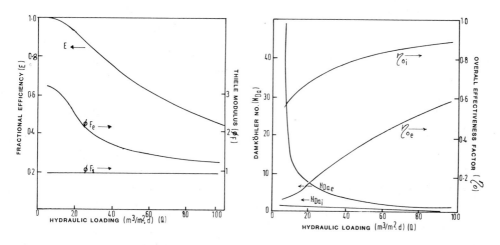

Fig. 5.1.4   Variation in Parameters to Maintain 90 Percent Efficiency

288

With increase in recycle ratio and correspondingly total flowrate the
inlet and outlet overall effectiveness factors tend towards each other as do
the Damköhler Numbers. The movement is generally from high mass transfer
limitations at zero recycle to smaller resistance to mass transport at high
recycle rates. Although the influent concentration is decreasing, the mass
transfer coefficient is increasing for increase in recycle rate. The product
of these coefficients decreases, thereby increasing the Damköhler Number.
For outlet conditions, the effluent concentration is constant, but mass transfer
coefficient increases with recycle ratio. This causes the Damköhler Number
to decrease. With modified Thiele modulus remaining substantially constant,
the Sherwood Number, through the mass transfer coefficient increases as the
recycle ratio. This then causes the overall effectiveness factor to slightly
increase.

5.1.1.5   Variation of Profiles with Specific Surface Area

For otherwise constant conditions to those of Table 5.1, the specific
surface area of media was varied. The major effect is that of an increase
in mass transfer coefficient as the specific surface area decreases, and a
decrease in the quantity of available bios for biochemical reaction.

The lefthand parts of Figures 5.1.2 and 5.1.5 can be interpreted
similarly as in Sec. 5.1.1.2. An increase in Reynolds Number causes a decrease
in modified Thiele modulus and increase in fractional effluent. It can also
be shown that $\log(C_e/C_i)\alpha(-A_v)$, especially over the higher range of specific
surface area.

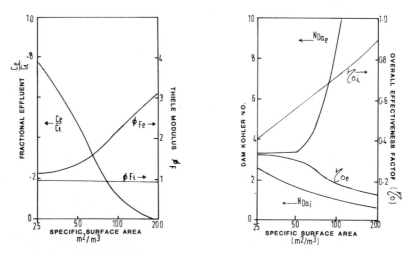

Fig. 5.1.5 Variation of profiles with Media Specific Surface Area

The righthand part of Figure 5.1.5 above shows that for a high specific surface area, the effluent is controlled by mass transfer limitations since the mass transfer coefficient is low and "food" concentration very low. This leads to a high Damköhler Number. With decrease in specific surface area, the overall effectiveness factors rapidly tend towards each other principally due to the modified Thiele moduli through the increase in effluent concentration.

5.1.1.6   Effect of Void Fraction on Profiles

In Sec. 4.4.3 it was noted that the Kataoka approach gave the best mass transfer coefficient estimate. By re-arranging these equations:

$$K_L \ \alpha \ \frac{(1-\varepsilon)^P}{\varepsilon^{2/3}} \qquad \text{for } N_{Re}' < 10 \ , \quad p = 1/3$$

$$N_{Re}' > 10 \ , \quad p = 1/6$$

for otherwise constant conditions of A_v, Q.

Different manufacture of media can produce packings of the same specific surface area but quite different void space. Smaller void fractions imply higher mass transfer coefficients and corresponding better fractional efficiencies. This would have to be balanced against lower upper irrigation rates and poorer ventilation through the filter. Figure 5.1.6 shows the trends in profiles for change in void fraction.

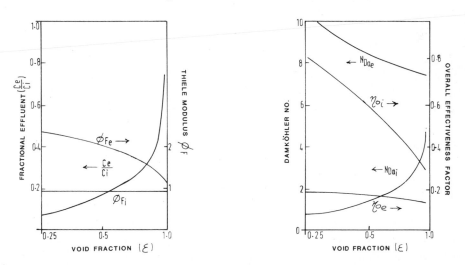

Fig. 5.1.6   Variation of Profiles with Media Void Fraction

Fractional effluent becomes much higher as the void fraction tends to unity. Overall effectiveness factors and modified Thiele modulus both drop off for increase in void fraction. The inlet and exit Damköhler Number tend to move towards each other.

All these changes correspond to a decrease in mass transfer coefficient and an increase in effluent concentration with increase in void fraction.

5.1.1.7 Effect of Packed Depth on Profiles

For otherwise constant conditions the simulation allowed the depth of packing to increase. With mass transfer coefficient constant the dominant effect was in exponential decrease in fractional effluent. Figure 5.1.7 shows the trends in parameters with packed depth. The results can be closely correlated by $\log(C_e/C_i)\ \alpha(-Z)$ over most of the range of depth. As the effluent concentration decreases, the overall effectiveness factor decreases since the modified Thiele modulus and Damköhler Number increases. The same discussion as in Sec. 5.1.1.1 applies here as to an increase in mass transfer resistance.

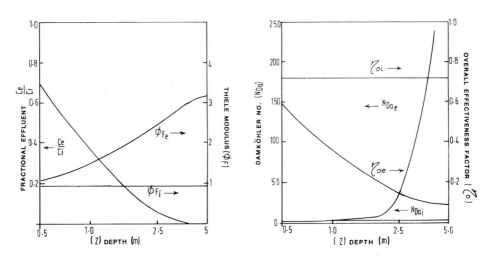

Fig. 5.1.7 Variation of Profiles with Packed Depth of Media

5.1.1.8 Effect of Diffusivity on Parameters

Estimates of effective diffusivity and the corresponding Schmidt Number are difficult to evaluate, but an approximation may be made by assuming that the effective diffusivity tends to the molecular diffusivity for a particular "food" compound. Table 5.2 provides estimates of diffusivity and Schmidt number for a number of compounds (see Mitchell, 1978).

TABLE 5.2 Diffusivity and Schmidt Number of Various Compounds

COMPOUND	D_e at $20^\circ C$ m^2/h	N_{Sc}
Ammonium nitrate	6.1×10^{-6}	590
Urea	4.4	825
Glycelglycine	2.4	1500
Glucose	2.1	1750
Raffinose	1.4	2500

The choice by Grady-Lim of $D_e = 1.39 \times 10^{-5}$ m^2/h gives $N_{Sc} = 259$, which is quite unrealistic in a practical situation. Figure 5.1.8 shows the effect a range in effective diffusivity has on the usual parameters. Since $K_L \propto D_e^{2/3}$, then an increase in diffusivity increases the mass transfer coefficient which causes a decrease in modified Thiele modulus and Damköhler Number. Fractional effluent decreases, which combined with the other parameters causes an increase in overall effectiveness factors.

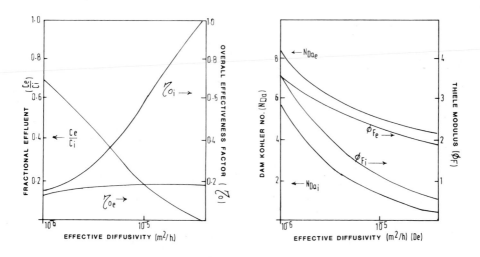

Fig. 5.1.8 Variation of Profiles with Diffusivity

5.1.1.9   Variation of Parameters with Maximum Specific Removal Rate

The maximum specific "food" removal rate, r_{max} in Eq. 4.3.5 is one of the most dominant factors which affects all parameters. Unfortunately, in the Grady-Lim model r_{max} and to a lesser extent maximum specific growth rate, μ_{max}, have to be *estimated* from external experimentation. The Damköhler Number and modified Thiele|modulus are both directly proportional to r_{max}.

Typical estimates of r_{max} are given in Table 5.3 for a number of "food" substances.

TABLE 5.3   Maximum Specific Removal Rates

"FOOD" SUBSTANCE	r_{max} kg/m^2h
Higher alcohols, glycols	1×10^{-3}
Phenols	2
Dairy, poultry wastes	3
Distillery wastes	4
Potato wastes	5
Frozen food, nylon wastes	8

Figure 5.1.9 shows the effect of maximum specific "food" removal rate on the usual parameter.

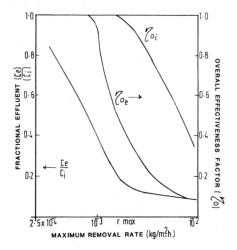 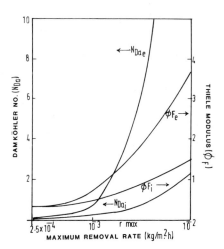

Fig. 5.1.9   Variation of Profiles with Maximum Removal Rate

The fractional effluent and correspondingly the overall effectiveness factors decrease sharply with increase in r_{max}. At low values of r_{max}, there is little resistance to mass transfer, but above about $r_{max} = 1 \times 10^{-3}$ kg/m^2h, Damköhler Number and modified Thiele modulus rapidly increase. This indicates an increasing resistance to mass transfer relative to the biochemical rate term.

The maximum specific growth rate, μ_{max} has little effect on the parameters. An increase in μ_{max} only slightly decreases fractional effluent and overall effectiveness factor. The modified Thiele modulus and Damköhler Number are only marginally increased which may be interpreted to be a slight shift towards mass transfer resistance.

5.1.1.10 Effect of Monod Saturation Coefficient on Parameters

The Monod saturation coefficient, K_s, together with r_{max} and μ_{max} are the real unknown factors when applied to trickling filter simulations. At the best, K_s would be an experimentally fitted coefficient. If Monod saturation coefficients are to be estimated by analogy to activated sludge systems, then residence time of effluent, age of biofilm and the particular type of "food" substances in the influent would be a guide.

In this simulation here, the influent "food" concentration was constant at 0.3 kg/m^3 and K_s varied between 0.015 and 0.3 kg/m^3. Figure 5.1.10 shows the effect a change in K_s has on the usual parameters.

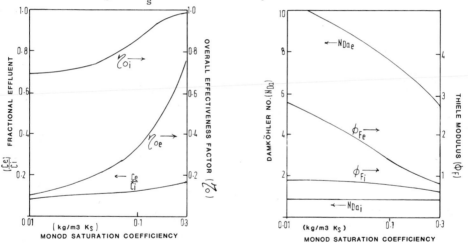

Fig. 5.1.10 Variation of Profiles with Monod Saturation Coefficient

In Eq. 4.3.5, K_s enters the equation to adjust the kinetics to between zero and first order, by the factor $C/(K_s+C)$.

When $K_s \ll C$ then the factor $C/(K_s+C) \to 1$, whereas if $K_s \gg C$ then the factor $C/(K_s+C) \to C/K_s$.

An increase in K_s only slightly increases fractional effluent but markedly alters most other parameters. The overall effectiveness factors are considerably increased implying reduced resistance to mass transfer. The modified Thiele modulus and Damköhler Number are both reduced approximately in proportion to the increase in Monod coefficient. Again this may be interpreted to be a reduction in resistance to mass transfer.

5.1.2 Interpretive Applications and Simulations using the Ames-Roberts Model

To illustrate the versatility of this model, a number of different examples are outlined in the following pages.

In most of these examples, no other model of the Eckenfelder type or Grady-Lim form can attempt to explain the reasons for change in trickling filter performance or satisfactorily predict operational behaviour.

The Eckenfelder type model incorporates two empirical coefficients while the Grady-Lim model contains at least six variables or factors which have to be estimated, excluding specific surface area and void fraction of media. Variants use ten independent biochemical coefficients, while the earlier McCarty (1976) model required an estimate of four factors. The Mutharasan (1978) model required an estimate of seven coefficients; Schroeder (1976) four factors, and Jennings (1976) model needs seven coefficients.   By comparison, the Ames-Roberts model requires knowledge of only the "food" Schmidt Number and an esti-mate of the biochemical term, both of which can be determined in most situations.

This concept of overall mass transfer coefficient incorporated within the model can adequately describe:

1. the variation of trickling filter efficiency with level of dissolved oxygen in the effluent due to $(KX/\alpha ...X)$,

2. the maturing of a trickling filter from start up, $(KX/\alpha ...\alpha)$,

3. decrease in performance and then recovery after a shock organic load or low concentration of toxic material, $(KX/\alpha ... X, \alpha)$,

4. the decrease in performance of each stage in a multistage combination of trickling filter units, $(KX/\alpha , N_{Sc})$,

5. performance of nitrification or denitrification units, $(KX/\alpha , N_{Sc})$,

6. the estimation of Eckenfelder type k, coefficient, $(KX/\alpha , N_{Sc})$,

7. the effect of temperature and flowrate on trickling filter performance, $(KX/\alpha , N_{Sc})$,

The following examples highlight the Ames-Roberts Model capabilities.

5.1.2-1 Variation of Efficiency with Oxygen Concentration

In work of the Water Research Centre (1965) conducted on an experimental filter, it was found that filter efficiency increased by only about 8 percent when the aerating gas composition changed from 2 to 20.6 percent oxygen.

A hypothetical situation will illustrate this finding, by application of the reciprocal series equation.

Assume that the kinetic-biological term is the major factor involved and that $[K/\alpha]$ is constant. The only change in the term $[KX/\alpha]$ is a 10-fold

increase in dissolved oxygen concentration [X] of the liquid trickling over
the packing due to Henry's Law of gas solubility.

Stone packing – size range: 51–102mm; packed depth: 2m

Hydraulic flowrate = $14.4^3/m^2d$ @ 20C

From Table 3.1, $A_V = 56.2 m^2/m^3$; $\varepsilon = 0.39$

From Table 2.1, M.W.R. = $3.5–5.4 m^3/m^2d$; U.I.R. = $51–104 m^3/m^2d$

which indicates that the hydraulic flow rate is in the accepted range of
operation.

By "a priori", the biochemical term, $[KX/\alpha] = 1310$

and the "food" Schmidt Number, $N_{Sc} = 1500$

Now $N'_{Re} = Q/\mu \cdot A_V(1-\varepsilon)$ $\qquad\qquad$ $\mu = 3.6 kg/m^2h$ (1 centipoise)

$$A_V = 56.2 m^2 m^3$$

$$\varepsilon = 0.39$$

$N'_{Re} = 600/3.6 \times 56.2 \times 0.61 = 4.86$ $\qquad$ $Q = 14.4 \times 10^3/24 kg/m^2h = 600$

$j_L = \dfrac{1.85}{N'_{Re}2/3}$ for $N'_{Re} < 10$ $\qquad$ $j_L = \dfrac{1.85}{4.862^{2/3}} = 0.645$

and $j_L = (\dfrac{1-\varepsilon}{\varepsilon})1/3 \cdot \dfrac{K_L N_{Sc}2/3}{Q/\varepsilon}$

re-arranging

$$K_L = (\dfrac{\varepsilon}{1-\varepsilon})^{1/3} (Q/\varepsilon) \dfrac{1}{N_{Sc}2/3} \times j_L$$

$$K_L = (\dfrac{0.39}{0.61})^{1/3} \cdot \dfrac{600}{0.39} \cdot \dfrac{1}{1500^{2/3}} \times 0.645 = 6.52 \ kg/m^2h$$

Now $\dfrac{1}{K_m} = \dfrac{1}{K_L \cdot A_V} + \dfrac{\alpha}{KX}$

$\dfrac{1}{K_m} = \dfrac{1}{6.52 \times 56.2} + \dfrac{1}{1310}$ $\quad \therefore \quad K_m = 286.3 \ kg/m^3h$

$\dfrac{C_e}{C_i} \cong \exp(-\dfrac{K_m Z}{Q})$

$\dfrac{C_e}{C_i} = \exp(-\dfrac{286.3 \times 2}{600}) = 0.385$

That is, removal efficiency across the trickling filter is $\cong$ 61.5 percent.

Now, due to a 10-fold increase in dissolved oxygen concentration $[KX/\alpha] \rightarrow 13100$.

Then $\dfrac{1}{K_m} = \dfrac{1}{6.52 \times 56.2} + \dfrac{1}{13100}$

$\qquad K_m = 356.5 \ kg/m^3h$

$\qquad \dfrac{C_e}{C_i} \cong \exp(-\dfrac{356.5 \times 2}{600}) = 0.305$

Or, removal efficiency across the tower $\cong$ 69.5 percent.

This is an improvement of 8 percent in efficiency.

5.1.2-2 A "Maturing" Trickling Filter

A new high rate trickling filter after initial start up, has the media surface rapidly covered with active bios.   An estimate can be made of performance of the "maturing" filter unit up to its steady-state efficiency of 70 percent in 16 weeks time.

Interpretation

After the media becomes covered with active bios (say 2 weeks) the liquid phase mass transfer coefficient is constant. The biochemical rate coefficient and dissolved oxygen concentration are also essentially independant of time. This means that the specific adsorption coefficient is the dominant variable in the whole process.

Unit Data: Packing - Filterpak I $A_v = 120 m^2/m^3$, $\varepsilon = 0.93$

"food" $N_{Sc} = 1800$; Packed depth = 5m; Hydraulic rate = $42 m^3/m^2 d \equiv 1750 kg/m^2 h$

From Table 2.1, M.W.R. = $30 m^3/m^2 d$

so the irrigation rate is satisfactory to completely wet the packing.

$$N_{Re}' = \frac{Q}{\mu A_v (1-\varepsilon)} = \frac{1750}{3.6 \times 120 \times 0.07} = 57.9$$

since $N_{Re}' > 10$, use the turbulent correlation

$$j_L = \frac{1.15}{N_{Re}'^{\frac{1}{2}}} = \frac{1.15}{57.9^{\frac{1}{2}}} = \underline{0.1511}$$

On "maturing", $\dfrac{C_e}{C_i} = \exp\left(-\dfrac{K_m Z}{Q}\cdot\right)$

$$0.30 = \exp\left(-\frac{K_m\ 5.0}{1750}\right)$$

$$K_m = \underline{421.4}\ kg/m^3 h$$

$$K_L = j_L \cdot \frac{Q}{\varepsilon} / (N_{Sc}^{2/3} \cdot (\frac{1-\varepsilon}{\varepsilon})^{1/3})$$

$$K_L = 0.1511 \cdot \frac{1750}{0.93} / (1800^{2/3} \cdot (\frac{0.07}{0.93})^{1/3}) = 4.55\ kg/m^3 h$$

$$\frac{1}{K_m} = \frac{1}{K_L \cdot A_v} + \frac{\alpha}{KX}$$

$$\frac{1}{421.4} = \frac{1}{4.55 \times 120} + \frac{\alpha}{KX}$$

$$[\frac{KX}{\alpha}] = \underline{1847}\ kg/m^3 h$$

"a priori" $[KX] = 1.385 \times 10^{-4}$ kg/m^{3}h substantially constant, so that $[\alpha] = 7.5 \times 10^{-8}$, on maturation.

An hypothetical growth curve can then be developed, relating specific adsorption coefficient $[\alpha]$ and elapsed time. Again, by the use of the series rate equation, the overall transfer rate and subsequently the removal efficiency estimated, as is shown below.

Parameter	Weeks Since Commencement						
	0	2	4	8	12	14	16
$[\alpha]$	–	–	6.15×10^{-6}	5.75×10^{-7}	1.34×10^{-7}	8.9×10^{-8}	7.5×10^{-8}
$[KX/\alpha]$	–	–	22.5	240.9	1034	1556	1847
K_m	–	–	21.64	167.2	357.3	404.2	421.4
C_e/C_i	1.0	1.0	0.94	0.62	0.36	0.315	0.30
Efficiency %	0	0	6	38	64	68.5	70

In the initial period, the specific adsorption coefficient and correspondingly, the very thin bios have few active sites for the "food" to be assimilated.
As time goes by, a thicker film of biomass results in a maximum of steady state
active sites.

5.1.2-3 Effect of Shock Organic Loading

An excellent example of the robustness of a high rate filter to shock load
is given by Hemming (1971). A milk processing plant effluent passes to a
FLOCOR unit, treating $50 m^3$/day of whey liquid waste of average influent
concentration, 0.96 kg/m^3 BOD_S. A shock load of $3 m^3$ of whey concentrate is
discharged to the balancing tank which causes the tower performance to drop
from 92 percent to 20 percent over the next 15 minutes. After 3 hours shock
delay time, the Unit commences to recover.

Unit Data: FLOCOR packing, $A_v = 95 m^2/m^3$; $\varepsilon = 0.96$
Packed depth $= 7.25m$; Hydraulic flowrate $= 35.2 m^3/m^2 d \equiv 1468$ kg/m^2h
From Table 2.1 M.W.R. $= 37 m^3/m^2 d$
which indicates that the tower is not quite "wetted".

Applying the effective wetted area equation of Sec.2.2, gives
$A_e = 94 m^2/m^3$.

Before the shock load, $C_e/C_i = 0.08$

i.e. $\dfrac{C_e}{C_i} \cong \exp\left(-\dfrac{K_m Z}{Q}\right)$

$0.08 \cong \exp\left(-\dfrac{K_m \cdot 7.25}{1468}\right)$

so that $K_m \cong 511.4$ kg/m^3h
When $\mu = 3.6$kg/m^2h

$N'_{Re} = \dfrac{Q}{\mu A_v (1-\varepsilon)} = \dfrac{1468}{3.6 \times 94 \cdot 0.04} = \underline{108.5}$

For $N_{Re} > 100$

$j_L = \dfrac{1.15}{N'^{\frac{1}{2}}_{Re}} = \dfrac{1.15}{108.5^{\frac{1}{2}}} = \underline{0.1104}$

By "a priori" the "food" $N_{Sc} = 690$, and

$K_L = \left(\dfrac{\varepsilon}{1-\varepsilon}\right)^{1/3} \cdot \dfrac{Q}{\varepsilon} \cdot \dfrac{1}{N_{Sc}^{2/3}} \cdot j_L$

$$K_L = (\frac{0.96}{0.04})^{1/3} \cdot \frac{1468}{0.96} \cdot \frac{1}{690^2/3} \cdot 0.1104 = \underline{6.24} \ kg/m^2h$$

Now $\frac{1}{K_m} = \frac{1}{K_L A_v} + \frac{\alpha}{KX}$

$$\frac{1}{511.4} = \frac{1}{6.24 \times 94} + \frac{\alpha}{KX}$$

or $[\frac{KX}{\alpha}] = 4000 \ kg/m^2h$

The dominant parameters affected by gross food concentration is the specific adsorption coefficient $[\alpha]$, being partly "blinded" - losing active sites and the dissolved oxygen concentration $[X]$, which markedly decreases. Mass transfer coefficient as such, is not significantly affected so that the mass transfer terms remain substantially constant.

On being shock loaded, the efficiency drops to 20 percent

i.e. $\frac{C_e}{C_i} = 0.80 \cong exp \ (- \frac{K_m Z}{Q}) = exp \ (- \frac{K_m \cdot 7.25}{1468})$

$K_m \cong \underline{45.2} \ kg/m^3h$

By repeated application of the reciprocal series equation the Kinetic-biological term is modified to accommodate change in $[X/\alpha]$ with time, as shown below:

Basis $\frac{1}{K_m} = \frac{1}{K_L A_v} + \frac{\alpha}{KX}$ with $K_L A_v$ and K, constant

$\frac{1}{45.2} = \frac{1}{586.6} + \frac{\alpha}{KX}$ from which $[KX/\alpha] = 49.0$

	Elapsed Time from initial recovery (hours)				
	0	4	8	12	16
Kinetic-Biological term $[KX/\alpha]$	49.0	86	320	3340	4000
Overall coefficient K_m	45.2	75.0	207	499	511.4
C_e/C_i		0.80	0.69	0.36	0.085	0.08
Efficiency %		20	31	64	91.5	92

The computed estimates closely match the actual response, shown for comparison in Figure 5.1.2-1.

Consider three identical high rate trickling filters in series with inter-stage settlement of sloughed bios. Each tower has the same hydraulic loading and operate at similar temperatures. Laboratory treatability studies show that the kinetic-biological rate is expected to decrease by a factor of two (2) between each stage of treatment. Provide a design estimate of stage efficiency.

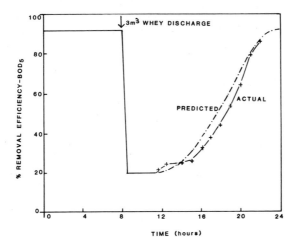

Fig. 5.1.2-1 Effect of Change in Treatability within Multistage Units

Unit Data: Packing SURFPAC: $A_v = 82m^2/m^3$; $\varepsilon = 0.94$

Packed depth = 6m ; Hydraulic rate = $24m^3/m^2d \equiv$ M.W.R. $\equiv 1000$ kg/m^2h

From laboratory treatability studies, at the individual stages,

$[\frac{KX}{\alpha}]_1 = 1000$; $[\frac{KX}{\alpha}]_2 = 500$; $[\frac{KX}{\alpha}]_3 = 250$ kg/m^3h

and $N_{Sc_1} = 1800$; $N_{Sc_2} = 1600$; $N_{Sc_3} = 1400$ (molecular weight or size of 'food' molecules becomes smaller on passage through each unit).

$$N_{Re}' = \frac{Q}{\mu A_v(1-\varepsilon)} = \frac{1000}{3.6 \times 82 \times 0.06} = 56.6$$

Since $N_{Re}' > 10$ use turbulent correlation

$$j_L = \frac{1.15}{N_{Re}^{1/2}} = \frac{1.15}{56.5^{1/2}} = \underline{0.1530}$$

	Stage 1	Stage 2	Stage 3
$K_L = j_L \frac{Q}{\varepsilon}/(N_{Sc}^{2/3}(\frac{1-\varepsilon}{\varepsilon})^{1/3})$	$0.1530 \times \frac{1000}{0.94}/(1800^{2/3}(\frac{0.06}{0.94})^{1/3})$	$1600^{2/3}$	$1400^{2/3}$
	$\underline{2.753}$ kg/m^2h	$\underline{2.978}$	$\underline{3.255}$
$\frac{1}{K_m} = \frac{1}{K_L A_v} + \frac{\alpha}{KX}$	$\frac{1}{2.753 \times 82} + \frac{1}{1000}$	$\frac{1}{2.978 \times 82} + \frac{1}{500}$	$\frac{1}{3.255 \times 82} + \frac{1}{250}$
K_m	$\underline{184.2}$	$\underline{164.1}$	$\underline{129.1}$
$\frac{C_e}{C_i} \cong \exp(-\frac{K_m Z}{Q})$	$\underline{0.331}$	$\underline{0.374}$	$\underline{0.461}$
Efficiency %	66.9	62.6	53.9

5.1.2-5   Denitrification in a Flooded Filter

For such a situation, influent soluble ammonium and/or nitrate/nitrite ions flow through the packed column with liquid and bios exchanging material. A continuous biochemical reaction occurs within the bios surface layers.

For nitrification, it is assumed that dissolved oxygen is in excess while for denitrification, a soluble carbon source is in excess, neither being rate controlling limitations.

The British Water Research Centre (1974) provides details of pilot plant performance at set flow rates and temperatures for the denitrification process. For example: 25mm pebbles, $A_v = 194 m^2/m^3$; $\varepsilon = 0.385$; $Z = 0.42m$; $Q = 183 kg/m^2 h$ for ammonium nitrate, $N_{Sc} = 590$; $\Theta = 1.065$

"a priori" $[\frac{KX}{\alpha}] = 2690$

By calculation, $N_{Re}' = 0.426$; $j_L = 3.267$; $K_L = 18.89$

$$\frac{1}{K_m} = \frac{1}{18.89 \times 194} + \frac{1}{2690}$$

$K_m = \underline{1551.3}\ kg/m^3 h$

so that $\dfrac{C_e}{C_i} \cong \exp\left(- K_m Z\, \Theta^{\Delta T}/Q\right) = \exp\left(- 3.5604 \cdot 1.065^{\Delta T}\right)$

at 20°C $\dfrac{C_e}{C_i} = \exp\left(- 1551.3 \times 0.42/183\right) = 0.0284$

Applying the temperature correction to the exponential term at several lower temperatures gives the following predicted performance which are compared with those reported.

Temperature °C	8	12	17	20
Predicted Removal %	81.2	88.0	94.8	97.2
Actual %	80	90	90	98

5.1.2-6   Prediction of Eckenfelder Type, k_1 Coefficient

The basic Eckenfelder equation is

$$\frac{C_e}{C_i} = \exp\left(- k_1 Z/Q^n\right) \qquad\qquad (4.1.12)$$

Experimental data indicates that this first order rate coefficient is a function of specific surface area of media and/or exponent (n) contained within the eqn. (4.1.12) for similar feed material.

Balakrishnan (1969) proposed $k_1 \propto A_v^{0.64}$, while Lipták (1974) suggested $k_1 \propto 1/n$.

A different explanation can be offered through application of the Ames-Roberts equation, which in effect predicts the magnitude of the Eckenfelder coefficient.

Lipták (1974) presents a table containing k_1 coefficients for a variety of media all evaluated using sewage as the feed material. If media are selected for which specific surface areas are known and for hydraulic regimes well above minimum wetting rates, the following data can then illustrate the procedure.

Media	Specific Surface Area $A_v m^3/m^3$	Void Fraction ε	Hydraulic Rate Q $m^3/m^2 d$	k_1 d^{-1}	Exponent n
2½" – 4" Rock	42	0.35	33 – 100	0.869	0.49
1" – 3" Granite	55	0.45	5 – 16	0.987	0.40
SURFPAC	82	0.92	34 – 269	1.025	0.45
1½" Raschig Rings	128	0.90	12.6 – 16	0.890	0.306

where $\dfrac{C_e}{C_i} = \exp(-k_1 Z/Q^n)$ with Q, Z in S.I. units.

The accompanying tabular computations summarise the estimates.

Media	N_{Re}'	j_L	k_L	Range in Biochemical Term $[\dfrac{KX}{\alpha}]$	k_M	Predicted k_1
2½ – 4" Rock	14.0	0.319	7.77	305 – 685	158 – 221	0.642-0.892
	42.4	0.177	13.05	585 –1425	283 – 396	0.649-0.908
1 – 3" Granite	1.91	1.200	3.97	205 – 460	106 – 148	0.969-1.350
	6.12	0.553	5.85	305 – 680	156 – 218	0.709-0.991
SURFPAC	48.0	0.166	4.15	530 –2000	207 – 291	0.714-1.004
	38.1	0.0589	11.70	1505 –15000	586 – 902	0.647-0.996
1½ Raschig Rings	11.4	0.365	3.38	410 – 910	211 – 293	0.873-1.21
	14.5	0.312	3.66	445 – 990	228 – 318	0.799-1.11

In summary, then,

Media	Measured k_1 (d^{-1})	Predicted k_1 (d^{-1})
2½ – 4" Rock	0.869	0.642 – 0.908
1 – 3" Granite	0.987	0.709 – 1.35
SURFPAC	1.025	0.647 – 1.004
1½" Raschig Rings	0.890	0.799 – 1.21

Despite the wide range in specific surface area of media, void fraction and flowrate, the spread in predicted Eckenfelder coefficients clearly overlap the measured values when due allowance is made for accuracy in the published results.

5.1.2-7 Evaluation of Kinetic-Biological Term in Ames-Roberts Model

A typical stone packed trickling filter has the following attributes:

Packing: $A_V = 61m^2/m^3$, $\varepsilon = 0.39$; "Food" $N_{Sc} = 1500$; $Q = 400$ kg/ m²h

The reference temperature is 20°C, maximum temperature is 37.5°C

and Theta $(\Theta) = 1.050$

Now $N_{Re}' = \dfrac{Q}{\mu A_V (1-\varepsilon)} = \dfrac{400}{3.6 \times 61 \times 0.61} = 2.99$

Corrected for viscosity, density

N_{Re}' @ 37.5C $= 2.99 \times 1.4573 = 4.36$

Since $N_{Re}' < 10$

$j_L = \dfrac{1.85}{N_{Re}'^{2/3}}$

j_L @ 37.5C $= \dfrac{1.85}{4.36^{2/3}} = \underline{0.694}$

j_L @ 20C $= \dfrac{1.85}{2.99^{2/3}} = \underline{0.892}$

Corrected for viscosity, density, diffusivity

N_{Sc} @ 37.5 $= 1500 \times 0.4389 = \underline{685.3}$

$k_L = j_L \cdot \dfrac{Q}{\varepsilon} / (N_{Sc}^{2/3} (\dfrac{1-\varepsilon}{\varepsilon})^{1/3})$

k_L @ 20C $= 0.892 \times \dfrac{400}{0.39} / (1500^{2/3} (\dfrac{0.61}{0.39})^{1/3}) = \underline{6.02}$ kg/m²h

k_L @ 37.5C $= 0.694 \times \dfrac{400}{0.39} / (658.3^{2/3} (\dfrac{0.61}{0.39})^{1/3}) = \underline{8.10}$ kg/m²h

At the maximum temperature, $[\dfrac{KX}{\alpha}] \to \infty$, so that $K_m = k_L A_v$

maximum K_{mT} @ 37.5C $= 8.10 \times 61 = 494.3$ kg/m³h

At the reference temperature, 20C

$K_m = K_{mT}/1.05^{37.5-20}$

$K_m = 4943/1.05^{17.5} = \underline{210.5}$ kg/m³h

At any temperature, $[\alpha/KX] = \dfrac{1}{K_m} - \dfrac{1}{K_L A_v}$

$[\alpha/KX] = \dfrac{1}{210.5} - \dfrac{1}{6.02 \times 61}$

so that @ 20C, $[\dfrac{KX}{\alpha}] = 493.6$ kg/m³h

The unique kinetic-biological term coefficient for a particular Theta value can be evaluated from:

$C_B = \dfrac{[\dfrac{KX}{\alpha}] \, \varepsilon \, (\dfrac{1-\varepsilon}{\varepsilon})^{1/3} \cdot N_{Sc}^{2/3} \cdot N_{Re}'^{2/3 \text{ or } 1/2}}{Q \cdot A_v}$

$$C_B = \frac{493.6}{400 \times 61} \cdot 0.39 \cdot (\frac{0.61}{0.39})^{1/3} \cdot 1500^{2/3} \cdot 2.99^{2/3 \text{ or } 1/2}$$

$C_B = \underline{2.488}$ (for $\Theta = 1.05$) for laminar conditions.

By similar calculations, for turbulent conditions, $C_B = \underline{1.804}$

For given flow regime, laminar or turbulent, the unique coefficient is independant of packing characteristics (A_v, ε) and of "food" Schmidt Number. Table 5.4 provides values of this unique coefficient.

TABLE 5.4   Unique Biological Coefficient C_B for $T_{max} = 37.5°C$

Flow Regime	Theta (Θ)							
	1.03	1.04	1.05	1.06	1.07	1.08	1.09	1.10
Laminar	7.543	3.897	2.488	1.748	1.298	0.9976	0.7859	0.6303
Turbulent	6.784	2.987	1.804	1.233	0.8998	0.6842	0.5349	0.4266

For any given packing (A_v, ε), waste (N_{Sc}), appropriate Theta (Θ) for the bio-chemical process (aerobic oxidation, nitrification or denitrification) and flow-rate, the maximum possible value of biochemical term can be determined from:

$$[KX/\alpha] = \frac{C_B \cdot Q \cdot A_v}{\varepsilon} \cdot (\frac{\varepsilon}{1-\varepsilon})^{1/3} / (N_{Sc}^{2/3} \cdot N_{Re}^{2/3 \text{ or } 1/2})$$

Example of Application

Packing: $A_v = 56.2 \text{ m}^2/\text{m}^3$; $\varepsilon = 0.39$; "Food" $N_{Sc} = 1500$; $Q = 600 \text{ kg/m}^2\text{h}$
The reference maximum temperature is 37.5C, $\Theta = 1.04$
$C_B = 3.8975$ for $N_{Re} < 10$ ($N_{Re} = 4.861$)

$$[KX/\alpha] = \frac{3.8975 \cdot 600 \cdot 56.2}{0.39} \cdot (\frac{0.39}{0.61})^{1/3}/(1500^{2/3} \cdot 4.86^{2/3}) = \underline{772}$$

5.1.2-8   Effect of Temperature on Ames-Roberts Model

Most previous authors have adjusted their pseudo-first order rate coefficient by the Streeter-Phelps factor:

$\Theta_{20}^{(T-20)}$ (see Sec.4.13 and Eqn. 4.1.13)

For the Eckenfelder or Atkinson type design equations, the form becomes

$$\frac{C_e}{C_i} = \exp(-k_{20} \cdot \Theta_{20}^{\Delta T} \cdot Z/Q^n)$$

while from the Ames-Roberts form:

$$C_e = C_r + (C_i - C_r) \exp(-K_{mT} \cdot Z/Q)$$

where $\dfrac{K_{mT}}{K_{m20}} = \Theta^{\Delta T}$

K_{mT} = overall mass transfer coefficient at temperature T

K_{m20} = overall mass transfer coefficient at temperature 20°C

Θ = Streeter-Phelps type coefficient

ΔT = temperature difference about the reference 20°C

Example of Application

Eckenfelder (1962) provides data on the performance of a POLYGRID tower treating diluted paper pulp black liquor.   Two sets of results are given here for BOD removed at two flowrates and temperatures over the tower depth.

Data   POLYGRID: A_V = 82 m²/m³; ε = 0.94

Q_I = 3.67m³/m²h, 24°C "a priori" N_{Sc} = 1500

Q_{II} = 7.34m³/m²h, 35°C Θ = 1.03

$$[\tfrac{KX}{\alpha}]_I = 2880 \; ; \; [\tfrac{KX}{\alpha}]_{II} = 4075$$

	Case I	Case II
N_{Re}'	207.2	414.4
j_L	0.07989	0.05649
k_L	5.956	8.423
K_{m20}	417.6	590.56
$1.03^{\Delta T}$	1.255 (4°C)	1.5580 (15°C)
K_{mT}	470.0	920.1

Re-arranging the Ames-Roberts Eqn. 4.4.6

$$\frac{C_e}{C_i} = C_r' + \exp\,(-\,K_m Z/Q) \qquad\qquad (4.4.6(a))$$

where $C_r' = \dfrac{C_r}{C_i} (1 - \exp\,(-K_m Z/Q)$

and again "a priori", C_r' = -0.05 @ Q = 3.67m³/m²h

C_r' = 0.025 @ Q = 7.34 m³/m²h

Substitution of the above coefficients into Eqn.(4.4.6(a)) for depths between 1-6 metres, leads to fraction remaining, as shown here

Z = 1 metre, Q_I = 3.67m³/m²h; Z = 6 metres, Q_{II} = 7.34m³/m²h

$$\frac{C_e}{C_i} = -0.05 + \exp\,(-\frac{470.0\times1}{3670}) \; ; \; \frac{C_e}{C_i} = -0.025 + \exp\,(-\frac{920.1\times6}{7340})$$

$$= \underline{0.830} \qquad\qquad\qquad = \underline{0.446}$$

Similar calculations produced the following results:

Temp.	Depth (m)					
	1	2	3	4	5	6
24°C	0.830	0.724	0.631	0.549	0.477	0.414
35°C	0.857	0.753	0.662	0.581	0.509	0.446

These computed estimates are compared with Eckenfelder's data in Figure 5.1.2.2

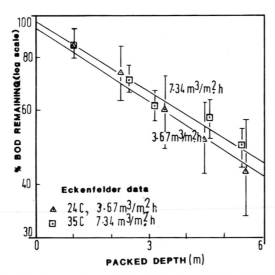

Fig. 5.1.2.2 Comparison of Predicted and Actual Data

5.1.2-9 An Elementary Design Example (Clark et al. (1971))

The raw wastewater flow from a municipality is 5676 m^3/d with an average BOD_5 strength of 180 mg/ℓ. Provide a design estimate of diameter and depth for a low rate trickling filter unit when the final effluent quality has to be better than 20 mg/ℓ BOD and BOD and organic loading not exceed 0.24 kg/m^3d.

Unit Data: Stone packing, A_v = 67 m^2/m^3, ε = 0.39
M.W.R. = 3.75 m^3/m^2d ≡ 156 kg/m^2h
"Food" N_{Sc} = 1750 and $[\frac{KX}{\alpha}]$ = 355 both "a priori"

Primary sedimentation - allow 35 percent BOD removal
Secondary settling - allow 7.5 percent BOD removal
Tower effluent = 21.6 mg/ℓ ; Tower influent = 117 mg/ℓ
C_e/C_i = 0.1846

$$N_{Re}' = \frac{Q}{\mu A_v (1-\varepsilon)} = \frac{Q}{3.6 \times 61 \times 0.61}$$

$N_{Re}' = 0.00747Q$ with $Q > 156$ kg/m²h and $N_{Re}' > \underline{1.17}$

$$j_L = \frac{1.85}{N_{Re}'^{2/3}}$$

$$j_L = \frac{1.85}{(0.00747Q)^{2/3}} \quad (1.671, \ N_{Re}' = 1.17)$$

$$K_L = j_L \cdot \frac{Q}{} / (N_{Sc}^{2/3} \cdot (\frac{1-}{})^{1/3})$$

$$K_L = j_L \cdot \frac{Q}{0.39} / (1750^{2/3} (\frac{0.61}{0.39})^{1/3})$$

$$K_L = 0.0152 j_L \cdot Q \quad (3.96 \ , \ N_{Re}' = 1.17)$$

$$\frac{1}{K_m} = \frac{1}{K_L A_v} + \frac{\alpha}{KX} \quad (143.8 \ , \ N_{Re}' = 1.17)$$

and $\dfrac{C_e}{C_i} = \exp\left(-\dfrac{K_m Z}{Q}\right)$

or $Z = -\log_e \left(\dfrac{C_e}{C_i}\right) \cdot \dfrac{Q}{K_m}$

$$Z = \frac{1.690}{K_m} \text{ metres} \quad (1.833\text{m} \ , \ N_{Re}' = 1.17)$$

Surface Area of Tower = 5676/(24 x 0.156) = <u>1516</u> m²

Volume of packing = 1.833 x 1516 = <u>2779</u> m³

Organic "food" to trickling filter = 5676 m³/d x 0.117 kg/m³ = 664 kg/d

Organic loading ≡ 664/2779 = <u>0.239</u> kg/m³d

Repeating the calculations for 3.84 m³/m²d hydraulic rate, produces the following results:

$N_{Re}' = 1.195$; $j_L = 1.643$; $K_L = 3.999$ kg/m²h

$K_m = 1.446$ kg/m³h ; $Z = 1.870$ m

Surface area = 1478m² ; Volume = 2764 m³

Organic loading = 0.24 kg/m³d

In summary, comparing these preliminary design results with those of Clark et al:

Hydraulic rate m³/m²d	3.84	3.59
Depth of packing (m)	1.87	1.75
Packed Volume (m³)	2764	2769
Surface Area (m²)	1478	1579

Clark et al.

5.1.2-10   Effect of Recycle

The general model described by Eqn. 4.3.5, the Ames-Roberts model described by Eqn. 4.4.6, and the Eckenfelder type Eqn. 4.1.12 each show the same trends on effect of recycle flow on effluent quality. Sec. 4.1.3 compared two empirical, multi-regression formulae and highlighted the different inter-pretations placed on recycle stream.

A perusal of the literature indicates complete lack of agreement on the effect of recycle flow on trickling filter system efficiency as noted in Sec. 2.3.   As discussed in Sec. 4.1.2, the Galler and Gotaas formula show that initially, recycle causes efficiency to increase, go through a maximum, then at high recycle cause efficiency to decrease.   The Hamman formula, however, shows that as recycle increases, efficiency continually increases.

The Eckenfelder type Eqn. 4.1.12 is simplest to demonstrate the effect of recycle flows on systems performance.

$$\text{Let } \frac{C_f}{C_i} = \exp\,(-A/Q^n) \tag{5.2.1}$$

where C_f = tower effluent concentration, kg/m^3

C_i = feed concentration to tower head, kg/m^3

$A = k_1 Z^m$ from Eqn. 4.1.12

$$\text{and } Q = F\,(1+\alpha) \tag{5.2.2}$$

$$C_i = (C_p + \alpha C_f)/(1+\alpha) \tag{5.2.3}$$

where F = flow rate coming to primary settling tank, $m^3/m^2 d$

C_p = raw feed concentration before mixing with recycled effluent, kg/m^3

substituting Eqn. 5.2.2, 3 into 5.2.1 and re-arranging:

$$1 - \frac{C_f}{C_p} = 1 - \frac{\exp\{-A/\,[F(1+\alpha)]^n\}}{1+\alpha\,-\,\alpha\,\exp\{-A/[F(1+\alpha)]^n\}} \tag{5.2.4}$$

which approximately describes the overall system efficiency. Differentiation of Eqn. 5.2.4 leads to a transcendental equation which has no unique analytic solution.   Numerical solution of Eqn. 5.2.4 or its differential form leads to complete understanding of the situation once appropriate values of (k_1, n, Z, Q) are decided upon from Table 4.1.2 and Sec. 2.2.

Figure 5.1.2-3 illustrates trends for the variation of recycle ratio from which the following generalisations can be made:

1.   for low to intermediate rate stone packed filters, any recycle is bene-ficial if the system efficiency is less than about 60 percent or if efficiency is less than 80 percent, α must exceed 6 : 1 to be advantageous;

2.   for intermediate to high rate stone or medium rate synthetic packed filters, any recycle is beneficial if the system efficiency is less than about 80 percent or if efficiency is less than about 90 percent, α must exceed 4 : 1 to be advantageous;

3.   for high rate synthetic packed filters, any recycle is beneficial.

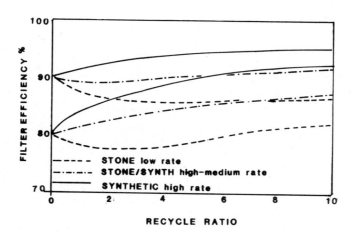

Fig. 5.1.2.3   Effect of Recycle on Filter Performance

The Grady-Lim simulation outlined in Sec. 5.1.1 and illustrated in Figure 5.1.4 shows a similar trend to that of a low rate filter. Overall performance decreases up to a recycle ratio of about 4 : 1, then commences to improve again. Thus, theoretical predictions from the Grady-Lim or Ames-Roberts models lead to the same interpretation of effect of recycle on trickling filter performance.

5.1.3   Sensitivity of Models to Parameter Changes

The general Grady-Lim model reduces to the Ames-Roberts system given the simplifications noted above, although the interpretation is somewhat changed. Disregarding the kinetic parameters inherent in the Grady-Lim approach, the remaining important factors which would need to be considered in any design estimate or simulation are:

1.   temperature, Theta (Θ) and Schmidt Number values for the appropriate biochemical process, and

2.   media characteristics of specific surface area and void fraction.

Typical packings and flowrates were used to assess the model sensitivity, as follows,

a low rate stone media: $A_v = 61m^2/m^3$, $\varepsilon = 0.35$, $Q = 15m^3/m^2d$

and a high rate FLOCOR media: $A_v = 95m^2/m^3$, $\varepsilon = 0.96$, $Q = 40m^3/m^2d$

$\Theta = 1.050$ and Schmidt Number 1500, Temperature = 20°C.

Depth of packing was that necessary to give a fractional effluent, $C_e/C_i = 0.20$, for either of the media.

A one percent increase in factors allowed the change in performance to be noted. Table 5.3 provides a summary of parameter sensitivity where this is defined as:

$$\text{Parameter Sensitivity} = \frac{(\frac{C_e}{C_i})+1\% - (\frac{C_e}{C_i})}{\text{Parameter}} \quad \text{where } \Delta \equiv +1\% \text{ change}$$

TABLE 5.3   Sensitivity of Model Parameters

Parameter	Low Rate - Stone	High Rate - FLOCOR
Temperature	- 0.0157	- 0.0157
Theta	+ 5.59	+ 5.59
Schmidt Number (diffusivity)	+ 0.000143	+ 0.000143
Specific surface area	- 0.00871	- 0.00504
Void Fraction	+ 0.778	+ 1.78

A + sign signifies a decrease in trickling filter efficiency; the dominant parameters are seen to be Theta or effect of temperature on the process (aerobic, anaerobic, nitrification or denitrification) and void fraction of the particular media when covered with bios and irrigated with effluent.

Realistic estimates of the possible errors involved in parameter estimation lead to the following deviations about $C_e/C_i = 0.20$, as shown in Table 5.4,

TABLE 5.4   Estimates of Parameter Errors

Parameter	Error	Low Rate - Stone	High Rate - FLOCOR
Temperature	±1°C	0.20∓0.016	0.20∓0.016
Theta	±0.01	±0.056	±0.056
Schmidt Number	±250	±0.036	±0.036
Specific Surface Area			
Random media	±5m²/m³	∓0.044	
Modular	±1		∓0.005
Void fraction			
Random media	±0.05	±0.039	
Modular	±0.01		±0.018

5.2   Dynamic Simulation of the Trickling Filter System

Steady state determinations for the operation of trickling filter systems under varying conditions can never fully explain major difficulties and unstable performance which may be encountered in full scale facilities. Wastewater treatment plants operate under unusually difficult conditions with regard to large fluctuations of influent concentration and flowrate. Himmelblau (1975) concluded from a stochastic analysis that these two variables were the most significant in altering system performance. Any close examination of entire process operation must recognise the dynamic nature of the coupled primary settling tank, trickling filter unit and secondary settling tank involving the recycle loop if a specified final effluent quality is desired to be maintained.

For *activated sludge* systems, control can be achieved by sludge recycle involving feed-forward or feed-backward control strategies; see Roberts (1978), Grady et al. (1978). This development of the late 1970's and early 1980's was made possible from the examination and subsequent simulation of dynamic response curves obtained from laboratory – pilot and fullscale activated sludge systems.

No significant publications have appeared concerning the modelling of an entire trickling filter system or attempts at simple effluent recycle control strategies. The possible explanation is that of deteriorating performance as recycle ratio is increased, see Sec. 5.1.2.10 and Fig. 5.1.2.3.

5.2.1. Modelling the Trickling Filter System

As pointed out by Grady et al. (1978), it is seldom possible to describe complex physico-chemical and biochemical processes in exact mathematical terms. The usual approach is to use approximating techniques to characterise dynamic process responses. By using pulse testing it has been found that complex processes can be satisfactorily described by models incorporating first or second order over-damped dynamics plus a dead time element.

In the simple simulation to be described, only the substrate "food" concentration will be investigated. Biological solids in the primary and secondary settling tanks will be disregarded as will sloughed bios intermittently escaping from the packed tower. The principal parameter to be considered in the system is the "food" concentration on its passage through each unit as a function of time.

Figure 1.2.1 is the trickling filter system to be modelled with recycled effluent passing to the *inlet* of the primary sedimentation tank or mixed with primary effluent *at the head* of the trickling filter.

BOD is anticipated to be lost in both settling tanks as explained in Sec. 3.5 and illustrated in Fig. 3.3. The amount of BOD removed is a function of influent residence time.

Instantaneous performance of the three units which make up the complete system were taken to be residence times for the settling tanks and biochemical shock delay time for the trickling filter unit.

For deviations away from steady state behaviour, exponental decay functions were considered most appropriate to simulate actual change in "food" concentration with time, as was shown in Sec. 5.1.2-3.

Current simulation techniques use digital computers and software packages such as IBM-CSMP or ICL-SLAM to mathematically describe the entire process and numerically solve a system of possibly coupled differential equations. As previously mentioned, the computer used here was a VAX 11-7/80 and the software package was an in-house *Program for Analogue Modelling*, running in double precision which is a simpler version of SLAM.

5.2.1.1 Modelling the Settling Tanks

The simplest empirical equation to describe performance as in Fig.3.3 was of the form

$$\text{Fractional Efficiency} = 1 - \exp\{-[A_1 + B_1 \log_e (T_{R_1})]\} \tag{5.2.1}$$

where $A_1 = 0.3147$ for primary settling

$\qquad A_1 = 0.1054 \qquad$ secondary settling

$\qquad B_1 = 0.1580$ for primary settling

$\qquad B_1 = 0.06715 \qquad$ secondary settling

$\qquad T_{R_1} =$ residence time for the settling tank $\qquad$ being proportional to inverse of flowrate.

Dynamic response of either settling tank, to describe an exponential decay was of the form;

$$\frac{dc}{dt} = A_2 . B_2 . \exp[B_2(t - T_{R_1})] \tag{5.2.2}$$

where $A_2 = C_t - C_o$

$\qquad C_o =$ initial "food" concentration at zero time

$\qquad C \ \ =$ instantaneous "food" concentration at time $= t$

$\qquad B_2 = \log_e(A_2)/T_{R_1}$

Integration of Eqn. 5.2.2 leads to the typical S shaped exponential decay curve of the form;

$$C_t = C_o + A_2 . \exp[B_2 . (t - T_R)] \tag{5.2.3}$$

5.2.1.2 Modelling the Trickling Filter Unit

The simplest form of the Ames-Roberts model is

$$C_e = C_i . \exp(-K_m . A/Q) \tag{5.2.4}$$

where K_m is evaluated as described in Sec. 5.1.2.

Dynamic response of the unit is again described by an exponential decay function similar to Eqn. 5.2.2 and its integration, Eqn. 5.2.3. The biochemical delay time was considered to be inversely proportional to square root of flow-rate.

5.2.1.3 Simulation of the Combined Units

All the elements of Sec. 5.2.1 are combined as a PAM program, detailed in Fig. 5.2.1. Each segment is self-explanatory.

The problem posed was a trickling filter unit packed to a depth of 7.25 metres with media characteristics; $A_V = 95m^2/m^3$, $\varepsilon = 0.94$. Flowrate is such that

for the primary settling tank, $T_R = 6$ hours

secondary settling tank, $T_R = 2$ hours and

biochemical shock delay time of $T_R = 2$ hours.

For the settling tank, the residence times are inversely proportional to flowrate, while for the trickling filter unit, inversely proportional to square root flow rate.

Feed composition to the filter unit depended upon the recycle ratio and location of mixing point − before or after the primary settling tank.

Two simple studies were investigated, involving:

(a)   recycle of final effluent back to the primary settling tank and

(b)   recycle of final effluent back to the head of the filter unit, after the primary settling tank.

These two recycle strategies were investigated for a 50% increase in organic loading by two different means:

(i)   constant influent flowrate but 50% increase in feed concentration, and

(ii)   constant feed concentration but 50% increase in influent flowrate.

Case 1

The feed to the primary settling tank initially was 181.82 mg/ℓ BOD and was upset by a 50 percent increase to 272.73 mg/ℓ BOD after six hours operation.

Initial performance was as below

Raw influent   181.82 mg/ℓ BOD ≡ 0.903 kg/m³ day
after primary settlement (6 hours) 100.00 mg/ℓ BOD ≡ 0.497 kg/m³ day
after trickling filter   20.00 mg/ℓ BOD
after secondary settlement (2 hours)   17.18 mg/ℓ BOD

Overall BOD removal   =   90.55%

```
    FLOCOR.PAM

      PROGRAM AMES
CONSTANT AV=95.0,E=0.94,DEPTH=7.25,ANSC=1500.0
CONSTANT START=6.0,FINAL=24.0
INPUT  TPRIM,TFIL,TSEC,ALFEED,HIFEED,Q
INPUT CPZERO,CFZERO,CSZERO
PROCEDURAL (ANRE,AJ,AZ=Q,AV,E)
      ANRE=Q/(3.6*AV*(1.-E))
      IF(ANRE.LT.33.0) AJ=1.85
      IF(ANRE.GT.33.0) AJ=1.15
      IF(ANRE.LT.33.0) AZ=2./3.
      IF(ANRE.GT.33.0) AZ=0.5
END
      AJL=AJ/ANRE**AZ
      AKL=AJL*Q/E/(ANSC**(2./3.)*((1.-E)/E)**(1./3.))
PROCEDURAL(CB=ANRE)
      IF(ANRE.LT.33.0)  CB=7.543
      IF(ANRE.GT.33.0)  CB=6.7844
END
      BIOL=CB*Q*AV/E*((E/(1.-E))**(1./3.))/(ANSC**(2./3.)*ANRE**AZ)
      RECIPR=1./(AKL*AV)+1./BIOL
      AKM=1./RECIPR
DERIVATIVE(FEED,FILEF,CSEC=TIME,START,PRIMEF,AKM,DEPTH,Q,ACTEFF,TSEC)
      IF(TIME.LT.START)  FEED=ALFEED
      IF(TIME.GE.START) FEED=HIFEED
      FILEF=PRIMEF*EXP(-AKM*DEPTH/Q)
      CSEC=ACTEFF*EXP(-0.10536-0.067151*LOG(TSEC))
END
DERIVATIVE(PRIMS,A1,A2,B1,B2=TPRIM,CPZERO,FILEF,CFZERO,TFIL)
      PRIMS=FEED*EXP(-0.3147-0.1580*LOG(TPRIM))
      A1=PRIMS-CPZERO
      B1=LOG(ABS(A1))/TPRIM
      A2=FILEF-CFZERO
      B2=LOG(ABS(A2))/TFIL
END
DERIVATIVE(A3,B3=CSEC,CSZERO,TSEC)
      A3=CSEC-CSZERO
      B3=LOG(ABS(A3))/TSEC
END
DERIVATIVE(DIBDT,DCBDT=A1,B1,TIME,START,A2,B2,TPRIM)
      DIBDT=A1*B1*EXP(-B1*(TIME-START))
      IF(TIME.LE.START)  DIBDT=0.0
      DCBDT=A2*B2*EXP(-B2*(TIME-(TPRIM+START)))
      IF(TIME.LE.(START+TPRIM))  DCBDT=0.0
END
DERIVATIVE(DEBDT=A3,B3,TIME,START,TPRIM,TFIL)
      DEBDT=A3*B3*EXP(-B3*(TIME-(TPRIM+START+TFIL)))
      IF(TIME.LE.(START+TPRIM+TFIL))  DEBDT-0.0
END
      PRIMEF=INTGRL(DIBDT,CPZERO)
      ACTEFF=INTGRL(DCBDT,CFZERO)
      FINEFF=INTGRL(DEBDT,CSZERO)
CINTERVAL  1.0
  INTGSTEPS  5
OUTPUT  TIME,FEED,PRIMEF,ACTEFF,FINEFF
TERMINATE(TIME.GE.FINAL)
REPEAT
END
```

Fig. 5.2.1 Details of the PAM program

After the upset, at infinite time:

Raw influent   272.73 mg/ℓ BOD ≡ 1.354 kg/m³ day
after primary settlement   150.0   mg/ℓ BOD ≡ 0.745 kg/m³ day
after trickling filter   30.0   mg/ℓ BOD
after secondary settlement   25.7   mg/ℓ BOD

Overall BOD removal   =   90.55%

Case 2

       The feed to the primary settling tank was constant at 181.82 mg/ℓ BOD. After six hours operation, the influent flowrate was increased by 50 percent.

       Following the upset and maintaining the same residence times, at infinite time:

Raw influent   181.82 mg/ℓ BOD ≡ 1.354 kg/m³day
after primary sedimentation (6 hours) 100.00 mg/ℓ BOD ≡ 0.745 kg/m³day
after trickling filter   26.88 mg/ℓ BOD
after secondary settlement (2 hours)   23.09 mg/ℓ BOD

Overall BOD removal   =   87.30%

       Four complete simulations were investigated:

1.   a 50 percent feed concentration increase with recycle back to mix with influent going to the primary settling tank,

2.   a 50 percent feed concentration increase with recycle back to mix with primary effluent going to the head of the trickling filter,

3.   a 50 percent flowrate increase with recycle back to mix with influent going to the primary settling tank, and

4.   a 50 percent flowrate increase with recycle back to mix with primary influent going to the head of the trickling filter.

       Recycle ratios were between zero and 4 : 1 while residence times were held between "best" and "worst" conditions.

       "Best" residence times envisaged constant six hours holding time for primary sedimentation and two hours for secondary settlement, irrespective of influent flowrate or recycle ratio. These conditions can be achieved practically by flow equalisation in primary sedimentation and floating head offtakes in variable volume conical or pyrimidal secondary settling tanks.

       "Worst" residence times would be encountered where there is no provision for flow equalisation in primary sedimentation and constant volume secondary settling tanks which would result in residence times decreasing with increase in flowrate.

A typical output is shown below:

Time (hours)	Feed	Primary Effluent (mg/ℓ)	Filter Effluent	Final Effluent
0	181.2	100.0	20.0	17.1
2.0	181.2	100.0	20.0	17.1
4.0	181.2	100.0	20.0	17.1
6.0	272.7	100.0	20.0	17.1
8.0	272.7	136.4	20.0	17.1
10.0	272.7	146.3	20.0	17.1
12.0	272.7	149.0	20.0	17.1
14.0	272.7	149.7	28.8	17.1
16.0	272.7	149.9	29.7	24.1
18.0	272.7	149.9	29.8	25.0
20.0	272.7	150.0	29.8	25.1
22.0	272.7	150.0	29.8	25.1
24.0	272.7	150.0	29.8	25.1

Thirty-eight simulation runs were made which are summarised in Table 5.2.1. The first set of results are for a 50 percent increase in influent feed concentration, no recycled effluent and noting the effect of increase in influent flowrate.

As previously outlined in Case 1, 2 situations, at *infinite* time the final effluents for + 50 percent increase in influent concentration or flowrate leads to:

for + 50% feed conc:- 25.77 mg/ℓ compared to 25.2 mg/ℓ in the simulation

for + 50% flowrate:- 23.09 mg/ℓ compared to 22.6 mg/ℓ in the simulation.

TABLE 5.2.1   Summary of Simulation Runs - Final Effluent Concentrations

Flowrate	Best T_{rs}	Worst T_{rs}	
36m³/m²d	25.2mg/ℓ	25.2mg/ℓ	+ 50% Feed concentration
72	40.5	47.4	no recycle.
108	50.1	64.2	

Recycle Ratio	Best T_{rs}	Worst T_{rs}	
0	25.2mg/ℓ	25.2mg/ℓ	+ 50% Feed concentration
1	21.7	25.3	Qi = 36m³m²d
2	18.8	24.0	Recycle back to mix with influent
4	17.4	22.0	*before* primary sedimentation.
8	16.0	20.2	

Recycle Ratio	Best T_{rs}	Worst T_{rs}	
0	25.2mg/ℓ	25.2mg/ℓ	+ 50% Feed concentration
1	22.5	23.4	Qi = 36m³/m²d
2	20.3	21.6	Recycle back to mix with primary
4	16.4	19.4	effluent *after* primary settlement.
6	15.8	19.0	

continued/..

Recycle Ratio	Best T_{rs}	Worst T_{rs}
0	22.6mg/ℓ	24.5mg/ℓ
1	18.1	23.3
2	16.1	21.8
4	13.9	19.9
8	13.6	18.6

+ 50% influent flowrate
C_i = 181.81 mg/ℓ BOD
Recycle back to mix with influent
before primary sedimentation.

Recycle Ratio	Best T_{rs}	Worst T_{rs}
0	22.6mg/ℓ	24.5mg/ℓ
1	19.4	21.7
2	18.0	19.9
4	17.2	18.8

+ 50% influent flowrate
C_i = 181.82 mg/ℓ BOD
Recycle back to mix with primary
effluent *after* primary settlement.

The empirical exponential decay functions incorporating the delay times for each plant unit, primary sedimentation, trickling filter and secondary settling tank are the cause for discrepancies between *infinite* time results and *real time* performance.

Fig.5.2.2 shows a typical simple output for the situation for a 50 percent feed change and no recycle.

Time (hours)	Feed	Primary Effluent (mg/ℓ)	Filter Effluent	Final Effluent
0	181.8	100.0	20.0	17.1
2.0	181.8	100.0	20.0	17.1
4.0	181.8	100.0	20.0	17.1
6.0	272.7	100.0	20.0	17.1
8.0	272.7	136.4	20.0	17.1
10.0	272.7	146.3	20.0	17.1
12.0	272.7	249.0	20.0	17.1
14.0	272.7	149.7	28.8	17.1
16.0	272.7	149.7	29.7	24.1
18.0	272.7	149.9	29.8	25.0
20.0	272.7	150.0	29.8	25.1
22.0	272.7	150.0	29.8	25.1
24.0	272.7	150.0	29.8	25.1

Fig.5.2.2   Typical PAM simulation output

The influent concentration increases after six hours, but the primary settling tank effluent takes another six hours to approach steady state. Because of the biochemical shock delay time involving the trickling filter unit, the filter effluent takes two hours to approach steady state. A residence time of two hours in the final settling tank causes the final effluent passing from the system to take six hours to approach steady state. Overall the delay is approximately twelve hours between feed concentration change in influent to final effluent reaching a near constant concentration.

Fig.5.2.3 shows a simple lineprinter output for the situation of a 50 percent flowrate increase and no recycle but with the best residence times for primary and secondary settlement.

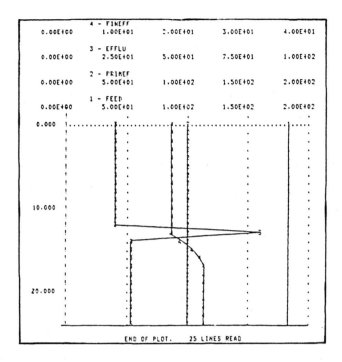

Fig.5.2.3   PAM simulation output for best residence times

The influent concentration remains constant throughout the simulation and with the best residence times for primary and secondary settlement, the primary effluent also remains constant, independent of flowrate. Trickling filter effluent is shock delayed by about two hours so that it takes approximately three and a half hours to approach steady state. As before, the final settling tank takes about five hours to approach steady state.   Overall, the delay in final effluent response to reach a near constant concentration is about eleven hours.

To show the effect of recycle back to mix with primary effluent, a simple output is given in Fig. 5.2.4.   This illustrates a 50 percent feed concentration increase, best residence times for primary and secondary settlement and a recycle ratio of 1:1.

After six hours steady state operation, the influent concentration increases.   Recycled final effluent is returned to mix with primary settling tank effluent and this is pumped to the trickling filter.   After a biochemical shock response, the filter effluent recovers to steady out while the final effluent coming from the secondary settling tank takes about five hours to reach near steady state.   The overall delay time is about twelve hours in

318

in alteration in feed concentration and recycle ratio to near constant final effluent concentration.

The effect of a 50 percent increase in influent flowrate for otherwise constant feed concentration is shown on Fig. 5.2.5. Worst residence times were involved, with recycle back to mix with influent before the primary settling tank.

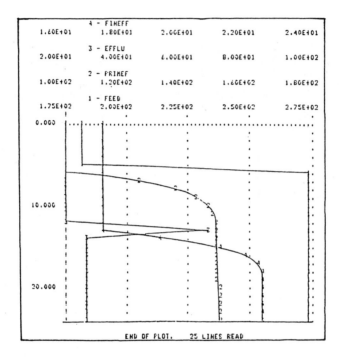

Fig.5.2.4 PAM output plot for recycle back after primary settling

In this situation the influent is constant for six hours, then a change in flowrate and a recycle ratio of 1:1 is imposed, mixing final effluent with raw influent. Primary effluent coming from the primary settling tank takes about three hours to reach steady state. After a biochemical shock response of near two hours, the trickling filter effluent recovers rapidly. The final effluent coming from the secondary settling tank takes about three hours to reach near steady state, while the overall delay is about six hours from step change in flowrate to near constant final effluent concentration.

These simulation results of Table 5.2.1 are shown for comparison in Fig.5.2.6. An influent *concentration* change together with recycle ratio and a mixing point is marginally worse than an influent *flowrate* change. The 'desired' concentration is that of the initial steady state response before a

step change.   Only by increasing the residence times, especially in the primary
settling tank, would recycle match desired performance when coping with a step
concentration increase.   If organic load increase is principally due to flowrate, then recycle back to mix with influent *prior* the primary settling tank
would seem to be the preferred route.

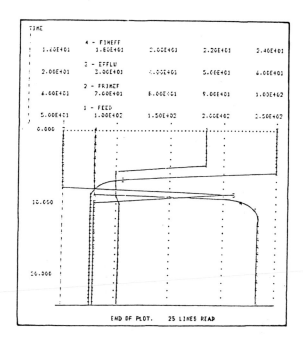

Fig.5.2.5   PAM output plot involving worst residence times

However, if organic load increase is principally due to feed concentration, recycle back to mix with primary effluent *coming from* the primary
settling tank would seem the better route.   With constant residence time in
the primary settling tank, independent of recycle flowrate, the fraction of
BOD removed remains constant.   This primary effluent when blended with a
relatively high recycle flow rate results in good quality final effluent.

An *increase* in residence times would move all the curves of Fig.5.2.6
downward to the left while a *decrease* in biological shock delay response time
would push all curves marginally upward to the right.

These findings which have been arrived at from a simple dynamic
simulation, confirm and amplify on the A.S.C.E. Manual of Practice No.36
(1977).   In part the Manual notes "fluctuations in the organic loading applied
to the filter are dampened by (a number of recycle) schemes.   The extent of
such dampening varies with the duration of organic load in the raw wastewater

320

and with the time the recirculated flow is retained in the settling tanks
through which it passes.   For stone medium trickling filters, the amount of
recirculation, not the system for obtaining it, is the important factor."

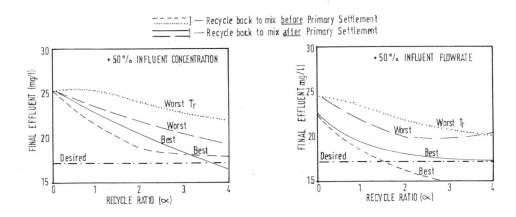

Fig.5.2.6   Predicted effect of recycle ratio on system performance

6.   SUMMARY

Elements of the information contained in Section 1-3 concerning operating and design parameters have been utilised in some of the mathematical
models outlined in Section 4.   The general model developed by Grady and Lim
may be useful in interpreting process performance.   This chemical engineering
science approach takes into account intrinsic biochemical rate of reaction in
the biofilm, diffusion in the film, liquid phase mass transfer and Monod type
kinetic coefficients, together with media characteristics.

From the model calculations it is possible to determine whether mass
transfer or biochemical rate is the controlling rate mechanism.   This Grady-
Lim model requires six coefficients to be independently determined for a
particular influent and the hydrodynamic conditions involved in trickle flow
over a particular media.

Recent research indicates that a surface reaction is involved with
liquid phase mass transfer from the bulk liquid into the surface bios.
This simplification of the Grady-Lim approach leads to a model which can be
developed from a quite different hypothesis;   the Ames-Roberts two phase film
model, requiring only two factors to be determined.   Physical chemical data
provide an estimate of the "food" Schmidt Number and the biochemical term can
be calculated from a knowledge of the maximum temperature of biological
activity.

It has been shown in Section 5 that the Ames-Roberts model is capable of accurately describing the process under steady state or dynamic conditions and make useful interpretations of change in performance.   The riddle of effect of recycle in filter operation has been clearly explained using this model, especially the role of the primary settling tank, by means of dynamic simulation.

Simple control of the trickling filter system by pumping back final effluent to mix with primary settling tank effluent at a relatively high recycle ratio seems the best of a number of strategies.

7. REFERENCES

Amado M.A., (1964). Analysis of BOD reduction in trickling filters, Master's Thesis, Cornell University.
Ames W.F., (1962). Transient operation of trickling filters. J.San.Eng.Div. ASCE, 88, 21-38.
American Soc.Civil Eng., (1977). Wastewater Treatment Plant Design. Manual of Practice No.36. Chapter 15, Biological Filters pp.283-320.
Arora H.C. and Chattopadhya S.N., (1980). Anaerobic contact filter processes for the treatment of vegetable effluents. J.Wat.Poll.Control, 79, 501-506.
Atkinson B., (1974). Biochemical Reactors. Pion Press, London, pp.80-88, 168-175.
Balakrishnan S., (1969). Organic removal by a selected trickling filter media. Water and Wastes Eng., 6, A22-25.
Barth, E.F., (1964). Effect of a mixture of heavy metals on sewage treatment processes. Proc.19th Ind. Waste Conf., Purdue University, pp.616-635.
Baumann E.R. and Stracke R.J., (1975). Biological treatment of a toxic industrial waste. Proc. 30th Ind. Waste Conf., Purdue University, pp.1131-1160.
Beg S.A., (1982). Inhibition of nitrification by arsenic, chromium and fluoride. J.Wat.Poll.Control Fed., 54, 482-488.
Benefield L.D. and Randall C.W., (1980). Biological Process Design for Wastewater Treatment. Prentice-Hall Inc., Eaglewood Cliffs, pp.24-30, 62-74, 87-98, 391-454.
Blain A.W. and McDonnell H.J., (1965). Discussion of Galler and Gottas paper. J.San.Eng.Div.ASCE, 91, SA4 57-61.
Blersch H.C., (1980). Conventional biological filters at ultra high loadings. Proc.Wat.Tech., 12, 243-249.
Brown G.G. (Editor), (1950). Unit Operations. John Wiley & Sons. pp.210-228.
Bruce A.M., (1970). Some factors affecting the efficiency of high rate biological filters. Proc.6th Int.Conf.Wat.Poll.Res. II, 1-8.
Bruce A.M. and Merkens J.C., (1973). Further studies of partial treatment by high rate biological filtration. J.Wat.Poll.Control, 72, 499-527.
Castaldi F.J. and Malina J.F., (1982). Velocity dependant reaction rates in a slime reactor. J.Wat.Poll.Control Fed., 54, 261-269.
Clark J.W. (Editor), (1971). Water Supply and Pollution Control. 2nd Ed., International Textbook Co., New York. pp.345-350, 379, 484-506.
Cook E.E. and Herning L.P., (1978). Shock load attenuation trickling filters. J.Envir.Eng.Div.ASCE., 104, 461-469.
Dennis N.D. and Jennett J.C., (1974). Pharmaceutical waste treatment with an anaerobic filter. Proc. 29th Ind.Waste Conf., Purdue University, pp.36-43.
Dosh T., (1974). Carbon oxidation-nitrification in synthetic media. J.Wat.Poll. Control Fed., 46, 2327-2339.

Eckenfelder W.W. and Barnhart E.L., (1962). Treatment of pulp and paper waste for high rate filtration using plastic media. Proc. 17th Ind.Waste Conf., Purdue University, pp.105-115.

Eckenfelder W.W., (1961), Trickling filter design and performance. J.San.Eng. Div.ASCE, 87, SA6, 33-45.

Eckenfelder W.W. (Editor), (1971). Design Guides for Biological Wastewater Treatment Processes. U.S. Envir.Prot.Agency, Washington, 11011ESQ 08/71, p.163.

Ellis K.V., (1980). The tertiary treatment of sewage. Effl. & Wat.Treat. J., 20, 527-537.

Elmaleh S., (1978). Biological filtration through a packed bed. Wat.Res. 12, 41-46.

Fink D.J., (1973). Effectiveness factor calculations for immobilized enzyme catalysts. J. Biotech & Bioeng., 15, 878-888.

Frechter A. (Editor), (1982). Advances in Biochemical Engineering, Vol.24. Springer-Verlag, Berlin, Chapter 2, Reaction Engineering Parameters for immobilised biocatalysts, pp.39-71.

Galler W.S. and Gotaas H.B., (1964). Analysis of biological filter variables. J.San.Eng.Div.ASCE, 90, 59-79.

Genung R.K., (1980). Energy conservation and scaleup studies for a wastewater treatment system based on a fixed film anaerobic bioreactor. Biotech. & Bioeng. Symp., No.10. John Wiley & Sons, New York, pp.295-316.

Grady C.P.L. and Lim H.C., et al. (1978). Automatic control of the activated sludge process. Wat.Res., 12, 81-99.

Grady C.P.L. and Lim H.C., (1980). Biological Wastewater Treatment. Marcel Dekker Inc., New York, pp.509-583, 715-753.

Hammam S., (1968). An investigation into the performance of a trickling filter plant. Stuttg.Ber.SiedWasserw. No.36.

Hemming M.L., (1971). Biological treatment in the dairy industry. Envir.Poll. Management, October. p24-29.

Howland W.E., (1953). Effect of temperature on sewage treatment processes. Sew. & Ind. Wastes., 25, 161-170.

Howland W.E., (1958). Flow over porous media as in a trickling filter. Proc.12th Ind.Waste Conf., Purdue University, pp.435-465.

Howland W.E., (1960). Kinetics in trickling filters. 3rd Annual Conference on Biological Waste Treatment. New York, pp.233-248.

Himmelblau D. and Mistry K.J., (1975). Stochastic analysis of trickling filters. J. Envir. Eng. Div. ASCE., 101, 330-350.

Hirose T. and Mori Y., (1976). Liquid to particle mass transfer in a fixed bed reactor. J.Chem.Eng., Japan, 9, 220-225.

Hoehn R.C. and Ray A.D., (1973). Effects of thickness on bacterial film. J.Wat.Poll.Control Fed., 45, 2302-2315.

Imhoff K., (1971). Disposal of Sewage and other Waterborne Wastes. 2nd Ed. Butterworths & Co., London, p.204.

Jackson S. and Brown V.M., (1970). Effect of toxic wastes on treatment processes and watercourses. J.Wat.Poll.Control, 69, 292-313.

Jenkins S.H. (Editor), (1973). Aspects of Environmental Protection. William Clowes and Sons, London, p.38.

Jennings P.A., (1976). Theoretical model for a submerged biological filter. J.Biotech. & Bioeng., 18, 1249-1273.

Kataoka T. and Yoshida H., (1972). Mass transfer in laminar region between liquid and packing material surface in the packed bed. J.Chem.Eng.Japan, 5, 132-136.

Keffer C.E. and Meisel J., (1952). Remodelling trickling filters exceeded expectations. Wat. & Sew. Works, 100, 277-279.

Kennedy K.L. and Vanden Berg L., (1982). Stability and performance of an anaerobic fixed film reactor. Wat.Res., 16, 1391-1398.

Kong M.F. and Yang P.Y., (1979). Practical design equation for trickling filter process. J.Biotech. & Bioeng., 21, 417-431.

Kornegay B.H. and Andrews J.F., (1968). Kinetics of fixed film biological reactors. J.Wat.Poll.Control Fed., 40, 460-468.

La Motta E.J., (1976). External mass transfer in a biological film reactor. ibid 18, 1359-1370.

La Motta E.J., (1976). Kinetics of growth and substrate uptake in a biological film system. J.Appl.Envir.Micros., 31, 286-293.

La Motta E.J., (1976). Internal diffusion and reaction in biological films. J.Envir.Sc.Tech., 10, 765-769.

Liptake B.G., (1974). Environmental Engineers Handbook, Vol.1. Chilton Book Co., Radnor, pp.439-444, 615-633, 823-835.

McCarty P.L. and Rittman B.E., (1981). Substrate flux into biofilms of any thickness. J.Envir.Eng.Div.,ASCE., 107, 831-849.

Mitchell R. (Editor), (1978). Water Pollution Microbiology, Vol.2. John Wiley & Son, New York, Chapter 4, Biofilm Kinetics, pp.71-109.

Miyauchi T., (1971). Film coefficients of mass transfer of dilute sphere packed beds in low flowrate regime. J.Chem.Eng.Japan, 4, 238-245.

Miyauchi T. and Nomura T., (1972). Liquid film mass transfer coefficient for packed beds in the low Reynolds number region. Int.Chem.Eng.J., 12, 360-366.

Moodie S.P. and Greenfield P.F., (1979). Substrate removal kinetics in trickling filters. J.Wat.Poll.Control Fed., 51, 1063-1067.

Mosey F.E., (1978). Anaerobic filtration: a biological treatment process for warm industrial effluent. J.Wat.Poll.Control, 77, 370-378.

Mutharasan R., (1978). An approximate solution to the theoretical model of a submerged biological filter. J.Biotech. & Bioeng., 20, 151-156.

National Research Council, (1946). Sewage treatment at military installations. Sew.Works J., 18, 787-1028.

Onda K.E., (1967). Wetted surface of packing material. Kagaku Kogaku 31, 126-129.

Quirk T.K., (1972). Scale-up and process design techniques for fixed film biological reactors. Wat.Res., 6, 1333-3160.

Rincke G., (1967). Neure gesichtspunkte zur Abwasserreinigung mit Tropfkorpern. DasGas und Wasserfach., 108, 667-673.

Roberts J., (1973). Towards a better understanding of high rate biological film flow reactor theory. Wat.Res., 7, 1561-1588.

Roberts J., (1973). A postulated model for high rate biological film flow reactors. 1st Aust.Conf.Heat & Mass Transfer. Sec.6., p57-64.

Roberts J., (1975). Fixed film reactor performance and simulation for nitrification-denitrification of organic wastes. 3rd Aust.Conf.Heat & Mass Transfer. Sec.T, p.1-18.

Roberts J., (1978). Dynamic performance of slurry biological reactor systems. Univ. Newcastle, N.S.W., ISBN 0725902469, 44 pages.

Sachs E.F., (1982). Pharmaceutical waste treatment by anaerobic filter. J.Envir.Eng.Div.ASCE., 108, 297-314.

Sarner E., (1981). Removal of dissolved and particulate organic matter in high rate trickling filters. Wat.Res., 15, 671-678.

Schroeder E.D. and Tchobanoglous G., (1975). Another look at the NRC formula. Wat. & Sew. Works, 122(7), 58-60.

Schroeder E.D. and Tchobanoglous G., (1976). Mass transfer limitations on trickling filter design. J.Wat.Poll.Control Fed., 48, 771-775.

Schulze K.L., (1960). Load and efficiency of trickling filters. J.Wat.Poll. Control Fed., 32, 245-261.

Shriver L.E. and Bowers D.M., (1975). Operational practices to upgrade filter plant performance. ibid 47, 2640-2651.

Stones T., (1980). The oxygen demand of sewage. Effl. & Wat.Treat.J., 20, 437-440.

Truesdale G.A. and Eden G.E., (1963). Comparison of media for percolating filters. Inst. Public Health Eng.J., 62, 283-302.

Upadhyay S.N. and Tripathi G., (1975). Liquid phase mass transfer in fixed and fluidised beds of large particles. J.Chem.Eng.Data, 20, 20-26.

Vanden Berg L. and Kennedy K.L., 91982). Comparison of intermittent and continuous loading of stationary fixed film reactors for methane production. J.Chem.Tech.Biotech., 32, 427-432.

Velz C.J., (1948). A basic law for the performance of biological filters.
    Sew.Works J., 20, 607-617.
Water Pollution Research, (1965). Oxygen requirements of percolating filters.
    British Gov., HMSO, p51.
Water Research Centre, (1974). Removal of nitrate from sewage effluents by
    biological methods. Notes on Water Pollution, No.66, HMSO.
Water Research Centre, (1978). The design of percolating filters and rotary
    biological contractors. British Gov., HMSO, p.44.
Webster I.A., (1981). Criteria for the prediction of diffusional control
    within whole cells and cell flocs., J.Chem.Tech.Biotech., 31, 178-182
Williamson K. and McCarty P.L., (1976). A model of substrate utilisation by
    bacterial films. J.Wat.Poll.Control Fed., 48, 9-24.
Williamson K. and Meunier A.D., (1981). Packed bed biofilm reactors. Part I,
    Simplified Model; Part II, Design. J.Envir.Eng.Div.ASCE., 107, 307-337.
Winkler M.A., (1981). Biological Treatment of Wastewater. John Wiley & Sons,
    New York. Chapter 6, Biological film systems. pp.31-37, 164-210, 226-269.
Witt E.R., (1979). Full scale anaerobic filter treats high strength wastes.
    Proc.34th Ind.Waste Conf., Purdue University, pp.229-234.

A LOADING MODEL FOR BIOLOGICAL ATTACHED GROWTH REACTORS - DEVELOPMENT AND APPLICATION

JAN A. OLESZKIEWICZ
Department of Civil Engineering
University of Manitoba
Winnipeg, Canada, R3T 2N2

1. INTRODUCTION

Standard engineering design textbooks such as Metcalf & Eddy (1979) as well as more specialized books on biological treatment (Benefield, et al., 1980) appear at a loss when it comes to a unified trickling filter design formula. Frequently, as many as three formulas are recommended; the NRC (1946), Eckenfelder's (1961) and Galler-Gotas (1969) equations have traditionally been chosen. The statistically developed NRC formula appears to be the preferred choice among designers.

Table 1 shows a comparison of expected performance of these popular models in various effluent recycle conditions. The differences in volumes calculated by these formulas for the same conditions reach 600%. The conditions are flow of 3800 m^3/d, influent BOD = 100 mg/l, removal efficiency E = 80%. The

TABLE 1

Comparison of trickling filter volumes

Recycle Ratio	Volume in (m^3) According to Formula:		
N	NRC (1946)	ECKENFELDER (1963)	GALLER & GOTAAS (1964)
0	1480	2150	10,000
1	930	560	850
3	670	150	160

recycle effects predicted by these formulas are not modelled similarly. The practice indicates improvement usually proportional to the change of concentration of influent BOD_5 when treating strong industrial wastewater. With usually dilute sewage these differences could show perhaps a 20-25% improvement.

Currently in research, as well as in engineering there is lack of agreement on effects of the fundamental variables and parameters in biological filtration, such as the effects of hydraulic load, effects of recirculation, filter depth and organic loading. Presently existing models usually fail to represent the changes effected by variations in these parameters. There is evidence available in literature to both prove and disprove the benefits of recirculation, the detrimental effects of increased hydraulic loading, and increased concentration. These differences may be due to different unit processes responsible for removal in various conditions. Thus, the transfer of data from one plant to another is not valid, particularly when treating industrial wastewater. The large amount of operational data gathered recently for municipal treatment plants has also shown a wide spread of removals - in spite of similarity of operational conditions (Pierce, 1978). Further modelling problems are due to the divergence of prevailing chemical reactor theory and actual performance of biological reactors, as noted by Busch (1984).

1.1 Aim and scope

The development and application of the design formula relating removal efficiency to a function of volumetric organic loading applied will be presented. The design procedure based on the formula includes the following parameters: hydraulic load (Q), depth (H), specific surface area (A) and influent concentration (S_0). Application of the formula to a variety of biofilters, both aerobic and anaerobic, will demonstrate that removal efficiency is proportional to the exposed surface area of the slime and that the differences in performance of, say, fluidized bed and stationary bed biofilters are due to the order of magnitude difference in the surface area, A.

No attempt at development of mechanistic mass transfer relationships will be made here although a brief discussion will be given. The developed model and design procedure assumes that in every case treatability studies are made. If literature data are used then only very similar wastewaters, concentrations and media types can be compared. As shown by many full scale failures, literature-based design cannot predict the performance of a biofilter in a range of loading conditions.

2. MECHANISTIC VERSUS EMPIRICAL MODELS

Industrial waste data were found to be poorly modelled by traditional empirical formulas and various pseudo-mechanistic correlations were used. Schulze's (1957) depth related model, still used by some media manufacturers, is one example.

The search for mechanistic models was based on actual transport and conversion phenomena into and in the biofilm. Usually modelled separately, under an assumption that the given modelled activity is rate determining, the following were included:

1. The kinetics of the biochemical reactions which reduce the biochemical oxygen demand in the wastewater;
2. The transfer of oxygen from the gas phase to and through the liquid phase to the bacterial sites on the media;
3. The transfer of organic compounds and nutrients from the liquid to the bacterial slime on the trickling filter media.

A thorough review of mechanistic models was given recently by Grady (1982). It suffices to say that the models deal with definition of the hydraulic regime in the wastewater-air and wastewater-slime interfaces; with definition of qualitative and quantitative stratification of the biomass in the reactor; flux of electron acceptors and donors; and reaction mechanisms within the slime.

The problem is further complicated by 1) the nature of organic substrate - its polarity and ameanability to sorption, its rate of biodegradation and rate of degradation of metabolic products of primary breakdown; 2) the heterogeneous nature of the biofilter in conditions of recycle and resulting stratification in biomass diversity and viability; 3) changes of the flux of substrate (electron donors) and oxygen (principal electron acceptor in aerobic, high surface irrigation rate biofilters) due to stochastic variability of the incoming load; 4) biofilter tower dispersion characteristics - i.e. effects of internal media configuration as well as the specific surface area $A(m^2/m^3)$; 5) various media and wastewater factors affecting the attachment and performance of the biological slime (Eighmy, et al., 1983 and Pringle et al., 1983); 6) changing role of adsorption and absorption mechanisms versus the biodegradation mechanisms in the biofilm. In studies with industrial effluents containing biodegradable solids, they were frequently recovered virtually unchanged in sloughed biofilm (Wheatley, 1979).

Few of these factors are taken into consideration in mechanistic models derived from studies on defined surface area, well described slime thickness and monosubstrates that are unlike the real wastes (e.g. glucose). To complicate this, it has been demonstrated in full scale that slime thickness and partitioning into anaerobic and aerobic layers are not "following" the hypothesis. The fungal slimes developed on industrial wastewater, some 3-5 mm thick, were mostly aerobic with very thin anaerobic sections (Oleszkiewicz & Eckenfelder, 1974; Williams, I.L., 1979).

This changing performance with variation in the hydraulic load has been illustrated recently by Muslu (1983). He has shown applicability of various models to the same filter as the hydraulic regime changed from laminar to completely mixed (turbulent).

In summary, the three-phase flow problem in chemical reactors, usually involving one or a few exactly defined monosubstrates and catalysts, still remains only vaguely defined in mechanistic terms. The biochemical reactions in the slime, are inherently more complex in terms of bulk mass transfer and micro-scale enzyme kinetics. Chances are, that the exact explanation of the fixed film biological system will continue to defy modelling attempts, directing designers to predictions based on (alas!) literature reference data or, preferably, pilot studies.

All these factors, coupled with difficulty in modelling three-phase systems even in purely chemical-physical operations, have led many (e.g. Sykes, 1984) to conclude that mechanistic models usually cannot be verified and that each treatment facility is unique and should be modelled through regression.

The model presented here is based on similar contention.

3. HYDRAULIC REGIME

Some of these variables are quantified by dimensionless numbers such as Peclet number (Pe) which is equal to the product of Reynolds number (Re) and Schmidt number (Sc). The Reynolds number defines the hydraulic regime in the thickness of laminar sublayer at slime-liquid interface. Schmidt number defines diffusivity of the organic substrate. Unfortunately the thickness of the liquid sublayer formed on the slime surface is dependent not only on the hydraulic regime but also on the Schmidt number of the quantity being transported.

The matter of adequate mechanistic presentation of the biofilter performance under transient conditions is further complicated by drastic changes in resistance to the flux of electron donors and electron acceptors at changing hydraulic loading. The Reynolds number above 4 to 25 are considered to represent the so called wavy or rippled flow with back-mixing and that, from the standpoint of mass transfer denotes turbulent flows (Popadic, private communication, 1974). This means that when the flow rate drops to Re below 4 to 25 the resistance to mass transfer increases proportionally to the ratio of turbulent to laminar diffusivity coefficients.

These hydraulic regime variations will influence accordingly the thickness of slime layer (by attrition and/or unloading) and hence its activity and depth of substrate penetration.

4. DEVELOPMENT OF THE LOADING MODEL

The model has been derived based on dimensional analysis, tests in over thirty curve fitting exercises and expanded into a Maclaurin series to show how it relates to the variables modelled by other empirical formulations (e.g. ICI Load-Efficiency Correlation). The load model differs from other empirical or statistically derived models in one respect: it does not define directly the "treatability" of the wastewater treated. Rather, it lumps together the effects on effluent quality of all parameters considered pertinent, i.e. influent substrate concentrations, wetting rate or hydraulic loading, depth and specific surface area of the media after it is coated with slime.

4.1 Definition of variables through dimensional analysis

In a study of pharmaceutical wastewater treatment a set of numbers from several runs consistently defied correlation by zero, first or second order models (Oleszkiewicz and Eckenfelder, 1974). Dimensional anlaysis was then employed. Mass transfer approach was applied, neglecting the effects of surface factors. From the functional relationship relating the height of transfer unit (HTU) and other variables involved, two dimensionless numbers appeared: the Schmidt and Reynolds numbers. As fundamental research into film flow has shown that liquid film wave inception threshold begins at Re $\geq$ 20 (Popadic, private communication, 1974) the traditionally assumed laminarity of flow was rejected. The conditions were assumed to be quasi-turbulent from the standpoint of mass transfer. Thus the Schmidt number, which is used to describe laminar diffusivity of substrate (or molecular properties), was assumed to lose its meaning. Finally, it was decided to use a modified Peclet number equal to the tower velocity-depth product divided by tower dispersion number.

Subsequently dimensional analysis on macro-parameters was performed using effluent concentration, applied load, specific surface area, hydraulic radius of the biofilter, and viscosity. Utilizing the Buckingham π method of analysis dimensionless groups were obtained, and then a graph of S_e versus L was made, showing a curvilinear relationship, with the points of all data pools spread around the curve. A plot of log S_e versus $1/L$ yielded the desired straight line. Based on graphical interpretation a model of the form:

$$S_e/S_o = e^{-K/L} = \exp(-K/L) \tag{1}$$

was found to be the most applicable to the case investigated (Oleszkiewicz & Eckenfelder, 1974).

4.2 Model development and practical modifications

Historically a variety of pseudo-first order models have been used relating removal (S_e/S_0) to the exponent of (-KH), similarly to Velz's depth-related removal concept. More recently formulas of the type:

$$S_e/S_0 = \exp(-Kt) \tag{2}$$

were introduced. This was a major step forward as hydraulic residence time, t, is a function of depth, flow rate and media characteristics.

The most comprehensive biochemical kinetics modelling approach has been offered by Fair & Geyer (1954):

$$dS/dt = K (S_e/S_0)^n S_e \tag{3}$$

Fair and Geyer have named the "k" factor a proportionality coefficient, rather than a treatability rate constant.

Grau & Dohanyos (1970) have further simplified this expression, assuming n = 1 in the plug flow activated sludge system. This resulted in, the now widely used equation for activated sludge:

$$S_e/S_0 = \exp(-Kt/S_0) \tag{4}$$

Taking into account that activated sludge food/microorganism ratio (F/M) is expressed as:

$$S_0/(X_v.t) \; ; \; (kg/kg.d) \tag{5}$$

and $K = k\,X_v$, the plug flow activated sludge system may be modelled by

$$S_e/S_0 = \exp(-k/(F/M)) = e^{-k/L'}, \tag{6}$$

where L' is applied load to the unit of biomass.

The biomass concentration (X_v) in the trickling filter is directly proportional to the specific surface area (A). The thickness of the film will vary with depth, with the slime surface hydraulic load (m^3/m^2 slime area.d) and temperature. There are a variety of techniques that allow measurement of the film thickness (Kristensen & Christensen, 1983; Williamson & McCarty, 1976). However, the design may be based on the determination of specific surface area of slimed media operating (pilot studies) under expected full scale conditions. Thus, K = kA and is expressed as ($kg/m^2.d$) (m^2/m^3), i.e. ($kg/m^3.d$). The units for F/M in trickling filters are ($kg/m^2.d$) rather than (kg/kg.d) as in Eq. 5 for activated sludge.

It appears then, that activated sludge and trickling filter performance can be modelled by similar kinetic expressions of pseudo-first order type. Both expressions use organic loading applied to the biomass in reactor as the major parameter determining effluent quality.

The volumetric load to the filter is expressed as

$$L = (Q \cdot S_0)/H \; ; \; (kg/m^3.d) \tag{7}$$

Since the volumetric removal coefficient $K = kA$ then the model (Eq. 1) takes the expanded form:

$$S_e/S_0 = \exp(-kAH/Q.S_0) \tag{8}$$

When recycle is practiced the value of S_0 has to be modified to include dilution by effluent:

$$S_a = (S_0 + NS_e)/(1 + N) \tag{9}$$

where N is the recycle ratio and S_a is the concentration of mixed raw wastes and recycle stream.

4.2.1 Similarity to the Eckenfelder (1961) Model

The model in the form of Eq. 8, first published by Oleszkiewicz & Eckenfelder (1974), differs from the Eckenfelder (1961) model as it includes influent concentration (S_0) effects. The comparison, given by Oleszkiewicz (1980), has shown significantly improved correlation of data from experiments where influent concentration varied. Eckenfelder (1961) has related removal to hydraulic residence time (HRT) expressed as a function of ratio of depth to flow (H^m/Q^n). The exponents modify the flow retention characteristics i.e. HRT. In this respect however they cannot be treated as specific for a given media, since this characteristic will change with the media being slimed and subject to varying hydraulic load.

In the Oleszkiewicz & Eckenfelder (1974) load model (Eq. 1) all these variables are practically incorporated in the $K = kA$ value. Effective specific surface area A is determined after sliming. The value of k ($kg/m^2.d$) then represents the actual load effects on the surface of slimed media. The value of K in turn, incorporates slime performance and the media configuration as well as its slimed area.

4.3 Empirical predecessors of the model

The majority of published practical and theoretical divagations regard the concept of surface organic load as the determining factor in filter performance. This has been expressed by Fair & Geyer (1954) and Jank & Drynan (1973) as well as in the design models introduced in Europe by Ganczarczyk (1957), Imhoff and Rincke (1967). The practical design models for laboratory or pilot data correlation used predominantly by media manufacturers (e.g. ICI-Flocor) have the form:

$$\text{Removal Efficiency } E = (S_0 - S_e)/S_0 = K/L \tag{10}$$

The K coefficient is not considered as "treatability" coefficient but proportionality constant. Out of a variety of equations of the type $y = a-bx$, presented by various authors, Rinke's (1967) formula, based on performance of 25 installations gained wide acceptance:

$$E(\%) = 93 - 0.017.L \tag{11}$$

where $L = BOD_5$ load $(kg/m^3.d)$

A direct correlation exists between all of these equations and the proposed Oleszkiewicz & Eckenfelder's model (eq. 1). Expansion of equation 1 into a MacLaurin series yields:

$$S_e/S_0 = \exp(-K/L) = 1 - K/L + (K/L)^2/2! - .... \tag{12}$$

which for small values of K/L, i.e. low removal rate coefficients - K, or high organic loads - L, reduces to:

$$S_e/S_0 = 1 - K/L \tag{13}$$

This form, after transformation is identical to eq. 10.

4.4 Temperature effects

The value of k $(kg/m^2.d)$ is affected by a variety of factors of which temperature effects are the most significant. The magnitude of response is entirely dependent on the system studied. In conventional high rate aerobic trickling filters the response may be simplistically approximated by

$$k_T = k_{20}\theta^{\Delta T} \tag{14}$$

where is assumed 1.020 unless experimentally determined to be different. In anaerobic systems with short solids retention times (SRT) in the order of 20 to 30 days this effect may be even more pronounced - i.e. the value of theta much greater - in the order of 1.10 - 1.25. In upflow fluidized or expanded bed anaerobic reactors characterized by very long SRT values (in excess of 100 days) and very thin, active slimes theta may be in the order of 1.05 to 1.10. This compensating effect of long biomass retention (SRT) on the temperature response has been demonstrated by successful operation at low temperatures of both the moving (expanded) bed reactors by Kelly & Switzenbaum (1983) and for a hybrid upflow biofilter with considerable biomass retained in the interstices, as shown by Oleszkiewicz & Koziarski (1982). It should be noted that kinetic response to temperature variation modelled by van't Hoff-Arrhenius-type formula is a great simplification. The biofilm temperature response is quite complex (Honda & Matsumoto, 1983). Both the biofilm yield coefficient and the endogenous respiration coefficient, increase with temperature. Oxygen solubility and transfer decreases at the same time. Both phenomena affect slime thickness and sloughing. Finally the value of temperature correction coefficient is constant in a narrow temperature range. Usually the effects are more pronounced at lower than at higher temperature ranges. In anaerobic conditions, the exact temperature response is very hard to predict as evidenced by diverging results on fluidized bed treatment of whey wastes reported by Switzenbaum & Jewell (1980) and Boeming & Larsen

(1982). One of the plausible explanations for the deviation of biofilter behavior from the Q_{10} rule is that acclimation conditions affect the temperature response. This was noted by Rusten (1984) who found $\theta = 1.016$ for his aerobic upflow biofilters.

5. GRAPHICAL DATA INTERPRETATION

Studies conducted on 6 parallel biofilters by Bruce & Merkens (1970, 1973) in Stevenage and data from parallel studies of sewage from Derby and Cheltenham (Joslin, 1970) on high rate experimental filters will be used here. Table 2 shows the condition for the 6 filters in Stevenage, numbered as in Fig. 1.

TABLE 2

Parameters of 6 filters in Stevenage

Filter No.	Medium	$A(m^2/m^3)$	L-Range (kg/m^3d)
1	Slag	40	0.58 - 1.61
2	Basalt	40	1.30 - 4.48
3	Flocor	85	1.71 - 4.49
4	Crinkle Close Surfpac	187	1.71 - 4.50
5	Cloisonyle	220	1.54 - 3.30
6	Surfpac	82	1.60 - 1.71

At Derby the influent concentration varied from 168 to 510 mg/l BOD_5, with a mean of 332 mg/l. At Cheltenham mean influent BOD_5 was 220 mg/l. The media used was of the same type and had the following theoretical specific surface areas: filter 1-15 cm slag .33 m^2/m^3; 2-10 cm slag, 49 m^2/m^3; 3-Flocor, 85 m^2/m^3; 4-Crinkle-close Surfpac, 180 m^2/m^3.

Fig. 1 illustrates the data from Stevenage plotted in accordance with eq. 1. Settled sewage was used in these experiments. On the basis of obtained results Bruce & Merkens have found that practical specific surface area of slimed Cloisonyle media was only 145 m^2/m^3 (No. 5). This is confirmed by correlation of specific surface area versus K $(kg/m^3.d)$ in Fig. 3a.

Fig. 2 shows Derby and Cheltenham data. Derby sewage was more concentrated and contained a high proportion of industrial effluents. Cheltenham sewage was primarily of domestic origin. The filters were operated without recycle.

Fig. 3b shows that C.C. Surfpac specific surface area does not give the expected K in the higher load regime. This is due to sliming and bridging resulting in decrease of the active exposed surface below the theoretical 180 m^2/m^3. In Cheltenham (Fig. 3c) the C.C. Surfpac combination used had 154 m^2/m^3 active surface area.

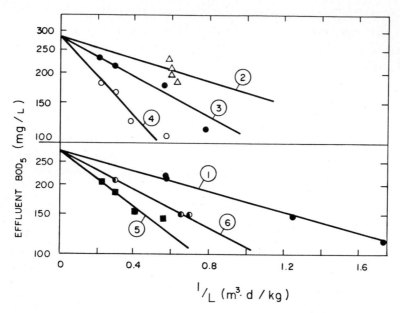

FIG. I   PARALLEL TRICKLING FILTERS TREATING SEWAGE IN STEVENAGE

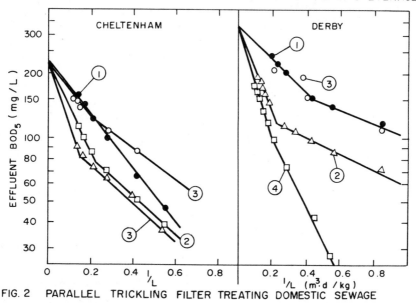

FIG. 2   PARALLEL TRICKLING FILTER TREATING DOMESTIC SEWAGE

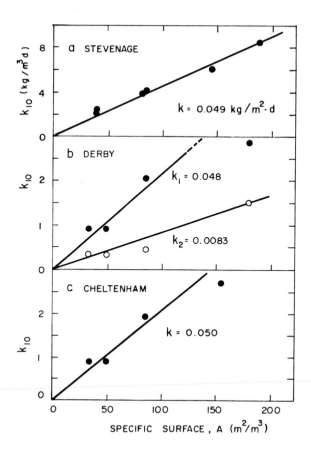

FIG. 3 CALCULATION OF SURFACE REMOVAL COEFFICIENT
FOR DOMESTIC SEWAGE TREATED ON DIFFERENT
MEDIA

The values of K are calculated as K_{10} = slope/0.434. The slope is calculated as equal to $(\log S_1/S_2)/(L_2-L_1)$.

Fig. 3 allows direct calculation of the surface load removal coefficient k $(kg/m^2.d)$. The coefficient is considered independent of media size, shape and configuration after the slime is fully developed. The high specific surface area media are usually exposing smaller effective slime surface in conditions of high organic loading and easily removable substrates, i.e. when slime build-up is heavy. Fig. 3b supports this contention. For high loads – the K_{10} versus A correlation shows that the CC Surfpac point at A = 175 m^2/m^3 is off the extrapolated line for the upper limbs (i.e. high loads) portions of the correlation in Fig. 2b. At lower loads, below 4 $kg/m^3.d$, the superiority of that media became apparent and this is reflected in the lower line in Fig. 3b with the slope k_2 = 0.0083 kg/m^2.d. The media has less slime (total mass/volume) and thus more area is exposed to BOD mass transfer. Thus all points are on the same line for lower limbs of S_e versus 1/L curves (Fig. 2b).

6.   EXAMPLES OF APPLICATION TO FIXED FILM REACTORS

6.1   Roughing aerobic biofilters
6.1.1 Downflow filters

Aerobic biofilter is perhaps the most efficient aerobic roughing pretreatment reactor available and is inherently less energy intensive than suspended growth systems (Eckenfelder, 1982). Oorthuys (1983) cites Dutch sewage treatment plants saving up to 60% of energy costs by using biofiltration pretreatment instead of activated sludge. The fixed film reactors that are used here are: 1) oriented plastic media down-flow trickling filter with recycle and 2) upflow packed bed reactor with cocurrent aeration. The aerobic, upflow fluidized bed reactor has not as yet been demonstrated to achieve effective oxygen transfer at very high organic loadings (i.e. above 8 to 10 kg BOD/m^3.d).

There are several examples of oriented plastic media roughing pretreatment data interpretation in the literature (Oleszkiewicz & Eckenfelder, 1974) and in Table 3. It would be informative to observe performance of a roughing random media biofilter. An example in Fig. 4 is taken from Oleszkiewicz (1981). With the loads reaching in excess of 20 kg BOD$_5$/m^3.d the experimental brushwood biofilter yielded 50 to 75% removals.

6.1.2 Upflow packed bed reactor

The packed bed reactor (PBR) differs significantly from down flow trickling filters. The upflow configuration and co-current air injection as well as presence of suspended biomass in the media interstices creates conditions of a hybrid fixed-film and suspended growth reactors. The load model (Eq. 1), S_e/S_0 = exp (-K/L) does not work as effectively here, as the volumetric removal rates are also dependent on the amount of viable biomass in the suspended form. A similar situation is experienced in upflow anaerobic fixed film reactors where it will be demonstrated that the correlation should involve total viable solids retention time (SRT). This should be more effective than simplified F/M ratio where M (microorganism concentration) is assumed directly related to surface available, A.

Fig. 5 illustrates a correlation obtained for a PBR treating benzol plant effluent wastewater. The reactor was packed with 35 mm plastic pall rings

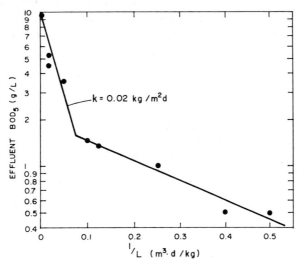

FIG. 4 PIGGERY MANURE BIOFILTRATION THROUGH BRUSHWOOD MEDIA

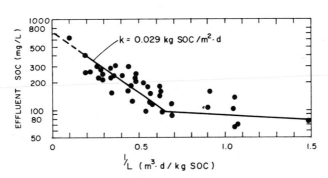

FIG. 5 AEROBIC PBR TREATING BENZOL WASTEWATER

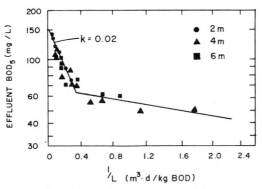

FIG. 6 SEWAGE TREATMENT ON PLASTIC MEDIA FILTERS
WITH DIFFERENT DEPTH

(Olthof & Oleszkiewicz, 1983). The final design conditions selected were SOC (soluble organic carbon) load of 2.2 kg/m^3.d yielding 57% SOC removal. The corresponding COD load was 7.5 kg O$_2$/m^{3}d yielding 67% COD removal and over 99.9% removal of phenols. The inflection point at L = 1.62 kg SOC/m^3.d (100 lbs/1000 CF.d) was selected as cut-off point for a cost effective application of the system.

Similar analysis was applied to data reported by Rusten (1984) for a PBR packed with PLASdek 140 m^2/m^3 media, treating municipal sewage. The distinct cut-off point was found at 5.6 kg COD/m^{3}d. Decrease of the load below 5.6 did little to improve the removal efficiency.

6.2 High rate aerobic trickling filters with and without nitrification

The high rate biofilter's performance was considered, by some, to be dependent on depth, by others, on retention time. Fig. 6 and reference (Oleszkiewicz, 1980) show that the depth or time dependence is of lesser importance than the volumetric load. The correlation is for three depths: H = 2, 4, and 6 m. All are within conventionally used design depths. The data comes from the work of Audoin (1970). For the slimed specific surface area of 145 m^2/m^3 the removal rate coefficient is k$_1$ = 0.02 kg/m^2.d. The curve fit appears satisfactory with an index of determination ID = 0.76 for the high rate (high load) portion of the curve.

Attached growth systems have recently received increased attention in nitrification applications. This is due to the ability to retain fragile nitrifying organisms on the media surface, to any desired extent of SRT, as opposed to activated sludge systems where this is not achieved easily. Usually, very fine plastic media is used (i.e. A > 120 m^2/m^3) and an example of such an application is presented in Fig. 7. Raw data was taken from the work of Huang et al. (1982). Effluent from an existing activated sludge tank was fed to two 55-liter biotowers in series, packed with plastic media of A = 140 m^2/m^3. The feed was enriched with ammonium chloride and sodium bicarbonate. Fig. 7 illustrates performance of the first nitrifying filter. The insert shows the same data points plotted against S$_e$/S$_o$ ratio - which yielded very similar k value as the plot of S$_e$ versus 1/L.

6.3 Anoxic and anaerobic biofiltration

The application of the presented model to correlation of low loading denitrifying (anoxygenic) biofilters is illustrated in Fig. 8. Full scale study data of a polishing denitrifying biofilter treating secondary effluent (high rate activated sludge) from an industrial scale piggery (Oleszkiewicz & Koziarski, 1981) was used. The biofilter was operated in a down flow mode

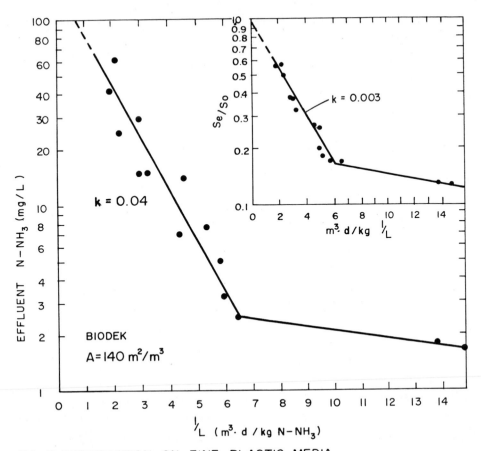

FIG. 7 NITRIFICATION ON FINE PLASTIC MEDIA

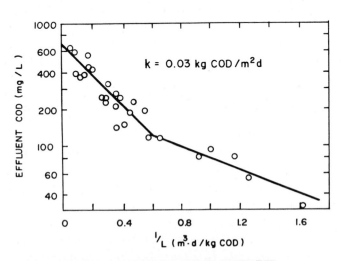

FIG. 8 ANOXIC DOWNFLOW COKE BIOFILTER

and was packed with coke. No carbon source was added to the wastewater.

An example of an upflow anaerobic biofilter data correlation is presented in Fig. 9, taken from Oleszkiewicz and Koziarski (1982). An upflow 2 m high laboratory filter was packed with 25 mm polypropylene spheres. The filter had developed very high suspended biomass in the interstices, amounting to almost 200 g/l of solids (expressed as non-filtered COD) at the bottom and gradually decreasing to 2 g/l at the top. The influent concentration was 10 g/l COD_{nf}. In order to account for suspended biomass a modified form of the load model was applied:

$$S_e/S_o = \exp(-K'/SRT) \tag{15}$$

In an upflow filter, with considerable amount of suspended growth, the removal has to be related to the residence time of solids in both attached and suspended state. The problem lies in difficulty of measuring the SRT in an actively operating anaerobic fixed growth reactor, without disrupting the process. In this case the bio-filter was drained and solids were measured in the liquid. Attached growth was determined by averaging the amount of slime scraped off several spheres. This was done before changing the loading.

Several load-efficiency correlations for anaerobic filters have been recently presented by Oleszkiewicz (1981). Data reported by Benjamin et al. (1981) for an upflow biofilter, packed with plastic cylindrical media, treating sulfite evaporator condensate is shown in Fig. 10. Influent raw wastes had COD of 5,000 mg/l and $BOD_{5,nf}$ of 3,000 mg/l. Recycle ratio was 4 to 15. Assuming A = 125 m^2/m^3 the removal coefficient for this easily degradable substrate is 0.056 $kg/m^2.d$. Eq. 1 was used also to fit data from the anaerobic fluidized bed reactors. Fig. 11 illustrates data collected by Switzenbaum & Danskin (1981) from a laboratory fluidized bed biofilter packed with aluminum oxide particles. The influent concentration was 10 g/l of COD (non-filtered) composed of whey wastes. The hydraulic residence time varied between 4 and 27 hours. The ratio of non-filtered and filtered COD was 1.2 and the process temperature was 25 to 31°C.

6.4 Removal coefficient

The volumetric removal coefficient K is equal to K = kA. The specific removal coefficient k ($kg/m^2.d$) should be independent of the type of media. Media configuration and liquid retention-mass transfer characteristics are incorporated in the effective specific surface area A, determined after the biofilter is in full operation. This determination can be made with the use of non-degradable fluorescent tracers or radioisotopes. Examples of removal coefficients obtained from a variety of biofiltration processes are presented in Table 3. The values are for the high rate portion of the

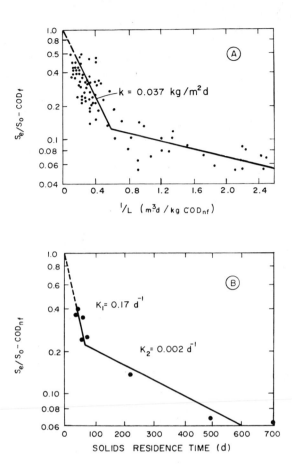

FIG. 9 ANAEROBIC UPFLOW BIOFILTRATION OF
PIGGERY WASTES

log S_e/S_0 versus $1/L$ curve.

7. CONVERGENCE OF ACTIVATED SLUDGE AND TRICKLING FILTER DESIGN APPROACH

Analysing Eqs. 4, 5 and 6 it appears that the plug flow activated sludge model is basically similar to the proposed model eq. 1. The models for the two systems would be identical if time dependence was accepted as the decisive factors as in Eq. 2.

The hydraulic residence time (HRT) in biofilter modelling is considered irrelevant by the author. This is due to the fact that HRT/SRT ratio is very low in attached growth reactors. In plastic media aerobic trickling filter HRT may be equal to 5 to 15 min while SRT is equal to 10+ days. In anaerobic biofilters the ratio of HRT/SRT is in the order of 1/100 or less for stationary media and 1/100 to 1/300 for fluidized bed systems. This is the single most significant difference between highly loaded homogeneous

TABLE 3   Removal coefficients for various wastewaters and reactors

Wastewater	Media	Average S_a(mg/l)	k(kg/m^2.d)	Data Base	Comments
A.	**TRICKLING FILTERS** (downflow, except as noted)				
Phenolic	Koroseal	340	0.021	BOD	
Domestic	6 different media	280	0.049	BOD	
Domestic	8 different media	200-520	0.045-0.05	BOD	
Domestic	Cloisonyle	266	0.02	BOD	
Kraft mill	Koroseal	210	0.016	BOD	
Potato	Flocor	500	0.051	BOD	
Bottling	Flocor	2,000	0.054	BOD	Series of 3
Benzol plant	Flexiring	435	0.029	SOC	Upflow packed bed
Domestic	Plasdek	300	0.033	COD	Upflow packed bed
Hardboard	Surfpac	443	0.007	BOD	
Pulp and paper	Surfpac	280	0.016	BOD	
Piggery	Brushwood	15,000	0.02	BOD	
Cannery + distillery	Coke	600	0.029	BOD	
Cannery	Coke	500	0.017	BOD	
Yeast	Coke	1,000	0.007	BOD	
Domestic	Biodek	50	0.0035	N-NH$_3$	Nitrifying filter
B.	**ANAEROBIC & ANOXIC BIOFILTERS** (upflow, except as noted)				
Piggery	20 mm spheres	10,000	0.037	COD	Stationary
Synthetic: carbohydrates & acetic acid	rock	2,000	0.060	COD	Stationary
Whey	0.4 mm	10,000*	0.020**	COD	Expanded
Whey	0.4 mm	5,000 to 20,000* (COD nf)	0.017**	COD	Expanded
Whey	0.4 mm	5,000 to 20,000* (COD nf)	0.007**	COD	Expanded
Piggery Wastes	Coke	395	0.032	COD	Denitrifying; downflow

* Value of S_o rather than S_a. Very high recycle rates were used (in excess of 20 to 50 N)

** Assumed 2000 m^2/m^3 specific surface area, based on 0.2 - 0.4 mm Al$_2$O$_3$ particle size.

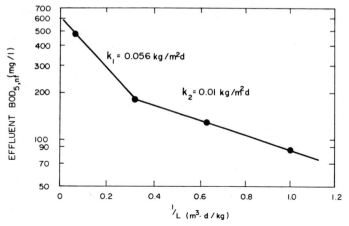

FIG. 10   ANAEROBIC BIOFILTRATION OF SEC.

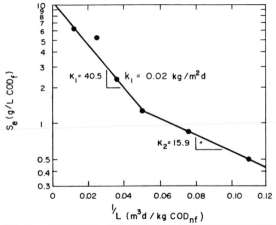

FIG. 11   ANAEROBIC EXPANDED BED TREATMENT OF WHEY WASTEWATER

stirred reactors such as activated sludge or anaerobic contact reactor and heterogeneous reactors such as biofilters.

The models of substrate removal kinetics converge only when Eq. 6 is taken into account.   F/M ratio in suspended growth, homogeneous systems means the mass of substrate applied to mass of bacteria in suspension (kg BOD/kg MLVSS.d).   The F/M for attached growth system means mass of BOD applied to the unit of slime area (kg BOD/m^2.d).   Excess sludge production kinetics has also been shown to be analogous in both systems.

The following factors affect the mass balance of excess slime: build-up due to influent suspended solids which are not degraded in the process; increase in biological volatile solids (VS) due to cell synthesis;   decrease in biological VS due to endogenous respiration;   decrease of sludge mass due to solids lost in the effluent.

To simplify, let us assume that influent is totally soluble and final clarification is 100% efficient then the resulting equation is:

$$\Delta X_V = aS_r - bX_V \tag{16}$$

where:

a - sludge or slime synthesis (yield) coefficient, (kg VSS produced/kg organics removed); and b - sludge endogenous respiration coefficient (kg VSS oxidized/kg MLVSS.d);

and where it is assumed that X_V represents 100% biodegradable fraction.

The design values for a and b coefficients are obtained from the plot of net sludge production versus activated sludge loading. In case of trickling filters - with proper hydraulic regime i.e. where adequate sloughing is maintained the active sludge is retained in the media in the form of biological slime. The sludge removed from the final clarifier should then be considered as excess sludge i.e. ΔX_V, proportional to the load removed S_r and to the active slime mass. The latter is directly proportional to the specific surface area A. It should be stressed here that sludge recycle, sometimes advocated, is without substantiation and may only be used for start-up. In active biological filter, sludge that is removed (sloughed), has little remaining treatment potential.

Results of studies on six parallel biofilters published by Bruce and Merkens (1970, 1973) have been utilized to show that sludge production in trickling filters can be accurately related to the applied and removed organic loadings. From the published data on average values of unit BOD removal and average quantity of wet film the values of kg BOD removed per kg of wet film-day were calculated and plotted in Fig. 12. The resulting coefficients are a = 0.59 kg TS/kg BOD removed and b = 0.016 d^{-1}.

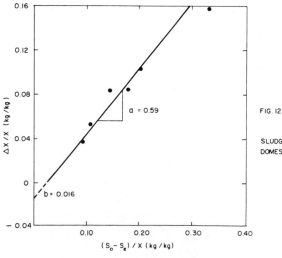

FIG. 12

SLUDGE PRODUCTION FOR A TRICKLING FILTER TREATING DOMESTIC SEWAGE

Rusten (1984) reported the sludge yield from an upflow, aerated biofilter (PBR), treating sewage to be 0.40 g TSS/g COD removed, at surface load of 25 g COD/m^2.d. This mass balance approach works in long-term studies with properly adjusted hydraulic loading, leading to even slime sloughing (Moodie & Kavanagh, 1981). Data is needed also to prove that the unit oxygen uptake rates and consumption by biofilter slimes can be used in a manner similar to activated sludge oxygen requirements formula. The oxygen information would be essential in scale-up work determining the porosity of the media and its ventilation requirements.

Oxygen uptake formula for activated sludge is:

$$O_2 \text{ uptake} = a'S_r - b'X \tag{17}$$

It was possible to measure the oxygen uptake coefficient (a') to be equal to a' = 0.05 to 0.6 g O_2/g COD removed, for pharmaceutical wastes, increasing with the recycle rate (Oleszkiewicz & Eckenfelder, 1974). The values of endogeneous respiration coefficients were not determined directly, because the study did not include slime weight determinations. It was found, however, that the auto-oxidation coefficient (b') is directly proportional to the biological activity of the slime, and increased whenever slime exhibited the increase of oxygen consumption. For the pharmaceutical waste studied, the value of b' increased with the increasing recycling until a plateau was reached at N = 400%, which is coincidental with other findings on optimizing the recycling rate in this particular study. Oxygen measurements in the trickling filter slime should be included, both in the studies and in routine plant performance monitoring. They are easily performed in micro and macroscale with the use of removable slimed slides.

8. DISCUSSION

The shortcomings of the presented model (Eqs. 1 and 8) are numerous. The formulas do not account for stratification of slime in the reactor along depth or do not model the various mass transfer limitations that may exist in the microcosm of slime-liquid-gas interfaces. It may also be inferred that the effects on effluent quality will be identical for a tall narrow tower under high surface hydraulic loading (m^3/m^2.d) as for a pancake-thin filter, provided that the organic loads L (kg/m^3.d) stay the same.

Finally, in a number of cases, the relationship of S_e/S_o versus 1/L in semi-log scale resulted in two distinct straight lines (sharp change of slope) with two significantly different values of k.

The idea behind the presentation of this model, however, was not to give a mechanistic representation of all mechanisms of the three phase biochemical reactor. The author is of the opinion that one mechanistic model uniting all the variables and including various environmental conditions will not be found - at least not soon. The presented volumetric load model (Eq. 1) is a method of correlating laboratory, pilot or full scale plant data that works in more cases than other, more refined approaches.

All current models can probably be led to the point of bankruptcy by setting demanding boundary conditions or even by simply comparing their results side by side. In some cases conflicting results can be obtained by different models for the same set of data points (Oleszkiewicz & Eckenfelder, 1974; Oleszkiewicz, 1982). It is postulated that fixed film biological reactors: aerobic and particularly anaerobic, should never be designed without treatability studies. This is because: 1) there are no identical industrial wastes; and municipal sewage becomes too complex in composition to regard it as "typical"; 2) the fixed film reactor performs as a heterogeneous or surface reactor. In most cases of stationary media, the aerobic and anaerobic layers of slime are in transient state depending on the hydraulic regime and slime thickness, media configuration and substrate diffusivity. These in turn have to be optimized experimentally, particularly when recycle is involved in order to: dilute a strong waste; activate the very top layers in the aerobic filters; create more of a complete mixing condition for the system; or to keep fluidized media in suspension. Interpretation through Eq. 1 frequently yields a sharp change of direction, resulting in a much lower removal rate at low loadings. This can be interpreted as a lower biodegradability of metabolic by-products of degradation reactions of the primary substrate. In other words the effluent BOD or COD from highly loaded biofilters contains significantly higher proportion of influent BOD or COD, than in case of low-loaded biofilters. Another explanation is that high organic loads are associated with higher hydraulic loads and this may induce better (more turbulent) mass transfer conditions. When higher concentrations are the cause of high organic load, then recycle is usually applied and this improves the wetting rate and mixing regime in the slime-wastewater interface.

The point of occurrence of this change of slope may then be taken as cut-off for further load decrease to achieve a desired effluent concentration. Significantly reduced removal rate does not warrant cost-effective operation. The bioreactor should then be operated at the high removal rate section of the curve. Further removal should be sought in a follow-up stage.

The interpretation of this point in the load model, Eq. 1, has been addressed by Beccari & Rozzi (1979) and commented on by Marais (personal communication, 1981). Beccari & Rozzi (1979) have suggested a concept of a multisubstrate model where the ratio of effluent to influent concentration is equal to the sum of removals of individual substrate components:

$$S_e/S_o = \sum_{i=1}^{N} \alpha_1 e^{-K_i/L} \tag{18}$$

When N = 1 this coincides with Eq. 1. The model has been further refined by introducing a term of non-biodegradable substrate fraction - β . For a case where there are two limbs of the standard plot of log S_e/S_o versus 1/L, such as in Figs. 4 through 10, the Oleszkiewicz & Eckenfelder Eq. 1 modifies to:

$$S_e/S_o = \alpha_1 e^{-K_1/L} + \alpha_2 e^{-K_2/L} + \beta \tag{19}$$

For the data of Audoin (1970), Fig. 6, Beccari & Rozzi (1979) have graphically determined: α_1 = 0.526, α_2 = 0.315, β = 0.159 and K_1 = 27.2, K_2 = 4.13 kg/m^3.d.

Marais (1981) has separately indicated the applicability of the two-substrates-concept for activated sludge. He has divided the influent into two sharply different substrates easily biodegradable soluble and slowly degradable particulate.

The difficulty in unified, continuous modelling is in the fallacy of considering effluent quality as "fraction remaining". It is shown to be an incorrect approach (Vaughan & Holder, 1984) by researchers working with easily degradable substrates.

It may be inferred here that the basic model of S_e/S_o = exp (-K/L) can take the form of a straight line, or a form of two lines with a sharp change of slope. The straight line is obtained when the loads (that is load L with its effects on F/M and SRT) are such that there is enough of original substrate (both soluble and particulate) remaining in the effluent.

A line with a point of sharp change of direction is formed when: a) there is considerable difference in degradation rates between raw waste components, e.g. soluble and particulate slowly degradable component; or b) when the load and raw substrate applied result in, a more or less, complete primary degradation into metabolites that have significantly slower rates of degradation than the original substrate. The two conditions leading to the change of direction are: low organic load (L, low F/M, long SRT) and high dilution through recycle. The recycle results in an almost completely mixed system, rather than plug flow, the latter being approximated in a downflow no-recycle systems. In a completely mixed system the concentration gradient - the driving force - of metabolites may be so low that they are hardly removed at all, as the biomass is selectively addressing the higher concentration of the raw influent substrate component.

Conversely, in such a situation it appears best to cut-off the removal in the first stage at 25% above the direction change-causing-load. The treatment should be continued in the second stage, where the biomass will adjust to the presence of metabolites alone, preferably in the plug flow mode.

Such staging procedure will optimize (i.e, minimize) the total volume required to achieve the desired effluent quality.

Based on results of biofiltration of similar wastewaters it may appear that removal efficiency coefficient K is dependent on the surface area. This contention infers that the performance of a biofilter is related more to the magnitude of $A(m^2/m^3)$, than to the type of media (e.g. plastic versus stone). Full scale data alternately verify and defy that. Pullen (1977), for example, has found that there are no benefits to be gained by replacing the stone media with the plastic media, unless the loads and resulting excess sludge production are high. Numerous other examples can be cited where not only type of media material, but shape and surface preparation have played a significant role (Roe, 1982; v. d. Berg & Kennedy, 1982; Ryall, 1982).

When two side-by-side media are compared in a downflow trickling filter the one with higher hydraulic retention time (HRT) and more tortuous path of flow will perform better at lower loads (if both filters have same A). At higher loads, when excess slime may begin to build up in the media, the other filter, with less tortuous liquid path and higher volume of open interstices, will be superior.

These factors will tend to confuse the interpretation of removal data. It is then important to conduct experimental work in full range of loads, to capture the external conditions. Larger scale pilot experiments are necessary to test media performance in the as-installed conditions.

9. EXPERIMENTAL AND DESIGN PROCEDURES
9.1 Design basis

Major drawback of trickling filter design is lack of standardized experimental testing procedures. This, coupled with the lack of agreement on data interpretation, has directed designers towards suspended growth systems.

In order to obtain adequate scale-up information large laboratory or pilot scale studies are mandatory, particularly when industrial wastes are treated. Predictions based on currently existing models are not possible. The model, Eq. 1 & 8, is specific for a given media-wastewater biodegradability-load combination. Data from studies of similar combinations can be extrapolated with caution. However, as shown by the wide range of results in Figs. 1 & 2, obtained in large parallel pilot reactors, the advantage of one media over another can only be determined in actual testing.

The philosophy behind the proposed experimental approach is as follows:

1) most of the predictive models have been shown to be inaccurate to the point of very significant deviation from the cost-effective design;

2) literature on full scale installations (e.g. Pierce, 1978) frequently shows no correlation between organic load, depth or hydraulic load and effluent quality for similar sewage;

3) the response to load has been shown to change in an abrupt way (the "inflection" point in S_e/S_o verus $1/L$ plot) that has to be determined experimentally;

4) it is mandatory to know the limitation of the fixed film bioreactor, i.e. the performance in the extremal conditions. This will allow for proper staging of reactors in series to arrive at a minimum total volume. (The total volume of two reactors in series is always smaller than the volume of one reactor achieving the same effluent quality, as series approximates plug flow conditions.

5) the performance at transient extremal conditions is necessary to design the hydraulic and organic load equalization facilities.

The predesign studies should take account of the following:

1) The laboratory fixed film should be packed with the same media that will be used in the full scale installation.

2) The results from the oriented type planar media, used in the pilot studies should not be extrapolated to a full scale installation packed with random media and vice versa.

3) Winter and summer conditions should be simulated by proper temperature control. Studies may be conducted at laboratory room temperature but the wastewater temperature should be controlled by refrigeration or heating tapes. It is preferable to simulate winter conditions by running the tests outside the laboratory facilities, as long as the scale-down of lab-units will not distort the actual temperature response of the full scale towers.

4) The wastewater should be fully equalized and pretreated with proper pH, nutrient and microelements content adjusted.

5) The dimension of depth H of the tower, should not be distorted by modelling.

6) The data collected should include influent and effluent parameters, flows, temperature, film thickness, and weight (g/m^2), sludge production and slime activity (oxygen uptake or unit methanogenic activity).

It is important to investigate the full range of organic and hydraulic conditions, with the recirculation ratio varying between 0 and 5, for aerobic trickling filters. The increasing recycle brings the biofilter, i.e. the heterogeneous reactor closer to pseudo-homogeneous conditions. For fluidized bed reactors the recycle is determined by the fluidization requirements.

A plot similar to Fig. 13-B should be used. This is true for both aerobic (Tanaka et al., 1981) and anaerobic (Boening & Larsen, 1982) reactors. The recycle ratio here may vary from N = 5 to 50+.

The design steps may be summarized as follows (Fig. 13):

1. Find the maximum load corresponding to the desired effluent quality S_e (Fig. 13a). If a change of direction point is obtained then both rate coefficients should be calculated and used in designing the filter. Higher rate (larger K) should be used to design the first stage.

2. Find the optimum recirculation ratio of N = Q_{rec}/Q_{raw} from Fig. 13b. (Examples - Oleszkiewicz, 1980, 1981, Tanaka et al., 1981.)

3. Determine the optimum biofilter height based on: a) construction site constraints; b) optimization of plan area to height ratio (structural). For fluidized bed a plot of effects of expansion will be helpful; c) an

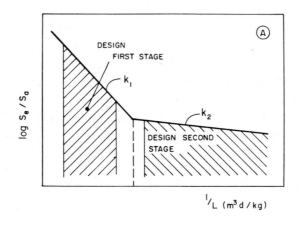

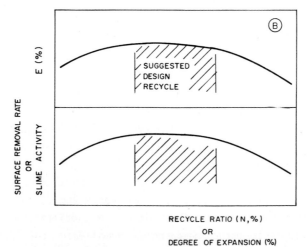

FIG. 13 BIOFILTRATION DESIGN STEPS

optimization analysis with the variables: S_e, H, L_{max}, N_{opt}. The liquid retention characteristics of the slimed surface should be taken into account: the random stone packing can be stacked to a much smaller depth than an oriented plastic media packing.

9.2 Experimental set-up

For oriented, planar media, smaller scale aerobic biofiltration studies can be conducted in a plane model of height equal to the expected full scale height. An example of the inclined plane biotower is given by Oleszkiewicz & Eckenfelder (1974). Careful scale-up procedures are needed here however if full scale media has a different liquid retention characteristics. In such cases larger pilot scale reactors should be used with actual packing in place. In an actual full scale study, reported by Oleszkiewicz & Eckenfelder (1974), six stage trickling filters in series, packed with expanded corrugated horizontal plastic media and oeprated without recycle (total of 36 meters of tower depth in series with five intermediate pumping stations) have been shown to be equivalent to one 6 m tall oriented vertical plastic media biofilter operated with 350% recycle. This study of the full scale pharmaceutical effluent treatment represents a failure of design when no treatability studies are performed and when media characteristics and recycle interrelationship is not taken into account.

Examples of pilot scale reactors are given by Bruce & Merkens (1970) and Gray & Learner (1983, 1984). The latter describe a pilot plant built of 0.6 m long sections of 1.6 m diameter concrete manhole sections. The 90 mm thick walls provide insulation and allow stacking to desired heights. Two different media can be packed in the same filter effected through a dividing wall. This allows side by side comparison of various media.

Anaerobic attached growth reactors can be modelled in pilot scale by insulating large diameter plastic pipes or fiberglass cast reactors molded to specification. Examples of pilot scale anaerobic biofilters have been given by DLA (1981) and Dahab & Young (1981). Full scale type plastic packing media can then be used as it is easily cut to desired shape.

Fluidised bed reactors are usually tested in small scale of 1 to 10 litres. It is essential to study hydraulics of these beds in a larger scale, to assure mixing and turbulence and contact opportunity expected of these reactors (Cooper & Atkinson, 1981).

10. SUMMARY AND CONCLUSIONS

The organic volumetric load model derived by Oleszkiewicz & Eckenfelder (1974) has been found applicable to a wide variety of experimental data from both aerobic and anaerobic attached growth reactors. In its basic form it resembles a black box formula:

$$S_e/S_a = e^{-K/L} \tag{1}$$

Closer analysis shows that it includes the most pertinent phenomenological variables:

$$S_e/S_a = \exp\ (kAH/QS_a)\quad;\ or \tag{8}$$
$$S_e/S_a = \exp\ (-k/(F/M)) \tag{6}$$

with the value of k changing with temperature according to the experimentally defined parameters in eq. 14.

In the form of Eq. 6 the model shows a direct parallel to the activated sludge (F/M) design concept. In fact the model - if it fits the data at hand states explicitly the direct relationship between removal efficiency and the ratio of substrate (F) to exposed microbial slime mass (M) ratio. Slime mass is assumed proportional to surface area A. For upflow attached growth hybrid reactors where considerable amount of solids are present in the interstices, as a suspension, the removal is proportional to the solids retentions time (SRT)

$$S_e/S_a = e^{-K'/SRT} \tag{15}$$

In most cases the basic load exhibits discontinuity when plotted as in Figs. 4 through 10. There are two explanations for this: 1) influent substrate is composed of soluble-easily-degradable organics and of slowly-degradable-insoluble organics; 2) the loads applied are low and the metabolites of primary degradation are degrading at a much slower rate. A mathematical representation has been given in Eq. 19.

Analysis of data on excess slime production proves that sludge production (yield) should also be monitored in fixed film reactors:

$$NET\ SLUDGE\ YIELD = YIELD - ENDOGENEOUS\ RESPIRATION;\ \Delta X = aS_r - bX \tag{16}$$

Based on similar oxygen kinetics (Eq. 17) it is postulated that the basic kinetic expressions: 1) substrate removal in response to load per biomass (X or A); 2) excess solids production; 3) oxygen uptake; should be used in a fashion similar to interpreting data for aerobic activated sludge. In both cases similar measurements should be taken during experimental treatability work.

Eq. 8 includes all variables pertinent to the performance evaluation study. Depth, influent hydraulic load, influent concentration and effective slimed area all combine to determine the substrate removal efficiency. The model, not inadvertently, states that their combined effect, expressed as load, matters more than the relative magnitude of individual parameters. It cannot hold in extreme conditions, say for very high "beanstick" towers or low "pancake" filters. For: normally used tower sizes (H = 3 to 10 m); hydraulic rates above the minimum wetting rate, and below excessive shearing rates; concentrations made compatible with filter capacity for mass transfer and unloading of produced solids through recycle; the main determining factor is the load applied to the slimed surface. Table 3 shows that specific removal coefficients, k, vary in the range of 0.007 to 0.05 $kg/m^2.d$. The fact that one reactor achieves 80% removal at 10 $kg/m^3.d$ load (e.g. anaerobic fluidized bed) while a trickling filter may require a load well below 1.0 $kg/m^3.d$ to achieve the same removal may be explained by differences in specific surface areas, e.g. 1500 m^2/m^3 versus 100 m^2/m^3.

An optimum design for the fixed film reactor would be to operate at an average daily load of about 80% of the maximum load on the slime still giving the desired effluent quality. Design for loads closer to the inflection point on the S_e/S_o vs 1/L curve is counter productive. Cost effectiveness consideration suggests a two stage design: 1st stage high-rate, high K, about 80% of the maximum load; 2nd stage - optimized for lower influent concentration and lower K value. Such reactor series is closer to a plug flow system and yields better performance due to higher concentration gradients than in one reactor only.

The design should invariably involve treatability study in a model with undistorted depth and media configuration. All variables needed to determine: the kinetics of substrate removal; oxygen uptake or methanogenic activity; sludge production; should be monitored at various recycle rates. Interpretation of the data could follow the procedures outlined in Fig. 13.

The procedure lends itself to selection of stages in the biological treatment process. The performance of two biofilters in series, with individual recycle, will exceed that of one biofilter of the same volume. Conversely, two stages in series will usually cost less, at the same effluent quality requirements, if the difference in removal rate coefficients, k_1 and k_2, from Fig. 13, is significant.

The need for experimental data cannot be overemphasized. The failure of a system of six biofilters in series, each 6 meters tall, described in 9.2, is the result of design without treatability studies. Attached growth reactors are proving to be very well suited to a variety of effluents, both in aerobic and in anaerobic mode. They are more resistant to shock loads, toxic impulses and temperature variations. Their behavior and performance, however, is harder to model and predict due to the complexity of the three phase flow and the diversity of conditions in the slime. The proposed Eq. 1 interprets the gross performance of such reactors, using engineering parameters of depth, slimed specific surface area of the reactor, hydraulic load and influent concentration.

11. NOMENCLATURE

a	sludge yield coefficient (MM^{-1})
a'	oxygen uptake coefficient per unit of biomass (MM^{-1})
A	specific surface area of the media; (L^2L^{-3})
b	autolysis or endogeneous respiration coefficient; (t^{-1})
b'	autoxidation coefficient; (t^{-1})
BOD	five day biochemical oxygen demand
COD	chemical oxygen demand
E	removal efficiency equal to $(S_0 - S_e)/S_0$; (%)
F/M	food to micro-organisms ratio, usually expressed per mass of organisms; $(MM^{-1}t^{-1})$
H	reactor height; (L)
HRT	hydraulic residence time; (t)
K	volumetric removal coefficient; $(ML^{-3}t^{-1})$
k	specific (area) removal coefficient; $(ML^{-2}t^{-1})$
L	volumetric organic loading; $(ML^{-3}t^{-1})$
N	recycle ratio equal to the ratio of recycled flow to the raw wastewater flow; (-) or (%)
Q	hydraulic loading of the cross-sectional area, in units of velocity; $(L^3L^{-2}t^{-1})$
S_a	mixed influent concentration, $S_a = (S_0 + NS_e)/(1+N)$; (ML^{-3})
S_e	effluent concentration; (ML^{-3})
S_0	raw influent concentration; (ML^{-3})
SRT	solid residence time (t)
T	temperature; (T); $\Delta T = T-20(°C)$
t	time (t)
X	mass of biological slime or solids, usually considered equal to volatile suspended solids; not necessarily equivalent to viable cell mass; (M), (ML^{-3})
α_1, α_2	constants; (-)
β	non-biodegradable fraction; (-)
ΔX	excess sludge growth (yield); (Mt^{-1})
π	dimensionless groups in Buckingham analysis
θ	temperature correction coefficient (-)

REFERENCES

Audoin, L., 1970. The use of plastic medium for trickling filters treating domestic sewage. Proceed. 5th Int'l. Confer. Water Pollut. Research, San Francisco 11-16.

Beccari, M., Rozzi, A., 1979. Analysis of rational approaches for trickling filter design. Environ. Protect. Engin. 5, 4, 353-364.

Benefield, L.D., Randall, C.W., 1980. Biological process design for wastewater treatment. Prentice-Hall, Inc., Englewood Cliffs.

Benjamin, M.M., Ferguson, J.F., Buggins, M.E., 1981. Treatment of sulfite evoporator condensate with an anaerobic reactor. Proceed. Technic. Assoc. Pulp, Paper Ind., 307-316.

Berg, v.d., B., Kennedy, K.J., 1982. Performance of anaerobic downflow fixed film reactors. Proceed. First Int'l Confer. on Fixed-Film Biological Processes, Kings Island, 1456-1475.

Boening, P.H., Larsen, V.F., 1982. Anaerobic fluidized bed whey treatment. Biotechnology & Bioengineer, 24, 2539-2556.

Bruce, A.M., Merkens, J.C., 1970. Recent studies of high rate biological filtration. Water Pollut. Control, 69, 113-148.

Bruce, A.M., Merkens, J.C., 1973. Further studies on partial treatment of sewage by high rate biological filtration. Water Pollut. Control, 72, 499-523.

Busch, A.W., 1984. Chemical reactor design theory and biological treatment of industrial wastes - is there a gap? Journal Water Pollut. Control Fed., 56, 215-218.

Cooper, P.F., Atkinson, B. (Editors), 1981. Biological fluidized bed treatment of water and wastewater. Ellis Horwood Ltd. Chichester, GB, 412 p.

Dahab, M.F., Young, J.C., 1981. Energy recovery from alcohol stillage using anaerobic filters. Proceed. 3rd Symposium on Biotechn. in Energy Product. and Conservation, Gatlingburg, TN, May 12-15.

DLA, 1981. Anaerobic waste treatment and energy recovery. Seminar sponsored by DLA, Inc., Consult. Engineers, Pittsburgh, PA, 300 p.

Eckenfelder, W.W., 1961. Trickling filter design and performance. Journ. Sanit. Engin. Div. ASCE, 87, SA 4, 33-45.

Eckenfelder, W.W., 1982. Workshop on research needs for fixed-film biological wastewater treatment. Proceed. 1st Intern. Conf. on Fixed-Film Biological Processes. Kings Island, OH., 1806-1899, April 20-23.

Eighny, T.T., Maratea, D., Bishop, P.L., 1983. Electron microscopic examination of wastewater biofilm formation and structural components. Applied & Environm. Microbiol., 45, 6, 1921-1931.

Ganczarczyk, J., 1957. Obliczanie wysoko-obciazonych zloz biologicznych metoda dopuszczalnych obciazen. (In Polish) Gaz, Woda & Techn. Sanitar., 31, 374.

Galler, W.S., Gotaas, H.B., 1964. Analysis of biological filter variables. Journ. Sanit. Engin. Div. ASCE, 90, SA6, 59-79.

Grady, C.P., 1982. Modelling of biological fixed films - a state-of-the-art review. Proceed. First Intern. Conf. on Fixed-Film Biological Processes. April 20-23, 1982, Kings Island, pp. 344-404.

Grau, P., Dohanyos, M., 1970. Substratova kinetika aktivovanego kalu. (In Czech) Rada B, Vodni hospodarstvi II, 298-305, Praha.

Gray, N.F., Learner, M.A., 1983. A pilot scale percolating filter for use in sewage treatment studies. Water Research, 17, 3, 249-253.

Gray, N.F., Learner, M.A., 1984. Comparative pilot scale investigation into up rating the performance of percolating filters by partial medium replacement. Water Research, 18, 409-422.

Honda, Y., Matsumoto, J., 1983. The effect of temperature on the growth of microbial film in a trickling filter. Water Research, 17, 4, 375-382.

Huang, J.M., Wu, Y.C., 1982. Molof, A.: Nitrified secondary treatment effluent by plastic-media trickling filter. Proceed. 1st Intern. Conf. on Fixed-Film Biological Processes, Kings Island, OH.

Jank, B.E., Drynan, W.R., 1973. Substrate removal mechanism of trickling filters. Journ. Environm. Engrg. Div., ASCE, 99, EE3, 187-204.

Joslin, J.R., 1970. High rate biological filtration - a comparative assessment. Joint Meeting E.W. Midland Branches, Burton-on-Trent (GB), Manuscript unpublished.

Kelly, C.R., Switzenbaum, M.S., 1983. Temperature and nutrient effects on the anaerobic expanded bed treating a high strength waste. Proceed. 38th Industr. Waste Confer., Purdue Univ. May 9-13.

Kristensen, G.H., Christensen, F.R., 1983. Application of cryo-cut method for measurements of biofilm thickness. Water Research, 16, 12, 1619-1622.

356

Marais, G.V.R., 1981. Personal communication. (13 April, 1981 letter). Univ. of Cape Town, Rondebosch, SA.

Metcalf-Eddy, 1981. Wastewater Engineering, McGraw-Hill Co., New York.

Moodie, S.P., Kavanagh, B.V., 1981. Production and conditioning of sludges from a high-rate biological filter. Wat. Pollut. Control, 80, 674-681.

Muslu, Y., 1983. Dispersion in the inclined plane model of trickling filters. Water Research, 17, 105-116.

NRC, National Research Council., 1946. Sewage treatment at military installation. Sewage Works Journ., 18, 787-1028.

Oleszkiewicz, J.A., Eckenfelder, W.W., 1974. Mechanism of substrate removal in high-rate plastic media trickling filters. Techn. Report No. 33, EWRE, Vanderbilt Univ. Press, 277 p. 1974.

Oleszkiewicz, J.A., 1980. Biofiltration design based on volumetric loading. Journ. Water Pollut. Control Fed., 52, 12, 2906-2913.

Oleszkiewicz, J.A., 1981. Aerobic and anaerobic biofiltration of agricultural effluents. Agricult. Wastes, 3, 285-296, 1981.

Oleszkiewicz, J.A., Koziarski, S., 1982. Low temperature anaerobic biofiltration in upflow reactors. Journ. Water Pollut. Control Fed., 54, 11, 1465-1471.

Olthof, M., Oleszkiewicz, J.A., 1983. Benzol plant wastewater treatment in a packed bed reactor. Proceed. 37th Industrial Waste Conf., Purdue Univ., 1982, Ann Arbor Scie. Publ., 519-525.

Oorthuys, F.M.L., 1983. Wastewater treatment and energy conservation in The Netherlands. Land & Water Intern'l., 51, 11-16.

Parsons, J.M., 1981. Operation and performance of roughing filters - a case study. Williamsburg Plant, Hampton Road Sanit. District, Virginia, April, 1981.

Pierce, D.M., 1978. Upgrading trickling filters. Techn. Report EPA 430/9-78-004.

Popadic, V., 1974. Falling film research work. Personal communications. Univ. of Kentucky, Lexington, Ky.

Pringle, J.H. Fletcher, M., 1983. Influence of substratum wettability on attachment of freshwater bacteria to solid surfaces. Applied & Environm. Microbiol., 45, 3, 811-817.

Pullen, K.G., 1977. Trials on the operation of biological filters. Water Pollut. Control, 76, 75-85.

Rincke, G., 1967. Gesichtspunkte zur Abwassereinigung mit Tropfkorpern. Das Gas-und Wasserfach, 108, 24, 667-73, 1967.

Roe, S., 1982. Miscellaneous data on pilot tests. Personal communication with Munters Co., Florida.

Rusten, Bjorn, 1984. Wastewater treatment with aerated submerged biological filters. Journ. Water Pollut. Control Fed. 56, 424-431.

Ryall, R.W., 1982. Miscellaneous data on pilot and full scale installations. Personal communication with B.F. Goodrich of Akron, Ohio.

Schulze, K.L., 1957. Experimental vertical screen trickling filter. Sewage & Industr. Wastes, 29, 4, 458-68.

Switzenbaum, M.S., Jewell, W.J., 1980. Anaerobic attached-film expanded bed reactor treatment. Journal Water Pollut. Control Fed., 52, 1953-1965.

Sykes, R.M., 1984. Indeterminancy in mechanistic biological models. Journ. Water Pollut. Control Fed., 56, 209-214.

Tanaka, H., Uzman, S., Dunn, I.J., 1981. Kinetics of nitrification using a fluidized sand bed reactor with attached growth. Biotechnol. & Bioengin., 23, 1683-1702.

Vaughan, G.M., Holder, G.A., 1984. Substrate removal in the trickling filter process. Journ. Water Pollut. Control Fed., 56, 417-423.

Wheatley, A.D., 1979. Discussion of "Plastic-media biological filters". Water Pollut. Control, 78, 371-381.

Williams, I.L., 1979. Discussion of "Plastic-media biological filters", by Porter, K.E., Smith, E., Water Pollut. Control, 78, 371-381.

Williamson, K., McCarty, P.L., 1976. Verification studies of the biofilm model for bacterial substrate utilization. Journ. Water Pollut. Control Fed., 48, 2, 281-296.

MATHEMATICAL MODELS FOR THE OXYGEN TRANSFER PROCESSES IN A ROTATING BIOLOGICAL CONTACTOR

Tomonori Matsuo and Kazuo Yamamoto

Department of Urban Engineering, Faculty of Engineering, The University of Tokyo,7-3-1 Hongo, Bunkyo-ku, Tokyo 113, Japan

This chapter consists of two parts: Part A presents a mathematical model of the physical oxygen transfer in rotating disk contactor and Part B describes a mathematical model of oxygen transfer in rotating biological contactor with substrate and oxygen consumption in the biofilm. The contents of Part B have been drawn mainly from manuscripts submitted to the International Seminar on Rotating Biological Discs, October 6-8, 1983, Fellbach, West-Germany.

CONTENTS

PART A : THE PHYSICAL OXYGEN TRANSFER IN A ROTATING DISK CONTACTOR

A.1 Introduction

In the studies of the physical transfer mechanism of oxygen from air to water by a rotating disk gas-liquid contactor, Yamane and Yoshida (1972), Bintanja et al. (1975), and Zeevalkink et al. (1979) suggested that the oxygen transfer process in the water film on the disk should affect the overall performance of the process. On the other hand, Ouano (1981) assumed the significance of the oxygen transfer through the agitated water surface in addition to the oxygen transfer in the water film. The agitated water surface certainly should have an effect in the oxygen transfer mechanism by the   rotating disk contactor, but the authors have obtained experimental data which show the importance of oxygen transfer through the water film on the disk at the usual depths of disk submergence and rotational speeds.

Regarding the model of oxygen transfer through the water film, Zeevalkink et al. (1979) showed that the model proposed by Yamane and Yoshida (1972) and Bintanja et al. (1975) gave incorrect results   at   intermediate   rotational   velocities   and   shallow submergence of the disks. They suggested that the difference between experimental and theoretical values should be explained by the incomplete mixing of the water film with the bulk water in the container during a revolution.

In this study, further theoretical and experimental analyses have   been   conducted   to develop   a mathematical   model   for establishing the physical oxygen transfer scale-up mechanism in the rotating biological disk contactor. In the model, the interchange coefficient for large-scale mixing induced by the rotating disk has been incorporated into the basic models of Yamane and Yoshida (1972), Bintanja et al. (1975) and Zeevalkink et al. (1979).

A.2 Observation of Flow Profile in the Bulk Water

The flow profile induced by the rotating disk in the bulk water was observed by using tracer techniques and an 8 mm cine camera.

Neutrally buoyant small particles (1.19 mm in diameter) and condensed milk with dye, which was applied to the disk surface, were used as the tracers for visualizing the flow profile which appeared near the surface of the rotating disk. General characteristics of the flow profile are described below.

At the points on the line AC in Fig. 1, the points of submergence, the condensed milk applied to the disk was not disturbed very much. It was assumed that the water film submerged smoothly into the bulk water and went on without being disturbed during the submergence, and that the mass transfer between the water film and the bulk water was controlled by a molecular diffusion process.

However, at the points on the line BC in Fig. 1, the points of emergence, the formation of a strong eddy current and the apparent mixing of the water film with the bulk water were observed. The schematic flow profile at these points is shown in Fig. 2; it is analogous to the flow profile proposed by Zeevalkink et al. (1978) for the flat plate withdrawn with constant velocity.

The following conclusions have been drawn from these observations.

Mixing by the eddy at the points of emergence is thought to be the main process which proceeds the mass transfer between the water film and the bulk water. The water film is not completely mixed with the bulk water as shown by Zeevalkink et al. (1978). In general, the preciseness of a model might be determined by the accuracy of a mathematical evaluation of the mixing mechanism at the emergence points of the rotating disk.

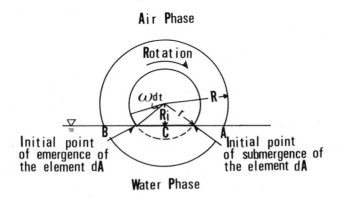

Fig. A.1. Diagram of a rotating disk

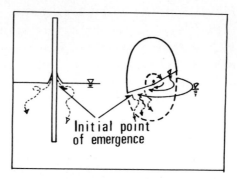

Fig. A.2. Schematic representation of strong eddy flow

A.3 Modelling of Physical Oxygen Transfer

To develop a mathematical model for the physical oxygen mass transfer from air into the bulk water, it is reasonable to divide the overall process into two steps: the first is the oxygen absorption process from air into the water film attached to the disk, and the second is oxygen transfer from the water film into the bulk water. In the first step, a molecular diffusion process controls the transfer of oxygen into the water film during the disk rotation through the air. In the second step, while the disk submerges into the bulk water, oxygen absorbed in the water film is transferred from the film to the bulk water, not only by molecular diffusion through the boundary layer which is assumed to be formed outside the water film, but also by the large scale mixing observed at the points of emergence.

Although these two oxygen transfer processes in the bulk water should be evaluated separately in the mathematical formulae, they may be expressed in a unified overall interchange coefficient, f, which is defined as the ratio of the volume of the water film interchanged by the bulk water to that of the water film attached to the disk just before submerging into the bulk water.

A.3.1 Mathematical Model and Solution

The mathematical model and its solution presented in this analysis have been developed with the following assumptions:
1. only the diffusion processes of a vertical direction to the disk surface are considered and those of radial and circumferential directions are neglected;
2. the water film is fixed to the disk surface throughout its rotation except at the points of emergence where large scale mixing

is observed;

3. the thickness of the water film is uniform over the disk surface; and

4. the distribution of the concentration of dissolved oxygen in the water film is uniform at the points of emergence owing to large-scale mixing and turbulence.

The mathematical models to describe the two steps in the physical oxygen transfer from air into the bulk water are shown below.

Step 1: Oxygen transfer in the air phase

A small surface element, dA, on the disk as schematically shown in Fig. 1, is expressed in Eq. 1.

$$dA = \omega r dr dt \tag{1}$$

where,

  dA = area of the small surface element,

  ω = angular velocity,

  r = radial distance from the disk axis to the points of submergebce and emergence, and

  t = time.

The retention time of the surface element rotating in the air is determined by Eq. 2.

$$t_a = \frac{2(\pi-\varphi)}{\omega} \tag{2}$$

where,

  $\varphi = \cos^{-1}(R_i / r)$,

  t_a = retention time of the element dA in the air, and

  R_i = distance between bulk water surface and disk axis.

The basic differential equation for the molecular diffusion process from air into the water film on the disk is given by Eq. 3.

$$\partial C / \partial t_a' = D_{cw} (\partial^2 C / \partial y^2) \tag{3}$$

where,

  C = dissolved oxygen concentration in the water film,

  D_{cw} = molecular diffusivity of oxygen in water,

  y = distance from the disk surface, and

  t_a' = travelling time of the element dA in the air.

The initial and boundary conditions are given as follows :

$$t_a' = 0, \quad 0 \leq y \leq \delta, \quad C = C_o$$

$$t_a' > 0, \quad y = 0, \quad \partial C / \partial y = 0 \qquad\qquad (4)$$

$$t_a' > 0, \quad y = \delta, \quad C = C^*$$

where,

C_o = initial dissolved oxygen concentration in the water film at the points of emergence,

C^* = saturation concentration of dissolved oxygen, and

δ = thickness of the water film on the disk.

The solution of Eq. 3 is given by Eq. 5.

$$C - C_o = (C^* - C_o) \sum_{n=0}^{\infty} (-1)^n \left[\mathrm{erfc}\left(\frac{(2n + 1)\delta - y}{2\sqrt{D_{cw}t_a'}}\right)\right.$$

$$\left. + \; \mathrm{erfc}\left(\frac{(2n + 1)\delta + y}{2\sqrt{D_{cw}t_a'}}\right)\right] \qquad\qquad (5)$$

where,

$$\mathrm{erfc}(x) = \frac{2}{\sqrt{\pi}} \int_x^\infty e^{-\xi^2} d\xi$$

$$= \text{complementary error function of } x.$$

The time-dependent oxygen flux from the air, N_t, can be calculated by differentiating Eq. 5 as:

$$N_t = D_{cw}\left(\frac{\partial C}{\partial y}\right)_{y=\delta}$$

$$= (C^* - C_o)\sqrt{\frac{D_{cw}}{\pi t_a'}}\left[1 + 2 \sum_{n=1}^{\infty} (-1)^n \exp\left\{-\left(\frac{n\delta}{D_{cw}t_a'}\right)^2\right\}\right] \qquad (6)$$

This solution was introduced by Yamane and Yoshida (1972) into the analysis of the oxygen transfer mechanism of a rotating disk gas-liquid contactor.

The average oxygen transfer coefficient for the element dA is given by:

$$k_L = \frac{1}{C^* - C_o} \cdot \frac{1}{t_a} \cdot \int_0^{t_a} N_t \, dt_a'$$

$$= 2\sqrt{\frac{D_{cw}}{\pi t_a}}\left[1 + 2\sqrt{\pi} \sum_{n=1}^{\infty} (-1)^n \; \mathrm{ierfc}\left(\frac{n\delta}{\sqrt{D_{cw}t_a}}\right)\right] \qquad (7)$$

where,

k_L = average oxygen transfer coefficient for the
element dA, and

$ierfc(x) = \int_x^\infty erfc(\xi)d\xi$
= integral complementary error function of x.

Step 2: Oxygen Transfer in the Bulk Water

The following equation describes the oxygen mass balance in the element dA during rotation time in the air.

$$dA\delta C_i(t) = dA\delta C_o(t - t_a) + dA\int_0^{t_a} N_t dt_a' \tag{8}$$

where C_i and C_o are the initial concentrations of dissolved oxygen in the water film at the point of submergence and at the point of emergence for the small surface element concerned, respectively; both concentrations are time-dependent variables.

In the bulk water, the oxygen mass balance may be expressed by the following equation:

$$dA\delta C_o(t) = (1 - f)dA\delta C_i(t - t_w) + fdA\delta C_b(t) \tag{9}$$

where,

f = overall oxygen mass interchange ratio coefficient,
C_b = dissolved oxygen concentration in the bulk water, and
t_w = retention time of the element dA in the bulk water.

The first term on the right hand side of Eq. 9 represents the residual oxygen mass which is retained in the water film without interchange to the bulk water, and the second term represents the oxygen mass which, in turn, is added to the water film from the bulk water.

From Eqs. 7, 8 and 9, the following equation is derived.

$$C_o(t) = (1 - f)\left[C_o(t - T) + \frac{t_a}{\delta} k_L\{C^* - C_o(t - T)\}\right]$$

$$+ fC_b(t) \tag{10}$$

where,

T = time per rotation.

Using a dimensionless parameter, $k' = k_L/(\delta/t_a)$, Eq. 10 can be rewritten in the form of Eq. 11.

$$C_o(t) = (1 - f)\{(1 - k')C_o(t - T) + k'C^*\} + fC_b(t). \tag{11}$$

Since the time for one rotation, T, is normally much less than the time constant of $(k_L a)^{-1}$, it is reasonably assumed that $C_b(t)$ is approximately equal to $C_b(t-T)$. Thus the following approximate expressions are assumed.

$$C_b(t) \cong C_b(t - T) \tag{12}$$

$$C_o(t) \cong C_o(t - T) \tag{13}$$

$$C_i(t) \cong C_i(t - T) \tag{14}$$

Considering these approximate relationships, Eq. 11 can be rewritten as Eq. 15.

$$C_o(t) = \frac{(1 - f)k'C^* + fC_b(t)}{(1 - f)k' + f} \tag{15}$$

From Eqs. 7, 8, 9 and 14, the relationships of Eq. 16 are obtained.

$$f\delta(C_i - C_b) = \int_0^{t_a} N_t dt'_a = t_a k_L (C^* - C_o) \tag{16}$$

The total oxygen mass balance in the bulk water is expressed in Eq. 17.

$$V \frac{dC_b}{dt} = \int_{R_i}^R f\delta(C_i - C_b)\omega r dr \tag{17}$$

where,

V = net volume of the bulk water, and

R = radius of the disk.

Substitution of Eqs. 2 and 16 into Eq. 17 yields:

$$V \frac{dC_b}{dt} = \int_{R_i}^R t_a k_L (C^* - C_o)\omega r dr$$

$$= \int_{R_i}^R 2k_L (C^* - C_o)(\pi - \varphi)r dr . \tag{18}$$

Assuming that the overall oxygen mass interchange ratio, f, is independent of radial distance, r, and that K_L and $\bar{C}_o$, which are defined as the mean values of k_L and C_o respectively for the entire radial distance, can be evaluated by similar forms as Eq. 7 and Eq. 15, the following equations are obtained.

$$K_L = 2 \frac{\sqrt{D_{cw}}}{\sqrt{\pi \bar{t}_a}} \left\{ 1 + 2\sqrt{\pi} \sum_{n=1}^\infty (-1)^n \text{ierfc}\left(\frac{n\delta}{\sqrt{D_{cw}\bar{t}_a}}\right) \right\} \tag{19}$$

$$\bar{C}_0 = \frac{(1 - f)kC^* + fC_b}{(1 - f)k + f} \tag{20}$$

where,

$\bar{t}_a$ = mean retention time in the air, and

$$k = K_L/(\delta/\bar{t}_a) \ . \tag{21}$$

$$V \frac{dC_b}{dt} = K_L(C^* - \bar{C}_0)\int_{R_i}^{R} 2(\pi - \varphi)rdr$$

$$= K_L A_a(C^* - \bar{C}_0) \tag{22}$$

where,

A_a = wetted area of the disk exposed to air.

Substituting Eq. 20 into Eq. 22, the overall expression of the mass transfer from the air to the bulk water can be derived as Eq. 23.

$$V \frac{dC_b}{dt} = \frac{f}{(1 - f)k + f} K_L A_a(C^* - C_b) \tag{23}$$

Considering the actual procedure to measure the oxygen transfer efficiency, the oxygen transfer phenomena can be evaluated in the overall expression as shown in Eq. 24.

$$\frac{dC_b}{dt} = K_L^* a(C^* - C_b) \tag{24}$$

where,

$a = A_c/V$ = specific gas-liquid contact area based on the total contact area between air and water,

A_c = total contact area between air and water, i.e. the sum of the A_a and the water surface of the bulk water, A_f, and

K_L^* = overall oxygen transfer coefficient.

Since the space between disks is usually much smaller than the radius of the disk, the surface area of the disk exposed to the air, A_a, is assumed to be almost same as A_c. By comparing Eq. 23 with

Eq. 24, the following relationship is obtained:

$$K_L^* = \frac{f}{(1 - f)k + f} K_L .$$ (25)

In the case where the water film is completely mixed with the bulk water, as assumed by Yamane and Yoshida (1972) and Zeevalkink et al. (1978), f in Eq. 25 is equal to unity and the equation is reduced to the original form given by these researchers. Under the conditions that f equals unity and the water film is entirely saturated with oxygen at the points of submergence, Eq. 17 can be rewritten as Eq. 26.

$$\frac{dC_b}{dt} = \frac{\delta}{\bar{t}_a} \cdot \frac{A_a}{V}(C^* - C_b)$$ (26)

In this case, the following relationships are obtained:

$$K_L^* = K_L = \delta / \bar{t}_a, \text{ and}$$

$$k = K_L\bar{t}_a/ \delta = 1.0.$$

Zeevalkink et al. (1979) showed that K_L approaches $\delta / \bar{t}_a$ at low rotational speeds and long retention time in the air ($\delta / \sqrt{D_{cw}\bar{t}_a} < 0.8$). Where K_L falls in the range $0 < K_L \leq \delta / \bar{t}_a$, k represents the dimensionless parameter designating the oxygen saturation level in the water film; the range of k is $0 < k \leq 1.0$.

A.3.2 Modelling of the Overall Mass Interchange Ratio
       Coefficient, f, in the Bulk Water

As mentioned previously, it is assumed that there are two succeeding oxygen transfer processes from the water film to the bulk water. The first one is the molecular diffusion process from the fixed water film on the disk to the bulk water during rotation in the bulk water. The second one is the large-scale turbulent mixing process between the water film and the bulk water at the points of emergence.

The overall mass interchange ratio coefficient, f, should be expressed in a unified form of the two hypothetical interchange ratio coefficients, f_1 and f_2, which correspond to the transfer processes of molecular diffusion and turbulent mixing, respectively. Here, f is formulated by using f_1 and f_2 as:

$$f = f_1 + (1 - f_1) f_2 .$$ (27)

In order to mathematically evaluate f_1, it is necessary to solve the partial differential equation (28), under the assumed initial and boundary conditions given as follows:

$$\frac{\partial C}{\partial t} = D_{cw} \frac{\partial^2 C}{\partial y^2} \qquad (28)$$

$$t = 0 \qquad 0 \leq y < \delta, \quad C = C_i$$
$$t = 0 \qquad y > \delta, \quad C = C_b \qquad\qquad (29)$$
$$t > 0 \qquad y = \infty, \quad C = C_b, \text{ and } y = 0, \frac{\partial C}{\partial y} = 0 .$$

The solution is:

$$C(y,t) = \frac{1}{2}(C_i - C_b)\{ \text{erf}(\frac{\delta - y}{2\sqrt{D_{cw}t}}) + \text{erf}(\frac{\delta + y}{2\sqrt{D_{cw}t}})\} + C_b \qquad (30)$$

where,

$$\text{erf}(x) = \frac{2}{\sqrt{\pi}} \int_0^x e^{-\xi^2} d\xi \quad \text{which is the error function of } x.$$

The interchange ratio coefficient, f_1, should then satisfy the relationship of Eq. 31 and should be expressed in Eq. 32 by integrating the right hand side of Eq. 31.

$$(1 - f_1)C_i + f_1 C_b = \frac{1}{\delta} \int_0^\delta C(y,\bar{t}_w) dy \qquad (31)$$

$$f_1 = \frac{\sqrt{D_{cw}\bar{t}_w}}{\delta} \{ \frac{1}{\sqrt{\pi}} - \text{ierfc}(\delta/\sqrt{D_{cw}\bar{t}_w})\} \qquad (32)$$

It is, however, more difficult to mathematically evaluate f_2. Therefore, it is necessary to return to the results of tracer observations described previously.

With regard to the large-scale water motion which appeared near the bulk water surface where the disk is withdrawn, Zeevalkink et al. (1978) discussed velocity profile of the backward flow to the bulk water. There has, however, been no attempt to relate this large scale motion to the mixing mechanism between the water film on the disk and the bulk water.

The basic assumption to formulate the large scale mass interchange ratio coefficient, f_2, is stated below.

The boundary layer flow near the rotating disk loses its kinematic energy at the stagnation points defined by Zeevalkink et al. (1978), close to the water surface. Part of the dissipated energy causes mixing of the water mass between the boundary layer and the water film. It is necessary to distinguish between the boundary layer which belongs to the bulk water and the water film in the boundary layer. The boundary layer used here means the boundary layer determined by Schlichting (1968) for a completely submerged rotating disk. Its thickness, L, is determined by Eq. 33 and Eq. 34 for the laminar condition and the turbulent condition, respectively.

Laminar condition: $L = \alpha_1' \sqrt{\nu/\omega}$ (33)

Turbulent condition: $L = \alpha_2' R^{3/5} (\nu/\omega)^{1/5}$ (34)

where,

α_1', α_2' = constants, and

ν = kinematic viscosity of water.

On the other hand, the thickness of the water film was given theoretically by Zeevalkink et al. (1978) as:

$$\delta = \frac{4}{15} \sqrt{\frac{2\nu}{g}} \sqrt{\omega R \sin\phi}$$ (35)

where,

g = acceleration of gravity, and

$\phi = \arccos(R_i/R)$.

The first approach or the first assumption is to formulate a relationship between f_2 and hydraulic properties concerning the mixing. One typical hydraulic property is the energy dissipation rate per unit mass of water at the relevant water surface. The assumed energy dissipation rate, ε, can be defined by Eq. 36 and formulated as Eq. 37.

$$\varepsilon \propto \frac{\left\{\begin{array}{l}\text{Kinematic energy of the boundary layer}\\ \text{just crossing the bulk water surface}\\ \text{in a vertical direction to the bulk water}\\ \text{surface per unit time}\end{array}\right.}{\left\{\begin{array}{l}\text{The mass of the water film crossing the}\\ \text{bulk water surface per unit time}\end{array}\right.}$$ (36)

$$\varepsilon = \frac{\beta_1 \rho (R\omega\sin\phi)^2 (\omega R^2 \sin^2\phi) L}{\rho(\omega R^2 \sin^2\phi)\delta} = \beta_1 (R^2\omega^2 L \sin^2\phi)/\delta$$ (37)

where,

β_1 = Proportional constant,

ρ = density of water,

L = thickness of the boundary layer, and

δ = thickness of the water film.

The first hypothetical relationship between f_2 and ε is linear as shown in Eq. 38.

$$f_2 = \beta\varepsilon$$ (38)

By substituting Eqs. 33, 34, 35 and 37 into Eq. 38, the latter can be rewritten as Eqs. 39 and 40 for laminar and turbulent conditions, respectively.

Laminar condition: $\qquad f_2 = \alpha_1 \omega R^{1.5} (\sin\phi)^{1.5}$ (39)

Turbulent condition: $\qquad f_2 = \alpha_t \nu^{-0.3} \omega^{1.3} R^{2.1} (\sin\phi)^{1.5}$ (40)

where, $\qquad \alpha_1 = \dfrac{15\sqrt{2g}}{8} \beta_1 \beta \alpha_1'$, and

$$\alpha_t = \dfrac{15\sqrt{2g}}{8} \beta_1 \beta \alpha_2'.$$

These are coefficients which should be experimentally determined, and should have a unique value for each specific rotating disk.

It should be noted that f_2 is valid only in the range of $0 \le f_2 \le 1$.

It is emphasized that the assumptions made in the theoretical part of the present study should be examined or verified by experiment. Before proceeding to experimental examination, a term from Eq. 19 needs to be quantitatively evaluated, i.e. the mean air retention time, $\bar{t}_a$.

Yamane and Yoshida (1972) showed that $\bar{t}_a$ could be evaluated by Eq. 41, using the relationship of $\bar{k}_L \approx 2\sqrt{D_{cw}/\pi\bar{t}_a}$ under the condition of $\delta / \sqrt{D_{cw}\bar{t}_a} \ge 1.7$.

$$(\tilde{\omega}/60) \cdot \bar{t}_a = \frac{\left[\int_H^1 \zeta\{1 - \frac{1}{\pi}\cos^{-1}(H/\zeta)\}^{3/2} d\zeta \right]^2}{\left[\int_H^1 \zeta\{1 - \frac{1}{\pi}\cos^{-1}(H/\zeta)\} d\zeta \right]^2}$$ (41)

where,

$H = R_i/R$ = dimensionless distance between the bulk water surface and disk axis.

Bintanja et al. (1975) gave Eqs. 42 and 43 to evaluate $\bar{t}_a$, corresponding to the case of $\delta / \sqrt{D_{cw}\bar{t}_a} \ge 1.7$ and $\delta / \sqrt{D_{cw}\bar{t}_a} < 0.8$, respectively.

$$(\tilde{\omega}/60)\bar{t}_a = \frac{\left[\int_H^1 \zeta\{1 - \frac{1}{\pi}\cos^{-1}(H/\zeta)\} d\zeta \right]^2}{\left[\int_H^1 \zeta\{1 - \frac{1}{\pi}\cos^{-1}(H/\zeta)\}^{1/2} d\zeta \right]^2}$$ (42)

$$(\tilde{\omega}/60)\bar{t}_a = \frac{\int_H^1 \zeta\{1 - \frac{1}{\pi}\cos^{-1}(H/\zeta)\} d\zeta}{\int_H^1 \zeta d\zeta}$$ (43)

TABLE A.1. Numerical comparison of Eqs. 41, 42, and 43

H	\multicolumn{3}{c}{$(\tilde{\omega}/60)\cdot\bar{t}_a$}
	Eq. 41	Eq. 42	Eq. 43
0	0.5 100 %	0.5 100 %	0.5 100 %
0.1	0.561 100 %	0.560 99.8%	0.559 99.3%
0.2	0.615 100 %	0.612 99.5%	0.611 99.3%
0.3	0.662 100 %	0.658 99.4%	0.657 99.2%
0.4	0.704 100 %	0.701 99.6 %	0.699 99.3%
0.5	0.743 100 %	0.740 99.6 %	0.739 99.5%
0.6	0.780 100 %	0.778 99.7 %	0.777 99.6%
0.7	0.818 100 %	0.816 99.8%	0.815 99.6 %
0.8	0.856 100 %	0.855 99.9 %	0.855 99.9%
0.9	0.902 100 %	0.901 99.9%	0.901 99.9%
1.0	1.0 100 %	1.0 100 %	1.0 100 %

A numerical comparison of Eqs. 41, 42 and 43 are shown in Table 1. The results show that there is little difference in these evaluations.   In the present analysis, the equation presented by Yamane and Yoshida, Eq. 41, was used as the standard method for the evaluation of $\bar{t}_a$. The mean retention time in the bulk water, $\bar{t}_w$, is determined by subtracting $\bar{t}_a$ from T, which represents a period of one rotation, as follows:

$$\bar{t}_w = T - \bar{t}_a . \tag{44}$$

For the numerical evaluation of the molecular diffusivity of oxygen in water, the relationship of Eq. 45 was used (Hirayama et al., 1982):

$$D_{cw} = 7.87 \times 10^{-10} (T_K/\mu) \tag{45}$$

where,

T_k = absolute temperature, and
μ = viscocity of water.

A. 4   Verification of the Model

A.4.1   Experimental Equipment and Method

    Four types of single-stage rotating disk contactors were used. Their dimensions and other specifications are shown in Table 2. Symbols L and S refer to the difference in the flat disk size; L corresponds to 90 cm in diameter and S to 30 cm. Symbols U and B denote different shaped container bottoms and clearances between the disk and the bottom; U has a cylindrical bottom and smaller clearance and B has a flat bottom and larger clearance. Polyvinyl chloride (PVC) was used as disk material for S-type contactors. Painted waterproof plywood was used for both L- and S-type contactors.

TABLE A.2.   Dimensions and other specifications of rotating disk contactors

Contactor type	LU	SU	LB	SB
Diameter of disk , cm	90	30	90	30
Thickness of disk , cm	0.9	0.1-0.9	0.9	0.1-0.5
Mutual distance of disks , cm	2.0-13.5	1.5-4.5	4.5	1.5
Number of disks	2 – 6	7 – 18	6	18
Net volume of container, L	106-115	12.9-14.0	164	24.1
Length of container , cm	36.5	40.0	36.5	40.0
Width of container , cm	99.9	33.4	100	40.0
Total contact area air water, A_c, m^2	1.83-4.76	0.70-1.59	4.82	1.63
of which contributed by the disks, A_a, m^2	1.44-4.44	0.58-1.47	4.44	1.47
by the bulk water surface , A_f, m^2	0.35-0.32	0.12-0.12	0.38	0.16

U-Type   B-Type

As expected, the following results show that there was no difference between materials. Each disk was designed to be removed from a shaft to change the number of disks and their spaces. The submergence depth of disk was changed by altering the volume of the bulk water and hence the water surface level.

Deoxygenation of the bulk water was carried out by bubbling nitrogen gas, or by adding sodium sulphite and cobalt chloride as a catalyst. Almost the same results were obtained by both deoxygenation methods. The increase in dissolved oxygen concentration in the bulk water was continuously measured with an oxygen probe during aeration by disk rotation.

Since it was confirmed that the recorder gave a linear reading with dissolved oxygen concentration, even in the case of the lowest rotational speed, values of the overall volumetric oxygen transfer coefficient, K_L^*a, were calculated from the recorder readings by using the logarithmic difference method and the moment method (Hashimoto et al., 1970).

Values of the specific contact surface area, a, were calculated by dividing the total contact surface area, A_c, by the net volume of the bulk water, V. A_c is the sum of the wetted surface area of the disks exposed to the air, A_a, and the free surface area of the bulk water, A_f. In most cases, A_f is almost negligible as shown in Table 2. The values of K_L^*, the overall oxygen mass transfer coefficient, were determined by dividing the values of measured K_L^*a by a.

A.4.2 Results and Discussion

The effect of different disk materials on K_L^* was examined by using the same size disks in the same container. The relationship between the rotational speed and K_L^* is shown in Fig. 3; it is obvious that there is no effect due to the difference between the PVC disk and the plywood disk on the oxygen mass transfer coefficient.

The effect of different bottom configurations of the container on K_L^* was examined. As shown in Fig. 4, the values of K_L^* obtained in the cylindrical bottom container are as almost same as those obtained in the flat bottom container.

The effect of varying the spacing between disks on K_L^* was also examined. There is no significant difference in the K_L^* values, as shown in Figs. 5 and 6. The lines in these figures represent three theoretical relationships of Eq. 25 for the laminar model, that for the turbulent model and Eq. 19.

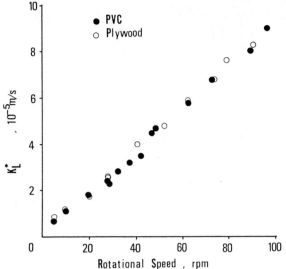

Fig. A.3. Effect of disk material on the overall oxygen
transfer coefficient, K_L^*. The SU-type contactor was used
and the experimental conditions were as R = 15 cm, H = 0.14,
d_i= 4.5 cm, and water temperature (w.t.) = 25.9 $\pm$ 0.7 °C.

Contactor type	d_i (cm)	w.t. (°C)	
△ LU	4.5	23.4-25.6	R=45, H=0.14
□ LB	4.5	24.4-26.4	
■ SB	1.5	23.3-26.4	R=15, H=0.14
▲ SU	1.5	24.7-26.2	

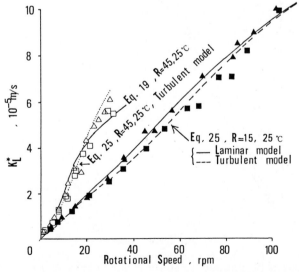

Fig. A.4. Effect of bottom configuration of the container
on the overall oxygen transfer coefficient, K_L^*.

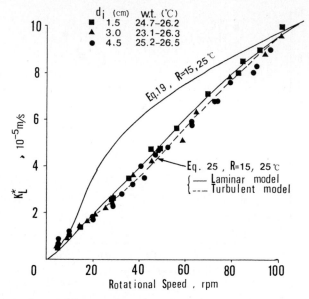

Fig. A.5. Effect of space between the disks, d_i, on the overall oxygen transfer coefficient K_L^*. The SU-type contactor was used with the disk radius of 15 cm and H = 0.14.

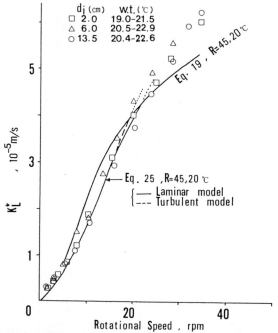

Fig. A.6. Effect of space between the disks, d_i, on the overall oxygen transfer coefficient K_L^*. The LU-type contactor was used with the disk radius of 45 cm and H = 0.14.

Considering all these effects on K_L^*, it was concluded that the mixing between the water film and the bulk water was not significantly affected by macroscopic flow characteristics in the contactor.

The relationship between the overall mass interchange ratio coefficient, f, and the disk rotational speed is shown in Fig. 7. The theoretical values of f were determined by using the relationships of Eqs. 27, 32, 39 and 40. Every variable and constant in Eq. 32 was theoretically determined by another equations, for instance Eqs. 35, 41, 44 and 45. The values of the coefficient α_1 and α_t in Eqs. 39 and 40 were determined by the experimental data plotted in Fig. 7. Theoretically, only one set of experimental conditions of ω, R, H and water temperature at the critical state of $f = f_2 = 1.0$ should be needed to determine the values of α_1 and α_t. Once these values are determined, they should have a universal characteristic for all experimental conditions.

Although there remain somewhat uncertain points in the determination of the critical state, where $f = f_2 = 1.0$, a linear extrapolation method was applied to the data, in which the rotational speeds were higher than 80 rpm. From the linear extrapolation, it was determined that $f = 1.0$ at $\omega = 110$ rpm in the case of R = 15 cm and H = 0.14. The values of α_1 and α_t were then calculated to be 1.52×10^{-3} ($cm^{-3/2}$ s) and 3.48×10^{-5} ($cm^{-3/2}$ s) using Eqs. 39 and 40, respectively, at the water temperature of 25 °C. The standard temperature of 25 °C was chosen from the range of 23.1 °C to 26.5 °C at which all experiments were carried out.

As stated above, once the values of α_1 and α_t had been determined, the theoretical prediction of the overall mass interchange ratio coefficient, f, was made by Eq. 27 in which the rotational speed and diameter of the disk were variable. The comparison between the theoretical values of f and the experimental results are also shown in Fig. 7. It was observed that the turbulent model generally agreed with experimental data better than the laminar model.

The comparison of theoretical values of K_L^*, which were determined by Eqs. 19, 25 for the laminar model and 25 for the turbulent model, with experimental data are shown in Figs. 3, 4, 5, 6 and 8. In the case of Fig. 8, the depth of disk submergence was varied. From these figures, it is clear that the fitness of the theoretical values from Eq. 25 to the experimental data is quite good, while those from Eq. 19 extreamly disagreed with the measured

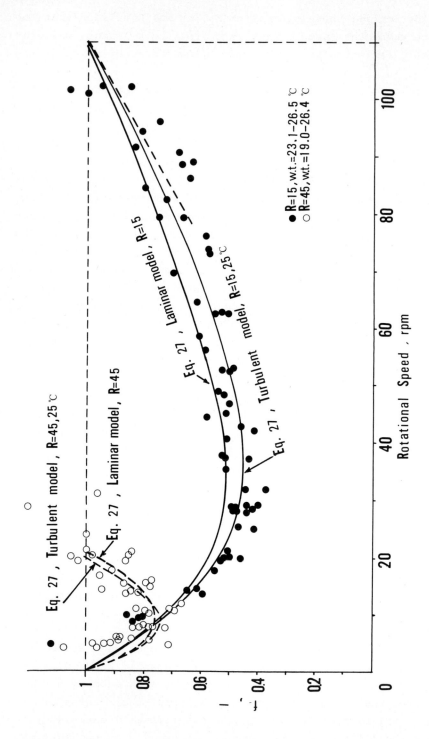

Fig. A.7. Relationship between the overall mass interchange ratio coefficient, f, and the disk rotational speed.

values.   The agreement shows that the mathematical model developed
in the present study has a reasonable basis and wide applicability.

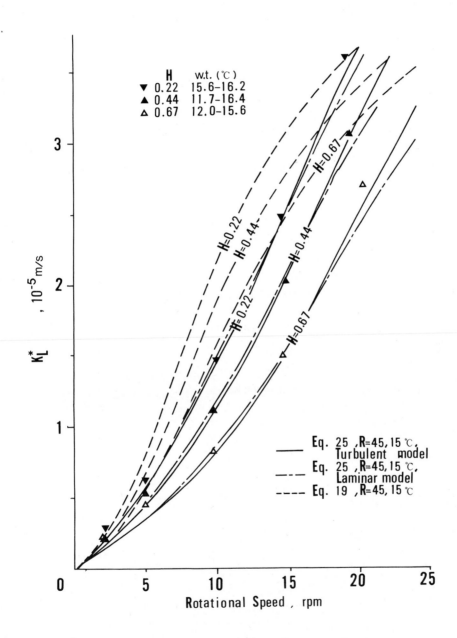

Fig. A.8. Effect of the depth of disk submergence on the
           overall oxygen transfer coefficient, K_L^*. The LU-
           type contactor was used with the disk rasdius of
           45 cm.

A.4.3 Application of the Turbulent Model in the Design of Rotating
Contactors

With the validity of the relationship of Eq. 25 proven, future applications of this relationship should be valuable in solving scale-up problems, estimation of aeration capacity and the design of rotating contactors. With respect to aeration capacity, K_L^*a should be quantitatively evaluated rather than K_L^*. However, considering the experimental results which showed that K_L^* is not dependent on the space between disks and the configuration of the container, and the fact that the specific contact area, a, can be designed independently of the disk size, the dependence of K_L^* on the disk size is an essential factor to scale up the aeration ability.

Examples of the analytical solutions of Eq. 25 for turbulent model are shown in Figs. 9 and 10, in which the relationships between disk size and K_L^* are examined under the constant rotational speeds and peripheral velocities. The following conclusions have been drawn from the results of Figs. 9 and 10:

1. K_L^* increased as disk radius increased at constant rotational speed, but decreased as disk radius increased at constant peripheral velocity;

2. at low rotational speeds, it was effective to use large size disk to achieve a large K_L^*; and

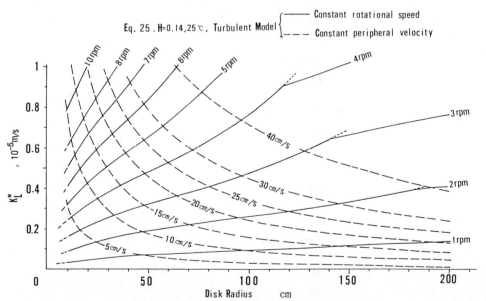

Fig. A.9. Relationship between disk radius and K_L^*
(in the case of $\tilde{\omega}$ < 10 rpm)

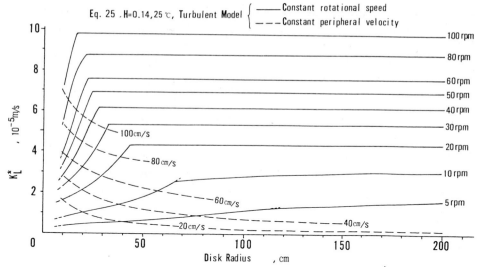

Fig. A.10. Relationship between disk radius and K_L^* (in the case of $5 < \tilde{\omega} < 100$ rpm)

3 K_L^* was almost independent on the disk size at high rotational speeds.

A. 5. Conclusions

In the present study, the physical oxygen mass transfer mechanism in the rotating disk gas-liquid contactor was investigated theoretically and experimentally. The main conclusions are summarized below.

1. It is reasonable to build a model of the overall oxygen mass transfer process from the air into the bulk water by dividing into two steps , one for the air phase and the other for the water phase.

2. In general, under the lower rotaional speeds, the overall oxygen transfer process is controlled by the transfer in the water phase, i.e. the mixing process. On the other hand, under the higher rotaional speeds, it is controlled by the transfer process in the air.

3. Introduction of the concept of mass interchange ratio coefficient and quantitative evaluations of f, f_1 and f_2 were very effective in the analyses of oxygen mass transfer.

4. The overall oxygen mass transfer coefficient, K_L^*, evaluated by Eq. 25 proved to be valid for various experimental conditions, such as various disk sizes, submergence depths of the disk and rotational speeds.

5. It is very important to know which process controls the rate of oxygen transfer in order to scale up the aeration capacity.

REFERENCES

Bintanja H.H.J., van der Erve J.J.V.M. & Boelhouwer C., 1975. Oxygen transfer in a rotating disc treatment plant. Water Research 9, 1147-1153.

Hashimoto S. & Fujita M., 1970. Theoretical study of overall volumetric oxygen transfer coefficient in wastewater treatment. Water Purification & Liquid Wastes Treatment 11, 25-38, (in Japanese).

Hirayama K. & Matsuo T., 1982. Study on oxygen transfer models. Wat. Sci. Tech. 14, 321-329.

Ouano E. A. R., 1981. Response to the comment by J. A. Zeevalkink on "Oxygen mass transfer scale up rotating biological filters" by Ouano (Water Research 12, 1005-1008 (1978)). Water Research 15, 1129-1130.

Schlichting H., 1968. Boundary-Layer Theory (6th ed.). McGraw-Hill , New York.

Yamane T. & Yoshida F., 1972. Absorption in a rotating-disk gas-liquid contactor. Journal of Chemical Engineering Japan 5, 381-385.

Zeevalkink J.A., Kelderman P. & Boelhouwer C., 1978. Liquid film thickness in a rotating disc gas-liquid contactor. Water Research 12, 577-581.

Zeevalkink J.A., Kelderman P., Visser D.C. & Boelhouwer C., 1979. Physical mass transfer in a rotating disc gas-liquid contactor. Water Research 13, 913-919.

SYMBOLS

a = Specific gas-liquid contact area $\qquad$ (cm^2/cm^3)

A_a = Wetted area of the disks exposed to air $\qquad$ (cm^2)

A_c = Total gas-liquid contact area $\qquad$ (cm^2)

A_f = Surface area of the bulk water $\qquad$ (cm^2)

C = Dissolved oxygen concentration in the water film $\qquad$ (mg/l)

C^* = Saturation concentration of dissolved oxygen $\qquad$ (mg/l)

C_i, C_o = Initial concentrations of dissolved oxygen in the water film at the points of submergence and at the points of emergence, respectively $\qquad$ (mg/l)

C_b = Dissolved oxygen concentration in the bulk water $\qquad$ (mg/l)

$\bar{C}_o$ = Mean of C_o for entire radial distance $\qquad$ (mg/l)

d_i = Space between disks $\qquad$ (cm)

D_{cw} = Molecular diffusivity of oxygen in water $\qquad$ (cm^2/s)

f = Overall oxygen mass interchange ratio coefficient $\qquad$ (-)

f_1 = Interchange ratio coefficient by molecular diffusion $\qquad$ (-)

f_2 = Interchange ratio coefficient by turbulent mixing $\qquad$ (-)

g = Acceleration of gravity $\qquad$ (cm/s^2)

$H = R_i/R$ = Dimensionless distance between the bulk water
surface and the disk axis $\qquad$ (-)

k = Dimensionless parameter defined by Eq. 21 $\qquad$ (-)

k_L = Average oxygen transfer coefficient for the element dA $\qquad$ (cm/s)

K_L = Oxygen transfer coefficient (mean of k_L for entire
contact area of the disk) $\qquad$ (cm/s)

K_L^* = Overall oxygen transfer coefficient $\qquad$ (cm/s)

L = Thickness of the boundary layer $\qquad$ (cm)

N_t = Oxygen flux from the air into the water film $\qquad$ (μg/(s$\cdot$cm^2))

r = Radial distance from the disk axis to the points
of submergence and emergence $\qquad$ (cm)

R = Radius of the disk $\qquad$ (cm)

R_1 = Distance between the bulk water surface and the disk axis $\quad$ (cm)

t_a, t_w = Retention times of the element dA in the air
and in the bulk water, respectively $\qquad$ (s)

$\bar{t}_a$ = Mean air retention time (Mean of t_a for entire contact
area of the disk) $\qquad$ (s)

$\bar{t}_w = T - \bar{t}_a$ = Mean retention time in the bulk water $\qquad$ (s)

T = Time per rotation $\qquad$ (s)

T_k = Absolute temperature $\qquad$ (K)

V = Net volume of water in the container $\qquad$ (cm^3)

y = Distance from the disk surface $\qquad$ (cm)

α_1', α_2' = Proportional constants
in Eqs. 33 and 34, respectively $\qquad$ (-)

α_1, α_t = Proportional constants
in Eqs. 39 and 40, respectively $\qquad$ (cm$^{-3/2}$s)

β = Proportional constant in Eq. 38 $\qquad$ (s^2/cm^2)

β_1 = Proportional constant in Eq. 37 $\qquad$ (-)

δ = Thickness of the water film on the disk $\qquad$ (cm)

ϵ = Energy dissipation rate defined by Eq. 36 $\qquad$ (cm^2/s^2)

μ = Viscosity of water $\qquad$ (g/(s$\cdot$cm))

ν = Kinematic viscosity of water $\qquad$ (cm^2/s)

ρ = Density of water $\qquad\qquad$ (g/cm^3)

φ = arccos(R_i/r)

Φ = arccos(R_i/R)

ω = Angular velocity $\qquad\qquad$ (rad/s)

$\bar{\omega}$ = Rotational speed $\qquad\qquad$ (rpm)

$$erf(x) = \frac{2}{\sqrt{\pi}} \int_0^x exp(-\xi^2)d\xi$$

$\qquad$ = Error function of x

$$erfc(x) = \frac{2}{\sqrt{\pi}} \int_x^\infty exp(-\xi^2)d\xi$$

$\qquad$ = Complementary error function of x

$$ierfc(x) = \int_x^\infty erfc(\xi)d\xi$$

$\qquad$ = Integral complementary error function of x

PART B: UNSTEADY-STATE OXYGEN UPTAKE AND SUBSTRATE
          UTILIZATION IN THE BIOFILM

B.1   Introduction

   A normal aerobic rotating biological contactor (RBC) has disks
which are half submerged in the bulk water . The biological film
fixed on the disk surface is rotated alternately between the air and
the bulk water.   In Part A of the present study, the authors
developed a mathematical model of physical oxygen mass transfer in
the rotating-disk contactor by introducing the concept of the mass
interchange ratio coefficient, f. However, it should be noted that
the physical oxygen transfer model developed in Part A can not be
directly applied to the actual biological rotating disk process,
since oxygen consumption in the biological film on the rotating disk
surface must first be taken into account. It should be emphasized
that oxygen consumption and substrate utilization in the biofilm
should finally control the overall performance of the rotating disk
biological contactor.

   There are several general models for steady-state biofilm
kinetics (Atkinson and Davies ,1974 ; Rittmann and McCarty,
1978,1981 ; Harremoës, 1977). Applications of the steady-state
models for submerged and/or half submerged rotating disk processes
were   made by Grieves (1972), Harremoës (1978), Watanabe et
al.(1982a),and Shieh (1982).   However, the steady-state models
cannot be unconditionally applied to the half submerged rotating
disk process, because the biofilm fixed on the disk experiences
cyclic changes of environment i.e. between the air phase and the
water phase. For example, the supply of substrate into the biofilm
is limited in the air and the concentration of the substrate varies
with the travelling time in the air. In order to analyse the mass
transfer mechanisms concerning the inside of the biofilm on the
rotating disk, it should be noted that the mechanism be essentially
described as an unsteady-state diffusion and reaction process.

   Famularo et al. (1978) calculated the time-dependent BOD
(Biochemical Oxygen Demand) and DO (Dissolved Oxygen) concentration
profiles inside the biofilm on the disk during a revolution by
dividing the biofilm into eight stationary sectors on its surface
and into four layers in its depth. Watanabe et al. (1982b) gave the

numerical simulation of the unsteady-state mass transfer for the nitrification process. Harremoës and Gonenc (1983) introduced a correction factor into their steady-state model to allow for the deviation due to unsteady-state. There has been, however, no analytical approach to the unsteady-state mass transfer in the rotating biological disk process.

In Part B of the present study, a new model has been developed to predict the oxygen consumption and ammonium nitrogen removal for the nitrification process in a rotating biological contactor. An attempt to obtain the analytical solutions of the unsteady-state mass transfer model will be shown with some experimental results which support the solutions. Effects of rotational speed, disk diameter and dissolved oxygen concentration in the bulk water on the ammonium nitrogen removal rate will be also examined.

B.2   Modelling of Unsteady-State Mass Transfer into the Biological Film

To develop a mathematical model of the unsteady-state mass transfer for nitrification in a rotating disk biological contactor, two simplified cases are assumed: the case of oxygen limitation so that nitrification is limited by lack of oxygen, and the case of substrate limitation in which the nitrification is limited by ammonium nitrogen. Further assumptions are listed below.

1. A uniform biofilm is attached to a disk. The penetration depth of the limiting substance into the biofilm is small enough that the thickness of the biofilm is assumed to be infinite.

2. A uniform water film is attached to the biofilm. The thickness of this water film has the same value as that attached to a bald disk without biofilm.

3. The reaction rate of nitrification in the biofilm is first order.

4. There is no accumulation of nitrite, so that the ammonium oxidation process limits the total oxidation process of nitrification.

5. Only the diffusion process in a vertical direction to the disk surface is considered; the diffusion process in radial and circumferential directions are neglected.

6. At the initial points of disk submergence (see Fig. 1), the attached water film is not disturbed and smoothly goes into the bulk water. At the initial points of disk emergence, however, the attached water film is disturbed by the turbulent motion of

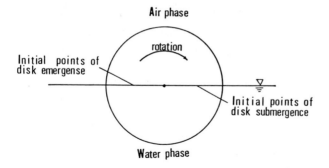

Fig. B.1.   Diagram of rotating disk

(A) Oxygen limitation

(1) Air phase   (2) Water phase

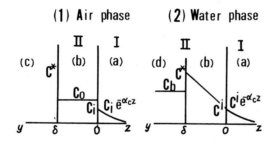

(B) Substrate limitation

(3) Water phase   (4) Air phase

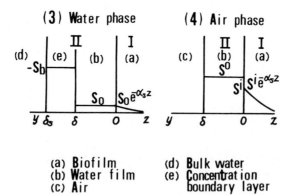

(a) Biofilm   (d) Bulk water
(b) Water film   (e) Concentration
(c) Air   boundary layer

Fig. B.2.   Assumed concentration of limiting substance
   at the initial points of disk submergence and disk
   emergence.

the flow near the disk, and the dissolved oxygen (DO) concentration is unified in the water film.

7. Regarding initial conditions, the concentration of the limiting substance decreases exponentially from the biofilm surface to its bottom and has a linear profile in the water film at both initial points of disk submergence and emergence (see Fig. 2).

These assumptions are discussed later.

B.2.1 Differential Equations and Solutions

B.2.1.1 Oxygen Limitation

B.2.1.1.1 Air Phase (Case 1)

Oxygen is transferred from the air into the biofilm through the water film in the air phase. Since no biological reaction takes place in the water film, only molecular diffusion needs to be considered. In the biofilm, however, biological reactions and molecular diffusion occur simultaneously. Thus, differential equations and initial and boundary conditions in the biofilm and water film are given as follows:

I. Biofilm
$$\frac{\partial C_I}{\partial t} = D_{cb}\frac{\partial^2 C_I}{\partial z^2} - \mu_c C_I \tag{1}$$

$$t = 0, \ z \geq 0, \ C_I = C_i \exp(-\alpha_c z)$$
$$t > 0, \ z = \infty, \ C_I = 0 \tag{2}$$

II. Liquid Film
$$\frac{\partial C_{II}}{\partial t} = D_{cw}\frac{\partial^2 C_{II}}{\partial y^2} \tag{3}$$

$$t = 0, \ \delta \geq y \geq 0, \ C_{II} = C_o$$
$$t > 0, \ y = \delta, \ C_{II} = C^* \tag{4}$$

$$t > 0, \ y = z = 0, \ C_I = C_{II} \ , \ D_{cb}\frac{\partial C_I}{\partial z} = - D_{cw}\frac{\partial C_{II}}{\partial y} \tag{5}$$

where,

C_I and C_{II} = dissolved oxygen concentrations in the biofilm and bulk water, respectively,

C_i = initial dissolved oxygen concentrarion at the biofilm surface,

C^* = saturation concentration of dissolved oxygen,

D_{cb} and D_{cw} = molecular diffusivities of oxygen in biofilm and water, respectively, y, z = distance from biofilm surface,

t = time,

δ = thickness of the water film,

μ_c = reaction rate constant, and

$\alpha_c = \sqrt{\mu_c/D_{cb}}$ = constant.

The above equations can be solved by using the Laplace transform method. The solution of Eq. 1 is given by:

$$C_I = (C_i - \frac{\eta_c^2 \alpha_c \delta}{1 + \eta_c^2 \alpha_c \delta} \cdot C^*) \cdot \exp(-\alpha_c z)$$

$$+ \int_0^t \{(C^* - C_o)\,\mathrm{erfc}(\frac{\delta}{2\sqrt{D_{cw}}(t-\xi)})$$

$$+ C_o - C_i\} \frac{z}{2\sqrt{\pi D_{cb}}\xi} \exp(-\mu_c \xi - \frac{z^2}{4D_{cb}\xi})d\xi$$

$$+ (C^* - C_o)\{ \frac{2}{\pi} \cdot \int_0^{\beta_c} \frac{\sin(\delta\xi/\sqrt{D_{cw}})}{\xi} \exp(-\xi^2 t - \frac{z^2}{4D_{cb}\xi})d\xi \cdot$$

$$+ \int_{\beta_c}^{\infty} \psi_1 \cdot \frac{\sin(\delta\xi/\sqrt{D_{cw}})}{\xi} \exp(-\xi^2 t) \cdot d\xi \} \tag{6}$$

where,

$$\eta_c = \sqrt{D_{cb}/D_{cw}} = \text{constant},$$

$$\beta_c = \sqrt{\mu_c} = \text{constant},$$

$$\psi_1 = -\frac{2}{\pi} \cdot \frac{1}{\{\xi^2 \cos^2(P) + \eta_c^2(\xi^2 - \mu_c)\sin^2(P)\}(C^* - C_o)} \cdot \left[(C^* - C_o) \cdot \right.$$

$$\cdot (\eta_c \sqrt{\xi^2 - \mu_c} - \xi)\{\xi\cos(P)\cos(P+Q) - \eta_c\sqrt{\xi^2 - \mu_c} \cdot \sin(P)\sin(P+Q)\}$$

$$+ (C_o - C_i)\eta_c\sqrt{\xi^2 - \mu_c}\{\xi\cos(P)\cos(Q) - \eta_c\sqrt{\xi^2 - \mu_c} \cdot \sin(P)\sin(Q)\}$$

$$\left. - \eta_c\sqrt{\mu_c} \cdot C_i\{\xi\cos(P)\sin(Q) + \eta_c\sqrt{\xi^2 - \mu_c} \cdot \sin(P)\cos(Q)\} \right],$$

$$P = \delta\xi/\sqrt{D_{cw}} \quad, \text{ and}$$

$$Q = z\sqrt{\xi^2 - \mu_c}/\sqrt{D_{cb}} \quad.$$

The solution of Eq. 3 is given by:

$$C_{II} = C_o + (C^* - C_o) \mathrm{erfc}(\frac{\delta - y}{2\sqrt{D_{cw}}\,t}) - \frac{\eta_c^2 \alpha_c (\delta - y)}{1 + \eta_c^2 \alpha_c \delta} \cdot C^*$$

$$+ (C^* - C_o) \cdot \left[\frac{2}{\pi} \int_0^{\beta_c} \sin(\frac{(\delta - y)\xi}{\sqrt{D_{cw}}}) \cdot \frac{\exp(-\xi^2 t)d\xi}{\xi} \right.$$

$$\left. + \int_{\beta_c}^{\infty} \psi_2 \sin(\frac{(\delta - y)}{\sqrt{D_{cw}}}) \cdot \frac{\exp(-\xi^2 t)d\xi}{\xi} \right] \tag{7}$$

where,

$$\psi_2 = -\frac{2}{\pi} \left[\frac{(\eta_c\sqrt{\xi^2 - \mu_c} - \xi)\{\xi\cos^2(P) - \eta_c\sqrt{\xi^2 - \mu_c}\cdot\sin^2(P)\}}{\xi^2\cos^2(P) + \eta_c(\xi^2 - \mu_c)\sin^2(P)} \right.$$

$$\left. + \frac{\eta_c\sqrt{\xi^2 - \mu_c}\{\xi(C_o - C_i)\cos(P) - \eta_c C_i \sqrt{\mu_c}\sin(P)\}}{\{\xi^2\cos^2(P) + \eta_c(\xi^2 - \mu_c)\sin^2(P)\}(C^* - C_o)} \right], \text{ and}$$

$$\mathrm{erfc}(x) = (2/\sqrt{\pi})\int_x^{\infty} \exp(-\xi^2)\,d\xi.$$

The average oxygen flux from the air into the liquid film $(\overline{N_{ca}})$ is defined as:

$$\overline{N_{ca}} = \frac{1}{\overline{t}_a} \int_0^{\overline{t}_a} D_{cw}(\frac{\partial C_{II}}{\partial y})_{y=\delta}\,dt \tag{8}$$

where, $\overline{t}_a$ = mean air retention time.

Substitution of Eq. 7 into Eq. 8 and integration of Eq. 8 yields:

$$\overline{N_{ca}} = \frac{\eta_c^2 \alpha_c D_{cw}}{1 + \eta_c^2 \alpha_c \delta} \cdot C^*$$

$$+ 2(C^* - C_o)\frac{\sqrt{D_{cw}}}{\sqrt{\pi \overline{t}_a}} \left[\mathrm{erfc}(\sqrt{\mu_c \overline{t}_a}) + \frac{1 - \exp(-\mu_c \overline{t}_a)}{\sqrt{\pi \mu_c \overline{t}_a}} \right]$$

$$- (C^* - C_o)\sqrt{D_{cw}} \cdot \frac{1}{\overline{t}_a} \cdot \int_{\beta_c}^{\infty} \psi_2 \frac{1 - \exp(-\xi^2 \overline{t}_a)}{\xi^2}\,d\xi. \tag{9}$$

The first term on the right hand side of Eq. 9 is called the steady-state term in this paper, because Eq. 9 becomes:

$$\overline{N}_{ca} = \frac{\eta_c^2 \alpha_c D_{cw}}{1 + \eta_c^2 \alpha_c \delta} \cdot C^* \quad , \quad as \quad t_a \to \infty \ . \tag{10}$$

However, other terms on the right hand side of Eq. 9 are called the unsteady-state terms which represent the contribution of unsteady-state oxygen transfer from the air. Since the last term on the right hand side of Eq. 9 is usually smaller than the other terms, this equation is reduced to:

$$\overline{N}_{ca} = \frac{\eta_c^2 \alpha_c D_{cw}}{1 + \eta_c^2 \alpha_c \delta} \cdot C^*$$

$$+ 2(C^* - C_o)\frac{\sqrt{D_{cw}}}{\sqrt{\pi t_a}} \left[\text{erfc}(\sqrt{\mu_c \overline{t}_a}) + \frac{1 - \exp(-\mu_c \overline{t}_a)}{\sqrt{\pi \mu_c \overline{t}_a}} \right] . \tag{11}$$

B.2.1.2 Water Phase (Case 2)

In the water phase, excluding the points of disk emergence, it is assumed that oxygen is transferred between the bulk water and biofilm through the attached water film, which has the same thickness, δ, as in the air phase.

The differential equation for the biofilm is also given by Eq. 1, but with the following initial and boundary conditions:

$$t = 0, \ z \geq 0, \ C_I = C^i \exp(-\alpha_c z)$$
$$t > 0, \ z = \infty, \ C_I = 0 \tag{12}$$

where,

C^i = dissolved oxygen concentration in the biofilm surface.

The differential equation for the water film and bulk water is also described by Eq. 3. The initial and boundary conditions are:

$$t = 0, \ \delta \geq y \geq 0, \ C_{II} = \varkappa_2 y + C^i$$

$$t = 0, \qquad y \geq 0, \ C_{II} = C_b \tag{13}$$

$$t > 0, \qquad y = \infty, \ C_{II} = C_b$$

where,

$$\varkappa_2 = (C^* - C^i)/\delta \quad , \ and \tag{14}$$

C_b = dissolved oxygen concentration in the bulk water.

Another boundary condition for a smooth connection of C_I and C_{II} at their interface is given by Eq. 5. The solution of C_I is given by:

$$C_I = (c^i - \frac{\varkappa_2}{n_c^2 \alpha_c}) \exp(-\alpha_c z)$$

$$+ (c^* - C_b) \left[\int_0^{\beta_c} \varphi_{11} \exp(-\xi^2 t - \frac{z\sqrt{\mu_c - \xi^2}}{\sqrt{D_{cb}}}) \cdot d\xi \right.$$

$$\left. + \int_{\beta_c}^{\infty} \varphi_{12} \exp(-\xi^2 t) \cdot d\xi \right] \tag{15}$$

where,

$$\varphi_{11} = \frac{2}{\pi} \cdot \frac{\varkappa_2\sqrt{D_{cw}} - n_c\sqrt{\mu_c - \xi^2} \cdot (c^* - C_b)}{\{\xi^2 + n_c^2(\mu_c - \xi^2)\}(c^* - C_b)} \cdot \cos(\frac{\delta\xi}{\sqrt{D_{cw}}})$$

$$+ \frac{2}{\pi} \cdot \frac{\varkappa_2\sqrt{D_{cb}}(\mu_c - \xi^2) + (c^* - C_b)\xi^2}{\xi\{\xi^2 + n_c^2(\mu_c - \xi^2)\}(c^* - C_b)} \cdot \sin(\frac{\delta\xi}{\sqrt{D_{cw}}}) \quad , \text{ and}$$

$$\varphi_{12} = \frac{2}{\pi} \cdot \frac{\varkappa_2\sqrt{D_{cw}}\cos(\frac{n_c\delta\xi + z\sqrt{\xi^2 - 1}_c}{\sqrt{D_{cb}}}) + (c^* - C_b)\xi\sin(\frac{n_c\delta\xi + z\sqrt{\xi^2 - 1}_c}{\sqrt{D_{cb}}})}{\xi(\xi + n_c\sqrt{\xi^2 - \mu_c})(c^* - C_b)} \cdot$$

Although the exact solution of C_{II} is given by Eq. 16, an approximate solution in series expansion, as given by Eq. 17, is useful to calculate the oxygen flux from the bulk water:

$$C_{II} = c^i - \frac{\varkappa_2}{n_c^2 \alpha_c} + (c^* - C_b) \left[\int_0^{\beta_c} \varphi_{21}\{\cos(\frac{y\xi}{\sqrt{D_{cw}}}) \right.$$

$$+ \frac{n_c\sqrt{\mu_c - \xi^2}}{\xi}\sin(\frac{y\xi}{\sqrt{D_{cw}}})\} \exp(-\xi^2 t)d\xi + \int_{\beta_c}^{\infty} \{\varphi_{22}\cos(\frac{y\xi}{\sqrt{D_{cw}}})$$

$$\left. + \frac{n_c\sqrt{\xi^2 - \mu_c}}{\xi} \cdot \varphi_{23} \cdot \sin(\frac{y\xi}{\sqrt{D_{cw}}})\}\exp(-\xi^2 t)d\xi \right] \tag{16}$$

where, $\quad \varphi_{21} = \frac{2}{\pi} \cdot \frac{\varkappa_2\sqrt{D_{cw}} - n_c\sqrt{\mu_c - \xi^2} \cdot (c^* - C_b)}{\{\xi^2 + n_c^2(\mu_c - \xi^2)\}(c^* - C_b)} \cdot \cos(\frac{\delta\xi}{\sqrt{D_{cw}}})$

$$+ \frac{2}{\pi} \cdot \frac{\kappa_2\sqrt{D_{cb}}(\mu_c - \xi^2) + (C^* - C_b)\xi^2}{\xi\{\xi^2 + \eta_c^2(\mu_c - \xi^2)\}(C^* - C_b)} \cdot \sin(\frac{\delta\xi}{\sqrt{D_{cw}}}) \quad ,$$

$$\varphi_{22} = \frac{2\left[\kappa_2\sqrt{D_{cw}}\cos(\frac{\delta\xi}{\sqrt{D_{cw}}}) + (C^* - C_b)\xi\sin(\frac{\delta\xi}{\sqrt{D_{cw}}})\right]}{\pi\xi\{\xi + \eta_c\sqrt{\xi^2 - \mu_c}\}(C^* - C_b)} \quad , \text{ and}$$

$$\varphi_{23} = \frac{2\left[\kappa_2\sqrt{D_{cw}}\sin(\frac{\delta\xi}{\sqrt{D_{cw}}}) - (C^* - C_b)\xi\cos(\frac{\delta\xi}{\sqrt{D_{cw}}})\right]}{\pi\xi\{\xi + \eta_c\sqrt{\xi^2 - \mu_c}\}(C^* - C_b)} \quad .$$

$$\begin{array}{ll} C_{II} \ (0 \le y \le \delta) \\ \\ C_{II} \ (\delta \le y) \end{array} = \begin{array}{l} \kappa_2 y + C^i \\ \\ C_b \end{array}$$

$$- (C^* - C_b)\left[\pm\frac{1}{2}\text{erfc}(\frac{\delta - y}{2\sqrt{D_{cw}t}}) + \eta_c\sum_{n=1}^{\infty}\Lambda\cdot\overset{(2n)}{\text{ierfc}}(\frac{\delta + y}{2\sqrt{D_{cw}t}})\right]$$

$$- \kappa_2\sqrt{D_{cw}t}\left[\text{ierfc}(\frac{\delta - y}{2\sqrt{D_{cw}t}}) + 2\eta_c\sum_{n=1}^{\infty}\Lambda\cdot\overset{(2n+1)}{\text{ierfc}}(\frac{\delta + y}{2\sqrt{D_{cw}t}})\right] \quad (17)$$

where,
$$\Lambda = (-1)^n \cdot f(n+1)(4\mu_c t)^n \quad , \tag{18}$$

$$f(n) = \Gamma(n - \frac{1}{2})/(2\sqrt{\pi}\cdot\Gamma(n + 1)), \tag{19}$$

$$\Gamma(x) = \int_0^{\infty} e^{-\xi} \cdot \xi^{x-1} \, d\xi \quad ,$$

$$\overset{(1)}{\text{ierfc}}(x) = \text{ierfc}(x) = \int_x^{\infty} \text{erfc}(\xi) d\xi \quad , \text{ and}$$

$$\overset{(n)}{\text{ierfc}}(x) = \int_x^{\infty} \overset{(n-1)}{\text{ierfc}}(\xi) d\xi \quad (n \ge 2).$$

When $\eta_c = 1$, Eq. 17 also gives the exact solution and is identical to Eq. 16. Table 1 shows comparisons between the calculated results from Eq. 16 and those from Eq. 17, in which parameter values were arbitrarily chosen. It is clear that, in Table 1, Eq. 17 is identical to Eq. 16 in the case of $\eta_c = 1$ and that Eq. 17 gives a good approximation in the case of $\eta_c \simeq 1$. By using Eq. 17, the average oxygen flux from the bulk water into the water film ($\overline{N_{cw}}$) is calculated as:

$$\overline{N_{cw}} = \frac{1}{\overline{t}_w} \int_0^{\overline{t}_w} D_{cw}(\frac{\partial C_{II}}{\partial y})_{y=\delta} \, dt$$

$$= s_1 \frac{D_{cw}\varkappa_2}{2} - s_2 \frac{\sqrt{D_{cw}}}{\sqrt{\pi \overline{t}_w}} \cdot (C^* - C_b) \tag{20}$$

where,

$$s_1 = 1 + 8\eta_c \cdot \sum_{n=1}^{\infty} (-1)^n f(n+1)(4\mu_c \overline{t}_w)^{n} \cdot \overset{(2n+2)}{ierfc}(\delta/\sqrt{D_{cw}\overline{t}_w}), \tag{21}$$

$$s_2 = 1 - 2\sqrt{\pi} \cdot \eta_c \cdot \sum_{n=1}^{\infty} (-1)^n f(n+1)(4\mu_c \overline{t}_w)^{n} \cdot \overset{(2n+1)}{ierfc}(\delta/\sqrt{D_{cw}\overline{t}_w}), \tag{22}$$

and $\overline{t}_w$ = mean residence time in the water phase.

TABLE B.1. Comparisons between calculated results from Eqs. 16 and 17

$t / \overline{t}_w$	y / δ	η_c	C_{II} / C_b Eq. 16	C_{II} / C_b Eq. 17	$\dfrac{Eq.\ 17}{Eq.\ 16}$
1.0	0.0	1.0	0.4143	0.4143	1.00
	0.2		0.4235	0.4235	1.00
	0.4		0.4776	0.4777	1.00
	1.0		0.6231	0.6231	1.00
	4.0		0.9697	0.9697	1.00
	8.0		1.000	1.000	1.00
0.0313	0.0	0.92	1.016	0.995	0.979
	0.2		1.136	1.125	0.990
	0.4		1.234	1.233	0.999
	1.0		1.251	1.259	1.01
	4.0		0.999	1.000	1.00
	8.0		1.000	1.000	1.00
0.5	0.0	0.92	0.593	0.607	1.02
	0.4		0.670	0.682	1.02
	1.0		0.823	0.830	1.01
	4.0		1.010	1.010	1.00
	8.0		1.000	1.000	1.00
1.0	0.0	0.92	0.439	0.455	1.04
	0.4		0.499	0.512	1.03
	1.0		0.641	0.651	1.02
	4.0		0.974	0.975	1.00
	8.0		1.000	1.000	1.00

Parameter values used

$\tilde{\omega} = 10$ rpm, $R = 13$ cm, $H = 0.1$, $C^* = 9.1$ mg/l,

$\mu_c = 0.8$ s^{-1}, $C_b = 5$ mg/l, $C_o = 3$ mg/l,

W. T. =. 17.5 °C, $\varkappa_2 = \dfrac{\eta_c^2 \alpha_c C^*}{1 + \eta_c^2 \alpha_c \delta}$

In order to evaluate the mixing effect on mass transfer between the water film and bulk water at the points of disk emergence, the average oxygen concentration in the water film at the points of disk emergence, C_{av}, is defined as:

$$C_{av} = \frac{1}{\delta} \int_0^\delta C_{II}(y, \bar{t}_w)\, dy. \tag{23}$$

Combining Eqs. 17 and 23 yields:

$$C_{av} = F \cdot C_b + (G - F) \cdot C^* \tag{24}$$

where,

$$F = \frac{\sqrt{D_{cw}\bar{t}_w}}{\delta} \left[\frac{1}{\sqrt{\pi}} - \text{ierfc}\left(\frac{\delta}{2\sqrt{D_{cw}\bar{t}_w}}\right) + 2\eta_c \cdot \sum_{n=1}^\infty (-1)^n f(n+1)(4\mu_c\bar{t}_w)^n \cdot \right.$$

$$\left. \cdot \left\{ \text{ierfc}^{2n+1}\left(\frac{\delta}{2\sqrt{D_{cw}\bar{t}_w}}\right) - \text{ierfc}^{(2n+1)}\left(\frac{\delta}{\sqrt{D_{cw}\bar{t}_w}}\right) \right\} \right], \quad \text{and} \tag{25}$$

$$G = \frac{C^* + C^i}{2C^*} - \frac{2\mu_2 D_{cw}\bar{t}_w}{\delta\, C^*} \cdot \left[\frac{1}{4} - \text{ierfc}^{(2)}\left(\frac{\delta}{2\sqrt{D_{cw}\bar{t}_w}}\right) \right.$$

$$\left. + 2\eta_c \cdot \sum_{n=1}^\infty (-1)^n f(n+1)(4\mu_c\bar{t}_w)^n \cdot \left\{ \text{ierfc}^{(2n+2)}\left(\frac{\delta}{2\sqrt{D_{cw}\bar{t}_w}}\right) - \text{ierfc}^{(2n+2)}\left(\frac{\delta}{\sqrt{D_{cw}\bar{t}_w}}\right) \right\} \right] \tag{26}$$

In Part A it was shown that the mixing of the water film with the bulk water occurs at the points of disk emergence and that the interchange ratio coefficient, f_2, which is defined as the ratio of the volume of the water film interchanged with the bulk water to the volume of the whole water film at those points, is successful in describing the mixing mechanism. By using the interchange ratio coefficient, f_2, C_o can be calculated as:

$$C_o = (1 - f_2) \cdot C_{av} + f_2 \cdot C_b. \tag{27}$$

Total oxygen balance in the bulk water is:

$$V \frac{dC_b}{dt} = \upsilon(C_{in} - C_b) + \left[-\overline{N_{cw}} + f_2(C_{av} - C_b) \cdot \frac{1}{\bar{t}_w} \right] \cdot A_w^e \tag{28}$$

where,

V = net volume of the container,

υ = flow rate,

C_{in} = influent dissolved oxygen concentration, and

A_w^e = effective disk surface area submerged in the bulk water, which is given by Eq. 29.

$$A_w^e = a_f \cdot A_w \tag{29}$$

where,

a_f = area factor, and

A_w = disk surface area submerged in the bulk water.

Assuming that steady-state conditions prevail in the bulk water , $V(dC_b)/dt$ becomes zero. By combining Eqs. 20, 24 and 28, C_b is finally obtained as:

$$C_b = \frac{\upsilon C_{in} + \{s_2\sqrt{D_{cw}/(\pi\bar{t}_w)} - s_1 D_{cw}^\kappa 2/(2C^*) + f_2\delta(G-F)/\bar{t}_w\}A_w^e C^*}{\upsilon + \{s_2\sqrt{D_{cw}/(\pi\bar{t}_w)} + f_2\delta(1-F)/\bar{t}_w\}\cdot A_w^e} . \tag{30}$$

The oxygen transferred into the water film during one rotational cycle should be entirely consumed by nitrifiers in the biofilm. Therefore, the total oxygen consumption in the biofilm per rotation, M_c (µg), is given by:

$$M_c = \{\ \bar{N}_{ca}\bar{t}_a\ +\ \bar{N}_{cw}\ \bar{t}_w + f_2(C_b - C_{av})\delta\ \}\cdot a_f A_T(1 - H^2) \tag{31}$$

where, A_T = total disk surface area.

By using the ratio of oxygen consumption to the consumption of ammonium nitrogen, k, which is defined in Eq. 32, the ammonium nitrogen removal rate based on unit disk area, r_s (g·day^{-1}·m^{-2}), is calculated by Eq. 33.

$$k = \frac{(\text{Oxygen consumption rate})}{(\text{Ammonium nitrogen consumption rate})} \tag{32}$$

$$r_s = \frac{60\cdot24\cdot\tilde{\omega}\cdot10^{-6}\cdot M_c}{10^{-4}\cdot A_T}$$

$$= 14.4(a_f\tilde{\omega}/k)[\bar{N}_{ca}\bar{t}_a + \bar{N}_{cw}\bar{t}_w + f_2(C_b - C_a)\delta](1 - H^2) \tag{33}$$

where, H = dimensionless distance between water surface and
disk axis; R_i/R.

Although different equations for the mean air retention time have been proposed by various researchers (Yamane et al., 1972; Bintanja et al., 1975), they give almost the same values, as previously shown in Part A. According to Yamane and Yoshida (1972), the mean air retention time, $\bar{t}_a$, can be evaluated by Eq. 34.

$$\bar{t}_a = [\int_H^1 \zeta \{1 - \tfrac{1}{\pi}arccos(\tfrac{H}{\zeta})\}^{1.5} d\zeta]^2 / [\tilde{\omega}/60 \int_H^1 \zeta\{1 - \tfrac{1}{\pi}arccos(\tfrac{H}{\zeta})\} d\zeta]^2 \qquad (34)$$

The mean retention time in the water phase is calculated by:

$$\bar{t}_w = T - \bar{t}_a \qquad (35)$$

where,   T = time per rotation.

The mean thickness of the water film, δ, which was theoretically given by Zeevalkink et al. (1978), is:

$$\delta = \frac{4}{15} \sqrt{\frac{2\nu \cdot \omega R sin(\Phi)}{g}} \qquad (36)$$

where,
  $sin(\Phi) = \sqrt{1 - (R_i/R)^2}$,
  R = disk radius,
  R_i = distance between the water surface and disk axis,
  g = acceleration of gravity,
  ν = kinematic viscosity of water, and
  ω = angular velocity.

It has been shown in Part A that the interchange ratio coefficient, f_2, could be expressed by:

$$f_2 = 3.48 \times 10^{-5} \nu^{-0.3} \omega^{1.3} R^{2.1} (sin(\Phi))^{1.5} , \quad (0 \le f_2 \le 1) , \qquad (37)$$

The molecular diffusivity of oxygen in water is given by:

$$D_{cw} = 7.87 \cdot 10^{-10} (T_k/\mu) \qquad (38)$$

where,

T_k = absolute temperature, and

μ = viscosity of water.

B.2.1.2   Substrate Limitation

B.2.1.2.1   Water Phase (Case 3)

The limiting substrate, which in this case is the ammonium ion, is supplied from the bulk water into the biofilm; it is not inversely transferred into the bulk water because the substrate concentration in the bulk water is always higher than those in the water film and the biofilm. This situation is quite similar to the case of oxygen transfer in the air phase under oxygen limitation (Case 1).   In case 3, a concentration boundary layer is introduced, as shown in Fig. 2, in order to compare the solutions with those obtained in the Case 1. Thus, the differential equations and the initial and boundary conditions are:

I.   Biofilm
$$\frac{\partial S_I}{\partial t} = D_{sb} \frac{\partial^2 S_I}{\partial z^2} - \mu_s S_I \tag{39}$$

$$
\begin{aligned}
&t = 0, \; z \geq 0, \; S_I = S_o \exp(-\alpha_s z) \\
&t > 0, \; z = \infty, \; S_I = 0
\end{aligned}
\tag{40}
$$

II. Liquid Film and Concentration Boundary Layer
$$\frac{\partial S_{II}}{\partial t} = D_{sw} \frac{\partial^2 S_{II}}{\partial y^2} \tag{41}$$

$$
\begin{aligned}
&t = 0, \; \delta \geq y \geq 0, \; S_{II} = S_o \\
&t = 0, \; \delta_s \geq y \geq \delta, \; S_{II} = S_b \\
&t > 0, \quad\quad y = \delta_s, \; S_{II} = S_b
\end{aligned}
\tag{42}
$$

$$t > 0, \; y = z = 0, \; S_I = S_{II} \; , \; D_{sb} \frac{\partial S_I}{\partial z} = -D_{sw} \frac{\partial S_{II}}{\partial y} \tag{43}$$

where,

S_I and S_{II} = ammonium nitrogen concentrations in the biofilm and the water film and concentration boundary layer, respectively,

S_o = initial ammonium nitrogen concentration in the

water film,

S_b = ammonium nitrogen concentration in the bulk water,

D_{sb} and D_{sw} = molecular diffusivities of ammonium ion in the biofilm and the water film, respectively,

δ_s = thickness of the concentration boundary layer,

μ_s = rate constant, and

$\alpha_s = \sqrt{\mu_s/D_{sb}}$ = constant.

The substrate concentration at any depth and time could be found by solving the above equations. However, only the S_{II} solution is required to calculate the substrate flux from the bulk water. The solution of Eq. 41 is given by:

$$\frac{S_{II} \; (0 \le y \le \delta)}{S_{II} \; (\delta \le y \le \delta_s)} = \frac{S_o}{S_b} - \frac{\eta_s^2 \alpha_s (\delta_s - y)}{1 + \eta_s^2 \alpha_s \delta_s} \cdot S_b$$

$$+ \frac{S_b - S_o}{2} \left[\; erfc\left(\frac{2\delta_s - \delta - y}{2\sqrt{D_{sw}t}} \right) \; \pm \; erfc\left(\frac{|\delta - y|}{2\sqrt{D_{sw}t}} \right) \; \right]$$

$$+ (S_b - S_o) \cdot \frac{2}{\pi} \int_0^{\beta_s} \sin\left(\frac{(\delta_s - y)\xi}{\sqrt{D_{sw}}} \right) \cos\left(\frac{(\delta_s - \delta)\xi}{\sqrt{D_{sw}}} \right) \cdot \frac{\exp(-\xi^2 t)d\xi}{\xi}$$

$$+ (S_b - S_o) \cdot \int_{\beta_s}^{\infty} \psi_s \sin\left(\frac{(\delta_s - y)\xi}{\sqrt{D_{sw}}} \right) \cdot \frac{\exp(-\xi^2 t)d\xi}{\xi} \tag{44}$$

where,

$\eta_s = \sqrt{D_{sb}/D_{sw}}$ = constant,

$\beta_s = \sqrt{\mu_s}$ = constant,

$$\psi_s = -\frac{2}{\pi} \left(\frac{(\eta_s\sqrt{\xi^2 - \mu_s} - \xi)\{\xi\cos^2(J) - \eta_s\sqrt{\xi^2 - \mu_s}\cdot\sin^2(J)\}}{\xi^2\cos^2(J) + \eta_s^2(\xi^2 - \mu_s)\sin^2(J)} \cdot \cos\left(\frac{(\delta_s - \delta)\xi}{\sqrt{D_{sw}}}\right) \right.$$

$$\left. - \frac{\eta_s^2\sqrt{\xi^2 - \mu_s}\cdot\sqrt{\mu_s}\cdot S_o\cdot\sin(J)}{\{\xi^2\cos^2(J) + \eta_s^2(\xi^2 - \mu_s)\sin^2(J)\}(S_b - S_o)} \right), \text{ and}$$

$J = \delta_s\xi/\sqrt{D_{sw}}$.

Since the last integral term on the right hand side of Eq. 44 is usually smaller than the other integral term, as illustrated in

Table 2, the average substrate flux from the bulk water into the water film, $\overline{N}_{sw}$, is obtained by using Eq. 44 and by neglecting the last term as:

$$\overline{N}_{sw} = \frac{\eta_s^2 \alpha_s D_{sw}}{1 + \eta_s^2 \alpha_s \delta} \cdot S_b + (S_b - S_o)\frac{\sqrt{D_{sw}}}{\sqrt{\pi \overline{t}_w}} \left\{ 1 + \sqrt{\pi}\,\text{ierfc}\left(\frac{\delta_s - \delta}{\sqrt{D_{sw}\overline{t}_w}} \right) \right\}$$

$$- 2(S_b - S_o)\sqrt{D_{sw}} \cdot \frac{1}{\pi \overline{t}_w} \cdot \int_0^{\beta_s} \frac{1 - \exp(-\xi^2 \overline{t}_w)}{\xi^2} \cos^2\left(\frac{\delta_s - \delta}{\sqrt{D_{sw}}}\xi\right) d\xi . \quad (45)$$

The average ammonium nitrogen concentration in the water film at the points of disk emergence, S_{av}, is defined as:

$$S_{av} = \frac{1}{\delta} \int_0^{\delta} S_{II}(y, \overline{t}_w)\, dy . \quad (46)$$

TABLE B.2.   Comparisons between the two integral terms in Eq. 44

Rotational Speed (rpm)	Ratio of the calculated value of the first integral term* to that of the second integral term** in Eq. 44
5	1.19×10^{-4}
10	1.35×10^{-3}
15	1.76×10^{-3}
20	4.85×10^{-3}

* First integral term

$$= (S_b - S_o) \cdot \frac{2}{\pi} \cdot \int_0^{\beta_s} \frac{\cos\{\xi(\delta_s - \delta)/\sqrt{D_{sw}}\} \cdot \sin\{\xi(\delta_s - \delta)/\sqrt{D_{sw}}\}}{\xi} \cdot e^{-\xi^2 \overline{t}_w} d\xi$$

** Second integral term

$$= (S_b - S_o) \cdot \int_{\beta_s}^{\infty} \frac{\Psi_s \sin\{\xi(\delta_s - \delta)/\sqrt{D_{sw}}\}}{\xi} \cdot e^{-\xi^2 \overline{t}_w} d\xi$$

Parameter values used

$R = 13$ cm, $H = 0.1$, $\mu_s = 0.188$ s^{-1}, $\eta_s = 0.92$,

$S_b = 20$ mg/l,   W.T. $= 17.5$ °C.

By using Eqs. 45 and 46, the ammonium nitrogen removal rate, r_s, in the case of substrate limitation is finally obtained as:

$$r_s = 14.4 \cdot a_f \tilde{\omega} \cdot \{\overline{N_{sw} \bar{t}_w} + f_2(S_b - S_{av})\delta\}(1 - H^2). \qquad (47)$$

B.2.1.2.2 Air Phase (Case 4)

The differential equations for the biofilm and water film have already been given by Eqs. 39 and 41, respectively; the initial and boundary conditions are given by Eqs. 43, 48 and 49.

$$\begin{array}{l} t = 0, \; z \geq 0, \; S_I = S^i \exp(-\alpha_s z) \\ t > 0, \; z = \infty, \; S_I = 0 \end{array} \qquad (48)$$

$$\begin{array}{l} t = 0, \; \delta \geq y \geq 0, \; S_{II} = S^o \\ t > 0, \quad y = \delta \; , \; \partial S_{II}/\partial y = 0 \end{array} \qquad (49)$$

where,
S^i and S^o = initial concentrations of ammonium nitrogen in the biofilm surface and in the water film, respectively.

Solving these equations yields:

$$S_{II} = \frac{2}{\pi} \cdot \int_{\beta_s}^{\infty} \frac{(S^o - S^i)\xi\sin(P) + \beta_s S^i\cos(P)}{n_s^2\xi(\xi^2 - \mu_s)\cos^2(P) + \xi^3\sin^2(P)} \cdot$$

$$\cdot n_s\sqrt{\xi^2 - \mu_s} \; \cos(\frac{(\delta - y)\xi}{\sqrt{D_{sw}}})\exp(-\xi^2 t) \; d\xi. \qquad (50)$$

Levich (1962) gave the thickness of the concentration boundary layer in the case of the steady-state diffusion of reactant from a fully-submerged disk surface. Levich's result, however, cannot be directly applied to the case of a half-submerged rotating disk, because the thickness of the boundary layer increases as it travels from the points of disk submergence to the points of disk emergence; the thicness is not constant. Although there is turbulent motion at the emergence points, the flow near the rotating disk would be laminar in most of the water phase. In the analogy of the flow near a half-endless flat plate (Schlichting, 1968), the thickness of the boundary layer at the emergence points can be evaluated by Eq. 51, if the effect of the turbulent motion on the thickness is neglected.

$$\delta_v = 5.0\sqrt{\nu/\omega}\sqrt{2\pi\bar{t}_w/T} = 12.5\sqrt{t_w/T}\sqrt{\nu/\omega} \tag{51}$$

Bird et al. (1960) showed the relationship between the thickness of boundary layer and that of concentration boundary layer in the case of the diffusion of reactant from a half-endless flat plat in a flow as follows:

$$\delta_s/\delta_v = Sc^{-1/3} = (D_{sw}/\nu)^{1/3} \tag{52}$$

where, Sc = Schmidt number.

Combining Eqs. 51 and 52 yields:

$$\delta_s = 12.5\sqrt{t_w/T} \cdot (D_{sw}/\nu)^{1/3}\sqrt{\nu/\omega} . \tag{53}$$

The thickness of the boundary layer given by Eq. 51 would be greatest over the submerged disk and the concentration of the substrate would not change outside the hypothetical layer with the constant thickness given by Eq. 53. Thus, the concentration boundary layer, which appears in the boundary conditions of Eq. 42, can be evaluated by Eq. 53.

According to Williamson and McCarty (1976), the value of the molecular diffusivity of ammonium ions in water at 20 °C is 1.5 cm^2/day. By using this value, the temperature dependence of the molecular diffusivity of ammonium ions can be formulated in a similar way to that of oxygen.

$$D_{sw} = 5.92 \times 10^{-10}(T_k/\mu) . \tag{54}$$

B.3. Calculation Procedure

B.3.1 Oxygen Limitation

To calculate the ammonium nitrogen removal rate in the case of oxygen limitation by using Eq. 33, it is necessary to determine the values of c^i, c_{av} and c_o. If it reaches steady-state at the end of the air phase, c^i becomes:

$$c^i = c^*/(1 + \eta_c^2\alpha_c\delta) . \tag{55}$$

It can be shown that the value of c^i is almost equal to the steady-state value which is given by Eq. 55. The ammonium nitrogen removal rate can, therefore, be calculated by the following procedure:

1. calculate c^i from Eq. 55 as its initial value,
2. calculate C_{av}, C_o and $\varkappa_2$ from Eqs. 23, 27 and 14, respectively,
3. substitute $y = 0$ and $t = \bar{t}_a$ into Eq. 7 and calculate $C_{II}(0,\bar{t}_a)$; then let the new c^i equal $C_{II}(0,\bar{t}_a)$,
4. repeat steps 2 and 3, until adequate convergence is achieved, and
5. calculate r_s from Eq. 33.

   Steps 3 and 4 can be omitted in normal conditions.

B.3.2 Substrate Limitation

   Similarly, the ammonium nitrogen removal rate in the case of substrate limitation can be calculated by the following procedure:

1. let $S_o = 0$ as its initial value,
2. calculate S_{av} from Eq. 46,
3. using Eq. 46, calculate s^o as:

$$s^o = f_2 s_b + (1 - f_2) \cdot S_{av} , \tag{56}$$

4. calculate s^i by substituting $y = 0$ and $t = \bar{t}_w$ into Eq. 44,
5. using Eq. 50, let the new $S_o = \{S_{II}(0,\bar{t}_a) + S_{II}(\delta,\bar{t}_a)\}/2$,
6. repeat the steps 2 to 5, untill adequate convergence is achieved, and
7. calculate the ammonium nitrogen removal rate from Eq. 47.

   Steps 3 to 6 can be omitted in normal conditions.

B.4 Experimental Evaluation

   Single stage laboratory scale RBC units were used. A flow diagram is shown in Fig. 3. Each has the same dimensions and specifications as given in Table 3. The composition of influent synthetic wastewater is given in Table 4. The major components of the influent were ammonium chloride and sodium bicarbonate which would allow the growth of nitrifying bacteria. No organic carbon was added. Before starting the experiment, biofilm was grown on the disk for an acclimatization period of more than 6 months until nitrification had occurred.

   The experimental conditions are summarized in Table 5. Influent dissolved oxygen was often supersaturated. The reaction rate constant, μ_c, and area factor, a_f, should be estimated from experimental data. Other parameters were fixed, as shown in Table 6 in accordance with experimental conditions. The repeating steps

TABLE B.3.   Dimensions and specifications of RBC unit

Number of disks	10	
Disk diameter	26	cm
Thickness of disk	0.1	cm
Disk material	PVC	
Space between disks	1.5	cm
Disk surface area submerged (A_w)	4670	cm^2
Ratio of submerged surface area to total disk surface area (A_w/A_T)	44	%
Net volume of tank (V)	7.2	L
Ratio of tank volume to disk surface area (V/A_T)	0.0068	m^3/m^2
Dimensionless distance between water surface and axis (H)	0.1	

TABLE B.4.   Composition of influent synthetic wastewater

Chemicals	mg/l	Chemicals	mg/l
NH_4Cl	150 - 225	$NaHCO_3$	500 - 750
K_2HPO_4	22.1	KH_2PO_4	14.1
$Na_2HPO_4 \cdot 12H_2O$	10.5	$MgSO_4 \cdot 7H_2O$	25.3
$MnSO_4 \cdot 4H_2O$	2.3	$FeCl_3 \cdot 6H_2O$	0.12
$CaCl_2 \cdot 2H_2O$	2.5		

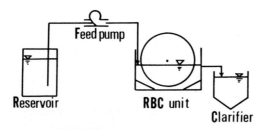

Fig. B.3.   Flow diagram of RBC unit

TABLE B.5.   Experimental conditions

Atmospheric temperature	20 ± 0.5 °C
Water temperature	17.5 ± 1.5 °C
pH	7.3 ± 0.3
Flow rate (υ)	1.9 ± 0.3 cm^3/s
Influent DO concentration (C_{in})	$9.5 \pm 1.5^*$ mg/l

Ammonium nitrogen concentration
in bulk liquid (S_b) at
rotational speeds of

$\tilde{\omega} \leq 5$ rpm	$S_b = 20 \pm 4$ mg/l
5 rpm $< \tilde{\omega} \leq 20$ rpm	$S_b > 14$
$\tilde{\omega} > 20$ rpm	$S_b > 6$

* The influent DO concentration was changed in the experiment to
  examine the effect of influent DO concentration (Fig. B. 11)

TABLE B.6.   Parameters for comparing analytical solutions
             with experimental data

W.T.	17.5 °C	A_w	4670 cm^2	υ	1.9 cm^3/s
C_{in}	9.5 mg/l	C^*	9.1 mg/l	k	$4.25^{(*)}$
H	0.1	$\eta_c^2 = \eta_s^2$	$0.85^{(**)}$	($\eta_c = \eta_s = 0.92$)

(*) The value of 4.25 was calculated from the stoichiometrics based
    on thermodynamic cell yields for Nitrosomonas and Nitrobacter.
    (Ito et al., 1980)
(**) The value of 0.85 was given by Williamson and McCarty (1976).

in the above mentioned procedures for calculating the ammonium
nitrogen removal rate could be omitted under these experimental
conditions.

Figure 4 shows a comparison between analytical solutions and
experimental data. At rotational speeds higher than 5 rpm, the
experimental bulk DO concentration was in agreement with its
theoretical lines (see Fig. 4a) and the experimental ammonium
nitrogen removal rate agreed well with Eq. 33 (see Fig. 4b), when
the value of μ_c was 0.8 sec^{-1} and a_f was chosen between 1.2 and
1.6. However, Eqs. 30 and 33 were not in agreement with the
experimental data when the rotational speed was lower than 5 rpm.
This suggests that the nitrification was not limited by oxygen at

404

the lower rotational speeds. Broken lines in Fig. 4b represent the theoretical values given by Eq. 47 for substrate limitation, using the same estimated values of μ_c and a_f. The ammonium nitrogen concentration in the bulk water was 20 ± 4 mg/l at rotational speeds lower than 5 rpm. The broken line with a parameter of $S_b = 20$ was in agreement with experimental data at rotational speeds lower than 5 rpm.

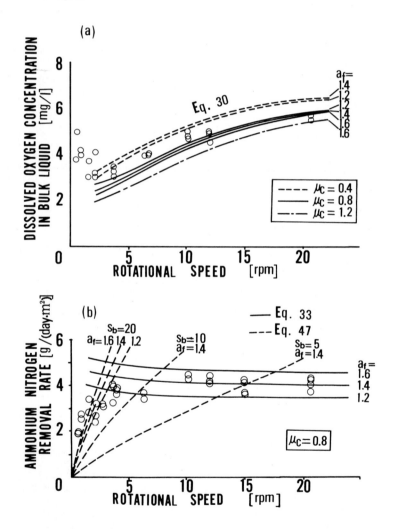

Fig. B.4. Comparison between analytical solutions and experimental data. The circles (o) represent experimental results and the lines represent analytical results. Parameters used for analytical results are shown in Table B.6.

B.5 Discussion

B.5.1 Assumption Propriety

Figs. 5 and 6 show calculated examples of DO concentration profiles under oxygen limitation in the air phase and the water phase, respectively. The DO concentration in the biofilm decreases exponentially from the biofilm surface. The DO concentration in the water film has a linear profile through most part of the air phase (Fig. 5); it almost reaches a steady-state at the end of the air phase. The DO concentration in the biofilm also decreases exponentially from the biofilm surface in the water phase (Fig. 6). Thus, it can be concluded that an exponential decrease of the DO concentration in the biofilm is reasonable for an initial condition. These figures also show that the DO concentration in the biofilm decreases sharply from the biofilm surface and the concentration almost reaches zero within the entire biofilm thickness about 1 mm in this experiment. The assumption of infinite biofilm can therefore be justified.

A microbial film fixed on the disk has a much higher sludge density than conventional activated sludge. The amount of limiting substance per unit of biomass would be significant for the biological reaction in the biofilm. Thus, the specific growth rate of ammonium oxidizing bacteria could be expressed by a Contois-type equation (Contois, 1959) for oxygen limitation.

$$\frac{1}{X} \cdot \frac{d\,X}{d\,t} = \frac{\mu_m \cdot C}{KX_T + C} = \frac{\mu_m \cdot C/X_T}{K + C/X_T} \tag{57}$$

where,

X and X_T = biomass densities of ammonium oxidizing
   bacteria and total biomass density, respectively,
μ_m = maximum specific growth rate of ammonium oxidizing
   bacteria,
K = dimensionless constant, and
C = DO concentration.

If $KX_T \gg C$, Eq. 57 becomes first order about C. It needs the order estimation of KX_T to discuss the propriety of the first order approximation of Eq. 57. Williamson and McCarty (1976) measured the density of biofilm of nitrifying bacteria and obtained its usual values of 50 to 80 mg/cm^3. Assuming $KX_T \gg C$, Eq. 57 becomes:

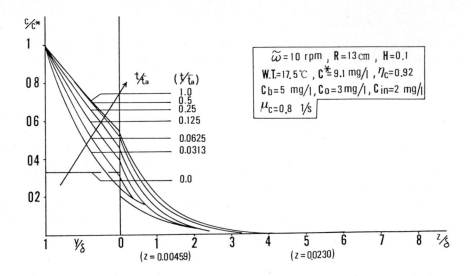

Fig. B.5. Example of dissolved oxygen concentration in the air phase. The liquid film thickness calculated from Eq. 36 was 45.9 μm. The DO concentration in the biofilm almost reaches zero at z = 0.023 cm (0.23 mm) in this figure.

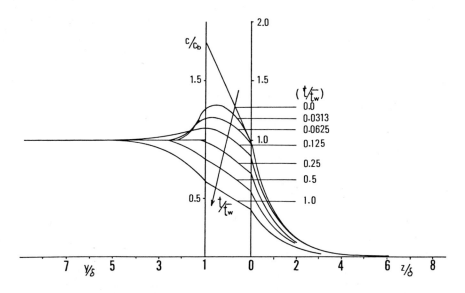

Fig. B.6. Example of dissolved oxygen concentration in the water phase. Parameters were given the same values as in Fig. B.5.

$$\frac{1}{X} \cdot \frac{dX}{dt} = (\mu_m/KX_T) \cdot C \ . \tag{58}$$

Assuming that the ammonium oxidizing bacteria had a density equal to nitrite oxidizing bacteria ($X_T = 2X$) in the biofilm, then

$$\mu_c C = (-\frac{dC}{dt})_{reaction} = \frac{k}{Y}(\frac{dX}{dt}) = (k\mu_m C)/(2YK) \tag{59}$$

where, Y is the growth yield of mass of ammonium oxidizing bacteria per mass of ammonium nitrogen utilized about 0.15 (Ito et al., 1980).

Rewriting Eq. 59 gives:

$$KX_T = (kX_T\mu_m)/(2Y\mu_c) \ . \tag{60}$$

According to Knowles et al. (1965), the maximum specific growth rate of ammonium oxidizing bacteria is about 0.6 day^{-1} at 17.5 °C. Substitution of $\mu_c = 0.8$ 1/s, $Y \simeq 0.15$, $k = 4.25$, $X_T = (50 \sim 80) \times 10^3$ and $\mu_m = 0.6/24/3600$ into Eq. 60 yields:

$$KX_T \simeq 4.25 \cdot (50 \sim 80) \cdot 10^3 \cdot (0.6/24/3600)/(2 \cdot 0.15 \cdot 0.8)$$

$$\simeq 6 \sim 10 \ mg/l \ .$$

It can be inferred that oxygen concentration in the biofilm would be much less than 6 $\sim$ 10 mg/l, therefore, $KX_T \gg C$ would be valid.

B.5.2 Effect of the Bulk Ammonium Nitrogen Concentration

Fig. 7 shows the relationship between the bulk ammonium nitrogen concentration, S_b, and the ammonium nitrogen removal rate, r_s. The experimental data in this figure were obtained from the authors' previous studies (Ito et al., 1980, 1981), in which the same experimental apparatus was used for nitrification and operated at the disk rotational speeds of between 13 and 15 rpm. It was concluded, in Fig. 7, that nitrification was limited by ammonium nitrogen at bulk ammonium nitrogen concentrations of below 5 mg/l and limited by oxygen at concentrations above 5 mg/l; thus 5 mg/l was the critical value at which the limiting substance changed from ammonium nitrogen to oxygen. Under substrate limitation, the ammonium nitrogen removal rate increased as the bulk ammonium nitrogen concentration increased. However, no increase in the removal rate was caused by an increase in the bulk ammonium nitrogen concentration under oxygen limitation.

It can be predicted from the analytical results shown in Fig. 4 that the critical value of the bulk ammonium nitrogen concentration increases as the rotational speed decreases. This is because, at low rotational speeds, the substrate may be consumed during a long retention time in the air phase despite a high concentration of the bulk substrate.

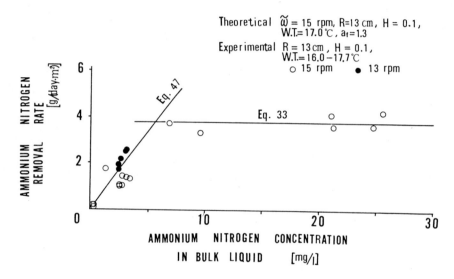

Fig. B.7. Relationship between the bulk ammonium nitrogen concentration and the ammonium nitrogen removal rate. The experimental data were obtained by Ito et al. (1980, 1981).

B.5.3 Oxygen Flux in the Case of Oxygen Limitation

The ratio of the mass of oxygen transferred into the biofilm from the air per rotation to the total oxygen mass transferred into the biofilm was calculated by using Eqs. 11 and 31 (see Fig. 8). It can be concluded that more than 90 % of the total oxygen mass transferred into the biofilm was from the air. Thus it is important to examine the content of oxygen flux from the air. Fig. 9 shows the effect of rotational speed on the ratio of the steady-state flux to the total flux from the air. Although the steady-state flux was dominant in this figure, the ratio of the steady-state flux gradually decreased as rotational speed increased.

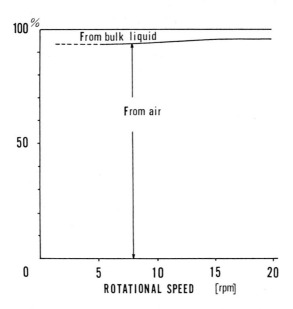

Fig. B.8. Effect of rotational speed on the ratio of the
mass of oxygen transferred into the biofilm from the air
per rotation to the total oxygen mass transferred into
the biofilm, with parameters of $\mu_C = 0.8$ and R = 13.
Other parameters are given in Table B.6.

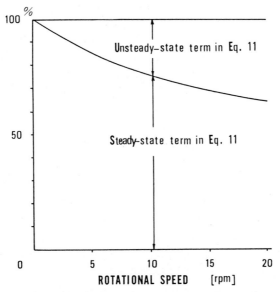

Fig. B.9. Effect of rotational speed on the ratio of the
steady-state oxygen flux to the whole flux from the air
under oxygen limitation, with parameters of $\mu_C = 0.8$ and
R = 13. Other parameters are given in Table B.6.

B.5.4 Effect of Rotational Speed

It can be concluded from Fig. 4 that the reaction mechanism changed from substrate limitation (ammonium limitation) to oxygen limitation at about 5 rpm. Also nitrification was limited by oxygen and the increase in rotational speed had no effect on the ammonium nitrogen removal rate at rotational speeds higher than 5 rpm. The DO concentration in the bulk water, however, increased with higher rotational speeds under oxygen limitation, because the bulk water was aerated as rotational speed increased. It can also be concluded that nitrification was limited by ammonium at rotational speeds lower than 5 rpm and that the increase in rotational speed led to the increase in the ammonium nitrogen removal rate.

It has already been shown that rotational speed had no effect on the ammonium nitrogen removal rate under oxygen limitation. This phenomenon is explained below. The steady-state flux is given by Eq. 10, and the increase in the thickness of the water film, δ, causes the decrease in the steady-state flux. δ is proportional to the square root of the rotational speed as given in Eq. 36. It is clear from Eqs. 10 and 36 that the increase in rotational speed causes the decrease in the steady-state flux. However, the increase in rotational speed causes the increase in the unsteady-state flux (see Fig. 9). The decrease in the steady-state flux may be compensated by the increase in the unsteady-state flux and a small change of the ammonium nitrogen removal rate is expected with an increase in rotational speed.

B.5.5 Substrate Flux in the Case of Substrate Limitation

Fig. 10 shows the ratio of the steady-state flux of ammonium nitrogen to the total flux into the biofilm. In this case, the steady state flux was not dominant, unlike the results shown in Fig. 8. This is because the thickness of the concentration boundary layer was much larger than that of the water film and the mass of substrate which initially existed within the concentration boundary layer is not negligible in comparison with the substrate mass transferred from the outside of the layer.

It was shown in Fig. 4 that the rotational speed had a large effect on the ammonium nitrogen removal rate under substrate limitation. This result is explained below. The decrease in the thickness of the concentration boundary layer causes the increase of the steady-state flux. The thickness is inversely proportional to the square root of the rotational speed. Therefore, the steady-state flux increases with the increase in the rotational speed. The unsteady-state flux also increases with rotational speed. Hence, the rotational speed has a large effect on the ammonium nitrogen removal rate.

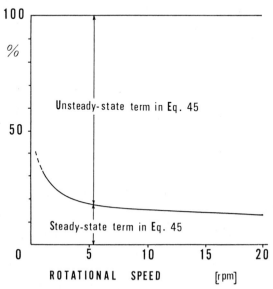

Fig. B.10 Effect of rotational speed on the ratio of the steady-state substrate flux to the total flux from the bulk water under substrate limitation, with parameters of μ_s= 0.188, R = 13 and S_b= 20. Other parameters are given in Table B.6.

B.5.6 Effect of Influent DO Concentration and Disk Size

The effect of influent DO concentration on the bulk DO concentration and ammonium nitrogen removal rate, under oxygen limitation, was also examined (see Fig. 11). Both the analytical solutions and experimental data in Fig. 11 show that the decrease in influent DO caused the decrease in the bulk DO, but had no effect on the ammonium nitrogen removal rate under oxygen limitation.

The effect of disk size on the ammonium nitrogen removal rate is shown in Fig. 12. As disk size increases, the maximum removal rate decreases. This figure suggests that results obtained from a small scale disk operating in a laboratory could never be obtained by a large disk operating in a full scale plant. Full scale RBCs which operate with rotational speeds of 1 - 2 rpm are easily under substrate limitation. If the peripheral velocity is chosen as a scale-up criteria and a small disk is operated with relatively high rotational speed, the results would be obtained under oxygen limitation and quite different from those obtained in the large scale disk operation. It should be obvious that peripheral velocity cannot be adopted as a scale-up criteria.

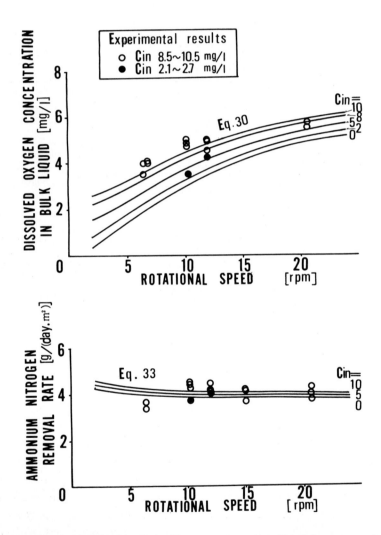

Fig. B.11. Effect of influent DO concentration on ammonium nitrogen removal rate and bulk DO concentration under oxygen limitation, with $\mu_c = 0.8$ and $a_f = 1.4$. Other parameters, except C_{in}, are given in Table B.6.

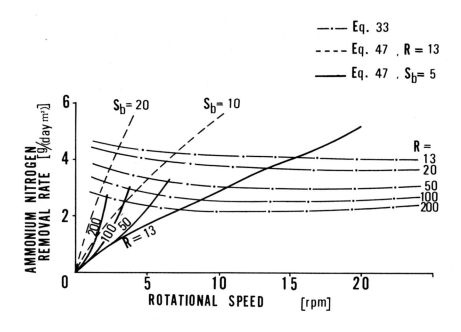

Fig. B.12. Effect of disk size on ammonium nitrogen
removal rate, with $\mu_C = 0.8$, $a_f = 1.4$ and hydraulic
loading of 0.155 $m^3/(day\ m^2)$. Other parameters,
except A_w and υ, are given in Table B.6.

B.6 Conclusions

In Part B of the present study, the unsteady-state oxygen
uptake and the substrate utilization for the nitrification process
in a rotating disk biological contactor were investigated
theoretically and experimentally. The analytical solutions of
differential equations of unsteady-state mass transfer were given
for both oxygen limitation and substrate limitation. The removal
rate of ammonium nitrogen was evaluated from the solutions and then
compared with experimental data. The main conclusions are
summarized below.

1. The unsteady-state model developed in the present study was
proved to be very effective in the analyses of the reaction
mechanism in the biofilm.

2. The critical value of disk rotational speed, at which the
reaction mechanism changed from ammonium limitation (substrate
limitation) to oxygen limitation, can be determined from the

414

model. In the present study, it was about 5 rpm.

3. At rotational speeds above 5 rpm, nitrification was limited by oxygen supply and further increases in speed had no effect on the ammonium nitrogen removal rate. The dissolved oxygen concentration in the bulk liquid, however, increased as rotational speed increased due to aeration of the bulk liquid.

4. It was estimated from the unsteady-state analytical solutions that more than 90 % of the oxygen used in the biofilm was supplied from the air, under the oxygen limitation, and that the maximum ammonium nitrogen removal rate decreased as disk diameter increased.

REFEFENCES

Atkinson B. & Davies I. J., 1974. The overall rate of substrate uptake (reaction) by microbial films. Part I - A biological rate equation. Trans. Instn Chem. Engrs 52, 248-259.

Bintanja H. H. J., van der Erve J. J. V. M. & Boelhouwer C., 1975. Oxygen transfer in a rotating disc treatment plant. Water Research 9, 1147-1153.

Bird R. B., Stewart W. E. & Lightfoot E. N., 1960. Transport Phenomena. John Wiley & Sons, New York.

Contois D. E., 1959. Kinetics of bacterial growth: Relationship between population density and specific growth rate of continuous cultures. J. Gen. Microbiol. 21, 41-50.

Famularo J., Mueller J. A. & Mulligan T., 1978. Application of mass transfer to rotating biological contactors. Journal WPCF 50, 653-671.

Grieves C. G., 1972. Dynamic and steady models for the rotating biological disc reactor. Ph.D. thesis, Dept. of Environ. Syst. Eng., Clemson Univ., USA.

Harremoës P., 1977. Harf-order reactions in biofilm and filter kinetics. Vatten 33, 122-143.

Harremoës P., 1978. Biofilm kinetics. In Water Pollution Microbiology. R. Mitchell (ed.), Vol.2, 82-109, John Wiley & Sons, New York.

Harremoës P. & Gonenc I. E., 1983. The application of biofilm kinetics to rotating biological contactors. Preprint for EWPCA/IAWPRC International Seminar on Rotating Biological Discs, Fellbach, West Germany, Oct. 6-8, 1983.

Ito K. & Matsuo T., 1980. The effect of organic loading on

nitrification in RBC wastewater treatment processes. Proceedings of First National Symposium on RBC Technology held at Champion, Pennsylvania, USA, Feb. 4-6, 1980, Vol.2, 1165-1175.

Ito K. & Matsuo T., 1981. Effect of amino acids on nitrification and denitrification in a rotating biological contactor. Proceedings of the 36th Annual Conference of JSCE, 2, 197-198. (in Japanese)

Knowles G., Downing A. L. & Barrett M. J., 1965. Determination of kinetic constants for nitrifying bacteria in mixed culture, with the aid of an electronic computer. J. Gen. Microbiol. 38, 263-278.

Levich V. G., 1962. Physicochemical Hydrodynamics. Prentice-Hall, New York.

Rittmann B. E. & McCarty P. L., 1978. Variable-order model of bacterial-film kinetics. Jour. of the Environmental Engineering Division, ASCE 104, 889-900.

Rittmann B. E. & McCarty P. L., 1981. Substrate flux into biofilms of any thickness. Jour. of the Environmental Engineering Division, ASCE 107, 831-849.

Schlichting H., 1968. Boundary-Layer Theory (6th ed.), McGraw-Hill, New York.

Shieh W.K., 1982. Mass transfer in a rotating biological contactor. Water Research 16, 1071-1074.

Watanabe Y. & Thanantaseth C., 1982a. Combined carbon oxidation-nitrification process in a rotating biological contactor. Jour. Japan Sewage Works Association 19, No. 221, 52-62.

Watanabe Y., Nishidome K. & Ishiguro M., 1982b. Simulation of nitrification process in a rotating biological contactor. Jour. Japan Sewage Works Association 19, No. 223, 30-39.

Williamson K. & McCarty P. L., 1976. Verification studies of the biofilm model for bacterial substrate utilization. J. WPCF 48, 281-296.

Yamane T. & Yoshida F., 1972. Absorption in a rotating-disk gas-liquid contactor. Journal of Chemical Engineering of Japan 5, 381-385.

Zeevalkink J. A., Kelderman P. & Boelhouwer C., 1978. Liquid film thickness in a rotating disc gas-liquid contactor. Water Research 12, 577-581.

SYMBOLS

A_T	= total disk surface area	(cm^2)
A_w	= submerged area of disk surface	(cm^2)
A_w^e	= effective disk surface area submerged $= a_f \cdot A_w$	(cm^2)
a_f	= area factor	$(-)$
C	= concentration of dissolved oxygen (DO)	(mg/l)
C^*	= saturated DO concentration	(mg/l)
C_b	= DO concentration in bulk water	(mg/l)
C_{av}	= average DO concentration in water film	(mg/l)
C_{in}	= initial DO concentration	(mg/l)
c^o, c^i, c_o, c_i	= initial value of C	(mg/l)
D_{cb}	= molecular diffusivity of oxygen in biofilm	(cm^2/s)
D_{cw}	= molecular diffusivity of oxygen in water	(cm^2/s)
D_{sb}	= molecular diffusivity of ammonium ion in biofilm	(cm^2/s)
D_{sw}	= molecular diffusivity of ammonium ion in water	(cm^2/s)
f_2	= interchange ratio coefficient	$(-)$
g	= acceleration of gravity	(cm/s^2)
H	= dimensionless distance between water surface and disk axis $= R_i/R$	$(-)$
J	= dimensionless parameter in Eq. 44 $= \delta_s \xi / \sqrt{D_{sw}}$	$(-)$
K	= dimensionless constant in Eq. 57	$(-)$
k	= ratio of mass of oxygen consumed to that of ammonium nitrogen consumed	$(-)$
$\overline{N}_{ca}$	= average oxygen flux from the air	$(\mu g/(s \cdot cm^2))$
$\overline{N}_{cw}$	= average oxygen flux from bulk water	$(\mu g/(s \cdot cm^2))$
$\overline{N}_{sa}$	= average ammonium nitrogen flux from the air	$(\mu g/(s \cdot cm^2))$
$\overline{N}_{sw}$	= average ammonium nitrogen flux from bulk water	$(\mu g/(s \cdot cm^2))$
P	= dimensionless parameter in Eq. 6 $= \delta \xi / \sqrt{D_{cw}}$	$(-)$
Q	= dimensionless parameter in Eq. 6 $= z\sqrt{\xi^2 - \mu_c} / \sqrt{D_{cb}}$	$(-)$
R	= radius of the disk	(cm)
R_i	= distance between water surface and disk axis	(cm)
r_s	= removal rate of ammonium nitrogen	$(g/(day \cdot m^2))$
S	= ammonium nitrogen concentration	(mg/l)
S_b	= ammonium nitrogen concentration in the bulk water	(mg/l)
S_{av}	= average ammonium nitrogen concentration in water film at the points of disk emergence	(mg/l)
s^o, s^i, s_o	= initial value of S	(mg/l)
Sc	= Schmidt number	$(-)$
s_1	= dimensionless parameter defined by Eq. 21	$(-)$
s_2	= dimensionless parameter defined by Eq. 22	$(-)$

T = time per one rotation   (s)

T_k = absolute temperature   (K)

$\bar{t}_a$ = mean retention time in the air phase   (s)

$\bar{t}_w$ = mean residence time in the water phase   (s)

V = net volume of container   (cm^3)

X = biomass density of ammonium oxidizing bacteria   (mg/1)

X_T = total biomass density   (mg/1)

Y = growth yield of mass of ammonium oxidizing bacteria
per mass of ammonium nitrogen utilized   (-)

y,z = distance from biofilm surface   (cm)

α_c = constant = $\sqrt{\mu_c/D_{cb}}$ (cm^{-1})

α_s = constant = $\sqrt{\mu_s/D_{sb}}$ (cm^{-1})

β_c = constant = $\sqrt{\mu_c}$ ($s^{-1/2}$)

β_s = constant = $\sqrt{\mu_s}$ ($s^{-1/2}$)

δ = thickness of water film   (cm)

δ_s = thickness of concentration boundary layer   (cm)

δ_v = thickness of boundary layer at the points of emergence   (cm)

$\varkappa_2$ = parameter defined by Eq. 14   ($\mu g/cm^4$)

Λ = dimensionless parameter defined by Eq. 18   (-)

η_c = constant = $\sqrt{D_{cb}/D_{cw}}$ (-)

η_s = constant = $\sqrt{D_{sb}/D_{sw}}$ (-)

μ = viscosity of water   (g/(s·cm))

μ_c = reaction rate constant   (s^{-1})

$\mu_s = \mu_c/k$ = reaction rate constant   (s^{-1})

μ_m = maximum specific growth rate of
ammonium oxidizing bacteria   (s^{-1})

ν = kinematic viscosity of water   (cm^2/s)

υ = flow rate   (cm^3/s)

ω = angular velocity   (rad/s)

$\tilde{\omega}$ = rotational speed   (rpm)

Φ = arccos(R_i/R)

$$\text{erfc}(x) = \frac{2}{\sqrt{\pi}} \int_x^\infty \exp(-\xi^2)d\xi$$

$$\overset{(1)}{\text{ierfc}}(x) = \text{ierfc}(x) = \int_x^\infty \text{erfc}(\xi)d\xi$$

$$\overset{(n)}{\text{ierfc}}(x) = \int_x^\infty \overset{(n-1)}{\text{ierfc}}(\xi)d\xi \qquad (n \geq 2)$$

$$\Gamma(x) = \int_0^\infty e^{-\xi}\xi^{x-1}d\xi$$

MATHEMATICAL MODELLING OF NITRIFICATION AND
DENITRIFICATION IN ROTATING BIOLOGICAL CONTACTORS

YOSHIMASA WATANABE,Department of Civil Engineering
Miyazaki University,Japan

1 Introduction
2 Steady State Biofilm Kinetics
3 Liquid Film Diffusion
4 Biological Nitrification Rate
5 Kinetic Analysis of Nitrification
6 Combined Organic Oxidation and Nitrification
7 Kinetic Analysis of Denitrification
8 Simultaneous Nitrification and Denitrification
9 Case Study
10 Summary and Conclusions

10.1 INTRODUCTION

In recent years,many reports have been published about fixed
film biological reactors. From the theoretical point of view,the
process of diffusion of substrates to the biofilm and that of diffusion with simultaneous biochemical reactions within the biofilm
must be taken into account in developing the biofilm kinetics:
some investigators have done this(La Motta,1976a,1976b;Williamson
and McCarty,1976;Harremoës,1976;Rittman and McCarty,1978;Riemer and
Harremoës,1978;Famularo,et al,1978). The Rotating Biological Conta-
ctor(RBC) is one application of a fixed-film biological reactor. It
has been pointed out that this process is very effective,in particular,for biological nitrification and denitrification(Ishiguro,et
al,1977a,1977b;Pretorius,1974;Davies and Pretorius,1975;Weng and
Molof,1974;Cheung and Krauth,1980). Watanabe,et al (1978a,1980a)
have also developed steady state biofilm kinetics and applied it
to the denitrification and nitrification in the RBC. The proposed
kinetics has been described as the process of molecular diffusion
with a simultaneous zero order biochemical reaction within the biofilm. It has been demonstrated that this would be applicable to anaerobic denitrification in a submerged RBC. The proposed kinetics
has also adequately explained most of the experimental data of nitrification in a partially submerged RBC in which the biofilm

420

alternately rotates into air and water. The partially submerged
RBC has no steady state substrate concentration profile within the
biofilm,even though the substrate concentration in the bulk water
is the steady state. Rittman and McCarty(1980) stated that the rotation of disks was fast enough that the substrate concentration
within the biofilm should not change significantly from the steady
state profile. However,there has been no evidence for this opinion.
Watanabe,et al(1982a) reported the results of a computer simulation
of nitrification designed to find out the reasoning behind the application of the steady state biofilm kinetics to the nitrification
process in a partially submerged RBC. Watanabe and Thanantaseth(
1982b)has also developed a mathematical equation which predicts the
ammonia removal in combined organic oxidation and nitrification in the
same kind of RBC. In this chapter,the development of a mathematical
model of the biofilm kinetics and its application to nitrification
and denitrification processes in the RBC are summarized. All substrate flux have been expressed on the submerged disk area basis.

10.2 STEADY STATE BIOFILM KINETICS

The clearest evidence for external mass transport and the necessity for its inclusion comes from microprobe measurements of the
dissolved oxygen profile up to and through a biofilm. The data from
Chen and Bungay(19 81) is shown in Fig.10.1 ,which clearly demonstrates
that the oxygen concentration at the biofilm:liquid interface can
be appreciably less than that in the bulk liquid. Therefore,it is

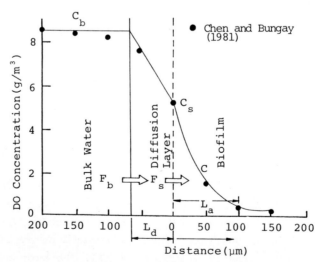

Fig.10.1 Biofilm system for kinetic analysis

convenient to divide the biofilm system into three major strata,i.e. bulk liquid,biofilm and the diffusion layer between the bulk liquid and the biofilm,as illustrated in Fig. 10.1. Watanabe(1984b) rearrenged a biofilm kinetics in order to calculate the flux of the rate limiting substrate,as described below. The biofilm kinetics has been developed under the following assumptions:

1. The bulk liquid is completely mixed;
2. Molecular diffusion of substrates occurs through the diffusion layer; and
3. Molecular diffusion of substrates with a simultaneous zero order biochemical reaction occurs within the biofilm.

The substrate flux through the diffusion layer,F_b,is given as follows:

$$F_b = (D_w/L_d)(C_b - C_s) \qquad (10.1)$$

where D_w=molecular diffusion coefficient of the rate limiting substrate,L_d=diffusion layer thickness,C_b=concentration of the rate limiting substrate in the bulk liquid,and C_s=concentration of the rate limiting substrate at the biofilm surface.

Substrate flux at the biofilm surface,F_s,can be formulated by solving the material balance equation within the biofilm:

$$D_f(\partial^2 C/\partial Z^2) - \partial C/\partial t - R = 0 \qquad (10.2)$$

where D_f=intrinsic molecular diffusion coefficient of the rate limiting substrate within the biofilm,C=concentration of the rate limiting substrate within the biofilm,Z=biofilm depth,and R=intrinsic biochemical reaction rate of the biofilm.

In a steady state,Eq.(10.2) is simplified as Eq.(10.3):

$$D_f(d^2 C/dZ^2) = R \qquad (10.3)$$

For the rate limiting substrate,boundary conditions for solving Eq.(10.3) are given as follows:

$$C = C_s \qquad \text{at} \quad Z = 0$$
$$dC/dZ = 0 \text{ and } C = 0 \quad \text{at} \quad Z = L_a \qquad (10.4)$$

where L_a=penetration depth of the rate limiting substrate,i.e.,the effective biofilm depth.

Solving Eq.(10.3) under the above boundary conditions gives the following intrinsic concentration profile of the rate limiting substrate:

$$C = (R/2D_f)Z^2 - \sqrt{2RC_s/D_f}\, Z + C_s \tag{10.5}$$

Then, the substrate flux at the biofilm surface is given as:

$$F_s = -D_f(dC/dZ)_{Z=0} = \sqrt{2D_f RC_s} \tag{10.6}$$

F_b is equal to F_s when the bulk substrate concentration is the steady state. Therefore, Eq. (10.7) is obtained in such a case, if D_f is expressed as αD_w, where α has a constant value of less than unity.

$$F_b = (D_w/L_d)(C_b - C_s) = \sqrt{2\alpha D_w RC_s} \tag{10.7}$$

Then, C_s is calculated from the known values of D_w, α, C_b, L_d and R as follows:

$$C_s = (2C_b + \lambda - \sqrt{\lambda^2 + 4\lambda C_b})/2 \tag{10.8}$$

$$\lambda = (2\alpha RL_d^2)/D_w \tag{10.9}$$

Through these kinetic considerations, it can be concluded that the bulk reaction rate, F_b, at a given value of C_b is described by the biofilm reaction characteristics such as α, D_w, R and L_d.

A list of aqueous and biofilm diffusion coefficients reported by Williamson and McCarty (1976) is presented in Table 10.1.

Table 10.1 Aqueous and biofilm diffusion coefficients at 20 °C reported by Williamson and McCarty (1976)

Biofilm Culture	Species	D_w (cm^2/d)	D_f (cm^2/d)	$\alpha = D_f/D_w$
Nitrobacter	NH_4^+	1.5	1.2	0.80
	NO_2^-	1.4	1.2	0.86
	NO_3^-	1.4	1.3	0.93
	O_2	2.6	2.2	0.85
Nitrosomonas-	NH_4^+	1.5	1.3	0.87
Nitrobactor	NO_2^-	1.4	1.2	0.86
	NO_3^-	1.4	1.4	1.00
	O_2	2.6	2.2	0.85

D_w = Aqueous diffusion coefficient
D_f = Biofilm diffusion coefficient

Table 10.2   Zero-order intrinsic biochemical reaction rates reported in literatures

Electron donor	Electron acceptor	Biofilm Characteristics	$R(g/m^3h)$ based on electron donor	Temp. (°C)	Reference
$C_6H_{12}O_6$	O_2	mixed culture	7.5×10^3	20	Onuma (1982)
$C_6H_{12}O_6$	O_2	mixed culture	6.0×10^4	20	Harremoës (1977)
$C_6H_{12}O_6$	NO_3^-	mixed culture	3.6×10^3	20	Mamiya (1983)
$C_5H_{10}O_6$	O_2	mixed culture	3.0×10^4	23	Watanabe, et al (1982b)
$HOCH_2CH_2OH$	NO_3^-	denitrifying bacteria	2.4×10^3	13	Mamiya (1983)
NH_4^+	O_2	nitrifying bacteria	7.3×10^3	28	Watanabe, et al (1980b)
CH_3OH	NO_3^-	denitrifying bacteria	3.8×10^3	20	Mamiya (1983)

Table 10.2 contains a list of zero order intrinsic biochemical reaction rates also reported in the literatures.

The flux of non rate limiting substrate, F_b^*, is given by Eq. (10.10)

$$F_b^* = F_b (\nu^* MW^* / \nu MW) \qquad (10.10)$$

where ν^*, ν=stoichiometric reaction coefficients for non rate limiting and rate limiting substrates, respectively, and MW^*, MW=molecular weight of non rate limiting substrate and rate limiting substrate, respectively.

Williamson and McCarty (1976) proposed the relationship to predict whether the biofilm reaction is limited by the electron acceptor or electron donor for a general metabolic reaction within a biofilm, as expressed by Eq. (10.11).

$$\nu_d D + \nu_a A + \text{Growth Requirements} \longrightarrow \text{End Products + Cells} \qquad (10.11)$$

where ν_d, ν_a=stoichiometric reaction coefficients for the electron donor (D) and the electron acceptor (A), respectively.

In terms of the bulk liquid concentrations, the electron acceptor is rate limiting in case of $D_{fa}/D_{fd}=D_{wa}/D_{wd}$ when:

$$C_{ba} < (D_{fd} \nu_a MW_a / D_{fa} \nu_d MW_d) C_{bd} \qquad (10.12)$$

where D_{fa}, D_{fd}=biofilm diffusion coefficient of electron acceptor and electron donor, respectively, D_{wa}, D_{wd}=aqueous diffusion coefficient of electron acceptor and electron donor, respectively, C_{ba}, C_{bd}=bulk concentration of electron acceptor and electron donor, respectively,

MW_a, MW_d=molecular weight of electron acceptor and electron donor, respectively.

10.3 LIQUID FILM DIFFUSION

Liquid film diffusion offers the diffusional resistance from the bulk liquid to the biofilm surface. The significance of the liquid film diffusion is expressed by the thickness of the diffusion layer, L_d,which is a physical characteristics related to the hydrodynamics in the reactor. Direct measurement of the diffusion layer thickness is very complicated in the RBCs. Indirect estimation of the diffusion layer thickness was carried out(Watanabe,et al,1980b).Fig.10.2 shows the relationship between the diffusion layer thickness and the disk rotational speed obtained by the following two indirect methods.

10.3.1 Estimation of the Diffusion Layer Thickness by Batch Experiment

In the batch biofilm reactor,the rate of ammonia reduction by biological nitrification is expressed by Eq.(10.13):

$$d(V_b C_{bA})/dt = -A\ K_d(C_{bA} - C_{sA}) \qquad (10.13)$$

where V_b=bulk water volume,C_{bA}=bulk ammonia concentration,t=time, A=submerged biofilm surface area,C_{sA}=ammonia concentration at the biofilm surface,$K_d=D_A/L_d$,D_A=molecular diffusion coefficient of ammnia.

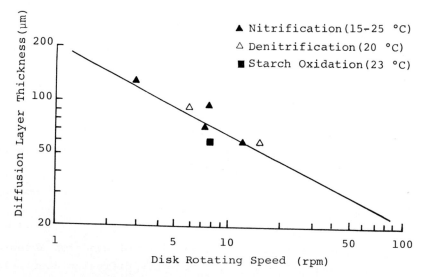

Fig.10.2 Relationship between disk rotating speed and diffusion layer thickness

The diffusion of ammonia to the biofilm surface is rate limiting at an extremely low C_{bA}. Under this situation, Eq. (10.13) is simplified to Eq. (10.14):

$$d(V_b C_{bA})/dt = -AK_d C_{bA} \qquad (10.14)$$

Integrating Eq. (10.14) under a constant V_b and $C_{bA} = C_{bA}^0$ at $t=0$ gives,

$$K_d = D_A/L_d = (V_b/A)\{ln(C_{bA}^0/C_{bA})/t\} \qquad (10.15)$$

where t is the elapsed time after Eq. (10.14) becomes valid.

Fig.10.3 is one of the experimental results obtained by the batch nitrification experiment. It shows that the diffusion of ammonia to the biofilm surface becomes rate limiting, i.e., Eq. (10.14) is valid, in a C_{bA} range of below 2 g/m^3. From the data shown in Fig. 10.3, the value of K_d is calculated by Eq. (10.15) at about 1.0×10^{-1} m/h. Then, L_d is evaluated as follows: $L_d = 7.5 \times 10^{-5}$ m, because $D_A = 7.5 \times 10^{-6}$ m^2/h. Table 10.3 shows the summary of experimental results.

Table 10.3   Determined values of diffusion layer thickness by batch nitrification experiment

N(rpm)	T(°C)	K_d(m/h)	D_A (m^2/h)	L_d(μm)
3	28.5	6.0×10^{-2}	7.5×10^{-6}	125
7.5	28.5	1.0×10^{-1}	7.5×10^{-6}	75
7.5	15	5.0×10^{-2}	5.0×10^{-6}	100
12.5	25	1.3×10^{-1}	7.0×10^{-6}	55

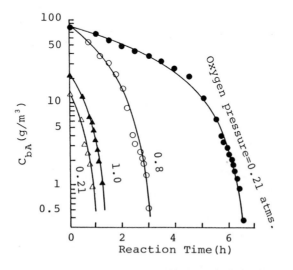

Fig. 10.3   Ammonia reduction in batch experiment

10.3.2 Estimation of the Diffusion Layer Thickness by Steady State Kinetics

Fig. 10.4 shows the relationship between bulk nitrate concentration and nitrate flux obtained in the denitrification experiment by a submerged RBC. According to the steady state biofilm kinetics, nitrate flux, F_{bN}, is expressed by the first order kinetics in an extremely low bulk nitrate concentration, C_{bN}:

$$F_{bN} = K_d C_{bN} \qquad (10.16)$$

where $K_d = D_N / L_d$, D_N = molecular diffusion coefficient of nitrate.

Therefore, the value of K_d can be determined from the slope of the straight line shown in Fig. 10.4 as follows:

$$K_d = F_{bN} / C_{bN} = 1.0 \times 10^{-1} \text{ m/h}$$

Then, L_d is estimated at 60 μm.

Fig. 10.5 shows the relationship between bulk COD and COD flux obtained in the starch oxidation experiment by a partially submerged RBC. In a similar way, L_d is estimated at 63 μm.

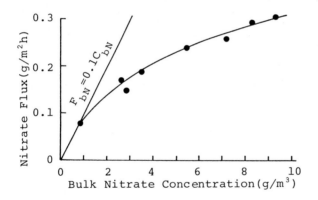

Fig. 10.4 Determination of K_d from denitrification data at 20 °C

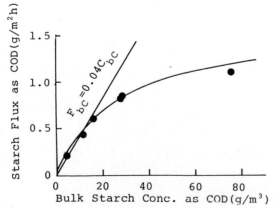

Fig. 10.5 Determination of K_d from starch oxidation data at 23 °C

10. 4 BIOLOGICAL NITRIFICATION RATE

10.4.1 Biological Nitrification Rate in Suspended Culture

In order to apply the kinetic model to biological nitrification in
an RBC, it is necessary to confirm that the biological nitrification
reaction itself is a zero order reaction. Many investigators (e.g.
Wong and Loehr, 1975) have reported that ammonia oxidation is a zero
order reaction. Watanabe, et al (1980a) carried out the batch experiment using suspended nitrifying bacteria to seek information on (a)
order of the reaction, (b) effect of microbial concentration, (c) effect
of dissolved oxygen concentration and aeration intensity, and (d) temperature effect. Fig. 10.6 shows one of the experimental results obtained in the batch experiment using the suspended nitrifying bacteria. A mixed culture of Nitrosomonas sp. and Nitrobacter sp. isolated
from activated sludge was used. This shows biological nitrification
(ammonia oxidation) is a zero order reaction and the accumulation of
nitrite is not remarkable in the pH range of 7 to 8. Fig. 10.7 shows
that the dissolved oxygen concentrtion is kept nearly constant during the ammonia oxidation. This also verifies ammonia oxidation to
be a zero order reaction, because the oxygen consumption rate must be
constant, regardless of the ammonia concentration, in order to keep
the dissolved oxygen concentration constant under a fixed aeration
intensity.

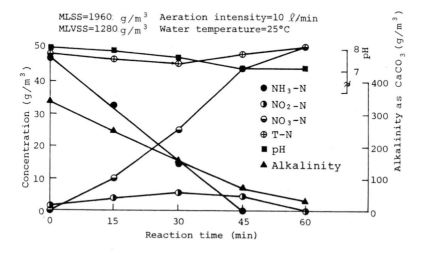

Fig. 10. 6 Nitrification rate in suspended culture

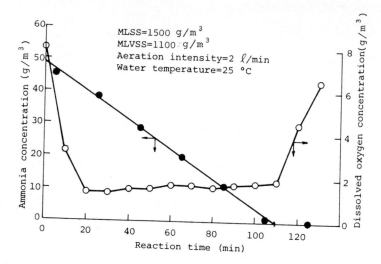

Fig. 10.7　Ammonia and DO concentrations in suspeded culture

Fig. 10.8 shows the relationship among the nitrifying bacteria concentration, aeration intensity and nitrification rate constant. It indicates that the nitrification rate constant is a function of nitrifying bacteria concentration and aeration intensity. At a fixed biomass concentration, the average floc size decreases and dissolved oxygen concentration increases as the aeration intensity increases. At a fixed aeration intensity, the flocculation rate and the the average floc size increase with the increase of biomass concentration. Therefore, the active part of the biomass increses as aeration intensity increases and biomass concentration decreases.

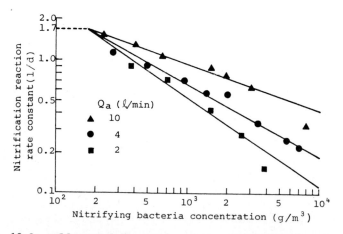

Fig. 10.8　Effect of bacteria conc. and aeration intensity on nitrification rate in suspended culture

However, as shown in Fig.10.8, the influence of biomass concentration on the nitrification rate constant diminishes at a biomass concentration of about 200 g/m^3, and the value of the nitrification rate constant converges to about 1.7 1/d. From these experimental results, ammonia oxidation rate in an active biomass is verified to be constant regardless of ammonia concentration, i.e., a zero order reaction. The effect of temperature on the nitrification reaction was evaluated by a series of experiments in which biomass concentration and aeration intensity were fixed at nearly 500 g/m^3 and 10 ℓ/min, respectively. The temperature was 10°C, 15°C, 20°C and 25°C. In every experiment, the nitrification reaction was confirmed to be a zero order reaction. Fig.10.9 is the Phelps plot of the nitrification rate constant and Eq.(10.17) is given for the temperature dependency of it.

$$r_{n,T} = r_{n,20}(1.15)^{T-20} \qquad (10.17)$$

where $r_{n,T}$=nitrification rate constant in suspended culture at T°C.

10.4.2 Intrinsic Nitrification Rate of Biofilm

In a ammonia rate limitation, steady state ammonia flux is given by Eq.(10.18), as explained in previous section.

$$F_{bA} = F_{sA} = \sqrt{2D_A R_n C_{sA}} \qquad (10.18)$$

where F_{bA}=ammonia flux through the diffusion layer, F_{sA}=ammonia flux at the biofilm surface, R_n=intrinsic nitrification rate of biofilm, defined as the intrinsic ammonia oxidation rate of biofilm.

When F_{bA}, C_{bA}, D_A and L_d are known, C_{sA} is calculated by Eq.(10.19), as derived from Eq.(10.7).

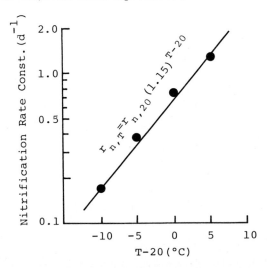

Fig.10.9 Temperature dependency of nitrification rate constant in suspended culture

$$C_{sA} = C_{bA} - (F_{bA}L_d/D_A) \tag{10.19}$$

As shown later,nitrification in a submerged RBC becomes ammonia limiting at a low bulk ammonia concentration,and Eq.(10.18) is available in such a case. Watanabe,et al(1982a) carried out the nitrification experiment using an RBC with hood to maintain an oxygen-rich gaseous phase. Fig.10.10 shows experimental result under the following experimental conditions: disk rotating speed=7.5 rpm,water temperature=28.5°C,and partial pressure of oxygen=0.8 atms. The exposed time of biofilm to high oxygen pressure was not enough to activate the nitrifying bacteria in the deeper part of biofilm, so,the experimental data are not completely followed by the half order plots as predicted by Eq.(10.18). Therefore,R_n has been estimated from the straight line in Fig. 10.10 as follows:

$$D_A = 7.5 \times 10^{-6} \ m^2$$

$$F_{bA} = 3.3 \times 10^{-1} \ g/m^2h$$

$$R_n = (F_{bA})^2/(2D_A C_{sA}) = 7.3 \times 10^3 \ g/m^3h$$

Under oxygen limitation,the ammonia flux is given as follows:

$$F_{bA} = \sqrt{2 D_O R_{on} C_{so}/a_n} \tag{10.20}$$

where D_O =molecular diffusion coefficient of oxygen,R_{on} =intrinsic oxygen uptake rate of the biofilm,C_{so} =dissolved oxygen concentration at the biofilm surface,a_n =oxygen requirement of the nitrification.

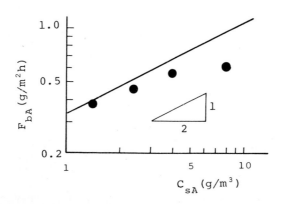

Fig.10.10 Determination of intrinsic nitrification rate based on half order kinetics

Therefore, the temperature dependency of the intrinsic nitrification rate can be expressed as follows:

$$F_{bA,T}/F_{bA,20} = \sqrt{(D_{O,T}/D_{O,20})(R_{n,T}/R_{n,20})(C_{so,T}/C_{so,20})} \qquad (10.21)$$

where $F_{bA,T}$=ammoia flux at T°C, $D_{O,T}$ =molecular diffusion coefficient of oxygen at T°C ,$C_{so,T}$=dissolved oxygen concentration at the bio-film surface at 20°C, $R_{n,T}$=intrinsic nitrification rate at 20°C.

Fig. 10.11 shows the temperature dependency of the ammonia flux obtained under oxygen limitation. The straight line in Fig. 10.11 is expressed as follows:

$$F_{bA,T}/F_{bA,20} = (1.04)^{T-20} \qquad (10.22)$$

The temperature dependency of the molecular diffusion coefficient is approximately written as follows:

$$D_{O,T}/D_{O,20} = (1.027)^{T-20} \qquad (10.23)$$

The following relationship has been found from the simulated data in the nitrification in a partially submerged RBC, as explained later.

$$C_{so,T}/C_{so,20} \cong C^*_{O,T}/C^*_{O,20} = (0.98)^{T-20} \qquad (10.24)$$

Introducing Eqs. (10.22), (10.23) and (10.24) into Eq. (10.21) gives the temperature dependency of the intrinsic nitrification rate as,

$$R_{n,T}/R_{n,20} = (1.07)^{T-20} \qquad (10.25)$$

The temperature coefficient of the nitrification rate for the suspended culture has been determined as 1.15 (Fig. 10.9). Therefore,

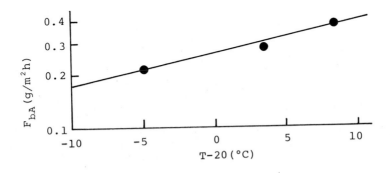

Fig. 10.11   Temperature dependency of ammonia flux in oxygen rate limitation

it is concluded that the intrinsic nitrification rate of biofilm
is less influenced by the bulk water temperature than is the nitrification rate in the suspended culture. The same conclusion has been
given for the temperature dependency of the denitrification rates.
As the intrinsic nitrification rate at 28.5°C has been estimated at
7.3 x 10^3 g/m³h, it is given by Eq.(10.26) at any bulk water temperature of 15 to 30°C.

$$R_{n,T} = (4.1 \times 10^3)(1.07)^{T-20} \qquad (10.26)$$

10.5 KINETIC ANALYSIS OF NITRIFICATION IN A PARTIALLY SUBMERGED RBC

10.5.1 Computer Simulation of Nitrification

10.5.1.1 Model Developement

The biofilm attached to a partially submerged RBC rotates alternately into air and water. During the rotation in the air phase, oxygen
is supplied to the biofilm from the air, but there is no ammonia
transport to the biofilm. During the rotation in the water phase,
ammonia diffuses into the biofilm from the bulk water. Watanabe, et
al(1982a, 1982c) have developed a computer simulation model to identify the change in the ammonia and dissolved oxygen concentration
profiles in the biofilm with the rotation of it. The simulation
model is based on the assumptions made for the developement of the
steady state biofilm kinetics. The basic equation of the simulation
model is the Fick's Second Law of Diffusion:

$$\partial C_A/\partial t = D_A(\partial^2 C_A/\partial z^2) \qquad (10.27)$$

where C_A=intrinsic ammonia concentration in the biofilm.

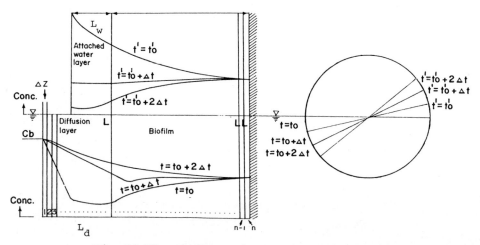

Fig. 10.12   Biofilm system division for the simulation

Eq. (10.27) has been directly applied to the diffusion process in the attached water film and the diffusion layer, but a biochemical reaction term has to be added in order to take into account the biological substrate uptake in the biofilm:

$$\partial C_A/\partial t = D_A(\partial^2 C_A/\partial Z^2) - R_n \qquad (10.28)$$

To obtain the solutions for Eqs. (10.27) and (10.28), the submerged ratio of the disks was assumed to be 0.5 and the disk surface was divided into 800 small sectors each with an area equql to ΔA. The biofilm, the attached water film, and the diffusion layer were divided into sub-layers, each with a thickness of ΔZ. Fig. 10.12 illustrates the system division for the computer simulation.

The difference form of Eq. (10.28) is

$$C_A(n+1,i) = K\{C_A(n,i-1) - 2C_A(n,i) + C_A(n,i+1)\}$$
$$+ C_A(n,i) - R_n\Delta t \qquad (10.29)$$
$$K = D_A\Delta t/(\Delta Z)^2 \qquad (10.30)$$

where subscript n refers to the number of Δt times; and subscript i refers to the concentration reference planes.

Ammonia flux to the elemental biofilm at any time during immersion can be obtained as:

$$F_{bA}(n) = (D_A/\Delta Z)\{C_A(n,1) - C_A^*(n,2)\} \qquad (10.31)$$

The average ammonia flux for all elemental biofilms in the water phase is

$$F_{bA} = (1/t)\sum_{n=1}^{400}\Delta t\, F_{bA}(n)$$
$$= (D_A/400\Delta Z)\sum_{n=1}^{400}\{C_A(n,1) - C_A(n,2)\} \qquad (10.32)$$

F_{bA} is equal to the ammonia flux determined by the experiment. The same procedure was followed for the calculation of the dissolved oxygen concentration in the biofilm system.

10.5.1.2 Analysis of Simulation Results

The diffusion layer thickness (Fig. 10.2), the intrinsic nitrification rate (Eq. 10.26), and the relationship between the bulk steady state concentrations of ammonia and dissolved oxygen (Fig. 10.13) have been obtained in previous studies (Watanabe, et al, 1980b). The thickness of the attached water film, L_w, was changed to match the simulated and experimental results. For the simulation conditions shown in Table 10.4, L_w was estimated at about 50 μm to give the best fit in the comparison between the simulated and experimentally determined ammnia flux. Bintanja et al(1975) proposed the following equation to calculate the attached water film thickness:

$$L_w = 0.93 \sqrt{\mu N r/g\rho} \qquad (10.33)$$

where μ=kinematic viscosity of water,N=disk rotational speed,r=disk diameter,g=gravitational acceleration and ρ=water density.

For the simulation conditions shown in Table 10.4,L_w is calculated at about 40 μm by Eq.(10.33).

Figs. 10.14 and 10.15 show the changes which occured in the simulated concentrations of ammonia and dissolved oxygen in the elemental biofilm with varying retention times for the air and water phases. The concentration profile changes,depending on the retention time in each phase,even that for the steady state condition of bulk ammonia and dissolved oxygen concentrations.

Table 10.4 Simulation conditions

Parameter	Run N-3	Run N-2
Biofilm thickness(μm)	900	900
Attached-water film thickness(μm)	50	50
Diffusion layer thickness(μm)	80	80
Intrinsic nitrification rate($g/m^3 h$)	7300	5200
Water temperature(°C)	28.5	23.5
Disk rotational speed(rpm)	7.5	7.5
Diffusion coefficient of NH_4^+	2.0 cm^2/day	1.8 cm^2/day
Diffusion coefficient of O_2	2.4 cm^2/day	2.1 cm^2/day

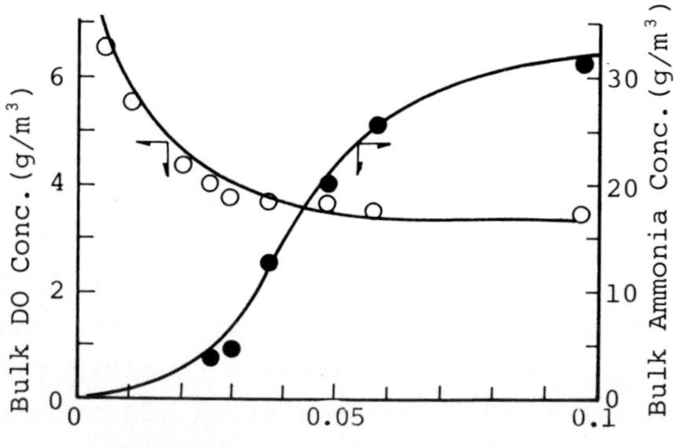

Fig. 10.13 Relationship between bulk ammonia and DO concentrations

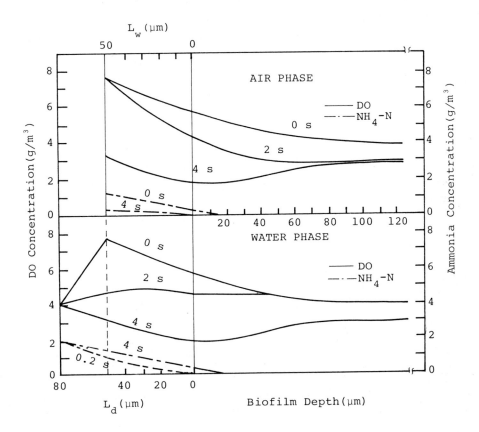

Fig. 10.14 Simulated changes of ammonia and DO concentration
profiles in air and water phases in ammonia rate
limitation

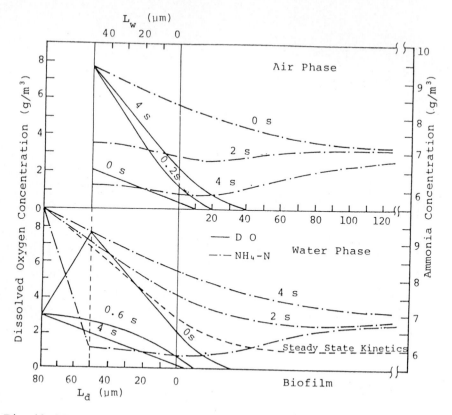

Fig.10.15 Simulated changes of ammonia and DO concentration
profiles in air and water phases in oxygen limitation

The ammonia flux to the elemental biofilm was calculated by Eq.(10.
31) and is shown in Fig.10.16 as a function of the retention time
during the water phase. At a fixed retention time, the ammonia flux
is almost constant in an ammonia concentration of more than 3.5
g/m^3 but increases with the bulk ammonia concentration of less than
3.5 g/m^3. The average flux for the elemental biofilms rotating
through the bulk water, which is the experimentally determined ammonia flux, was calculated by Eq.(10.32) and is shown in Fig.10.16.
Fig.10.17 shows the comparison between the simulated average ammonia flux and the experimentally determined ammonia flux. The simulated results confirm that the ammonia flux to the biofilm rotating
through the water phase would be equal to the average ammonia flux
of each elemental biofilm.

10.5.1.3 Effects of Simulation Parameters on RBC Performance

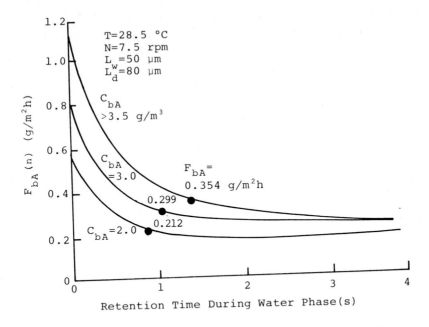

Fig.10.16 Ammonia flux as a function of retention time during water phase

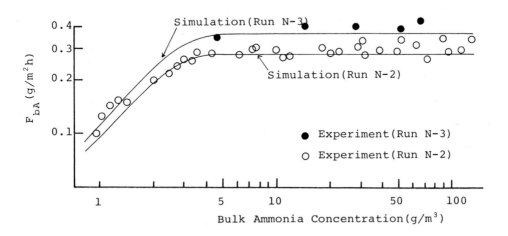

Fig.10.17 Comparison of experimentally determined and simulated flux

438

The average ammonia flux of the elemental biofilms rotating in the water phase is written as the ammonia flux in the discussion of the simulated results. Fig. 10.18 shows the effect of L_w on the ammonia flux with oxygen rate limitation. Fig. 10.19 shows the effect of L_d on the ammonia flux with ammonia rate limitation. Fig. 10.20 shows the effect of bulk dissolved oxygen concentration on the ammonia flux with oxygen rate limitation. It demonstrates that the RBCs have a good performance even in a low dissolved oxygen level. This comes from the fact that the RBCs take most of oxygen during the rotation in the air phase. The partial pressure of oxygen in the air phase, P_o, significantly influences the RBC performance with oxygen rate limitation. Fig. 10.21 shows the relationship between the ammonia

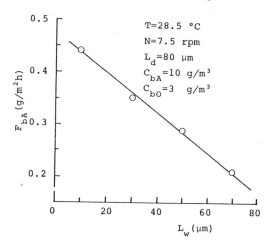

Fig. 10.18   Effect of L_w on ammonia flux in oxygen limitation

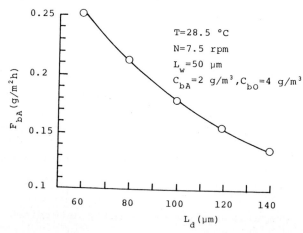

Fig. 10.19   Effect of L_d on ammonia flux in ammonia limitation

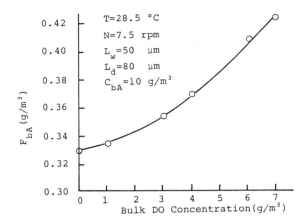

Fig.10.20   Effect of bulk DO conc. on ammonia flux in
            oxygen limitation

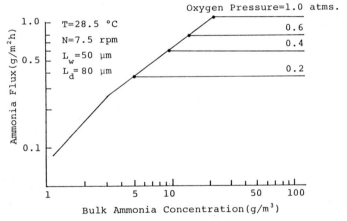

Fig.10.21   Effect of oxygen pressure on ammonia flux

flux and bulk ammonia concentration with varying partial pressures of
oxygen in the air phase. Comparison of the simulated ammonia flux
under oxygen rate limitation with the experimental data obtained by
Pretorius(1974) and Watanabe et al(1980a) has been made,as shown in
Fig.10.22. Pretorius(1974) supplied pure oxygen to maintain a required oxygen pressure for 36 hrs but Watanabe,et al(1980a) for 5 to
10 hrs. Exposure time to a high oxygen pressure in the later experiment was not enough to grow the nitrifying bacteria in the deeper
parts of the biofilm. This is the reason why the ammonia flux obtained in the later research was less than the simulated flux.

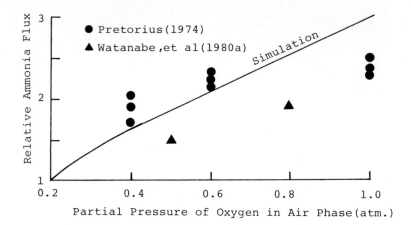

Fig. 10.22 Effect of oxygen pressure on ammonia flux
in oxygen limitation

10.5.2 Application of Steady State Kinetics

As shown in Fig. 10.14, it has been demonstrated that under ammonia limitation, the ammonia concentration in the biofilm would reach the steady state profile in a short time after the biofilm enters the water phase. Fig. 10.15 also demonstrates that under oxygen limitation, the dissolved oxygen concentration in the biofilm would reach the steady state profile in a short time after the biofilm enters air or water phase. Fig. 10.23 shows the contribution of unsteady state oxygen flux during the air phase at varing disk rotational speeds. Simulation was done at a fixed disk pheripheral speed, i.e., a fixed L_w. Fig. 10.23 demonstrates that unsteady state diffusion of oxygen is almost negligible in the usual disk rotational speed. Based on these evidences, it can be concluded that the steady state biofilm kinetics is applicable to predict the approximate nitrification rate in a partially submerged RBC. Experimental data shown in Fig. 10.24 are used to confirm the availability of the steady state biofilm kinetics in a partially submerged RBC.

Under ammonia limitation, the ammonia flux, F_{bA}, can be calculated by Eqs. (10.7), (10.8) and (10.9). For Run N-2 in Fig. 10.24, F_{bA} under ammonia limitation is calculated as follows:

Calculation conditions; $C_{bA}= 5$ g/m^3, L_d=100 um = 10^{-4} m, $D_A = 5.0$ x 10^{-6} m^2/h (α is assumed to be unity), R_n=2.9 x 10^3 g/m^{3}h.

$$\lambda = (2R_n L_d^2)/D_A = 12.3 \text{ g/m}^3$$

$$C_{sA}= (2C_{bA} + \lambda - \sqrt{\lambda^2 + 4\lambda C_{bA}})/2 = 1.2 \text{ g/m}^3$$

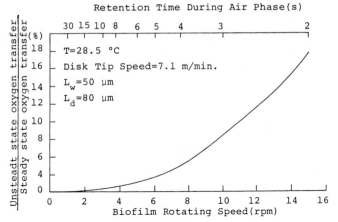

Fig.10.23 Contribution of unsteady state oxygen transfer in oxygen limitation

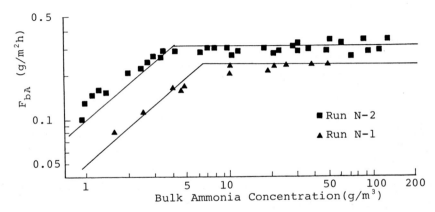

Fig.10.24 Kinetic Analysis of nitrification data in partially submerged RBC

$$F_{bA} = (D_A/L_d)(C_{bA} - C_{sA}) = 0.19 \text{ g/m}^2\text{h}$$

F_{bA} in ammonia limitation can be calculated in a similar way. The oblique solid line in Fig. 10.24 was calculated, as explained above.

Under oxygen limitation, F_{bA} can be calculated as follows: From Eqs.(10.7),(10.8) and (10.9), the following equations are derived for calculating the oxygen flux to the biofilm rotating in the air and water phases:

Oxygen flux to the biofilm rotating in the air phase, F_{aO}:

$$F_{aO} = D_O(C_O^* - C_{so,a})/L_w = \sqrt{2D_O R_{on} C_{so,a}} \qquad (10.34)$$

$$C_{so,a} = (2C_O^* + \lambda_a - \sqrt{\lambda_a^2 + 4\lambda_a C_O^*})/2 \qquad (10.35)$$

$$\lambda_a = (2R_{on}L_w^2)/D_O \qquad (10.36)$$

where D_O =molecular diffusion coefficient of oxygen,C_O^*=saturation concentration of dissolved oxygen,$C_{so,a}$ =dissolved oxygen concentration at the biofilm surface during the air phase,R_{on}=intrinsic oxygen consumption rate of the biofilm,L_w=thickness of the attached water film.

Oxygen flux to the biofilm rotating in the water phase,F_{wO} :

$$F_{wO} = D_O(C_{bo} - C_{so,w})/L_d = \sqrt{2D_O R_{on}C_{so,w}} \qquad (10.37)$$

$$C_{so,w} = (2C_{bo} + \lambda_w - \sqrt{\lambda_w^2 + 4\lambda_w C_{bo}})/2 \qquad (10.38)$$

$$\lambda_w = (2R_{on}L_d^2)/D_O \qquad (10.39)$$

where C_{bo}=bulk dissolved oxygen concentration,$C_{so,w}$=dissolved oxygen concentration at the biofilm surface during the water phase,L_d =thickness of the diffusion layer.

When an RBC,with a submerged ratio of 0.5,is operated under oxygen limitation,the ammonia flux is given by Eq.(10.40),which comes from the concept expressed by Eq.(10.10).

$$F_{bA} = (R_n/R_{on})(F_{aO} + F_{wO}) = (F_{aO} + F_{wO})/4.33 \qquad (10.40)$$

For Run N-2 in Fig.10.24,F_{bA} under oxygen limitation,is calculated using the following values: L_d=8 x 10^{-5} m,L_w=5 x 10^{-5} m, D_O=8.8 x 10^{-6} m^2/h,C_O^*=8.5 g/m^3,C_{bo}=3.0 g/m^3,R_{on}=4.33 R_n=2.3 x $10^4 g/m^3 h$,where it is assumed that 4.33 g of molecular oxygen is consumed to oxidize 1 g of ammonia to nitrate through nitrite.

Calculated values for Run N-2 are as follows:

$$F_{aO} = 1.02 \ g/m^2h$$

$$F_{wO} = 0.31 \ g/m^2h$$

$$F_{bA} = 0.31 \ g/m^2h$$

Similarly,the following values are obtained for Run N-1 under oxygen limitation:

$$F_{aO} = 0.84 \ g/m^2h$$

$$F_{wO} = 0.20 \ g/m^2h$$

$$F_{bA} = 0.24 \ g/m^2h$$

The horizontal solid lines in Fig.10.24 indicate the calculated F_{bA}. Comparison between the calculated and experimental F_{bA} confirms the

availability of the steady state biofilm kinetics in a partially
submerged RBC.

10.6 COMBINED ORGANIC OXIDATION AND NITRIFICATION

Many kinds of wastewaters such as domestic sewage contain both
soluble organic matter and ammonia. Therefore,the process of molecular diffusion with simultaneous organic oxidation and nitrification occurs in the biofilm used to treat such wastewaters. These two
biochemical reactions influence each other,since both are aerobic
reactions. A mathematical modelling of simultaneous organic oxidation and nitrification developed by Watanabe,et al(1982b,1984a) is
presented in this section.

10.6.1 Mathematical Analysis of Experimental Data

Three completely mixed flow type units with disk diameter of 30 cm
were used. Units 1,2 and 3 consist of 13,15 and 10 disks,respecti-
vely,mounted 2 cm apart on a horizontal shaft. The experimental
variables were (a)hydraulic loading,(b)influent concentration of
organic matter,and (c)type of organic matter(glucose,starch,or ethylene glycol). The disk rotational speed was fixed at 8.5 rpm. Influent ammonia concentration was 45 g/m^3. The experimental conditions are summerized in Table 10.5.

  Reduction of ammonia in simultaneous organic oxidation and nitrification consists of reduction due to nitrification and cell synthesis. Eq.(10.41) shows the ammonia flux in the simultaneous organic
oxidation and nitrification.

Table 10.5   Experimental conditions

Run No.	Organic matter	Influent conc.	Hydraulic loading	Water temp.
1	Glucose	36 g/m^3	2.5-7.6L/m^2h	27 °C
2	Glucose	75	2.5-10.6	25
3	Glucose	180	2.0-7.1	25
4	Starch	233	1.7-7.6	25
5	Starch	57-490	2.6	23
6	Ethylene glycol	30-200	5.7	25
7	Ethylene glycol	60-190	3.6	25

$$F_{bA} = F'_{nA} + F_{gA} \qquad (10.41)$$

where F'_{nA} =ammonia flux due to the nitrification, F_{gA}=ammonia flux due to the cell synthesis.

Ammonia flux in a completely mixed flow RBC is calculated as :

$$F_{bA} = Q(C_{iA} - C_{eA})/A \qquad (10.42)$$

where Q=flow rate, C_{iA} and C_{eA}=influent and effluent ammonia concentrations, respectively.

The mass balance of the oxidized nitrogen in such an RBC is

$$V_b(dC_{bN}/dt) = QC_{iN} - QC_{eN} + F'_{nA}A \qquad (10.43)$$

where C_{bN}=bulk oxidized nitrogen concentration, C_{iN} and C_{eN}=influent and effluent oxidized nitrogen concentrations, respectively.

If the influent contains no oxidized nitrogen, Eq.(10.43) becomes Eq.(10.44) at steady state.

$$C_{eN} = F'_{nA}A/Q \qquad (10.44)$$

Assuming that the ratio of oxidized organic matter to ammonia flux for the cell synthesis is constant, i.e., $F_{gA}=kF'_{bC}$, Eq.(10.45) is obtained by combining Eqs.(10.41) and (10.44).

$$(C_{eN}/F_{bA})/(A/Q) = 1 - k(F'_{bC}/F_{bA}) \qquad (10.45)$$

where F'_{bC}=organic matter flux in combined organic oxidation and nitrification.

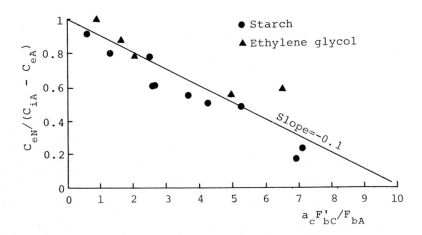

Fig.10.25 Determination of ammonia amount utilized for cell synthesis of heterotrophic bacteria

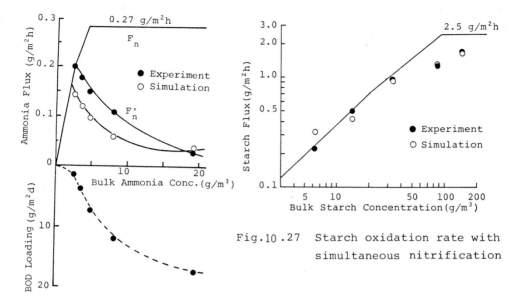

Fig.10.27   Starch oxidation rate with
simultaneous nitrification

Fig.10.26   Effect of organic oxidation
on nitrification rate

Combining Eqs.(10.42) and (10.45) gives Eq.(10.46).

$$C_{eN}/(C_{iA} - C_{eA}) = 1 - (k/a_c)(a_c F'_{bC}/F_{bA}) \qquad (10.46)$$

where C_{iA}=influent ammonia concentration,C_{eA}=effluent ammonia concentration,a_c=oxygen requirement of biological organic oxidation.

Experimental data in Runs 4,5,6 and 7 were plotted as shown in Fig.
10.25. Based on Fig.10.25,the relationship among F_{bA},F'_{nA} and F'_{bC} is
given by Eq.(10.47).

$$F_{bA} = F'_{nA} + F_{gA} = F'_{nA} + 0.1a_c F'_{bC} \qquad (10.47)$$

a_c was experimentally determined as 0.71 g O_2/g oxidized glucose,
0.55 g O_2/g oxidized starch,0.63 g O_2/g oxidized ethylene glycol(
Watanabe et al,1982b). Heukelekian et al(1955) reported 0.50 to
0.74 g O_2/g oxidized glucose,and 0.22 to 0.68 g O_2/g oxidized
starch.

Fig.10.26 shows the effect of organic oxidation on the relationship between the ammonia flux and bulk ammonia concentration in

Run 5. F'_{nA} was calculated using the experimental values of F_{bA} and F'_{bC}. In Eq.(10.47),a_c was assumed as 0.55 g O_2/g oxidized starch. Fig.10.27 shows the relationship between the starch flux and bulk starch concentration in Run 5. The solid line in Fig.10.27 shows the same relationship for pure starch oxidation which has been calculated by the following mathematical analysis.

For the mathematical analysis of simultaneous organic oxidation and nitrification,Watanabe and Thanantaseth(1982b) have made the following hypothesis: " the intrinsic oxygen uptake rate of the biofilm has a constant value at a fixed temperature,independent of the composition of the aerobic bacteria". This hypothesis can be mathematically expressed by Eq.(10.48).

$$R_o = a_n R_n = a_n R'_n + a_c R'_c = a_c R_c \qquad (10.48)$$

where R_o=intrinsic oxygen uptake rate of the biofilm, and R'_n,R'_c= intrinsic nitrification and organic oxidation rates of the biofilm in simultaneous organic oxidation and nitrification.

From the data of the nitrification experiment,Watanabe,et al(1982c) obtained an intrinsic nitrification rate for pure nitrification, R_n,as 5.2×10^3 g/m^{3}h at 23.5°C. Therefore,the intrinsic starch oxidation rate for pure starch oxidation,R_c,at the same temperature is calculated as follows:

$$R_c = (a_n R_n)/a_c = (4.33 \times 5.2 \times 10^3)/0.55 = 4.1 \times 10^4 g/m^3h$$

The molecular diffusion coefficient of starch has been calculated as 2.4×10^{-6} m^2/h by the Wilk and Chang's equation(Welty et al, 1969).

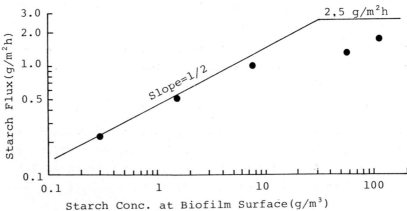

Fig. 10.28   Relationship between starch flux and starch conc.
at biofilm surface

Under starch rate limitation, the starch flux for pure starch oxidation is expressed as follows:

$$F_{bC} = \sqrt{2D_c R_c C_{sC}} \qquad\qquad (10.49)$$

where D_c=molecular diffusion coefficient of the starch within the biofilm, C_{sC}=starch concentration at the biofilm surface.

Graphically described, Eq.(10.49) is given by an oblique line having a slope of 1/2 in Fig.10.28, if R_c and D_c are 4.1 x 10^4 g/m^3h and 2.4 x 10^{-6} m^2/h, respectively. The data in Fig. 10.28 were obtained by converting the measured bulk starch concentration to the starch concentration at the biofilm surface using Eq.(10.19). The relationship between the bulk starch concentration and starch flux has been calculated using Eq.(10.7) for starch rate limitation and Eq.(10.50) for oxygen rate limitation(see Eq.(10.40)).

$$F_{bC} = (R_c/R_{oc})(F_{aO} + F_{wO}) = (F_{aO} + F_{wO})/a_c \qquad (10.50)$$

where R_{oc}=intrinsic oxygen uptake rate in pure starch oxidation.

The following numerical values have been used to calculate F_{aO} and F_{wO} by Eqs.(10.34) and (10.37). D_o= 8.8 x 10^{-6} m^2/h, R_{oc}=$a_c R_c$= 2.2 x 10^4 g/m^3h, L_w=5.0 x 10^{-5} m, C_O^* =8.5 g/m^3, L_d=8 x 10^{-5} m and C_{bo}=3g/m^3. The starch flux in the oxygen limitation has been calculated as follows:

$$F_{bC} = F_O/a_c =(F_{aO} + F_{wO})/a_c = 2.5 \text{ g/}m^2h$$

The bulk starch concentration at the transition from the starch to the oxygen rate limitation has been calculated using the steady state biofilm kinetics as 94 g/m^3. Then, the starch concentration at the biofilm surface in such a case has also been calculated as 32 g/m^3. The calculated relationship between F_{bC} and bulk starch concentration, C_{bC}, can be shown graphically by the solid line in Fig.10.27, which demonstrates that the reduction of the starch flux due to simultaneous nitrification is almost negligible until the bulk starch concentration reaches about 40 g/m^3 (corresponding BOD is about 20 g/m^3), but it becomes remarkable in the bulk starch concentration of more than 40 g/m^3 because the inner part of the aerobic biofilm consists of both heterotrophic and nitrifying bacteria.

Based on Eqs.(10.34) and (10.37), and the hypothesis expressed by Eq.(10.48), the following hypothesis has been derived: " the amount of oxygen supplied to the biofilm is the same under a fixed operating condition, independent of the composition of aerobic bacteria". This hypothesis is mathematically expressed as follows :

$$F_O = a_n F_{nA} = a_n F'_{nA} + a_c F'_{bC} = a_c F_{bC} \qquad (10.51)$$

where F_{nA}, F_{bC}=ammonia and organic matter flux in pure nitrification and pure starch oxidation, respectively.

Dimensionless form of Eq.(10.51) is

$$F'_{nA}/F_{nA} = 1 - (1/a_n)(a_c F'_{bC}/F_{nA}) = 1 - 0.23(a_c F'_{bC}/F_{nA}) \qquad (10.52)$$

Introducing Eq.(10.47) into Eq.(10.52) gives

$$F_{bA}/F_{nA} = 1 - 0.13(a_c F'_{bC}/F_{nA}) \qquad (10.53)$$

The experimental data in Runs 1 to 7 were plotted as the relationship between F'_{nA}/F_{nA} and $a_c F'_{bC}/F_{nA}$, as shown in Fig.10.29. This figure gives the experimental verification of Eq.(10.51).

10.6.2 Computer Simulation

Computer simulation was carried out to determine the change in the ammonia, starch and oxygen concentration profiles in the biofilm attached to a partially submerged RBC. The basic equation for the simulation is Eq.(10.54).

$$\partial C_i/\partial t = D_i(\partial^2 C_i/\partial z^2) - R_i \qquad (10.54)$$

where C_i, D_i=concentration and molecular diffusion coefficient of the substrate i, respectively, R_i=intrinsic reaction rate of the substrate i.

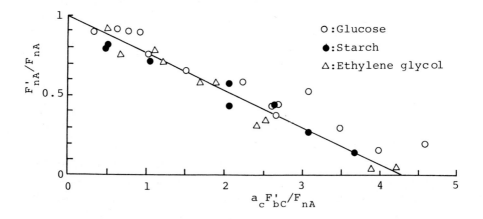

Fig.10.29   Experimental verification of Eq.(10.51)

The simulation procedure has already been presented in previous section. In a biofilm, with simultaneous organic oxidation and nitrification, the activity of the nitrifying bacteria would gradually increase with the depth of aerobic biofilm, while that of the heterotrophic bacteria would decrease with the depth of aerobic biofilm, because of the difference in the specific growth rates of both bacteria. Watanabe, et al (1984a) have proposed the emprical formulae, such as Eqs. (10.55) to (10.57), to express the above phenomenon, which also satisfy the concept expressed by Eq. (10.48).

Intrinsic nitrification rate:

$$R_n'/R_n = \gamma \qquad\qquad \text{at } 0 < Z < Z_0$$
$$R_n'/R_n = (Z - Z_0)/(K_n + (Z - Z_0)) \quad \text{at } Z \geq Z_0$$
$$\text{(10.55)}$$

Intrinsic organic oxidation rate:

$$R_c'/R_c = 1 - \gamma \qquad\qquad \text{at } 0 < Z < Z_0 \qquad \text{(10.56)}$$
$$R_c'/R_c = K_n/(K_n + (Z - Z_0)) \qquad \text{at } Z \geq Z_0$$

Intrinsic ammonia removal rate:

$$R_a'/R_a = (Z + 0.1K_n a_n)/(K_n + Z) \qquad\qquad \text{(10.57)}$$

where Z_0=numerical constant, $R_n'/R_n = R_c'/R_c = 1/2$ at $Z=Z_0 + K_n$, γ=constant, R_n', R_c' and R_a'=intrinsic nitrification, organic oxidation and ammonia removal rates in combined organic oxidation and nitrification, respectively, R_n, R_c and R_a=above mentioned rates in pure nitrification and organic oxidation, respectively.
Fig. 10.30 illustrates the concept expressed by Eqs. (10.55) and (10.56).

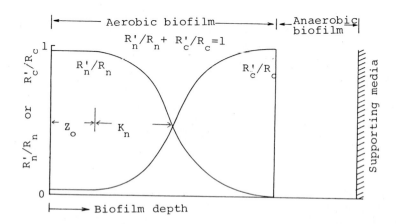

Fig. 10.30 Intrinsic organic oxidation and nitrification rates

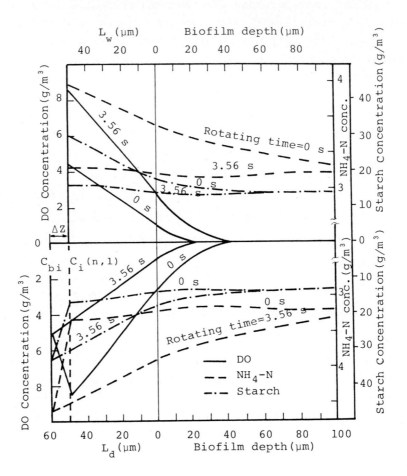

Fig. 10.31 Simulated profiles of starch, ammonia and DO
concentrations (Run 5)

451

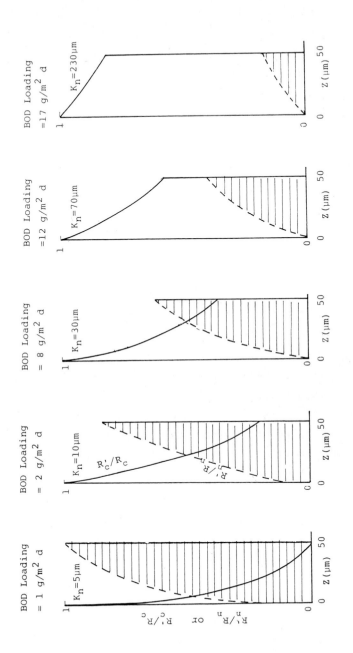

Fig. 10.32 Estimated profile of intrinsic organic oxidation and nitrification rates ($\gamma=0$)

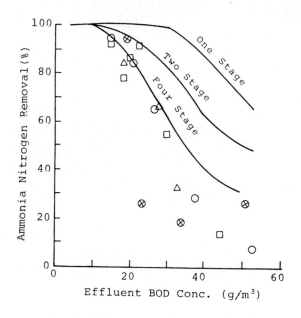

(N=2.0 rpm, T=17-19 °C)
(N=4.5-5.0 rpm, T=12-19 °C)
(N=3.2 rpm, T=14-20 °C)
(N=5.0 rpm, T=4-10 °C)

Fig.10.33 Comparison of experimental and calculated
ammonia removals in combined organic oxidation
and nitrification

Table 10.6   Numerical constants for the simulation

Biofilm Thickness	900 μm
Attached-water layer thickness	50 μm
Diffusion layer thickness	70 μm
Max.intrinsic starch oxidation rate	4.1×10^4 g/m^{3}h
Max.intrinsic nitrification rate	5.2×10^3 g/m^{3}h
Diffusion coefficient of oxygen	8.8×10^{-6} m^2/h
Diffusion coefficient of starch	2.4×10^{-6} m^2/h
Diffusion coefficient of ammonia	7.5×10^{-6} m^2/h
Oxygen requirement of starch oxidation	0.55 gO$_2$/g oxidized starch
Oxygen requirement of nitrification	4.33 gO$_2$/g oxidized ammonia
Disk rotating speed	8.5 rpm

The simulation was conducted for the experimental conditions in Run 5. Table 10.6 shows the numerical values in the simulation. The constant, K_n, was changed to match the simulated results with the experimental results. Fig.10.31 shows the simulated profiles of starch, ammonia, and dissolved oxygen in Run 5.

Using the simulated substrate profiles, the flux of substrate i can be calculated by Eq.(10.58).

$$F_i = (D_i/N \, \Delta Z) \sum_{n=1}^{N} (C_{bi} - C_i(n,1)) \qquad (10.58)$$

where the submerged biofilm surface is divided into N elemental surfaces, and the biofilm depth is divided into sub-layers, each of them ΔZ thick. N and ΔZ were 400 and 10 μm, respectively. Comparison of the simulated and experimentally determined ammonia flux is shown in Fig.10.26. Fig.10.27 also gives the same comparison in starch oxidation. Fig.10.32 shows the intrinsic profiles of organic oxidation and nitrification rates found out by the simulation. The calculated relationship between the effluent BOD and the ammonia percent removal for a multi-stage, completely mixed flow RBC is shown in Fig.10.33 along with USEPA(1971) data. The calculation was made for an influent ammonia concentration of 30 g/m^3 and an influent BOD of 150 g/m^3. USEPA data were collected in a two stage RBC whose flow direction was parallel to the rotating shaft. The average influent Kjeldahl nitrogen concentration and BOD were 29 g/m^3 and 147 g/m^3, respectively.

10.7 KINETIC ANALYSIS OF DENITRIFICATION IN A SUBMERGED RBC

The submerged RBC has been successfully applied to an anaerobic denitrification(Davies and Pretorius,1975; Ishiguro and watanabe, 1977b). An. experimental verification of the applicability of the steady state biofilm kinetics to the same process has been done(Watanabe et al,1978a), as explained below.

Fig.10.34 shows the relationship between the nitrate flux and the bulk nitrate concentration under nitrate rate limitation. The diffusion layer thickness in Runs D-1 and D-2 was estimated at about 90 μm and 60 μm,respectively. At a water temperature of 20°C,the molecular diffusion coefficient of nitrate ion is about 6×10^{-6} m²/h. Using these values,the nitrate concentration at the biofilm surface, C_{sN},is calculated by Eq.(10.59),as derived from Eq.(10.7).

$$C_{sN} = C_{bN} - (F_{bN}L_d/D_N) \qquad (10.59)$$

where C_{bN}=bulk nitrate concentration,F_{bN}=nitrate flux,D_N=molecular diffusion coefficient of nitrate ion.

Fig.10.35 was thus obtained. It demonstrates that the steady state biofilm kinetics is applicable to the denitrification in a submerged RBC. The intrinsic denitrification rate at 20°C has been estimated at about 1.5×10^3 g/m³h,from the slope of the straight line in Fig.10.35. The solid lines in Fig.10.34 were calculated by the steady state biofilm kinetics,using the determined values of L_d and R_d

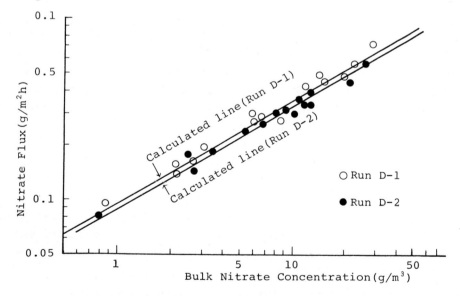

Fig.10.34 Logarithmic plots of nitrate flux and
bulk nitrate concentration in submerged RBC

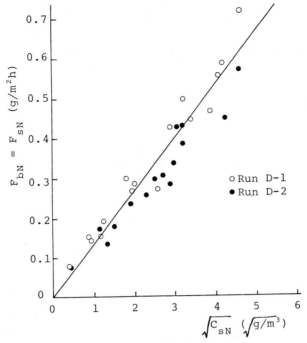

Fig. 10.35 Verification of half order kinetics

The temperature dependency of the intrinsic denitrification rate was determined, using the nitrate flux under nitrate limitation (Watanabe, 1984b). With this condition, the nitrate flux is expressed as follows:

$$F_{bN} = \sqrt{2D_N R_d C_{sN}} \qquad (10.60)$$

where R_d = intrinsic denitrification rate.

Therefore, the intrinsic denitrification rate is expressed as,

$$R_d = (F_{bN})^2 / 2D_N C_{sN} \qquad (10.61)$$

In the denitrification experiment with a submerged RBC, F_{bN} and C_{bN} were measured at varying temperatures (Zaitsu and Koresawa, 1976). For example, at the water temperature of 10°C, F_{bN} was 0.096 g/m²h in the bulk nitrate concentration of 3.6 g/m³. D_N and L_d at this experimental condition are 4.5 x 10^{-6} m²/h and 9 x 10^{-5} m, respectively. Therefore, the intrinsic denitrification rate at 10°C is calculated as follows: 6.4 x 10^2 g/m³h.

In the same manner, the intrinsic denitrification rates at 15, 20 and 25°C have been determined as 1.1 x 10^3, 1.5 x 10^3 and 2.0 x 10^3 g/m³h, respectively. Fig. 10.36 shows the temperature dependency of the intrinsic denitrification rate, and it is expressed as follows:

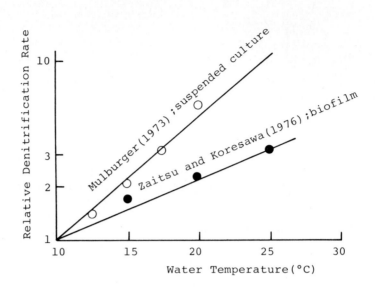

Fig. 10.36   Temperature dependency of denitrification rate

$$R_{d,T} = (1.5 \times 10^3)(1.07)^{T-20} \qquad\qquad (10.62)$$

where $R_{d,T}$=intrinsic denitrification rate at T °C.

From the data of Mulburger(1973),the temperature coefficient for the suspended denitrifying bacteria becomes 1.17. Fig. 10.36 demonstrates that the intrinsic denitrification rate is less influenced by the temperature than is the denitrification rate in suspended culture. This is the same conclusion obtained for the nitrification rate.

10.8 SIMULTANEOUS NITRIFICATION AND DENITRIFICATION IN A PARTIALLY SUBMERGED RBC

10.8.1 Experimental Data on Simultaneous Nitrification and Denitrification

In a partially submerged RBC,oxygen seldom penetrates into the deepest part of the biofilm,and the penetration depth of oxygen changes with the rotation of the biofilm. Therefore,the facultative denitrifying bacteria can convert the oxidized nitrogen such as nitrite and nitrate into the gaseous nitrogen by the biological denitrification reaction,if they exist in the atmosphere where the oxidized nitrogen and the organic matter coexist. Masuda et al(1982) have confirmed this phenomenon of the simultaneous nitrification

and denitrification(referred to as SND) by measuring the nitrogen
gas content in the air phase of a closed type submerged RBC.
Fig. 10.37 shows the increase of nitrogen and unknown(maybe ,N_2O)
gases in the air phase of such an RBC. The nitrogen removal efficiency by the SND depends on the partial pressure of oxygen in the
air phase,temperature,ammonia and organic loadings,retention time,
and type of the organic sources. Fig.10.38 shows the effect of the
ammonia loading,and methanol to ammonia concentration in the influent on the nitrogen removal by the SND. Fig. 10.39 shows the effect
of the partial pressure of oxygen on the removal of ammonia and nitrogen by the SND.

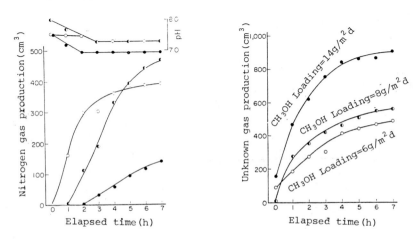

Fig. 10.37 Gas production with the increase of elapsed time

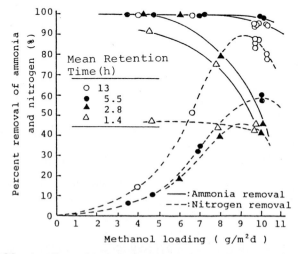

Fig.10.38   Effect of methanol loading on simultaneous nitrification
and denitrification

458

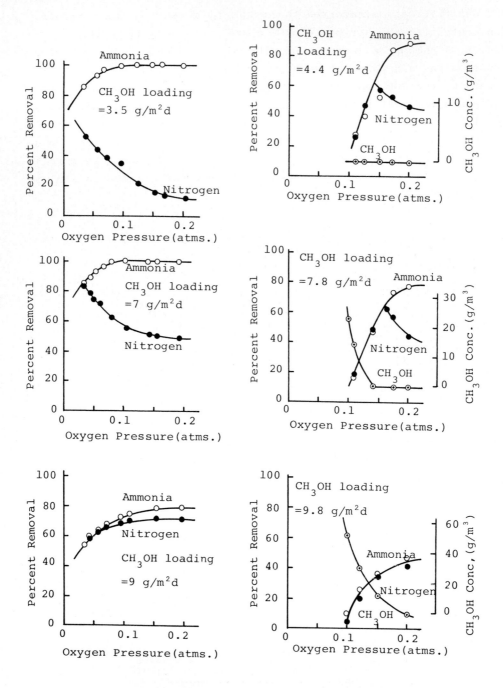

Fig. 10.39   Effect of oxygen pressure on simultaneous
nitrification and denitrification

10.8.2 Computer Simulation of the SND

Computer simulation was carried out to determine the relationship between the partial pressure of oxygen in the air phase, and the ammonia and nitrogen removal rates (Watanabe, et al, 1984 a). The simulation model has been the same as that used for the simulation of simultaneous organic oxidation and nitrification. However, the equations describing the diffusion of organic matter and nitrate with the simultaneous zero order intrinsic denitrification rate are also included in the simulation model for the SND. The intrinsic denitrification rate in the organic source of methanol is expressed as,

$$R_d' = R_d \quad \text{and} \quad R_c'' = 2.5 R_d' \tag{10.63}$$

where R_d' =intrinsic denitrification rate in the simultaneous nitrification and denitrification, R_c'' =intrinsic organic oxidation rate due to the nitrification reaction.

Eq.(10.63) means that the facultative denitrifying bacteria exist in the same activity as that in pure denitrification, and 2.5 grams of methanol is oxidized per unit gram of denitrified nitrogen. The oxygen penetration depth in the biofilm changes with the rotation of the biofilm. This penetration forms a zone within the biofilm where aerobic and anaerobic atmospheres appear alternately. The facultative denitrifying bacteria act as aerobic heterotrophic bacteria in the aerobic atomosphere and anaerobic denitrifying bacteria in the anaerobic atmosphere.

The reproduction of the experimental data by the simulation was made for the data shown in Table 10.7. Table 10.8 shows the numerical constants used in the simulation.

Table 10.7 Experimental data for the simulation

	P_O(atm.)	C_{ba} (g/m^3)	C_{bn} (g/m^3)	C_{bm} (g/m^3)	C_{bo} (g/m^3)
Run 1	0.21	0.5	22.0	1.3	1.0
Run 2	0.18	0.5	20.0	1.3	0.9
Run 3	0.11	7.0	0.2	8.5	0.1
Run 4	0.08	20.0	0.1	56.0	0.1

P_O: Partial pressure of oxygen in the air phase.
C_{ba}: Bulk ammonia concentration.
C_{bn}: Bulk nitrate concentration.
C_{bm}: Bulk methanol concentration.
C_{bo}: Bulk DO concentration.

460

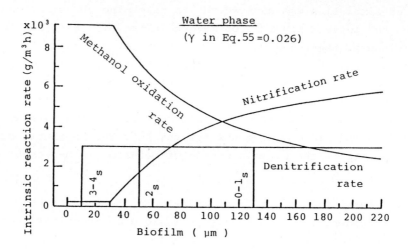

Fig.10.40   Profiles of intrinsic reaction rates
            during water phase

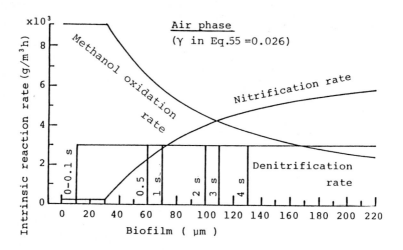

Fig.10.41   Profiles of intrinsic reaction rates
            during air phase

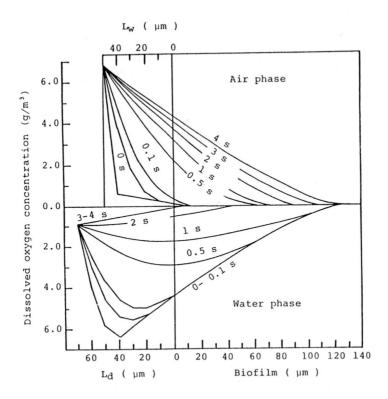

Fig.10.42 Simulated profile of DO in simultaneous
nitrification and denitrification(Run 2)

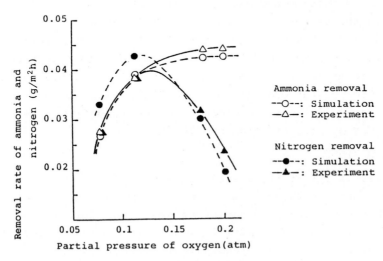

Fig.10.43 Comparison of experimental and simulated rates
of nitrification and denitrification

Table 10.8   Simulation conditions for SND

Biofilm thickness	900 μm
Attached-water layer thickness	50 μm
Diffusion layer thickness	70 μm
Intrinsic nitrification rate	8000 g/m^3h
Intrinsic denitrification rate	3000 g/m^3h
Intrinsic methanol oxidation rate	9000 g/m^3h
Diffusion coefficient of ammonia	9.0 x 10^{-6} m^2/h
Diffusion coefficient of oxygen	10.4 x $10^{-6}m^2/h$
Diffusion coefficient of methanol	7.5 x 10^{-6} m^2/h
Diffusion coefficient of nitrate	7.2 x 10^{-6} m^2/h
Disk rotating speed	7.5 rpm

The data in Table 10.7 were obtained by an experiment in which the vent holes in the cover of the closed type RBC were closed. Steady state operation at an oxygen pressure of 0.21 atms had been completed before the vent holes were closed. The partial pressure of oxygen decreased with the elapsed time after the vent holes were closed. It took about 6 hours to decrease the partial pressure of oxygen from 0.21 to 0.08 atms. Therefore, the profiles of the intrinsic organic oxidation and nitrification rates were fixed in the simulation, because 6 hours would be too short a time to significantly change the profiles. Figs. 10.40 and 10.41 show the profiles of intrinsic methanol oxidation and nitrification rates which gave the best fit for the simulated and experimental results. The simulation determined the values of K_n and Z_o to be 70 μm and 30 μm, respectively. Fig. 10.42 shows the simulated dissolved oxygen profile. Considering the simulated profiles shown in Fig. 10.42, the profile of intrinsic denitrification rate is shown in Figs. 10.40 and 10.41 as a function of the biofilm rotating time during the air and water phases. Fig. 10.43 shows the comparison of the experimental and simulated rates of nitrification and denitrification.

 10.9 CASE STUDY

Groundwater or infiltrating surface water moving through solid wastes can produce the leachate. The biochemical characteristics of the leachate may depend on the structure of the landfill, material to be buried, age of the landfill, amount of rainfall, weather and so on. The Haginodai sanitary landfill, Miyazaki City, Japan has started in 1968 and every kind of municipal solid wastes including garbage was buried there until 1974. Since then, only noncombustible refuse has been buried while combustible materials are burned in a newly

constructed incinerator. The area and the volume of this landfill
are about 6.2 ha and 400,000 m³,respectively. About 150 tons of
refuse is buried every day. Miyazaki City is located in the most
southern part of Kyushu island and has semi-tropical climate with
annual rainfall of about 2,500 mm. The leachate from the Haginodai
sanitary landfill enters a small river from which irrigation water
has been drawn for the rice fields. As the leachate has been rich
in ammonia,bio-stimulation of the rice plant occurred and the nei-
ghbouring farmers called for the nitrogen removal from the leachate
to prevent this. In order to resolve the problem,Miyazaki Sanitary
Authority planned to construct a leachate treatment plant for the
nitrogen removal. As a result,an RBC plant was constructed. The
plant consists of nitrification RBC ,denitrification RBC ,reaera-
tion RBC and sedimentation tank,as shown in Fig.10.44. Each RBC,
which is divided into four stages,has corrugated plate disks with
diameter of 3.6 m. Submerged ratio of nitrification RBCs is 0.4.Total
disk surface area of nitrification RBC,denitrification RBC and reaeration RBC is 18,680 , 8,490 and 1,380 m²,respectively. The disk
rotational speed is 1.6 rpm in nitrification RBC and denitrification RBC. The main characteristics of the leachate are as follows:
pH=7.6 to 7.9, alkalinity=800 to 1,100 g/m³, ammonia concentration
=100 to 150 g/m³, organic carbon concentration=170 to 400 g/m³,
BOD=10 to 20 g/m³, DO=4 to 6 g/m³. Operation of the plant began in
October,1976. Fig.10.45 shows the operational data during one year

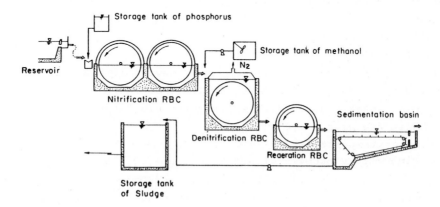

Fig.10.44 Flow sheet of Haginodai RBC plant

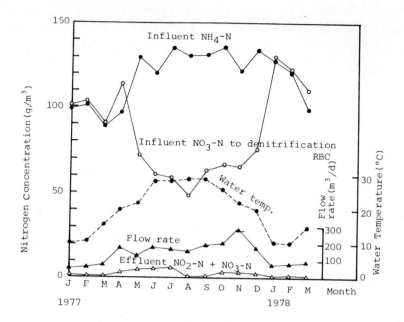

Fig.10.45 Operation results of Haginodai RBC plant

which were taken once a week as a grab sample.The monthly data in
Fig.10.45 are the averaged weekly data.

Simultaneous nitrification and denitrification is quite active
when the influent leachate temperature is more than 20°C. Nitrogen
removal percentage by the SND is mainly influenced by the hydraulic
loading and the influent leachate temperature as shown in Fig.10.46.
Fig.10.47 shows the typical profiles of the treated water qualities
in nitrification and denitrification RBCs.

Kinetic analysis of the nitrification in the first stage of nitrification RBC ,which was operated under oxygen limitation,has been
carried out under the following conditions:

$\quad$ disk rotational speed,N = 1.6 rpm = 0.027 rps

$\quad$ disk radius,r = 180 cm

$\quad$ water temperature,T = 20°C

$\quad$ kinematic viscosity of water,μ= 1.0 x 10^{-2} g/cm s

$\quad$ attached-water film thickness,L_w = 6.2 x 10^{-5} m (Eq.(10.33))

$\quad$ diffusion layer thickness,L_d = 1.5 x 10^{-4} m (Fig.10.2)

$\quad$ saturation DO concentration,C_o^* = 8.2 g/m^3

$\quad$ intrinsic nitrification rate,R_n = 5.8 x 10^3 g/m^{3}h (Eq.(10.26))

$\quad$ bulk DO concentration,C_{bo} = 0.7 g/m^3

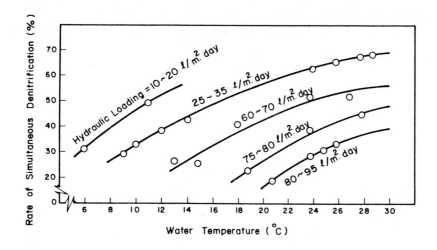

Fig.10.46   Temperature effect on simultaneous
            nitrification and denitrification

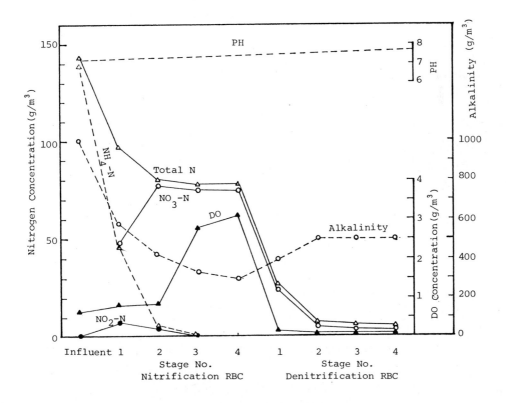

Fig.10.47 Profiles of treated water quality

intrinsic oxygen uptake rate, $R_{on} = a_n R_n = 2.5 \times 10^4$ g/m^{3}h
molecular diffusion coefficient of ammonia, $D_A = 9.0 \times 10^{-6}$ m^2/h

Calculation of ammonia flux, F_{bA} can be done as follows:

$\lambda_a = 21.4$ g/m^3 by Eq. (10.36)

$C_{so,a} = 1.9$ g/m^3 by Eq. (10.35)

$F_{aO} = 0.91$ g/m^{2}h   by Eq. (10.34)

$\lambda_w = 125$ g/m^3 by Eq. (10.39)

$C_{so,w} = 0$ g/m^3 by Eq. (10.38)

$F_{wO} = 0.042$ g/m^{2}h   by Eq. (10.37)

$F_{bA} = (F_{aO} + F_{wO})/a_n = 0.22$ g/m^{2}h   by Eq. (10.40)

Calculation of effluent ammonia concentration can be done as follows:

As the flow direction of RBC is perpendicular to the disk surface, the flow pattern is not completely mixed flow. However, F_{bA} is written by Eq. (10.64) in a reactor with any flow pattern if F_{bA} is independent of the bulk ammonia concentration.

$$F_{bA} = (Q/A)(C_{iA} - C_{eA}) \tag{10.64}$$

Numerical values of the parameters in the above Equation are as follows:

Flow rate, $Q = 107$ m^3/day

Submerged biofilm surface area, $A = 4670 \times 0.4 = 1,868$ m^2

Influent ammonia concentration, $C_{iA} = 140$ g/m^3

Ammonia flux, $F_{bA} = 0.22$ g/m^{2}h

Effluent ammonia concentration, C_{eA} can be calculated as follows:

$$C_{eA} = C_{iA} - (AF_{bA}/Q) = 140 - 92 = 48 \text{ g/m}^3$$

The calculated value of F_{bA} is very close to the experimentally determined value of 45 g/m^3.

10.10 SUMMARY AND CONCLUSIONS

A steady state biofilm kinetics is described as the process of molecular diffusion with a simultaneous zero order biochemical reaction within the biofilm. According to the proposed kinetics, the flux of the rate limiting substrate to the biofilm can be determined by the molecular diffusion coefficient of the substrate,intrinsic biochemical reaction rate,and the diffusion layer thickness. The intrinsic nitrification and denitrification rates have been experimentally determined,and it is concluded that they are less influenced by the temperature than are the nitrification and denitrification rates in the suspended culture. The diffusion layer thickness in the RBC would be inversely proportional to the contactor rotating speed,and also depends on the contactor configuration. The proposed kinetics is applicable to the denitrification in a submerged RBC. However,it is not applicable to the biochemical process in a partially submerged RBC which has no steady state substrate profile within the biofilm,since the biofilm alternately rotates from water to air.

The computer simulation of the nitrification in such an RBC was carried out,which produced an average ammonia flux for the elemental biofilms rotating in the water phase that is almost equal to the ammonia flux obtained by the experiment in the same process. The simulation result also demonstrates the following points:

1. Under ammonia limitation,the ammonia concentration in the biofilm would reach the steady state profile in a short time after the biofilm enters the water phase, and

2. Under oxygen limitation,the oxygen concentration in the biofilm would reach the steady state profile in a short time after the biofilm enters the air or water phase .

Based on the above evidences,it is concluded that the proposed kinetics can be applicable to the nitrification in a partially submerged RBC when the nitrification process is under ammonia limitation,and the following equations are available when it is under oxygen limitation:

$$F_{bA} = (F_{aO} + F_{wO})/4.33$$

$$F_{aO} = D_O (C_O^* - C_{so,a})/L_w$$

$$F_{wO} = D_O (C_{bo} - C_{so,w})/L_d$$

Comparison between the calculated and experimental F_{bA} confirmed the availability of the above equations.

Based on the hypothesis stating that the oxygen flux to the biofilm is almost the same under fixed operating conditions,independent of the aerobic bacteria composition,the following equa-

468

tion has been proposed for the ammonia flux due to nitrification in the combined organic oxidation and nitrification in a partially submerged RBC:

$$F'_{nA}/F_{nA} = 1 - (a_c/a_n)(F'_{bC}/F_{nA})$$

The experimental verification of the above equation was also demonstrated.

The phenomenon of simultaneous nitrification and denitrification in an RBC was mentioned and a computer simulation model of the same phenomenon was presented. The simulation model is the same as that used for simultaneous nitrification and denitrification, except for the introduction of the action of the facultative denitrifying bacteria. The simulated profile of dissolved oxygen concentration within the biofilm shows that the denitrification would occur mainly in a zone formed within the biofilm where aerobic and anaerobic atmospheres alternately appear. The depth of such a zone depends on the factors such as the loadings of organic matter and ammonia, water temperature, and partial pressure of oxygen in the air phase. Comparing the simulated and experimental results of simultaneous nitrification and denitrification, the proposed simulation model was found to be almost satisfactory.

The operational data of the leachate treatment plant, which consists of nitrification and denitrification RBCs were presented, and the kinetic analysis of the nitrification RBC was carried out.

REFERENCES

Bintanja,H.H.J.,et al(1975) Oxygen Transfer in a Rotating Disk Treatment Plant,Water Research,Vol.9

Chen,Y.S. and Bungay,H.R.(1981) Microelectrode Study of Oxygen Transfer in Trickling Filter Slimes,Biotech. and Bioeng.,Vol.23

Cheung,P.S. and Krauth,K.(1980) The Effect of Nitrate Concentration and Detention Period on Biological Denitrification in Rotating Disk System,Journal of Water Pollution Control Federation,Vol.52

Davies,T.R. and Pretorius,W.A.(1975) Denitrification with Bacterial Disk Unit,Water Research,Vol.9

Famularo,J.,Mueller,J.A. and Mulligan,T.(1978) Application of Mass Transfer to Rotating Biological Contactors,Journal of Water pollution Control Federation,Vol.50,No.4

Harremoës,P.(1976) The Significance of Pore Diffusion to Filter Denitrification, Journal of Water Pollution Control Federation,Vol.48,No.2

Harremoës,P.(1977) Half-Order Reactions in Biofilm and Filter Kinetics,Vatten 33

Harremoës,P. and Gonenc,I.E.(1983) The Application of Biofilm Kinetics to Rotating Biological Contactors,International Seminar on Rotating Biological Disks,Stuttgart,Oct. 6-8

Heukelekian,H. and Rand,M.C.(1955) Biological Oxygen Demand of Pure Organic Compounds,Sewage and Industrial Wastes,27

Ishiguro,M.,Watanabe,Y. and Masuda,S.(1977a) Advanced Wastewater Treatment by Rotating Biological Disk Unit(11),Journal of Japan Sewage Works Association,Vol.14,No.152

Ishiguro,M.,Watanabe,Y. and Masuda,S.(1977b) Advanced Wastewater by Rotating Biological Disk Unit(111),Journal of Japan Sewage Works Association,Vol.14,No.161

La Motta,E.J.(1976a) Internal Diffusion and Reaction in Biological Films,Environmental Science & Technology,Vol.10

La Motta,E.J.(1976b) External Mass Transfer in a Biological Reactor Biochemistry & Bioengineering,Vol.18

Levich,V.G.(1962) Physicochemical Hydrodynamics,Printice-Hall Inc.

Mamiya,T.(1983) Effect of Organic Compounds Characteristics on Simultaneous Nitrification and Denitrification,Master's Thesis, Dept. of Civil Engineering,Miyazaki University,Japan

Masuda,S.,Watanabe,Y. and Ishiguro,M.(1982) Simultaneous Nitrification and Denitrification in a Rotating Biological Contactor, Proceedings of the First International Conference on Fixed-Film Biological Processes

Masuda,S.,Watanabe,Y. and Ishiguro,M.(1983) Mathematical Model of Simultaneous Nitrification and Denitrification in a Rotating Biological Contactor,Proceedings of the 38th Conference of Japan Society of Civil Engineers

McCarty,P.L.,et al(1969) Biological Denitrification of Wastewater by Addition of Organic Materials,Proceedings of the 24th Industrial Waste Conference,Purdue University

Mulbarger,M.C.(1973) Nitrification and Denitrification Facilities Wastewater Treatment,Technology Transfer Seminar Publication,USEPA

Onuma,M. and Omura,T. (1982) Mass-Transfer Characteristics within Microbial Systems,Water Technology and Science,Vol.14

470

Pretorius,W.A.(1974) Nitrification on the Rotating Disk Unit, Progress in Water Technology,Vol.8

Riemer,M. and Harremoës,P.(1978) Multi-Component Diffusion in Denitrifying Biofilm,Progress in Water Technology,Vol.10,Nos.5/6

Rittman,B.E. and McCarty,P.L.(1978) Variable-Order Model of Bacterial-Film Kinetics,Journal of the Environmental Engineering Division, ASCE,EE5

Rittman,B.E. and McCarty,P.L.(1980) Design of Fixed-Film Processes with Steady State Biofilm Model,Progress in Water Technology,Vol.12

USEPA(1971) Application of Rotating Disc Process to Municipal Wastewater Treatment,Water Pollution Control Reaearch Series 17050 DAM 11/71

USEPA(1973) Nitrification and Denitrification Facilities Wastewater Treatment,Technology Transfer Seminar Publication,EPA-625/4-73-004a

Watanabe,Y. and Ishiguro,M.(1978a) Denitrification Kinetics in a Submerged Rotating Biological Disk Unit,Progress in Water Technology,Vol.10,Nos.5/6

Watanabe,Y.,Ishiguro,M. and Nishidome,K.(1978b) Kinetic Analysis of Denitrification by Rotating Bio-Disk Unit,Proceedings of the Japan Society of Civil Engineers,No.276

Watanabe,Y.(1979) Nitrogen Removal from Sanitary Landfill Leachate: a Case Study in Japan, Proceedings of the 2nd Regional Seminar on Solid Waste Management

Watanabe,Y.,Ishiguro,M. and Nishidome,K.(1980a) Nitrification Kinetics in a Rotating Biological Disk Reactor,Water Technology and Science,Vol.12

Watanabe,Y.,Ishiguro,M. and Nishidome,K.(1980b) Purification Mechanism of Rotating Biological Contactors(11),Journal of Japan Sewage Works Association,Vol.17,No.195

Watanabe,Y,Bravo,H.E. and Nishidome,K.(1982a) Simulation of Nitrification and Its Dynamics in a Rotating Biological Contactor, Water Technology and Science,Vol.14

Watanabe,Y. and Tanantaseth,C.(1982b) Combined Organic Oxidation and Nitrification Process in Rotating Biological Contactors,Journal of Japan Sewage Works Association,Vol.19,No.221

Watanabe,Y.,Nishidome,K. and Ishiguro,M.(1982c) Simulation of Nitrification Process in Rotating Biological Contactors,Journal of

Japan Sewage Works Association,Vol.19,No.223

Watanabe,Y,.Masuda,S.,Nishidome,K. and Wantawin,C.(1984a) Mathematical Model of Simultaneous Organic Oxidation,Nitrification,and Denitrification in Rotating Biological Contactors,Water Technology and Science,Vol.16

Watanabe,Y.(1984b) Unpublished paper

Welty,J.R.,et al(1969) Fundamentals of Momentum,Heat and Mass Transfer,John Wiley & Sons Inc.

Weng,C.M. and Molof,A.H.(1974) Nitrification in the Biological Fixed-Film Rotating Disk System,Journal of Water Pollution Control Federation,Vol.45

Wezernak,C.T. and Cannon,J.J.(1967) Oxygen Nitrogen Relationship in Autotrophic Nitrification,Applied Mocrobiology,Vol.15,No.5

Williamson,K. and McCarty,P.L.(1976) A Model of Substrate Utilization by Bacterial Films,Journal of Water Pollution Control Federation,Vol.48,No.1

Wong-Chong,G.M. and Loehr,R.C.(1975) The Kinetics of Microbial Nitrification,Water Research,Vol.9

Zaitsu,Y. and Koresawa,Y.(1976) A Fundamental Study on Biological Denitrification in a Submerged Rotating Biological Contactor, Graduation Thesis,Dept. of Civil Engineering,Miyazaki University, Japan

MATHEMATICAL MODELLING OF BIOLOGICAL PACKED AND FLUIDIZED BED REACTORS

S.W. HERMANOWICZ[*] and J.J. GANCZARCZYK
University of Toronto, Toronto, Canada, M5S 1A4

(*) present address: Department of Civil Engineering,
                     University of California, Berkeley, CA 94720, U.S.A.

1. INTRODUCTION

Some biological fixed-film processes have been long established in the practice of wastewater treatment; others are relatively new in this field. The fixed-film processes are usually classified into the following categories:

- packed beds;

- two-phase fluidized beds (liquid-solids);

- three-phase fluidized beds (liquid-gas-solids);

- rotating biological contactors (RBC);

- trickling filters.

The first two systems contain only liquid and solid phases. In the latter two (RBC and trickling filters) three phases: gas, liquid, and solid co-exist in the reactor, and the presence of all these components is essential for the proper functioning of the system. Three-phase fluidized beds can be placed, from the technological standpoint, in a border area between two and three phase systems although, strictly speaking, three physical phases are present in the reactor. However, if the volume of the gas phase is small relatively to the volume of liquid and solids, the effects of the gas phase can be neglected, and the system can be considered as an essentially two-phase reactor. This situation is likely to occur when the gas is produced in biochemical reactions in the system (like in denitrification or anaerobic digestion) and its flow is small compared with the flow of the liquid. Even then, some phenomena characteristic for three-phase systems may occur (e.g. flotation).

In this work only the modelling of two-phase reactors (liquid-solids) will be discussed. In these systems the solid phase consists of two components: biomass and carrier particles. The carrier can be inert (like sand or plastic), can be active as a sorbent (like activated carbon or other sorptive materials), or even can be chemically active. The latter possibility was studied recently by Kowalski and Lewandowski (1983) in a nitrifying bed packed with crushed marble. The interactions between sorption, desorption, some chemical processes and biological removal are very complex. Many different phenomena may control the process during the operation. Modelling of those interactions is beyond the scope of this work.

Microorganisms in a fixed-film reactor are either attached to the surface of a carrier or are suspended in liquid phase in the form of flocs or pellets. It is possible to eliminate carrier particles and cultivate microorganisms in the form of bacterial pellets forming a sludge blanket in the reactor (Klapwijk et al., 1979; Lettinga et al., 1980). The use of bacterial cells or enzymes artificially immobilized on the surface of a carrier has been practiced for fermentation processes but only recently this technique was applied for wastewater treatment (Vossoughi et al., 1982; Kokufuta et al., 1982). The applications of fixed-film processes presented above show that a fixed-film biological reactor can be operated in a variety of forms. Some of them, like fluidized biological beds, can be even considered as modifications of the activated sludge process (Ganczarczyk, 1983). There are, however, some basic principles which apply to all of these variants. The similarities and the differences among various applications are discussed in the next part of this work.

2. PACKED AND FLUIDIZED BEDS - SIMILARITIES AND DIFFERENCES

A basic common characteristics of all fixed-film reactors discussed in this work is the presence of two physical phases in the reactor. The liquid phase contains soluble or particulate substrates and products which are formed inside the solid phase. Transport phenomena play an important role in an overall substrate removal and product formation. The overall removal process of soluble substrates is composed of the following steps:
- diffusion of substrates from the bulk of liquid through a liquid film around the particle to the liquid/biofilm interface;
- diffusion of substrates within the biofilm and simultaneous reaction;
- diffusion of reaction products from the place of formation through the biofilm towards its surface;
- diffusion of products through the liquid film into the bulk of solution.

The particulate substrates must undergo hydrolysis outside the biofilm before the diffusion can take place. The mechanism of this process is not understood well enough to warrant any modelling attempt. Since the removal of pollutants (substrates) constitutes a major concern in the wastewater treatment practice, the first two steps are probably most important for modelling purposes. However, if an out-diffusion of products is slower than their formation, accumulated product may inhibit microorganisms and adversely affect the removal rate. Diffusional restrictions in the substrate transfer may result in a limited substrate availability in deeper layers of the biofilm and affect its physical structure.

In all fixed-film reactors biomass is mostly attached to or immobilized at the surface of a carrier. Therefore, microorganisms are **essentially** contained within the reactor and only a small fraction of them is carried away with flowing liquid. Even in a sludge blanket reactor, where no carrier is used, the hydraulic flux (liquid velocity) is maintained in such a range as to allow microbial aggregates to remain inside the reactor. Therefore, long biomass retention times and a high concentration of biomass can be achieved in the fixed-film reactors.

The reactor containing a biological bed is usually designed in a tubular form and mixing conditions depend on a number of factors like diameter, height, size of carrier particles, liquid velocity, etc. Mixing, however, is not complete and a concentration gradient usually exists within the bed. At the inlet, where the concentration of substrate is higher, the removal rate is also increased and biomass growth is promoted. At the outlet, the removal rate is generally lower due to lower substrate concentration and bacterial growth is less intense. This phenomenon was observed both in packed beds (Faup et al., 1982) and in fluidized bed when the intensity of solids mixing was small (Hermanowicz, 1982). These features, common to packed and fluidized systems, rationalize the use of similar modelling methods for both types. However, there are some important differences between the two types of the reactors. The main difference between a packed bed and a fluidized bed lies in the mobility of the solid particles. In the fluidized systems, the particles are "suspended" in a stream of liquid and are able to move with regard to each other and to the walls of the reactor. If the particles are of different size and/or density (e.g. as a result of biofilm growth), they tend to separate, the larger and lighter accumulating at the top of the bed. This phenomenon is counteracted by solids mixing in the fluidized reactor. The movement of solids also alters the mixing pattern of the liquid phase in the fluidized bed as compared with the packed bed. Ngian and Martin (1980a) postulated that solids mixing may result in microorganisms being periodically exposed to different environmental conditions (e.g. aerobic/anaerobic, low/high substrate concentrations, etc.). This may affect the metabolism of microorganisms, but at present, is only marginally understood.

Biomass removal from the reactor may occur in several ways. Some of the removal mechanisms are common for both types of systems. They include microbial death and lysis, and hydraulic sloughing, although the intensity of the removal may differ in both systems. In a fluidized bed, biomass is also removed by "mechanical" attrition as a result of more or less vigorous contacts between the particles. The dynamics of biofilm development was recently discussed by Characklis (1981) and Characklis and Trulear (1982). In fluidized beds an intentional and controlled removal of excess biomass is also possible. For this purpose, a part of the particles covered with the biofilm is removed from the reactor, usually from the top of the bed, separated from the biomass, and cleaned particles are returned to the reactor. This type of cleaning creates a recycle flow of solids through the reactor which can significantly alter biomass distribution and the efficiency of the process (Bousfield and Hermanowicz, 1984).

The type and size of carrier used in packed and fluidized beds is probably the most visible difference between these two types. Particles used in packed beds are generally larger, often specially designed to achieve larger porosity of the bed and prevent clogging. In a fluidized bed finer media are used to reduce the required liquid velocity (fluidization velocity) and to increase the specific surface area available for bacterial growth. Clogging in these beds is prevented by their expansion during the operation. The difference in size and type of medium is of minor importance in the modelling since the basic principles of the process are not altered by the choice of medium. This choice, though, may affect the practical efficiency of the process and its operation.

3. BASIC STRUCTURE OF A MODEL

Three main parts can be distinguished in a model of a fixed-film biological bed. They are:
- a reactor;
- substrates uptake and products formation;
- biomass development.

Each of the three parts is composed of several elements which are listed in Fig. 1. The elements related to fluidized beds only are marked with an asterisk. In Fig. 1 the interactions are indicated only between the main parts of the model. This does not imply that any of the main parts represents a closed system. On the contrary, the elements of each part do interact with the elements of the other parts, e.g., the growth of biomass affects substrate removal as well as biomass distribution and solids segregation. The main parts shown in Fig. 1 should be thus regarded as an organizational framework facilitating the development and description of the model. Each of the

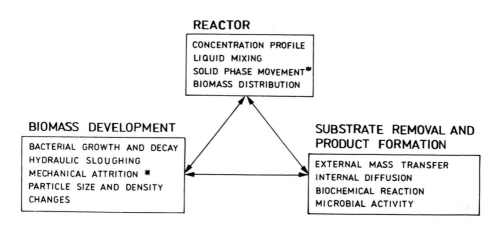

Fig. 1. Basic structure of a model

elements listed in Fig. 1 can be treated dynamically as changing in time or assumed to be in a steady state (either in static conditions or dynamic equilibrium).

4. DESCRIPTION OF MODEL ELEMENTS

4.1 Reactor Analysis: Liquid Phase

One of the elements of a biological packed or fluidized bed model is a description of the reactor. The bed is placed in a tubular reactor with a distribution system and a liquid collection system. The reactor can be operated as a downflow packed bed or as an upflow packed or fluidized bed. A schematic representation of downflow and upflow reactors is shown in Fig. 2. Above the bed packing a free zone is usually provided to allow the bed to expand during its operation (fluidized beds) or during its backwashing (packed and occasionally fluidized beds). This part of the reactor is quite often neglected in various analyses, despite the fact that its volume can be significant and may play a considerable role in the dynamics of the whole system.

A large number of various models of packed and fluidized beds has been developed, mainly in the field of chemical engineering. Despite this considerable effort, the modelling of these types of reactors still creates a great deal of controversy and many questions remain unanswered. Two types of the reactor models will be presented in this work. At this moment, only the liquid phase in the reactor will be considered. Modelling of solid phase movement in fluidized beds will follow. These two models: dispersion model and tanks-in-series model, though most common in the literature and probably most

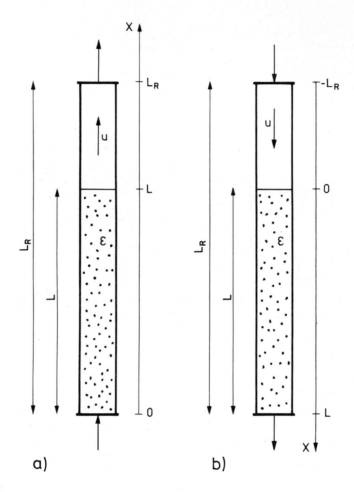

Fig. 2. Schematic representation of a reactor
a - upflow reactor (packed or fluidized)
b - downflow reactor (packed)

useful for a sanitary engineer, do not pretend to cover all aspects of reactor modelling. For further details, the reader is referred to chemical engineering literature; a particularly good introduction to these problems is presented by Froment and Bischoff (1979) and a more detailed treatment can be found in the work of Wen and Fan (1975).

The first type of model discussed in this work is usually called the "plug flow with axial dispersion" or "Fickian" model. For the purpose of this model it is assumed that the liquid phase flows through a pseudo-homogeneous medium of porosity ε. The flow is unidimensional (along Ox axis) with the velocity $u = Q/A_r$, where Q is a volumetric flow rate, and A_r is a cross section area of the empty reactor. The flow is isothermal and concentration changes do not alter the density of the liquid. Then, the mass balance at particular cross-section of the reactor is given by the following equation:

$$\varepsilon \cdot \partial c / \partial t + u \cdot \partial c / \partial x - D_e \cdot \varepsilon \cdot \partial^2 c / \partial x^2 + r_{vr} = 0 \qquad (1)$$

where c is a concentration of substrate or product in the bulk of liquid. The first term in this equation describes an accumulation of the substrate, the second-term — convective transport with the stream of liquid, and the last term — production or removal rate. The third term, with the second-order derivative, is the most important and the most controversial in this model. It was introduced to describe the observed dispersion of tracer during the flow through the bed. This approach bears clear resemblance to molecular diffusion described by Fick's law and the mathematical structure of Eq. 1 is identical to the well-known diffusion formulations. An effective dispersion coefficient D_e is substituted here for molecular diffusivity. It can also be shown (Froment and Bischoff, 1979), that the dispersion of liquid in a laminar flow through a tubular reactor caused by the velocity profile can be well described by Eq. 1. In this case:

$$D_e = D_m + u^2 \cdot d_c^2 / (192 \cdot D_m) \qquad (2)$$

where D_m is the molecular diffusivity and d_c is the reactor diameter. Similar correlations are also available for turbulent flows (Himmelblau and Bischoff, 1968; Wen and Fan, 1975). Eq. 1 can be thus applied to the free zone above the bed with $\varepsilon = 1$ and $r_{vr} = 0$. The concept of effective dispersion has also been applied to the flow of liquid through porous media in a packed bed (see Froment and Bischoff, 1979) and to the flow through fluidized beds (Potter, 1971). A generalized correlation between D_e and the parameters of the system was developed in the following form for packed and fluidized beds by Chung and Wen (1968):

$$\varepsilon \cdot Pe / \Gamma = 0.20 + Re^{0.48} \qquad (3)$$

where $Pe = u \cdot d_p / D_e$ is Peclet number, $Re = u \cdot d_p / \nu$ is Reynolds number, d_p is the diameter of solid particles, ν is kinematic viscosity of the liquid phase; Γ equals 1 for packed beds and for fluidized beds is defined as the ratio of the minimum fluidization velocity to the velocity u. The data used for the development of this correlation were considerably scattered and, thus, Eq. 3 should be used with a certain amount of caution.

Despite the wide use of Eq. 1 in the literature for flow modelling, the dispersion model has been severely criticized. The model, due to its diffusion-like second-order term, predicts backmixing of the liquid phase in the reactor. This backmixing can be expected in fluidized beds where it is induced by solids mixing. In packed beds, however, the experiments by Hiby (1963) showed virtually no backmixing even at high Reynolds numbers. The existence of backmixing in tubular reactors has been questioned by Deckwer and Mahlmann (1976). Additionally, Eq. 1, being of parabolic type, predicts an infinite speed of propagation of any concentration change which is clearly not possible. Sundaresan, Amundson and Aris (1980) presented an excellent discussion of the deficiencies of the dispersion model. They attempted to construct a new model which would satisfy the conditions of mass conservation, finite propagation speed, no backmixing and correct asymptotic form. They examined some new hyperbolic-type differential models and found that no second-order differential equation and probably no finite-order equation can satisfy all conditions. From their analysis "it appears that no model has an outstanding advantage over any other" since all of them are good approximations for large space and time variables. They concluded that the dispersion model is quite adequate for "present design purposes" although it is "substantially deficient in details".

To complete the model, Eq. 1 must be accompanied by a set of initial and boundary conditions. The initial conditions describe a concentration profile at an initial moment $t = 0$ and have the following form:

$$\text{for} \quad t = 0 : \qquad c(x,t) = c_{in}(x) \tag{4}$$

The initial conditions are not required if a steady state is assumed. In this case, $\partial c/\partial t = 0$ and Eq. 1 transforms into an ordinary second-order differential equation. Since the dispersion model consists of a second-order differential equation two boundary conditions are required. The first proposition of the boundary conditions was presented by Danckwerts (1953) in the following form:

$$c_o = c(0^+) - \frac{D_e}{u} \cdot \frac{dc}{dt}\bigg|_{0^+} \tag{5a}$$

$$\frac{dc}{dx}\bigg|_{L^-} = 0 \tag{5b}$$

These conditions, proposed for steady-state model, were further developed by Wehner and Wilhelm (1956). They considered a more complex reactor with fore and after sections (Fig. 3) with the dispersion coefficients D_{e1} and D_{e2}, respectively. They showed that, at a steady state, Eq. 5b can be applied to the end of the reactive zone. They also postulated that Eq. 5a is only a

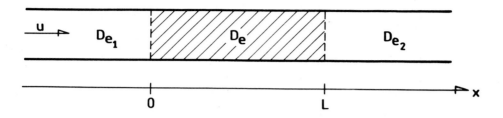

Fig. 3. Reactor with fore and after sections

special case of a more general equation involving the dispersion coefficient in the fore section, D_{el}. Their general conditions were further used for transient states (Van Cauwenberghe, 1966) and form a second set of generally used boundary conditions (see Fig. 3):

$$c(-\infty, t) \;=\; \text{finite} \tag{6a}$$

$$c(0^-, t) \;=\; c(0^+, t) \tag{6b}$$

$$c(0^-, t) - \frac{D_{el}}{u} \cdot \frac{\partial c}{\partial x}\Big|_{0^-} \;=\; c(0^+, t) - \frac{D_e}{u} \cdot \frac{\partial c}{\partial x}\Big|_{0^+} \tag{6c}$$

$$c(L^-, t) \;=\; c(L^+, t) \tag{6d}$$

$$c(L^-, t) - \frac{D_e}{u} \cdot \frac{\partial c}{\partial x}\Big|_{L^-} \;=\; c(L^+, t) - \frac{D_{e2}}{u} \cdot \frac{\partial c}{\partial x}\Big|_{L^+} \tag{6e}$$

$$c(+\infty, t) \;=\; \text{finite} \tag{6f}$$

It should be noted that both types of conditions (Eqs. 5 and 6) were derived in a speculative way and only few experimental studies were reported (Deckwer and Mahlmann, 1974, 1976). From the practical point of view, the mathematical subtleties and the choice of boundary conditions are probably not very important since they are only an approximation of real inlet and outlet conditions. The influence of the boundary conditions on the solution of Eq. 1 is limited only to large D_e (precisely to small Bodenstein numbers $Bo = u \cdot L/D_e$) and in that case the high intensity of mixing is not well described by the diffusion-like mechanism. Transfer functions $T(s)$ corresponding to two sets of boundary conditions were plotted in Fig. 4 for $Bo = 5$ and $Bo = 20$ (for non-reactive system, $r_{vr} = 0$, and real arguments, only). For large values of the Bodenstein number the transfer functions are very close indicating a similar response of the reactor despite different boundary conditions.

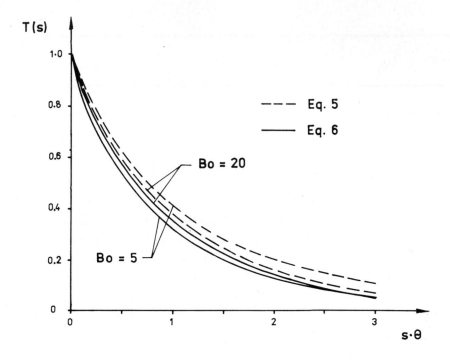

Fig. 4. Transfer functions T(s) for plug flow reactor with axial
dispersion with different boundary conditions

Mathematical problems of the dispersion models are largely avoided in a second class of reactor models. In this class, the reactor is described as a series of completely mixed tanks. A schematic representation of such a model is presented in Fig. 5. It consists of a series of n completely mixed tanks (reactors) each of the volume v_i (i=1,...,n). In general, two streams flow through the reactor: forward and backward. Thus, two streams are originated in each tank and two are received. If the total net flow through the reactor is Q, the forward flow from the i-th tank is $Q_{F,i}$ and the backward flow is $Q_{B,i}$ then at a constant volume of the i-th tank, we have:

$$Q_{F,i-1} - Q_{B,i} = Q_{F,i} - Q_{B,i+1} = Q$$

Denoting $Q_{B,i}/Q = q_i$ we obtain $Q_{F,i} = (1+ q_{i+1}) \cdot Q$ and $Q_{B,i} = q_i \cdot Q$

Usually it is assummed that all tanks have the same volume $v = V/n$, where V is the total volume of the reactor and that the backflow is also the same ($q_i = q$). Then the mass balance equation has the following form:

- for the first tank

$$V/(n \cdot Q) \cdot dc_1/dt = c_o + q \cdot c_2 - (1+q) \cdot c_1 - r_{vr,1} \cdot V/(n \cdot Q) \qquad (7a)$$

- for the i-th tank

$$V/(n \cdot Q) \cdot dc_i/dt = (1+q) \cdot c_{i-1} - (1+2 \cdot q) \cdot c_i + q \cdot c_{i+1} - r_{vr,i} \cdot V/(n \cdot Q) \qquad (7b)$$

- for the last n-th tank

$$V/(n \cdot Q) \cdot dc_n/dt = (1+q) \cdot c_{n-1} - (1+q) \cdot c_n - r_{vr,n} \cdot V/(n \cdot Q) \qquad (7c)$$

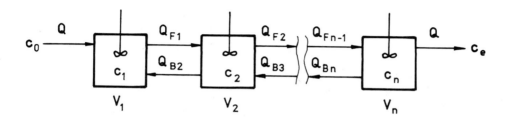

Fig. 5. Reactor model - a series of completely mixed tanks

The equations of this model are indeed simpler than those of the dispersion model. Initial and boundary conditions are also simplified. At the initial moment, $t = 0$, concentrations in all tanks must be specified and as a boundary conditions the influent concentration function $c_o(t)$ has to be known. The model described by Eqs. 7 (a-c) predicts both backmixing and dispersion of the liquid phase. The propagation speed of a signal is infinite like in the dispersion model. These two models give very similar responses. Although there is no strict way to compare both models, equating the variances of the impulse responses can give an idea about the corresponding values of the parameters. Froment and Bischoff (1979) showed that the equality of variances (second moments) yields the following approximate relation:

$$n/(1 + 2 \cdot q) \cong (Bo + 1)/2 \qquad (8)$$

This equation indicates one of the drawbacks of the tank-in-series model. It includes two adjustable parameters, n and q, one more than the dispersion model. The number of tanks, n, is an arbitrary parameter and can only be determined experimentally from tracer analysis. The number of parameters can be reduced by assuming q = 0. The model then transforms into the simple tank-in-series model which does not predict any backmixing. Despite these problems, the tank-in-series model has been widely used and its detailed treatment was provided by Roemer and Durbin (1967). A modification of this model, predicting a finite propagation speed, was presented by Sundaresan, Amundson and Aris (1980). They incorporated a time delay between subsequent tanks. Physically, it means that the completely mixed tanks are separated by plug flow sections with the detention time Δt (Fig. 6). The equations describing the model are formed from Eqs. 7 (a-c) and for the i-th tank are as follows:

$$v_i/Q \cdot dc_i/dt = (1+q) \cdot c_{i-1}[t-\Delta t] \quad - \quad (1+2 \cdot q) \; c_i[t] \quad +$$

$$+ \; q \cdot c_{i+1}[t+\Delta t] \quad - \quad r_{vr,i} \cdot v_i/Q \tag{9}$$

and $\quad V = n \cdot v_i + 2(n-1) \cdot Q \cdot \Delta t$

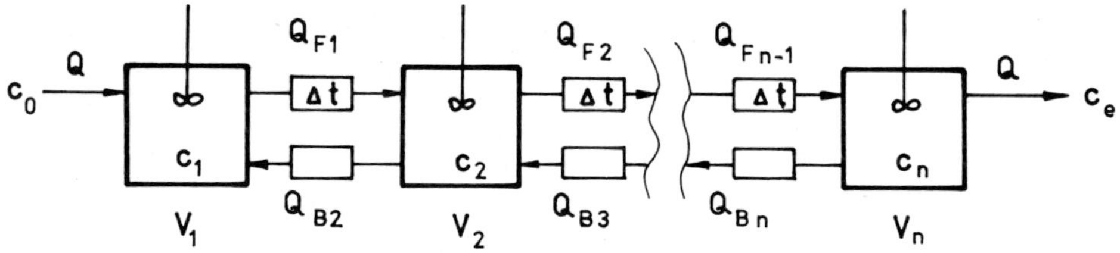

Fig. 6. Reactor model - a series of completely mixed tanks with plug flow delay sections

The modification proposed by Sundaresan and co-workers rectifies the violation of the law of physics (infinite propagation speed) but at a cost of introducing one more parameter Δt. Some other models of this type can be developed combining completely mixed tanks, plug flow or dispersive sections, and backward flows. In the authors' opinion, shared with other researchers (Froment and Bischoff, 1979; Sundaresan et al., 1980), none of the models has any clear a priori advantage over the others and whenever possible the choice should be made based on tracer analysis and mathematical handling of the system. It is sometimes believed that the mathematical complexity of the dispersion model gives an advantage from the computational point of view to the tank-in-series model. This may not be true in all cases. The results of tracer analysis of a fluidized bed done by Hermanowicz (1982) indicated that the bed could be described equally well as a series of completely mixed tanks and as a dispersive flow. The number of tanks required, and thus the number of differential equations to be solved, was in a range of 20 to 37. On the other hand, the second-order differential equation, similar to Eq. 1, describing the same reactor, was effectively transformed by using the orthogonal collocation method (Lee, 1968) into a set of 6 ordinary differential equations.

4.2 Reactor Analysis: Solid Phase

The main difference between packed beds and fluidized beds is the mobility of particles in the latter. In a packed bed, the particles remain in the same place and the only change in the reactor hydraulics is caused by porosity (voids) changes. They are mostly a result of biofilm growth and removal (decay, sloughing, etc.). For spherical particles in a packed bed, the biofilm thickness, δ, is related to the actual bed porosity, ε, and the initial porosity of clean bed, ε_o, by an equation:

$$1 - \varepsilon = (1 - \varepsilon_o)(1 + 2 \cdot \delta/d_s)^3 \qquad (10)$$

In a fluidized bed the situation is more complex. The growth of the microorganisms in the form of biofilm changes the size of a carrier particle and its average density. The density of a biocoated particle is expressed in the following form:

$$\rho_p = \rho_b + (\rho_s - \rho_b) \cdot (d_s/d_p)^3 \qquad (11)$$

$$d_p = d_s + 2 \cdot \delta$$

where d_p is diameter of a biocoated particle, d_s – diameter of clean particle, δ – biofilm thickness, ρ_b – wet biofilm density and ρ_s – solids density. Both Eq. 10 and Eq. 11 assume that the biofilm covers uniformly a spherical particle. The changes in particle diameter and density affects the expansion of the fluidized bed. Among several methods of calculating the expansion, the method of Zaki and Richardson (e.g. Richardson, 1971) has been extensively used. In this method, the porosity of the bed is calculated from the following formulae:

$$u/u_i = \varepsilon^n \qquad (12a)$$

$$u_i/u_t = 10^{-d_p/d_c} \qquad (12b)$$

$$n = (4.4 + 18 \cdot d_p/d_c) \cdot Re_t^{-0.1} \quad \text{for} \quad 1 < Re_t = u_t \cdot d_p / \nu < 200 \tag{12c}$$

where u_t is a terminal free fall velocity of the particles of diameter d_p, and d_c is the diameter of the reactor. Cleasby and Fan (1981) extended this method to non-spherical particles by modifying Eq. 12c as follows:

$$n' = n \cdot \psi^{(-2.942 \, \cdot \psi^{0.884} \cdot Re_t^{-0.363})} \tag{12d}$$

and the substitution of n' for n in Eq. 12a. ψ is the sphericity of the particles. Smith and co-workers (1978) found that the Richardson and Zaki equation could be used for fluidization of microbial aggregates like flocs of yeast and fungal molds. They estimated the value of n from the experiments and did not calculate the particle terminal velocity in their analysis. Ngian and Martin (1980b) reported significant discrepancies between the predicted expansion, using Eqs. 12 a–c, and the observed data for a fluidized bed with biocoated particles. They calculated the terminal velocity, u_t, from the formula:

$$u_t = \left[\frac{4 \cdot g \cdot (\rho_p - \rho_1)}{3 \cdot C_D \cdot \rho_1} \cdot d_p \right]^{1/2} \tag{13}$$

and the drag coefficient, C_D, from the following equation:

$$C_D = 18.5 \cdot Re_t^{-0.6} \tag{14}$$

The above equation was developed for smooth, rigid particles unlike the particles coated with biofilm. It was reported by Hermanowicz (1982) and by Hermanowicz and Ganczarczyk (1983) that the settling velocity of biocoated particles differs significantly from that calculated with Eqs. 13 and 14. This difference was attributed to higher drag forces exerted by liquid on biocoated particles due to the roughness of their surface. A new formula for the drag coefficient, C_D, was developed:

$$C_D = 17.1 \cdot Re_t^{-0.47} \tag{15}$$

which yields higher values of C_D than those calculated from Eq. 14. The new formula was developed on the basis of terminal settling velocity measurements of a limited number of particles with the biofilm composed mainly of nitrifying microorganisms and, at present, should be treated only as a tentative one. It is probably valid for the intermediate range of Reynolds numbers and at $Re = 1.9$ coincides with the drag coefficient formula for laminar flow $C_D = 24/Re$. The use of terminal velocity calculated with Eq. 15 in the Zaki and Richardson method (Eqs. 12) gave a good approximation of bed expansion (Hermanowicz and Ganczarczyk, 1983). In a biologically active fluidized bed the particles are, in general, not uniform and exhibit a gradation in size and density which leads to different settling velocities, u_t. The expansion of such a non-uniform bed can be calculated as if they were the sum of m single species beds (Epstein et al., 1981):

$$1 - \varepsilon = \left[\sum_1^m \frac{\alpha_i}{1 - \varepsilon_i} \right]^{-1} \qquad (16)$$

where α_i is the volume fraction of species i in the total volume of solids ($\Sigma \alpha_i = 1$) and ε_i is the porosity or liquid hold up when species i is fluidized alone with the same superficial velocity, u. The bed containing particles of different size and/or density can be considered, as a whole, to be in dynamic equilibrium, with the expansion expressed by Eq. 16. An individual particle, however, is subject to unbalanced forces causing the particle to move. This movement is a result of two counteracting phenomena: segregation of solids and mixing. Segregation of solids in a liquid fluidized bed was recently analyzed by Van Duijn and Rietema (1982) under the assumption that mixing can be neglected and the particles are free to move separately. From this theoretical analysis supported by some experimental results they drew the following conclusions: "If the free falling velocity of the heavier component in the fluid is equal to or higher than that of the lighter one, the heavier component always moves downwards. If the free falling velocity of the heavier component is smaller, the bed porosity and the ratio of the particle sizes of both solids determine if the heavier component moves upwards or downwards." Van Duijn and Rietema also proposed a method for determination which solid in a binary mixture will move upwards and which will move downwards. This method can be extended in a straight-forward manner for mixtures of more components. If there is no mixing in the bed, the solids will strafify into separate layers. The bulk density in these layers, $\bar{\rho}_i$, can be calculated from:

$$\bar{\rho}_i = \varepsilon_i \cdot \rho_1 + (1 - \varepsilon_i) \cdot \rho_{pi} \qquad (17)$$

where ε_i is the porosity of the layer containing only component i at the superficial velocity u. The porosity ε_i can be found from Eqs. 12 or by other suitable methods. Van Duijn and Rietema (1982) postulated that the segregated bed is stable if:

$$d \bar{\rho}/dx < 0 \qquad (18)$$

where $\bar{\rho}$ is the bulk density of the bed at the height x from the bottom. In

488

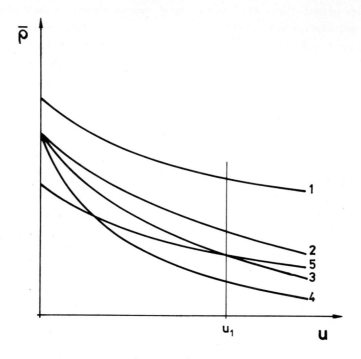

Fig. 7. Fluidized bed bulk density as a function of liquid velocity

Fig. 7, bulk densities $\bar{\rho}_i$ are plotted as functions of u. It can be noted that the intersection of a density curve with the $\bar{\rho}$-axis (ordinate) is equal to the particle density ρ_{pi} corresponding to the solid component i. Obviously, the density curves have a physical meaning only for $u_{mf,i} < u < u_{t,i}$.

The plots in Fig. 7 have the following interpretation. Component 1 consists of heavy particles with a large free fall velocity and thus remains at the bottom of the bed at all velocities u. The family of curves 2,3 and 4 represents the components of the same density but of decreasing diameter ($dp2 > dp3 > dp4$). Component 5 has lower density but higher free fall velocity than components 3 and 4. Therefore, at low velocities, u, it moves to the top of the bed but at higher velocities (and porosities) it is found between component 3 and 2. At the superficial velocity u_1 the bulk density of component 3 is equal to that of component 5 and both components become mixed. At the superficial velocity u_1 the bed will stratify as follows: from the top – component 4, components 3 and 5 mixed, component 2 and component 1.

The segregation mechanism proposed by Van Duijn and Rietema assumes that the particles can move freely in the bed and the mixing of solids is negligible. At low expansions, commonly used for biological beds, a kind of structure can be formed in the bed as observed by Kmiec (1978). He also noted that for porosity $\varepsilon < 0.55$ the movement of particles was restricted to their neighborhood. A similar feature was observed in a biological bed operated at the porosity of about 0.55 - 0.6 where the particles with the thickest biofilm (hence the largest diameter and lowest density) were found in the middle of the bed (Hermanowicz, 1982). This phenomenon was somewhat similar to the clogging of packed beds in the regions of high substrate concentration and high biomass growth indicating that the free movement of particles was restrained. On the other hand, in the same bed a range of particle diameters was observed at each depth instead of total segregation. This was caused by local mixing existing even at low porosities. The significance of mixing increases at higher porosities. A model of particle mixing was proposed by Kennedy and Bretton (1966) and extended by Burovoi and Ibraev (1971). They postulated a diffusional mechanism of solid mixing described by Fick's law. The diffusional flux of i-th component is described by:

$$D_{si} \cdot dc_{pi}/dx$$

where D_{si} is the diffusivity of the i-th component and c_{pi} its concentration at the height x from the bottom of the bed. At a steady state, in dynamic equilibrium this flux is equal to the segregation flux and the equation describing particles mixing has the following form:

$$D_{si} \cdot dc_{pi}/dx = c_{pi} \cdot (u - u_{pi})/ \varepsilon \qquad (19)$$

The right hand side of Eq. 19 represents the segregation flux, as proposed by Kennedy and Bretton (1966), in which u_{pi} is the velocity of the liquid phase required to achieve the porosity ε in a bed containing the i-th component only. The porosity is a function of the height and can be related to concentrations c_{pi} with the help of Eq. 16. Yutani and co-workers (1982) reported the values of particles diffusivity for water fluidization of glass beads ($d_p = 0.042 - 0.072$ cm; $u = 1.5 - 5$ cm/s) in a range of 0.3 to 4 cm^2/s. A similar range of values was reported by Wojcik (1976) who examined mixing of glass, aluminum and polystyrene particles fluidized with water. The size of the particles was in a range of 0.17 to 0.3 cm. The diffusivity of the particles was correlated with the parameters of the system in the following dimensionless form (Wojcik, 1976):

$$D_s/\nu = 0.335 \, (\varepsilon - \varepsilon_{mf}) \cdot \varepsilon^{-1} \cdot Re^{1.32} \cdot (L/d_p)^{0.33} \cdot \phi^{2.5} \qquad (20)$$

where L is the height of the bed and Re = $u \cdot d_p/\nu$. Since this correlation was also developed for non-spherical particles the shape factor ϕ (reciprocal of the sphericity) is involved. The correlation was reported to cover the following range of parameters:

$$0.12 < (\varepsilon - \varepsilon_{mf})/\varepsilon < 0.57$$
$$5 < Re < 300$$
$$29 < L/d_p < 520$$
$$1 < \phi < 2.5$$

If the bed contains n fractions (components) of different density and/or diameter, then the solution of a set of n differential equations yields a description of particles distribution in the bed. It is particularly convenient to express concentrations c_{pi} in terms of volume fraction of the bed occupied by the particles of component i since the porosity is then $\varepsilon = 1 - \Sigma c_{pi}$ and the velocity u_{pi} can be back-calculated from Eqs. 12 or equivalent. Wojcik (1976) analyzed a set of Eqs. 19 and noted that the set was stable if boundary conditions were specified for the top of the bed.

In a biological bed, the size and density of the particles change in time as a result of biomass growth and removal. Even if a short-term equilibrium is assumed, the calculations have to be repeated in short-time intervals when a new particle distribution is achieved. The effects of biofilm development (growth, decay, detachment) are further discussed in Section 6. The problem of uneven concentration of biomass along the bed was generally neglected in the models of biological fluidized beds presented in the literature. Only a few models include some elements of solid mixing and segregation influenced by biomass growth. Eggers and Terlouw (1979) used an experimentally determined biomass concentration profile in a fluidized bed for their steady-state calculations. Experimental data were also used as initial conditions in a dynamic model of nitrifying fluidized beds (Hermanowicz and Ganczarczyk, 1984) in which provision was made for porosity changes due to biomass growth. Rittmann (1982) assumed a completely mixed fluidized bed (with regard to the biomass concentration) and a segregated packed bed.

An interesting approach was presented by Andrews and Tien (1982) who represented the solid phase of the reactor in a form of a matrix. A row of the matrix described the distribution of particle size at a certain height in the bed and each column showed the distribution of particles of a given size along the bed. They proposed to update the matrix in the process of simulation after every time step in accordance with biomass growth or decay. They employed their concept for complete solids mixing and complete solids stratification and concluded that neither complete mixing nor stratification could occur in the reality. The question of biomass distribution and particle motion seems quite important as indicated by the simulation results by Rittman (1982) and Bousfield and Hermanowicz (1984) who reported a significant influence of biomass profile on the efficiency of the process. The phenomena of solids mixing and segregation should be more studies and incorporated into models of biological fluidized beds.

5. SUBSTRATE REMOVAL

Substrate(s) removal and product(s) formation take place in a biofilm growing on the surface of carrier particles. The amount of suspended microorganisms in a packed or fluidized bed which do not form aggregates or film is usually small and will not be considered in the present work. Obviously, a biofilm consists of a complex structure of viable microorganisms, dead cells and other components. Its thickness can vary substantially over a very short distance in the reactor and its composition may be different at various depths. To make mathematical modelling possible some simplifying assumptions must be made. It is postulated that the local intrinsic rate of biochemical reactions in the biofilm is described by the same equations as those governing the suspended-growth culture. It is also assumed that the properties of the biofilm (like density, diffusivity, etc.) are uniform, at least locally. Some of models (e.g. Grasmick et al., 1979) divide the biofilm into two layers: an active one penetrated by substrates, and the deeper one in which the microbial activity is restrained by the lack of one or more substrates. The inactive layer may differ from the active one not only in removal rate but also in density, decay rate, structure and ability to withstand shearing forces. These phenomena are, at present, far from being well understood; the current point-of-view on the subject was presented by Characklis (1981). Recently, Bryers (1984) discussed the development of a biofilm composed of several bacterial species utilizing the same substrate in a continuous fermenter showing that microbial species can coexist in the biofilm despite their different growth rates.

Since biochemical reactions occur inside the microbial film, the substrate(s) must penetrate into the biofilm and the product(s) must be transported outside. This transport is a two-step process involving the transport through a liquid film layer around the surface of the biofilm and the transport inside the biofilm combined with simultaneous reaction. The mechanism is relatively well understood only for soluble substrates. Unfortunately, a considerable fraction of organic substances in wastewater is present in a form of suspended or colloidal matter. The utilization of this form of substrates requires its solubilization with the help of extracellular enzymes and further penetration of solubilized compounds into microbial cells.

Dold, Ekama and Marais (1980) incorporated the particulate form of substrates into their model of the activated sludge process and it seems that these concepts could be applied to fixed-film processes. However, better understanding of the mechanism of suspended and colloidal matter entrapment in a biological bed is required before any modelling attempts can be made. The present work is limited to the removal of soluble substrates. Their transport appears to be essentially of diffusional nature but some other mechanisms may also be involved, particularly inside the biofilm. They may include electrostatic forces caused by the fixed charges in the biofilm as well as active biological transport of substrates, and convective flow of liquid. These phenomena and their role are very poorly understood; some of them were briefly discussed by Harremoes (1978) in his review of biofilm kinetics. In the present work it will be assumed that the transport mechanisms can be described by Fick's diffusion equation.

5.1 Substrate Removal Inside the Biofilm

Diffusional substrate transport and removal in the biofilm can be described by the following equation:

$$\frac{\partial c}{\partial t} - \frac{D_{eff}}{z^s} \cdot \frac{\partial}{\partial z} (z^s \cdot \frac{\partial c}{\partial z}) + r_v = 0 \tag{21}$$

The first term of this equation describes accumulation of substrate at the position z in the biofilm, the second term denotes diffusional transport with an effective diffusivity D_{eff}, and the last term is substrate removal rate. The parameter s depends on the choice of coordinates; s = 0, 1, 2 for rectangular, cylindrical and spherical coordinates, respectively. The time scale of the diffusion transport is in the order of magnitude of δ^2/D_{eff} which for common values δ = 100 µm and D_{eff} = $2 \cdot 10^{-5}$ cm^2/s gives 5 s. This time is generally small compared to the time scale of concentration changes outside the biofilm in the reactor indicating that a steady state is approached rapidly in the biofilm and the accumulation term, $\partial c/\partial t$ can usually be taken as 0. This assumption simplifies the analysis and transforms Eq. 21 into a second-order ordinary differential equation:

$$\frac{D_{eff}}{z^s} \cdot \frac{d}{dz} (z^s \cdot \frac{dc}{dz}) = r_v \tag{22}$$

The removal rate, r_v, is the intrinsic removal rate of the substrate by the microorganisms of the biofilm. It can be expressed with a number of equations proposed by various researchers but the discussion of this subject is beyond the scope of the present work. Two formulae most popular in the literature will be used:

- n-th order reaction:

$$r_v = k_n \cdot X \cdot c^n \qquad \text{for} \quad 0 \leq n \leq 1 \qquad (23)$$

- Michaelis-Menten kinetics:

$$r_v = k_m \cdot X \cdot c / (c + K_s) \qquad (24)$$

In both expressions the reaction rate depends on the local concentration of the microorganisms, X, expressed usually, but not necessarily, as dry mass density. The concentration of the substrate at the surface of the biofilm is c_s and decreases as the substrate penetrates into the biofilm (Fig. 8). At the surface of the carrier there is no substrate transport through the interface between the carrier and the biofilm and the diffusional flux is equal zero. This situation leads to a set of boundary conditions for Eq. 22 in the following form:

$$- \text{for } z = z_p \qquad dc/dz = 0 \qquad (25a)$$

$$- \text{for } z = z_p + \delta \qquad c = c_s \qquad (25b)$$

If the removal rate is large, the concentration decreases rapidly and becomes very small in deeper layers of the biofilm. Based on this phenomenon some authors (Meunier and Williamson, 1981; Rittman and McCarty, 1978) distinguish between a "deep" and a "shallow" biofilm. In the shallow biofilm the concentration decreases and reaches zero at a certain point inside the biofilm, $z_p + \delta_o$ (Fig. 8). Therefore, a layer of biofilm of the thickness δ_o adjacent to the surface of the carrier is not penetrated by the substrate. In the deep biofilm, the substrate concentration is greater than zero in every point. In the case of a "deep" biofilm the boundary condition of Eq. 15a can be moved to the "division" point and expressed in the following form:

$$- \text{for } z = z_p + \delta_o \qquad c = 0 \quad \textbf{and} \quad dc/dz = 0 \qquad (25c)$$

An additional equality in Eq. 25c allows for determination of the division point position. The idea of the shallow film is very intuitive but it may lead, if carelessly applied, to a mathematical problem which is ill-posed. It can be shown that Eq. 25c can be fulfilled only if:

$$\text{for } c \longrightarrow 0 \qquad dr_v/dc \longrightarrow \infty \qquad (26)$$

494

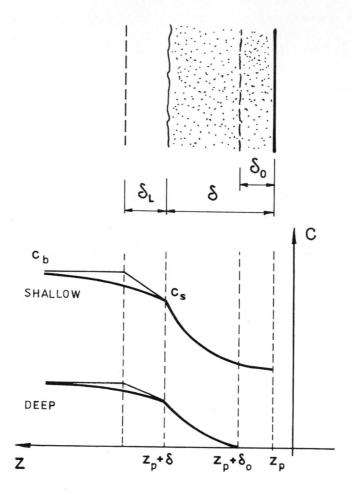

Fig. 8.   Substrate concentration profiles in biofilm

In other cases Eq. 25c can be fulfilled only for an infinite biofilm thickness
and thus normally a non-zero concentration profile is established throughout
the biofilm.   For the discussed reaction types (Eqs. 23 and 24), a shallow
biofilm may occur only for Eq.   23 with the restriction $0 \leq n < 1$.   The concept
of "deep" and "shallow" biofilm, however, is quite important from the practical
point-of-view since an increase of biofilm thickness beyond the critical
thickness of "deep" biofilm does not significantly increase the overall removal
rate.   It seems more appropriate to define "deep" biofilm based on the overall
reaction rate criterion as proposed by Howell and Atkinson (1976a) which will

be discussed later. Since in the analysis of a biological bed reactor its performance is described in terms of substrate concentration in the bulk of liquid, it is convenient to express the overall removal rate by a unit volume of the biofilm in the same terms. The overall removal rate by the biofilm, r_{vb}, is defined as:

$$r_{vb} = \frac{1}{V_b} \cdot \int_{V_b} r_v(c) \cdot dV_b \tag{27}$$

where V_b is the volume of biofilm. The overall removal rate is expressed as a product of an effectiveness factor, η, and the removal rate corresponding to the substrate concentration at the surface of the biofilm, c_s:

$$r_{vb} = \eta \cdot r_v(c_s) \tag{28}$$

The effectiveness factor is, in other words, a ratio of the actual removal rate by the biofilm to the removal rate exhibited by the same volume of biomass totally exposed to the concentration c_s. The value of the effectiveness factor is generally less than or equal to 1 since the reaction rate in deeper parts of biofilm is lower due to lower substrate concentration. In case of substrate inhibition, however, η can be greater than unity, as shown by Moo-Young and Kobayashi (1972). The effectiveness factor depends on biofilm thickness, substrate diffusivity and the intrinsic reaction rate and can be obtained from the solution of Eq. 22 with the help of Eq. 27. The overall removal rate by the biofilm, r_{vb} is related to the removal rate per unit volume of the bed through the following formula:

$$r_{vr} = r_{vb} \cdot V_b/V$$

Since the biofilm thickness in a reactor is generally small as compared with the radius of the carrier particles, the biofilm can be treated as a flat layer neglecting its curvature. This approximation simplifies the mathematical treatment and does not introduce any large error. (This point will be discussed later.) Therefore, the geometry parameter, s, in Eq. 22 assumes the value of 0 and Eq. 22 reduces to the form

$$D_{eff} \cdot d^2 c/dz^2 = r_v(c) \tag{29}$$

with the origin of the coordinates at the interface between the carrier and the biofilm ($z_p=0$).

If the reaction rate is expressed by Eq. 23 (n-th order reaction), then the dimensional analysis of Eq. 29 leads to a generalized Thiele modulus for this type of reaction:

$$\Phi_n^2 = \frac{(n+1) \cdot k_n \cdot X \cdot \delta^2 \cdot c_s^{(n-1)}}{2 \, D_{eff}} \qquad \text{for} \quad 0 \leq n \leq 1 \tag{30}$$

For the first-order reaction ($n=1$), the effectiveness factor has a generally known form:

$$\eta = \tanh(\Phi_n)/\Phi_n \tag{31}$$

As discussed previously, the concentration in the biofilm with the first-order reaction does not reach zero within a finite thickness and the biofilm, strictly speaking, remains shallow. In the case of lower reaction rate orders ($0 \leq n < 1$) a deep biofilm can be formed with the concentration equal to zero in the inner part. For the deep biofilm (and $0 \leq n < 1$) Eq. 29 can be easily be integrated (with boundary conditions of Eq. 25c) and the effectiveness factor is:

$$\eta = 1/\Phi_n \tag{32}$$

The deep biofilm exists for:

$$\Phi_n \geq (1+n)/(1-n) \tag{33}$$

For smaller Φ_n the biofilm becomes shallow and analytical solution of Eq. 29 exists only for $n=0$ (as well as for $n=1$, Eq. 31). For zero-order kinetics the rate of reaction does not depend on substrate concentration and since the shallow biofilm is fully penetrated by the substrate, the effectiveness factor is:

$$\eta = 1 \tag{34}$$

For other values of n ($0 < n < 1$) Eq. 29 was numerically solved using a Runge-Kutta method and the resulting effectiveness factor was plotted together with Eqs. 31, 32 and 34 in Fig. 9. The critical values of Φ_n for deep biofilm (Eq. 33) were also marked for various n.

If the intrinsic removal rate r_v is described by the Michaelis-Menten equation (Eq. 24), no analytical solution is known. Atkinson and Davies (1974) solved numerically this equation and calculated the values of the effectiveness factor. They approximated these values with the following analytical function:

$$\eta = 1 - \frac{\tanh(\Phi_m)}{\Phi_m} \cdot \left[\frac{\Phi_p}{\tanh(\Phi_p)} - 1 \right] \qquad \text{for } \Phi_p < 1 \qquad (35a)$$

$$\eta = \frac{1}{\Phi_p} - \frac{\tanh(\Phi_m)}{\Phi_m} \cdot \left[\frac{1}{\tanh(\Phi_p)} - 1 \right] \qquad \text{for } \Phi_p > 1 \qquad (35b)$$

where
$$\Phi_p = \Phi_m \cdot \tilde{c}_s \cdot [2 \, (1+\tilde{c}_s)^2 \cdot (\tilde{c}_s - \ln(1+\tilde{c}_s))]^{-1/2} \qquad (36)$$

and
$$\Phi_m = \left[\frac{k_m \cdot X \cdot \delta^2}{D_{eff} \cdot K_s} \right]^{1/2} \qquad \tilde{c}_s = c_s / K_s$$

(Φ_m is the Thiele modulus for the Michaelis–Menten reaction.)

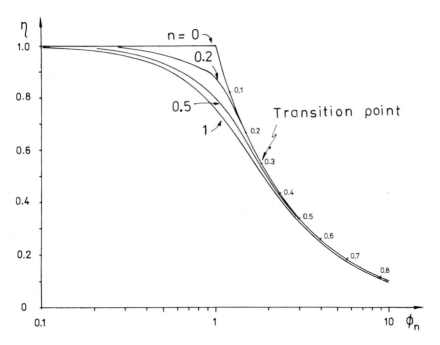

Fig. 9.   Effectiveness factor for n-th order kinetics in flat biofilm

If K_s increases and the removal rate approaches first order, the effectiveness factor for the Michaelis–Menten reaction (Eq. 35) coincides with the effectiveness factor expressed by Eq. 31. In a similar way, Eq. 35 approaches Eqs. 32 and 34 if K_s becomes small (zero-order reaction). The values of the effectiveness factor from Eq. 35 are shown in Fig. 10 as a function of Φ_p which

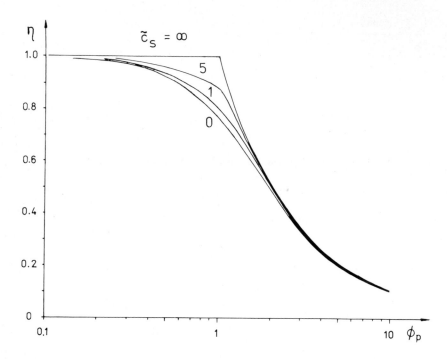

Fig. 10. Effectiveness factor for Michaelis–Menten kinetics in flat biofilm

is a modified Thiele modulus proposed by Atkinson and Davies (1974). It is
interesting to note that the values of the effectiveness factor for different
reaction kinetics as shown in Fig. 9 and 10 are very close for small and for
large values of the appropriate Thiele moduli. For small values corresponding
to thin biofilm, the diffusional limitations do not impair substrate transport
and the concentration throughout the biofilm (hence the reaction rate) is close
to that at the surface. Largest discrepancies caused by different kinetics
occur for values of the Thiele modulus around 1. Unfortunately, it can be
expected that biological beds will operate in this region of large
discrepancies where the form of reaction rate equation is important. The plots
of the effectiveness factor indicate that an increase of biofilm thickness
(increasing Thiele modulus) diminish the effectiveness factor. An overall
removal rate in the biofilm of thickness δ and surface area A is:

$$r_{vb} \cdot V_b = A \cdot \eta \cdot \delta \cdot r_v(c_s)$$

and the substrate flux into the biofilm, j, is equal to:

$$j \; = \; r_{vb} \cdot V_b/A \; = \; \eta \cdot \delta \cdot r_v(c_s) \tag{37}$$

Defining the dimensionless flux $\tilde{j}$ for the n-th order reaction as:

$$\tilde{j} \; = \; j \cdot \frac{n+1}{2 \cdot D_{eff} \cdot k_n \cdot X \cdot c_s^{(n+1)}}$$

we obtain with the help of Eq. 30 the following equation:

$$\tilde{j} \; = \; \Phi_n \cdot \eta \tag{38}$$

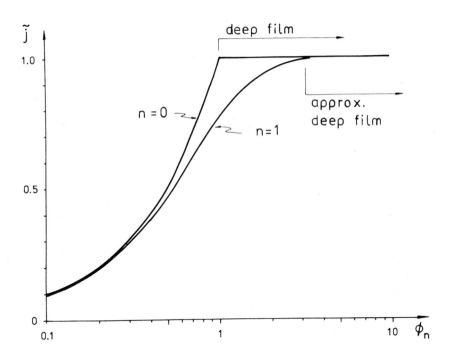

Fig. 11. Dimensionless substrate flux into flat biofilm

The values of $\tilde{j}$ are plotted in Fig. 11 for n=0 and n=1. With the increase of biofilm thickness (resulting in the increase of Φ_n) the total flux (and the substrate uptake) also increase but there is a limiting maximum value of $\tilde{j}$ corresponding to the infinite biofilm thickness. The value of $\tilde{j}$ can be used as a criterion for classification of biofilms into "deep" and "shallow" categories. The biofilm can be considered "deep" if the flux $\tilde{J}$ exceeds an arbitrary fraction of the maximum flux, say 99%. From the plots in Fig. 11, it can again be noted that the biofilm with zero-order kinetics can become exactly deep with the flux actually reaching its maximum value whereas for the biofilm with first-order reaction, the flux only approaches the maximum value.

The previous considerations of the reaction rate and effectiveness factor were limited to flat biofilms (s=0). In practice, this kind of biofilm is not common. Usually the biofilm is grown on the surface of carrier particles resembling spheres, though in some applications the inner surface of tubes or specially shaped plastic media are used for biofilm support. The effects of carrier surface and biofilm curvature were not studied extensively. The calculations of the effectiveness factor of a shell-like spherical biofilm were reported by Mulcahy and co-workers (1981) for zero and first-order kinetics. They found that if the effective biofilm thickness, δ_e, is substituted for δ in the Thiele modulus (Eq. 30), then the effectiveness factor can be expressed as a function of the (modified) Thiele modulus only and does not depend on particles geometry. The effective biofilm thickness, δ_e, is defined as a ratio of biofilm volume to its external surface area. The findings of Mulcahy and co-workers coincide with the statement of Atkinson and Davies (1974) and with the results of Horvath and Engasser (1973). The formulae proposed by Mulcahy and co-workers were obtained for d_s/d_p in a range of 0.1 to 0.9 in the following form:

- for zero-order reaction (n=0):

$$\eta = 1 \qquad\qquad \text{for } \Phi_n < 0.81$$

$$\eta = 0.8274 \cdot \Phi_n^{-0.9} \qquad\qquad \text{for } \Phi_n > 0.81 \tag{39}$$

- for first-order reaction (n=1):

$$\eta = \frac{\coth(3 \cdot \Phi_n)}{\Phi_n} - \frac{1}{3 \cdot \Phi_n^2} \tag{40}$$

Eqs. 39 and 40 are shown in Fig. 12 together with the values of the effectiveness factor for a flat biofilm (Eqs. 31, 32 and 34). The effectiveness factor for shell-like biofilms is smaller than that for flat biofilms at the corresponding Thiele modulus. For two biofilms with the same specific surface area (the same ratio of surface area to biofilm volume), the thickness of the flat biofilm is smaller than that of the shell-like biofilm, resulting in better penetration of the substrate and higher effectiveness.

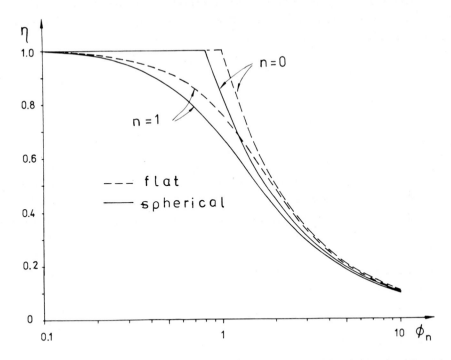

Fig. 12.   Effectiveness factor for flat and spherical biofilm

5.2 Double Substrate Limitation

In the previous discussion it was assumed that there was only one substrate which limited the reaction rate of the microorganisms.   In practical applications it is likely that the availability of more than one nutrient is restricted and a number of substrates may influence the reaction rate.   The phenomenon of multiple substrate limitations is far from being fully understood.   Since biochemical reactions involve an electron donor and acceptor it seems that these two substances are most likely to control the overall reaction rate.   In this case, the biochemical reaction can be written in the following general form:

$$\nu_a \cdot A + \nu_d \cdot D + \text{other nutrients} \longrightarrow \text{products} + \text{cells} + \text{energy} \qquad (41)$$

where A and D are electron acceptor and donor, respectively. This pair may include oxygen and organic substances, oxygen and ammonia, nitrate and methanol or any other two substances which limit the rate of reaction.   The rate of removal (uptake) of each of the two substrate can be expressed by the Michaelis-Menten equation (Eq. 24). Since the maximum removal rates of both substrates must be equal:

$$\nu_a \cdot MW_a \cdot k_{m,a} = \nu_d \cdot MW_d \cdot k_{m,d} \tag{42}$$

then the electron acceptor is **locally** limiting the overall reaction rate if:

$$c_a/K_{s,a} < c_d/K_{s,d} \tag{43}$$

MW in Eq. 42 is the molecular weight of the acceptor and the donor, respectively and the maximum removal rates k_m are expressed on mass basis. Eq. 43 constitutes a limitation condition at a particular position in the biofilm. The overall limitation condition depends also on the diffusivities of both substrates. The condition was developed for the case of zero-order kinetics (Williamson and McCarty, 1976a). The electron acceptor controls the overall reaction rate in the biofilm if:

$$c_a < \frac{D_{eff,d} \cdot \nu_a \cdot MW_a}{D_{eff,a} \cdot \nu_d \cdot MW_d} \cdot c_d \tag{44}$$

Howell and Atkinson (1976a) considered a more complex situation in which the intrinsic removal rate is a function of both substrate concentrations. They examined a double Michaelis-Menten kinetics:

$$r_v = k_m \cdot X \frac{c_a}{c_a + K_{s,a}} \cdot \frac{c_d}{c_d + K_{s,d}} \tag{45}$$

and the Bright and Appleby model:

$$r_v = k_m \cdot X \frac{c_a \cdot c_d}{c_a \cdot c_d + K_{s,a} \cdot c_d + K_{s,d} \cdot c_a} \tag{46}$$

and the single substrate Michaelis-Menten kinetics. For a "deep" biofilm (i.e. for large values of the Thiele modulus) the limitation conditions are approximately expressed by Eq. 44. Around the transition point from "shallow" to "deep" biofilm, the removal rate depends on both concentrations and the half-saturation constants K_s. The influence of both substrates in thinner biofilms could not be expressed in a simple form and required solving the diffusion equations.

5.3 Biokinetic and Diffusivity Parameters

The calculations of the effectiveness factor and the reaction rates require the values of several parameters. Generally, only some rough estimates are available and the values reported in the literature may vary considerably. A number of values of the biokinetic parameters and the diffusion coefficients were reported by Harremoes (1978), Williamson and McCarty (1976b), Atkinson and How (1974), Howell and Atkinson (1976a), Meunier and Williamson (1981). Based on these data, the following conclusions can be reached. The diffusivity in the biofilm varies form approximately 3 cm^2/d for simple molecules like O_2, 1.2 - 1.4 cm^2/d for NH_4^+ to 0.1 - 0.3 cm^2/d for more complex substrates like glucose or organic substances in sewage (as BOD). Generally, the values of diffusivity in the biofilm are in a range of 75-95% of the values for water although much lower values (7%) were also reported. Tomlinson and Snaddon (cited in Harremoes, 1978) found the diffusivity of oxygen in the biofilm much higher than in water (924%). Their claims were not confirmed by other authors; Ngian and Lin (1976) questioned the accuracy of determination of biofilm thickness in that case. It is, however, conceivable that an abnormally high diffusivity was caused by some active transport mechanisms which are not, at present, fully recognized. The maximum removal rate, k_o or k_m, as reported in the previous references, are in a range of 1 to 7.2 d^{-1} and the half-saturation (Michaelis-Menten) constants in a range of 0.06 to 0.5 mg/L for O_2, NH_4-N, NO_2-N and much higher (2.5 - 400 mg/L) for glucose and sewage BOD. Harremoes (1976) suggested that these high values might be caused by an erroneous interpretation of the diffusional limitations as a biochemical effect. The possibility of such misinterpretation was clearly demonstrated by Atkinson and Ur-Rahman (1979) and Shieh and LaMotta (1979) who took appropriate measures to minimize the diffusional effects or to include them in the analysis. Another source of variability of biokinetic parameters is probably related to physiological states of microbial cultures and their activity. These effects were investigated in a fluidized bed by Hermanowicz and Ganczarczyk (1984) and were recently reviewed by Daigger and Grady (1982).

The concentration of the microorganisms in the biofilm, X (or biofilm dry density), was estimated at 108 mg VS/cm^3 by Hermanowicz (1982) and in range 40-110 mg TS/cm^3 by Williamson and McCarty (1976b). These results coincide with the values reported by Shieh, Sutton and Kos (1979) from their own experiments and from the literature references. The range of these values was from 35 to 105 mg/cm^3 and usually around 75-90 mg/cm^3. The results of Hoehn (cited by Williamson and McCarty, 1976a) showed a dependence of concentration X on biofilm depth with a maximum value of 100 mg TS/cm^3 at a 200 μm depth. Mulcahy and LaMotta (cited in Shieh et al., 1979) reported also a decrease of microorganism concentration for the biofilms exceeding 300 μm. The results of Zelver and Trulear reported by Characklis (1981) indicated that biofilm density may increase with increasing turbulence of the liquid phase and increasing substrate loading. Increased turbulence may result in developing a denser, well-structured biofilm or in physical "squeezing" out water from the biofilm. Increased substrate loading may promote more intense growth of microorganisms resulting in a higher concentration of microorganisms rather than a thicker biofilm since substrate utilization is more efficient in a thinner film.

5.5. Liquid Film Transport

Substrate transport into the biofilm results in a decrease of substrate concentration in a liquid layer adjacent to the biofilm/liquid interface. The flux of chemical substance through this layer, j, is proportional to the difference between the concentrations in the bulk of liquid and at the biofilm surface, c_b and c_s, respectively (Fig. 8):

$$j = k_f \cdot (c_b - c_s) \tag{46}$$

with the mass transfer coefficient k_f. At steady-state conditions, the flux through the liquid layer is equal to the flux through the biofilm surface (Eq. 37) yielding the following equation:

$$k_f \cdot (c_b - c_s) = \eta \cdot \delta \cdot r_v(c_s) \tag{47}$$

This equation can be solved with regard to c_s and the overall removal rate r_{vb} can be determined (Eq. 37). The liquid film resistance to substrate transport lowers the surface concentration c_s and reduces the overall effectiveness factor, η_b, which is defined in the same way as η (Eq. 28) but is based on the substrate concentration in the bulk of liquid rather than that at the biofilm surface. This effectiveness factor is a function of the biokinetic parameters, substrate diffusivity and the biofilm thickness, like η, but also depends on the mass transfer coefficient k_f. For a flat biofilm η_b is expressed by the following formulae:

- for zero-order reaction (n=0):

$$\eta_b = 1 \qquad \text{for } \Phi_{n,b}^2 < M \tag{48}$$

$$\eta_b = \frac{Sh}{\Phi_{n,b}^2 + \Phi_{n,b} \cdot \sqrt{\Phi_{n,b}^2 + Sh^2}} \qquad \text{for } \Phi_{n,b}^2 > M \tag{48}$$

$$M = (1 + 2/Sh)^{-1}$$

- for first-order reaction (n=1):

$$\eta_b = \frac{\tanh(\Phi_{n,b})}{\Phi_{n,b}} \cdot \frac{Sh}{Sh + \Phi_{n,b} \cdot \tanh(\Phi_{n,b})} \tag{49}$$

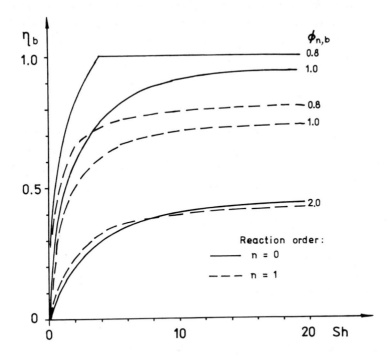

Fig. 13. Effects of external mass transfer on effectiveness factor for flat biofilm.

where $\Phi_{n,b}$ is the Thiele modulus expressed by Eq. 30 with the substitution of c_b for c_s and Sh is Sherwood number ($Sh = k_f \cdot \delta/D_{eff}$). The values of η_b are plotted in Fig. 13 as function of Sh. As the Sherwood number increases the liquid film resistance decreases and the effectiveness factor approaches the values shown in Fig. 9 for the corresponding values of the Thiele modulus. It is interesting to note that the overall effectiveness factor η_b for zero-order reaction (which accounts for liquid film resistance) can actually reach its limiting value of 1 for $\Phi_{n,b} < 1$. In that case, for sufficiently large Sh, the resistance of the liquid film does not influence biofilm efficiency.

The overall effectiveness factor for other types of reactions (n-th order, Michaelis-Menten) can be found from the following equation:

$$\eta_b = \eta \cdot r_v(c_s)/r_v(c_b) \tag{50}$$

where the concentration c_s is calculated from Eq. 47.

The transport of chemical compounds in the liquid layer near the biofilm surface has a rather complex mechanism involving molecular diffusion and micromixing of liquid. The change of concentration between the bulk of liquid and the biofilm is gradual; however, it is convenient for analytical purposes to introduce a concept of stagnant liquid film of thickness δ_L adjacent to the biofilm surface (see Fig. 8) which gives an equivalent resistance to the mass transfer coefficient k_f :

$$\delta_L = D_m/k_f \tag{51}$$

The thickness δ_L depends on the diffusivity of the species involved and on hydrodynamic conditions near the biofilm. A large number of experimental work has been devoted to the determination of the mass transfer coefficient in various systems and a number of formulae have been proposed. Most of these works are related to liquid films adjacent to surfaces other than biofilm. It can be expected that biofilm surface characteristics (roughness, deformations due to shear stress, extracellular polymers, etc.) may substantially change the liquid mass transfer in a way similar to the changes described by Characklis (1981) with regard to frictional forces and heat transfer. To account for these factors, Williamson and McCarty (1976a) divided the liquid film into two sublayers of thickness δ_{L1} and δ_{L2}, respectively. In their experiments with a nitrifying biofilm, they found that the sublayer closer to the biofilm had a constant thickness $\delta_{L2} = 56$ μm independently of liquid velocity. They believed that this layer resulted "from the uneven or sponge-like nature of the liquid-biofilm interface" but "whether such a layer exists in all biofilms is currently unknown." It should be noted that the biofilm used by Williamson and McCarty was formed by filtering a suspension of microbial cells rather than cultivating the microorganisms on some surface and might not be representative for other types of biofilm.

The mass transfer coefficient k_f for the flow through a porous bed depends on hydrodynamics of the system and is usually expressed in a dimensionless form:

$$J_D = \frac{k_f}{u} \left(\frac{\nu}{D_m}\right)^{2/3} = f(\varepsilon) \cdot Re^m \tag{52}$$

as a function of bed porosity ε and the Reynolds number $Re = u \cdot d_p/\nu$. If the existence of two liquid sublayers is assumed, following Williamson and McCarty, then the coefficient k_f in Eq. 52 is equal to D_m/δ_{L1} and the overall mass transfer coefficient (as used in Eqs. 46-49) would be

$$D_m/\delta_{L1} + D_m/\delta_{L2}$$

Few formulae for the J_D modulus listed below are plotted in Fig. 14 and 15.

- for fixed beds

$$J_D = 1.99 \cdot (1 - \varepsilon)^{0.5} \cdot Re^{-0.5} \qquad \text{Pohorecki and Wronski, 1977} \quad (53)$$

$$J_D = 1.625 \cdot Re^{-0.507} \qquad \text{for Re} < 120$$
$$\qquad\qquad\qquad\qquad\qquad\qquad\qquad\qquad\qquad \text{McCune and Wilhelm, 1949} \quad (54)$$
$$J_D = 0.687 \cdot Re^{-0.327} \qquad \text{for Re} > 120$$

- for fluidized beds

$$J_D = 0.0795 \cdot (1 - \varepsilon)^{0.345} \cdot (Re/Ga)^{-0.345} \qquad \text{Keinath and Weber, 1968} \quad (55)$$

$$J_D = 0.783 \cdot (1 + F(\varepsilon))^{0.54} \cdot Re^{-0.54} \qquad \text{Hermanowicz and Roman, 1980} \quad (56)$$

$$F(\varepsilon) = 2 \cdot (1 - \varepsilon) \cdot (2 - \varepsilon)^2 \cdot \varepsilon^{-2}$$

- for both types of beds

$$J_D = 0.813 \cdot \varepsilon^{-1.5} \cdot Re^{-0.5} \qquad \text{Snowdon and Turner, 1967} \quad (57)$$

$$J_D = 3.39 \cdot (1 - \varepsilon)^{0.75} \cdot \varepsilon^{-0.75} \cdot Re^{-0.5} \qquad \text{Rittman, 1982} \quad (58)$$

The values of J_D for fixed beds (Fig. 14) were calculated assuming $\varepsilon = 0.4$ and the values for fluidized beds (Fig. 15) with ε from the Zaki and Richardson method for biocoated particles (Eqs. 12, 13 and 15).

6. BIOFILM DEVELOPMENT

The amount of biomass in a fixed-film reactor, like in any biological system is a fundamental parameter affecting the performance of the reactor. However, unlike a suspended-growth system, not only the rate of biochemical reactions is influenced by the amount of microorganisms but also the physical characteristics of the system. In fixed beds, growing biomass reduces the porosity of the bed and leads to clogging. In fluidized beds, the clogging phenomenon is usually avoided if particle movement is not restricted. In general, biomass is not uniformly distributed in the reactor and local variations of the reaction rate and other characteristics may be significant.

The changes of porosity (liquid hold-up) in the packed bed are expressed by the following formula:

$$\varepsilon = \varepsilon_o - V_b/V \qquad (59)$$

where ε_o is the initial porosity of the clean bed, V_b is biomass volume and V denotes the volume of the bed. For spherical particles Eq. 59 is equivalent to

508

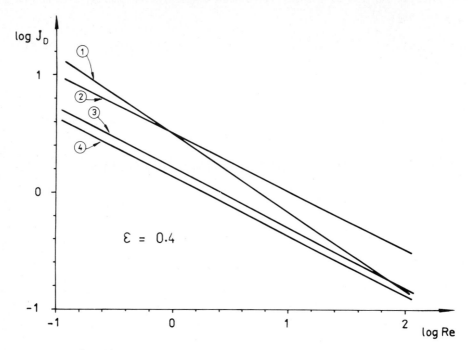

Fig. 14. External mass transfer modulus J_D for fixed beds.
1 – Eq. 58; 2 – Eq. 57; 3 – Eq. 54; 4 – Eq. 53

Eq. 10 presented previously. The effect of biomass growth on the porosity of a fluidized bed is more complicated. The porosity can be calculated using the Zaki-Richardson method modified for biocoated particles (Eqs. 12, 13 and 15). From a similar analysis reported by Andrews (1982), it appears that the increasing biomass volume may increase or decrease the porosity of the fluidized bed depending on a value of parameter B:

$$B = (\rho_b - \rho_1)/(\rho_s - \rho_1) \tag{60}$$

A key equation of Andrews' analysis (modified to conform with Eq. 15) has the following form:

$$\frac{u_t}{u_{to}} = \frac{(1 + B \cdot y)^{0.654}}{(1 + y)^{1/3}} \tag{61}$$

where u_t is the terminal free fall velocity of a biocoated particle, u_{to}

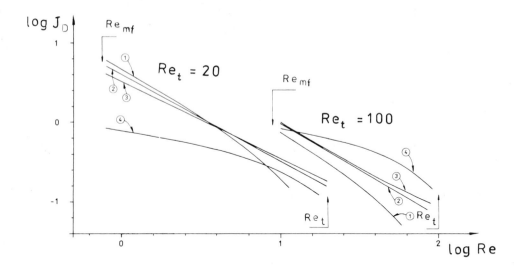

Fig. 15.   External mass transfer modulus J_D for fluidized beds.

1 - Eq. 58; 2 - Eq. 56; 3 - eq. 57; 4 - Eq. 55

denotes the terminal velocity for clean particle and y is a ratio of biomass volume to carrier particle volume (for spherical particles y = $(d_p{}^3-d_s{}^3)/d_s{}^3)$. The exponent 0.654 corresponds to Eq. 15 (0.654=1/(2-0.47)) and is valid for the same region of Re_t as Eq. 15.   Eq. 61 is plotted in Fig. 16.   For small values of B, typical for sand carrier particles, an increase of biomass volume (biofilm thickness) results in a decrease of the terminal velocity and the corresponding expansion of the bed according to Eq. 12.   If, however, the density of the carrier is close to that of the biofilm (i.e. B close to 1) biomass growth results in a higher terminal velocity. The expansion of the fluidized bed in such case may decrease and cause clogging. From this point of view, light support carrier particles like polypropylene and PVC, advocated by Atkinson and co-workers (1979, 1981) may not be advantageous.   It can be seen from Fig. 16 that for B in a range 0.3 - 0.4, the biomass growth, up to a rather large value, causes little change in the terminal velocity indicating that the expansion of the bed filled with particles of suitable density may remain relatively constant during its operation.

The growth of the biofilm may also affect, through size and density changes, the pattern of solids mixing/segregation in the fluidized bed.   These phenomena were discussed previously in this work.   Most of the presented conclusions are based on observations of biologically non-active systems.   It is currently not

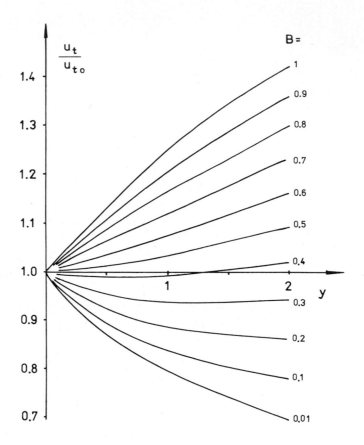

Fig. 16. Effects of biomass growth on terminal free fall velocity
of biocoated particles (after Andrews, 1982)

clear whether those conclusions also apply to fluidized beds with a biofilm.
It can be expected that the biological activity may alter surface
characteristics of the particles through excretion of biopolymers or other
compounds. It was observed (Hermanowicz, 1982) that, during an intense growth,
some biocoated particles formed aggregates containing up to 10-15 sand grains
enveloped in the biomass. The formation of such aggregates promoted solids
segregation in the reactor. In the same experiments, a number of particles
were "colonized" by stalked ciliates. These colonies with the extended
flexible branches up to 1 mm long, changed dramatically the hydraulic
properties of the particles.

In a theoretical analysis of solids mixing and stratification, Andrews
(1982) concluded that for B close to 1, as in sludge blanket reactors, a
complete stratification must occur in steady-state conditions. He also argued
that in a case of a large difference between biofilm density and carrier

density (large B) the solids in a fluidized bed become completely mixed at steady-state conditions. His conclusion was based on the following reasoning: As a substrate is being removed a concentration gradient is established in the bed with higher concentrations at the bottom near the inlet. Thus the rate of microbial growth is higher at the bottom of the bed and the particles become covered with a thicker biofilm. Such particles are lighter and move to the top of the bed where the microorganisms are exposed to a lower substrate concentration and their growth rate become smaller than the decay rate. Then, the biofilm becomes thinner, the density of the particles increases and the particles sink to the bottom of the bed. The results of this analysis of biofilm development by Andrews are not supported by experimental observations of distinct biomass profiles in fluidized bed (Shieh, 1981; Eggers and Terlouw, 1979; Hermanowicz, 1982).

The analysis presented by Andrews does not take into considerations some factors. First, it assumes that the rate of growth is greater than the rate of decay (or other kind of biomass removal) at the bottom of the bed and the opposite at the top with an equilibrium between growth and decay/removal only for the whole bed. It is conceivable that such equilibrium can be reached at every position in the bed eliminating the driving force for particles circulation. Second, intentional biomass wasting from a bed by removing particles from the top, cleaning them and recycling to the bottom imposes a flow of solids through the bed. Such flow can alter biomass distribution and process efficiency (Bousfield and Hermanowicz, 1984).

The phenomenon of bed stratification according to local biomass concentration was also observed in fixed beds (Rittmann and McCarty, 1980b; Faup et al., 1982) although its mechanism is different. Since substrate concentration at the inlet is higher than that in other parts of the bed, microbial growth at the inlet is most intense and leads to highest biomass concentration. The only mechanism of biomass transport in a fixed bed is that with liquid carrying suspended microorganisms and fragments of detached biofilm.

Biofilm development in a fixed-film reactor includes a number of processes:
- attachment of microorganisms to the surface of carrier particles;
- biomass growth and maintenance;
- microbial death and lysis;
- biofilm detachment.

A mass balance equation can be formulated for attached biomass at any place in the bed:

accumulation = growth + attachment - detachment - decay $\hspace{3cm}$ (62)

The present knowledge of those processes is limited.   In many cases only qualitative statements are possible.   In the authors' opinion, biofilm development constitutes the weakest point in the modelling of biological fixed-film reactors and more investigation is required to provide a better understanding and a quantitative description of the biofilm dynamics. Nonetheless, an attempt will be made to present here some principles of biofilm development.

Four basic processes shown in Eq. 62 contribute to biofilm development. The individual significance of each process is different in different stages of biofilm development.   Initially, a clean solid surface immersed even in a dilute solution of microbial nutrients (like natural waters) quickly adsorbs organic molecules (Marshall, 1978; Characklis, 1981).   Then, microorganisms suspended in the liquid are transported to the surface and adhere to it.   This initial adhesion is weak and reversible; the microorganism can easily be removed from the surface by rinsing.   The attached microorganisms subsequently become more strictly bonded to the surface with the aid of extracellular polymers (Marshall, 1978) forming a complex biofilm structure.   The rate of attachment initially controls the development of biomass but when a developed biofilm is established, its contribution to biofilm accumulation can be neglected unless suspended microorganisms are present in a very high concentration.

The metabolism of the biofilm microorganisms is regulated by the similar principles as that in suspended growth.   Substrate utilization yields energy required for cell growth, maintenance, and reproduction.   It is commonly assumed that the amount of biomass grown is proportional to the amount of substrate removed with the yield, Y, being the proportionality coefficient.   A more complex approach was presented among other by Powell (1967), Ollis (1977) and Fredrickson and Tsuchiya (1977) but it will not be discussed in this work. Thus, for a biofilm:

$$\text{rate of growth} = Y \cdot A \cdot j = Y \cdot A \cdot \delta \cdot r_v(c_b) \cdot \eta_b \tag{63}$$

where A denotes the surface area of the biofilm. Since the effectiveness factor η_b decreases with increasing biofilm thickness, the rate of growth declines as the biofilm grows.   This decline was experimentally observed by Trulear and Characklis (1982).   Grasmick and co-workers (1979) described the growth of biofilm in their model of a submerged biological filter as:

$$\text{rate of growth} = k_b \cdot m_a \tag{64}$$

where m_a is mass of active layer in the biofilm.   If a zero-order kinetics is assumed, Eq. 64 conforms with Eq. 63 since the effectiveness factor is proportional to the mass of active biolayer (assuming constant X). The solution of Eq. 64 leads, in absence of detachment and decay, to an expotential increase of biofilm thickness in the initial phase of growth followed by a linear increase of thickness.

The growth of microorganisms resulting from substrate utilization and the increase of biofilm thickness are controlled by microbial decay and biofilm detachment. The microbial decay includes all physiological processes which lead to a decline of microbial mass. Herbert (1958) postulated that an endogenous metabolism occurs at all phases of growth supplying energy for maintenance purposes. Pirt (1966) supposed that maintenance energy is derived directly from substrate utilization rather than through biomass synthesis and decay. The problems of maintenance and endogenous metabolism were discussed in detail by Powell (1967), and Fredrickson and Tsuchiya (1977), among others.

This discussion leads to a general conclusions that the decay rate is proportional to the mass of microorganisms, m_b:

$$\text{decay rate} = b \cdot m_b \qquad\qquad (65)$$

with b denoting the decay rate coefficient. Little is known about the specific value of the decay coefficient, b. Rittmann and McCarty (1980b) measured its value at 0.205 d^{-1} for microorganisms isolated from sewage and grown on sodium acetate and methylene chloride medium. Beck (1981) estimated b at 0.17-0.2 d^{-1} for nitrifying bacteria in activated sludge, while Stratton and McCarty (1967) proposed b = 0.05 d^{-1} for the nitrifying microorganisms in natural streams. Benefield and Randall (1980) recommended a value of 0.1 d^{-1} for activated sludge; b=0.24 d^{-1} was used by Dold, Ekama and Marais (1980) and b=0.044 d^{-1} was estimated by Gujer (1980).

As the biofilm grows some fragments of accumulated biomass detach from the carrier and are carried away with the flowing liquid. In a fluidized bed, the detachment can be caused by shear forces exerted by the flowing liquid and as a result of collisions of biocoated particles. In a packed bed, the liquid shear forces are alone responsible for biofilm detachment. It is likely that the formation of a substrate-deficient layer in a deep biofilm may affect the structure of the biofilm matrix and decrease its resistance to shear forces. This mechanism may be particularly important if gaseous products are formed within the biofilm (nitrogen gas in denitrification, carbon dioxide and methane in anaerobic fermentation). Harremoes, Jansen and Kristensen (1980) reported that nitrogen bubbles formation weakened markedly the biofilm and enhanced its sloughing. Howell and Atkinson (1976b) presented a model of biofilm sloughing in a trickling filter in which sloughing occurs when the substrate concentration at the biofilm/carrier interface falls below a critical value. Since substrate penetration in the biofilm depends on its depth, the rate of biofilm detachment will also depend on biofilm thickness. The rate of detachment depends as well on the surface area of biofilm. The product of the surface area and the biofilm thickness is proportional to the mass of biofilm indicating that the rate of detachment could be related to the mass of microorganism. Trulear and Characklis (1982) reported that the rate of detachment was indeed an increasing function of biofilm mass. Thus, the detachment rate can be expressed as:

$$\text{detachment rate} = k_d \cdot m_b \tag{66}$$

The values of k_d estimated from the data of Trulear and Characklis for a biofilm cultivated in an annular reactor with glucose as limiting nutrient were in a range of 0.9 to 1.8 d^{-1}. Eq. 66 was also used by Hermanowicz and Ganczarczyk (1984) in a model of a nitrifying fluidized bed although it was possible to estimate only an overall biomass removal coefficient describing both biomass decay and detachment. The values of this coefficient (equal to b + k_d) were from 0.04 d^{-1} to 0.17 d^{-1}.

The velocity of liquid, or more precisely shear stress, affects the rate of detachment. Trulear and Characklis (1982) reported that the detachment rate in the annular reactor increased with increasing liquid velocity. It is likely that a similar effect would be observed in a packed bed where growing biomass fills void spaces and increases the interstitial liquid velocity. In a fluidized bed, however, Hermanowicz and Ganczarczyk (1984) found lower values of the biomass removal coefficient, b + k_d, for higher liquid velocities. At a lower superficial liquid velocity, the porosity of the fluidized bed is also lower resulting in lower liquid shear forces but the rate of collision between the particles is increased. Therefore, the lower values of the biomass removal coefficient at low liquid velocities, if confirmed in other studies, would indicate that the detachment of biofilm in fluidized beds is mainly caused by particle collision rather than liquid shear.

The rates of individual processes affecting biofilm development (Eqs. 63 – 66) combined into Eq. 62 yield a biofilm development equation:

$$dm_b/dt = Y \cdot A \cdot \delta \cdot \eta_b \cdot r_v(c_b) - (b + k_d) \cdot m_b \tag{67}$$

This equation describes the changes of biofilm at a given place in the bed. It does not include any changes caused by the movement of biocoated particles due to mixing and segregation. These effects were discussed in previous sections of this work. Since the present knowledge of mixing and segregation is quite limited, the modelling of biomass development in a fluidized bed is probably restricted to two extreme cases: complete solids mixing and complete segregation. The behavior of a real fluidized bed lies somewhere between these two extremes.

An analysis of Eq. 67 originally reported by Rittmann and McCarty (1980a) leads to a conclusion that a steady-state biofilm ($dm_b/dt=0$) cannot exist at substrate concentrations below a critical value. In a limiting case of very thin biofilm $\eta_b=1$ and at steady state, Eq. 67 reduces to:

$$Y A \cdot \delta \cdot r_v(c_b) = (b + k_d) \cdot m_b \qquad (68)$$

Since, for the thin biofilms $m_b = A \cdot \delta \cdot X$, regardless of particle geometry, the following condition for the critical substrate concentration, $c_{b,min}$ is valid:

$$Y \cdot r_v(c_{b,min}) = (b + k_d) \cdot X \qquad (69)$$

The minimum substrate concentration supporting the existence of biofilm, $c_{b,min}$, can be calculated from Eq. 69. This minimum concentration is applicable to steady state conditions only. Such conditions can be achieved in a packed bed with a constant influent concentration as demonstrated by Rittman and McCarty (1980b). In a fluidized bed mixing results in individual biocoated particles being exposed to varying substrate concentration and not achieving a steady state. The bed may be in a dynamic equilibrium with individual particles moving between the inlet region of high concentration and biofilm growth and the outlet region of low substrate concentration and biofilm decay. This oscillation between biofilm growth and decay allows the concentration of substrate to fall below $c_{b,min}$ as postulated by Rittman (1982).

7. SUMMARY

In this work some elements of the modelling of fixed-film reactors were presented. Since major developments in mathematical description of biofilm related phenomena have been achieved only in the last few years, the task of biofilm reactor modelling is still far from completion. Nonetheless, it seems possible to propose a general framework for such models. In authors' opinion, the following problems should be considered in a model of a fixed-film reactor:
- reactor characteristics (flow regime, mixing, concentration profile);
- substrate utilization in the biofilm (effectiveness factor, formation of substrate-deficient layers, external mass transport, growth requirements);
- biofilm formation and development (rate of growth, rate of detachment, particle density changes and resulting changes of bed porosity and expansion, biofilm thickness control).

At present, several limitations of accurate modelling exist in all of the above listed areas. Despite a large number of studies of chemical reactors, a description of liquid mixing in packed and fluidized beds is not well developed. It may be, thus, reasonable in some cases to check the prediction of a particular reactor model with those obtained for a completely mixed reactor or a plug flow model (with regard to liquid phase) as the two extremes of the flow regime. If the plug flow model with axial dispersion is used, the choice must be made among boundary conditions, although their influence is probably small.

The effects of biomass growth on reactor behavior cannot be underestimated, particularly for fluidized beds. The growth of biomass on the surface of carrier particles affects the porosity of the bed. In packed beds the biomass simply fills the void space but in fluidized beds the mechanism is more complex. The growth of biomass changes the density of the particles, hence the expansion of the bed. It has also been recognized that a microbial coating on the particles increases the drag forces exerted by flowing liquid although the quantitative data are limited. The changes of particle density in the fluidized bed, combined with the increased drag forces, affect the solids mixing pattern. These effects were not yet well investigated.

Since a concentration gradient usually exists in the reactor, it is likely that the growth of biomass and biofilm thickness also vary along the bed affecting local substrate removal rate. The distribution of biomass in the reactor can be fairly easily described in a packed bed. In a fluidized bed, however, the biofilm non-uniformity induces particles segregation which, together with mixing, substrate removal, and resulting microbial growth and biofilm detachment, form a complex mechanism regulating biomass distribution in the bed. At present the lack of full understanding of this mechanism and the mathematical complications practically limit the modelling to simplest cases of either complete solids mixing or total stratification.

The kinetics of soluble substrate removal in a microbial film has been thoroughly investigated in a theoretical way. As a result of this analysis, it is now possible to calculate the effectiveness factor and determine the overall removal rate for a flat or spherical biofilm, including external mass transport limitations and for a number of types of microbial kinetics. With the advances of numerical techniques, the computations can easily be performed even for a complicated biofilm geometry or microbial kinetics. For these computations, several assumptions are made and a number of parameters is required. It is generally assumed that the intrinsic kinetics of substrate utilization in the biofilm can be described in the same form as that for suspended-growth cultures. A variety of equations has been proposed among which the n-th order and the Michaelis-Menten kinetics are most widely used. While these equations may quite well describe a steady state substrate removal, a new approach must be developed to represent dynamic situations. Even at a steady state the values of biokinetics parameters have to be determined for each microbial population and type of substrate. Molecular diffusivity for simple substrates (like glucose, oxygen, ammonium, etc.) has been estimated but the values for many other compounds present in wastewaters are not available. Modelling of complex substrates removal in the biofilm presents a great problem, since the principles of microbial kinetics for such substrates are not yet established.

The physical structure of the biofilm is not well understood despite its importance in the modelling. The concentration of microorganisms in the biofilm, or biofilm density, has been roughly estimated but the effects of substrate loading, biofilm thickness and shear stress on the microbial concentration are not well-known. Similar problems are related to the estimation of biofilm detachment rate which controls the thickness of the biofilm. The rate of detachment can be presently estimated within the order of magnitude, but the effects of the specific factors (liquid velocity, bed expansion, substrate concentration, etc.) cannot be quantitatively evaluated.

Despite considerable advances in recent years in the modelling of the fixed-film reactors, an "open field" still remains for many investigations, theoretical and experimental, before a model fully representing this complex process can be proposed. Nonetheless, the present "deficient" models can also yield valuable results useful for the design purposes and for planning of future studies.

8. REFERENCES

Andrews, G.F., 1982. Fluidized-Bed Fermenters: A Steady State Analysis. Biotechn. Bioengineering, 24: 2013-2030.
Andrews, G.F. and Tien, C., 1979. The Expansion of a Fluidized Bed Containing Biomass. A.I.Ch.E. Journal, 25: 720-723.
Andrews, G.F. and Tien, C., 1982. An Analysis of Bacterial Growth in Fluidized-Bed Adsorption Column. A.I.Ch.E. Journal, 28: 182-190.
Atkinson, B., Black, G.M., Lewis, P.J. and Pinches, A., 1979. Biological Particles of Given Size, Shape, and Density for Use in Biological Reactors. Biotech. Bioengineering, 21: 193-203.
Atkinson, B., Black, G.M. and Pinches, A., 1981. The Characteristics of Solid Supports and Biomass Support Particles When Used in Fluidized Beds. In: P.F. Cooper and B. Atkinson (Editors), Biological Fluidized Bed Treatment of Water and Wastewater. WRC & Ellis Horwood, Chichester, U.K., pp. 75-106.
Atkinson, B. and Davies, I.J., 1974. The Overall Rate of Substrate Uptake (Reaction) by Microbial Films. Part I. A Biological Rate Equation. Trans. Instn. Chem. Eng., 52: 248-259.
Atkinson, B. and How, S.Y., 1974. The Overall Rate of Substrate Uptake (Reaction) by Microbial Films, Part II. Effect of Concentration and Thickness with Mixed Microbial Films. Trans. Instn. Chem. Eng., 52: 260-268.

Atkinson, B. and Ur-Rahman, F., 1979. Effect of Diffusion Limitations and Floc Size Distributions on Fermenter Performance and the Interpretation of Experimental Data. Biotech. Bioengineering, 21: 221-251.

Beck, M.B., 1981. Operational Estimation and Prediction of Nitrification Dynamics in the Activated Sludge Process. Water Research, 15: 1313-1330.

Benefield, L.D. and Randall, C.W., 1980. Biological Process Design for Wastewater Treatment. Prentice-Hall, Englewood Cliffs, N.J.

Bousfield, D.W. and Hermanowicz, S.W., 1984. Biomass Distribution in a Biological Fluidized Bed Reactor. Presented at 2nd Intnl. Conf. on Fixed-Film Biological Processes, July 10, 1984, Arlington, Va.

Bryers, J.D., 1984. Biofilm Formation and Chemostat Dynamics: Pure and Mixed Culture Considerations. Biotechn. Bioengineering, 26: 948-958

Burovoi, I.A. and Ibraev, A.H., 1971. Matematiceska Model' Processa Peremesivania i Separacii Polidispersnogo Materiala v Kipascem Sloe. (A Mathematical Model of Mixing and Separation of Polydispersed Material in a Fluidized Bed. In Russian). Izvest. Vyssh. Uceb. Zaved. Cvet. Met., 14: No. 1, 136-140.

Characklis, W.G., 1981. Fouling Biofilm Development: A Process Analysis. Biotech. Bioengineering, 23: 1923-1960.

Characklis, W.G., Trulear, M.G., Bryers, J.D. and Zelver, N., 1982. Dynamics of Biofilm Processes: Methods. Water Research, 16: 1207-1216.

Chung, S.F. and Wen, C.Y., 1968. Longitudinal Dispersion of Liquid Flowing Through Fixed and Fluidized Beds. A.I.Ch.E. Journal, 14: 857- 866.

Cleasby, J.L. and Fan, F., 1981. Predicting Fluidization and Expansion of Filter Media. Journal Env. Engng. Div. A.S.C.E., 107: EE3, 455-471.

Daigger, G.T. and Grady, C.P.L., 1982. The Dynamics of Microbial Growth on Soluble Substrates. A Unifying Theory. Water Research, 16: 365-382.

Danckwerts, P.V., 1953. Continous Flow Systems. Distribution of Residence Times. Chem. Engng. Sci., 2: 1-9.

Deckwer, W.D., Mahlman, E.A., 1974. Dispersed Flow Reactors with Sections of Different Properties. Adv. Chem. Ser., 133: 334-344.

Deckwer, W.D. and Mahlman, E.A., 1976. Boundary Conditions of Liquid Phase Reactors with Axial Dispersion. Chem. Engng. Journal, 11: 19-25

Dold, P.L., Ekama, G.A. and Marais, G.V.R., 1980. A General Model for the Activated Sludge Process. Prog. Water Technology, 12: 47-77.

Eggers, E. and Terlouw, T., 1979. Biological Denitrification in a Fluidized Bed with Sand as Carrier Material. Water Research, 13: 1077-1090.

Epstein, N., Leclair, B.P. and Pruden, B.B., 1981. Liquid Fluidization of Binary Particle Mixtures. Part 1. Overall Bed Expansion. Chem. Engng. Sci., 36: 1803-1809.

Faup, G.M., Leprince, A. and Pannier, M., 1982. Biological Nitrification in an Upflow Fixed Bed Reactor. Water Sci. Technology, 14: 795-810.

Fredrickson, A.G. and Tsuchiya, H.M., 1977. Microbial Kinetics and Dynamics. In: L. Lapidus and N.R. Amundson (Editors), Chemical Reactor Theory. A Review. Prentice Hall, Englewood Cliffs, pp. 405-483.

Froment, G.F. and Bischoff, K.B., 1979. Chemical Reactor Analysis and Design. Wiley, New York.

Ganczarczyk, J.J., 1983. Activated Sludge Process: Theory and Practice. Dekker, New York and Basel, 270 pp.

Grasmick, A., Elmaleh, S. and Ben Aim, R., 1979. Theorie de l'Epuration par Filtration Biologique Immergee. Water Research, 13: 1137-1147.

Gujer, W., 1980. The Effect of Particulate Organic Material on Activated Sludge Yield and Oxygen Requirement. Prog. Water Technology, 12: 79-95.

Harremoes, P., 1976. The Significance of Pore Diffusion to Filter Denitrification. Journal Water Poll. Control Fed., 48: 377-389.

Harremoes, P., 1978. Biofilm Kinetics. In: Water Pollution Microbiology. Wiley, New York, Vol. 2, pp. 71-109.

Harremoes, P., Jansen, J. and Kristensen, G.H., 1980. Practical Problems Related to Nitrogen Bubble Formation in Fixed Film Reactors. Prog. Water Technology, 12: 253-269.

Herbert, D., 1958. Some Principles of Continuous Culture. In: G. Tunevall (Editor), Recent Progress in Microbiology. Almqvist and Wiksell, Stockholm.

Hermanowicz, S.W., 1982. Dynamics of Nitrification in a Fluidized Bed Reactor. Ph.D. Thesis, Department of Civil Engineering, University of Toronto.

Hermanowicz, S.W. and Ganczarczyk, J.J., 1983. Some Fluidization Characteristics of Biological Beds. Biotech. Bioengineering, 25: 1321-1330.

Hermanowicz, S.W. and Ganczarczyk, J.J., 1984. Dynamics of Nitrification in a Biological Fluidized Bed Reactor. Water Sci. Technology, 17: 351-366.

Hermanowicz, S.W. and Roman, M., 1980. A Comparison of Packed-Bed and Expanded-Bed Adsorption Systems. In: L. Pawlowski (Editor), Physiochemical Methods for Water and Wastewater Treatment. Pergamon Press, Oxford, pp. 141-152.

Hiby, J.W., 1963. Longitudinal and Transverse Mixing During Single-Phase Flow Through Granular Beds. In: Interaction Between Fluids and Particles. Institute Chemical Engineering, London, pp. 312-321.

Himmelblau, D.M. and Bischoff, K.B., 1968. Process Analysis and Simulation. Wiley, New York.

Horvath, C. and Engasser, J.M., 1973. Pellicular Heterogeneous Catalysts. A Theoretical Study of the Advantages of Shell Structured Immobilized Enzyme Particles. Ind. Engng. Chem. Fundam., 12: 229-235.

Howell, J.A. and Atkinson, B., 1976a. Influence of Oxygen and Substrate Concentrations on the Ideal Film Thickness and the Maximum Overall Substrate Uptake Rate in Microbial Film Fermenters. Biotech. Bioengineering, 18: 15-35.

Howell, J.A. and Atkinson, B., 1976b. Sloughing of Microbial Film in Trickling Filters. Water Research, 10: 307-315.

Keinath, T.M. and Weber, W.J., 1968. A Predictive Model for the Design of Fluidized-Bed Adsorbers. Journal Water Poll. Control Fed., 40: 741-759.

Kennedy, S.C. and Bretton, R.H., 1966. Axial Dispersion of Spheres Fluidized with Liquid. A.I.Ch.E. Journal, 12: 24-30.

Klapwijk, A., Jol, C. and Donker, H., 1979. The Application of an Upflow Reactor in the Denitrification Step of Biological Sewage Purification. Water Research, 13: 1009-1020.

Kmiec, A., 1978. Particle Distributions and Dynamics of Particle Movement in Solid-Liquid Fluidized Beds. Chem. Engng. (Lausanne), 15: 1-12.

Kokufuta, E., Matsumoto, W. and Nakamura, I., 1982. Immobilization of Nitrosomonas europaea Cells with Polyelectrolyte Complex. Biotech. Bioengineering, 24: 1591-1603.

Kowalski, E. and Lewandowski, Z., 1983. Nitrification Process in a Packed Bed Reactor with a Chemically Active Bed. Water Research, 17: 157-160.

Lee, E.S., 1968. Quasilinearization and Invariant Imbedding. Academic Press, New York.

Lettinga, G., Velsen, A., Hobma, S.W., deZeeuw, W. and Klapwijk, A., 1980. Use of the Upflow Sludge Blanket Reactor Concept for Biological Wastewater Treatment, Especially for Anaerobic Treatment. Biotech. Bioengineering, 22: 699-734.

Marshall, K.C., 1978. The Effects of Surfaces on Microbial Activity. In: Water Pollution Microbiology, Wiley, New York, Vol. 2., pp. 51-70.

McCune, L. and Wilhelm, R., 1949. Mass and Momentum Transfer in Solid-Liquid System - Fixed and Fluidized Beds. Ind. Engng. Chem., 41: 1124-1136.

Meunier, A.D. and Williamson, K.J., 1981. Packed Bed Biofilm Reactors: Simplified Model. Journal Env. Engng. Div. A.S.C.E., 107: EE2, 307-317.

Moo-Young, M. and Kobayashi, T., 1972. Effectiveness Factors for Immobilized-enzyme Reactions. Canadian Journal Chem. Engng., 50: 162-167.

Mulcahy, L.T., Shieh, W.K. and LaMotta, E.J., 1981. Simplified Mathematical Models for a Fluidized Bed Biofilm Reactor. A.I.Ch.E. Symp. Ser., 77: (209) 273-294.

Ngian, K.F. and Lin, S.H., 1976. Diffusion Coefficients of Oxygen in Microbial Aggregates. Biotech. Bioengineering, 18: 1623-1627.

Ngian, K.F. and Martin, W.R.B., 1980a. Biologically Active Fluidized Bed: Mechanistic Considerations. Biotechn. Bioengineering, 22: 1007-1014.

Ngian, K.F. and Martin, W.R.B., 1980b. Bed Expansion Characteristics of Liquid Fluidized particles with Attached Microbial Growth. Biotech. Bioengineering, 22: 1843-1856.

Ollis, D.F., 1977. Biological Reactor Systems. In: L. Lapidus and N. Amundson (Editors), Chemical Reactor Theory. A Review. Prentice Hall, Englewood Cliffs, pp. 484-531.

Pirt, S.J., 1961. The Maintenance Requirement of Bacteria in Growing Cultures. Proc. Royal Society (London), B163: 224-231.

Pohorecki, R. and Wronski, S., 1977. Kinetyka i termodynamika procesow inzynierii chemicznej. (Kinetics and Thermodynamics of Chemical Engineering Processes. In Polish). Wyd. Nauk. Tech., Warsaw.

Potter, O.E., 1971. Mixing. In: J.F. Davidson and D. Harrison (Editors), Fluidization, Academic Press, London, pp. 293-381.

Powell, E.O., 1967. The Growth Rate of Microorganisms as a Function of Substrate Concentration. In: E.O. Powell (Editor), Microbial Physiology and Continuous Culture. Her Majesty's Stationery Office. pp. 34-55.

Richardson, J.F., 1971. Incipient Fluidization and Particulate Systems. In: J.F. Davidson and D. Harrison (Editors), Fluidization. Academic Press, London, pp. 25-64.

Rittmann, B.E., 1982. Comparative Performance of Biofilm Reactor Types. Biotech. Bioengineering, 24: 1341-1370.

Rittmann, B.E. and McCarty, P.L., 1978. Variable-Order Model of Bacterial-Film Kinetics. Journal Env. Engng. Div. A.S.C.E., 104: EE5, 889-900.

Rittmann, B.E. and McCarty, P.L., 1980a. Model of Steady-State Biofilm Kinetics. Biotech. Bioengineering, 22: 2343-2357.

Rittmann, B.E. and McCarty, P.L., 1980b. Evaluation of Steady-State Biofilm Kinetics. Biotech. Bioengineering, 22: 2359-2373.

Roemer, M.H. and Durbin, L.D., 2967. Transient Response and Moments Analysis of Backflow Cell Model for Flow Systems with Longitudinal Mixing. Ind. Engng. Chem. Fundam., 6: 120-123.

Shieh, W.K., 1981. Comments on Biologically Active Fluidized Bed. Biotech. Bioengineering, 23; 2145-2147.

Shieh, W.K. and LaMotta, E.J., 1979. The Intrinsic Kinetics of Nitrification in a Continuous Flow Suspended Growth Reactor. Water Research, 13: 1273-1279.

Shieh, W.K., Sutton, P.M. and Kos, P., 1981. Oxitron System Fluidized Bed Wastewater Treatment Process: Predicting Reactor Biomass Concentration. Journal Water Poll. Control Fed., 53: 1574-1584.

Smith, E.L., James, A. and Fidgett, M., 1978. Fluidization of Microbial Aggregates In Tower Fermenters. In: J.F. Davidson and D.L. Keairns (Editors), Fluidization. Proc. 2nd Engineering Foundation Conference, Cambridge, U.K., Cambridge University Press.

Snowdon, C.B. and Turner, J.C.R., 1967. Mass Transfer in Liquid Fluidized Bed of Ion Exchange Resin Beads. Proc. International Symposium on Fluidization, pp. 599-608.

Stratton, F.E. and McCarty, P.L., 1967. Prediction of Nitrification Effects on the Dissolved Oxygen Balance of Streams., Env. Sci. Tech., 1: 405-410.

Sundaresan, S., Amundson, N.R. and Aris, R., 1980. Observations of Fixed-Bed Dispersion Models: The Role of the Interstitial Fluid. A.I.Ch.E. Journal, 26: 529-536.

Trulear, M.G. and Characklis, W.G., 1982. Dynamics of Biofilm Processes. Journal Water Poll. Control Fed., 54: 1288-1301.

Van Cauwenberghe, A.R., 1966. Further Note on Danckwerts' Boundary Conditions for Flow Reactors. Chem. Engng. Sci., 21: 201-205.

Van Duijn, G. and Rietema, K., 1982. Segregation of Liquid-Fluidized Solids. Chem. Engng. Sci., 37: 727-733.

Vossoughi, M., Laroche, M., Navarro, J.M., Faup, G. and Leprince, A., 1982. Continuous Denitrification by Immobilized Cells. Water Research, 16: 995-1002.

Wen, C.Y. and Fan, L.T., 1975. Models for Flow Systems and Chemical Reactors. Dekker, New York.

Wehner, J.F. and Wilhelm, R.H., 1956. Boundary Conditions of Flow Reactor. Chem. Engng. Sci., 6: 89–93.

Williamson, K. and McCarty, P.L., 1976a. A Model of Substrate Utilization by Bacterial Films. Journal Water Poll. Control Fed., 48: 9–24.

Williamson, K. and McCarty, P.L., 1976b. Verification Studies of the Biofilm Model for Bacterial Substrate Utilization. Journal Water Poll. Control Fed., 48: 281–296.

Wojcik, M., 1976. Badanie dyspersji osiowej w zlozu fluidalnym dla ukladu ciecz-cialo stale. (Investigations of Axial Dispersion in a Fluidized Bed for a Liquid-Solids Systems. In Polish). D.Sc. Thesis, Warsaw Technical University.

Yutani, N., Otake, N., Too, J.R. and Fan, L.T., 1982. Estimation of the Particle Diffusivity in a Liquid-Solids Fluidized Bed Based on a Stochastic Model. Chem. Engng. Sci., 37: 1079–1085.

9. NOTATIONS

Dimensions

L — length

M — mass

T — time

Symbols

A — biofilm surface area, L^2;

A_r — cross section area of reactor, L^2;

b — biomass decay rate coefficient, T^{-1};

B — density parameter (in Eq. 61), dimensionless;

$Bo = u \cdot L/D_e$ — Bodenstein number, dimensionless;

c — substrate concentration, ML^{-3};

c_a — electron acceptor concentration, ML^{-3};

c_b — substrate concentration in the bulk of liquid, ML^{-3};

$c_{b,min}$ — minimum substrate concentration in the bulk of liquid supporting the existence of biofilm, ML^{-3};

c_d — electron donor concentration, ML^{-3};

C_D — drag coefficient, dimensionless;

c_i — substrate concentration in the i-th tank, ML^{-3};

$c_{in}(x)$ — initial substrate concentration in the reactor, ML^{-3};

c_{pi} — concentration of solids species i, $L^3 L^{-3}$;

c_s — substrate concentration at biofilm surface, ML^{-3};

$\tilde{c}_s = c_s/K_s$ — concentration parameter, dimensionless;

c_o — substrate influent concentration, ML^{-3};

d_c — reactor column diameter, L;

D_e — axial liquid dispersion coefficient, $L^2 T^{-1}$;

D_{e1}, D_{e2} — liquid dispersion coefficients, fore and after sections, respectively, $L^2 T^{-1}$;

D_{eff} — effective substrate diffusivity in biofilm, $L^2 T^{-1}$;

$D_{eff,a}$, $D_{eff,d}$ – effective diffusivity in biofilm of electron acceptor and donor, respectively, L^2T^{-1}.

D_m – molecular diffusivity, L^2T^{-1};

d_p – particle diameter, L;

d_s – biofilm support (carrier) particle diameter, L;

D_s – solids dispersion coefficient, L^2T^{-1};

D_{si} – solids dispersion coefficient of species i, L^2T^{-1};

g – gravity acceleration, LT^{-2};

$Ga = d_p^3 \cdot (\rho_p - \rho_l) \cdot g / (\nu^2 \cdot \rho_l)$ – Galileo number, dimensionless;

j – substrate flux into biofilm, ML^2T^{-1};

$\tilde{J}$ – substrate flux (in Eq. 38), dimensionless;

J_D – external mass transfer modulus (in Eq. 52), dimensionless;

k_b – biofilm growth rate (in Eq. 64), T^{-1};

k_d – biofilm detachment rate, T^{-1};

k_f – external mass transfer coefficient, LT^{-1};

k_m – maximum substrate utilization rate (Michaelis-Menten kinetics), T^{-1};

$k_{m,a}$, $k_{m,d}$ – maximum utilization rate for electron acceptor and donor, respectively, T^{-1};

k_n – n-th order substrate utilization rate coefficient, $(ML^{-3})^{1-n} T^{-1}$;

K_s – half-saturation (Michaelis-Menten) constant, ML^{-3};

$K_{s,a}$ $K_{s,d}$ – half-saturation constants for electron acceptor and donor, respectively, ML^{-3};

k_o – zero-order rate coefficient, $ML^{-3}T^{-1}$;

L – bed length, L;

L_r – reactor length, L;

m_a – mass of active biofilm layer, M;

m_b – total mass of biofilm, M;

MW – molecular weight;

n – order of reaction, dimensionless;

  – exponent in Eq. 12, dimensionless;

  – number of tanks in tanks-in-series model, dimensionless;

$Pe = u \cdot d_p / D_e$ – Peclet number, dimensionless;

Q – volumetric flow rate, L^3T^{-1};

$Q_{B,i}$ – backward flow rate from i-th tank, L^3T^{-1};

$Q_{F,i}$ – forward flow rate from i-th tank, L^3T^{-1};

$q_i = Q_{B,i}/Q$ – tank-in-series model parameter, dimensionless;

$Re = u \cdot d_p / \nu$ – Reynolds number, dimensionless;

$Re_{mf} = u_{mf} \cdot d_p / \nu$ – Reynolds number corresponding to minimum fluidization velocity, dimensionless;

$Re_t = u_t \cdot d_p / \nu$ – Reynolds number corresponding to terminal free fall velocity, dimensionless;

r_v – intrinsic substrate utilization (uptake) rate in biofilm, $ML^{-3}T^{-1}$;

r_{vb} – overall substrate utilization rate per unit volume of biofilm, $ML^{-3}T^{-1}$;

r_{vr} – substrate utilization rate per unit volume of the bed, $ML^{-3}T^{-1}$;

s – geometry factor, s = 0, 1, 2 for flat, cylindrical, spherical coordinates, respectively, dimensionless;

$Sh = k_f \cdot \delta / D_{eff}$ – Sherwood number, dimensionless;

t – time, T;

$T(s)$ – reactor transfer function, Laplace transform of impulse response of reactor as function of complex variable s;

u – liquid superficial velocity, LT^{-1};

u_{mf} – minimum fluidization velocity, LT^{-1};

u_{pi} – liquid velocity required to fluidize species i at porosity ε, LT^{-1};

u_t – terminal free fall velocity, LT^{-1};

u_{to} – terminal free fall velocity of a clean particle, LT^{-1};

V – total bed volume, L^3;

V_b – biofilm volume, L^3;

v_i – volume of i-th tank, L^3;

x – coordinate (distance) along reactor axis, L;

X – microbial concentration in biofilm, biofilm dry density, ML^{-3};

y – ratio of biofilm volume to support carrier volume, dimensionless;

Y – biomass yield coefficient, MM^{-1};

z – position (coordinate) inside biofilm, L;

z_p – position (coordinate) of biofilm/carrier interface, L;

α_i – volume fraction of solids species i, dimensionless;

Γ – parameter in (Eq. 3), 1 for packed beds, Re_{mf}/Re for fluidized beds, dimensionless;

δ – biofilm thickness, L;

δ_e – effective biofilm thickness, ratio of biofilm volume to biofilm surface area, L;

δ_L – liquid film thickness, L;

δ_{L1}, δ_{L2} – sublayer thickness in liquid film, L;

δ_o – inactive sublayer thickness of biofilm, L;

Δt – time delay in tank-in-series model, T;

ε – bed porosity, liquid hold-up, dimensionless;

ε_i – bed porosity at separate fluidization of species i, dimensionless;

ε_{mf} – minimum fluidization porosity, dimensionless;

ε_o – initial porosity of bed of clean particles, dimensionless;

η – effectiveness factor, dimensionless;

η_b – effectiveness factor based on bulk substrate concentration dimensionless;

ν – liquid kinematic viscosity, L^2T^{-1};

$\bar{\rho}$ – bed bulk density, ML^{-3};

ρ_b – wet biofilm density, ML^{-3};

$\bar{\rho}_i$ – bulk density of bed containing species i, ML^{-3};

ρ_l – liquid density, ML^{-3};

ρ_p – particle density (average), ML^{-3};

ρ_{pi} – particle density of species i, ML^{-3};

ρ_s – biofilm support (carrier) density, ML^{-3};

ϕ – particle shape factor (reciprocal of sphericity), dimensionless;

Φ_m – Thiele modulus for Michaelis-Menten kinetics (in Eq. 36), dimensionless;

Φ_n – Thiele modulus for n-th order kinetics (in Eq. 30), dimensionless;

$\Phi_{n,b}$ – Thiele modulus for n-th order kinetics based on bulk substrate concentration, dimensionless;

Φ_p – modified Thiele modulus for Michaelis-Menten kinetics (in Eq. 36), dimensionless;

ψ – particle sphericity, dimensionless.

MATHEMATICAL MODELS FOR BIOLOGICAL AEROBIC FLUIDIZED BED REACTORS

S. ELMALEH and A. GRASMICK
Laboratoire de Génie Chimique
Université des Sciences et Techniques du Languedoc
34060 Montpellier Cedex - France

1. Introduction
2. Oxygen transfer in a fluidized bed
3. Modelling of aerobic biological fluidized reactors
4. Experimental validation and use of the model
5. Conclusions
6. Notations
7. References.

1. INTRODUCTION

Aerobic biological fluidized reactors are used for organic carbon abatement and nitrification of ammonia containing water. In spite of a high interest in anaerobic fluidized reactors which, at laboratory scale, are able to purify influent containing low levels of organic carbon (Jewell, 1981), it is likely that aerobic fluidized reactors will continue to be developed for a large range of inlet concentrations.

In the first congress entirely devoted to biological fluidized bed treatment of water and wastewater held in Manchester, it was said that "extensive use must be made of modelling for the evaluation of process concepts in a predictive sense, prior to the construction of laboratory and pilot-scale devices" (Atkinson, 1981). Considering the data obtained from experimental studies about biological fluidized bed technology, parameters to be dealt with include volumetric reaction rates, purification efficiency and retention times linked only by empirical correlations.

On the other hand, it is known that biological processes are not simple to handle mathematically ; as a consequence, most models introduce a large number of parameters, too many to be of any use to the process engineer. It is not uncommon to find from literature a model where five or six parameters characterize the analysed system ; moreover, in some cases, all the parameters are not fully independent.

In many engineering problems, a precise model is not essential but rather a calculation method of sufficient accuracy where the influence of each process variable is easily derived.

First, this chapter will consider the requirements resulting from oxygen transfer followed by the development of the model of pollution conversion by a fluidized biofilm. After having given some experimental evidence of the validity of the model, a design method will be proposed and an example case studied.

2.   OXYGEN TRANSFER IN A FLUIDIZED BED

Oxygen can be supplied to a biological fluidized system by one of the following (Elmaleh, 1982) :

1. gas injection in the reactor, i.e. the reactor is a three-phase fluidized bed (Chatib et al., 1981) ;
2. external air or oxygen supply by an oxygenation device situated on a recycle line (Sutton et al., 1981 ; Sehic, 1981) ; and
3. injection of an oxygen liberating chemical, e.g. hydrogen peroxide (Yahi et al., 1982).

The effectiveness of oxygen transfer greatly influences the economics of the whole process.

2.1.   Oxygen transfer characteristics of three-phase fluidized beds

There has been growing interest in the use of three-phase fluidized beds for simultaneously contacting a gas, a liquid and a solid in many fields of chemical engineering (Østergaard, 1971 ; Shah, 1979). In all applications an important aspect of contacting is the dispersion of the gas within the reactor.

2.1.1. General character of gas dispersion

2.1.1.1. Low gas rates

At relatively low gas rates, a three-phase fluidized bed can be most readily envisaged as a bed of solids which initially is fluidized by liquid alone and then has the gas introduced. The gas dispersion in a bed of small particles can be quite different for a bed composed of large particles, e.g. 4-6 mm diameter.

The characteristic differences in behaviour between large and small particles are clearly seen at low gas rates. Large particles, with high inertia, can deform and split any gas bubble entering the bed (Fig. 1a). The result is a dispersion of relatively small gas bubbles of diameter 2 mm or less. Small particles, e.g. 2mm, however, are apparently unable to penetrate large bubbles in opposition to surface tension forces. Most of the gas passes through the bed as large, spherical, cap-type bubbles which tend to coalesce as they rise through the bed (Fig. 1b).

For biological fluidized bed reactors, because of importance of aeration, beds of relatively large particles giving good gas dispersions of small diameter bubbles are likely to be of greater interest. Moreover, in biological systems, the surface tension is likely to be considerably less than for pure water which tends to enhance gas dispersion by favouring the break-up of bubbles.

2.1.1.2. High gas rates

Three-phase fluidized beds with relatively high gas rates, e.g. gas superficial velocity of 10 cm/s or more, can best be thought of as starting with a steady gas flow through a fixed bed of particles and then gradually increasing the liquid flow rate (Lee and Al-Dabbagh, 1978 ; Begovich and Watson, 1978). As it has been observed for gas-liquid flow through fixed beds (Shah, 1979), the flow within the bed is not steady and over a wide range of flow rates, pulsations occur in pressure drop and in the local gas-liquid ratio.

Even with large particles, at high gas rates, large bubbles or slugs are produced and the whole bed passes into a stage described as churn-turbulent. Packets of particles are thrown about violently and the noise of particles clashing together increases. It is thought that in some regions of the bed, the particles become defluidized and can form a more rigid kind of structure bridging the roof of large bubbles (Fig. 1c). The violent movements associated with the churn-turbulent regime could well be a disadvantage when it is necessary to retain a microbial film on the external surface of the particles.

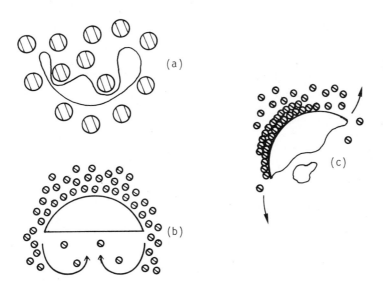

Fig. 1. Bubble-particle interaction in a three-phase fluidized bed (Lee and Buckley, 1981). (a) Large particles - low gas rates. (b) Small particles - low gas rates. (c) High gas rates.

2.1.2. Gas-liquid mass transfer

The advantage of using relatively large diameter particles in a three-phase fluidized bed is that the small gas bubbles produced give a large interfacial area for mass transfer. Measurements for 6 mm diameter glass beads and distilled water of the overall mass transfer coefficient, K_La, are given in Fig. 2. They show a strong dependence on gas velocity before reaching a plateau. On the other hand, the superficial liquid velocity has no significant effect on the steady value. It is interesting to compare mass transfer coefficient values in beds of small and large particles (Table 1).

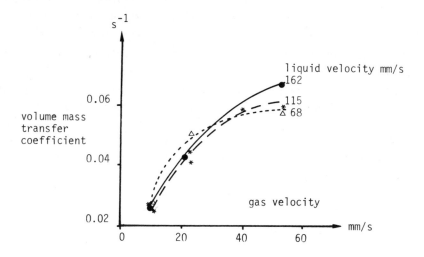

Fig. 2. Overall oxygen transfer coefficient (Ostergaard and Fosbol, 1972).

TABLE 1.

Oxygen transfer coefficient in a three-phase fluidized bed .

Particles	Liquid velocity u_L (m/s)	Gas velocity u_G (m/s)	$\dfrac{u_L}{u_G}$	Oxygen transfer coefficient K_La (s^{-1})	References
6 mm glass beads	0.162	0.04	4	0.06	Ostergaard and Fosbol (1972)
4-6 mm gravel particles	0.088	0.04	2.2	0.09	Elmaleh (1981)
0.6-1 mm crushed brick granules	0.0083	0.0088	0.94	0.007	Chatib et al. (1981)

2.1.3. Conclusions

Fluidized beds of relatively large particles such as 6 mm and 4 mm diameter glass beads have good aeration characteristics especially when the surface tension of the water is reduced by surface active agents. Nevertheless, the fluidization of such particles requires an important power dissipation and, in most cases, the whole process will not be economically viable. The only exception occurs when the support medium density is low enough to compensate the large size effect ; the density can even be lower than the water density. Atkinson et al. (1981) have proposed such particles which are large enough to provide a high gas dispersion and light enough to be of low energy requirement.

However, fluidized beds of small particles of classic materials, e.g. sand, do not present good oxygen transfer characteristics and the process is drastically limited by aeration (Grasmick et al., 1981).

2.2.   Oxygen supply to a two-phase biological fluidized reactor

Oxygen is transferred to the reactor by an external device situated on the recycle line (Fig. 3). Liquid oxygen can be provided through an aeration cone (Jeris et al., 1981) or dissolved in the feed of the reactor in a downflow bubble contactor (Cooper and Wheeldon, 1981) or by countercurrent contact with saturated air through sieve plates in a tower (Sehic, 1981). It is also possible to feed the reactor with an oxygen liberating chemical, e.g. hydrogen peroxide (Grigoropoulou, 1980 ; Yahi et al., 1982).

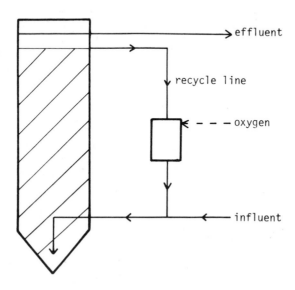

Fig. 3. Flowsheet of an aerobic biological fluidized reactor.

Note that the maximum attainable dissolved oxygen concentration using air is about 10 mg/l at 20°C. With pure oxygen, 25 to 40 mg/l can be reached. This means that, considering a gross stoichiometry for BOD elimination of about 1 kg of oxygen needed for 1 kg of BOD, high values of recycle ratio are required, i.e. from 10 to 100. The energy demand for oxygen transfer can then be conside- rable but lower than in a three-phase fluidized bed of sand particles.

For technical and economic reasons, it is thought that the aerobic fluidized reactor will develop principally as a two-phase reactor where oxygen supply will be regulated according to the inlet pollution mass flowrate. The following model will essentially treat this case.

3. MODELLING OF AEROBIC BIOLOGICAL FLUIDIZED REACTORS
3.1. Fundamentals
3.1.1. What is a biofilm ?

A biofilm results from the fixation of micro-organisms to a solid material (Zobell, 1939). Microscopic observation shows an agglomerate of various micro-organisms included in a gelatinous matrix (Fig. 4). This heterogeneous population should hold different transport and metabolic properties (Aleman and Veil, 1981). Another factor of heterogeneity is the oxygen gradient which, depending on oxygen concentration, induces a development of aerobic, facultati- ve or anaerobic organisms. The organisms can hold different physical properties, e.g. filamentous and flocculating bacteria (Roques, 1980 ; Grady, 1982).

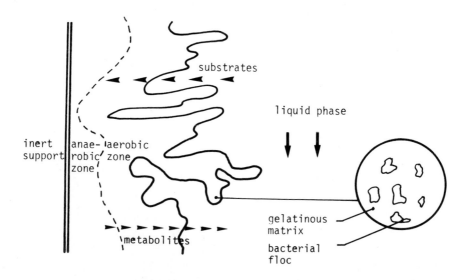

Fig. 4. Schematic cross-section of a biofilm.

Some special devices have been tested to develop a relatively controlled biofilm in order to measure characteristic parameters of transport and reaction:

1. The plate reactor (Maier et al., 1967 ; Atkinson et al., 1967) is an inclined plate on which a submerged biofilm is developed. The liquid flow is well approximated by plug-flow and the parameter values are distributed along the plate (Fig. 5).

2. The rotating cylinder (Tomlinson and Snaddon, 1966 ; Hoehn and Ray, 1973) is an inclined tube rotating around its axis (Fig. 5). The tube is partially filled with liquid and so the biofilm is alternately submerged or aerated.

3. The rotating reactor is composed of two coaxial cylinders (Kornegay and Andrews, 1969 ; La Motta, 1974 ; Jansen and Kristensen, 1980 ; Grasmick, 1982). The rotation of the external cylinder provides a perfect mixing of the liquid phase. The biofilm develops on the internal surface of both cylinders (Fig. 5).

3.1.1.1. Biofilm thickness

Considering the complexity of the biofilm structure and the heterogeneity of its surface, it is particularly difficult to define and measure a biofilm thickness. The experimental measurement can be made directly or can involve such sophisticated procedure as congelation of the whole reacting medium. Accuracy on values of thickness less than 100 µm is 20 % but on values of about 2 mm, discrepancy can reach 300 % (Grasmick, 1982).

Depending on hydrodynamic conditions, Atkinson and Fowler (1974) found values between 0.07 and 4 mm. When the film is mechanically or hydraulically controlled, its thickness does not exceed 0.2 mm (Grasmick, 1982).

3.1.1.2. Biofilm density

It has been verified that the biofilm density depends on its thickness (Hoehn and Ray 1973). The different steps of the development of an aerobic biofilm can be described as follows (Fig. 6) :

a   - initially the biofilm is composed of only some aerobic bacteria included in a gelatinous matrix, i.e. the density is low,

b-c - aerobic micro-organisms grow rapidly and the density is a growing function of the thickness,

d   - as oxygen depletion occurs in the biofilm, an anaerobic zone appears near the support,

e   - anaerobic and facultative bacteria grow near the support as aerobes decay causing decreasing density,

f   - an equilibrium between aerobes and anaerobes is reached : density is stabilized, and

g   - the equilibrium is maintained until substrates are exhausted in the deeper zone ; then, anaerobes themselves begin to decay ; finally, a part of the biofilm flows away.

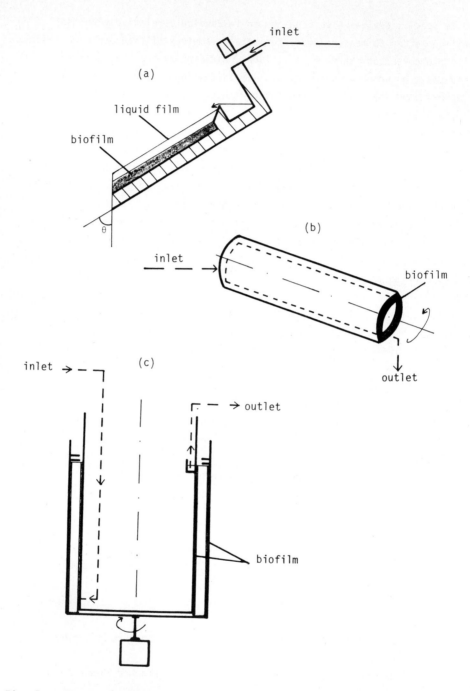

Fig. 5. Laboratory-scale reactors. (a) Plate reactor. (b) Rotating cylinder. (c) Rotating reactor.

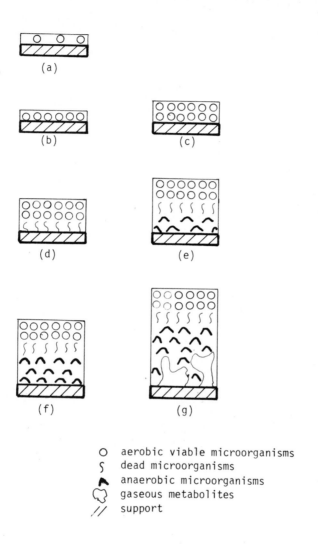

(a)

(b)   (c)

(d)   (e)

(f)   (g)

O   aerobic viable microorganisms
S   dead microorganisms
▲   anaerobic microorganisms
ⓢ   gaseous metabolites
//   support

Fig. 6. Transient biofilm.

3.1.1.3. Transport and reaction within a biofilm

In spite of the biofilm heterogeneity, most of the models assume that substrates are transported by molecular diffusion and, therefore, that an effective diffusivity is a characteristic constant of the system (Atkinson and Fowler, 1974 ; Harremoes et al., 1975 ; Harremoes, 1975, 1976, 1978 ; La Motta, 1976 ; Williamson and MacCarty, 1976 ; Grasmick et al., 1979, 1981 ; Rittman and Mac Carty,1981). In many works, values of effective diffusivity are assumed to be the same as that in pure water. They have been measured in some cases on special reactors such as the rotating reactor or calculated after adjustment of a model (Table 2).

TABLE 2.

Effective diffusivity in pure water and in biofilms

Pollution	Substrate	Conditions	Temperature ($°C$)	Effective diffusivity 10^{-10} (m2/s)	Reference
pure water	O_2		20	15	
milk	O_2	fixed bed	20	3 - 9	Grasmick et al. (1980)
beef extract	O_2	fixed bed	20	8 - 10	Grasmick et al. (1980)
wastewater	O_2	rotating cylinder	20	12 - 17	Tomlinson and Snaddon (1966)
glucose	O_2		20	4 - 20	Matson and Charaklis (1976)
nitrogen compounds	O_2	pure culture *Nitrobacter* *Nitrobacter* + *Nitrosomonas*		25	Williamson and Mac Carty (1976)
methanol	methanol	rotating reactor	22	20 - 50	Jansen and Kristensen (1980)
glucose	glucose	rotating reactor	22	1 - 5	La Motta (1976)
milk	TOC	rotating reactor	25	6 - 50	Grasmick (1982)

Grasmick (1982) has extensively applied the concept of effective diffusivity on biofilms developed on a rotating reactor. The results show that the concept of a constant effective diffusivity is not always consistent and that some convection through the biofilm must interfere ; this phenomenon can compare to some extent with convection through porous catalysts (Rodrigues, 1982).

The reaction rate in a biofilm is based on the concept of limiting substrate. This section will consider only the overall reaction of aerobic wastewater treatment which can be written, depending on the process, in the form :

$$\nu_{TOC} TOC + \nu_{O_2} O_2 \longrightarrow \text{metabolites}$$

(organic carbon conversion)

$$\nu_N NH_4^+ + \nu_{O2} O_2 \longrightarrow \text{metabolites}$$

(nitrification)

Therefore, the limiting substrate will be oxygen, organic carbon or ammonia.

The intrinsic reaction rate of a limiting substrate follows, depending on the authors, a Monod-type equation, a first order equation or a zero order equation. In wastewater treatment, it has been.experimentally shown that the best approximation is zero order (La Motta, 1976 ; Riemer and Harremoës, 1978 ; Grasmick, 1982). Depending on the penetration into the biofilm, the apparent reaction rate will, therefore, follow zero order kinetics for full penetration and half-order kinetics for partial penetration. Some values of the reaction rate coefficient for the intrinsic kinetics or the apparent kinetics are reported in Table 3.

Grasmick (1982) has observed that the intrinsic zero order rate is not constant for all biofilm thicknesses : k_o values measured in a rotating reactor decrease when biofilm thickness increases. This comes from the difficulty of defining a biofilm thickness and an exchange area ; the model is probably better for small biofilms, i.e. thickness less than 300 μm.

3.1.1.4. Liquid-biofilm mass transfer

Before any reaction inside the biofilm, the substrates are to be transferred from the bulk liquid to the solid phase. The existence of a liquid-biofilm mass transfer resistance has been shown by experimental runs using the rotating reactor. When all other variables are maintained steady except the rotating speed which is slowly increased, the conversion reaches a plateau starting at a linear velocity of about 1 m/s (La Motta, 1976b ; Grasmick, 1982). Chen and Bungay

TABLE 3.

Intrinsic zero order and apparent half-order kinetic coefficients.

Pollution	Limiting substrate	Conditions	Temperature (°C)	Intrinsic zero order rate per unit biofilm volume $(kg/m3.s \ 10^{-3})$	Apparent half-order rate coefficient per unit biofilm volume $(kg^{1/2}_m{}^{-1/2}_s{}^{-1} \ 10^{-5})$	Apparent half-order rate coefficient per unit reactor volume $(kg^{1/2}_m{}^{-3/2}_s{}^{-1} \ 10^{-3})$	Reference
milk	O_2	fixed bed	20		0.12		Grasmick et al. (1980)
beef extract	O_2	fixed bed	20		0.32		Grasmick et al. (1980)
methanol	methanol	rotating reactor	22		0.16 - 0.19	0.38 - 1.58	Jansen and Kristensen (1980)
milk	TOC	rotating reactor	20	0.27 - 0.59	0.083- 0.18		Grasmick (1982)

(1981) clearly showed that oxygen concentration in the interface is significantly less than in the bulk flowing phase (Fig. 7). Moreover, Grasmick (1982) showed that the liquid-solid mass transfer coefficient is 10^3 higher when rotating speed passes from 60 rpm to 180 rpm ; such an increase for a relatively low increment in the energy input cannot only be due to a decrease of the boundary layer. It might be attributed to an increase of the mass transfer area and, probably, to a modification of the exchange surface itself due to a drastic variation of the physical properties of the biofilm. However, in some reactors, the error committed when this resistance is neglected does not exceed 7 % (Shieh, 1980). Howell and Atkinson (1976) think that the phenomenon is negligible in most cases.

In the following model development a liquid-biofilm mass transfer resistance will be introduced and then the conditions in which its influence can be neglected will be determined.

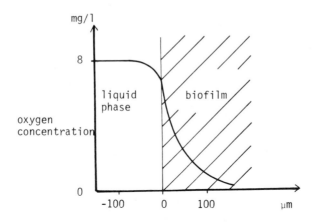

Fig. 7. Oxygen concentration profile.

3.1.2. Reactor structure

Overall kinetics in the case of chemical reaction mass transport interaction has been extensively studied in the field of heterogeneous catalysis. In biochemical engineering, it is usual to couple diffusion limitation to zero order, first order or Monod-type kinetics. It has been shown that the effectiveness factor is relatively insensitive to both reaction order and particle geometry (Weisz, 1973).

A biological fluidized reactor is often viewed as spherical particles coated with a biofilm suspended by upflowing water. Generally a segregation occurs, the apparent particles density decreasing as the biofilm thickness increases. In this work, it is considered that a fluidized reactor is equivalent for its conversion to a plate support coated with a biofilm of uniform thickness. The relationship between this plate reactor and a fluidized bed of spherical particles will be established later.

3.1.3. Basic assumptions

The model is based on the following assumptions :

1. the biofilm is homogeneous ;
2. all the substrates are diffusion-transported within the biofilm ;
3. the intrinsic reaction rate of the limiting substrate follows zero order kinetics ;
4. the liquid-solid mass transfer resistance is well described by a phenomenological equation ;
5. the reactor is considered as an inert plate coated with a biofilm ; and
6. steady-state operation.

3.2. Diffusion, reaction and liquid-solid mass transfer resistance

3.2.1. Transport through the biofilm

The mass balance for the limiting substrate in a volume element of the biofilm (Fig. 8) is :

$$D \frac{d^2C}{dy^2} - k_o H (y - y^*) = 0 \tag{1}$$

where, D = effective diffusivity of the substrate in the biofilm

k_o = intrinsic zero order reaction rate per unit volume of biofilm

C = concentration within the biofilm at the position y

y^* = position where both the concentration and its space derivative are zero, and

H = Heaviside function.

By the introduction of dimensionless variables normalized by the biofilm thickness and the bulk concentration : $x = \frac{y}{\delta}$ and $f = \frac{C}{C_B}$, then :

$$\frac{d^2f}{dx^2} = \Phi^2 \tag{2}$$

where Φ = Thiele modulus referred to the bulk concentration :

$$\Phi = \delta \left(\frac{k_o}{DC_B} \right)^{1/2} \tag{3}$$

As there is experimental evidence that most biofilm reactors work in a diffusional regime with partial penetration of substrate, the boundary conditions are :

$$x = x^*, \ f = 0 \ \text{and} \ \left(\frac{df}{dx} \right)_{x=x^*} = 0$$

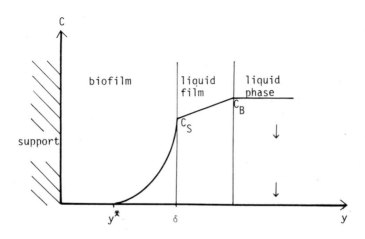

Fig. 8. Limiting substrate concentration profile with liquid film resistance.

3.2.2. Liquid-solid mass transfer

The mass flux through the interface is given by :

$$J = - K_f (C_B - C_S) \tag{4}$$

where, K_f = mass transfer coefficient through a transition zone, i.e. a liquid film, and

C_S = concentration at the interface.

At $y = \delta$, the equality of fluxes is :

$$K_f (C_B - C_S) = D (\frac{dC}{dy})_{y=\delta} \tag{5}$$

which is re-written in dimensionless form :

$$B_i (1 - f_{x=1}) = (\frac{df}{dx})_{x=1} \tag{6}$$

where, $B_i = \dfrac{K_f \delta}{D}$ = mass Biot number which compares the liquid film conductance, K_f, to the biofilm conductance, $\dfrac{D}{\delta}$.

3.2.3. Effectiveness factor

The effectiveness factor, η , is defined by :

$$r_a = \eta k_o \tag{7}$$

where, r_a = apparent reaction rate per unit volume of biofilm.

For zero order kinetics, the effectiveness factor is simply the ratio of biofilm volume with substrate concentration greater than zero to the total biofilm volume.

The integration of Eq. 2 gives :

$$\frac{df}{dx} = \phi^2 (x - x^*) \tag{8}$$

At the interface   liquid-biofilm :

$$(\frac{df}{dx})_{x=1} = \phi^2 (1 - x^*) = \phi^2 \eta \tag{9}$$

Combining with the preceding relationships, the dimensionless concentration at the interface is :

$$f_{x=1} = \frac{1}{2} (\eta \phi)^{1/2} \tag{10}$$

And then the effectiveness factor is calculated by :

$$\eta = (\frac{2}{\phi^2} + \frac{1}{B_i^2})^{1/2} - \frac{1}{B_i} \tag{11}$$

The effectiveness factor is plotted as a function of the Thiele modulus with the Biot number as a parameter in Fig. 9. Two regimes are observed :

a) $\eta = 1$: rate is controlled by reaction only ; and

b) $\eta < 1$: rate is controlled by mass transport phenomena and reaction.

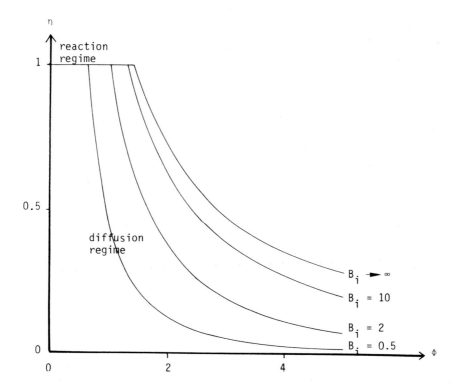

Fig. 9. Effect of Thiele modulus on effectiveness factor.

The plot of η vs B_i with the Thiele modulus as a parameter shows that the liquid-biofilm resistance is negligible when B_i value is greater than 10 (Fig. 10). This case will now be examined.

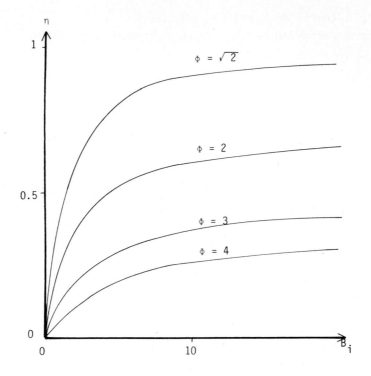

Fig. 10. Effect of Biot number on effectiveness factor.

3.3. Diffusion and reaction without any liquid-solid resistance

3.3.1. Transport through the biofilm

In such a system (Fig. 11), the mass balance is the same as in the preceeding case, i.e. : $\dfrac{d^2f}{dx^2} = \phi^2$

The boundary conditions are changed in :

$$x = x^* \; : \; f^* = 0 \text{ and } \left(\frac{df}{dx} \right)_{x=x^*} = 0 \tag{12}$$

$$x = \delta \; : \; f = 1 \tag{13}$$

The integration of this new differential system gives :

$$\frac{df}{dx} = \phi^2 (x - x^*) \tag{14}$$

and

$$f = \frac{\Phi^2}{2} \left(x - x^* \right)^2 \tag{15}$$

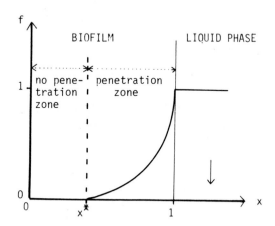

Fig. 11. Limiting substrate concentration profile without any liquid film resistance.

The concentration profile through the biofilm is plotted in Fig. 11.

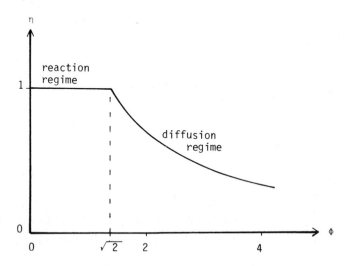

Fig. 12. Effect of Thiele modulus on effectiveness factor without any liquid film resistance.

3.3.2. Effectiveness factor

The effectiveness factor is easily derived from Eq. 11 where B_i is made infinite :

$$\eta = \frac{\sqrt{2}}{\Phi} \tag{16}$$

The plot of η vs Φ illustrates again two regimes (Fig. 12).
a) $\eta = 1$: rate is only controlled by reaction ; and
b) $\eta < 1$: rate is controlled by reaction and diffusion.

3.4. Biomass hold-up - Plate reactor and fluidized reactor

In a plate reactor coated with a biofilm, biomass hold-up, ε_b, is given to a good degree of approximation by $a_b\delta$ where a_b is the specific interfacial area of the biofilm. The volume of the liquid phase being the same as in the fluidized reactor :

$$a_b = \frac{1 - \varepsilon}{d_p + \delta}$$

where, ε = bed porosity, and
d_p = characteristic thickness of the plate support
From this :

$$\varepsilon_b = \frac{\dfrac{\delta}{d_p}}{1 + \dfrac{\delta}{d_p}} \, (1 - \varepsilon) \tag{17}$$

The characteristic length, d_p, is calculated by writing that the clean plate reactor and the clean fluidized bed of spherical particles have an identical supporting area which leads to :

$$d_p = \frac{d}{6} \tag{18}$$

where, d = mean clean particle diameter.

Another relationship between the plate reactor and the fluidized reactor is derived by writing that the volumetric rate of reaction is the same in both of the contactors and is expressed by the following equation :

$$(\eta\varepsilon_b)_{\text{plate reactor}} = (\eta\varepsilon_{bF})_{\text{fluidized reactor}}.$$

In the case where liquid film resistance is negligible, also considering that the effectiveness factor is relatively insensitive to biofilm geometry (Weisz, 1973), combination with Eq. 16 gives an approximate relationship :

$$\frac{\varepsilon_b}{\delta} = \frac{\varepsilon_{b_F}}{\sigma} \tag{19}$$

where, ε_{b_F} = biomass hold-up in the fluidized reactor or spherical particles coated with a biofilm of uniform thickness, σ. Shieh et al. (1981) have found that ε_{b_F} is given by :

$$\varepsilon_{b_F} = (1 - \varepsilon)\left[1 - \frac{1}{(1 + \frac{2\sigma}{d})^3}\right] \tag{20}$$

By combining Eqs. 19 and 20, a relationship between δ and σ is derived, thus :

$$\frac{\delta}{\sigma} = \frac{1}{1 - \frac{1}{(1+2\frac{\sigma}{d})^3}} - \frac{1}{6\frac{\sigma}{d}} \tag{21}$$

The latter equation is represented graphically in Fig. 13 which shows that δ and σ are of the same order of magnitude. Although this relation has been established between systems without liquid film transfer resistance, its validity is assumed for any case.

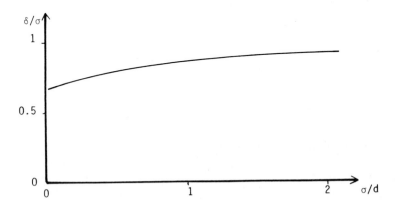

Fig. 13. δ/σ vs σ/d

3.5. Conversion in ideal reactors

Information thus obtained can be used to calculate the conversion of substrate or reactor efficiency. Consider two extreme flow situations : continuous stirred tank reactor (CSTR) and plug-flow reactor.

3.5.1. Conversion in CSTR and plug-flow reactors

The mass balances over ideal reactors are :

$$\text{for a CSTR} \qquad : C_i = C_e + k_o \, \varepsilon_b \, \eta \, \tau \tag{22}$$

$$\text{for a plug-flow reactor} : dC_B + k_o \, \varepsilon_b \, \eta \, d\tau = 0 \tag{23}$$

where, C_i and C_e = inlet and outlet concentrations respectively and
τ = space time calculated for an empty reactor.

3.5.1.1. With liquid-solid resistance

Taking into account Eqs. 3 and 11 and noting that in a CSTR $C_e = C_B$, some algebraic manipulations of the mass balance (Eq. 22) produces :

$$E_A = \left[\alpha^2 \left(1 + \frac{1}{N_f}\right)^2 + 2\alpha \right]^{1/2} - \alpha \left(1 + \frac{1}{N_f}\right) \tag{24}$$

The conversion, E_A, is expressed as a function of two parameters :
. the parameter α is given by :

$$\alpha = \frac{\varepsilon_b^{\,2} \, \tau^2 \, k_o D}{C_i \, \delta^2} \tag{25}$$

α itself is the product of two dimensionless numbers :

$$\alpha = N_r \, N_b \tag{26}$$

where, N_r = the number of reaction units or Damkholer number :

$$N_r = \frac{k_o \, \varepsilon_b \, \tau}{C_i} \tag{27}$$

and N_b = the number of diffusion units for the biofilm :

$$N_b = \frac{D \, \varepsilon_b \, \tau}{\delta^2} \tag{28}$$

. The parameter N_f is the number of transfer units through the liquid film defined by :

$$N_f = \frac{K_f \, \varepsilon_b \, \tau}{\delta} \tag{29}$$

In the same manner, integration of the mass balance (Eq. 23) gives the following equation :

$$\left[\frac{2}{\alpha}\frac{C_e}{C_i}+\frac{1}{N_f^2}\right]^{1/2}-\left[\frac{2}{\alpha}+\frac{1}{N_f^2}\right]^{1/2}+\frac{1}{N_f}Log\frac{\left[\frac{2}{\alpha}\frac{C_e}{C_i}+\frac{1}{N_f^2}\right]^{1/2}-\frac{1}{N_f}}{\left[\frac{2}{\alpha}+\frac{1}{N_f^2}\right]^{1/2}-\frac{1}{N_f}}=-1 \qquad (30)$$

Eq. 30 is an implicit relationship between conversion and the two parameters α and N_f.

N.B. : Significance of N_r, N_b and N_f

1. In the case of full penetration of the biofilm and no liquid film resistance, the only limiting process is the zero order reaction. Conversion is the same in a CSTR and in a plug-flow reactor :

$$E_A = E_p = N_r \quad \text{when } \tau < \frac{C_i}{k_o\,\varepsilon_b}$$

and

$$E_A = E_p = 1 \quad \text{when } \tau > \frac{C_i}{k_o\,\varepsilon_b}.$$

2. The number of diffusion units compares the overall mass transfer coefficient to the volumetric convection flowrate in the bulk phase. For instance, N_b is defined by :

$$N_b = \frac{\frac{D}{\delta}\,A_b}{Q} \quad \text{where, } A_b = \text{interfacial area liquid-biofilm}$$

$$A_b = \frac{\varepsilon_b V}{\delta} \;.\; \text{Then, } N_b = \frac{D\,\varepsilon_b\,\tau}{\delta^2}$$

In the same way, the number of transfer units through the liquid film is :

$$N_f = \frac{K_f\,A_b}{Q}\;.$$

A plot of conversion in a CSTR and in a plug-flow reactor as a function of α, N_f being a parameter, is given in Fig. 14. A plug-flow reactor is always more efficient than a CSTR particularly for large values of N_f. The case of infinite N_f corresponds to a lack of liquid film resistance. However isoconversion curves in the (α, N_f) plane are more significant (Fig. 15). They show three regions :

548

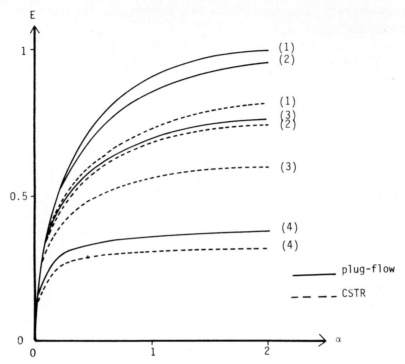

Fig. 14. Conversion in ideal reactors (with liquid film resistance)
(1) $N_f \to \infty$. (2) $N_f = 10$. (3) $N_f = 2$. (4) $N_f = 0.5$.

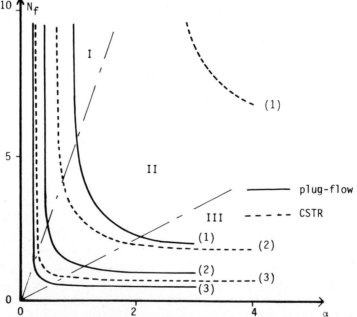

Fig. 15. Iso-conversion curves in ideal reactors (with liquid film resistance)
(1) $E = 0.8$. (2) $E = 0.6$ (3) $E = 0.4$.

1. In region I, the rate is controlled by diffusion and reaction only, i.e. there is no liquid film resistance and conversion is predicted by the limits of Eq. 24 or 30 when N_f is made infinite :

$$E_A = (\alpha^2 + 2 \alpha)^{1/2} - \alpha$$

$$E_p = (2 \alpha)^{1/2} - \frac{\alpha}{2}$$

2. In region II, the rate is controlled by liquid-film resistance diffusion through biofilm and reaction ; conversion is then predicted by Eqs. 24 or 30.
3. In region III, the rate is controlled only by liquid film mass transfer and it can then easily be shown that conversion is predicted by :

$$E_A = \frac{1}{1 + \frac{1}{N_f}} \quad \text{or } E_p = 1 - e^{-N_f}$$

If the biofilm is fully penetrated, which corresponds to $\phi < \left[\dfrac{2}{1 + \dfrac{2}{B_i}} \right]^{1/2}$, the conversion is simply $E_A = E_p = N_r$.

3.5.1.2. Without any liquid film resistance

It has just been demonstrated that when liquid-solid resistance is negligible, conversion is situated in region I of the (α, N_f) plane (Fig. 15). In the case of CSTR, by making N_f infinite in Eq. 24, then :

$$E_A = \left[\alpha (\alpha + 2) \right]^{1/2} - \alpha \tag{31}$$

For a plug-flow reactor, it is more convenient to consider again the mass balance (Eq. 23) associated with Eqs. 3 and 16. By integration :

$$\tau = \frac{\delta}{\varepsilon_b} \left(\frac{2}{k_o D} \right)^{1/2} \left(\sqrt{C_i} - \sqrt{C_e} \right) \tag{32}$$

This last relationship shows that there exists a critical space time for which the outlet concentration will be zero. This critical space time is :

$$\tau_m = \frac{\delta}{\varepsilon_b} \left(\frac{2 \, C_i}{k_o D} \right)^{1/2} \tag{33}$$

When $\tau < \tau_m$, the substrate conversion in a plug-flow reactor is calculated by re-writting Eq. 32 in the following form :

$$E_p = \sqrt{2\alpha} - \frac{\alpha}{2} \tag{34}$$

Eq. 34 shows the existence of a critical value $\alpha_m = 2$ for which $E_p = 1$.

The plug-flow reactor gives a better conversion than CSTR but the relative improvement does not exceed 20 % (Fig. 16), the value is easily calculated from Eqs. 31 and 34. In most cases therefore, as wastewater treatment units are generally calculated with a 20 % accuracy, the influence of the flow pattern can be neglected. Moreover, precise residence time distribution measurements (Riemer et al., 1980) are not, at this stage, required.

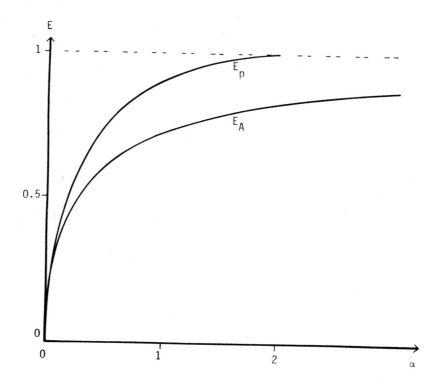

Fig. 16. Conversion in ideal reactors (without liquid film resistance).

To display both reaction and diffusion regimes, it is interesting to plot reactor efficiency against N_r with N_b as a parameter. Since in the reaction regime $E = N_r$ for both types of reactor, no solution exists to the left of that straight line (Fig. 17) ; then the mathematical solution for the diffusional regime is only valid for $N_r \geqslant \dfrac{2\,N_b}{1 + 2\,N_b}$ in the case of a CSTR and for $N_f \geqslant \dfrac{2\,N_b}{(1 + \dfrac{N_b}{2})^2}$ in the case of a plug-flow reactor.

These discontinuities in the E vs N_r curves while passing from reaction to diffusion regime have been already noted by Harremoës (1975).

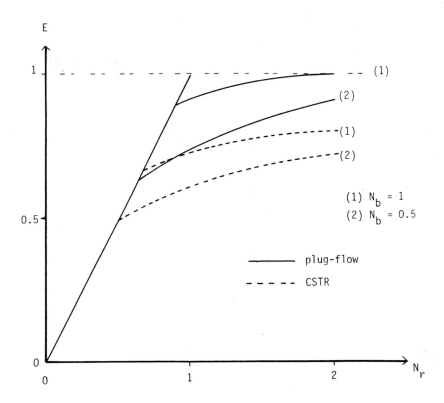

Fig. 17. Effect of the number of reaction units on conversion in ideal reactors.

3.5.2. Effect of space time and inlet concentration on reactor efficiency (no liquid film transfer resistance)

The relations obtained to express conversion in ideal reactors where liquid film transfer is negligible will be now exploited to derive the effect of space time and inlet concentration on the conversion. It will be assumed that biofilm thickness remains steady in the range of variation of τ and C_i. This assumption will be justified later when considering variations with upflow superficial velocity.

3.5.2.1. Effect of space time

The mass balance over a CSTR can be re-written in the following from :

$$C_i = C_e + \varepsilon_b \tau \; \frac{\sqrt{2} \, k_o \, D}{\delta} \; \sqrt{C_e} \qquad (35)$$

From Eq. 35, a simple algebraic manipulation gives :

$$\frac{E_A}{\sqrt{1 - E_A}} = 2 \, \frac{\tau}{\tau_m} \tag{36}$$

Similarly, the following relationship is derived from Eq. 41 for a plug-flow reactor :

$$\sqrt{1 - E_p} = 1 - \frac{\tau}{\tau_m} \tag{37}$$

The last two relationships show that a plot of :

$$\frac{E_A}{\sqrt{1 - E_A}} \quad \text{or} \quad \sqrt{1 - E_p} \quad \text{against } \tau \quad \text{will give a straight line} : \tau_m \text{ is a constant}$$

when space time only is varied. It has the significance of a critical space time in a plug-flow reactor. It should be noticed that $\tau_m/\sqrt{2}$ is the geometric mean between the time constant for the reaction and the time constant for the diffusion :

$$\frac{\tau_m}{\sqrt{2}} = \sqrt{\tau_r \, \tau_b} \tag{38}$$

Indeed, in a batch reactor where the reaction is the sole phenomenon, the time constant is :

$$\tau_r = \frac{C_i}{k_o \, \varepsilon_b} \tag{39}$$

It can be also checked that in a batch reactor, where diffusion through the biofilm induces the only consumption effect, the time constant is :

$$\tau_b = \frac{\delta^2}{D \, \varepsilon_b} \tag{40}$$

3.5.2.2. Effect of inlet concentration

The mass balance over a CSTR can also be written in the following way :

$$\frac{1 - E_A}{E_A^2} = \frac{1}{4} \, \frac{C_i}{C_m} \tag{41}$$

where, C_m = a constant concentration when C_i only is varied :

$$C_m = \frac{\varepsilon_b^2 \, \tau^2 \, k_o D}{2 \, \delta^2} \tag{42}$$

In the same way, the corresponding relationship for a plug-flow reactor is :

$$\sqrt{1 - E_p} = 1 - \frac{\sqrt{C_m}}{\sqrt{C_i}} \tag{43}$$

C_m appears then as a critical inlet concentration for which all the substrate is converted in a plug-flow reactor. A relationship equivalent to Eq. 43 is :

$$\left(\frac{1}{1 - \sqrt{1 - E_p}} \right)^2 = \frac{C_i}{C_m} \tag{44}$$

Eq. 41 and 44 show that plots of $\dfrac{1 - E_A}{E_A^2}$ and $\left(\dfrac{1}{1 - \sqrt{1 - E_p}} \right)^2$ against C_i

are straight lines passing through the origin.

3.6. Effect of recycle

Most biological fluidized reactors are operated with a recycle flow which helps fluidization and provides a convenient way of cleaning the support and transferring oxygen. The model's predictions will now be determined when a fraction of the effluent is recirculated. This approach will be used only where liquid-solid resistance is neglected.

3.6.1. Effect of recycle on a CSTR

It is well known that recirculation has no effect on conversion in a CSTR. This can easily be verified from an overall mass balance.

3.6.2. Effect of recycle on a plug-flow reactor

It is assumed here that the recycle does not affect the flow pattern in the reactor. Then, the mass balance over an element volume is :

$$dC_B + \frac{k_o \, \varepsilon_b}{(1+R) \, \delta} \left(\frac{2 \, DC_B}{k_o} \right)^{1/2} d\tau = 0 \tag{45}$$

The integration of Eq. 45 gives :

$$\sqrt{C_1} = \sqrt{C_e} + \frac{\varepsilon_b}{(1+R) \, \delta} \left[\frac{k_o \, D}{2} \right]^{1/2} \tau \tag{46}$$

where, C_1 = inlet concentration into the reactor :

$$C_1 = \frac{C_i + R \, C_e}{1 + R} \tag{47}$$

The combination of Eqs. 46 and 47 leads to a second degree algebraic equation, the sole acceptable solution of which allows the calculation of the conversion :

$$E = \left[\frac{R}{1+R}\alpha^2 + 2\alpha\right]^{1/2} - \frac{1+2R}{1+R}\frac{\alpha}{2} \tag{48}$$

It can easily be checked that when there is no recycle, the conversion in a once-through plug-flow reactor is $E_p = \sqrt{2\alpha} - \frac{\alpha}{2}$ and when R is infinite, the conversion in a CSTR is $E_A = \sqrt{\alpha(\alpha+2)} - \alpha$.

It may be noted that Eq. 48 is valid only for α less than the critical value 2. For a given α, conversion is a decreasing function of recycle ratio tending asymptotically to the corresponding conversion value in a CSTR (Fig. 18). The aim of recirculation therefore cannot be an improvement of conversion but essentially to provide a sufficient oxygen mass flow rate for pollution abatement.

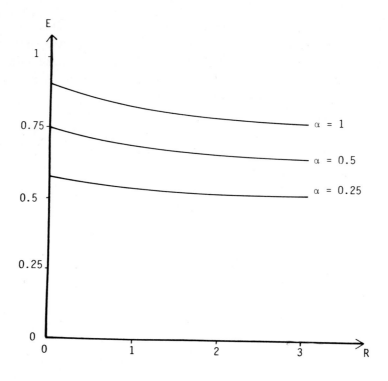

Fig. 18. Effect of recycle on conversion in a plug-flow reactor.

3.6.2.2. Effect of space time on a recycled plug-flow reactor.

A relationship equivalent to Eq. 42 having similar applications is derived from the mass balance (Eq. 45) :

$$(1 + R) \left[\frac{1 + R (1 - E)}{1 + R} \right]^{1/2} - \sqrt{1 - E} = \frac{\tau}{\tau_m} \tag{49}$$

When space time only is varied, a plot of

$$(1 + R) \left[\frac{1 + R (1 - E)}{1 + R} \right]^{1/2} - \sqrt{1 - E} \quad \text{against } \tau \text{ is a straight line}$$

passing through the origin with a slope of $1/\tau_m$.

3.6.2.2. Effect of inlet concentration on a recycled plug-flow reactor

Similarly, when all other conditions are maintained steady, variations of inlet concentration can be taken into account through the following equation :

$$\frac{(1 + R)^2}{\left[\frac{1 + R (1 - E)}{1 + R} \right]^{1/2} - \sqrt{1 - E}} = \frac{C_i}{C_m} \tag{50}$$

$$\frac{(1 + R)^2}{\left[\frac{1 + R (1 - E)}{1 + R} \right]^{1/2} - \sqrt{1 - E}} \quad \text{vs } C_i \text{ is a straight line passing through}$$

the origin with a slope of $1/C_m$.

3.7. Conversion of a non-limiting substrate

A biological oxidation can be schematically represented by the overall chemical reaction :

$$\upsilon_A \text{ A} + \upsilon_B \text{ B} + \ldots \xrightarrow{\text{micro-organisms}} \text{cells} + \text{metabolites}$$

where, A and B = respectively the electron donor and the electron acceptor, and υ_A and υ_B are stoichiometric coefficients.

The flux of A is linked to the flux of B by (Frank-Kamenetskii, 1955) :

$$\frac{J_A}{\upsilon_A M_A} = \frac{J_B}{\upsilon_B M_B} \tag{51}$$

where, M_A and M_B = respectively the molar mass of A and B.

This relationship shows that the flux of the electron donor is proportional to the flux of the electron acceptor. It will, therefore, be assumed that the consumption rate of a non limiting substrate is proportional to the consumption rate of the limiting substrate :

$$r_S = \lambda \, r_a \tag{52}$$

One can then easily verify that conversion of a non-limiting substrate is linked to conversion of the limiting substrate by :

$$E_S = \beta \, E \tag{53}$$

$$\text{where,} \quad \beta = \frac{\lambda \, C_i}{S_i}, \text{ and} \tag{54}$$

S_i = the inlet non-limiting substrate concentration.

For instance, conversion of a non-limiting substrate in a CSTR where the liquid-solid resistance is negligible is given by :

$$E_{SA} = \beta \left[\sqrt{\alpha(\alpha + 2)} - \alpha \right] \tag{55}$$

where α is calculated for the limiting substrate.

3.8.   How to calculate a biological fluidized bed reactor

The equations just derived link the conversion to physico-reactional varia-bles and operating variables. Now the relationship between the state of fluidization and the biomass hold-up will be established.

3.8.1.   Fluidization and biomass hold-up

The bed voidage, ε , can be estimated from the correlation proposed by Wen and Yu (1966) as follows. Based on the analysis of forces acting on the suspending particles, a relationship between the drag coefficient, γ_D, of a constituent particle in a multiple particle suspension and the drag coefficient, γ_{DS}, of a single particle in an infinite expanse of fluid is established :

$$\gamma_D = \gamma_{DS} \, \varepsilon^{-4.7} \tag{56}$$

By combination of Eq. 56 with the correlation for γ_{DS} (Grace and Clift, 1974 ; Schiller and Naumann, 1935) and the pressure drop of the bed, $\Delta P = (\rho_S - \rho_L) \, g \, (1 - \varepsilon) \, L$, the following equation for the bed voidage of a bubbleless fluidized bed is obtained :

$$\varepsilon = \left[\frac{18 \, R_e + 2.7 \, R_e^{1.69}}{Ar} \right]^{1/4.7} \tag{57}$$

where, R_e = Reynolds number for the particle :

$$R_e = \frac{\rho_L \, u \, (d + 2 \, \sigma)}{\mu} \tag{58}$$

and A_r is the Archimedes number :

$$A_r = \frac{(d + 2 \, \sigma)^3 (\rho_b - \rho_L) \, \rho_L \, g}{\mu^2} \tag{59}$$

where, μ = dynamic viscosity of the flowing liquid, and

ρ_b = density of a spherical particle.

$$\rho_b = \frac{\rho_S + \left[(1 + \frac{2\sigma}{d})^3 - 1 \right] \rho_w}{(1 + \frac{2\sigma}{d})^3} \tag{60}$$

where, ρ_S = density of the inert supporting material and,

ρ_w = density of drained wet biomass.

It should be remembered that the biomass hold-up is given by Eq. 20 :

$$\varepsilon_{b_F} = (1 - \varepsilon) \left[1 - \frac{1}{(1 + \frac{2\sigma}{d})^3} \right]$$

Examination of Eqs. 20 and 57 reveals that the biomass concentration in the fluidized bed reactor is mainly a function of the biofilm thickness and of the hydraulic characteristics of the system, provided that the characteristics of the biofilm such as density and moisture content are not affected by the mode of operation. Thus, the conversion in the fluidized bed reactor is affected by a number of variables that are under the control of the design engineer. These variables include the expanded bed height, amount of clean media used, size of the media, biofilm thickness and reactor cross-sectional area. The biofilm thickness can also be controlled by the design engineer. For a given media, the equilibrium biofilm thickness is dependent on superficial upflow velocity, expanded bed height and media volume.

The effect of superficial upflow velocity, u, on the biomass hold-up is viewed graphically in Fig. 19.

The decrease in ε_{bF} with increasing superficial upflow velocity is much more significant when the biofilm is thicker. For a given media size, there exists one u value at which two different biofilm thicknesses will yield the same hold-up in the reactor.

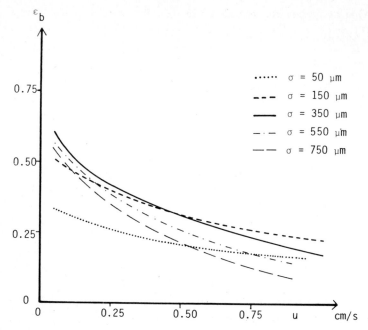

Fig. 19. Effect of superficial upflow velocity on biomass hold-up (d = 400 μm).

The effect of biofilm thickness is shown in Fig. 20. A plateau can be observed in the range of 200 μm < σ < 500μm and low values of superficial upflow velocity.

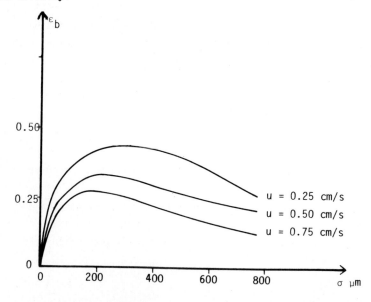

Fig. 20. Effect of biofilm thickness on biomass hold-up (d = 400 μm).

Fig. 21 presents the biomass hold-up as a function of the media size for a range of superficial upflow velocities at a given biofilm thickness. The curves possess a similar trend, i.e. the biomass hold-up increases with increasing media size until a maximum value is reached. The media size that results in the maximum hold-up depends on both the superficial upflow velocity and the biofilm thickness. There is a possibility of carry-over (decrease of hold-up to 0) at smaller media size when a certain critical thickness is reached.

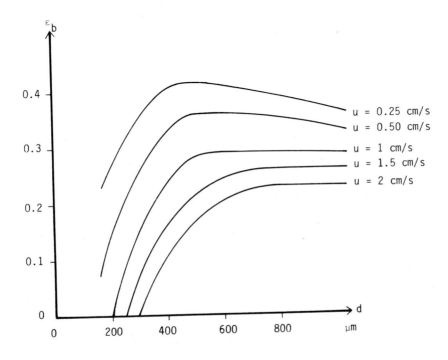

Fig. 21. Effect of support particle diameter on biomass hold-up (σ = 150 μm).

The expansion of a biological fluidized bed calculated from the clean fixed bed is given in Fig. 22 against superficial upflow velocity at constant biofilm thickness. A tremendous increase can be noted between 150 μm and 350 μm thickness.

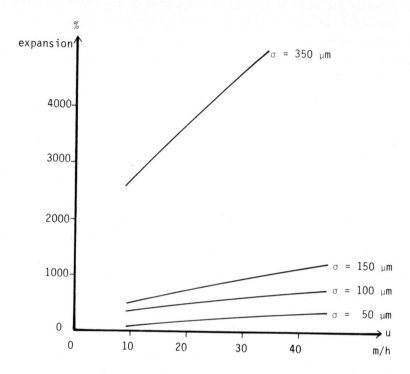

Fig. 22. Effect of superficial upflow velocity and biofilm thickness on bed expansion (d = 400 μm).

3.8.2. Reactor calculation

Because of the low difference between predicted conversion in plug-flow and CSTR reactor, it will be assumed for the design that an aerobic fluidized reactor is perfectly mixed. This assumption is still justified by the high recycle ratio required for oxygen transfer. The liquid film resistance is taken to be negligible.

When the performance equation of a CSTR is associated with the model of biomass hold-up in a fluidized bed of spherical particles, all the variables are determined when only two of them are fixed, i.e. the whole model has two independent variables. For instance, biofilm thickness and superficial upflow velocity can be used as design variables bearing in mind that biofilm thickness is controlled by the bed expansion of the allowed bed height.

The whole reactor calculation is detailed in Table 4.

TABLE 4

Reactor calculation.

DATA	Q, C_i E	influent characteristics and required conversion
	R	required recycle ratio for oxygen transfer
	d, ρ_s	media size and density
	k_o, D, ρ_w	characteristics of the biofilm
	μ, ρ_L	dynamic viscosity and density of wastewater

ENTRY→ σ give a value to biofilm thickness around a spherical particle

DATA

$$d_b = d + 2\sigma$$ bioparticle diameter

$$\delta = \frac{\sigma}{1 - (1 + \frac{2\sigma}{d})^{-3}} - \frac{d}{6}$$ biofilm thickness in the plate reactor (Eq. 21)

$$\rho_b = \frac{\rho_s + \left[(1 + 2\frac{\sigma}{d}) - 1\right]^3 \rho_w}{(1 + 2\frac{\sigma}{d})^3}$$ bioparticle density

ENTRY→ u   give a value to superficial upflow velocity

DATA

$$R_e = \frac{\rho_L \, u \, d_b}{\mu}$$ Reynolds number

$$A_r = \frac{d_b^3 \, \rho_L \, (\rho_b - \rho_L) \, g}{\mu^2}$$ Archimedes number

$$\varepsilon = \left(\frac{18 \, R_e + 2.7 \, R_e^{1.69}}{A_r}\right)^{\frac{1}{4.7}}$$ bed porosity (Eq. 57)

$$\varepsilon_{b_F} = (1 - \varepsilon)\left[1 - \frac{1}{(1 + 2\frac{\sigma}{d})^3}\right]$$ biomass hold-up in the fluidized reactor (Eq. 20)

$$\varepsilon_b = \frac{6(1 - \varepsilon)\frac{\delta}{d}}{1 + 6\frac{\delta}{d}}$$ biomass hold-up in the plate reactor (Eq. 17).

.../...

562

TABLE 4

(continued)

TEST

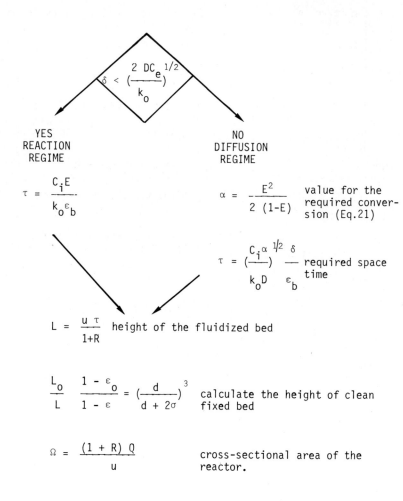

$$\delta < \left(\frac{2\ DC_e}{k_o}\right)^{1/2}$$

YES
REACTION
REGIME

NO
DIFFUSION
REGIME

$$\tau = \frac{C_i E}{k_o \varepsilon_b}$$

$$\alpha = \frac{E^2}{2\ (1-E)} \quad \text{value for the required conversion (Eq.21)}$$

$$\tau = \left(\frac{C_i}{k_o D}\right)^{\alpha\ 1/2} \frac{\delta}{\varepsilon_b} \quad \text{required space time}$$

$$L = \frac{u\ \tau}{1+R} \quad \text{height of the fluidized bed}$$

$$\frac{L_o}{L} \quad \frac{1-\varepsilon_o}{1-\varepsilon} = \left(\frac{d}{d+2\sigma}\right)^3 \quad \text{calculate the height of clean fixed bed}$$

$$\Omega = \frac{(1+R)\ Q}{u} \quad \text{cross-sectional area of the reactor.}$$

3.9. Conclusions

1. In this approach, the real reactor is replaced by a plate reactor on which a plane biofilm is grown. The relationships between both of the system are obtained by writing the equality of :

   a) volume occupied by the liquid phase ;

   b) clean surface area ; and

   c) apparent volumetric reaction rate.

2. The model is then based on a zero order intrinsic reaction of a limiting substrate and a diffusional mass transport through the biofilm. Two cases are then to be considered depending on the importance of resistance to mass transfer from the bulk of the liquid to the biofilm :

   a) if this resistance has to be taken into account, the model is fully characterized by two parameters, namely α which combines reaction and diffusion, and N_f which represents liquid film mass transfer ; and

   b) when this resistance is negligible, the model is characterized only by the parameter α .

3. When comparing conversions in a continuous stirred tank reactor and a plug-flow reactor, they are of the same order of magnitude ; the error made by neglecting the nature of the fluid flow does not exceed 20 % which is acceptable in the field of biological treatment.

4. The model that neglects liquid film mass transfer provides a very easy method to check its own validity when only space time or inlet concentration is changed ; experimental results can be plotted in such a way that straight lines are obtained (table 5).

5. Biomass hold-up in a fluidized bed of spherical particles can be predicted from a given biomass thickness from the hydraulic characteristics of the system.

6. An aerobic biological fluidized reactor is fully characterized when biofilm thickness and superficial upflow velocity are given.

4.   EXPERIMENTAL VALIDATION AND USE OF THE MODEL

It is not intended to give an extensive description of experimental work but only to illustrate some examples from experimental essays and show how the model can be fitted and how it can help in unit design.

4.1.   Some examples of experimental validation

4.1.1.   Liquid film mass transfer resistance

The influence of the liquid film mass transfer resistance can be easily observed using the cylindrical rotating reactor where runs are carried out at different rotating speeds (Fig. 5). In Fig. 23, $\dfrac{1 - E_A}{E_A^2}$ vs C_i has been

TABLE 5

Inlet concentration and space time effect -
Plot of straight lines.

Variable	CSTR	Plug-flow reactor	Plug-flow reactor with recycle
space time	$\dfrac{E_A}{\sqrt{1 - E_A}} = 2\,\dfrac{\tau}{\tau_m}$	$\sqrt{1 - E_p} = 1 - \dfrac{\tau}{\tau_m}$	$(1+R)\left[\sqrt{\dfrac{1 + R(1 - E)}{1 + R}} - \sqrt{1 - E}\,\right] = \dfrac{\tau}{\tau_m}$
inlet concentration	$\dfrac{1 - E_A}{E_A^{\,2}} = \dfrac{1}{4}\,\dfrac{C_i}{C_m}$	$\left(\dfrac{1}{1 - \sqrt{1 - E_p}}\right)^2 = \dfrac{C_i}{C_m}$	$\dfrac{(1 + R)^2}{\sqrt{\dfrac{1 + R(1 - E)}{1 + R}} - \sqrt{1 - E}} = \dfrac{C_i}{C_m}$

plotted for various rotational speeds. Line 3 is in agreement with Eq. 41 which predicts a straight line when liquid film resistance is negligible ; at lower rotating speeds, the liquid film resistance becomes more important and curves 1 and 2 are not linear. Since α is proportional to $1/C_i$, from Eq. 24 it can be seen that :

$$\lim_{C_i \to 0} E_A = \frac{1}{1 + \frac{1}{N_f}} \tag{61}$$

and then for $C_i = 0$, $\quad \dfrac{1 - E_A}{E_A^2} = \dfrac{1}{N_f (1 + N_f)}\quad$ as shown in Fig. 23.

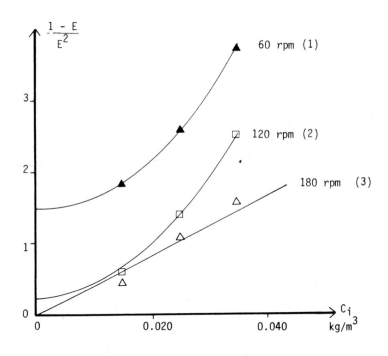

Fig. 23. Rotating biological reactor. Effect of rotational speed (τ = 0.23 h) ; limiting substrate : organic carbon. Experimental points are plotted on theoretical curves (Grasmick, 1982).

566

Another example is given by the work of Chatib et al. (1981) on a three-phase fluidized bed reactor. It has been found that organic carbon conversion is independent on the liquid superficial velocity until a break point after which it falls away as shown in Fig. 24. The plateau in the efficiency may be explained by considering the liquid film mass transfer as the controlling mechanism.

In spite of this experimental evidence, it can be taken for granted that in most cases, the calculations which neglect the liquid film resistance are accurate enough for design purposes.

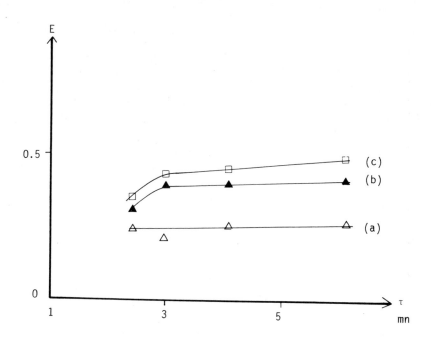

Fig. 24. Conversion in a three-phase fluidized bed for organic carbon removal (Chatib et al., 1981). Gas superficial velocity : (a) 12.7 m/h ; (b) 25.4 m/h ; (c) 38.2 m/h.

4.1.2. Use of straight lines for space time or inlet concentration variation

The equations recalled in Table 5 are valid when space time or inlet concentration only are varied. Note that in an aerobic biological fluidized reactor, where high values of recycle ratio are in use, the superficial upflow velocity is nearly independent of space time, and therefore biofilm thickness and biomass hold-up remain almost constant (Fig. 19).

Figs.25 a and b show the straights lines obtained when plotting $E_A/\sqrt{1-E_A}$ vs τ and $(1-E_A)/E_A^2$ vs C_i for data obtained on a pilot fluidized reactor treating synthetic sewage water made with molasses added to urban wastewater. The recycle flow rate is 4 m^3/h while influent flow rate is varied from 0.02 to 0.09 m^3/h and influent BOD$_5$ concentration is varied from 1 to 2 kg/m^3. From the slopes of the straight lines, the critical space time and the critical inlet concentration are calculated : τ_m = 6.4 h and C_m = 0.93 kg/m^3.

Another example is given by data on nitrification of secondary effluent enriched by ammonia in a pilot-scale fluidized reactor run by Dearborn Environmental Consulting Services (1980). Those data are plotted in the straight line fashion for various values of space time and inlet concentration (Figs. 26 a and b). The consistency of the method is checked by plotting τ_m against $\sqrt{C_i}$ and C_m against τ^2; in conformity with τ_m and C_m definitions these plots are also straight lines (Figs. 27 a and b) :

$$\tau_m = \frac{\delta}{\varepsilon_b}\left[\frac{2\,C_i}{k_o\,D}\right]^{1/2} = \left[\frac{C_i}{\kappa}\right]^{1/2} \tag{62}$$

$$C_m = \frac{\varepsilon_b^2\,\tau^2\,k_o D}{2\delta^2} = \kappa\,\tau^2 \tag{63}$$

$$\text{where,} \quad \kappa = \frac{\varepsilon_b^2 k_o D}{2\delta^2}$$

can be calculated using one or other of the straight lines of Fig. 27 which gives the same result, i.e. κ = 5.48 mg.l^{-3}h^{-2}.

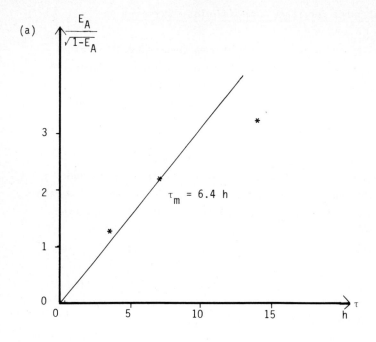

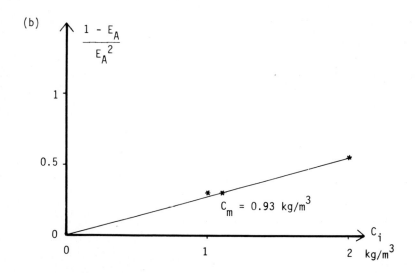

Fig. 25. Pilot plant data plotted on straight lines : limiting substrate: organic carbon ; fluidized sand : 400 - 600 μm ; reactor cross-sectional area : 0.1 m2. Data from Grasmick (1982).

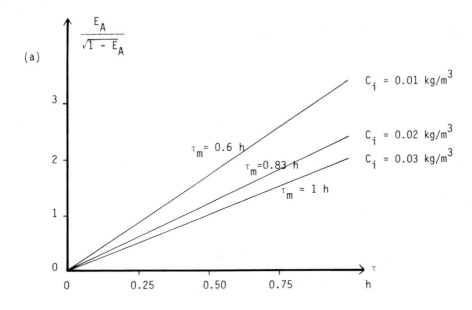

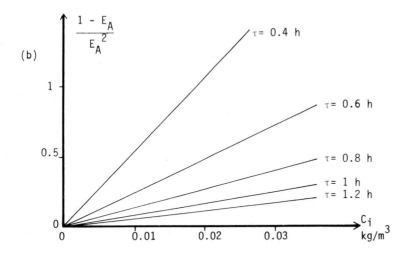

Fig. 26. Pilot plant data plotted on straight lines: limiting substrate :
ammonia ; fluidized quartzite : 480 μm ; reactor diameter : 28.9 cm
(Dearborn Environmental, 1980).

(a)

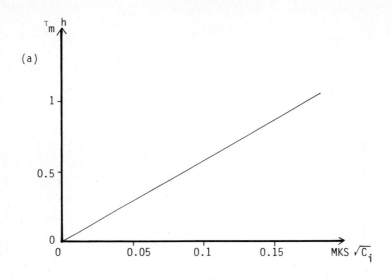

(b)

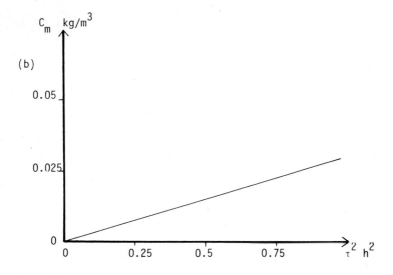

Fig. 27. τ_m vs $\sqrt{C_i}$ and C_m vs τ^2 from Figure 26.

4.2.   Case study
4.2.1. Data

A wastewater containing 1 kg BOD_5/m^3 is to be treated in an aerobic biological fluidized reactor. The support media is sand of density 2.5 10^3 kg/m^3
and mean diameter 400 μm. Pure oxygen is transferred to the recycle line
through a bicone device able to deliver 40 g/m^3 . Experimental measurements
on a rotating cylindrical reactor have given the following values of the
intrinsic zero order rate and the substrate diffusivity :

$$k_o = 5.10^{-7} g.cm^{-3}. s^{-1} \text{ and } D = 5.10^{-6} cm^2. s^{-1}$$

1. Calculate the recycle rate required by stoichiometry for a conversion of
   0.95.
2. How does space time depend on upflow superficial velocity and biofilm
   thickness ?
3. Calculate the height of the fluidized bed.
4. What is the cross-sectional area needed to treat a volumetric flowrate
   of 10 m^3/h ?
5. Calculate the power needed to fluidize the bioparticles.

4.2.2. Reactor calculation
1. Write the equality between the required oxygen mass flowrate and the
   delivered oxygen mass flowrate :

$$Y_{O_2} C_i E = R C_{O_2} \qquad (64)$$

where, Y_{O_2} = mass of oxygen required for abatement of 1 kg BOD_5 :
            Y_{O_2} = 0.8 kg O_2/kg BOD_5 and,
        C_{O_2} = oxygen concentration delivered in the liquid phase by the
            bicone device.
The recycle ratio is then :

$$R = \frac{0.8 \times 1 \times 0.95}{0.04} = 19$$

This high R value completely justifies the assumption of perfect mixing
of the liquid phase and the use of the CSTR equations.
2. The plate reactor equivalent to the fluidized bed of spherical sand particles has a characteristic length $d_p = \frac{d}{6} = 67$ μm.

All the computational steps detailed in Table 4 are followed in a computer programme where the entry data are values of superficial upflow velocity and biofilm thickness.

The effect of superficial upflow velocity on space time is shown Fig. 28. Variation of space time is linear for values of biofilm thickness less than the critical value which corresponds to the transition from reaction to diffusion regime. For higher values of σ , space time increases with a parabolic shape.

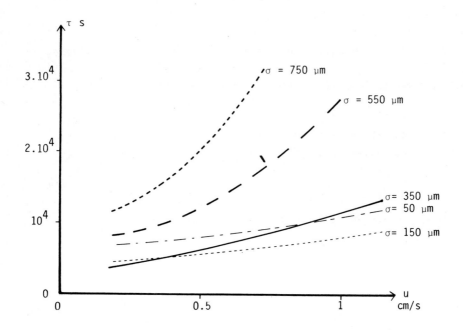

Fig. 28. Effect of superficial upflow velocity on space time at constant biofilm thickness (d = 400μm).

The effect of superficial upflow velocity on space time at constant biomass hold-up is viewed in Fig. 29. The different points of the curves are plotted for increasing values of biofilm thickness.

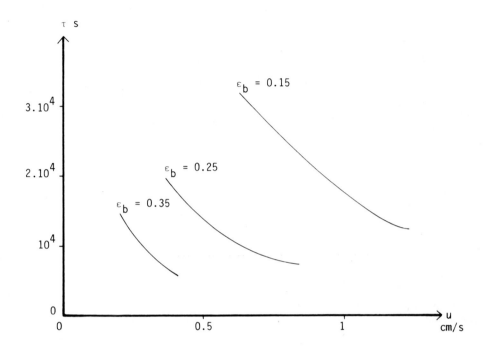

Fig. 29. Effect of superficial upflow velocity on space time at constant biomass hold-up (d = 400 μm).

If space time is plotted against biofilm thickness at constant superficial upflow velocity, the optimum value of biofilm thickness appears to be the maximum penetration depth allowed by reaction and diffusion (Fig. 30). Note that only little data is known about fixed biomass growth and related sludge production : it is most probable that the sludge production is maximum when the biofilm is at reaction regime which implies a minimum endogenous decay.

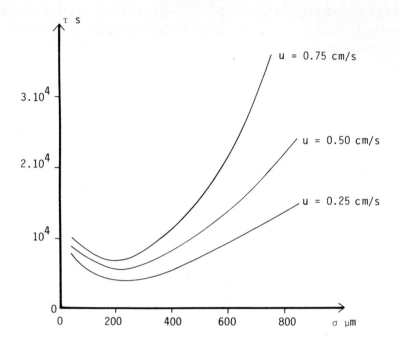

Fig. 30. Effect of biofilm thickness on space time at constant superficial upflow velocity (d = 400 μm).

3. The height of the fluidized bed is given by the following equation :

$$L = \frac{u\tau}{1+R} \tag{65}$$

L depends on u and σ as illustrated in Fig. 31. The optimum biofilm thickness again appears to be the transition value between reaction and diffusion regime.

The incipient fluidization of clean sand is obtained for a superficial upflow velocity of 25 x 10^{-4} m/s. To avoid hydraulic and fluid distribution problems, it is usual to maintain a superficial upflow velocity between 0.005 and 0.01m/s and a biofilm thickness between 50 and 400μm.

There is of course an infinity of reactors able to give an answer to the problem but an economic optimization is needed.

Taking u = 0.01 m/s and the biofilm thickness value giving a minimum height
σ = 200 μm, the height of the fluidized bed will be L = 4.20 m (Fig. 31).

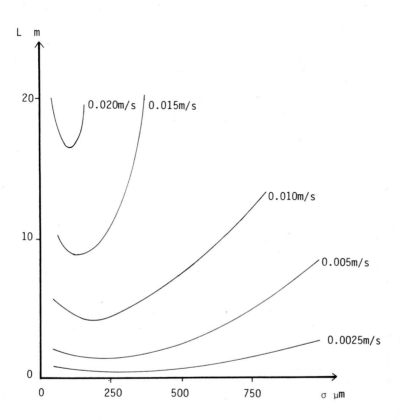

Fig. 31. Effect of biofilm thickness on fluidized bed length at constant
superficial upflow velocity (d = 400 μm).

4. The cross-sectional area of the reactor is calculated by the following
equation :

$$\Omega = \frac{(1 + R)\ Q}{u} \tag{66}$$

with the given numerical values :

$$\Omega = \frac{20 \times 10}{3600 \times 0.01} = 5.6 \ m^2$$

5. The power needed for the fluidization of the bioparticles is :

$$P = (1 + R) Q \ \Delta P \tag{67}$$

where, ΔP = pressure drop through the reactor :

$$\Delta P = (\rho_b - \rho_L) \ g \ L \ (1 - \varepsilon) \tag{68}$$

P is a decreasing function of σ at a constant superficial upflow velocity (Fig. 32). For the designed unit, P = 115 W which is a very low energy requirement.

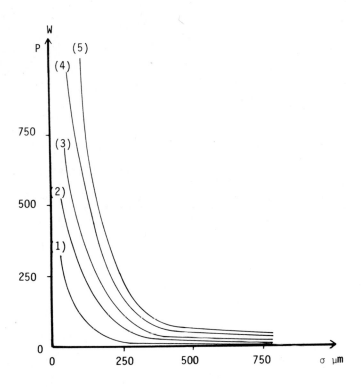

Fig. 32. Required power for fluidization against biofilm thickness at constant superficial upflow velocity (d = 400 μm). (1) u = 0.0025 m/s. (2) u = 0.005 m/s. (3) u : 0.010 m/s. (4) u = 0.015 m/s. (5) u = 0.020 m/s.

5.   CONCLUSIONS

The model has shown to be accurate enough in fitting various experimental
data and could help in unit design. Further research is needed :
1. to tabulate transfer and reaction constants, i.e. diffusivity and zero
   order reaction rate, for most types of wastewater ;
2. to check the model predictions in units designed by the proposed method.

6.   NOTATIONS

a	specific area	L^{-1}
a_b	biofilm specific area	L^{-1}
A_b	biofilm interfacial area	L^2
A_r	Archimedes number	
B_i	Biot number	
BOD	Biochemical oxygen demand	ML^{-3}
C	limiting substrate concentration	ML^{-3}
C_i	inlet concentration	ML^{-3}
C_e	outlet concentration	ML^{-3}
C_m	critical concentration	ML^{-3}
C_B	concentration in the bulk liquid phase	ML^{-3}
C_s	concentration at the interface biofilm-liquid	ML^{-3}
COD	chemical oxygen demand	ML^{-3}
d	mean diameter of spherical support material	L
d_b	bioparticle diameter	L
d_p	plate reactor characteristic length	L
D	diffusivity	L^2T^{-1}
$E = 1 - \dfrac{C_i}{C_e}$ conversion		
E_A	CSTR conversion	
E_p	plug-flow reactor conversion	
E_S	conversion of a non-limiting substrate	
$f = \dfrac{C}{C_B}$ dimensionless concentration		
g	gravitational acceleration	LT^{-2}
H	Heaviside function	
J	mass flux	$ML^{-2}T^{-1}$
k_o	intrinsic zero order reaction rate	$ML^{-3}T^{-1}$
K_f	liquid-biofilm overall mass transfer	LT^{-1}
K_1	gas-liquid overall mass transfer coefficient	LT^{-1}
L	height	L

P	power	ML^2T^{-3}
Q	volumetric flowrate	L^3T^{-1}
r_a	apparent reaction rate of a limiting substrate per unit biofilm volume	$ML^{-3}T^{-1}$
r_S	apparent reaction rate of a non-limiting substrate per unit biofilm volume	$ML^{-3}T^{-1}$
R	recycle ratio	
R_e	Reynolds number	
S_i	inlet concentration of a non-limiting substrate	ML^{-3}
TOC	total organic carbon	ML^{-3}
u	superficial upflow velocity	LT^{-1}
V	volume	L^3
$x = \frac{y}{\delta}$	dimensionless abscissa	
x^*	dimensionless substrate penetration	
y	abscissa	L
y^*	substrate penetration	L

Greek letters

α	$= N_r\,N_b$	
α_m	critical value of α	
β	$= \dfrac{\lambda C_i}{S_i}$	
γ_D	drag coefficient of a particle	
γ_{DS}	drag coefficient of a single particle in an infinite expanse of fluid	
δ	biofilm thickness in the plate reactor	L
ϵ	porosity	
ϵ_o	porosity of clean fixed bed	
ϵ_b	biomass hold-up in the plate reactor	
ϵ_{bF}	biomass hold-up in the fluidized reactor	
Φ	Thiele modulus	
η	effectiveness factor	
λ	coefficient of proportionality between r_S and r_a	
μ	dynamic viscosity	$ML^{-1}T^{-1}$
υ	stoichiometric coefficient	
ρ_b	bioparticle density	ML^{-3}
ρ_L	wastewater density	ML^{-3}
ρ_s	solid support density	ML^{-3}

ρ_w	drained wet biomass density	ML^{-3}
σ	biofilm thickness around a spherical particle	L
τ	space time	T
τ_m	critical space time	T
Ω	reactor cross-sectional area	L^2

7 REFERENCES

Aleman, J.E. and Veil, J.A., 1981. Observations of attached biofilm composition and structure using scanning electron microscopy. Annual Meeting of the A.I.Ch.E., New Orleans, La.

Atkinson, B., Swilley, E.L., Bush, A.W. and Williams D.A., 1967. Kinetic, mass transfer and organisms growth in a biological film reactor. Trans. Inst. of Chem. Eng., 45, 257.

Atkinson, B. and Fowler, H.W., 1974. The significance of microbial film in fermenters. Advances in Biochemical Engineering, 3, 221.

Atkinson B., 1974. Biochemical Reactors, Pion Press, London.

Atkinson, B., Black, G.M. and Pinches, A., 1981. The characteristics of solid support particles when used in fluidised beds, in Biological Fluidised Bed Treatment of Water and Wastewater. Cooper, P.F. and Atkinson, B., Editors, Ellis Horwood Limited, Chichester.

Atkinson, B., 1981. Immobilised biomass-a basis for process development in wastewater treatment in Biological Fluidised Bed Treatment of Water and Wastewater. Cooper, P.F. and Atkinson, B., Editors, Ellis Horwood Limited, Chichester.

Begovich, J.M. and Watson, J.G., 1978. Hydrodynamic characteristics of three-phase fluidised beds, in Fluidisation edited by Davidson, J.F. and Keairns, D.L., Cambridge University Press, Cambridge.

Buckley, P.S., 1976. Gas dispersions in three-phase fluidised beds. Ph. D. Thesis, University of Wales.

Chatib, B., Grasmick, A., Elmaleh, S. and Ben Aim, R., 1981. Biological wastewater treatment in a three-phase fluidised bed reactor, in Biological Fluidised Bed Treatment of Water and Wastewater. Cooper, P.F. and Atkinson, B., Editors, Ellis Horwood Limited, Chichester.

Chen, Y.S. and Bungay, H.R., 1981. Microelectrode studies of oxygen transfer in trickling filter slimes. Biotech. Bioeng. 23, 781.

Cooper, P.F. and Wheeldon, P.H.V., 1981. Complete treatment of sewage in a two-fluidised bed system, in Biological Fluidised Bed Treatment of Water and Wastewater. Cooper, P.F. and Atkinson, B., Editors, Ellis Horwood Limited, Chichester.

Dearborn Environmental Consulting Services, Missisauga, Ontario, 1980. Pilot-scale assessement of the biological fluidized bed process for municipal wastewater treatment ; Information and Communications Centre, Canada Mortgage and Housing Corporation, Ottawa, Ontario.

Elmaleh, S., 1981. Transfert d'oxygène en réacteur fluidisé triphasique, unpublished results.

Elmaleh, S., 1982. Les réacteurs à biomasse fixée, in Point sur l'Epuration et le Traitement des Effluents. Martin, G., Editor, Technique et Documentation, Lavoisier, Paris.

Frank-Kamenetskii, D.A., 1955. Diffusion and Exchange in Chemical Kinetics, Princeton Univ. Press.

Grace, J.R. and Clift, R., 1974. Chem. Eng. Sc., 29, 327.

Grady, C.P.L. Jr, 1982. Modelling of biological fixed films. A state of the art review, 1st Int. Conf. on fixed film biological processes, Kings Island Resort, Ont.

Grasmick, A., Elmaleh, S. and Ben Aim, R., 1979. Théorie de l'épuration par filtration biologique immergée. Wat. Res., 13, 1137.

Grasmick, A., Elmaleh, S. and Ben Aim, R., 1980. Etude expérimentale de la filtration biologique immergée. Wat. Res., 14, 613.

Grasmick, A., Chatib, B., Elmaleh, S. and Ben Aim, R., 1981. Epuration hydrocarbonée en couche fluidisée triphasique. Wat. Res., 15, 719.

Grasmick, A., 1982. Contribution à la modélisation des réacteurs à cellules immobilisées sur support granulaire en couche fixe ou fluidisée. Dr. Sc. Thesis, Toulouse.

Grigoropoulou, H., 1980. Etude de l'oxygénation d'un filtre immergé alimenté par le peroxyde d'hydrogène. Docteur-Ingénieur Thesis, Toulouse.

Harremoes, P. and Riemer, M., 1975. Pilot experiments on down filter denitrification. Conf. on Nitrogen as a Water Pollutant, Copenhagen.

Harremoes, P., 1975. The significance of pore diffusion to filter denitrification. Conf. on Nitrogen as a Water Pollutant, Copenhagen.

Harremoes, P., 1976. The significance of pore diffusion to filter denitrification. J.W.P.C.F., 48, 2.

Harremoes, P., 1978. Biofilm kinetics, in Water Pollution Microbiology, edited by R. Mitchell, J. Wiley.

Hoehn, R.C. and Ray, A.D., 1973. Effects of thickness on bacterial film. J.W.P.C.F., 45, 2302.

Howell, J.A. and Atkinson, B., 1976. Sloughing of microbial film in trickling filters. Wat. Res., 10, 307.

Jansen, J. and Kristensen, G.H., 1980. Fixed film kinetics. Denitrification in fixed films, Technical University of Denmark, Dept of Sanitary Engineering.

Jeris, J.S., Owens, R.W. and Flood, F., 1981. Secondary treatment of municipal wastewater with fluidised bed technology, in Biological Fluidised Bed Treatment of Water and Wastewater. Cooper, P.F. and Atkinson, B., Editors, Ellis Horwood Limited, Chichester.

Jewell, W.J., 1981. Development of the attached microbial film expanded-bed process for aerobic and anaerobic waste treatment in Biological Fluidised Bed Treatment of Water and Wastewater. Cooper, P.F., and Atkinson, B., Editors, Ellis Horwood Limited, Chichester.

Kornegay, B.H. and Andrews, J.F., 1969. Characteristics and kinetics of biological fixed film reactors. Dept. Env. System Eng., Clemson University.

La Motta, E.J., 1974. Evaluation of diffusional resistances in substrate utilisation by biological films. Ph. D. Thesis, Univ. of North Carolina, Chapel Hill.

La Motta, E.J., 1976. Internal diffusion and reaction in biological films. Env. Sc. and Technol., 10, 8, 765.

La Motta, E.J., 1976b. External mass transfer in a biological film reactor. Biotech. Bioengng., 18, 1359.

Lee, J.C. and Al Dabbagh, N., 1978. Three-phase fluidised beds : onset of fluidisation at high gas rates, in Fluidisation, edited by Davidson, J.F. and Keairns, D.L., Cambridge University Press, Cambridge.

Lee, J.C. and Buckley, P.S., 1981. Fluid mechanics and aeration of fluidised beds in Fluidisation, edited by Davidson, J.F. and Keairns, D.L., Cambridge University Press, Cambridge.

Maier, W.J., Behn, V.C. and Gates, C.D., 1967. Simulation of the trickling filter process. J. ASCE, SA4, 91.

Matson, J.V. and Charaklis, W.G., 1976. Diffusion into Microbial Aggregates, Dept of Civil Eng., Univ. of Houston.

Nutt, S.G., Stephenson, J.P. and Pries, J.H., 1981. Steady and non-steady state performance of the aerobic(oxygenic) biological fluidised bed in Biological Fluidised Bed Treatment of Water and Wastewater. Cooper, P.F., and Atkinson, B., Editors, Ellis Horwood Limited, Chichester.

Ostergaard, K., 1971. Fluidisation, edited by Davidson, J.F. and Harrison, D., Academic Press, New York.

Ostergaard, K. and Fosbol, P., 1972. Transfer of oxygen across the gas-liquid interface in gas-liquid fluidised beds. Chem. Eng. J. 3, 105.

Riemer, M., Kristensen, H. and Harremoes, P., 1980. Residence time distribution in submerged biofilters. Water Research, 14 (8), 949.

Riemer, M. and Harremoes, P., 1978. Multi-component diffusion in denitrifying biofilms. Prog. in Water Technology, 10, 149.

Rittman, B.E. and Mac Carty, P.L., 1981. Substrate flux into biofilms of any thickness. Journal of the Env. Eng. Div. ASCE.

Rodrigues, A.E., 1982. Non-linear adsorption, reaction, diffusion and intraparticle convection phenomena in flow systems, in Residence Time Distribution Theory in Chemical Engineering, edited by Petho, A. and Noble, R.D., Verlag Chemie, Weinheim.

Roques, H., 1980. Fondements théoriques du traitement biologique des eaux. Technique et Documentation.

Sehic, O.A., 1981. Fluidised sand recycle reactor for aerobic biological treatment of sewage, in Biological Fluidised Bed Treatment of Water and Wastewater. Cooper, P.F. and Atkinson, B., Editors, Ellis Horwood Limited, Chichester.

Shiller, L. and Naumann, A., 1935. Z. Ver. Dtsch. Ing., 77, 285.

Shah, Y.T., 1979. Gas-liquid-solid reactor design. Mc Graw Hill, New York.

Shieh, W.K., 1980. A suggested kinetic model for the fluidized bed biofilm reactor. Biotech. Bioeng., 22, 667.

Shieh, W.K., Sutton, P.M. and Kos, P., 1981. Predicting reactor biomass concentration in a fluidised bed system. J.W.P.C.F., 53 (11), 1574.

Sutton, P.M., Shieh, W.K., Kos, P. and Dunning, P.R., 1981. Dorr Oliver's Oxitron systems fluidised bed water and wastewater treatment process in Biological Fluidised Bed Treatment of Water and Wastewater. Cooper, P.F. and Atkinson, B., Editors, Ellis Horwood Limited, Chichester.

Tomlinson, T.G. and Snaddon, D.N., 1966. Biological oxidation of sewage by films of micro-organisms. Air Wat. Poll. 10, 865.

Wen, C.Y. and Yu, Y.H., 1966. Chem. Eng. Prog. Symp. Ser. 62, 100.

Weisz, P.W., 1973. Diffusion and chemical transformation : an interdisciplinary excursion. Science, 179, 433.

Williamson, K. and Mac Carty, P.L., 1976. Verification studies of the biofilm model for bacterial substrate utilization. J.W.P.C.F., 48, 281.

Yahi, H., Grasmick, A., Elmaleh, S. and Faup, G.M., 1982. Nitrification par réacteur à cellules immobilisées en couche fixe et oxygénation par le peroxyde d'hydrogène. Env. Technol. Letters, 3, 281.

Zobell, C.E., 1939. The role of bacteria in the fouling of submerged surfaces. Biological Bulletin, 77, 302.

MATHEMATICAL MODELLING OF THE ANAEROBIC DIGESTION PROCESS

Keisuke HANAKI[*], Tatsuya NOIKE[**] and Junichiro MATSUMOTO[***]

* Department of Urban Engineering, University of Tokyo, Hongo, Tokyo 113 , Japan; ** Sewage Works Division, Public Works Research Institute, Tsukuba, Ibaraki 305, Japan; *** Department of Civil Engineering, Tohoku University, Aoba, Sendai 980, Japan

CONTENTS

1. INTRODUCTION

The anaerobic digestion process stabilizes a wide variety of organic materials and concurrently produces methane from them. It has been employed for the treatment of sewage sludge, industrial wastes and agricultural wastes and, particularly in Japan, it has been applied to the treatment of night soil. This process is able to convert more kinds of organic matter to the useful fuel, methane, than does the ordinary application of the same method as described above. The shortage of fossil energy resources will require the cutting of energy consumption in sewage treatment plants in the near future. Anaerobic digestion will be more advantageous because it does not consume energy and, in some cases, can even generate electricity for supply within the treatment plant. The main disadvantage of the anaerobic digestion process is that it requires a long retention time, usually longer than ten days. The reasons for this are the slow growth rate of bacteria and poor process stability. Although the reaction which occurs in the anaerobic digestion process is widespread in the natural system, the sanitary engineer must control it and enhance its efficiency. The development and application of a mathematical model is a useful tool for improving this process. The steady state model helps the rational design of the process and the dynamic model enables the effective process control. Estimation of the effects of environmental factors is necessary for maintaining the stability of the process.

2. PROCESS DESCRIPTION

2.1 Principles of reactions
2.1.1 Overview of degradation process
Various kinds of organic materials, such as sewage sludge, municipal solid waste, industrial wastewater and night soil are degraded and ultimately converted into methane (CH_4) and carbon dioxide (CO_2) in the anaerobic digestion process. Fig. 1 illustrates the scheme of this degradation process. At first, complex materials (polysaccharides, proteins, neutral fats) are hydrolyzed into the component monomers (monosaccharides, amino acids, long-chain fatty acids) by extracellular enzymes (Step 1). These monomers are then fermented to intermediates such as volatile fatty acids (acetate, propionate, butyrate) and hydrogen gas (Step 2).

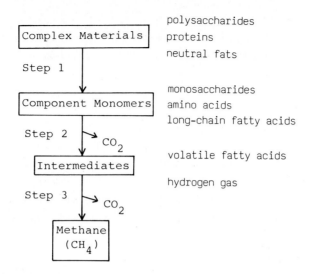

Fig. 1. Scheme of Degradation Process in Anaerobic Digestion

These intermediates are ultimately converted to methane (Step 3).

Several nomenclatures have been proposed for these three steps. The first, second and third steps are usually called hydrolysis, acidogenesis and methanogenesis, respectively, although the hydrolysis step is sometimes regarded as a part of the acidogenesis. Toerien and Hattingh (1969) called the first and the second steps the non-methanogenic phase and the third step, the methanogenic phase because compounds other than acids (hydrogen gas as an example) are also produced in the second step. Speece and McCarty (1962) called the first and the second steps the constant BOD phase and the third step, the reducing BOD phase because only the methane formation in the third step brings about the reduction of BOD or COD through the whole process. Liquefaction and gasification are relatively empirical terms, which originate from the observation that the sewage sludge in a solid state is liquefied in the first and the second steps, and soluble organic matters are gasified in the third step. In this chapter, the first and the second steps are combined and to be called the acidogenic phase, and the third step is called the methanogenic phase.

The reaction in the acidogenic phase are conducted by a group of bacteria called acidogenic bacteria, while a syntrophic consortium of methanogenic bacteria and hydrogen-producing acetogenic bacteria is responsible for the methanogenic phase.

2.1.2 Acidogenic bacteria

A group of acidogenic bacteria includes various kinds of bacteria which ferment organic matter and produce organic acids in anaerobic conditions. The number of strains of acidogenic bacteria which occur in the anaerobic digester is so large that computer-aided classifications of these bacteria have been tried (Toerien, 1970). The development of the anaerobic culture technique revealed that strict anaerobes composed most part of this group of bacteria (Toerien et al., 1967) although it had been previously believed that facultative bacteria played more important roles than strict anaerobes (McKinney, 1962). Only a small proportion of acidogenic bacteria can hydrolyze polymerized organic matter, while the remainder can degrade monosaccharides or amino acids which are liberated by hydrolysis. Carbohydrates are fermented via pyruvate mainly to acetate (CH_3COOH), propionate (CH_3CH_2COOH), butyrate ($CH_3CH_2CH_2COOH$), hydrogen gas (H_2), lactate ($CH_3CHOHCOOH$), formate (HCOOH), ethanol (CH_3CH_2OH) and carbon dioxide (CO_2). The distribution of the final products depends on the species of acidogenic bacteria and on the environmental conditions such as pH and temperature. Amino acids which derive from protein are degraded by Clostridium sp. (Siebelt and Toerien, 1969) mainly via the Stickland reaction (Nagase and Matsuo, 1982) and volatile fatty acids are produced. Neutral fat, which is a typical lipid in wastewater treatment, is hydrolyzed into glycerol and long-chain fatty acids. The former enters the degradation pathway of carbohydrates, but the latter are not degraded by acidogenic bacteria.

2.1.3 Methanogenic bacteria

The knowledge of the methanogenic bacteria has been advanced significantly in the last decade. Methanogenic bacteria form a special group of micro-organisms which have specific coenzyme, CoM, and electron carriers, F_{420}, F_{430}. A genetic analysis showed that methanogenic bacteria are quite distinct from other groups of bacteria (Fox et al., 1977). Methanogenic bacteria are strict anaerobes which are most fastidious to cultivation (Zeikus, 1977). Bergey's manual (Bryant, 1974) classified them into four genera according to their morphology, namely, Methanobacterium, Methanococcus, Methanosarcina and Methanospirillum. A new classification based on genetic characteristics was proposed recently as shown in Table 1 (Balch et al., 1979).

Table 1. New Taxonomy Proposed by Balch et al. (1979)

Order	Family	Genus
Methanobacteriales	Methanobacteriaceae	Methanobacterium Methanobrevibacter
Methanococcales	Methanococcaceae	Methanococcus
Methanomicrobiales	Methanomicrobiaceae	Methanomicrobium Methanogenium Methanospirillum
	Methanosarcinaceae	Methanosarcina

Almost all the species of methanogenic bacteria utilize hydrogen as a substrate, and some species can also utilize formate. Only a few species can utilize acetate as the sole carbon source in spite of the fact that acetate is a more important intermediate than hydrogen as a precursor of methane in the anaerobic digestion process. There might be some species of acetate-utilizing methanogenic bacteria which have not yet been isolated. Methanol is also utilized by Methanosarcina barkerii. Methanogenesis from four substrates are as follows:

Hydrogen: $4H_2 + CO_2 \longrightarrow CH_4 + 2H_2O$ (1)

Formate: $HCOO^- + \frac{1}{2} H_2O + \frac{1}{4} CO_2 \longrightarrow \frac{1}{4} CH_4 + HCO_3^-$ (2)

Acetate: $CH_3COO^- + H_2O \longrightarrow CH_4 + HCO_3^-$ (3)

Methanol: $CH_3OH \longrightarrow \frac{3}{4} CH_4 + \frac{1}{2} H_2O + \frac{1}{4} CO_2$ (4)

Although not only H_2 but also CO_2 is required in Eq. 1, CO_2 does not limit this reaction because an excessive amount of CO_2 or HCO_3^- exists in the anaerobic digesters. Formate is not usually converted to methane directly according to Eq. 2, but is degraded once into H_2 and CO_2 and then converted to methane in mixed culture systems (Zeikus, 1977; Pretorius, 1972). Methanol does not occur in wastewater treatment except for a particular industrial waste, namely pharmaceutical waste (Jennett and Dennis, 1975) or waste from pulp industry (Endo and Tohya, 1984). Therefore only methanogenesis from hydrogen and that from acetate are important in the ordinary anaerobic digestion.

2.1.4 Hydrogen-producing acetogenic bacteria

The existence of this group of bacteria was proposed recently by Bryant (1976, 1979). He found that the methane production from ethanol is performed by syntrophic association of methanogenic bacteria and a new type of bacteria (Bryant et al., 1967) while it had been believed that ethanol is converted to methane directly by one species of methanogenic bacteria "Methanobacterium omelianskii". Short-chain fatty acids such as propionate and butyrate are produced by the acidogenic bacteria and converted to hydrogen and acetate by hydrogen-producing acetogenic bacteria as follows:

Butyrate: $CH_3CH_2CH_2COO^- + 2H_2O \longrightarrow 2CH_3COO^- + 2H_2 + H^+$

$$\Delta G_o' = +48.1 \text{ kJ/mol (Thauer et al., 1977)} \qquad (5)$$

Propionate: $CH_3CH_2COO^- + 3H_2O \longrightarrow CH_3COO^- + 3H_2 + HCO_3^- + H^+$

$$\Delta G_o' = +76.1 \text{ kJ/mol (Thauer et al., 1977)} \qquad (6)$$

It was believed that Methanosarcina methanica and Methanobacterium propionicum converted butyrate and propionate to methane, respectively, but now the existence of these species of methanogenic bacteria is doubted. Long-chain fatty acids are also degraded to acetate and hydrogen via β-oxidation cycle (Jeris and McCarty, 1965) by hydrogen-producing acetogenic bacteria as follows:

Palmitate: $CH_3(CH_2)_{14}COO^- + 14H_2O \longrightarrow 8CH_3COO^- + 14H_2 + 7H^+$

$$\Delta G_o' = +345.6 \text{ kJ/mol (Brown, 1969)} \qquad (7)$$

Values of the standard free energy change, $\Delta G_o'$, in Eqs. 5 to 7 are positive. From the viewpoint of the thermodynamics, a negative value of free energy change is necessary for any reaction to proceed without input of external energy. This theory apparently suggests that hydrogen-producing acetogenic bacteria cannot obtain energy for growth from these reactions. However, the value of free energy change in the actual environment surrounding the bacteria, $\Delta G'$, is different from that of $\Delta G_o'$ and depends on the concentrations of substrates and products as follows:

$$\Delta G' = \Delta G_o' + RT_A \ln \frac{[P_1] \cdot [P_2] \cdots}{[S_1] \cdot [S_2] \cdots} \qquad (8)$$

where,

$\Delta G'$ = free energy change at pH = 7 (kJ/mol),

$\Delta G_o'$ = standard free energy change at pH = 7 (kJ/mol),

R = gas constant = 0.082 l·atm/mol·°K,

T_A = temperature (°K),

$[P_1]$, $[P_2]\cdots$ = product concentration (mol/l or atm), and

$[S_1]$, $[S_2]\cdots$ = substrate concentration (mol/l or atm).

Only low partial pressure of hydrogen can give negative values of $\Delta G'$ in Eqs. 5 to 7, because substrate concentration cannot be so high and acetate concentration is not so low in anaerobic digesters. Table 2 shows critical partial pressures of hydrogen, beyond which $\Delta G'$ becomes positive in these reactions under a hypothetical conditions of substrate and other compounds concentration. This calculation shows that extremely low partial pressure of hydrogen is essential for hydrogen-producing acetogenic bacteria although they themselves produce hydrogen. Kaspar and Wuhrmann (1978a) demonstrated experimentally that the hydrogen partial pressure higher than 5×10^{-3} atm ceased the degradation of propionate by hydrogen-producing acetogenic bacteria.

Hydrogen-utilizing methanogenic bacteria can serve such a thermodynamically favourable condition for hydrogen-producing acetogenic bacteria in anaerobic digesters, in other words, activity of hydrogen-producing acetogenic bacteria depends on the existence of methanogenic bacteria. Methanogenic bacteria receive hydrogen as a substrate from hydrogen-producing acetogenic bacteria. The interrelationship between these two groups of bacteria is called interspecies hydrogen transfer, which also exists between acidogenic bacteria and methanogenic bacteria. Acidogenic bacteria produce more hydrogen and acetate than propionate or lactate and obtain more energy under low hydrogen

Table 2. Critical Hydrogen Partial Pressure for Hydrogen-producing Acetogenic Bacteria

substrate	critical hydrogen partial pressure (atm)
n-butyrate	6.0×10^{-4}
propionate	1.0×10^{-4}
palmitate	4.4×10^{-4}

Hypothetical condition: concentrations of n-butyrate, propionate, acetate, palmitate and HCO_3^- are 600 mg/l as acetate, 600 mg/l as acetate, 600 mg/l, 1000 mg/l and 2000 mg/l as $CaCO_3$, respectively.

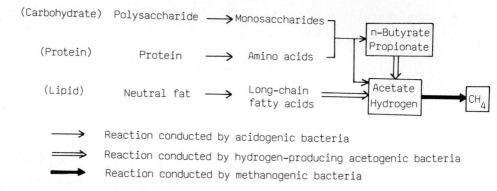

Fig. 2.   Degradation of Each Organic Component by Three Kinds of
          Bacteria in Anaerobic Digestion

partial pressure which is kept by methanogenic bacteria. The
interspecies hydrogen transfer is favourable but not essential for
acidogenic bacteria, while it is indispensable for hydrogen-
producing acetogenic bacteria.   Reactions conducted by hydrogen-
producing acetogenic bacteria should be included in the
methanogenic phase and not in the acidogenic phase because their
activity depends on methanogenic bacteria.

2.1.5 Summary

     Degradation of carbohydrates, proteins and lipids by the three
groups of bacteria is summarized in Fig. 2. Hydrogen, acetate,
propionate and n-butyrate are important as intermediates while
lactate and ethanol are less important in wastewater treatment.
About 70% of methane is produced via acetate in sewage sludge
digestion (Jeris and McCarty, 1965; Smith and Mah, 1966).

2.2   Digester operation

     Anaerobic digestion has been used in sewage treatment plants
since the 1920's. A classic type of digester operation does not
involve mixing or heating and requires a long retention time. A
conventional digester is operated with heating but without mixing,
where scum layer, supernatant, actively digesting sludge and
stabilized sludge are stratified. A high-rate digester is mixed
sufficiently to turn the whole digester into an active zone.
Table 3 shows operating conditions in these two types of digester
operation (Water Pollution Control Federation, 1968). The high-
rate digestion process is more efficient than the conventional one,

Table 3. Operation of Anaerobic Digestion

Parameter	Conventional	High-rate
loading rate (kgVS/m^{3}d)	0.48-1.12	1.6-6.4
Retention time (d)	30-60	15 or less
feed and withdrawal manner	intermittent	continuous or intermittent
mixing condition	stratification	homogeneity

although the former usually has to be followed by the second stage digester in order to separate the supernatant from the solid. Therefore, the total hydraulic retention time of the high-rate digestion is not much shorter than that of the conventional type.

Operation of the two phase digestion process is completely different from the operations described above. This process is based on the idea of separating the acidogenic reaction from the methanogenesis. Although several operations have been proposed for the separation of the two types of bacteria, the one proposed by Ghosh et al. (1975) seems to be most feasible. He proposed to operate the acidogenic reactor under a very short retention time (about one day), where methanogenic bacteria cannot keep their number and are washed out. As this process is relatively new, its advantages and disadvantages are yet to be made clear.

The digestion processes are classified into mesophilic and thermophilic digestion according to the temperature employed. The optimum range of temperature is believed to be about 30-40°C and about 50-60°C (McCarty, 1964a) for mesophilic and thermophilic digestion, respectively. The effects of temperature will be discussed in Section 4.1.

3. STEADY STATE MODELS

3.1 Formulation of models

Kinetics of the anaerobic digestion process is not a particular one and the universal model for biological treatment process can basically be applied. Anaerobic digestion has two phases, acidogenic and methanogenic phase as stated before, but a common type of model is employed to describe both of them. Material balance and rate equation of reaction are considered to formulate kinetic equations.

3.1.1 Material balance

Fig. 3 shows a scheme of the anaerobic digester, which is mixed completely and fed continuously without solid recycle. Material balance for the substrate gives

$$dS/dt = Q(S_1 - S_2)/V - r_v \qquad (9)$$

where,

S = substrate concentration in the reactor,
Q = flowrate,
S_1 = influent substrate concentration,
S_2 = effluent substrate concentration ($S_2 = S$),
V = reactor volume,
r_v = substrate utilization rate per volume, and
t = time.

As only the steady state condition is considered here, $dS/dt = 0$. Therefore,

$$r_v = (S_1 - S_2)D \qquad (10)$$

where,

$D = Q/V$ = dilution rate (reciprocal of hydraulic retention time, Θ).

Specific substrate utilization rate, r_X, is defined as follows:

$$r_X = r_v/X \qquad (11)$$

where,

X = biomass concentration in the reactor.

Eq. 10 is rewritten as

$$r_X = (S_1 - S_2)D/X \qquad (12)$$

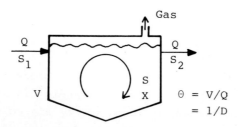

Fig. 3. Scheme of Completely Mixed Digester

Material balance for biomass gives the following equation:

$$dX/dt = -QX/V + \mu X \qquad (13)$$

where,

μ = specific biomass growth rate.

Biomass in the influent is ignored here. Under the steady state condition, Eq. 13 yields

$$\mu = D = 1/\theta \qquad (14)$$

The growth of micro-organisms is proportional to the substrate utilization, and biomass decay or endogenous metabolism should also be taken into account to describe the specific growth rate as follows:

$$\mu = Y_G r_X - b \qquad (15)$$

where,

Y_G = growth yield coefficient, and
b = specific biomass decay coefficient.

Combining Eqs. 14 and 15 gives

$$D = Y_G r_X - b \qquad (16)$$

or

$$r_X = \frac{D}{Y_G} + \frac{b}{Y_G} \qquad (17)$$

These coefficients can be evaluated by making laboratory-scale experiments. Chemostat-type reactor is usually operated under various retention times, in other words, various dilution rates. Influent substrate concentration, S_1, is controlled and effluent substrate concentration, S_2, and biomass concentration, X, are measured. Specific substrate utilization rates are calculated according to Eq. 12 and are plotted against dilution rate, D. The slope and the intercept of the regression line for these plots give Y_G and b according to Eq. 17.

Ghosh and Pohland (1974) introduced substrate utilization coefficients to describe a formation of end products based on Pirt's theory (Pirt, 1975). It was assumed that the substrate utilization is divided into three parts, namely one for the energy production of synthesis, one for the energy production of maintenance and one for assimilation to biomass, and that the end products are produced in the substrate utilization for energy production. The relationship between substrate utilization and biomass production is

$$S_1 - S_2 = (U_p + U_e + m\Theta)X \qquad (18)$$

where,

U_p = substrate utilization coefficient for synthesis,

U_e = substrate utilization coefficient for energy production, and

m = maintenance coefficient for micro-organism.

Eqs. 12 and 18 yield

$$r_X = (U_p + U_e)D + m \qquad (19)$$

That is

$$D = \frac{r_X}{U_p + U_e} - \frac{m}{U_p + U_e} \qquad (20)$$

By comparing Eq. 20 with Eq. 16, the following is obtained:

$$Y_G = 1/(U_p + U_e) \qquad (21)$$

$$b = mY_G \qquad (22)$$

The formation of the end products is expressed as

$$P = Y_P(U_e + m\Theta)X \qquad (23)$$

where,

P = concentration of products in effluent, and

Y_P = true product yield coefficient.

Utilization of end products is neglected here. The observed product yield coefficient, $Y_{P,obs}$, is defined as

$$Y_{P,obs} = \frac{P}{S_1 - S_2} \qquad (24)$$

and can be expressed as follows:

$$Y_{P,obs} = \frac{Y_p(U_e + m\Theta)}{U_p + U_e + m\Theta} \qquad (25)$$

These considerations are useful especially in the acidogenic phase where organic acids are the main end products. The evaluation of the organic acids production in the acidogenic phase is important because these acids become substrates in the methanogenic phase. The coefficients, U_p, U_e, Y_P, m are estimated experimentally using

Eqs. 20 and 23. Matsumoto et al. (1981) claimed that the substrate utilization coefficient for growth, U_p, can be calculated theoretically for a given substrate. Assuming that the average formula of anaerobic bacteria is $C_5H_9O_3N$ (Speece & McCarty, 1962), the carbon content of biomass is 45.8%. As the organic carbon in the substrate is assimilated and constitutes the new biomass, carbon balance gives the value of U_p, which is a ratio of the substrate utilized for biomass assimilation to biomass produced. For instance, when glucose, which has a carbon content of 40%, is used as a substrate, U_p can be calculated as

$$U_p = \frac{100/40.0}{100/45.8} = 1.15 \tag{26}$$

3.1.2 Rate equation

The substrate utilization rate depends on the substrate concentration and several types of equations have been proposed to describe the relationship. The most popular is the Monod-type equation which is written as follows:

$$r_X = \frac{S}{K_S + S} \; r_{X,max} \tag{27}$$

where,

$r_{X,max}$ = maximum specific substrate utilization rate,
K_S = saturation constant for substrate, and
S = substrate concentration in the reactor ($S = S_2$).

The value of r_X is half of $r_{X,max}$ when substrate concentration is K_S. This equation is quite similar to the equation proposed by Monod, which describes the relationship between the growth rate and the substrate concentration. One of the following three types of linear plot can be used to evaluate $r_{X,max}$ and K_S using experimental data of r_X and S.

$$\frac{1}{r_X} = \frac{K_S}{r_{X,max}} \frac{1}{S} + \frac{1}{r_{X,max}} \tag{28}$$

$$r_X = r_{X,max} - K_S \frac{r_X}{S} \tag{29}$$

$$\frac{S}{r_X} = \frac{S}{r_{X,max}} + \frac{K_S}{r_{X,max}} \tag{30}$$

These are called Lineweaver-Burk plot, Eadie plot and Hofstee plot, respectively. Maximum specific biomass growth rate, μ_{max}, is obtained by substituting $r_{X,max}$ for r_X in Eq. 15 as follows:

$$\mu_{max} = Y_G r_{X,max} - b \tag{31}$$

A critical retention time, Θ_C, is defined as the retention time below which substrate utilization does not occur because of the wash out of micro-organisms. By using Eq. 14, Θ_C is expressed as

$$\Theta_C = \frac{1}{\mu_{max}} \tag{32}$$

The substrate and the biomass concentrations in the effluent under a given dilution rate are expressed using these coefficients as follows:

$$S_2 = \frac{K_S(D + b)}{Y_G r_{X,max} - (D + b)} \tag{33}$$

$$X = \frac{Y_G D}{D + b} (S_1 - S_2) \tag{34}$$

A few other types of rate equation are also used to describe substrate utilization in the anaerobic digestion. The first order equation is sometimes used because of its simplicity when the overall anaerobic digestion process is described. A volumetric substrate utilization rate is

$$r_v = kS \tag{35}$$

where,

k = first order rate constant.

Replacing S in Eq. 27 with S/X gives Contois-type equation such as

$$r_X = \frac{(S/X)}{K_{S,C} + (S/X)} r_{X,max} \tag{36}$$

where,

$K_{S,C}$ = kinetic coefficient in Contois-type model.

This equation is effective when the ratio of substrate to biomass governs the substrate utilization rate.

Chen and Hashimoto (1978) proposed the following model.

$$\mu = \frac{S_2/S_1}{K + (1 - K)S_2/S_1} \mu_{max} \tag{37}$$

where,

K = dimensionless kinetic coefficient.

As they disregarded the term of biomass decay in Eq. 15 and let the biomass growth rate be proportional to the substrate utilization rate, Eq. 37 is equivalent to :

$$r_X = \frac{S_2/S_1}{K + (1 - K)S_2/S_1} r_{X,max} \tag{38}$$

Rearranging Eq. 37 using Eq. 32 yields

$$\frac{S_2}{S_1} = \frac{K}{\theta/\theta_c - 1 + K} \tag{39}$$

This equation suggests that S_2/S_1 is a function of hydraulic retention time, θ, only and effluent substrate concentration, S_2, depends on influent substrate concentration, S_1, while S_2 is independent of S_1 in Monod-type model as shown in Eq. 33. However, experimental data show that the kinetic coefficient, K, actually varies according to the influent substrate concentration (Chen and Hashimoto, 1978).

3.2 Application of models to the anaerobic digestion process

The description of the anaerobic digestion process using the formulated models and the estimation of kinetic coefficients are effective means to clarify the process characteristics and provide basic informations for the design criteria. The kinetics of acidogenesis and methanogenesis should be evaluated separately by using the two phase system since these two steps are distinct from each other, while the estimation of the overall kinetics of anaerobic digestion is also useful from a practical aspect. Some examples of the application of the model using the steady state data from the anaerobic chemostat reactor are presented and some comparisons of the kinetic coefficients are made here.

3.2.1 Acidogenic phase

Although a great variety of substrates is utilized in the acidogenic phase, kinetic coefficients were reported only for limited kinds of substrates. Glucose is probably the simplest substrate and supports the growths of various species of acidogenic bacteria. A procedure for the estimation of kinetic coefficients for acidogenesis from glucose using experimental data obtained by Matsumoto et al. (1981) is shown below as an example. A substrate which consisted of glucose, inorganic nitrogen, phosphate and

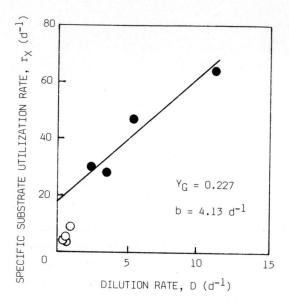

Fig. 4. Estimation of Y_G and b according to Eq. 17 for Acidogenesis from Glucose

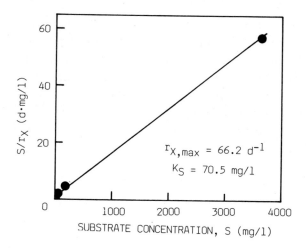

Fig. 5. Estimation of K_S and $r_{X,max}$ according to Eq. 30 for Acidogenesis from Glucose

several elements was fed continuously to a completely mixed reactor at 35°C and effluent glucose concentration and mixed liquor volatile suspended solids (MLVSS) were measured. A plot is made to evaluate Y_G and b according to Eq. 17 as shown in Fig. 4. The regression line gives $Y_G = 0.227$ and b $= 4.13$ d^{-1}. The data at a dilution rate less than 2 d^{-1} (expressed by open circles in Fig. 4)

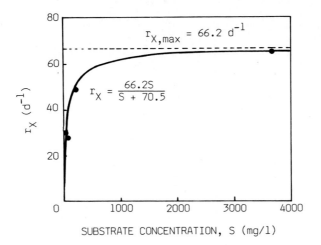

Fig. 6. Experimental Data and Calculated Curve of Monod-type Model

were omitted in calculating the regression line, because the activity of methanogenic bacteria as well as acidogenic bacteria was observed. Coefficients K_S and $r_{X,max}$ in a Monod-type model can be estimated according to Eq. 30 as shown in Fig. 5. The regression line gives an equation

$$r_X = \frac{66.2S}{S + 70.5} \tag{40}$$

where r_X and S are expressed as d^{-1} and mg/l, respectively. Fig. 6 illustrates the calculated curve (Eq. 40) and the experimental data. The substrate utilization coefficient, U_p, is 1.15 as already shown in Eq. 26 when glucose is the substrate. The substrate utilization rate for energy production, U_e, is calculated according to Eq. 21, which yields

$$U_e = \frac{1}{Y_G} - U_p$$

$$= 3.26 \tag{41}$$

The true product yield coefficient, Y_p, is estimated using Eq. 23, where the concentration of products, P, is represented by total organic acids (mg/l as CH_3COOH). The value of Y_p is estimated for each dilution rate and the average of Y_p is 0.33. Kinetic coefficients for glucose degradation obtained by Ghosh and Pohland (1974) are shown with the above coefficients in Table 4. They estimated the biomass concentration by measuring the dehydrogenase

Table 4. Kinetic Coefficients for Acidogenesis (35-37°C)

Substrate	Y_G	b d^{-1}	$r_{X,max}$ d^{-1}	K_S mg/l	θ_C d	substrate index	biomass estimation	reference
Glucose	0.227	4.13	66.2	70.5	0.092	glucose	MLVSS	Matsumoto et al. (1981)
Glucose	0.173	1.06*	180	22.5	0.033	glucose	dehydrogenase activity	Ghosh & Pohland (1974)
Soluble starch	0.350	3.61	37.5	591	0.105	carbohydrate	MLVSS	Yaguchi (1982)
	0.259	1.15	42.3	836	0.102	carbohydrate	organic-N	
Cellulose	0.16	——	10.7*	36800	0.584	cellulose	MLVSS	Ghosh & Klass (1978)
Waste activated sludge	0.40	0.32*	10.4*	26000	0.240	volatile solids	dehydrogenase activity	Ghosh et al. (1975)

* calculated value according to Eqs. 22 and 31.

activity which may have caused the difference between their coefficients and that of Matsumoto et al. (1981).

The degradation of polysaccharide such as cellulose and starch includes the hydrolysis step which limits more or less the substrate utilization rate. Kinetic coefficients for soluble starch are also shown in Table 4. The maximum specific substrate utilization rate is smaller than that of glucose and the saturation constant is about ten times the value for glucose. A larger saturation constant means a lower affinity of biomass for the substrate. When the substrate concentration is 600 mg/l, for example, specific substrate utilization rates are about 90 and 50% of the maximum value for glucose and soluble starch, respectively.

The saturation constant for cellulose is extremely high as shown in Table 4, which means that the acidogenesis from cellulose is a first order reaction within the usual range of substrate concentration. Endo (1980) showed that the Contois-type model (Eq. 36) described the cellulose degradation better than the Monod-type model and reported kinetic coefficients at 35°C; $r_{X,max} = 1.11$ d^{-1} and $K_{S,C} = 4.09$.

Besides synthetic wastes listed above, sewage sludge from actual treatment plants were used to evaluate the coefficients for acidogenesis. Eastman and Ferguson (1981) showed that the hydrolysis of raw primary sludge was a first order reaction and the kinetic constant was 3.0 d^{-1}. They reported that the lipid fraction of the sludge was not decreased during acidogenesis, which suggested that the degradation of long-chain fatty acids by hydrogen-producing acetogenic bacteria occurred in the methanogenic phase, as pointed out in Section 2.1. Ghosh et al. (1975) applied a Monod-type model to the acidogenesis from a waste activated sludge and obtained coefficients (Table 4). Values of $r_{X,max}$ and K_S are similar to those for cellulose.

3.2.2 Methanogenic phase

Acetate is the most important precursor in methanogenesis in the anaerobic digestion. Chang et al. (1982) carried out a study to examine the kinetics of methanogenesis from acetate. These researchers varied not only the dilution rate but also the influent acetate concentration in a chemostat-type experiment at 35°C. The effluent substrate concentration should be constant irrespective of the influent substrate concentration in a Monod-type model as shown in Eq. 33, but they reported that a high acetate loading rate

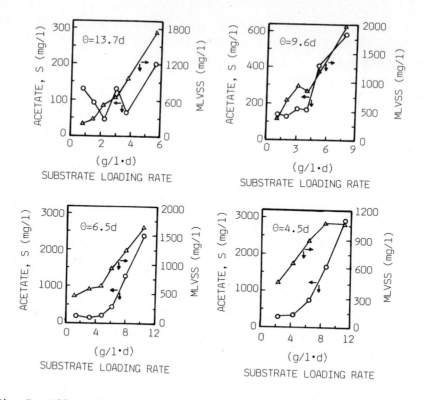

Fig. 7. Effect of Loading Rate on Effluent Acetate Concentration
in Methanogenesis from Acetate

brought about a high concentration of acetate in the effluent (Fig. 7). They also showed that a substrate loading rate higher than around 12 g/l·d caused a process failure, which was observed as a sudden drop of methane production and rapid accumulation of acetate. The growth yield coefficient, Y_G, and biomass decay coefficient, b, were determined for each influent substrate concentration by plotting r_X versus D according to Eq. 17. Fig. 8 shows a family of the regression lines. The slope of the line becomes high, which means small Y_G value, with the increase of S_1. The regression line for estimating $r_{X,max}$ and K_S is also influenced by S_1 (Fig. 9). The kinetic coefficients for different influent acetate concentrations are compared in Table 5. Y_G and b decrease, and $r_{X,max}$ and K_S increase by increasing the influent acetate concentration, while θ_C is comparatively constant. This result suggests that the Monod-type model cannot describe the methanogenesis from acetate perfectly. Kinetic coefficients obtained by Lawrence and McCarty (1969) are also shown in Table 5.

603

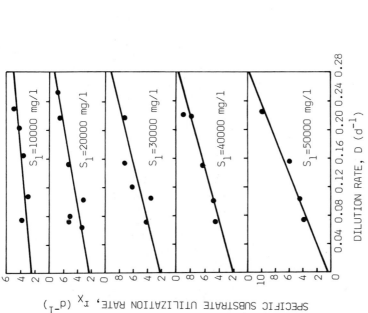

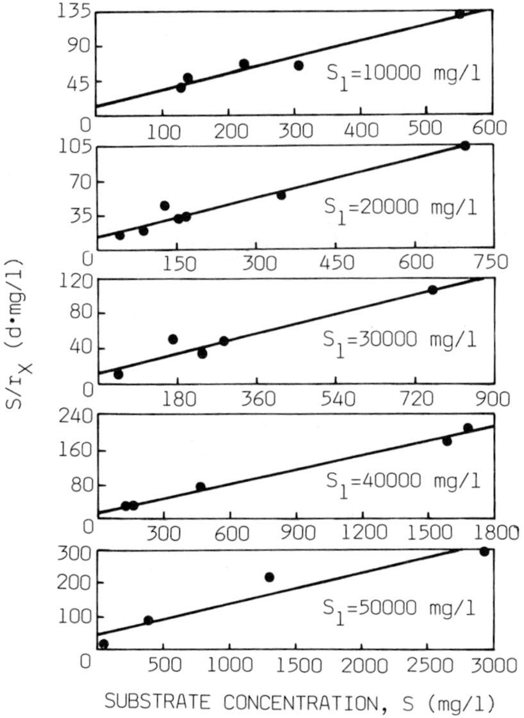

Fig. 9. Effect of Influent Acetate Concentration on $r_{X,max}$ and K_S

Fig. 8. Estimation of Y_G and b for Several Influent Acetate Concentrations

Table 5. Kinetics of Methanogenesis from Acetate (35°C)

S_1 mg/l	Y_G	b d^{-1}	$r_{X,max}$ d^{-1}	K_S mg/l	Θ_C d	substrate basis	biomass estimation	reference
10,000	0.115	0.283	4.72	53	3.85			
20,000	0.056	0.128	7.62	90	3.35			
30,000	0.040	0.090	8.02	101	4.33	acetic acid	MLVSS	Chang et al. (1982)
40,000	0.037	0.070	8.22	157	4.40			
50,000	0.025	0.010	10.91	508	3.81			
1,568	0.040	0.019	12.3	207	2.0	acetic acid	organic-N	Lawrence & McCarty (1969)
3,135	0.040	0.019	8.0	155	3.2			

Another important reaction is methane formation from hydrogen and carbon dioxide (Eq. 1). It is difficult to estimate the kinetics of this reaction by continuous-fed experiments because substrate must be supplied in the gas phase: a study made by Shea et al. (1968) is probably the only case and their results are shown in Table 6. The hydrogen partial pressure in a normal methanogenic digester must be lower than 10^{-4} atm to permit the degradation of propionate as discussed before (Table 2). The hydrogen partial pressures in a laboratory-scale digester and in an actual digester were reported to be about 5×10^{-5} atm (Hanaki et al., 1981) and less than 10^{-4} atm (Kasper & Wuhrmann, 1978b), respectively. These values are about 10^4 times smaller than K_S value of Shea et al. (1968), which means that

$$
\begin{aligned}
r_X &= \frac{S}{K_S + S}\, r_{X,max} \\
&= 10^{-4} \times r_{X,max} \\
&= 2.5\times10^{-3}\ (d^{-1})
\end{aligned}
\tag{42}
$$

However, r_X for methanogenesis from hydrogen in an actual digester is much larger than this calculated value. This comparison raises a suspicion that Shea et al. (1968) might have overestimated the K_S value or underestimated the $r_{X,max}$ value. As they obtained a fairly good regression line for the estimation of these coefficients, it is unlikely that incorrect coefficients are caused by the experimental error. A pure culture study of hydrogen-utilizing thermophilic methanogenic bacteria, <u>Methanobacterium thermoautotrophicum</u> gave $K_S = 0.2$ atm at 65°C (Schönheit et al., 1980) and Kaspar and Wuhrmann (1978b) reported $K_S = 0.105$ atm for a mixed culture. These considerations suggest that the Monod-type

Table 6. Kinetic Coefficients for Methanogenesis
from Hydrogen at 37°C (Shea et al., 1968)

Y_G	0.043
b	$-0.009\ d^{-1}$
$r_{X,max}$	$24.8\ d^{-1}$
K_S	569 mmHg or 0.749 atm
Θ_C	0.943 d

Utilized substrate and biomass are expressed as
COD and MLVSS, respectively.

equation should not be extrapolated to the lower range of substrate concentration when the kinetic coefficients have been estimated by the experiment in high substrate concentration. Kinetic studies under low hydrogen partial pressures may be necessary to characterize the hydrogen utilization in actual digesters. Hungate (1975) showed that K_S was 1 µM, approximately about 1.5×10^{-3} atm in a rumen where methane is produced only from hydrogen.

Fatty acids with a carbon chain longer than that of acetate are degraded by hydrogen-producing acetogenic bacteria which do not act without the presence of methanogenic bacteria. As an enrichment culture of only hydrogen-producing acetogenic bacteria cannot be obtained, the consortium composed of this type of bacteria, hydrogen-utilizing methanogenic bacteria and acetate-utilizing methanogenic bacteria is usually used to estimate kinetic coefficients; although the presence of acetate-utilizing methanogenic bacteria, however, is not indispensable for hydrogen-producing acetogenic bacteria. Kinetic coefficients for the degradations of n-butyrate and propionate, both of which are important intermediates in anaerobic digestion, are shown in Table 7. Chang (1982) did not distinguish the biomass of hydrogen-producing acetogenic bacteria from that of other bacteria so that the growth yield and specific utilization rate were based on biomass of the whole organisms which consist of hydrogen-producing acetogenic bacteria, hydrogen-utilizing methanogenic bacteria and acetate-utilizing methanogenic bacteria. Lawrence and McCarty (1969) introduced a substrate-specific fraction of microbial mass. They assumed that propionate or n-butyrate was degraded to acetate and methane by substrate-specific methanogenic bacteria which did not utilize acetate, and that the growth rate of each methanogenic bacteria was proportional to the methane production rate from each substrate.

The effluent of an acidogenic reactor contains acetate, propionate and n-butyrate, and flows into the methanogenic reactor in the two phase digestion process. Performance in the methanogenic phase can be predicted theoretically using the kinetic coefficients obtained for each volatile fatty acid. However, Chang (1982) made an experiment in which a mixture of acetate, propionate and n-butyrate was fed to a methanogenic reactor, and showed that its performance was better than had been predicted.

Long-chain fatty acids liberated by the hydrolysis of neutral fats are decomposed into acetate and hydrogen, and then converted to methane in the methanogenic phase. Novak and Carlson (1970)

Table 7. Kinetic Coefficients for Degradation of Fatty Acids Higher than Acetate (35-37°C)

Substrate	Y_G	b d^{-1}	$r_{X,max}$ d^{-1}	K_S mg/l	Θ_C d	substrate index	biomass estimation	reference
Propionate	0.043	0.092	6.46	15	5.38	COD	MLVSS	Chang (1982)
n-Butyrate	0.022	0.005	7.96	42	5.88			
Propionate	0.042	0.010	9.6	32	2.54*	fatty acid (as acetate)	organic-N	Lawrence & McCarty (1969)
n-Butyrate	0.047	0.027	15.6	5	1.42*			
Myristate (14:0)	average	average	0.95	105	10.6*	fatty acid	MLVSS minus insoluble fatty acid	Novak & Carlson (1970)
Palmitate (16:0)			1.0	143	10.0*			
Stearate (18:0)	0.10-	0.00-	0.77	417	13.4*			
Oleate (18:1)	0.12	0.02	4.0	3180	2.33*			
Linolate (18:2)			5.0	1816	1.85*			
Long-chain fatty acids in sewage sludge			6.67	2000	4.0	COD		O'Rourke (1968)

* calculated value according to Eqs. 31 and 32.

studied the degradation of various long-chain fatty acids, which often appear in the sewage sludge digestion (Heukelekian and Mueller, 1958). Novak and Carlson (1970) did not distinguish hydrogen-producing acetogenic bacteria which degraded long-chain fatty acids from methanogenic bacteria although they estimated the biomass by subtracting the insoluble fatty acids from the observed value of MLVSS. The kinetic coefficients obtained by these researchers and those obtained by O'Rourke (1968) for long-chain fatty acids in sewage sludge are shown in Table 7. The maximum specific substrate utilization rates, $r_{X,max}$, for oleate and linolate are much higher than that for saturated fatty acids, although saturation constants, K_S, for oleate and linolate are also high, suggesting that the substrate utilization rates for them are lower than the maximum values in an actual digester.

3.2.3 Comparison of kinetic coefficients

The comparison of kinetic coefficients between the acidogenic phase (Table 4) and methanogenic phase (Tables 5-7) reveals the differences between their characteristics. The growth yield coefficient, Y_G, for acidogenic bacteria is 0.16-0.40, while that for methanogenic bacteria and hydrogen-producing acetogenic bacteria is 0.02-0.12. This comparison suggests that acidogenic bacteria constitute the majority of the population in the anaerobic digestion process in which the acidogenic and methanogenic reactions simultaneously take place. The much larger maximum specific utilization rate, $r_{X,max}$, for the acidogenesis from glucose or starch compared to that for the methanogenic phase means that the reactions in methanogenic phase limit the overall rate of methane fermentation from soluble carbohydrates. On the other hand, $r_{X,max}$ for the acidogenesis from cellulose or sewage sludge is comparable to that for the methanogenic phase and the saturation constant for them is very high. These comparisons suggest that the hydrolysis of cellulose or sewage sludge rather than methanogenesis could be the rate-limiting step. Critical retention time, Θ_C, is an important parameter for the process design and operation and it gives the retention time required to maintain the biomass properly. Acidogenic bacteria need a much shorter retention time (0.03-0.6 day) than hydrogen-producing acetogenic bacteria and methanogenic bacteria (0.9-13 days). This difference enables the separation of the acidogenic phase from the methanogenic phase by controlling the retention time in a two phase digestion process.

The typical values of kinetic coefficients in an activated sludge process, which is a representative aerobic treatment system, are as follows: Y_G = 0.4 mgVSS/mgCOD, b = 0.06 d^{-1}, $r_{X,max}$ = 5.0 d^{-1}, K_S = 40 mg/l in 20°C (Metcalf & Eddy, Inc., 1979), and a calculation from these values gives θ_C = 0.52 d. The kinetic coefficients for the acidogenesis from glucose or starch are comparable to those for the aerobic system, while those for the methanogenic phase are not.

3.2.4 Overall kinetics of the anaerobic digestion

Although the anaerobic digestion is a dual phase process, kinetic coefficients for the overall process are sometimes evaluated from a relatively practical point of view. Table 8 shows some examples of the kinetic coefficients. The growth yield coefficient is high because the acidogenic bacteria contribute a large part of the biomass except for the case for amino acids or fatty acids. The maximum specific utilization rate is governed by the activity of the methanogenic bacteria and is underestimated because it is calculated by dividing the reaction rate of methanogenesis by the whole biomass which includes large amount of acidogenic bacteria.

The first order model was applied to describe the methane fermentation of several kinds of organic matter such as brewery by-product (Keenan and Kormi, 1977) and cow manure (Shelef et al., 1980) where biomass cannot be estimated. However, this model does not seem to describe the process very well.

Chen and Hashimoto (1978) applied their model (Eqs. 37-39) to the results which had been reported by several investigators and showed that the kinetic coefficient, K, depended on the influent substrate concentration. Hashimoto (1982) expressed the effect of influent substrate concentration, S_1, on K during the methane fermentation of beef-cattle manure as follows:

$$K = 0.8 + 0.016\exp(0.06S_1) \tag{43}$$

where,

K = dimensionless kinetic coefficient, and

S_1 = influent substrate concentration (kg VS/m^3).

The meaning of the kinetic coefficient, K, seems to be somewhat obscure in this model and the collection of K values for various kinds of substrate is necessary.

Table 8. Overall Kinetics of Anaerobic Digestion (34-37°C)

Substrate	Y_G	b d^{-1}	$r_{X,max}$ d^{-1}	K_S mg/l	Θ_C d	substrate index	biomass estimation	reference
Landfill leachate	0.10	0.006	5.88	4020	1.7	BOD	MLVSS	Cameron & Koch (1980)
Wine stillage	0.268	0.001	0.488	0.062	7.71*	carbon	carbon	Sanders (1977)
Glucose or starch	0.46	0.088						
Nutrient broth	0.076	0.014				COD	suspended solids	Speece & McCarty (1962)
Amino acids or fatty acids	0.054	0.038						
Brewery by-product	0.421	0.026				COD	VSS	Keenan & Kormi (1977)

* calculated value according to Eqs. 31 and 32.

4. ENVIRONMENTAL FACTORS AFFECTING MATHEMATICAL MODELLING

4.1 Temperature

The anaerobic digestion process is affected by temperature in the same way as all the other biological wastewater treatment processes. Many studies have been devoted to the temperature characteristics of the anaerobic digestion process. Fair and Moore (1932) rearranged the experimental results on effects of temperature on anaerobic digestion of sewage sludge in order to develop design criteria for a digestion tank. They showed that there are two significant temperature zones: (a) a zone of high temperatures in which thermophilic organisms are responsible for digestion, and (b) a zone of moderate temperatures in which mesophilic organisms are active.

For convenience, anaerobic digestion has been regarded as a dual step process composed of the acidogenic phase and the methanogenic phase, each of which is carried out by two physiologically distinct group of bacteria as stated in Section 2.1. It should be recognized, however, that in the digester there is a continuum of metabolic reactions probably brought about by a complex intimate association between hydrogen-producing acetogenic bacteria and methanogenic bacteria. Fair and Moore (1932), and Golueke (1958) investigated the effects of temperature on the overall anaerobic digestion. Zoetemeyer et al. (1982) and Endo et al. (1983) separated the acidogenic phase from the methanogenic phase using the difference of growth rate of populations responsible for acid and methane production in order to grasp the temperature characteristics of the acidogenic phase.

4.1.1 Kinetic models based on the effects of temperature

Topiwara and Sinclair (1971) introduced the effects of temperature into the Monod equation and the mass balance. They expressed maximum specific biomass growth rate, specific biomass decay coefficient, saturation constant and growth yield coefficient by temperature dependent functions, $\mu_{max}(T)$, $b(T)$, $K_S(T)$ and $Y_G(T)$. They investigated the relationship between kinetic coefficients and temperature on the basis of experimental results by pure culture of Aerobacter aerogenes and reported that μ_{max}, K_S and b were dependent on temperature, but Y_G was independent.

Chen et al.(1980) and Hashimoto (1982) showed that temperature influenced the coefficient μ_{max}, but not K in their model (Eq. 37) when beef cattle manure was applied to the anaerobic digestion process in the temperature range of 35-65°C.

Novak (1974) suggested that the effect of temperature on the kinetic coefficients in the Monod-type model (Eq. 27) could be estimated in the following manner:

$$r_{X,max}(T) = r_{X,max}(T_o) \exp\{C_1(T - T_o)\}$$ (44)

$$K_S(T) = K_S(T_o) \exp\{C_2(T - T_o)\}$$ (45)

where,

$r_{X,max}(T) = r_{X,max}$ at a temperature, T°C;

$r_{X,max}(T_o) = r_{X,max}$ at the reference temperature, T_o°C;

$K_S(T) = K_S$ at a temperature, T°C;

$K_S(T_o) = K_S$ at the reference temperature, T_o°C; and

C_1, C_2 = constants = to the slope of $\ln(r_{X,max})$ and $\ln(K_S)$ versus temperature lines, respectively. Combining Eqs. 27, 44 and 45 yields the following equation:

$$r_X(T) = \frac{S \cdot r_{X,max}(T_o) \exp\{C_1(T - T_o)\}}{K_S(T_o) \exp\{C_2(T - T_o)\} + S}$$ (46)

where,

$r_X(T)$ = specific substrate utilization rate at T°C.

The relationship between the rate constant and temperature can be expressed also by the Arrhenius equation.

$$\frac{d(\ln k)}{dT_A} = \frac{E_a}{RT_A^2}$$ (47)

where,

k = first order rate constant,

R = gas constant,

E_a = a constant for the reaction termed the activation energy, and

T_A = absolute temperature.

Integrating Eq. 47 gives

$$\ln k = -E_a/RT_A + \text{const}$$ (48)

A plot of ln k versus $1/T_A$ according to Eq. 48 gives a value of activation energy of which the biochemical meaning is not clear. Here, temperature dependencies of substrate utilization, microbial growth and acid fermentation rate are investigated using the following equation (Eq. 49) in which the activation energy is not introduced:

$$r(T_A) = C \exp(-u/T_A) \tag{49}$$

$$\ln r(T_A) = -u/T_A + C' \tag{50}$$

$$\ln\{r(T_{A,1})/r(T_{A,2})\} = -u(1/T_{A,1} - 1/T_{A,2}) \tag{51}$$

where,

$r(T_A)$ = reaction rate at T_A°K,

u = temperature dependency coefficient, and

C, C' = constants.

4.1.2 Effects of temperature on the acidogenic phase

Endo et al. (1983) carried out the continuous experiments on the effects of temperature on the acidogenic phase using an anaerobic chemostat reactor with glucose as a main substrate, retention time of one day, and temperatures ranging from 5 to 60°C at intervals of 5°C.   Fig. 10 shows average values of glucose concentration and soluble COD of the mixed liquor in reactors. Fig. 11 shows the steady state MLVSS and ammonium concentration and

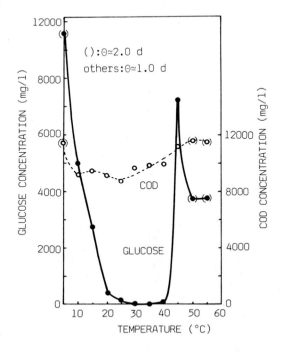

Fig. 10. Effect of Temperature on the Concentration of Glucose and Dissolved COD (Endo et al., 1983)

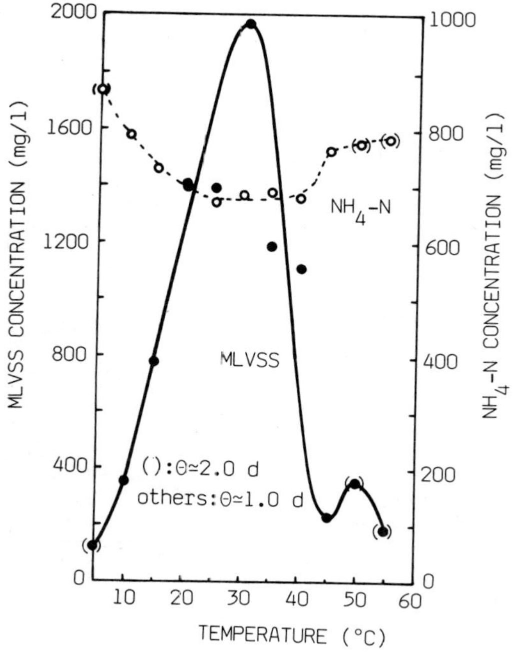

614

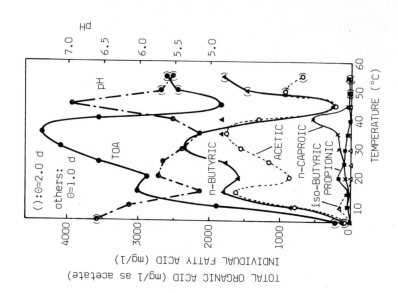

Fig. 12.   Effect of Temperature on pH and Distribution of Volatile Fatty Acid (Endo et al., 1983)

Fig. 11.   Effect of Temperature on the Concentration of MLVSS and NH$_4$-N (Endo et al., 1983)

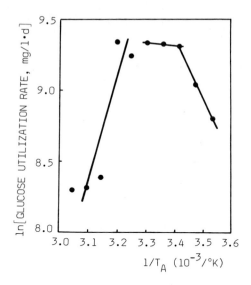

Fig. 13.   A Plot of $1/T_A$ versus Natural Log of Glucose
Utilization Rate (Endo et al., 1983)

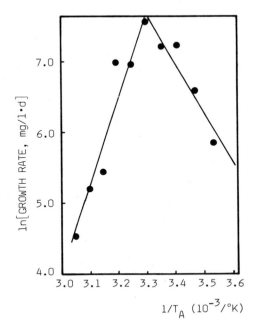

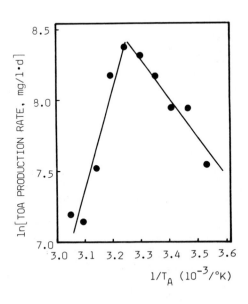

Fig. 14.   A Plot of $1/T_A$ versus
Natural Log of Growth Rate
(Endo et al., 1983)

Fig. 15.   A Plot of $1/T_A$ versus Nat-
ural Log of Total Organic
Acid Production Rate
(Endo et al., 1983)

Table 9. Temperature Dependent Coefficient, u, in Eq. 51. (Endo et al., 1983)

Reaction	Temperature range (°C)	u (°K)
Glucose consumption	10 - 20	4260
	35 - 55	-5900
Microbial growth	10 - 30	6950
	30 - 55	-13400
Total organic acid production	10 - 35	2750
	35 - 55	-6860

Fig. 12 shows the steady state total organic acid (TOA) and individual volatile fatty acids at each experimental temperature. The growth rate of acidogenic bacteria has a maximum at 30°C and TOA production rate has a maximum at 35°C. The 45°C temperature forms the boundary between mesophilic and thermophilic zones where the rates of acid production and microbial growth begin to decrease. Therefore, effects of temperature on substrate utilization, bacterial growth and acid production are alike in character.

Figs. 13 to 15 show plots of glucose utilization rate, growth rate of acidogenic bacteria and fermentation rates of organic acids versus $1/T_A$ according to Eq. 51. The results shown in these figures give temperature coefficients, u, in Table 9. Positive value of u means that the rate of substrate utilization, microbial growth or TOA production increases with the increase of temperature, and vice versa.

4.1.3   Effects of temperature on the methanogenic phase

Lawrence and McCarty (1969) reported that the coefficient K_S, but not $r_{X,max}$ varied with temperature (25-35°C) in the methanogenesis from acetate, and expressed the following equation similar to Eq. 51.

$$\log \left\{ \frac{K_S(T_{A,2})}{K_S(T_{A,1})} \right\} = 6980 \left\{ \frac{1}{T_{A,2}} - \frac{1}{T_{A,1}} \right\} \tag{52}$$

or

$$\ln \left\{ \frac{K_S(T_{A,2})}{K_S(T_{A,1})} \right\} = 16070 \left\{ \frac{1}{T_{A,2}} - \frac{1}{T_{A,1}} \right\} \tag{53}$$

where,

$K_S(T_{A,1})$, $K_S(T_{A,2})$ = K_S values at $T_{A,1}$ and $T_{A,2}$ °K, respectively.

O'Rourke (1968) made experiments in the temperature range of 20 to 35°C and described the effect of temperature on the degradation of lipid in sewage sludge by Eqs. 44 and 45 as follows:

$$r_{X,max}(T) = 6.67 \times 10^{0.015(T - 35)} \quad (d^{-1}) \tag{54}$$

$$K_S(T) = 2224 \times 10^{-0.046(T - 35)} \quad (mg/l) \tag{55}$$

or

$$r_{X,max}(T) = 6.67 \exp\{0.035(T - 35)\} \quad (d^{-1}) \tag{56}$$

$$K_S(T) = 2224 \exp\{-0.106(T - 35)\} \quad (mg/l) \tag{57}$$

These equations should not be extrapolated beyond the temperature range in which the experiments were carried out.

4.2 Volatile fatty acids and pH

The accumulation of volatile fatty acids and the drop of pH are usually observed simultaneously when the anaerobic digestion process becomes upset. The primary factors which hinder the activity of methanogenic bacteria are reported as follows:

1. high concentration of volatile fatty acids, higher than 2,000 mg/l as acetate (Buswell, 1947; Schlenz, 1947);
2. H^+ ions, higher than $10^{-6.5}$ mol/l, i.e. pH lower than 6.5 (McCarty and McKinney, 1961; Radhakrishnan et al., 1969; Dague et al., 1970); and
3. un-ionized volatile fatty acid, higher than 10 mg/l as acetate (Andrews, 1969; Kroeker et al., 1979).

The relationship among un-ionized and ionized volatile fatty acid and pH is shown as follows:

$$pH = pK_a + \log \frac{[S^-]}{[HS]} \tag{58}$$

where,

$pK_a = -\log K_a$,
K_a = dissociation constant of volatile fatty acid,
  $= 1.81 \times 10^{-5}$ (35°C) for acetate,
$[S^-]$ = ionized volatile fatty acid concentration, and
$[HS]$ = un-ionized volatile fatty acid concentration.

Fig. 16 shows the relationship among ionized and un-ionized volatile fatty acid and pH based on the experimental results which Chang et al. (1983) obtained from the continuous anaerobic digestion experiment using chemostat reactors with the substrate having acetate for its main ingredient. The circles and triangles

618

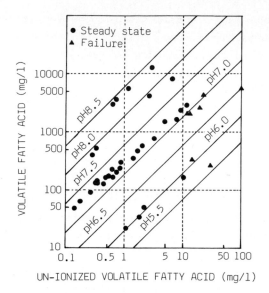

Fig. 16. Effect of Un-ionized Volatile Fatty Acid on
Process Stability (Chang et al., 1983)

on Fig. 16 show the experimental results at the steady state and
failure conditions, respectively. It is clear that a digester can
be operated at a steady state condition independently of the
residual volatile fatty acid concentration and pH value in the
digester if un-ionized volatile fatty acid is maintained at less
than 12 mg/l as acetate. In most case when methane production
stops, the high concentration of volatile fatty acid accumulates
and lower the pH value in the digester. The effects of pH on the
activity of methanogenic bacteria depend on un-ionized volatile
fatty acid concentration as mentioned before; the Monod-type
equation (Eq. 27) is corrected by introducing the pH influence as
follows:

$$r_X = \frac{[HS]}{K_S + [HS]} \, r_{X,max} \qquad (59)$$

Un-ionized volatile fatty acid at each pH value is calculated by
Eq. 58. Eq. 59 can be transformed to:

$$\frac{[HS]}{r_X} = \frac{[HS]}{r_{X,max}} + \frac{K_S}{r_{X,max}} \qquad (60)$$

A plot of $[HS]/r_X$ versus $[HS]$ results in the maximum specific
substrate utilization rate, $r_{X,max}$, and saturation constant, K_S, of

Table 10. Predominant Species of Methanogenic Bacteria in Reactors
at Different pH and Retention Times (Chang et al., 1983)

Retention time (d)	pH			
	6.0	7.2	7.9	8.4
shorter than 9.35	M S	M B	M B	M B
longer than 6.49	M S	M S	M B	M B

M S: Methanosarcina, M B: Methanobacterium

methanogenic bacteria.   As mentioned in Section 3.2, the metabolic
activity of methanogenic bacteria changes as the dilution rate and
substrate concentration of the influent change, so kinetic parameters, $r_{X,max}$ and K_S also change. Table 10 shows predominant
methanogenic bacteria in reactors at different pH and retention
time (Chang et al., 1983).   It was observed that Methanobacterium
sp. and Methanosarcina sp. were predominant in the ranges of high
and low pH values, respectively.   Therefore, it is necessary to
estimate $r_{X,max}$ and K_S for Methanobacterium sp. and Methanosarcina
sp. individually as shown in Figs. 17 and 18.   According to the
results shown in these figures, $r_{X,max}$ of Methanosarcina sp. was
larger than that of Methanobacterium sp., and the substrate
utilization ability of Methanosarcina sp. is superior to that of
Methanobacterium sp.   The K_S value is based on the un-ionized
volatile fatty acid here, and it can be expressed on the basis of
the total volatile fatty acid as follows:

$$K_{S,T} = \{1 + \frac{K_a}{[H^+]}\}K_{S,U}$$ (61)

where,

$K_{S,T}$ = saturation constant based on total volatile fatty acid,
and

$K_{S,U}$ = saturation constant based on un-ionized volatile fatty
acid.

4.3 Heavy metals toxicity

Heavy metals are the main toxic materials which exist in the
industrial wastewaters discharged into public sewerage sytems.   In
general, both the primary sludge and the waste activated sludge are
digested anaerobically, and the concentrations of heavy metals in
these sludges are less than a few hundred milligrams per liter.

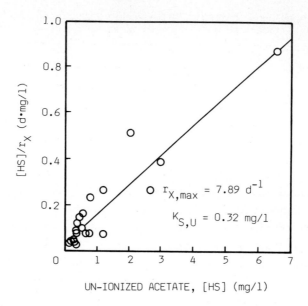

Fig. 17. Estimation of $r_{X,max}$ and $K_{S,U}$ for <u>Methanobacterium</u> sp.

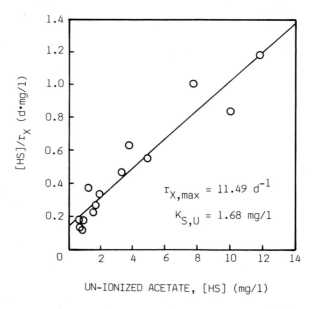

Fig. 18. Estimation of $r_{X,max}$ and $K_{S,U}$ for <u>Methanosarcina</u> sp.

These metals seem to have no marked effects on sludge digestion because the heavy metals are diluted with the existing mixed liquor in the digester and because the bacteria become acclimatized to toxic materials during a long digestion period. However, some accidents caused by the inflow of high concentration of heavy metals to digesters have been reported (Rudgal, 1946; Regan and Peters, 1970). It may be possible that the low concentration of heavy metals in raw sludge accumulates in the digesting sludge to become high concentrations during a long retention period. There have been only few experimental studies on the durability of digester to toxic materials. The study made by Matsumoto and Noike (1980) is presented below.

4.3.1 Inhibition rate

The slowest or rate-limiting step in anaerobic digestion is generally believed to be the methanogenesis. Methanogenesis is important since this is the major process by which wastes are stabilized in anaerobic digestion. Hence, it is considered suitable to examine inhibition at the methanogenic step in order to study the inhibition of anaerobic digestion by heavy metals. The inhibition rate defined by the following equation was introduced.

$$\text{Inhibition rate (i)} = 1 - \frac{\text{(Amount of methane gas from the digester affected by the heavy metal addition)}}{\text{(Normal amount of methane gas)}} \quad (62)$$

The amount of methane gas is used instead of the amount of total gas because the total gas from the digester is affected by the addition of heavy metal and contains CO_2 which results from the liberation of bicarbonate owing to the drop of pH.

4.3.2 Heavy metal concentration in the digester

The anaerobic digestion process is normally operated as a two stage system. The main reaction of anaerobic digestion is completed in the primary digester, and the solid and liquid phases are separated in the secondary digester. Therefore, the results of anaerobic digestion in the primary digester are considered to be responsible for the separation of solid and liquid phases and the qualities of the supernatant and the digested sludge in the secondary digester. It is important to study the effects of heavy metals on anaerobic digestion in the primary digester. Here, $Q(m^3/d)$ is the raw sludge volume daily fed to the digester, $M(mg/l)$

622

is the heavy metal concentration of the raw sludge and V (m^3) is the volume of the digester. In their study, the supernatant and the mixed liquor were drawn by turns every other day so that grams of heavy metal per one gram of volatile solids of the mixed liquor were held as constant as possible throughout the experimental periods. Assuming that the digester is completely mixed and that the supernatant does not contain heavy metal, the content of heavy metal (in grams) in the digester since the heavy metal addition started is calculated as follows:

1st day: MQ;

2nd day: $MQ\{1 + (1 - \frac{Q}{V})\}$;

3rd day: $MQ + MQ\{1 + (1 - \frac{Q}{V})\}$;

4th day: $MQ\{1 + (1 - \frac{Q}{V})\}\{1 + (1 - \frac{Q}{V})\}$;

$(2n - 1)$th day: $MQ + MQ\{1 + (1 - \frac{Q}{V})\} \cdot \dfrac{1 - (1 - \frac{Q}{V})^{n-1}}{1 - (1 - \frac{Q}{V})}$; and

$(2n)$th day: $MQ\{1 + (1 - \frac{Q}{V})\} \cdot \dfrac{1 - (1 - \frac{Q}{V})^{n}}{1 - (1 - \frac{Q}{V})}$

If C_V (mg/l) is the volatile solids concentration of the mixed liquor, the heavy metal concentrations of the mixed liquor on the $(2n-1)$th and the $(2n)$th days are shown as follows:

$(2n-1)$th day:

$$\frac{M}{C_V} \left[\frac{1}{\Theta} + (2 - \frac{1}{\Theta}) \cdot \{1 - (1 - \frac{1}{\Theta})^{n-1}\} \right] \quad (g/gVS) \tag{63}$$

and

$(2n)$th day: $\quad \dfrac{M}{C_V} (2 - \frac{1}{\Theta})\{1 - (1 - \frac{1}{\Theta})^{n}\} \quad (g/gVS) \tag{64}$

Table 11. The Copper Concentration of the Mixed Liquor on the 10th Day (Matsumoto & Noike, 1980)

	Digester number					
	No.1	No.2	No.3	No.4	No.5	No.6
Cu in raw sludge (mg/l)	250	500	1000	2000	4000	8000
Volatile Solids in mixed liquor (mg/l)	13700	14500	14500	15300	15200	17000
Calculated Cu concentration, C_{10}, (mg/gVS)	5.57	10.5	21.0	39.8	80.3	143.0
Measured Cu concentration, C'_{10}, (mg/gVS)	11.2	12.0	25.0	42.0	81.0	103.0
$C_{10} - C'_{10}$ (mg/gVS)	-5.63	-1.5	-4.0	-2.2	-0.7	+40.5

Table 11 shows the copper concentrations measured and calculated by Eq. 64 on the 10th day in the digester to which various concentrations of copper was added.

4.3.3 Effects of the heavy metal mixture addition

The terms summation, antagonism and synergism which have been used in pharmacology are applied to the results obtained from the heavy metal mixture addition experiment. When the inhibitors, I_1, I_2 .... and I_n act separately, the inhibition rates are described as i_1, i_2 .... and i_n, respectively. If all the inhibitors act simultaneously, the expected inhibition will be as follows (Webb, 1963):

$$\text{Summation:} \quad i_{1,2...n} = 1 - (1-i_1)(1-i_2)....(1-i_n)$$
$$\text{Antagonism:} \quad i_{1,2...n} < 1 - (1-i_1)(1-i_2)....(1-i_n) \qquad (65)$$
$$\text{Synergism:} \quad i_{1,2...n} > 1 - (1-i_1)(1-i_2)....(1-i_n)$$

Therefore, antagonism is commonly assumed to occur when all the inhibitors produce an effect that is less than expected on the basis of their individual actions, and synergism is deduced when the combined effect is greater than expected. Kugelman and McCarty (1965) investigated the effects of the combination of essential components like Na^+, K^+, Ca^{2+}, Mg^{2+} and NH_4^+ on anaerobic digestion, and reported that either antagonism or synergism occurred according to the combination of components and that the component concentration range in which antagonism occurred was

Table 12. The Inhibition in the Digester to which the Heavy Metal Mixture was Added (Matsumoto & Noike, 1980)

Heavy metals added (mg/l)	day	heavy metals concentration (mg/gVS)	$i_{1,2,3,4,5}$	$1-(1-i_1)\cdot(1-i_2)\cdots(1-i_5)$
1000	35th	4.9	0.301	0.087
		25.0	0.239	0.387
		32.0	0.243	0.387
		37.0	0.186	0.383
		46.7	0.195	0.403
		52.3	0.188	0.453
2000	15th	9.2	0.500	0.215
		31.5	0.610	0.412
		39.0	0.665	0.415
		58.0	0.698	0.512
		66.0	0.830	0.549
		68.0	0.990	0.533

narrower than that in which synergism occurs.

Table 12 shows the inhibition rate in the digester to which the mixture of five kinds of equivalent heavy metals, Cu, Zn, Ni, Cr, Cd were added. In the digester to which 1,000 mg/l of the heavy metal mixture was added, antagonism ocurred to diminish the inhibitory effects of individual heavy metals and the inhibition rate decreased as the heavy metal concentration of the mixed liquor increased. In the digester to which 2,000 mg/l of the heavy metal mixture was added, however, synergism occurred and the inhibition rate increased as the heavy metal concentration of the mixed liquor increased.

4.4 Inhibition by long-chain fatty acids

Long-chain fatty acids are the intermediates which occur during the degradation of lipids in anaerobic digestion as stated in Section 2.1, and are also present in the raw sewage sludge. The degradation of long-chain fatty acids contributes much to the methane production, since the lipid content in the sewage sludge is high, and a larger amount of methane is produced from these materials than from other components. Long-chain fatty acids have, however, toxic effects on many kinds of bacteria (Galbraith et al., 1971); inhibition in the anaerobic digestion process has also been reported (McCarty, 1964b).

Hanaki et al. (1981) studied to clarify the inhibitory effect of long-chain fatty acids on each reaction in the anaerobic digestion process. They made batch experiments and showed that the methanogenesis from acetate and β-oxidation of long-chain fatty acids and of n-butyrate were inhibited severely and that the methanogenesis from hydrogen was inhibited partially, while the acidogenesis from glucose was not inhibited by long-chain fatty acids of less than 1000 mg/l. They showed that the neutral fat in a whole milk was also inhibitory since it was easily hydrolyzed to long-chain fatty acids. They found that the addition of calcium chloride, but not of insoluble calcium carbonate, reduced the inhibitory effect of long-chain fatty acids by forming the insoluble salt of long-chain fatty acids.

Lipid content of sewage sludge was reported to be 28% of the total organic solids (Chynoweth and Mah, 1971) and its concentration in the feed sludge may be as high as 10,000 mg/l. Such a high concentration of long-chain fatty acids can cause the inhibition in the anaerobic digestion process; this inhibition does not usually occur because the digester is operated continuously which prevents their concentration from becoming high enough under the normal condition and allows methanogenic bacteria to become acclimatized to long-chain fatty acids (Hanaki et al., 1983). However, an overloading or a shock loading of the sewage sludge to the digester may provoke inhibition by long-chain fatty acids.

5. DYNAMIC MODELS AND PROCESS CONTROL

The application of results of the steady state models suggest that the retention time can be shorter than the typical design value (15 days) in the anaerobic digestion process. The shorter retention time makes digester smaller and saves the energy required for its heating. However, it increases the risk of process failure which is characterized by a drastic decrease in methane production and a drop of pH in the digester. It takes a long time for the digester to recover from the failure once it has taken place. Process failure is believed to be provoked by the fluctuation of the quality and/or quantity of influent or by receiving toxic materials. The steady state models cannot, of course, describe the process failure which occurs out with the steady state conditions. Dynamic models are required to describe process failure and to control the process for its prevention.

5.1 Description of the process failure

As process failure is characterized by the retardation of methanogenesis as the acidogenesis proceeds, equation for the retardation of methane production has to be formulated in order to describe the process failure. Graef and Andrews constructed a dynamic model which described the process failure (Andrews, 1969, 1978; Andrews and Graef, 1971; Graef and Andrews, 1974a,b). They assumed that process failure resulted from the inhibition of methanogenesis by un-ionized fatty acids or from the death of methanogenic bacteria caused by toxic materials. The former was formulated as follows:

$$\mu = \frac{\mu_{max}}{1 + K_S/[HS] + [HS]/K_{I,a}} \tag{66}$$

where,

$[HS]$ = un-ionized fatty acids, and

$K_{I,a}$ = inhibition coefficient of un-ionized fatty acid.

Un-ionized fatty acids concentration depends on both the total fatty acids concentration and pH, as mentioned in Section 4.2; the value of pH in the digester is also affected by CO_2 content in the gas phase and by alkalinity. The relationship among them is governed by the carbonate-bicarbonate buffer system. The following equation was formulated for the effect of toxic materials:

$$r_K = K_T \cdot T_X \qquad (67)$$

where,

r_K = rate of organism kill,

K_T = toxic rate constant, and

T_X = concentration of toxic material.

Other factors, such as the gas transfer rate, are also related to the condition of the digester. Graef and Andrews (1974a) con-structed a model system which comprised many equations describing the phenomena in the anaerobic digestion process (Fig. 19). They applied this model to the methanogenesis from acetate and made several computer simulations. An example of their simulation is shown in Fig. 20 which illustrates the responses of substrate concentration, methane production rate, CO_2 content in the gas phase and pH when influent acetate concentration was increased from 10,000 mg/l to 24,000, 33,000, 35,700 or 36,300 mg/l. The step increase of influent acetate to 36,300 mg/l resulted in process failure, while those up to 35,700 mg/l did not provoke failure and the process recovered of itself.

Hill and Barth (1977) developed a dynamic model which took account of the inhibition of both un-ionized ammonia and un-ionized fatty acids in the manner similar to Eq. 66. They expressed the specific biomass growth rate as

$$\mu = \frac{\mu_{max}}{1 + K_S/[HS] + [HS]/K_{I,a} + [NH_3]/K_{I,n}} \qquad (68)$$

where,

$[NH_3]$ = un-ionized ammonia, and

$K_{I,n}$ = inhibition coefficient of un-ionized ammonia.

Un-ionized ammonia is affected by pH in the same way as un-ionized fatty acid. They simulated the start-up period of animal waste digestion and made experiments to verify the model.

Carr and O'Donnell (1976) carried out laboratory scale experiments to examine the dynamics of methanogenesis from acetate.

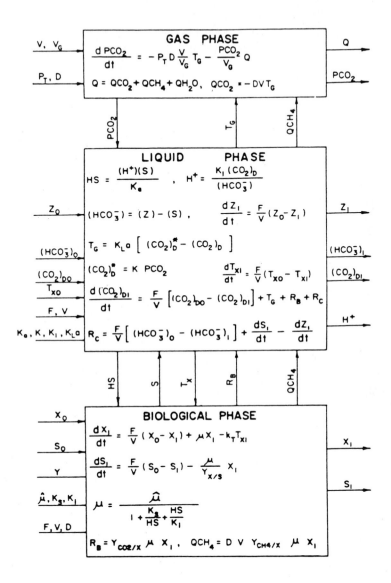

Fig. 19. Dynamic Model Proposed by Graef and Andrews (1974a)

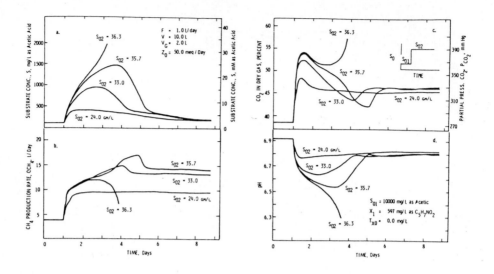

Fig. 20. Results of Simulation Using the Model of Graef and
Andrews (Graef and Andrews, 1974a)

They showed that the computer simulation using the model of Graef
and Andrews could predict the experimental data.

5.2 Process control strategy

One of the most important objects in the development of the
dynamic model is its application to the control of the anaerobic
digestion process. Graef and Andrews (1974a,b) discussed an
indicator which could warn of process failure. Their simulation
showed that methane production rate, substrate concentration, CO_2
content in dry gas and pH were the effective indices. They
estimated the effect of feedback control on process stability
against the failure. In their simulation, they used pH as the
feedback control signal and examined four control actions, namely,
gas scrubbing and recycle, base addition, organism recycle and flow
reduction. The gas scrubbing operation is expected to raise pH
by removing CO_2 in the gas phase.

Collins and Gilliland (1974) obtained transfer functions when
pH was regarded as the process output, and estimated the effect of
feedforward control with computer simulation using the model of
Graef and Andrews. They showed that predictive flow control gave a
good performance in controlling pH.

6. SUMMARY AND CONCLUSION

The importance and the advantage of anaerobic digestion are expected to receive a greater recognition in the future. This trend has been yielding the intensive research activity in recent years on various aspects of this process. The microbiology and the metabolic mechanism of the process have been clarified and the knowledge was used to improve mathematical modelling of the process. However, the design and operation of actual digesters have undergone little improvement over the last twenty years. The gap between the understanding of the process and the operation of digesters derives from the apprehension of process failure which is the characteristic phenomenon in the anaerobic digestion process. Process failure gives rise to a serious situation in which the digester loses its primary function for a long period of time, although the sludge treatment must continue by any means. There-fore, a large safety factor has been taken into account in the design and operation in order to prevent failure. An innovation which breaks this conventional concept has to be introduced. The most hopeful means of diminishing the safety factor is appropriate and effective process control, which enables higher efficiency than the present "high-rate" process. Both the steady state kinetic model and the dynamic model will contribute much to the innovation of the process, but the improvement of models and the accumulation of data of the kinetic coefficients are still required. Good maintenance of the digester, including proper heating, satisfactory mixing, keeping the anaerobic condition and cleaning of the accumu-lated sand, are also necessary before the successful application of the mathematical model since a primary assumption of the model is the maintenance of these physical conditions to a high standard.

SYMBOLS

b = specific biomass decay coefficient $\hspace{3cm}$ d^{-1}

C_1, C_2, C, C' = constants $\hspace{5cm}$ -

C_V = volatile solids concentration of mixed liquor $\hspace{1cm}$ mg/ℓ

$D = Q/V$ = dilution rate (reciprocal of hydraulic retention
$\hspace{2cm}$ time, θ) $\hspace{6cm}$ d^{-1}

E_a = a constant for the reaction termed the activation
$\hspace{1cm}$ energy $\hspace{7cm}$ kJ/mol

630

[HS] = un-ionized fatty acids   mg/ℓ

i = inhibition rate defined by Eq. 62   –

K = dimensionless kinetic coefficient in Eq. 37   –

K_a = dissociation constant of volatile fatty acid
    = 1.81×10^{-5} (35°C) for acetate

$K_{I,a}$ = inhibition coefficient of un-ionized fatty acid   mg/ℓ

$K_{I,n}$ = inhibition coefficient of un-ionized ammonia   mg/ℓ

K_S = saturation constant for substrate   mg/ℓ

$K_S(T)$ = K_S at a temperature, T°C   mg/ℓ

$K_S(T_o)$ = K_S at the reference temperature, T_o°C   mg/ℓ

$K_S(T_{A,1})$, $K_S(T_{A,2})$ = K_S values at $T_{A,1}$ and $T_{A,2}$ °K   mg/ℓ

$K_{S,C}$ = kinetic coefficient in Contois-type model   –

$K_{S,T}$ = saturation constant based on total volatile fatty acid   mg/ℓ

$K_{S,U}$ = saturation constant based on un-ionized volatile
        fatty acid   mg/ℓ

K_T = toxic rate constant d^{-1}

k = first order rate constant d^{-1}

M = heavy metal concentration of raw sludge   mg/ℓ

m = maintenance coefficient for micro-organism d^{-1}

[NH_3] = un-ionized ammonia   mg/ℓ

P = concentration of products in effluent   mg/ℓ

[P_1], [P_2]··· = product concentration   mol/ℓ or atm

pK_a = -log K_a

Q = flowrate   ℓ/d or m^3/d

R = gas constant   0.082 ℓ·atm/mol·°K

r_K = rate of organism kill   mg/ℓ·d

r_v = substrate utilization rate per volume   mg/ℓ·d

r_X = specific substrate utilization rate d^{-1}

$r_X(T)$ = specific substrate utilization rate at T°C d^{-1}

$r_{X,max}$ = maximum specific substrate utilization rate d^{-1}

$r_{X,max}(T)$ = $r_{X,max}$ at a temperature, T°C d^{-1}

$r_{X,max}(T_o)$ = $r_{X,max}$ at the reference temperature, T_o°C d^{-1}

$r(T_A)$ = reaction rate at T_A °K   mg/ℓ·d

S = substrate concentration in the reactor   mg/ℓ

S_1 = influent substrate concentration   mg/ℓ

S_2 = effluent substrate concentration   mg/ℓ

[S^-] = ionized volatile fatty acid concentration   mg/ℓ

[S_1], [S_2]··· = substrate concentration   mol/ℓ or atm

T = temperature   °C

T_A = temperature   °K

T_X = concentration of toxic material   mg/ℓ

t = time   d

U_e = substrate utilization coefficient for energy production   -

U_p = substrate utilization coefficient for synthesis   -

u = temperature dependency coefficient   °K

V = reactor volume   ℓ or m^3

X = biomass concentration in the reactor   mg/ℓ

Y_G = growth yield coefficient   -

Y_P = true product yield coefficient   -

$Y_{P,obs}$ = observed product yield coefficient   -

μ = specific biomass growth rate d^{-1}

μ_{max} = maximum specific biomass growth rate d^{-1}

Θ = hydraulic retention time   d

Θ_c = critical hydraulic retention time   d

$\Delta G_o'$ = standard free energy change at pH = 7   kJ/mol

$\Delta G'$ = free energy change at pH = 7   kJ/mol

REFERENCES

Andrews, J. F., 1969. Dynamic model of the anaerobic digestion process. Proc. Amer. Soc. Civil Eng., San. Eng. div., 95, SA1: 95-116.

Andrews, J. F., 1978. The development of a dynamic model and control strategies for the anaerobic digestion process. In: A. James (Editor), Mathematical Models in Water Pollution Control. John Wiley & Sons, pp. 281-302.

Andrews, J. F. and Graef, S. P., 1971. Dynamic modeling and simulation of the anaerobic digestion process. Adv. in Chemistry Series, 105: 126-162.

Balch, W. E., Fox, G. E., Magrum, L. J., Woese, C. R. and Wolfe, R. S., 1979. Methanogens: reevaluation of a unique biological group. Microbiol. Reviews, 43: 260-296.

Brown, H. D.(Ed.), 1969. Biochemical Microcalorimetry. Academic Press.

Bryant, M. P., 1974. Methane-producing bacteria. In: R. E. Buchanan and N. E. Gibbons (Editors), Bergey's Manual of Determinative Bacteriology (8th ed.), The Williams & Wilkins, Part 13.

Bryant, M. P., 1976. The microbiology of anaerobic degradation and methanogenesis with special reference to sewage. In: H. G. Schlegel and J. Barnea (Editors), Microbial Energy Conversion. Erich Goltze KG, Göttingen, pp. 107-117.

Bryant, M. P., 1979. Microbial methane production ———— Theoretical aspects. J. Animal Science, 48: 193-201.

Bryant, M. P., Wolin, E. A., Wolin, M. J. and Wolfe, R. S., 1967. Methanobacillus omelianskii, a symbiotic association of two species of bacteria. Archiv. für Mikrobiol., 59: 20-31.

Buswell, A. M., 1947. Important considerations in sludge digestion,
    Part II. Microbiology and theory of anaerobic digestion. Sewage
    Works Journal, 19: 28.
Cameron, R. D. and Koch, F. A., 1980. Trace metals and anaerobic
    digestion of leachate. J. Water Poll. Control Fed., 52: 282-292.
Carr, A. D. and O'Donnell, R. C., 1977. The dynamic behavior of an
    anaerobic digester. Progress in Water Technology, 9: 727-738.
Chang, J., 1982. Studies on methanogenic phase in anaerobic diges-
    tion. Doctoral thesis at Tohoku University, Japan, pp.200 (in
    Japanese).
Chang, J., Noike, T. and Matsumoto, J., 1982. Effect of retention
    time and feed substrate concentration on methanogenesis in anaero-
    bic digestion. Proc. Japan Society of Civil Engineers, No. 320:
    67-76 (in Japanese).
Chang, J., Noike, T. and Matsumoto, J., 1983. Effect of pH on
    methanogenesis in anaerobic digestion. Proc. Japan Society of
    Civil Engineers, No. 333: 101-108 (in Japanese).
Chen, Y. R. and Hashimoto, A. G., 1978. Kinetics of methane fermenta-
    tion. Biotechnol. Bioeng. Symposium, 8: 269-282.
Chen, Y. R., Varel, V. H. and Hashimoto, A. G., 1980. Effect of
    temperature on methane fermentation kinetics of beef-cattle manure.
    Biotechnol. Bioeng. Symposium, 10: 325-339.
Chynoweth, D. P. and Mah, R. A., 1971. Volatile acid formation in
    sludge digestion. Adv. in Chemistry Series, 105: 41-54.
Collins, A. S. and Gilliland, B. E., 1974. Control of anaerobic
    digestion process. Proc. Amer. Soc. Civil Eng., Env. Eng. div.,
    100, EE2: 487-505.
Dague, R. R., Hopkins R. L. and Tonn, R. W., 1970. Digestion
    fundamentals applied to digester recovery —— two case studies. J.
    Water Poll. Control Fed., 42: 1666-1675.
Eastman, J. A. and Ferguson, J. F., 1981. Solubilization of particu-
    late organic carbon during the acid phase of anaerobic digestion.
    J. Water Poll. Control Fed., 53: 352-366.
Endo, G., 1980. Studies on acidogenic phase in anaerobic digestion.
    Doctoral thesis at Tohoku University, Japan, 234pp. (in Japanese).
Endo, G. and Tohya, Y., 1984. Anaerobic biological decomposition of
    malodorous compounds in kraft pulping wastewater. Proc. First
    IAWPRC Symposium on Forest Industry Wastewaters (held at Tampere).
Endo, G., Noike, T. and Matsumoto, J., 1983. Effect of temperature
    and pH on acidogenic phase in anaerobic digestion. Proc. Japan
    Society of Civil Engineers, No. 330: 49-57 (in Japanese).
Fair, G. M. and Moore, E. W., 1932. Effect of temperature of incuba-
    tion upon the course of digestion. Sewage Works Journal, 4: 589.
Fox, G. E., Magrum, L. J., Balch, W. E., Wolfe, R. S. and Woese, C.
    R., 1977. Classification of methanogenic bacteria by 16S ribosomal
    RNA characterization. Proc. Natl. Acad. Sci. USA, 74: 4537-4541.
Galbraith, H., Miller, T. B., Paton, A. M. and Thompson, J. K., 1971.
    Antibacterial activity of long chain fatty acids and the reversal
    with calcium, magnesium, ergocalciferol and cholesterol. J. Appl.
    Bacteriol., 34: 803-813.
Ghosh, S., Conrad, J. R. and Klass, D. L., 1975. Anaerobic acidogene-
    sis of wastewater sludge. J. Water Poll. Control Fed., 47: 30-45.
Ghosh, S. and Klass, D. L., 1978. Two-phase anaerobic digestion.

Process Biochemistry, 13, No.4: 15-24.

Ghosh, S. and Pohland, F. G., 1974. Kinetics of substrate assimilation and product formation in anaerobic digestion. J. Water Poll. Control Fed., 46: 748-759.

Golueke, C. G., 1958. Temperature effects on anaerobic digestion of raw sewage sludge. Sewage and Industrial Wastes, 30: 1225-1232.

Graef, S. P. and Andrews, J. F., 1974a. Stability and control of anaerobic digestion. J. Water Poll. Control Fed., 46: 666-683.

Graef, S. P. and Andrews, J. F., 1974b. Mathematical modeling and control of anaerobic digestion. AIChE Symposium Series, 70, No. 136: 101-131.

Hanaki, K., Ishikawa, T. and Matsumoto, J., 1983. Inhibitory and stimulative effects of oleate on methanogenesis from acetate in anaerobic digestion. Technology Reports, Tohoku University, 48: 123-135.

Hanaki, K., Matsuo, T. and Nagase, M., 1981. Mechanism of inhibition caused by long-chain fatty acids in anaerobic digestion process. Biotechnol. Bioeng., 23: 1591-1610.

Hashimoto, A. G., 1982. Methane from cattle waste: Effects of temperature, hydraulic retention time, and influent substrate concentration on kinetic parameter (K). Biotechnol. Bioeng., 24: 2039-2052.

Heukelekian, H. and Mueller, P., 1958. Transformation of some lipids in anaerobic sludge digestion. Sewage and Industrial Wastes, 30: 1108-1120.

Hill, D. T. and Barth, C. L., 1977. A dynamic model for simulation of animal waste digestion. J. Water Poll. Control Fed., 49: 2129-2143.

Hungate, R. E., 1975. The rumen microbial ecosystem. Annual Rev. of Ecology and Systematics, 6: 39-66.

Jenett, J. C. and Dennis, N. D., 1975. Anaerobic filter treatment of pharmaceutical waste. J. Water Poll. Control Fed., 47: 104-121.

Jeris, J. S. and McCarty, P. L., 1965. The biochemistry of methane fermentation using C^{14} tracers. J. Water Poll. Control Fed., 37: 178-192.

Kasper, H. F. and Wuhrmann, K., 1978a. Product inhibition in sludge digestion. Microbial Ecology, 4: 241-248.

Kaspar, H. F. and Wuhrmann, K., 1978b. Kinetic parameters and relative turnovers of some important catabolic reactions in digesting sludge. Appl. Environ. Microbiol., 36: 1-7.

Keenan, J. D. and Kormi, I., 1977. Methane fermentation of brewery by-products. Biotechnol. Bioeng., 19: 867-878.

Kroeker, E. J., Schulte D. D., Sparling, A. B. and Lapp, H. M., 1979. Anaerobic treatment process stability. J. Water Poll. Control Fed., 51: 718-727.

Kugelmann, I. J. and McCarty, P. L., 1965. Cation toxicity and stimulation in anaerobic waste treatment. J. Water Poll. Control Fed., 37: 97-116.

Lawrence, A. W. and McCarty, P. L., 1969. Kinetics of methane fermentation in anaerobic treatment. J. Water Poll. Control Fed., 41: R1-R17.

Matsumoto, J. and Noike, T., 1980. Effects of heavy metals on anaerobic sludge digestion (V). Technology Reports, Tohoku University, 45: 57-69.

Matsumoto, J., Noike, T. and Endo, G., 1981. Microbial growth and organic acid formation in acidogenesis phase of anaerobic digestion. Technology Reports, Tohoku University, 46: 293-311.

McCarty, P. L., 1964a. Anaerobic waste treatment fundamentals Part 2 Environmental requirement and control. Public Works, 95, 10: 123-126.

McCarty, P. L., 1964b. Anaerobic waste treatment fundamentals Part 3 Toxic materials and their control. Public Works, 95, 11: 91-94.

McCarty, P. L. and McKinney, R. E., 1961. Volatile acid toxicity in anaerobic digestion. J. Water Poll. Control Fed., 33: 223-232.

McKinney R. E., 1962. Microbiology for Sanitary Engineers. McGraw-Hill, 293pp.

Metcalf & Eddy, Inc., 1979. Wastewater Engineering —— treatment, disposal, reuse ——. McGraw-Hill, 920pp.

Nagase, M. and Matsuo, T., 1982. Interactions between amino-acid-degrading bacteria and methanogenic bacteria in anaerobic digestion. Biotechnol. Bioeng., 24: 2227-2239.

Novak, J. T., 1974. Temperature-substrate interaction in biological treatment. J. Water Poll. Control Fed., 46: 1984-1994.

Novak, J. T. and Carlson, D. A., 1970. The kinetics of anaerobic long chain fatty acid degradation. J. Water Poll. Control Fed., 42: 1932-1943.

O'Rourke, J. T., 1968. Kinetics of anaerobic treatment at reduced temperatures. Thesis presented to Stanford University in partial fulfillment of the requirements for the degree of Doctor of Philosophy: cited by Lawrence, A. W., 1971. Application of process kinetics to design of anaerobic processes. Adv. in Chemistry Series, 105: 163-189.

Pirt, S. G., 1975. Principles of Microbe and Cell Cultivation. Blackwell Scientific Publications, 274pp.

Pretorius, W. A., 1972. The effect of formate on the growth of acetate utilizing methanogenic bacteria. Water Research, 6: 1213-1217.

Radhakrishnan, I., De, S. B. and Nath, B., 1969. Evaluation of loading parameters for anaerobic digestion of cane molasses distillery wastes. J. Water Poll. Control Fed., 41: R431-R440.

Regan, T. M. and Peters, M., 1970. Heavy metals in digesters: failure and cure. J. Water Poll. Control Fed., 42: 1832-1839.

Rudgal, H. T., 1946. Effect of copper-bearing wastes on sludge digestion. Sewage Works Journal, 18: 1130-1137.

Sanders, M. C., 1977. Determination of the kinetic constants for the anaerobic treatment of wine stillage. In: Proc. 7th Public Aspirations Realities Water Resour. Manage., Fed. Conv., Australian Water Wastewater Assoc.: 479-513.

Schlenz, H. E., 1947. Important considerations in sludge digestion, Part I. Practical aspects. Sewage Works Journal, 19: 19.

Schönheit, P., Moll, J. and Thauer, R. K., 1980. Growth parameters (K_s, μ_{max}, Y_s) of Methanobacterium thermoautotrophicum. Arch. Microbiol., 127: 59-65.

Shea, T. G., Pretorius, W. A., Cole, R. D. and Pearson, E. A., 1968. Kinetics of hydrogen assimilation in the methane fermentation. Water Research, 2: 833-848.

Shelef, G., Kimchie, S. and Grynberg, H., 1980. High-rate thermophilic anaerobic digestion of agricultural wastes. Biotechnol. Bioeng. Symposium, 10: 341-351.

Siebelt, M. L. and Toerien, D. F., 1969. The proteolytic bacteria present in the anaerobic digestion of raw sewage sludge. Water Research, 3: 241-250.

Smith, P. H. and Mah, R. A., 1966. Kinetics of acetate metabolism during sludge digestion. Appl. Microbiol., 14: 368-371.

Speece, R. E. and McCarty, P. L., 1962. Nutrient requirements and biological solids accumulation in anaerobic digestion. Proc. first Int. Conf. on Water Pollution Research, Vol. 2: 305-322.

Thauer, R. K., Jungermann, K. and Decker, K., 1977. Energy conservation in chemotrophic anaerobic bacteria. Bacteriol. Rev., 41: 100-180.

Toerien D. F., 1970. Population description of the non-methanogenic phase of anaerobic digestion —— II. Hierarchical classification of isolates. Water Research, 4: 285-303.

Toerien, D. F. and Hattingh, W. H. J., 1969. Anaerobic digestion —— I. The microbiology of anaerobic digestion. Water Research, 3: 385-416.

Toerien, D. F., Siebelt, M. L. and Hattingh, W. H. J., 1967. The bacterial nature of the acid-forming phase of anaerobic digestion. Water Research, 1: 497-507.

Topiwara, H. and Sinclair, C. G., 1971. Temperature relationship in continuous culture. Biotechnol. Bioeng., 13: 795.

Water Pollution Control Federation, 1968. Anaerobic Digestion. WPCF Manual of Practice, No. 16, 73pp.

Webb, 1963. Enzyme and Inhibitors I. Academic Press, 507pp.

Yaguchi, J., 1982. Anaerobic degradaion of carbohydrates. Master's thesis at Tohoku University, Japan, 130pp. (in Japanese).

Zeikus, J. G., 1977. The biology of methanogenic bacteria. Bacteriol. Rev., 41: 514-541.

Zoetemeyer, R. J., Arnoldy, P., Cohen, A. and Boelhouwer, C., 1982. Influence of temperature on the anaerobic acidification of glucose in a mixed culture forming part of a two-stage digestion process. Water Research, 16: 313-321.

MATHEMATICAL MODELS IN ANAEROBIC TREATMENT PROCESSES

Alberto Rozzi(+) and Roberto Passino(*)

(+) IRSA-CNR, Via de Blasio 5, 70123 Bari, Italy
(*) IRSA-CNR, Via Reno 1, 00198 Roma, Italy.

1. INTRODUCTION

The progress in wastewater treatment technology is expected to be characterized by a shift from conventional aerobic to anaerobic or combined anaerobic/aerobic processes.

Most of the progress is due to a better knowledge of the pertinent microbiology and biochemistry, as well as an improved capacity of process modeling.

In anaerobic treatment processes, the physico-chemical and the biological systems are linked by complex feedback loops and the behaviour of the digesters, especially in dynamic conditions, is not at all evident. Mathematical modeling is invaluable to explore the interactions between the physico-chemical environment and the microbial populations because it can be used to determine the individual and aggregated influence of each variable on the process. Models can be transformed into suitable programs in order to simulate any operational condition of anaerobic reactors on digital computers.

The usefulness of mathematical models was already apparent when methanogens were assumed to be the only rate-limiting anaerobic microorganisms and digesters were modelled as simple one population systems.

Models are now essential as more complex anaerobic systems are considered which take into account four or more bacterial populations. The complexity of the interactions in these systems is very high. For example, the concentration of some gaseous end-products (i.e.: H_2 and CO_2 indirectly through the pH), strongly affects the metabolic pathways and the growth rates of several anaerobic microorganisms.

Modeling studies are important also because real experiments on anaerobic processes are very lengthy, labour intensive and expensive. Simulation tests on computer may considerably decrease the risks of running useless experiments or of designing inefficient process control strategies.

This chapter is divided in two main sections: a) definition and analysis of the mathematical models of anaerobic processes and b) applications.

In the first section, the complexity of the anaerobic system is first reduced by defining and analysing separately its main components: the physico-chemical environment and the active biomass. The main parameters affecting the process are examined one by one. Therefore, in analysing the physico-chemical system, the conversion rate from substrate to end products, i.e. the biomass metabolism, is assumed to be unaffected by the changes of the environment. Special attention is devoted to the definition of physico-chemical equilibria and to liquid/gas interactions, the importance of which is frequently overlooked.

The biological system is then considered assuming constant physico-chemical conditions and kinetic models for the degradation of typical substrates, i.e. particulate suspended solids and glucose, are presented.

A general model of the anaerobic degradation of glucose which summarizes the interactions between the biomass and the physico-chemical environment is also outlined.

Finally, anaerobic biofilms adhering to fixed or mobile surfaces are considered. The main problems related to their modeling, such as the determination of biofilm thickness and of apparent kinetics, which depend on substrate diffusion, are also summarized.

In the second section of the chapter, applications of mathematical models to anaerobic processes are presented.

The most important application so far is the use of mathematical models to investigate anaerobic process stability and potential process control strategies, both during start-up and transient full load operation.

Models can also be applied to evaluate the most suitable anaerobic process as a function of waste characteristics, and, more generally, to assist engineers in process design and economical evaluation, as shown in the following paragraphs.

In the interpretation of the plotted results, functional analysis is applied whenever possible, in order to link graphical patterns with equations and to give a better insight of the investigated phenomena.

2. DEFINITION OF THE ANAEROBIC SYSTEM

As any biological system, a generalized anaerobic system can be visualized as a physico-chemical (P/C) system interacting with a biochemical (B/C) or biological one (Fig. 1).

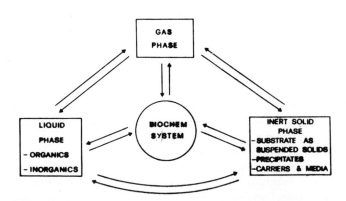

Fig. 1. Schematic representation of the interactions between the P/C and the B/C components in anaerobic systems.

The P/C system consists of the following phases: the gas, the liquid and the biologically inert fraction of solids. The B/C system is made of the microbial cells and of the related exoenzymes which act as a mass and energy tranfer unit.

Interactions between the P/C and the B/C systems are probably more complicated in anaerobic processes than in aerobic ones because of the stronger influence of CO_2/HCO_3^- equilibria in the former, due to much higher values of the CO_2 partial pressure (one or two orders of magnitude).

There are several important differences between the anaerobic and the aerobic biochemical systems other than the obvious one related to endogenous or external electron acceptors. Aerobic oxidation of organics to CO_2 and H_2O can be performed by many bacterial species while the anaerobic degradation of the same compounds to CO_2 and H_2O is carried out in series by different microbial populations each feeding on the metabolite(s) produced by the organisms 'upstream' in the food chain. Moreover the anaerobic degradation of carbohydrates (which usually are a major component in wastewaters) is shifted along different metabolic paths depending on the level of the H_2 partial pressure.

Last but not least, anaerobic processes are used to digest waste waters high in suspended solids, whereas in conventional aerobic processes most of the polluting load is related to soluble or/and colloidal compounds. The particulate suspended solids entering a biological reactor affect very strongly the composition of the mixed liquor sludge with respect to the fraction of active biomass, and this influence is usually understated in anaerobic digestion studies.

From the above remarks, it follows that the overall picture of anaerobic degradation is quite complicated and consequently a clear definition both of the P/C and B/C systems and of their complex interactions is essential for a better understanding and control of the anaerobic process.

For the sake of clarity, several simplified models related to specific P/C and B/C aspects are examined first independently, and then a model of a generalized anaerobic system is defined and its complex behaviour analyzed.

2.1 THE PHYSICO-CHEMICAL SYSTEM

2.1.1 Definition of the model.

A simple model has been considered in which the complex biochemical system has been substituted by an ideal catalyst (as defined below) which is part of the P/C system, in order to analyse separately the interactions between the main physico-chemical parameters (Rozzi, 1980). The anaerobic digester is then modelled as a simple catalytic reactor in which the substrate is converted to CH_4, CO_2, insoluble compounds (to take into account biomass production) and in some instances to acetic acid. This model does not take into account the formation of other volatile acids because their effect on the P/C system (on molar basis) is similar to acetic acid.

The following assumptions are made:
1. The digester is a completely stirred tank reactor (CSTR).
2. The substrate is made of simple carbohydrates.
3. The ideal catalyst converts the substrate at rates independent of

640

temperature and of pH variations.

4. The concentration of the catalyst is constant.

5. To simulate biomass syntesis, a fraction Y of the substrate is converted to insoluble compounds, which are discharged to the effluent.

6. Remaining substrate is converted to acetic acid.

7. Gas and liquid phases are in equilibrium conditions and steady state conditions are considered.

8. CH_4 solubility in water (%15 ppm vol.) is neglected.

9. Ammonium is assumed to behave as a strong cation, as most digesters operate at pH < 7.5 and its concentration is given as sodium.

10. Nutrient requirements for biomass syntesis are neglected.

11. Activity coefficients are considered equal to one irrespective of the ionic strength.

Because of assumptions 4 and 5, the proposed model behaves as a fluidized bed system holding a constant biomass. The implications related to the autocatalytic nature of microbiological reactors (Atkinson, 1974) are neglected.

Equilibria related to gas and liquid phases are considered first.

Because of assumption 8, the substrate converted to CH_4 is entirely discharged as gas while the substrate converted to CO_2 is partly evacuated as a gas and partly discharged through the liquid flow as inorganic carbon (mainly as bicarbonate and as dissolved CO_2). The simplified mass flow sheet is illustrated in Fig. 2. The fraction of substrate converted to CO_2 which remains in solution as inorganic carbon depends on the concentration of "available cations" in the influent. In this model available cations are Na^+ ions added to the influent as hydroxide or as bicarbonate.

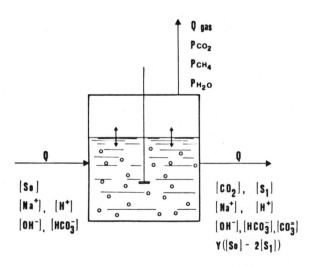

Fig. 2. Simplied flow sheet of the P/C system.

In real anaerobic systems, available cations are metal ions added or present in the feed in the above forms and/or balanced by organic and inorganic anions, such as CH_3COO^- and SO_4^{2-}, which are mainly converted to undissociated species discharged in gas form (CH_4, CO_2, H_2S) or NH_4^+ ions derived from the degradation of aminoacids. The influence of hydrogen sulfide and of ammonia on the P/C system is not considered here for sake of semplicity, but it could be easily included in the model (McCarty, 1964c), (Hill and Barth, 1977), (Ferguson et al., 1982), (Kroiss and Plahn-Wabnegg, 1983).

2.1.2 Equations describing the system.

Inorganic carbon (IC) mass balance in the liquid phase of the reactor is:

$$[IC] = [CO_2] + [HCO_3^-] + [CO_3^{2-}] \tag{1}$$

Combining the bicarbonate equilibrium equation

$$Ka1 = [H^+] \cdot [HCO_3^-]/[CO_2] \tag{2}$$

and Henry's law:

$$[CO_2] = KH \cdot PCO_2 \tag{3}$$

the following equation is obtained:

$$PCO_2 = \frac{[H^+] \cdot [HCO_3^-]}{Ka1 \cdot KH} \tag{4}$$

This equation, plotted for different pH values in Fig. 3, (McCarty, 1964b), is very useful in checking the consistency of pH, PCO_2 and bicarbonate alkalinity (BA), which are very important operational parameters in anaerobic treatment processes. It is advisable for digester operators to verify routinely analytical data on Fig. 3 in order to detect relevant errors (due, e.g., to pH electrode fouling, incorrect sampling procedures etc.).

Taking into account assumptions 4 and 5, the substrate mass balance (as moles of C) is given as:

$$Q \cdot [So] = \frac{(Qg-Qw) \cdot To}{22.4 \cdot T} + (Y \cdot ([So]-2 \cdot [S1]) + ([IC]+2 \cdot [S1])) \cdot Q \tag{5}$$

where Q is the influent flow rate, [So] the influent substrate concentration (as Total Organic Carbon, TOC) and [S1] the molar concentration of acetate + acetic acid in the effluent. The factor $To/(22.4 \cdot T)$ is used to convert gas volumes at temperature T to moles of carbon. Eq. (5) implies no inorganic carbon in the influent (i.e.: sodium is added as hydroxide and not as bicarbonate).

The gas volume balance is:

$$Qg - Qw = QCH_4 + QCO_2 \tag{6}$$

and from Dalton's law:

$$QCH_4 + QCO_2 = QCH_4 \frac{Pg - Pw}{Pg - Pw - PCO_2} \tag{7}$$

in which Pg, PCO_2, PCH_4 and Pw are the digester gas pressure, CO_2 and

642

CH$_4$ partial pressures and water saturation pressure respectively.

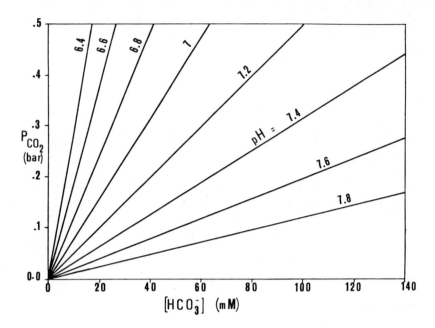

Fig. 3. CO$_2$ partial pressure as a function of bicarbonate alkalinity and pH.

Taking into account assumptions 5 and 6, substrate conversion to methane is given by:

$$\frac{Q_{CH_4} \cdot T_0}{22.4 \cdot T} = A \cdot ([S_0] - 2 \cdot [S_1]) \cdot (1-Y) \cdot Q \tag{8}$$

in which A is the ratio of TOC converted to methane over the TOC converted to methane and carbon dioxide (the latter both in the gas and in the effluent as IC). For simple carbohydrates such as glucose A = 0.5.

Substituting Eqs. (6,7) in Eq.(5), the following mass balance for Total Carbon is obtained:

$$(1-Y) \cdot ([S_0] - 2 \cdot [S_1]) \cdot (1 - A \frac{P_g - P_w}{P_g - P_w - P_{CO_2}}) = [IC] \tag{9}$$

Eq. (9) simply indicates that the substrate which is not converted to gas, to insoluble compounds (biomass) or to acetic acid, is discharged as inorganic carbon in the effluent stream.

Taking into account the dissociation equations for acetic acid and for carbonate, the water ionic product, the mass balance for acetate and acetic acid:

$$[S_1] = [HAc] + [Ac^-] \tag{10}$$

and the charge balance for the digester solution

$$[Na^+] + [H^+] = [HCO_3^-] + [Ac^-] + 2 \cdot [CO_3^{2-}] + [OH^-] \tag{11}$$

a system of 9 equations with 9 variables $[H^+]$, $[OH^-]$, $[CO_2]$, $[HCO_3^-]$, $[CO_3^{2-}]$, $[IC]$, $[Ac^-]$, $[HAc]$ and PCO_2 is defined, which can be easily solved by iterative procedures with currently available digital computers.

In the authors' experience the easiest way to solve the above system is to assume a tentative value for the pH (or H^+ concentration), solve eq. (9) which is a second degree equation in PCO_2 and check the charge balance, eq. (11). According to the inequality in eq. (11), adjust the pH value and iterate until the required accuracy is attained.

Sulfide, ammonia and phosphate equilibria could be easily added to the system of equations if required, without complicating the solution if the above procedure is used.

Another way to solve the system is the following. H^+ concentration is obtained from the charge balance for a tentative value of PCO_2, then eq. (9) is checked and if necessary a corrected value of PCO_2 is used. This second procedure is less convenient, because the addition of other ions in eq. (11) increases the number of roots. The solution and the discrimination of the correct root become quite complicated and moreover the computer program has to be substantially modified each time new ionic species are added to the system.

2.1.3 Process parameters and operating conditions for the model.
In order to show the influence on the anaerobic system of the physico-chemical parameters which can be controlled to some extent by digester operators (i.e.: substrate concentration, added alkalinity, temperature), a simple simulation study is carried out making use of equilibrium constants from the literature (Perry et al.,1963), (Weast, 1968) and assuming a mean growth yield factor for both acidifiers and methanogens $Y = .2$ (Henze and Harremoes, 1983). Gas pressure Pg in the digester is assumed to be constant and equal to 1 bar. The constants are reported in Table 1.

TABLE 1
Constants related to the P/C model.

T (ºC)	Ka1 (10^{-7} M)	Ka2 (10^{-11} M)	KAc (10^{-5} M)	KH (10^{-3} M/bar)	Pw (10^{-2} bar)
10	3.43	3.75	1.72	53	1.2
20	4.15	4.13	1.75	39.2	2.3
30	4.71	5.16	1.75	29.9	4.2
37	5.06	5.86	1.72	24.6	7
40	5.06	6.03	1.70	23.8	7.3
50	5.16	6.73	1.65	19.6	12.3
60	5.02	7.20	1.61	16.3	19.9

Substrate (as $C_6H_{12}O_6$) and cation (as Na^+) concentrations in the influent, in-reactor acetic+acetate concentration $[S1]$ and temperature T are assumed as independent variables. pH and CO_2 partial pressure PCO_2 are considered as dependent variables. Both parameters are currently used for process control. pH is also important because methanogens have a narrow optimum pH range, between 6 and 8 (Zehnder et al., 1982), for volatile acids concentrations of the order of 0.2 to 2 kg/m³ as HAc which are typical of real anaerobic systems (Kroeker et al.,1979). PCO₂ is a relevant parameter for the determination of biogas quality and of outcoming instability conditions (Graef and Andrews, 1974a).

Although PCO₂ changes can be ascribed to changes of the metabolic pathways related to the biochemical system, as it will be shown later, in this paragraph only variations of PCO₂ due to the P/C system are considered.

The model has first been used to simulate the operation of a mesophilic digester (T = 37 °C).

2.1.4 Influence of substrate and of cation concentrations.

pH and PCO₂ are calculated as functions of substrate and cation concentrations assuming that all the substrate is converted to CH_4 , CO_2 and into insoluble compounds to simulate biomass production. Hence [S1] = 0. Results of simulation runs are plotted in Fig. 4 as pH and PCO₂ vs. substrate concentration (given as M TOC) for different concentrations of cations (given as mM NaOH). The lower part of Fig. 4 clearly shows that the anaerobic degradation of carbohydrate solutions with little buffering potential is possible only at low pH values, which is unlikely. Fig. 4 also shows that, for the same cation concentration, the influence of substrate concentration on pH and on PCO₂ is much higher at low than at high TOC influent concentration. This difference is explained qualitatively if one considers that the amount of CO_2 which remains in solution as IC depends mainly on cation concentration (see Eq. (11)) which can be approximated as $[Na^+]$ = $[HCO_3^-]$, neglecting $[H^+]$, $[OH^-]$ and $[CO_3^{2-}]$ contributions (carbonate can be neglected because, if the pH is lower than 8, then $[CO_3^{2-}]/[IC]$ < .01). Hence, as the influent TOC concentration increases, keeping $[Na^+]$ constant, the fraction of CO_2 discharged as IC decreases, and PCO₂ will tend to the value:

$$PCO_2 = Pg \cdot (1 - A) \cdot (1 - \frac{Pw}{Pg}) \qquad (12)$$

As the influent TOC concentration decreases, the fraction of CO_2 discharged as IC increases and PCO₂ decreases appreciably.

Eq. (4) can be written as:

$$[HCO_3^-] = (10^{-pH}) \cdot PCO_2 \cdot Ka1 \cdot KH \qquad (13)$$

$[HCO_3^-]$ is practically constant, for a given cation concentration, and, from Eq. (13), the reversed trends of pH and of PCO₂ in Fig. 4 are easily explained.

In Fig. 5 the same data are plotted as pH and PCO₂ vs [NaOH] for different influent TOC concentrations. The lines [So] = .1 and .25 M TOC indicate that, if the cation concentration exceeds the molar concentration of substrate converted to CO_2 (taking into account TOC converted to "biomass"), there is a sharp rise of pH because $[Na^+]$ > $[HCO_3^-]$ and PCO₂ tends to zero. Fig. 5 also shows that the influence of cation concentration on pH and on PCO₂ decreases if the substrate concentration increases, as explained above.

If sodium cations are added in bicarbonate form instead of hydroxide, then,

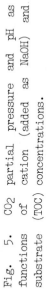

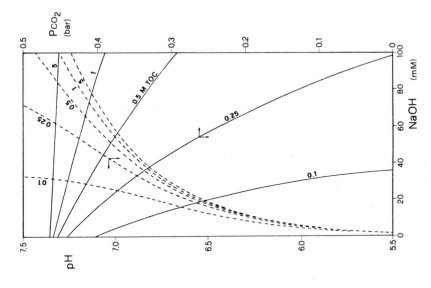

Fig. 5. CO$_2$ partial pressure and pH as functions of cation (added as NaOH) and substrate (TOC) concentrations.

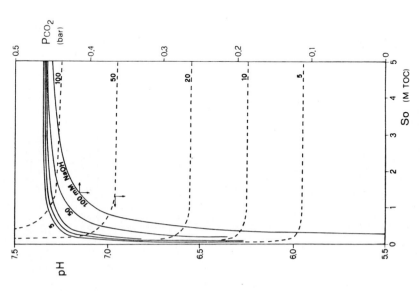

Fig. 4. CO$_2$ partial presure and pH as functions of substrate (TOC) and cation (added as NaOH) concentrations.

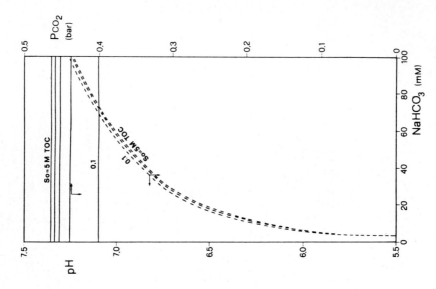

Fig. 7. CO_2 partial pressure and pH as functions of cation (as $NaHCO_3$) and substrate (TOC) concs.; [TOC] = .1, .25, .5, 1, 5M.

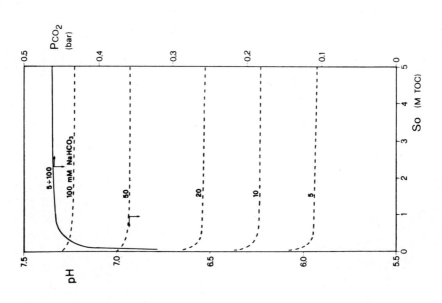

Fig. 6. CO_2 partial pressure and pH as functions of substrate (TOC) and cation (added as $NaHCO_3$) concentrations.

the LHS term of Eq. (5) must be written as Q·(So+BAo), to take into account the inorganic carbon flowing into the system as bicarbonate. Eq. (9) becomes:

$$(1-Y)\cdot\{[So]-2\cdot[S1]\}\cdot\{1-A\ \frac{Pg-Pw}{Pg-Pw-PCO_2}\} + [BAo] = [IC] \qquad (14)$$

For a digester model fed with a solution of glucose and sodium bicarbonate, the trends of pH and of PCO_2 substrate and bicarbonate concentrations as independent variables are shown in Figs. 6 and 7 respectively. Figs. 4 and 6 show, by comparison, that if cations are added as bicarbonate, the influence of cation concentration on PCO_2 is practically negligible, except at very low substrate concentration; pH lines are also flatter.

The above trends are easily explained if one considers that IC concentrations in the influent and in the effluent only differ by the amount of dissolved CO_2 and practically the whole CO_2 production is discharged into the gas except for very low substrate concentrations when the fraction of dissolved CO_2 to the total CO_2 produced becomes appreciable.

2.1.5 Influence of acetic acid concentration.

PCO_2 and pH were calculated as functions of influent cation concentration for different concentrations [S1] of acetic acid in the digester and for different TOC influent concentrations.

In Fig. 8 pH and PCO_2 are plotted vs. [NaOH] for acetic acid concentrations ranging from 0 to 50 mM HAc and assuming [So] = 0.5 M TOC in the influent. Fig. 8 shows that when acetic acid concentration increases, keeping [Na$^+$] constant (e.g.: from point 1 to 2), pH decreases while PCO_2 increases. This behaviour is explained considering that for pH >= 6.5 then [S1] $\simeq$ [Ac$^-$] and hence Eq. (11) can be approximated as:

$$[Na^+] = [HCO_3^-] + [S1] \qquad (15)$$

It follows that when [S1] increases, [HCO$_3^-$] decreases correspondingly. The fraction of substrate converted to CO_2, which is discharged as IC, decreases as well and consequently PCO_2 values rise. Due to the variations of [HCO$_3^-$] and of PCO_2, pH will decrease according to Eq. (13).

If the digester model operates at different TVA concentrations but at the same in-reactor bicarbonate concentration (points 1 and 3 in Fig. 8), approximately the same pH values (and PCO_2 values) are obtained. In fact PCO_2 in point 3 is slightly lower because, if [S1] increases, the fraction of substrate converted to CO_2 and the % of CO_2 in the gas both decrease. This PCO_2 variation accounts also for the slight pH increase between points 1 and 3 (see Eq. (4)).

In Fig. 9 pH and PCO_2 are plotted vs. [NaOH] for [S1] = 0 to 50 mM HAc, assuming influent concentration equal to 5 M TOC. The influence of cation concentration on pH is similar to the previous case plotted in Fig. 8 whereas the influence on PCO_2 is practically negligible because the substrate fraction converted to CO_2 and discharged as IC in the effluent is very low.

In Fig. 10 the same conditions as in the case described by Fig. 8 are assumed except that sodium is provided as bicarbonate instead of hydroxide. The PCO_2 lines have a much flatter trend for the same reasons presented for the discussion of Fig. 7.

648

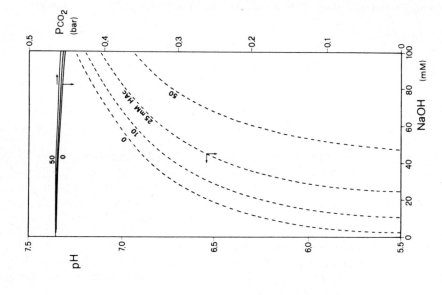

Fig. 9. CO_2 partial pressure and pH as functions of cation (added as NaOH) and acetic acid concentrations (So = 5 M).

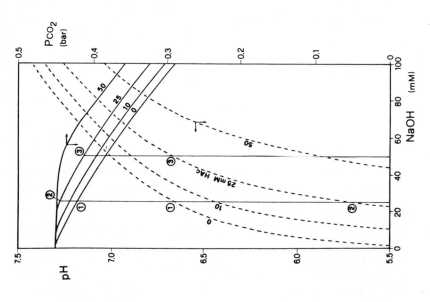

Fig. 8. CO_2 partial pressure and pH as functions of cation (added as NaOH) and acetic acid concentrations (So = 0.5 M).

2.1.6 Influence of temperature.

The model is used to simulate a digester operating within a temperature range from 10 to 60 ^{o}C. pH and PCO_2 are calculated as functions of substrate and of cation concentrations assuming negligible volatile acids contents in the effluent or [S1] = 0.

In Figs. 11 and 12 pH and PCO_2 are plotted vs. temperature T for different values of the influent TOC, assuming cation concentrations equal to 10 and 50 mM respectively. Henry's constant (in arbitrary units) and the partial pressure Pci:

$$Pci = (1 - A) \cdot (Pg - Pw) \tag{16}$$

are also plotted on Fig. 11. Pci is the limit value of PCO_2 if all the CO_2 produced were discharged in the gas flow (as if CO_2 were completely insoluble and IC = 0).

The influence of the independent variables on PCO_2 is shown in Figs. 11 and 12. If substrate and cation concentrations are both low, then PCO_2 is a strong function of Henry's law, as most of CO_2 produced is discharged as IC. This fact is evidenced in Fig. 11 which shows that PCO_2 line for [So] = 50 mM TOC has the reversed trend of Henry's law constant. For higher substrate and cation concentrations ([So] >= 0.5 M TOC; [NaOH] = 10 mM in Fig. 11. Any [So] and [NaOH] = 50 mM in Fig. 12) PCO_2 depends mainly on Pw, which influence is indicated by the trend of Pci. The above dependence can be explained as follows. For a given cation concentration, bicarbonate concentration is practically constant because of the charge balance (Eq. 14) and IC changes mainly depend on dissolved CO_2 concentration. In the case of low substrate and cation concentrations, the fraction of substrate converted to CO_2 which is discharged in the effluent is relevant and hence the influence of Henry's law on PCO_2 is important. If substrate and/or cation concentrations are high, the flow of dissolved CO_2 in the effluent is negligible if compared to the flow of CO_2 in the gas and/or to the flow of bicarbonate in the effluent.

The dependence of pH on temperature is more complex as it is related to the product of bicarbonate and of Henry' constant, as indicated by Eq. (13). For a given cation concentration, bicarbonate alkalinity remains constant while the product $PCO_2 \cdot Ka1 \cdot KH$ decreases with temperature unless both [So] and [Na$^+$] are very low. It follows that in most of the cases considered here pH increases with temperature.

The above parametric study shows the appreciable influence of physico-chemical parameters, such as the strength of the waste, the available cation concentration and the temperature, on the characteristics of both liquid effluent and gas for an idealised anaerobic system. Knowledge of the mechanisms which control the interactions between the P/C parameters is essential for a better understanding and control of the overall anaerobic process.

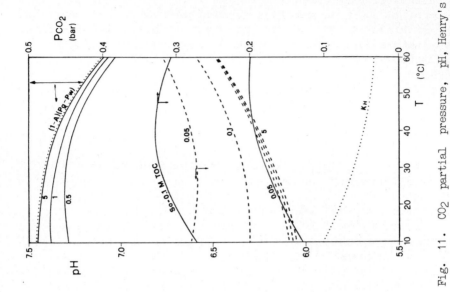

Fig. 11. CO_2 partial pressure, pH, Henry's constant KH as functions of temperature and substrate (TOC) concentration (for [NaOH] = 10 mM).

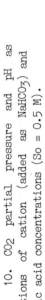

Fig. 10. CO_2 partial pressure and pH as functions of cation (added as $NaHCO_3$) and acetic acid concentrations (So = 0.5 M).

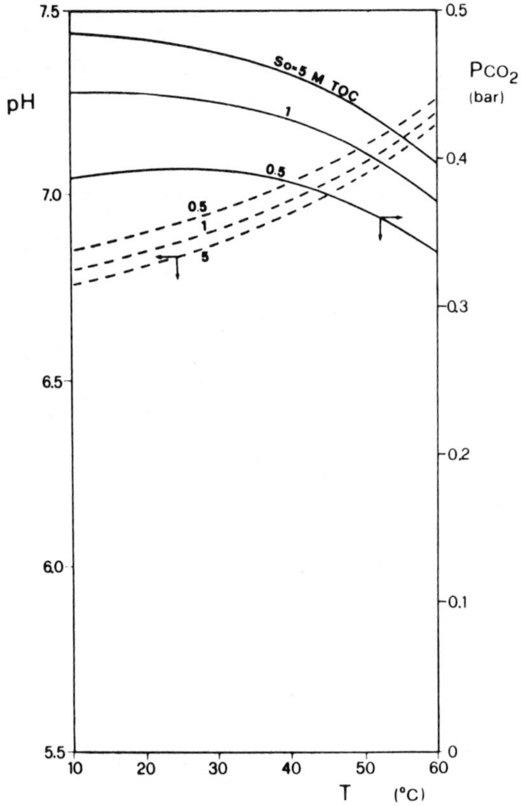

651

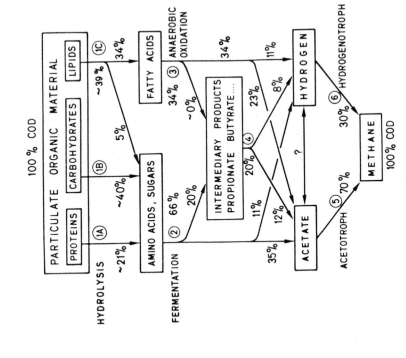

Fig. 13. Proposed reaction scheme for the anaerobic degradation of a complex substrate (From Gujer and Zehnder, 1983).

Fig. 12. CO$_2$ partial pressure and pH as functions of temperature and substrate (TOC) concentration (for [NaOH] = 50 mM).

2.2 THE BIOCHEMICAL SYSTEM

2.2.1 Definition of the system.

The biochemical system is made of the microorganisms and the related enzymes which degrade anaerobically the organic substances both as suspended solids and/or soluble and colloidal compounds. Within the anaerobic systems there are also autotrophic microorganisms which cannot be neglected, the CO_2 reducing methanogens which are extremely important because they control the concentration of H_2 in the anaerobic environment, and the sulfate-reducing bacteria which convert SO_4^{2-} to H_2S using H_2 or organic compounds (e.g.: propionic acid) as electron donors (mixotrophic microorganisms).

In Fig. 13 the biochemical system related to the anaerobic degradation of a complex substrate made of carbohydrates, proteins and lipids is shown (Gujer and Zehnder,1983).

2.2.2 Anaerobic Microorganisms.

The microoganisms which are involved in the conversion of organic matter to CH_4 and to CO_2 can be roughly classified according to their metabolic function.

The hydrolytic microorganisms convert complex organic polymers to monomers, the acidifiers convert the monomers to simpler compounds, mainly short chain alyphatic acids, other organic acids, alcools, H_2, CO_2, NH_3, H_2S. Two special groups of microorganisms, which are scarcely known, produce acetic acid: a) the Obligate Hydrogen Producing Acetogens (OHPA) convert organic compounds such as ethanol, (Bryant et al., 1967), lactate (Bryant et al., 1977), propionate and longer-chained fatty acids to acetate, H_2 and, in the case of odd-numbered carbon energy sources, to CO_2 (McInerney and Bryant, 1981),(McInerney et al., 1979); b) the homoacetogenic bacteria produce acetic acid and sometimes other acids from CO_2 and H_2 (Zeikus, 1980), (McInerney et al., 1980). The role of these latter bacteria is not quite clear. According to Zeikus (1980) they contribute to the stability of the overall conversion process by keeping, amongst others, the H_2 partial pressure within the digester low.

Methane is produced from acetic acid by the acetoclastic methanogens and from CO_2 and H_2 by the CO_2 reducing methanogens.

It is interesting to note that both acetic acid and methane can be produced by heterotrophic and autotrophic microorganisms.

Readers interested to chemical and thermodynamic aspects of anaerobic degradation of organic matter can refer to the literature on the subject, e.g.:McInerney and Bryant (1981) or Thauer et al. (1977). The tables published by the above authors list tens of reactions which indicate the complexity of possible metabolic pathways in anaerobic digestion, and hence the difficulties of modelling the biochemical system in anaerobic reactors.

In general the kinetic description of processes, both in chemical and in biochemical engineering, is less advanced than stoichiometry and energetics (Roels,1982). This is particularly true for anaerobic digestion. Recent reviews of the literature (McInerney and Bryant, 1981), (Henze and Harremoes, 1983), show that very few studies have been carried out on the kinetics of anaerobic microorganisms. Even relatively simple aspects such as substrate inhibition of acetoclastic methanogens have not been extensively investigated and kinetic equations proposed in the late sixties (like the Michaelis-Menten-Haldane equation, (Andrews, 1968)) have not yet been confirmed to the authors' knowledge since the pioneering studies by Andrews (1969).

On the other hand, the need for precise kinetic relationships is questionable because: a) it is quite difficult, while working on biological systems, to obtain accurate and reproducible results which could be used to define the kinetic equations; b) in the experimental systems utilized for these determinations the diffusion resistance around and within the bacterial aggregates should be negligible, otherwise the kinetic constants would be system dependent and should be termed instead as system coefficients. This latter situation is typical of most real processes except in some instances for fluidized beds; c) if the rate of substrate addition controls the specific growth rates, as it happens in most biological treatment systems, "the process is under transport rather than kinetic control. Therefore any of the multitude of equations proposed in the literature (Roels and Kossen, 1978) will be equally successful in the costruction of an unstructured model of batch, fed batch or continuous culture growth" (Roels, 1982).

According to several studies (Kirsh and Sykes, 1971) , (McCarty, 1971), in actual practice Monod kinetics have been found to apply with a reasonable approximation to anaerobic systems. As a general rule it must be remembered that the interpolation of experimental data which are more or less scattered (as customary with anaerobic digesters) is a very delicate operation because different kinetic equations can easily be "forced" through a set of data. For instance, satisfactory interpolation of experimental results related to first order kinetics can be obtained for the Monod equation if the interpolation procedure gives values of the saturation constant Ks which are much higher than the maximum experimental datum.

2.2.3 Degradation kinetics for anaerobic microorganisms.

The metabolism of hydrolytic microorganisms is relatively well known while the related degradation kinetics have not been extensively studied. Direct application of kinetic equations, derived for simple homogeneous systems (soluble substrate and enzyme), to complex heterogeneous systems such as the hydrolytic ones is objectionable because the same assumptions do not apply for both cases. Recent papers indicate that the knowledge of kinetics related to hydrolytic microorganisms is rather poor (Ghosh and Klass, 1978), (Eastman and Ferguson, 1981) and that even the kinetics related to degradation of suspended solids by simple enzymes is not well understood (Dixon and Webb, 1979), (Lee and Fan, 1982), (Lee et al., 1982).

In several cases the hydrolytic and the acidifying reactions are performed by the same microorganisms. As the hydrolysis is always rate-limiting with respect to acidification , in anaerobic degradation models it is equivalent to consider two distinct populations or only one.

Kinetics related to acidifiers and acetogenic (OHPA) bacteria are discussed in the paragraph related to glucose degradation.

Monod kinetics were found to be reasonably satisfactory for acetoclastic methanogens (Lawrence and McCarty, 1970) and were also proposed for CO_2 reducers (Mosey, 1983a)

2.2.4 Definition of the model.

More or less simplified models can be defined depending on the substrate composition (concentration, fractions of soluble and particulate substrate) and on the objective of the modeling study (e.g.: design of reactors in steady state operation, investigations on process stability in dynamic conditions).

The hydrolysis is normally rate-limiting if the substrate is in particulate form (e.g.: cellulose and other biopolymers). If soluble substrate is provided to the system, then the rate-limiting reactions can be the acetoclastic methanogenesis or, in some cases, the acetogenesis by the OHPA microorganisms, depending on the environmental conditions.

If steady state operating conditions are considered, then the anaerobic degradation of insoluble substrates can be modelled in a first approximation approach by a simple hydrolysis reactor. The picture is more complicated for digesters fed on soluble substrates, but in general the rate-limiting reaction is the acetoclastic methanogenesis.

Before defining a general model, some particular cases are examined.

2.2.5 Model for the anaerobic degradation of suspended solids.

Heterogeneous reaction systems, such as particulate substrates contacted with microbial cells and related enzymes, are more difficult to study than homogeneous systems in which both the substrate and the enzymes are assumed to be soluble. In the latter case all the substrate molecules are equally available (on a statistical basis) to the enzymatic attack, while for particulate substrates the position of the biopolymer molecules within the particle seems to affect their biodegradability, and consequently reaction rates also depend on the "history" of the particle to be degraded. These effects have been found even in relatively simple systems such as pure cellulose particles hydrolyzed by pure soluble enzymes (Lee and Fan, 1982).

Kinetic equations usually considered in biological treatment processes have been more or less successfully tested on the anaerobic degradation of suspended solids, e.g.: Monod equation (Ghosh and Klass, 1978), (Hobson, 1983), Haldane equation (Hill, 1983), first order with respect to the substrate (Pfeffer, 1968), (Kaspar, 1977), (Eastman and Ferguson, 1981), first order with respect to the substrate and 0.5 order with respect to the biomass (Rozzi and Verstraete, 1981), (Rozzi, 1984) first order with respect to substrate and biomass (Rozzi, 1982) and Contois equation (Chen and Hashimoto, 1980), (Hill, 1982).

In Table 2, the kinetic functions indicated above and the related effluent substrate concentrations (for CSTR) and microbial growth rates are listed. Kinetic equations and the related growth rates should always be examined jointly and their consistency with the behaviour of real microbial systems analysed. This cross-examination is performed very seldom and yields disturbing results for some kinetic equations.

For the first order kinetics with respect to the substrate only, a finite maximum value for μ does not exist, i.e.: for wash-out conditions, when $S \longrightarrow S_o$, then $\mu \longrightarrow \infty$. In spite of this major biological incongruity which should rule out the model on rational grounds, this equation is one of the most successfully applied to interpolate experimental data of suspended solids anaerobic degradation (municipal sludge, urban refuse slurries, etc.).

TABLE 2
Kinetic functions used in anaerobic digestion studies.

kinetic equation $\dfrac{dS}{dt} =$	effl. substrate conc. $S =$	growth rate $u =$	eq. N.
$\dfrac{K \cdot S \cdot X}{Ks+S}$	$\dfrac{Ks}{SRT \cdot \mu - 1}$	$\dfrac{\mu \cdot S}{Ks+S}$	(17)
$K \cdot S \cdot X$	$\dfrac{So}{SRT \cdot \mu}$	$\dfrac{\mu \cdot S}{So}$	(18)
$K \cdot S$	$\dfrac{So}{1+K \cdot SRT}$	$\dfrac{K \cdot S}{So-S}$	(19)
$\dfrac{Km \cdot S}{Kx \cdot X+S}$	$\dfrac{K \cdot So}{SRT \cdot \mu + K-1}$	$\dfrac{\mu \cdot S}{K \cdot So+(1-K) \cdot S}$	(20)

The above kinetic models were also compared, on the basis of the LYRS diagram proposed by Hawkes and Horton (1981), to correlate experimental data from the literature on sewage sludges digestion. In the LYRS diagram (Loading rate against the gas Yield for Retention time and volatile Solids contents), gas production (as m^3 gas/kg VS added) is plotted vs volumetric loading rates (as kg VS/$m^3 \cdot$d) as a grid of lines at constant influent substrate concentration and constant hydraulic retention times in once-through CSTR.

As a general rule, it would be better to make use of methane yields instead of gas yields, because COD mass balances are related to methane and biomass production only, and CO_2 is partly discharged in the effluent (see paragraph on P/C systems). LYRS diagrams are derived for the different kinetic models as follows. The relationship which gives the effluent substrate concentration (see Table 2) is entered into Eq. (21), given below, which is then solved for different values of So and HRT = SRT, and plotted as a function of the volumetric loading rate.

The methane yield My is given as:

$$My = A1 \frac{So-S}{So} \tag{21}$$

A1 is a methane to substrate conversion factor, which takes into account substrate converted to biomass.

The digester volumetric loading rate L is given as:

$$L = \frac{Q \cdot So}{V} = \frac{So}{HRT} \tag{22}$$

In Figs. 14 and 15 LYRS diagrams are plotted for first order kinetics with respect to the substrate only and with respect to both substrate and biomass concentration (Eqs. (19) and (18) in Table 2 respectively). Quite different trends are obtained. Contois equation gives a LYRS plot similar to Fig. 14 while the remaining kinetic equations listed in Table 2 give plots similar to Fig. 15 which compares favourably with the empirical diagram derived by Hawkes and Horton (1981). More complicated models which take into account the CO_2/HCO_3 equilibria could be used to calculate the gas yield from the methane production, but similar results would be obtained.

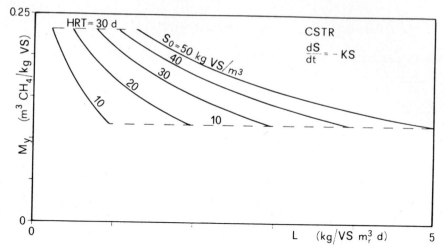

Fig. 14. Methane yield as a function of the volumetric loading rate (LYRS diagram) for first order kinetics with respect to substrate.

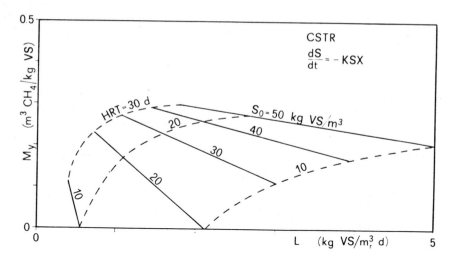

Fig. 15. Methane yield as a function of the volumetric loading rate (LYRS diagram) for first order kinetics with respect to substrate and to biomass.

2.2.6 Model for the anaerobic degradation of glucose.

The biochemical system related to the anaerobic degradation of glucose has been studied by Mosey who developed a cornerstone model (Mosey, 1983a, 1983b) which considers the following microbial populations: acid forming bacteria, acetogenic bacteria, acetoclastic and H_2 utilizing methanogens (CO_2 reducers). The microbial ecology of the model is illustrated in Fig. 16, which is a generally accepted flow-sheet, assuming that the conversion of glucose to lactic acid and ethanol is negligible.

The main assumptions of this model are:

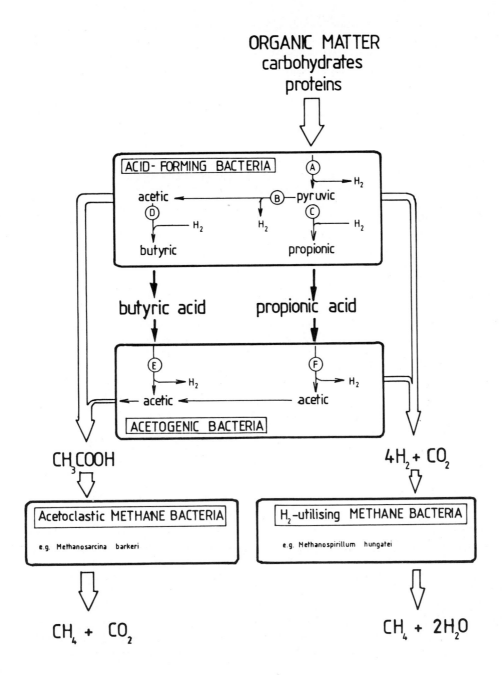

Fig. 16. Microbial ecology of the anaerobic digestion process. (From: Mosey, 1983a).

1. The relative availabilities of the reduced (NADH) and oxidised (NAD+) forms of the carrier NAD (Nicotinamide Adenine Dinucleotide) within the bacteria control both the overall rate of conversion down the metabolic pathway and the composition of the mixture of acids formed.

2. The bacteria are homeostatic with regard to their internal pH, which is assumed to be 7.

3. The bacteria are "transparent to H_2" so that the hydrogen reducing potential is transferred freely through the bacteria semi-permeable membrane (Mosey, 1983c).

4. The bacteria are "transparent to redox potential (ORP)" so that the ORP inside the cells is the same as the one of the growth medium.

5. The gas and liquid phases are in equilibrium conditions.

The oxidation state of the NAD carrier molecule can then be related to the partial pressure of H_2 in the gas (PH_2), by equating the oxidation/reduction reactions for NAD and hydrogen via their redox potential, which must be the same because of assumption 4, to yield the following equation (Mosey, 1983a):

$$\frac{[NADH]}{[NAD^+]} = 1500 \cdot PH_2 \tag{23}$$

The substitution of Eq. (23) into the rate equations for acid formation of acids from glucose, for OHPA and for CO_2 reducing methanogens provides the basis of the model, as shown in Table 3, in which the rate equations used by Mosey (1983a) are reported.

TABLE 3

Rate equations for the formation of VFA from glucose (Mosey, 1983a)

Reaction Product	Derived Equation	
Acetic Acid	$Rg \cdot (F_1H)^2 \cdot (1-2 \cdot F_2H)$	(24)
Propionic Acid	$Rg \cdot F_1H \cdot F_2H$	(25)
Butyric Acid	$Rg \cdot (F_1H)^2 \cdot F_2H$	(26)

Rg = maximum glucose uptake if hydrogen is not rate-limiting

$F_1H = 1/(1 + 1500 \cdot PH_2)$ (PH_2 in bars)   (27)

$F_2H = 1500 \cdot PH_2/(1 + 1500 \cdot PH_2)$ (28)

The function F_1H is a hyperbolic equation. Its concavity is upwards and its value tends to zero for high hydrogen concentrations. F_2H is an hyperbolic equation like Monod relationship (concavity downwards; maximum asymptotic value for high hydrogen concentrations). The trend of the fraction of acetic acid formation is controlled by F_1H and its value decreases continuously as PH_2 increases. The maximum value of rate formation for butyric acid is reached for lower hydrogen concentrations than for propionic acid because the factor $(F_1H)^2$ exerts a stronger damping action than F_1H.

Relative rates of acid formation as fractions of metabolized glucose are plotted in Fig. 17 as a function of hydrogen concentration in the biogas.

The above mathematical model is used to simulate steady state and dynamic performance of mesophilic (T=35 °C) completely mixed anaerobic reactors without solids recycle and fed on glucose solutions (So = 2 kg/m3).

For the dynamic model, a pH inhibition function for bacteria based on data from Clark and Speece (1971) is used.

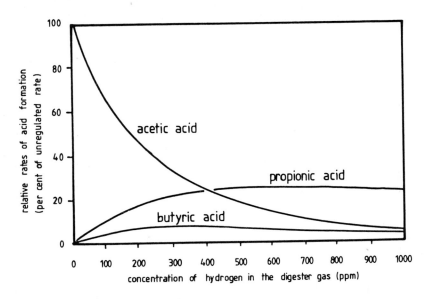

Fig. 17. Relative rates of acid formation from glucose as functions of H_2 concentration in the gas. (From Mosey, 1983a).

In Fig. 18 the data related to the steady state simulation are plotted as concentrations of volatile fatty acids in the effluent and of hydrogen in the gas vs the hydraulic (and solids) retention time.

The shape of hydrogen concentration curve in Fig. 18 is typical of the residual substrate concentration for wastewater processes controlled by Monod equation (Lawrence and McCarty, 1970). These kinetics imply that the effluent concentration is independent of the influent substrate concentration and depend only on the kinetic constants and on the residence time.

The shapes of the volatile acid concentration curves in Fig. 18 require some comments. For HRT < 3 days, OHPA bacteria and acetoclastic methanogens are washed out because their maximum growth rates are smaller than 0.33 d-1. The digester becomes an acidification reactor in which some methane is still produced by the CO_2 reducers (whose growth rates are relatively high). The concentrations of volatile acids are proportional to the relative rates of acid formation and hence are directly controlled by hydrogen concentration as already indicated in Fig. 17 (e.g.: low acetic acid production for high H_2 concentration). For hydraulic retention times higher than 3 days, the volatile acids concentrations decrease approximately as imposed by Monod kinetics considering each acid as substrate (Lawrence and McCarty, 1970).

For HRT values in the range from 10 to 40 days, which are typical of sewage sludge digesters, the model predicts hydrogen concentrations in the gas of 15–60 ppm which compare favourably with the range 40–80 ppm determined in full scale digesters (Mosey, 1983a). Better fitting between experimental data and model results can be obtained by selecting an appropriate value for the saturation constant Ks related to hydrogen (Mosey, 1983b).

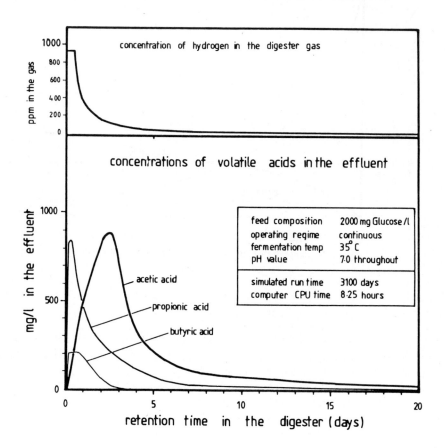

Fig. 18. Computer simulation of steady state operation of anaerobic digester model. (From Mosey, 1983a).

In Fig. 19 the computer simulation of a laboratory digester in dynamic conditions is shown. Organic overloading is induced by step increases in substrate concentration.

The surges of H_2 concentration in the gas, as shown in Fig. 19, are damped by the fairly large headspace of gas assumed for the model (equal to 2.24 times the liquid digestion volume).

It is interesting to note that the effect of the second step overload on process stability is much lower than the first one. The time interval between the two overload trials is too short (1.50 times the HRT) to permit the re-establishment of the original steady state conditions (and related distribution of the various microbial species in the biomass). Therefore the digester still retains most of the OHPA acetogenic microorganisms which grew up during the first overload.

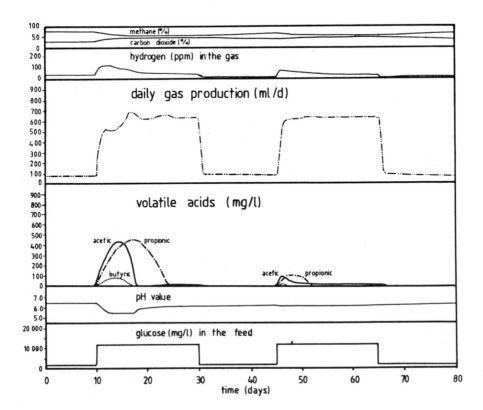

Fig. 19. Computer simulation of dynamic operation (step overload) of anaerobic digester model. (From Mosey, 1983a).

Similar trends related to volatile acids concentrations were obtained in experiments on chemostats temporarily operated in batch conditions and overloaded with concentrated glucose solutions (Cohen, 1982).

2.3 THE OVERALL P/C–B/C ANAEROBIC SYSTEM MODEL.

The principal interactions between the different anaerobic trophic groups are found in the acidification and methane generation steps of soluble substrates. The hydrolysis of complex substrates to monomers can be rate limiting (Pfeffer, 1968), (Hobson, 1983), but its dependence on the subsequent degradation steps is generally considered to be negligible because inhibition of the hydrolysis reaction occurs only at relatively high concentrations of those products which are normally well degraded by the acidifying and methanogenic bacteria.

The study of dynamic mathematical models of biological waste water treatment was started in the sixties when analog and digital computers became commercially available and made possible the solution of the complex systems of differential equations related to microbial conversion processes.

Andrews was the first to develop a very interesting dynamic model of the biochemical process linked to the physico-chemical one. This model "involves intricate interplays between physical, chemical and biological factors. It therefore exemplifies a synthesis of classical engineering skills with basic biological knowledge" (Bailey and Ollis, 1977) and it has been used as the basis for the majority of anaerobic mathematical modelling studies published in the last decade. The key factors of Andrews' dynamic model are:
1. An inhibition function to relate volatile acids concentration and specific growth rate for methane (acetoclastic) bacteria.
2. Consideration of the unionized (acetic) acid as the growth limiting substrate and inhibiting agent.
3. Consideration of a toxicity effect which increases the decay coefficient Kd proportionally to the concentration of the toxicant.
4. Consideration of the interactions which occur in and between the liquid, gas and biological phases of the digester.

The model is related to the anaerobic degradation of sewage sludge but, as hydrolysis and acidification were assumed not to be rate-limiting, a microbial population only is considered i.e. the methanogenic acetoclasts which produce most of the methane in conventional digesters (Jeris and McCarty, 1965).

The summary of the mathematical model is shown in Fig. 20 (Andrews, 1978).

The set of mathematical equations is solved on a hybrid computer. The high speed of analog computer permits days of digester operation to be contained within seconds without the numerical integration (and iteration) problems related to digital computers (Graef and Andrews, 1974a) In this model, gas and liquid phases are not assumed to be in equilibrium conditions and the net rate of carbon dioxide transfer between the gas and liquid can be expressed by the 'two film' theory as:

$$TG = KL \cdot a \cdot ([CO_2]' - [CO_2]) \qquad (29)$$

in which $KL \cdot a$ is the overall gas transfer rate and $[CO_2]'$ and $[CO_2]$ are respectively the actual and the equilibrium carbon dioxide concentrations in the liquid phase. Andrews' model does not take into account the complex interactions between the different microbial populations (i.e.: acidogens, OHPA and methanogens) which were discovered later, but it is still very useful to study: a) the dynamic behaviour of digesters for short periods of variable operating conditions; b) the control strategies to be developed for anaerobic process control; c) the effects of overloading on a digester.

Only a few authors consider the hydrolysis and the acidification steps in their models, among them: (Hill and Barth, 1977), (Installe' and Antunes, 1980) (Kleinstreuer and Poweigha, 1982), (Hill, 1983), while most papers on the subject investigate the dynamic behaviour of a biological system made of acetoclastic methanogens. Taking into account the recent discoveries on anaerobic metabolism, more complex models can now be defined and analysed.

The general anaerobic system defined here refers to the degradation of a simple exose (glucose) solution with sulphate and sodium ions. Six microbial populations are considered: a) acidogens which convert glucose to acetic, propionic and butyric acids; b) OHPA microorganisms which convert propionic

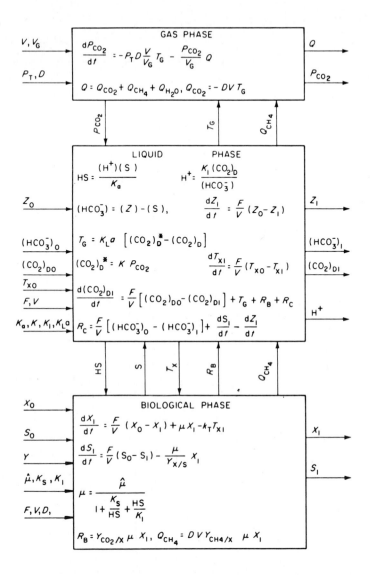

Fig. 20. Summary of mathematical model and information flow (from Andrews, 1978).

and butyric to acetic acid; c) acetoclastic methanogens; d) CO_2 reducing methanogens; e) homoacetogens which convert hydrogen and CO_2 to methane; f) sulphate reducing bacteria which use hydrogen or volatile acids (mainly propionic and butyric). The conversion of glucose to ethanol and lactic acid is not considered though it seems that it is not really negligible.

The general mass flow-sheet is shown in Fig. 21. The circles indicate the "biochemical nodes" in which the organic and inorganic compounds are

664

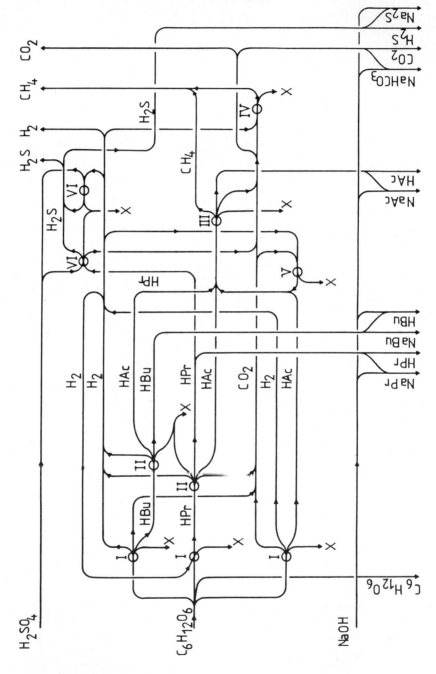

Fig. 21. Summary of the six populations mathematical model. I = Acidogens; II = OHPA; III = acetoclastic methanogens; IV = CO₂ reducing methanogens; V = homoacetogens; VI = sulphate reducers.

metabolized by the different groups of microorganisms. The diagram shows the remarkable complexity of a degradation system related to a simple substrate such as an exose which can be converted to CH_4 and CO_2 along different metabolic pathways.

It should be remembered that in the above system the microorganisms are considered as simple autocatalytic enzymes and that the higher complexity level related to the metabolism of each individual microbial species has been neglected.

The flow of metabolites along the different pathways is controlled by a complex set of constraints on the process variables. This set can be visualized as a 'control system' as follows: suitable sensors control, on a feedback mode, valves throttling the flow rates of the substrate and intermediate metabolites. According to present knowledge, the most important control variables in anaerobic processes are molecular hydrogen and proton concentrations.

2.4 MODELS FOR ANAEROBIC BIOFILM SYSTEMS (PACKED AND FLUIDIZED BEDS)

Throughout the above discussion on biological reaction kinetics, the diffusion resistances of reactants and of products in the liquid layer sourrounding the bacteria are implicitly considered to be negligible. In fact this assumption is basically correct only for well stirred biological systems made of small aggregates of bacteria (particles or thin films) and operating at relatively high substrate concentrations, because in these cases the effect on the kinetics of the diffusion resistance within the liquid concentration boundary layer and within the biomass can be neglected.

The basic concept that diffusional resistance has a significant effect on reaction phenomena in a dense matrix originates from the development of chemical engineering science during the sixties (Henze and Harremoes, 1983). This concept was first applied to biological fixed film reactors by Atkinson et al. (1967) and later it was developed for anaerobic systems too (Young, 1968), (Williamson and McCarty, 1976), (Rittmann and McCarty, 1980a and 1980b).

Most of the progress in this field has been achieved since the introduction of electronic computers, which made possible the numerical solution of the general set of differential equations representing the idealized conversion system, namely: substrate utilization, molecular diffusion and mass transport, and biological growth and decay.

The general model for anaerobic biofilms is conceptually similar to those derived for aerobic or anoxic wastewater attached growth systems. In the latter cases the electron donor and acceptor are always two different compounds and hence there can be two different cases of kinetic rate limitations. In anaerobic systems the electron donor and the electron acceptor can be different compounds or the same one depending on the microbial species (e.g.: CO_2 reducing methanogens and sulfobacteria on one hand and acetoclastic methanogens on the other hand).

The basic assumptions of a simplified steady state anaerobic biofilm model are:

1. The biofilm is homogeneous with respect to the different microbial populations.

2. The substrate is soluble.

3. The system is a differential reactor (bulk substrate concentration is constant).

4. The substrate diffuses through the liquid concentration boundary layer (or diffusion layer) and within the biofilm according to Fick's law. Its concentration varies only in the z direction which is normal to the surface of the film.

5. Reaction products diffuse through the biofilm as the substrate.

6. The biofilm thickness is determined by the flux of substrate into the biofilm and by the bacterial growth and decay rates, in other words, the total amount of biofilm mass is just equal to that which can be supported by the substrate flux (Rittmann and McCarty, 1980a).

7. The substrate utilization at any point of the biofilm is assumed to follow a kinetic equation of the following form:

$$\frac{dS}{dt} = X \cdot F(Sf) \tag{30}$$

where $F(Sf)$ is a function of the substrate concentration Sf in the biofilm. Because of assumption 4, transport of substrate within the biofilm is given by:

$$\frac{dS}{dt} = Df \frac{d^2Sf}{dz^2} \tag{31}$$

where Df is the molecular diffusivity of the substrate in the biofilm.

As the system is assumed to be in steady state conditions, Eqs. (30) and (31) yield the ordinary differential equation for substrate concentration within the biofilm:

$$Df \frac{d^2Sf}{dz^2} = X \cdot F(Sf) \tag{32}$$

Fick's law applied to the diffusion boundary layer gives the substrate flux J:

$$J = D \frac{dS}{dz} - \frac{S - Ss}{Ld} \tag{33}$$

Assuming a decay rate kd, the net growth of the bacterial mass within the differential volume dV = A·dz is given by:

$$dV \frac{dX}{dt} = Y \cdot F(Sf) \cdot X \cdot dV - kd \cdot X \cdot dV \tag{34}$$

The maximum conversion rate of a biofilm is reduced by the diffusional resistance which limits the availability of the substrate to the bacteria deep within the biofilm and also limits the release of the metabolic products.

The conceptual basis for biofilm model is shown in Fig. 22 which illustrates the three characteristic substrate concentration profiles (Rittmann and McCarty, 1980a). Within a deep biofilm (Rittmann and McCarty, 1980a), the substrate concentration decreases asymptotically to zero. The deeper portion of the film receives little or no substrate. The maximum substrate flux for the deep biofilm is limited by the substrate concentration Ss at the biofilm/liquid interface. This type of biofilm is defined by

Harremoes (1978) as partly penetrated.

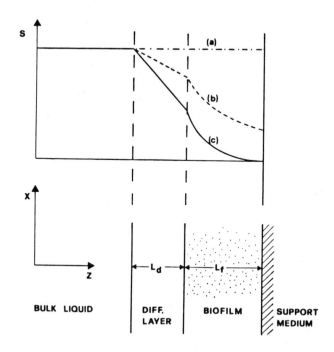

Fig. 22. Conceptual basis for biofilm model. (a): fully penetrated; (b): shallow; (c): deep. (Adapted from Rittmann and McCarty, 1980a).

All the cases in which the substrate concentration at the biofilm/attachement medium interface is $0 <= Sf <= Ss$ are defined as shallow or fully penetrated respectively by the authors quoted above. It should be noted that Rittmann and McCarty use the fully penetrated definition for the case of neglegible diffusion resistance within both the the diffusion layer and the biofilm ($Ss = S$, see case (a), Fig. 22). The different definitions of penetration are listed in Table 4.

TABLE 4
Definitions of penetration in biofilms.

Rittmann and McCarty (1980a)	Harremoes (1978)
Deep biofilm	Partly penetrated
Shallow biofilm	Fully penetrated
Fully penetrated	Fully penetrated ($Df \longrightarrow \infty$)

Diffusion resistance within the biofilm affects the conversion reaction rates. Overall or apparent kinetics are then different from intrinsic kinetics which refer to a sistem without diffusion resistance.

"In the case of zero order reaction, the reduced rate at partial penetration (Harremoes, 1978), as compared to full penetration of the biofilm (or shallow biofilm) is expressed by the conversion of the intrinsic zero order reaction rate to an overall half order reaction. In the case of first order reaction, the rate remains 1st order, but at a reduced rate. Monod kinetics provide nothing more than smooth transitions between different order and rates" (Henze and Harremoes, 1983).

The significance of diffusional resistance as a function of biofilm thickness and of bulk substrate concentration was evaluated by Henze and Harremoes (1983) assuming one rate limiting reaction (Monod) only for the anaerobic biofilm model. The kinetic constants and biofilm parameters are reported in Table 5 (Henze and Harremoes, 1983).

TABLE 5
Kinetic constants and biofilm parameters.

Biofilm concentration: $X = 50$ kg VSS/m^3
Substrate removal rate: $r = 2$ kg COD/(kg VSS·d)
Monod constant: $K_s = 0.2$ Kg COD/m^3
Biofilm diff. coeff.: $D_f = 10^{-4}$ m^2/d
Yield coefficient: $Y = 0.1$ kg VSS/kg COD

Results are plotted in Fig. 23. Transition from Monod to first order kinetics occurs for $S/K_s = 2$ which gives $S = .4$ kg COD/m^3 for $K_s = .2$ kg COD/m^3. Transition from bulk zero order to half order reaction occurs for $L_f^2 = 2 \cdot 10^{-6} \cdot S$ which gives a biofilm thickness L_f of 1mm for the above substrate concentration, as shown in Fig. 23.

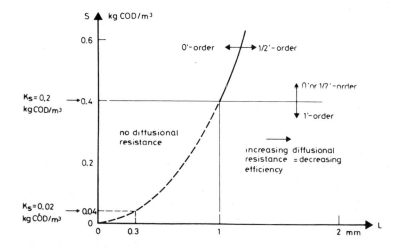

Fig. 23. Diffusional resistance in anaerobic biofilm. (From Henze and Harremoes, 1983).

Biofilm thickness in steady state conditions was evaluated by Rittmann and McCarty (1980a) who equated available and maintenance energy rates according to assumption 6 as:

$$Ld \cdot Kd \cdot X \ = \ J \cdot Y \qquad\qquad\qquad (35)$$

Since the model does not take into account biomass losses such as sloughing, it can overestimate the biofilm mass.

The above authors also determined the minimum substrate concentration which can maintain the thinner possible film, i.e. a monolayer of bacteria. Model results and experimental studies (Rittmann and McCarty, 1980b) showed that for acetoclastic methanogens the minimum bulk substrate concentration which can support a biofilm is very low. This finding supports the use of attached growth processes for the anaerobic treatment of low stregth wastewaters.

The above simplified model for a differential reactor does not take into account: a) the variations of concentration of the different bacterial populations along the biofilm depth; b) the complicated relationships between removal rates, diffusional build-up of acids within the biofilm and the related inhibiting effects on specific bacterial populations; c) the feed-back interactions between points a) and b); d) outdiffusion of bubbles of gases from the back of the biofilm and related interferences on the whole diffusional patterns; e) sloughing of the biofilm; f) adsorption and hydrolysis phenomena within the biofilm; g) the variation of substrate concentration along the axial digester coordinate which should be taken into account for plug-flow reators of finite length.

From the above considerations it is clear that anaerobic biofilm modelling provides a good example of the choice which faces the researcher trying to model complex real ecosystems: on one hand a simple rough model which can be easily related to operational parameters but which largely oversimplifies the real system and on the other hand more complex models that use parameters and constants which are very difficult or practically impossible to measure experimentally.

There are other factors which make quite difficult the modelling of fixed film systems: 1) real biofilm reactors are seldom pure attached growth systems. Upflow anaerobic filters are usually hybrid systems which contain appreciable amounts of suspended active biomass in the lower section (Young, 1983). This effect can also be appreciable in downflow anaerobic filters operating at HRT lower than the wash out retention times for suspended microorganisms; 2) mixing conditions in anaerobic filters can be highly variable and this fact strongly affects the substrate and metabolite concentration distributions along the filter height (Hall, 1982).

3. APPLICATIONS OF MATHEMATICAL MODELS

Applications of mathematical models to anaerobic processes are still scarce and, with a few exceptions, in a more or less embryonic stage.

The first, and up to now, the most important application concerns the basic modeling of the anaerobic P/C and B/C system in order to obtain a better understanding of the process and a better evaluation of the related control strategies, especially in dynamic conditions. Experiments performed on real digesters usually last very long (some 50 to 200 days) and simulations can be very useful in screening the probable "wrong" tests or the ineffective control strategies.

Mathematical models are also being applied to investigate the behaviour of anaerobic processes as a function of the wastewater characteristics. Here too they can be very useful to obtain a better understanding of the digester operation and to avoid time consuming trial and error experimental procedures.

Another major application concerns reactor design and cost analysis. A few studies only have been published on digester economics and reactor modelling while the related literature on aerobic processes is quite broad. This fact is surprising if one takes into account the potential economic advantages of anaerobic processes which allow the recovery of a valuable byproduct such as methane and consume much less mechanical energy than aerobic treatment systems.

3.1   INVESTIGATIONS ON PROCESS STABILITY AND CONTROL

Andrews was the first to use mathematical models to investigate stability and control problems in anaerobic digesters. His main objectives were: a) evaluation of process stability and b) definition of control strategies to keep stable process operation in "stressed" conditions (overloadings and start-up) (Andrews, 1969), (Graef and Andrews, 1974a and 1974b), (Andrews, 1975), (Andrews, 1978).

Andrews' model has been already described in the paragraph related to the overall anaerobic system and only the results related to process stability and control strategies are summarized here.

3.1.1 Overloads

Process behaviour for different cases of overloading (hydraulic, organic and toxic) is simulated for CSTR such as sludge digesters. The simulations provide an abundance of data for evaluating the stability indicators, i.e. the parameters most sensitive and best suited for detecting the onset of process instability.

The model is also used to formulate and evaluate strategies for digester control. Control systems include: "a) a control action, b) a feed-back control signal, and c) a controller mode. As with selecting stability indicators and evaluating factors enhancing stability, the combinations and alternatives in formulating control strategies were numerous. Thus computer simulations were used to sort the most favourable methods from those which were ineffective or of little use" (Graef and Andrews, 1974a).

As control process stability indicators, the following parameters are selected: a) rate of methane production, b) % CO_2 in the dry gas, c) total volatile acids concentration and d) pH, because they offer a great potential from the standpoint of indicating stability and because they can be readily measured in the field with currently available equipment and sensors.

In Fig. 24 results related to organic overloading (step increase of substrate concentration) are presented. Methanogenesis is assumed to be the rate-limiting reaction and the digester is simulated by a reactor fed on acetic acid.

The faster response after a step load is obtained from gas parameters. Indeed the solubility of gases in the liquid is relatively low (little hold-up

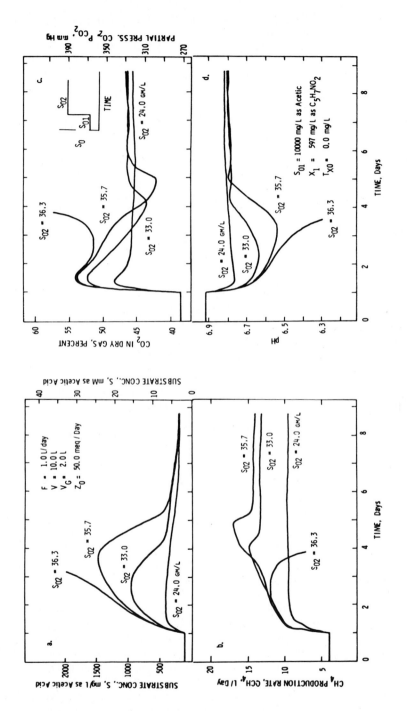

Fig. 24. Dynamic responses of anaerobic digester model to step changes in substrate concentration. (From Graef and Andrews, 1974a)

672

in the liquid phase) and the gas retention time in the digester is orders of magnitude smaller than the liquid one.

Graef and Andrews also investigated the effects of other operational parameters on process stability, e.g. bicarbonate alkalinity (BA) and solids retention time. They simulated organic overloading of a digester operating in steady state conditions by a step increase of the substrate concentration until the process failed (sour digester) and plotted the influent concentration (after the step load) required to cause failure as a function of steady state bicarbonate alkalinity and of hydraulic retention time. Simulation results, plotted in Fig. 25, confirm that process stability strongly depends on the bicarbonate alkalinity of the mixed liquor.

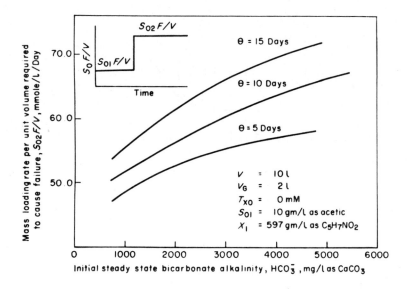

Fig. 25. Influence of hydraulic residence time θ and alkalinity on process stability. (From Andrews, 1978).

Mathematical models can be very useful to evaluate control strategies on digesters, as experimentation on anaerobic processes is extremely time-consuming. "Formulating suitable control strategies involves selecting feedback control variables and automatic control modes. A feedback control variable is a measurable process quantity that can serve as a guideline for adjusting the conditions in a process.(..omissis..). In selecting a control variable, it is important that the variable chosen be directly related to the cause of process instability as well as the corrective action that is is being exerted." (Graef and Andrews, 1974a). Automatic controller modes can be quite sophisticated, but for real process conditions the control system should be as simple as possible, such as "on-off" control.

Among the different control strategies defined by the above authors the most interesting one is gas scrubbing and recycle. In this control configuration a fraction of the biogas is scrubbed to remove CO_2 and then recycled to the digester. When the pH falls below a threshold value, gas flow

673

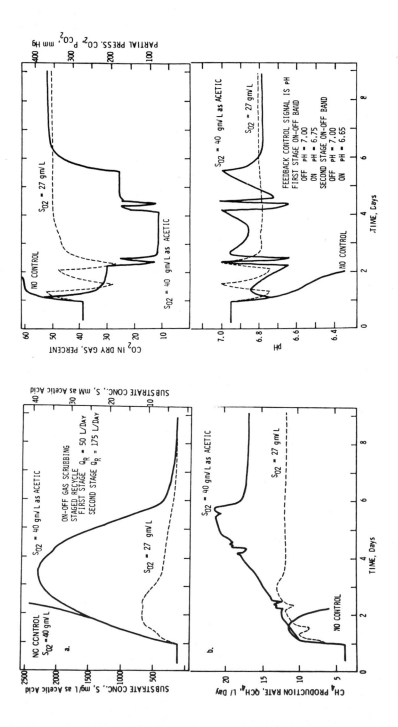

Fig. 26. Response of digester model to control by gas scrubbing and recycle. (From Andrews, 1978).

674

through the CO_2 scrubber is increased. The recycled gas, mainly made of methane, lowers the concentration of dissolved CO_2 in the mixed liquor, thereby increasing the pH, as indicated in Fig. 3. This subtle approach to pH control has several advantages relative to conventional techniques as it removes a weak acid from the system instead of adding a base. In this way the new pH conditions are obtained in a rather short time throughout the bulk of the liquid phase, while if a strong base solution is added, localized regions of very high pH may temporarily build up in the digester. Simulation results related to this control action, which has not been actuated so far, are reported in Fig. 26.

Making use of a model similar to the one defined by Andrews, Rozzi (1981) compared control strategies based on pH and bicarbonate alkalinity monitoring. Different assumptions were used only for the inhibitory function on methanogens (growth rate dependent on pH instead of Haldane equation) and for the interactions between the liquid and the gaseous phases (equilibrium conditions instead of a rate of CO_2 transfer defined by the two film theory).

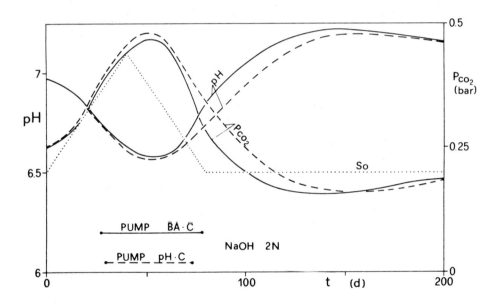

Fig. 27. Response of digester model to control by pH or bicarbonate sensors. Concentration of soda solution for process control: 2N.

In Fig. 27 the response of the system to a wedge shaped organic overload is shown. Similar control actions are obtained by monitoring bicarbonate alkalinity (BA) and pH. The segments in the lower part of the figure indicate the time interval in which the control pump adding the alkali is operating.

The superiority of either method depends on the sensitivity and on the accuracy of available probes. As pH is a logarithm on base 10 and bicarbonate alkalinity could be automatically monitored with fairly good accuracy, of the order of 10 % (Rozzi et al., 1983), it is likely that control systems based on the latter parameter would give better results.

More sophisticated models which take into account the complex interactions between the main anaerobic populations and the main role played by hydrogen partial pressure are being developed, (Mosey, 1983a-1983b), (Rozzi et al., 1984) Making use of a model based on thermodynamic and kinetic interactions of the final steps of anaerobic digestion, Gujer and Zehnder (1983) simulated the operation of a sewage sludge digester and suggested two monitoring parameters: a) hydrogen partial pressure, which indicates instantaneous process upsets and b) acetate concentration, giving an early warning of pH drop which strongly affects CO_2 reducing methanogens (Zehnder and Wuhrmann, 1977) and consequently disrupting the whole anaerobic system. Monitoring hydrogen concentration down to 10 ppm is now practicable (Fernandez and Mosey, 1983).

Simulation results related to organic overloading of a anaerobic contact reactor operating at low HRT (1d) (Rozzi et al., 1984) are plotted in Fig. 28. The biomass is assumed to degrade carbohydrates according to the model defined by Mosey (1983a). Substrate concentration was linearly increased from 10 to 30 mM glucose in 0.5 days. Hydrogen is the parameter which presents the fastest response to the organic overload, but then its concentration remains at high levels until propionate too is degraded (not shown in Fig. 28).

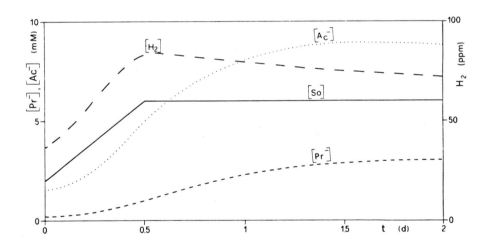

Fig. 28. Response of a four populations digester model to overloading. Acetic acid, propionic acid and hydrogen concentrations vs time.

New algorithms for anaerobic process control have been recently proposed (Dochain and Bastin, 1984), (Bastin and Dochain, 1984) both for mixed and separated (acidogenic and methanogenic phases) systems. An important feature of the proposed algorithms is that they do not require any specific analytical description of the microbial growth rates.

The above few examples show the potential of dynamic simulation in investigating anaerobic process overload. Taking into account the progress made in analytical instrumentation and in commercial computers it is expected

676

that in a very short time it will be possible to validate very complex models related to anaerobic process overloading.

3.1.2 Start-ups.

Start up is a very special case of controlled overloading, which has been little considered so far, both in real and simulation investigations, although it is a crucial phase in digester operation.

A few modeling studies have been published on start-up procedures (Andrews, 1978), (Installe' and Antunes, 1980), (Van den Heuvel and Zoetemeyer, 1982), and automatic start-up strategies are among the most interesting aspects of anaerobic digestion modeling presently under study in several research centers.

Simulations of start-up of a CSTR fed on sewage sludge (Andrews, 1978) indicated that operating conditions are improved by increasing the inoculum concentration and initial pH and that digester failure can be avoided by slowly increasing the digester loading up to the final value. These results are in agreement with experimental data on laboratory and full scale digesters.

Van den Heuvel and Zoetemeyer (1982) investigated the stability of a methane reactor both in steady state and in start-up conditions, making the following basic assumptions: a) acetoclastic methanogens to be the only

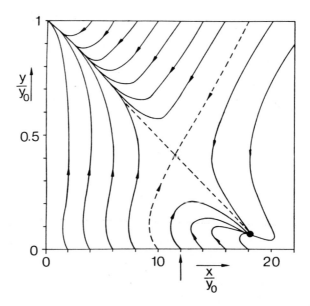

Fig. 29. Dimensionless biomass concentration vs dimensionless substrate concentration in multiple steady state region. Conditions: $y_o = S_o/K_s = 330$; $m/p = u /SRT = 1.68$; $p = HRT/SRT = 0.05$; $S_o = 10.7$ kg COD/m^3; HRT = 5 h. (From Van den Heuvel and Zoetemeyer, 1982).

microbial population in the reactor model and b) Haldane degradation kinetics for acetic acid. They derived adimensional equations to describe the time-dependent behaviour of the reactor and solved the coupled substrate/biomass equations system both in steady state and in dynamic conditions. They found that a well defined domaine of residence times exists with three solutions in the steady state: a high conversion, a low conversion and a wash-out state. The low conversion rate was shown to be unstable. The dynamic behaviour of the system was also investigated for any given initial condition, therefore including any possible start-up. The use of adimensional parameters is obviously very convenient for investigating a wide range of different cases.

In Fig. 29 an example is shown of the obtained results, for a given set of operational conditions. The abscissa and the ordonate are dimentionless biomass and substrate concentrations respectively. Fig. 29 shows that, if the biomass inoculum (seed sludge) is not large enough ($x/yo < 8$) then wash-out ($y/yo = 1$; $x/yo = 0$) is reached irrespectively of the initial start-up substrate concentration (ranging from the raw wastewater to infinitely diluted one, condition y/yo from 1 to 0).

Fabiani et al. (1983) carried out similar studies and compared state trajectories for CST reactors assuming Contois and Haldane kinetics for methanogens.

3.2 PROCESS SELECTION AS A FUNCTION OF WASTEWATER COMPOSITION.

Mathematical modeling in the area of anaerobic digestion has largely been restricted to process optimization and to stability simulation studies. However, taking advantage of the many recent advances in both anaerobic digester technology and microbiology, kinetic models can also be used for process selection purposes.

Making use of a simple kinetic model, Rozzi and Verstraete (1981) and De Baere et al. (1983) investigated the loading limitations of anaerobic digesters as a function of waste water composition (suspended solids and soluble+colloidal substrate fractions) and of solids retention times. The analysis carried out in the papers quoted above is summarized here.

The basic assumptions for the model are the following:
1. The substrate So is the sum of a fraction of volatile suspended solids Sao (expressed as dry weight, VSS) and a fraction of soluble + colloidal components Sbo (expressed as COD). In wastewaters a considerable part of VSS is non-biodegradable and it is accounted for as Fa·Sao.
2. Hydrolysis is the rate limiting step in the degradation of organic suspended solids VSS.
3. Two different kinetic equations are used to describe the degradation of VSS and of the soluble substrate (first order on the substrate and 0.5 order on the biomass equation for the VSS and Monod equation for the soluble substrate).
4. The digester is an anaerobic contact system which allows the independent control of the sludge residence time and of the hydraulic residence time.
5. The efficiency of the settler is assumed to be equal to 100% and independent of the organic and hydraulic loading rates.

The mean retention time of the VSS substrate particles within the digester is equal to the sludge retention time SRT and consequently the degradation of Sao must be referred to it. The concentration of VSS within the reactor derived from the suspended fraction of the substrate is Sa1. Concentrations of biomasses feeding respectively on suspended solids and on soluble substrate are Xa and Xb.

The concentration of total suspended solids within the digester is:

$$SS1 = Sa1 + (Xa + Xb) \tag{36}$$

and can be easily determined as functions of the kinetic and operational parameters (Rozzi and Verstraete, 1981).

A parametric study was carried out in order to define the dependence of Sa1, X, SS1 on HRT, SRT and on the volumetric loads LSa and LSb referred to particulate and soluble substrate respectively assuming the process operating conditions reported in Table 6. The kinetic constants and process parameters used in the simulation study are reported elsewhere (Rozzi and Verstraete, 1981).

TABLE 6

Process operating conditions.

	LSa $(\frac{kgVSS}{m^3d})$	LSb $(\frac{kgCOD}{m^3d})$	HRT (d)	Sao $(\frac{kgVSS}{m^3})$	Sbo $(\frac{kgCOD}{m^3})$
Case A	1	5	0.5	0.5	2.5
"	1	5	20	20	100
Case B	5	1	0.5	2.5	0.5
"	5	1	20	100	20
Case C	1	1	0.5	0.5	0.5
"	1	1	20	20	20
Case D	5	5	0.5	2.5	2.5
"	5	5	20	100	100

SRT range: 10 to 100 days.

In Fig. 30, total sludge concentration SS1 in the reactor (i.e.: biomass + inert solids) is plotted vs SRT for different loading conditions. From the figure it is apparent that the mixed liquor concentration SS1 strongly depends on the suspended solids loading LSa, while the soluble COD loading LSb has a much weaker effect and the related increase in sludge concentration is only due to biomass feeding on the soluble substrate. It is therefore clear that wastewaters containing appreciable concentrations of suspended solids can not be fed to digesters operating at high volumetric loading rates and/or at high SRT, because the mixed liquor concentration would be too high.

In Fig. 31 the ratio between biomass and total sludge concentrations vs SRT is plotted for cases A,B,C and D. The diagram shows that, if the substrate is mainly made of suspended solids (Case B), the fraction of biomass in the digester sludge is very low and it is not worth using the anaerobic contact process of other suspended growth process configurations operating at high SRT to recycle and/or keep into the reactors sludges mainly consisting of inert suspended solids. The situation is different for waste waters prevailingly

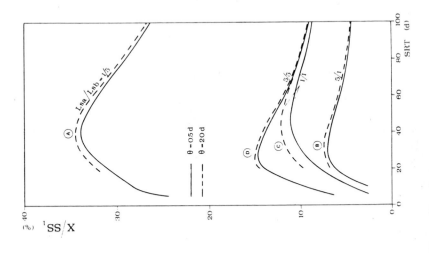

Fig. 31. Ratio of active biomass over sludge concentration vs HRT for different load conditions (cases A to D). (From Rozzi and Verstraete, 1981).

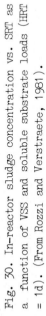

Fig. 30. In-reactor sludge concentration vs. SRT as a function of VSS and soluble substrate loads (HRT = 1d). (From Rozzi and Verstraete, 1981).

made of soluble pollutants (Case A). The absolute concentration of the two fractions (suspended solids and soluble COD) does not seem to play a major role provided that the settler operation is not influenced by the HRT (assumption 5). It follows that it is not the absolute concentration of SS in the waste which is important to determine the maximum loading rate of a digester but the ratio between the suspended solids and the soluble COD fractions.

In fact the interactions between the operating parameters in actual anaerobic processes are more complicated because: a) the hydraulic loading rate, i.e. the HRT, has a definite influence on the efficiency of the settler (and hence on the SRT) and b) the settling characteristics of the biomass and of the inert suspended solids in the waste may be very different and then the use of an average SRT for the sludge is meaningless. Nevertheless the principal relationships between the operational parameters, as shown by the simulations on the above model, have been observed in real digesters.

Similar results were obtained in other modeling studies in which the interactions between maximum suspended solids loading rates, the fraction of biomass in the sludge and the fraction of biodegradable suspended solids were investigated (Lettinga, 1980), (Henze and Harremoes, 1983).

3.3 PROCESS DESIGN AND ECONOMICAL EVALUATION.

Computer aided process design and economical evaluation are widely used in many branches of industrial technology, e.g. mechanical, electrical and chemical engineering.

As margins for profit are often quite small in commercial digesters, careful and detailed design by computer of anaerobic systems would be particularly helpful for the construction of more efficient plants. It is unlikely that in the near future the whole design procedure may be implemented by a computerized system. Certain decisions and choices are still better left to man, nevertheless most calculation procedures in the design of a digester can easily be entered in a suitable computer program at a very reasonable cost.

Procedures for designing and costing anaerobic systems are at present available (Hawkes and Rosser, 1983), (Rush et al., 1983).

Hawkes and Rosser (1983) analysed the complex process of digester systems design and selected the following typical elements for a design model: a) wastewater characteristics; b) type and basic sizing of the digester; c) estimated net gas production, i.e. the difference between the gas produced and the gas used for digester heating; d) product utilization, i.e. economical evaluation of biogas and electricity and e) economical analysis of the process as a whole, including capital and running costs. The above elements and their main interactions within the model are shown in Fig. 32. The blocks 2 (Digester) and 3 (Gas production) in Fig. 32 which are the real core of the anaerobic model can be dealt with by two different approaches:

The first approach attempts to build, using adequate microbial kinetics and data related to the physico-chemistry of the medium, a model of the process which can be made to describe and predict any operational state by selecting suitable coefficients and initial conditions. This approach, which is widely applied in chemical engineering, makes use of complicated models built from

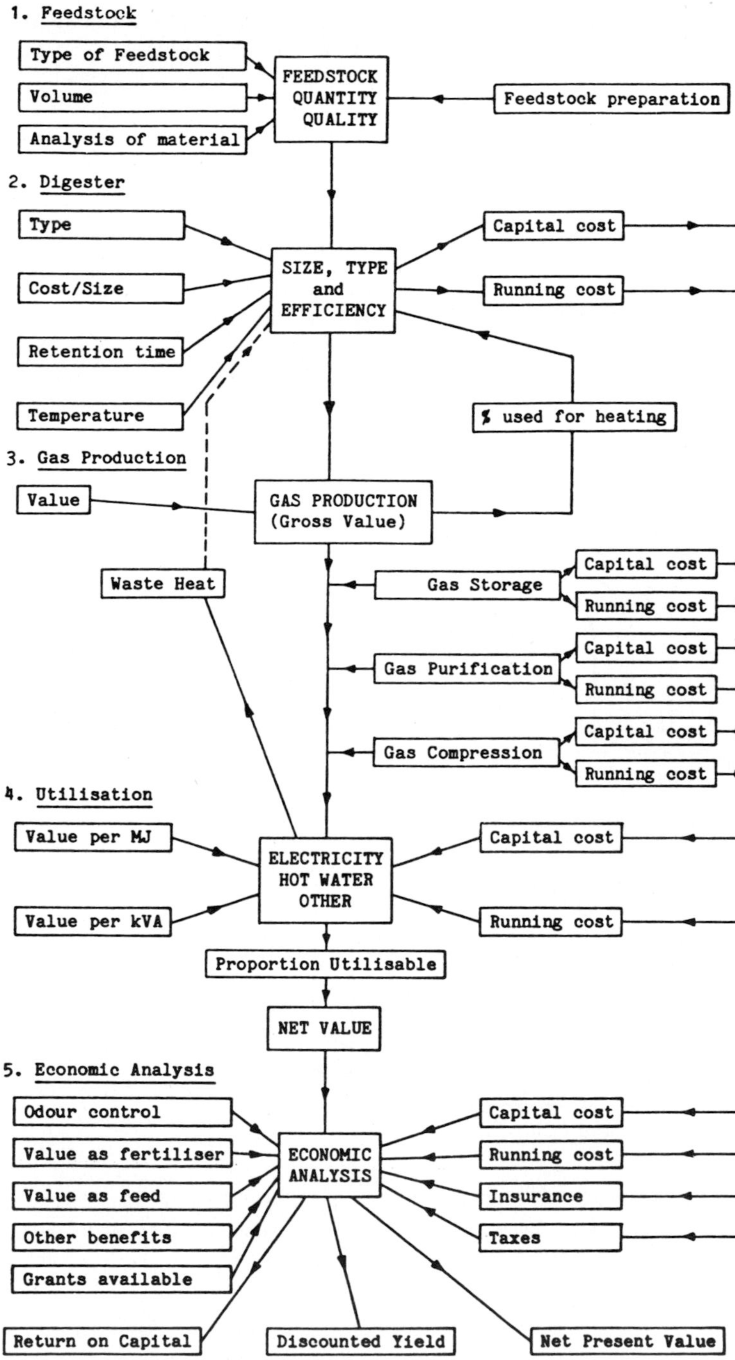

Fig. 32. Techno—economical digester model (From Hawkes and Rosser, 1983).

differential equations, and its potential applicability to biotechnology has been shown by recent studies (Roels and Kossen, 1978), (Roels, 1982), (Roels, 1983).

An alternative, and more pragmatic approach, does not seek generality but uses some arbitrary "best-fit" representation of some macroscopic aspect of the system behaviour, (Hawkes and Horton, 1981), (Hawkes and Rosser, 1983). This method involves a more direct use of the original experimental results within the model and uses operational parameters such as HRT and gas yields directly. A major limitation of this approach is that predictions can be made only for a system operating in similar conditions and with a similar waste to the actual system which was used to obtain the experimental data. The method ignores changes in microbial populations within the system which is treated as a "black-box" and only considers inputs and outputs over a given time period.

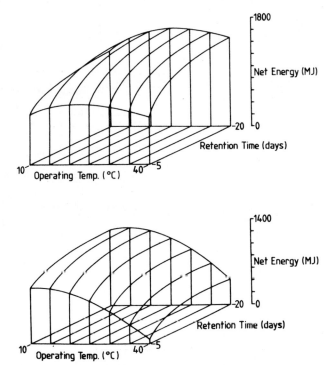

Fig. 33. Plots of predicted net energy in the digestion of pig slurry of two different concentrations (7% upper, 3.5% lower) generated by models based on published data. (From Hawkes and Rosser, 1983).

Energy and mass balances are equally important in anaerobic systems. Energy balance in some instances can be appreciably positive because the gas used to heat the digester is more than the gas produced by the system. If facilities are available near the digester to use the net energy directly as thermal energy or as electric power, a careful economic analysis of the anaerobic system can show when and how the digester can be profitable. Energy balances heavily depend on influent and on ambient temperatures. Predicted net energy derived from a pragmatic model (Hawkes and Rosser, 1983) as function of operating temperature, retention time and feedstock (pig slurry) concentration is plotted in Fig. 33. In this tridimentional graph it is interesting to note that, especially for the more diluted waste, the maximum net energy is not obtained for the maximum operating temperature which corresponds to the highest degradation rates, because the increased gas yields at the higher temperatures do not compensate the much higher energy requirements to heat the feed.

4. SUMMARY AND CONCLUSIONS

Anaerobic treatment processes are complex systems made of two main components: a physico-chemical system and a biological one. The first one is relatively simple and can be defined with a good approximation. The second one is quite complicated as it is made of several microbial populations which interact in feedback loops. The separated or joined dynamic behaviour of both components can be investigated either performing actual time consuming experiments or simulating the process by suitable mathematical models. The scopes of applying mathematical models to anaerobic processes are the same as for any other biological wastewater treatment process: a) basic studies on the interactions between the different microbial populations of the anaerobic ecosystem and validation of related hypotheses; b) investigations on process stability and on process control strategies; c) start-up procedures; d) design and economical analyses.

Economic balances can be much more favourable for anaerobic processes than for conventional aerobic ones because in the former case mechanical energy inputs are much lower and it is often possible to obtain a net energy output as methane. Therefore it is expected that in a near future a real optimization of anaerobic treatment will be required for commercial applications. This objective can be implemented only by a closer interaction between mathematical modeling and other science and technology branches, e.g. microbiology, biochemistry and engineering.

Acknowledgments.
Many thanks are due to Mr. Nicola Palmisano for preparing a taylor made computer word processor to edit this chapter and to Mr. Nicola Limoni for contributing to the preparation of the camera-ready manuscript.

684

REFERENCES

ANDREWS, J.F., 1968. A mathematical model of the continuous culture of microorganisms utilizing inhibitory substrates. Biotech. and Bioeng., 10: 707-723.

ANDREWS, J.F., 1969. Dynamic model of the anaerobic digestion process. J. San. Eng. Div. ASCE, 95, SA1: 95-116.

ANDREWS, J.F., 1971. Kinetic models of biological waste treatment. Biotech. and Bioeng. Symposium n. 2: 5-33.

ANDREWS, J.F., 1975. Mathematical models and control strategies for biological wastewater treatment processes. In: T.K. Keinath and M. Wanielista (Editors), Mathematical modeling for water pollution control processes. Ann Arbor Science, Ann Arbor, Michigan.

ANDREWS, J.F., 1978. The development of a dynamic model and control strategies for the anaerobic digestion process. In: A. James (Editor), Mathematical Models in Water Pollution Control. Wiley, Chichester.

ATKINSON, B., BUSH, A.W., SWILLEY, E.L. and WILLIAMS, D.A., 1967. Kinetics, mass tansfer and organism growth in biological film reactor. Trans. Inst. Chem. Engrs., 48: 245-264.

ATKINSON, B., 1974. Biochemical reactors. Pion, London.

BASTIN, G. and DOCHAIN, D., 1984. Stable adaptive controllers for waste treatment by anaerobic digestion. Proc. of the II International Workshop on Modelling and Control of Biological Wastewater Treatment, 24-26 September, at Firenze, Italy, Environm. Tech. Letters (in press).

BAILEY, J.E. and OLLIS, D.F., 1977. Biochemical Engineering Fundamentals, Mc Graw Hill, New York.

BRYANT, M.P., WOLIN, E.A., WOLIN, M.J. and WOLFE, R.S., 1967. Methanobacillus omelianskii, a symbiotic association of two species of bacteria. Arch. Mikrobiol., 59: 20-31.

BRYANT, M.P., CAMPBELL, L.L., REDDY, C.A. and CRABILL, M.R., 1977. Growth of desulfovibrio in lactate or ethanol media low in sulphate in association with H2-utilizing methanogenic bacteria. Appl. Environ. Microbiol., 33: 1162-1169.

CHEN, Y.R. and HASHIMOTO, A.G., 1980. Substrate utilization kinetic model for biological treatment process. Biotech. and Bioeng., 22: 2081-2095.

CLARK, R.H. and SPEECE, R.E., 1971. The pH Tolerance of Anaerobic Digestion. In: S.H. Jenkins (Editor), Advances in Water Pollution Research. 1: II27/1-II27/14.

COHEN, A., 1982. Optimization of anaerobic digestion of soluble carbohydrate containing wastewaters by phase separation, Ph. D. Thesis, Universiteit van Amsterdam.

DE BAERE, L., VERSTRATE, W. and ROZZI, A., 1983. Solubilization of particulate organic matter as rate-limiting step in anaerobic digestion loading. Trib. Cebedeau, 37 (484): 75-81.

DIXON, M. and WEBBS, E.C., 1979. Enzymes, pp. 78-79, Longman, London.

DOCHAIN, D. and BASTIN, G., 1984. Adaptive identification and control algorithms for nonlinear bacterial growth systems. Automatica, 20 (5): 621-634.

EASTMAN, J.A. and FERGUSON, J.F., 1981. Solubilization of particulate organic carbon during the acid phase of anaerobic digestion. J. Wat. Pollut. Control Fed., 53: 352-366.

FABIANI, P., HIRTSCH, D. and BOREL, J., 1983. A simple model of anaerobic digestion and its state trajectories properties. Poster Paper presented at the III Internat. Symposium on Anaerobic Digestion, 14-19 August, at Boston, Massachusetts.

FERGUSON, J.F., EIS, B.J. and BENJAMIN, M.M., 1982. Neutralization in anaerobic treatment of an acidic waste. In: Proc. IAWPRC Specialized Seminar "Anaerobic Treatment of Wastewater in fixed film reactors", 16-18 June, at Copenhagen, Danemark.

FERNANDEZ, X. and MOSEY, F.E., 1983. Monitoring of hydrogen concentrations in biogas. Paper presented at the European Symposium on Anaerobic Waste Water Treatment, 23-25 November, at Noordwijkerhout, Netherlands.

GRAEF, S.P. and ANDREWS, J.F., 1974a. Mathematical modeling and control of anaerobic digestion. WATER-73 AIChE Symposium Series N.136, 70: 101-130.

GRAEF, S.P. and ANDREWS, J.F., 1974b. Stability and control of anaerobic digestion. J. Wat. Pollut. Control Fed., 46: 666-683.

GHOSH, S., 1981. Kinetics of Acid-Phase Fermentation in Anaerobic Digestion. Biotech. and Bioeng. Symp. N.11, 301-313, Wiley, New York.

GHOSH, S. and KLASS, D.L., 1978. Two phase anaerobic digestion, Process Biochemistry, 13 (4): 15-24.

GUJER, W. and ZEHNDER, A.J.B., (1983). Conversion processes in anaerobic digestion. In: Proc. IAWPCR Specialized Seminar "Anaerobic Treatment of Wastewater in fixed film reactors", Copenhagen, 16-18 June 1982, at Copenhagen, Danemark. Water Science and Technology, 15 (8-9): 127-168.

HALL, E.R., 1982. Biomass retention and mixing characteristics in fixed-film and suspended growth anaerobic reactors. Proc. IAWPCR Specialized Seminar "Anarobic Treatment of Wastewater in fixed film reactors, 16-18 June, at Copenhagen, Danemark.

HARREMOES, P., 1978. Biofilm Kinetics, in: Water Pollution Microbiology, R. Mitchel (Editor), Wiley, New York.

HAWKES, D.L. and HORTON, H.R., 1981. Optimization of anaerobic digesters for maximum energy production. Studies in Environmental Science, 9: 131-142.

HAWKES, D.L. and ROSSER, B.L., 1983. Computer aided design of anaerobic digesters. Proc. III Internat. Symposium on Anaerobic Digestion, 14-19

August, at Boston, Massachusetts.

HENZE, M. and HARREMOES, P., 1983. Anaerobic treatment of wastewater in fixed film reactors – a literature review. In: Proc. IAWPCR Specialized Seminar "Anaerobic Treatment of Wastewater in fixed film reactors", 16–18 June 1982, at Copenhagen, Danemark. Water Science and Technology, 15 (8–9): 1–102.

HILL, D.T. and BARTH, C.L., 1977. A dynamic model for simulation of animal waste digestion. J. Wat. Pollut. Control Fed., 49: 2129–2143.

HILL, D.T., 1982. Design of digestion systems for maximum methane production. Transactions of the ASAE, 25 (3): 226–230.

HILL, D.T., 1983. Simplified Monod Kinetics of methane fermentation of animal wastes. Agricultural Wastes, 5: 1–16.

HOBSON, P.N., 1983. The kinetics of anaerobic digestion of farm wastes. J. Chem. Tech. Biotechnol.: 33B, 1–20.

INSTALLE', M. and ANTUNES, S., 1980. The mathematical modeling and control of anaerobic digestion processes. Inter-university course on Anaerobic Digeston sponsored by the Council of Europe. 8–18 December 1980, at Dijon, France. In: D.J. Picken, Fox, M.F. and R. Buvet (Editors), Biomethane: Production and Uses. Turret-Wheatland Ltd, Rickmansworth, UK, 1984.

JERIS, J.S. and McCARTY, P.L., 1965. The biochemistry of methane fermentation using C^{14} tracers. J. Wat. Pollut. Control Fed., 37: 178–192.

KASPAR, H.F., 1977. Untersuchungen sur Koppelung von Wasserstoof und Methanbildung in Faulschlamm. Ph. D. Thesis, Swiss Federal Institute of Technology, Zurich.

KIRSH, E.J. and SIKES, R.M., 1971. Anaerobic digestion in biological waste treatment. Prog. Ind. Microbiol., 9: 155–235.

KLEINSTREUER, C. and POWEIGHA, T., 1982. Dynamic simulator for anaerobic processes. Biotech. and Bioeng., 24: 1941–1951.

KROEKER, E.J., SCHULTE, D.D., SPARLING, A.B. and LAPP, H.M., 1979. Anaerobic treatment process stability. J. Wat. Pollut. Control Fed., 51: 718–727.

KROISS, H. and PLAHN-WABNEGG, F., 1983. Sulfite toxicity with anaerobic waste water treatment. Proc. European Symposium "Anaerobic Waste Water Treatment", 23–25 November, at Noordwijkerhout, Netherlands.

LAWRENCE, A.W. and McCARTY, P.L., 1970. Unified basis for biological treatment design and operation. J. San. Eng. Div. ASCE, 96: 757–778.

LEE, Y.H., and FAN, L.T., 1982. Kinetic studies of enzymatic hydrolysis of insoluble cellulose: analysis of the initial rates. Biotech. and Bioeng., 24: 2383–2406.

LEE, S.B., SHIN, H.S. and RYU, D.D.Y., 1982. Adsorption of Cellulase on Cellulose: Effect of Physicochemical Properties of Cellulose on Adsorption and Rate of Hydtolysis. Biotech. and Bioeng.: 24, 2137–2153.

LETTINGA, G., 1980. Anaerobic treatment of low strength wastes. Inter-university course on Anaerobic Digestion sponsored by the Council of Europe, 8–18 December, at Dijon, France. In: D.J. Picken, Fox, M.F. and R. Buvet (Editors), Biomethane: Production and Uses. Turret-Wheatland Ltd, Rickmansworth, UK, 1984.

LETTINGA, G., VAN DEN GEEST, A.T., HOBMA, S. and LAAN, J.V.D., 1979. Anaerobic treatment of methanolic wastes. Water Res., 13: 725–737.

LYND, L.H. and ZEIKUS, J.G., 1983. Metabolism of H2-CO2, Methanol and Glucose by Butyribacterium methylotrophicum. J. Bacter., 153: 1415–1423.

McCARTY, P.L., 1964a. Anaerobic waste treatment fundamentals. Part I. Chemistry and Microbiology. Public Works, 95 (9): 107–112

McCARTY, P.L., 1964b. Anaerobic waste treatment fundamentals. Part II. Environmental requirements and control. Public Works, 95 (10): 123–126.

McCARTY, P.L. 1964c. Anaerobic waste water fundamentals. Part III. Toxic materials and their control. Public Works, 95 (11): 91–94.

McCARTY, P.L., 1971. Energetics and kinetics of anaerobic treatment. Adv. Chem. Ser., 105: 91–107.

McINERNEY, M.J., BRYANT, M.P. and PFENNIG, N., 1979. Anaerobic bacterium that degrades fatty acids in syntrophic association with methanogens. Arch. Microbiol., 122: 129–135.

McINERNEY, M.J. and BRYANT, M.P., 1981. Review of Methane Fundamentals. In: D.L. Wise (Editor), Fuel Gas Production from Biomass, CRC Press Inc., Boca Raton, Florida.

MOSEY, F.E., 1983a. Mathematical modelling of the anaerobic digestion process: regulatory mechanisms for the formation of short chain volatile acids. Proc. IAWPR Specialized Seminar "Anaerobic Treatment of Wastewater in fixed film reactors", 16–18 June 1982, at Copenhagen, Danemark. Water Science and Technology, 15 (8–9): 209–231.

MOSEY, F.E., 1983b. Kinetic descriptions of anaerobic digestion, Proc. III International Symposium on Anaerobic Digestion, 14–19 August, at Boston, USA.

MOSEY, F.E., 1983c. Personal communication.

PFEFFER, J.T., 1968. Increased loadings on digesters with recycle of digester solids. J. Wat. Pollut. Control Fed., 40 (11): 1920–1933.

PERRY, R.H., CHILTON, C.H. and KIRKPATRICK, S.D. (Editors), 1963. Chemical Engineer's Handbook, Mc Graw Hill, New York.

RITTMANN, B.E. and McCARTY, P.L., 1980a. Model of Steady State Biofilm Kinetics. Biotech. and Bioeng., 22: 2343–2357.

RITTMANN, B.E. and McCARTY, P.L., 1980b. Evaluation of Steady-State Biofilm Kinetics. Biotech. and Bioeng., 22: 2359–2373.

ROELS, J.A., 1982. Mathematical models and the design of biochemical reactors.

688

J. Chem. Tech. Biotechnol., 32: 59-72.

ROELS, J.A., 1983. Energetics and kinetics in biotechnolgy. Elsevier Science Publishers, Amsterdam.

ROELS, J.A. and KOSSEN, N.W.F., 1978. On the modelling of microbial metabolism, Progress in Ind. Microbiol., 14: 95-203.

ROZZI, A., 1980. Physico-chemical equilibria in anaerobic digesters. Inter-university course on Anaerobic Digestion sponsored by the Council of Europe. 8-18 December, at Dijon, France. In:D.J. Picken, Fox, M.F. and R. Buvet (Editors), Biomethane: Production and Uses. Turret-Wheatland Ltd, Rickmansworth, UK, 1984.

ROZZI, A., 1981. Modeling and control of anaerobic digestion processes. Proc. of the I International Workshop on Modelling and Control of Biological Wastewater Treatment, 29-30 June, at Firenze, Italy. In: Transactions of The Institute of Measurement and Control (in press).

ROZZI, A., 1982. A rational approach to suspended solids degradation kinetics. Seminar held at the Universite' Catholique de Louvain, (unpubl.).

ROZZI, A. and VERSTRAETE, W., 1981. Calculation of active biomass and sludge production vs waste composition in anaerobic contact processes. Trib. Cebedeau, 34 (455): 421-427.

ROZZI, A., BRUNETTI, A., LABELLARTE, G. and LONGOBARDI, C., 1983. Anaerobic process control by automatic bicarbonate monitoring. Poster Paper presented at the III International Symposium on Anaerobic Digestion, 14-19 August, at Boston, USA.

ROZZI, A., MERLINI, A. and PASSINO, R., 1984. Development of a four population model of the anaerobic degradation of carbohydrates. Proc. II International Workshop on Modelling and Control of Biological Wastewater Treatment, 24-24 September, at Firenze, Italy. Environm. Tech. Letters (in press).

ROZZI, A., 1984. Confronto tra diversi modelli cinetici per processi biologici mediante il diagramma efficienza/carico. Ingegneria Ambientale (in Italian, in press).

RUSH, R.J., MONTHEITH, H.D. and STICKNEY, A.R., 1983. A systematic procedure for designing and costing anaerobic digestion facilities. Proc. III Internat. Symposium on Anaerobic Digestion, 14-19 August, at Boston, USA.

SHOBERT, S., 1977. Acetic acid from H_2 and CO_2. Formation of acetate by cell extracts of acetobacterium Woodii. Arch. Microbiol., 14: 143-148.

STUMM, W. and MORGAN, J.J., 1970. Aquatic Chemistry, Wiley Intescience, New York.

THAUER, R.K., JUGERMANN, K. and DECKER, K., 1977. Energy conservation in chemotrophic anaerobic bacteria. Bacteriol. Rev., 41: 100-180.

VAN DEN BERG, L. and KENNEDY, K.J., 1983. Comparison of advanced anaerobic reactors. Proc. III International Symposium on Anaerobic Digestion, 14-19 August, at Boston, 1983.

VAN DEN HEUVEL, J.C. and ZOETEMEYER, R.J., 1982. Stability of the methane reactor: a simple model including substrate inhibition and cell recycle. Process Biochemistry, 17 (3): 14–19.

WEAST, R.C. (Editor)., 1968. Handbook of Chemistry and Physics, The Chemical Rubber CO., Cleveland.

WILLAMSON, K. and McCARTY, P.L., 1976. A Model of Substrate Utilization by Bacterial Films, J. Wat. Pollut. Control Fed., 48: 9–24.

YOUNG, J.C., 1968. The anaerobic filter for waste treatment. Ph.D. thesis, Stanford University, Stanford, California.

YOUNG, J.C., 1983. The anaerobic filter. Past, present and future. Proc. III International Symposium on Anaerobic Digestion, 14–19 August, at Boston, USA.

ZEHNDER, A.J.B. and WUHRMANN, K., 1977. Physiology of a Methanobacterium strain AZ. Arch. Microbiol, 11: 199–205.

ZEHNDER, A.J.B., INGVORSEN, K. and MARTI, T., 1980. Microbiology of methane bacteria. In: Hughes, D. E. et al. (Editors), Anaerobic Digestion 1981, Elsevier Biomedical, Amsterdam.

ZEIKUS, G, 1980. Microbial populations in digesters. In: D.A. Stafford, B.I. Wheatley and D.E. Hughes (Editors), Anaerobic Digestion, Applied Science Publishers, Barking, Essex.

ZOETEMEYER, R.P., 1982. Acidogenesis of soluble carbohydrate–containing wastewaters. Ph. D. Thesis, Universiteit van Amsterdam.

LIST OF SYMBOLS

Symbol	Description	Units
A	constant (eq. 8)	(–)
A1	methane to substrate conversion factor	(m^3/kg)
$[Ac^-]$	acetate ion concentration	(mol/l = M)
[BA]	bicarbonate alkalinity	(M)
$[CO_3^{2-}]$	carbonate concentration	(M)
$[CO_2]$	concentration of disolved CO_2	(M)
D	substrate diffusivity in liquid	(m^2/s)
Df	substrate diffusivity in biofilm	(m^2/s)
[Hac]	free acetic acid concentration	(M)
$[HCO_3^-]$	bicarbonate concentration	(M)
HRT	Hydraulic Retention Time	(d)
[IC]	inorganic carbon concentration	(M)
J	substrate flux	$(kg/m^2 \cdot d)$
K	kinetic constant	
Ka1	bicarbonate equilibrium constant	(m^2)
Ka2	carbonate equilibrium constant	(m^2)
KAc	acetate equilibrium constant	(m^2)
kd	decay rate	(d^{-1})
KH	Henry's constant	(M/bar)
KL·a	overall gas transfer rate (eq. 29)	(d^{-2})

Ks	saturation constant (Monod)	(M)
Kx	Contois' constant (eq. 20)	(-)
L	volumetric load (eq. 22)	$(kg/m^3 \cdot d)$
Ld	diffusion boundary layer thickness	(mm)
Lf	biofilm tickness	(mm)
LSa	volumetric load referred to SS	$(kg/m^3 \cdot d)$
LSb	volumetric load referred to sol. COD	$(kg/m^3 \cdot d)$
My	Methane yield	(m^3/kg)
PCH_4	methane partial pressure	(bar)
PH_2	hydrogen partial pressure	(bar)
Pci	CO_2 partial pressure (eq. 16)	(bar)
PCO_2	CO_2 partial pressure	(bar)
Pg	digester gas pressure	(bar)
Pw	water vapour saturation pressure	(bar)
Q	feed flow rate	(m^3/d)
QCH_4	methane flow rate	(m^3/d)
QCO_2	gaseous CO_2 flow rate	(m^3/d)
Qg	biogas flow rate	(m^3/d)
Qw	water vapour flow rate (in biogas)	(m^3/d)
r	substrate removal rate	$(kgCOD/(kgVSS \cdot d))$
Rg	maximum glucose uptake	(M/d)
S	substrate concentration	(M)
So	substrate concentration (as TOC)	(M)
Sf	substrate concentration in biofilm	(M)
Ss	substrate concentration at the biofilm-diffusion boundary layer interface	(M)
S1	concentration of HAc + Ac$^-$ (as TOC)	(M)
SRT	Sludge Retention Time	(d)
T	temperature	(°C)
To	standard temperature (zero)	(°C)
[TOC]	Total Organic Carbon	(M)
V	volume	(m^3)
X	biomass concentration	(kg/m^3)
Y	growth yield factor for biomass	(d^{-1})
z	coordinate normal to biofilm surface	(m)
μ	biomass growth rate	(d^{-1})

NOTE: whenever possible, SI system has been used. In some cases the dimensions used by the authors of the papers quoted in the text have been kept.

MATHEMATICAL MODELLING OF ENERGY CONSUMPTION IN BIOLOGICAL WASTEWATER TREATMENT

Hiroshi Kubota and Koichi Fujie

Research Laboratory of Resources Utilization, Tokyo Institute of Technology

1   INTRODUCTION

In order to obtain high quality effluent from biological wastewater treatment plants, a considerable amount of energy is required. Although the configurations of typical modern aerobic treatment processes differ from one to the other, the necessary supply of oxygen from the atmosphere to the microorganisms, which are decomposing and assimilating organic matter in the wastewater, is the main cause of energy consumption.

Recently, owing to the increased need for energy conservation, a method to decrease requisite energy, while maintaining a sufficiently high quality of effluent wastewater from the treatment plant, is of great concern. The correct selection of the process and the rational design of the configuration and the dimensions of the plant, necessary to meet the quantity and quality of the raw wastewater, are essential in order to realize the optimal practices which minimize the energy required.

In order to assess the performance of each process, information is required about the quantitative relationships between performance of organic removal and the energy consumption under widely varied operating conditions of the process. For these reasons, the application of mathematical models is essential to depict the characteristics of the process as a function of operating parameters.

In this chapter, methods of mathematical modelling of the biological wastewater treatment processes proposed by the authors, and their application to the characterization of the process in relation with the power economy concept, which is defined as an amount of organic removal per unit power consumption in the process, are introduced.

2   DEFINITION OF POWER ECONOMY

The concept of power economy was first introduced for oxygen supply in the activated sludge wastewater treatment process and was defined as the amount of oxygen dissolved into the water in the aeration tank per unit power consumption. It has been used as an indicator in evaluating the performance of an air-liquid contacting facility. Hosono et al. (1980) have extended the concept of power economy to the trickling filter wastewater treatment process, in which case the power economy is defined as the amount of BOD removed per unit power consumption.

In the trickling filter and rotating biological contactor (RBC) processes, the amount of oxygen consumed can not be readily determined, while the amount of

BOD removed can be directly observed. By using the power economy concept based on the amount of BOD removed, the performance of different biological wastewater treatment processes, with respect to their power consumption, can be easily compared.

Under steady state operating conditions with constant inflow rate of wastewater Q, influent BOD concentration S_0, effluent BOD concentration S_e and fraction of BOD removal x, the amount of BOD removed W in a wastewater treatment unit is given as

$$W = Q (S_0 - S_e) = QS_0 x \qquad\qquad (.1)$$

If the power required is denoted by P_w, the power economy is evaluated by

$$W / P_w = QS_0 x / P_w \qquad\qquad (.2)$$

In cases, where there exists daily fluctuations in Q and S_0 the daily average power economy is given by

$$W / P_w = \frac{\int_0^{24} Q (S_0 - S_e) \, d\theta}{\int_0^{24} P_w \, d\theta} = \frac{\int_0^{24} QS_0 x \, d\theta}{\int_0^{24} P_w \, d\theta} \qquad\qquad (.3)$$

If the oxygen consumption rate is evaluated in the wastewater treatment unit, the power economy can also be evaluated by

$$W / P_w = Q_{O_2} / \tilde{Y}_0 P_w \qquad\qquad (.4)$$

where $\tilde{Y}_0$ is the overall oxygen yield factor, which is defined as the amount of oxygen consumed per unit amount of BOD removed. In aerobic biological wastewater treatment processes, oxygen is consumed by the bio-oxidation of organic substrate, the endogenous respiration of cells, and the nitrification of ammonia.

3   CONVENTIONAL ACTIVATED SLUDGE PROCESS

3.1   Oxygen transfer in aeration tank

(i)   Concept of diffuser performance   In the conventional activated sludge process energy is required mainly to supply oxygen to the aeration tanks. An increase in aeration efficiency is essential to save the cost of energy necessary to remove the BOD of wastewater.

In activated sludge aeration tanks, equipped with air diffusers as aeration devices, the air diffusers are located along the side walls near the bottom of the tank. Air bubbles generated from the diffusers exist in only a part of the tank, where oxygen is transfered from the air bubbles to liquid. In such tanks, as the liquid flows in a spiring motion with a relatively high velocity caused by the gas-lift action of the rising air bubble swarms, the dissolved oxygen in

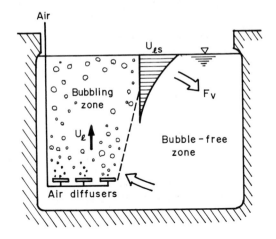

Air

$U_{\ell s}$

Bubbling zone

U_ℓ

Bubble-free zone

F_v

Air diffusers

Fig. .1 Schematic side view of conventional activated sludge aeration tank with spiral liquid circulation.

the bubbling zone moves to the bubble-free zone along with the spiral liquid flow, as schematically shown in Fig. .1. After designating the volumeric oxygen transfer coefficient and the cross-sectional area of the bubbling zone by K_{La}' and S_a', respectively, the total oxygen transfer rate Q_{O_2} in the aeration tank under the steady state condition is given by

$$Q_{O_2} = \int_0^h K_{La}' \ (\ C_0 z - C \ ) \ S_a' \ d\ell \qquad\qquad (\ .5)$$

where C_0 is the dissolved oxygen concentration in liquid saturated with pure oxygen gas, C is the dissolved oxygen concentration in liquid bulk, ℓ is distance from the air diffuser, h is depth of the diffuser in the tank, and z is mole fraction of oxygen in air bubbles.

The overall volumetric oxygen transfer coefficient K_{La} based on the total volume of aeration tank is usually defined as

$$Q_{O_2} = K_{La} \ (\ \overline{C}_s - \overline{C} \ ) \ V \qquad\qquad (\ .6)$$

where $\overline{C}_s$ and $\overline{C}$ are the average values of $C_0 z$ and C within the tank, respectively, and V is effective volume of the tank. Combining Eqs. (.5) and (.6), K_{La} is expressed as

$$K_{La} = \frac{\int_0^h K_{La}' \ (\ C_0 z - C \ ) \ S_a' d\ell}{(\ \overline{C}_s - \overline{C} \ ) \ V} \qquad\qquad (\ .7)$$

This shows clearly that K_{La} is dependent on the individual aeration tank, which has given values of V, S_a', and h, and does not show quantitatively the oxygen transfer performance of the air diffuser installed in each tank.

Instead of K_{La}, a new concept of diffuser performance as a measure of oxygen transfer was proposed by Kubota et al. (1979) and was modified by Sekizawa et al. (1984). The mathematical expression of the diffuser performance Φ is given

694

as

$$\Phi = \frac{G_v h}{K_{La} V} \cdot \frac{2P^*}{P_h\{1 + (P^*/P_h)^{1/2}\}} = \frac{\kappa}{12} \sqrt{\frac{\pi}{D} \frac{d_{BO}}{u_s}} \left(u_s + u_\ell \right) d_{BO} \tag{ .8}$$

To derive the equation, it was assumed that bubbles generated from the air diffuser located near the bottom of the tank ascend in the mixed liquor without
coalescence and disruption, and the Higbie's penetration theory was applied to
express the volumetric oxygen transfer coefficient K_{La}'.

Notations in Eq. (.8) are as follows : D is diffusivity of dissolved oxygen
in liquid, d_{BO} is average bubble diameter at the diffuser depth h, P^* is atmospheric pressure, P_h is pressure at the diffuser depth, U_ℓ is superficial liquid
velocity in the bubbling zone, U_s is slip velocity of air bubbles to liquid, κ
is a correction factor denoting deviation from the Higbie's theory (refer to
Sekizawa et al., 1984).

The diffuser performance Φ in Eq. (.8) was defined to being almost independent of the geometry and dimensions of aeration tanks and characterizes the oxygen transfer performance of the air diffusers, although the term U_ℓ in the right
hand side of Eq. (.8) is not only influenced by the geometry and dimensions of
the tanks but also by the arrangement of the diffusers.

(ii) <u>Determination of diffuser performance</u> The rate of oxygen consumption
Q_{O_2} under steady state operating condition is equal to the rate of oxygen transfer through the gas-liquid interface. For the aeration tank in which air is
fed into the mixed liquor through air diffusers

$$Q_{O_2} = \frac{G_v z_0 \eta}{v_m} = K_{La} (\bar{C}_s - \bar{C}) V \tag{ .9}$$

where z_0 is mole fraction of oxygen in atmosphere, and v_m is molar volume of
oxygen.

The relation between oxygen utilization η and the oxygen mole fraction in the
exhaust gas z_e is given by

$$\eta = \frac{z_0 - z_e}{(1 - \delta z_e) z_0} \tag{ .10}$$

Assuming that carbon dioxide, which mole fraction is z_c, is only a gaseous product of aerobic bio-oxidation, δ in Eq. (.10) is determined by

$$\delta = 1 - \frac{z_c (1 - z_0)}{z_0 - z_e - z_0 z_c} \tag{ .11}$$

Therefore, the value of η, evaluated from Eqs. (.10) and (.11) using the experimentally observed data of exhaust gas analysis, gives the value of K_{La} under practical operating conditions of the aeration tank. The diffuser performance Φ is then evaluated directly from the following relation, which is obtained by eliminating K_{La} from Eqs. (.8) and (.9).

$$\Phi = \frac{v_m h \ (\overline{C}_s - \overline{C})}{z_0 \ \eta} \ \frac{2P^*}{P_h \ 1 + (P^*/P_h)^{1/2}} \qquad (.12)$$

The exact value of η can be obtained by simultaneously measuring concentrations of oxygen and carbon dioxide in the exhaust gas from the aeration tank (Kubota et al., 1981).

(iii) <u>Spiral liquid flow rate and oxygen transfer</u> The rate of liquid flow caused by the air lift action of bubble swarms can quantitatively be represented by the liquid flow velocity at the liquid surface of the aeration tank $U_{\ell s}$. Hoshino et al. (1977b) discussed theoretically the mechanism of the air lift action in the aeration tank and proposed the following semi-empirical relation between $U_{\ell s}$ and the product of U_g and h, where U_g is superficial air velocity, that is, the air flow rate per unit sectional area of the tank, and h is diffuser depth in the tank.

$$U_{\ell s} = a \{ hU_g \ (h/H)^{1/2} (H/W)^{1/3} \}^m \qquad (.13)$$

Effects of configuration of tanks are corrected by terms of $(h/H)^{1/2}$ and $(H/W)^{1/3}$, where H and W are liquid depth and width of the tank, respectively. Values of empirical constants a and m in the equation are shown in TABLE .1. The comparison between observed values and the values calculated from Eq.(.13) of $U_{\ell s}$ is shown in Fig. .2.

It was found that the ascending liquid velocity in the bubbling zone U_ℓ, which affects the value of the air diffuser performance Φ, as seen in Eq. (.8),

TABLE .1
Values of constants a and m in Eq. (.13), where $\phi = hU_g (h/H)^{1/2} (H/W)^{1/3}$

Type of air diffuser	ϕ (cm^2/sec)	m	a
Fine bubble types *	$\phi \le 20$	0.64	7.0
	$\phi > 20$	0.46	12.0
Coarse bubble types **	$\phi \le 20$	0.78	3.5
	$\phi > 20$	0.56	4.9

* Porous plates and tubes.
** Perforated plates and tubes, single nozzles and others.

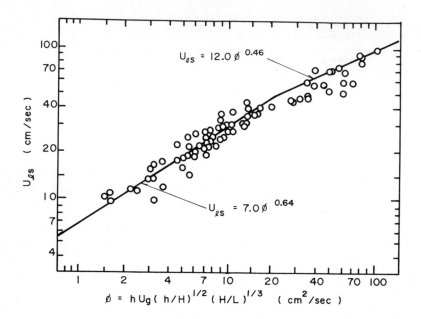

Fig. .2 Application of Eq. (.13) for observed liquid velocity at water surface $U_{\ell s}$ of conventional activated sludge aeration tanks

was about one-third of $U_{\ell s}$.

In aeration tanks, the dissolved oxygen in the bubbling zone moves to the bubble-free zone along with the spiral liquid flow. The volumetric liquid exchange coefficient E between the bubbling zone and the bubble-free zone is given by (Sekizawa and Kubota, 1979),

$$E = F_v/V = (H/2) \overline{U}_\ell /HL = \beta U_{\ell s}/2L \qquad (.14)$$

Here, F_v and $\overline{U}_\ell$ are liquid flow rate and average liquid flow velocity between two zones, respectively, V is effective volume of the tank, and β is the ratio of $U_{\ell s}$ to $\overline{U}_\ell$. In Eq. (.14), it was assumed that the upper half of the tank serves as liquid flow path from the bubbling zone to the bubble-free zone and the lower half of the tank serves as liquid flow path in the opposit direction.

The transfer rate of oxgyen from the bubbling zone to the bubble-free zone q_{O_2} is then given by

$$q_{O_2} = E (C' - C'') V \qquad (.15)$$

where C' and C'' are dissolved oxygen concentrations in the bubbling zone and the bubble-free zone, respectively.

Using these correlations the resistance of oxygen transfer between the bubbl-

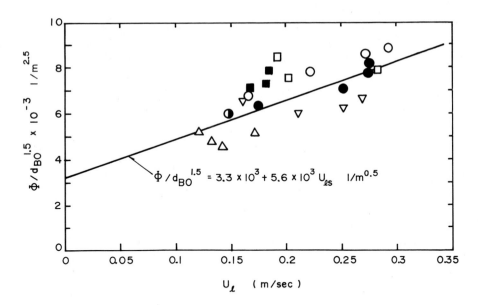

Fig. .3 Relations between $\Phi/d_{BO}^{1.5}$ and U_ℓ for data obtained at five munici-pal activated sludge wastewater treatment plants (Sekizawa et al., 1984)

ing zone and the bubble-free zone was evaluated and was found to be at most 5% of the resistance of oxygen transfer at the gas-liquid interface in the bubbling zone. In other words, the difference between dissolved oxygen concentrations C' and C" is usually negligible. That is, it is reasonable to assume a uniform distribution of the dissolved oxygen concentration in the conventional activa-ted sludge aeration tank.

(iv) <u>Air diffuser performance of various diffusers</u> From Eq. (.8), one can obtain

$$\frac{\Phi}{d_{BO}^{1.5}} = \frac{\kappa}{12}\sqrt{\frac{\pi}{D}}\sqrt{U_s} + \frac{\kappa}{12}\sqrt{\frac{\pi}{D}}\frac{1}{\sqrt{U_s}}U_\ell \qquad (\ .16)$$

For simultaneously observed values of Φ, d_{BO}, and U_ℓ under practical operating conditions of five municiple wastewater treatment plants, the plot of $\Phi/d_{BO}^{1.5}$ against U_ℓ is shown in Fig. .3. From this plot, values of U_s and κ were ob-tained of 20.3 cm/sec and 2.18, respectively.

Taking these values for U_s and κ, and the relation of $U_\ell = (1/3)U_{\ell s}$, and assuming average bubble diameter at the diffuser depth d_{BO} under practical aer-ation conditions, Eq. (.16) is converted to

$$\Phi/d_{BO}^{1.5} = 3.3 \times 10^3 + 5.6 \times 10^3 U_{\ell s} \qquad 1/m^{0.5} \qquad (\ .17)$$

where the dimensions of d_{BO}, $U_{\ell s}$ and Φ are to be m, m/sec, and 1/m, respectively.

If values of Φ can be evaluated from previously reported data of K_{La} or η in an aeration tank with a definite configuration and dimensions, in which $U_{\ell s}$ can be predicted form Eq. (.13), the value of characteristic bubble diameter of air diffusers installed in the tank d_{BO} can be evaluated from Eq. (.17).

For various types of air diffusers, typical values of d_{BO} obtained by the above procedure and the range of values of Φ for various air diffusers are summarized in Table .2. From the values of characteristic bubble diameters d_{BO}, the air diffusers used in conventional activated sludge aeration tanks are classified into three categories : very fine bubble diffuser, fine bubble diffuser, and coarse bubble diffuser.

3.2 Oxygen uptake and BOD removal

(i) Liquid flow model in aeration tank Activated sludge aeration tanks are usually long in the direction of liquid flow, and are sometimes separated into several stages by partition walls. Liquid mixing in the flow direction can be characterized by a combination of dispersion coefficients in each stage of the

TABLE .2

Values of diffuser performance Φ and characteristic bubble diameter d_{BO} of various air diffusers (Sekizawa et al., 1984)

Clasifi-cation	Air diffuser	Φ (m)	d_{BO} (mm)	Remarks
Very fine bubble	MPC nozzle	0.2 - 0.4	1.0	D
	Aquarator	0.5 - 0.8	2.2	I
Fine bubble	Brandol 60	0.8 - 1.6	2.5	I
	Ceramic plate	0.8 - 1.6	3.3	D
	Ceramic tube	1.1 - 1.6	3.2	I
	Toray aerator	1.3	3.3	I
	Plastic porous plate	1.3 - 1.7	3.5	I
	Plastic porous tube	1.0 - 1.5 1.6 - 1.9	3.1 4.1	D D
	Sanitarie Model D-24	1.7 - 2.0	3.8	I
	Jet aerator	1.7 - 1.8	4.0	I
	Line-mixer aerator	1.6 - 2.2	4.2	I
Coarse bubble	Wakatsuchi fine diffuser	2.3	5.3	D
	Perforated tube	2.2 - 3.5	5.5	I
	Sparger	3.1 - 3.7	6.0	I

Note : D in remarks indicates directly observed values and I indirectly evaluated ones from reported K_{La} and the corresponding aeration conditions.

tank E_z, and back flow rate through the partition wall F_b. The liquid mixing in the aeration tank is governed by spiral liquid circulation at right angle to the direction of liquid flow caused by the air lift action of rising bubble swarms in the tank. As mentioned above the spiral liquid circulation is characterized by the measurable velocity at the liquid surface $U_{\ell s}$, which can be predicted from Eq. (.13). The relation between E_z and $U_{\ell s}$ and that between F_b and $U_{\ell s}$ were given by Fujie et al.(1983b) as

$$E_z = 0.0115 \left\{ 1 + H/L \right\}^{-3} U_g^{-0.34} U_{\ell s} \left(H + W \right) \qquad m^2/hr \qquad (.18)$$

$$F_b = 0.045 U_{\ell s} A_b \qquad m^3/hr \qquad (.19)$$

where U_g is superficial air flow rate in the tank, H, W and L are liquid depth, width and length of the tank, respectively, and A_b is free area of the partition wall. The adaptability of Eq. (.18) for previously obtained data is shown in Fig. .4.

If one defines the overall dispersion coefficient through the tank as E_z, it can be related with E_z and $\gamma = F_b/F$, where F is net liquid flow rate, as follows (Fujie et al., 1983b) :

$$\frac{U_\ell L}{E_z} = U_\ell \sum^{i} \frac{\Delta \ell_i}{E_{zi}} + \left\{ \frac{n}{2(n - 1)^2} + \frac{\gamma}{n} \right\}^{-1} \qquad (.20)$$

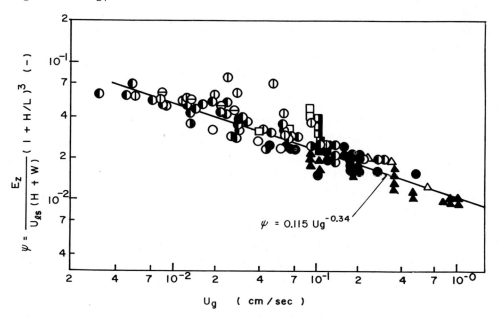

Fig. .4 Correlation of Eq. (.16) for longitudinal dispersion coefficients observed in aeration tanks with different configuration and dimensions (Fujie et al., 1983b)

where U_ℓ is superficial liquid velocity in the direction of liquid flow, E_{zi} is longitudinal dispersion coefficient, $\Delta\ell_i$ is distance between the partition walls, n is the number of stages, and subscript i denotes i-th stage.

The number of equivalent stages n' in the completely mixed tanks in the series model, which will be a more intuitive expression of the extent of longitudinal liquid mixing $\tilde{E}_z$, can be evaluated by

$$n' = \frac{L}{2 \tilde{E}_z U_\ell} \qquad\qquad (\ .21)$$

It was shown by Fujie et al. (1983b) that the values of n' for a five stage tank and a two stage tank were 1.92 and 1.08, respectively. It can be concluded that the single completely mixed flow model may reasonably be applied for the sake of simplicity to most activated sludge aeration tanks, in order to represent the liquid mixing.

(ii) <u>Relation between oxygen uptake rate and BOD removal rate</u> The relation between oxygen uptake rate $-r_{O_2}$ and BOD removal rate $-r_s$ per unit volume of the aeration tank is given by

$$-r_{O_2} = Y_0 (-r_s) + b X \qquad\qquad (\ .22)$$

where Y_0 is oxygen yield factor defined as mass of oxygen consumed per unit mass of BOD removal, b is endogenous respiration rate per unit mass of sludge, and X is sludge concentration. If completely mixed flow of the liquid in the aeration tank is assumed as described above, the BOD removal rate $-r_s$ in the tank is given by

$$-r_s = (S_0 - S_e) / \bar{\theta} \qquad\qquad (\ .23)$$

where S_0 and S_e are influent and effluent BOD concentrations, respectively, and $\bar{\theta}$ is hydraulic retention time in the tank.

As was shown by Kubota and Miyaji (1977), when the effluent BOD concentration falls less than a certain value, nitrification of ammonium nitrogen proceeds abruptly. Under such conditions the oxygen uptake due to nitrification must also be taken into account.

Therfore, an overall oxygen yield factor $\tilde{Y}_0$, which is defined as total mass of oxygen consumed per unit mass of BOD consumed, is represented by

$$\tilde{Y}_0 = Y_0 + \frac{b X \bar{\theta}}{(S_0 - S_e)} + \frac{Y_N(N_0 - N_e)}{(S_0 - S_e)} \qquad\qquad (\ .24)$$

where Y_N is oxygen yield factor of nitrification defined as mass of oxygen consumed per unit mass of ammonium nitrogen oxidized, and N_0 and N_e are influent and effluent ammonium nitrogen concentrations, respectively. The relation between the effluent BOD S_e and that of ammonium nitrogen N_e, when the nitrification was proceeding, was given by

$$N_e = \frac{K_N S_e}{\beta - S_e} \qquad (\ .25)$$

where constants K_N and β were determined empirically as 1.47 and 19.7 mg/ℓ, respectively (see Kubota and Miyaji, 1977). If the effluent BOD S_e is larger than 19.7 mg/ℓ, N_e should be set as N_0, that is the nitrification does not proceed, and therefore the third term of right hand side of Eq. (.24) will be negligible. The relation between the rate of BOD removal W and the rate of oxygen consumption Q_{O_2} in the activated sludge aeration tank is thus correlated by

$$W = (\ S_0 - S_e\ )Q = Q_{O_2} / \widetilde{Y}_0 \qquad (\ .26)$$

where Q is inflow rate of wastewater into the tank.

3.3 <u>Power economy</u>

(i) <u>Requisite energy of aeration</u> In aeration tanks with diffusers, energy is consumed to compress air supplied from the atmospheric to the pressure at the diffuser depth. Assuming adiabatic compression, the energy required to compress air with a volumetric feed rate G_v is expressed by

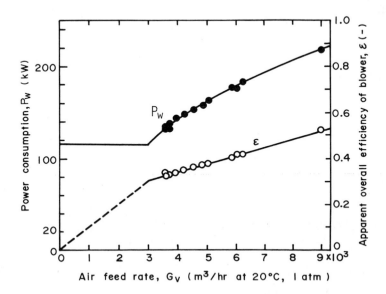

Fig. .5 An illustrative example of power consumption P_W and overall efficiency ε observed on the air blower in a municipal wastewater treatment plant. The efficiency of the blower includes that of the electric motor(Fujie et al,1983a).

$$P_W = \frac{G_v \, \gamma}{\varepsilon(\, \gamma - 1 \,)} (\, \frac{P_2}{P_1} \,)^{\frac{\gamma-1}{\gamma}} - 1 \qquad\qquad (\; .27)$$

where P_1 and P_2 are respective pressures at inlet and outlet of the air blower, γ is heat capacity ratio of air, given as 1.40, and ε is efficiency of the air blower.

The apparent value of ε is the product of the efficiencies of the electric motor and the blower. If the maximum air feed rate of the blowers satisfies the designed conditions of the opertation, which depends on the required oxygen consumption rate in the aeration tank, the air feed rate G_v can be controlled by manipulating the opening of the inlet valve of the blower until a critical value of G_{vc} is achieved. Below this, a reduction in G_v can not be achieved without surging of the blower. Under such conditions, the efficiency of the blower ε decreases drastically. Values of ε for the air blower used in a municipal wastewater treatment plant under varing conditions of G_v are illustratively shown in Fig. .5. It should be noted that the adequate selection of a blower which fits the oxygen consumption rate required in the aeration tank, is essential to reduce the requisite electric power of the air blower.

(ii) <u>Procedure of evaluation of power economy</u> In an aeration tank equipped with air diffusers having their characteristic bubble diameter d_{BO}, the evaluation of power economy is performed in the following porcedure :

1. Obtain values of Φ under varing aeration rate per unit volume of the tank G_v/V, using Eq. (.17) in which $u_{\ell s}$ is predicted by Eq. (.13) corresponding to the configuration and dimensions of the aeration tank.

2. For assigned values of the dissolved oxygen concentration $\overline{C}$ and the diffuser depth h, values of oxygen utilization η are calculated from Eq. (.12) and oxygen uptake rates Q_{O_2} are obtained from Eq. (.9) as a function of G_v/V.

3. If a BOD removal rate in the aeration tank W is assigned, the BOD in the tank, that is the effluent BOD S_e is obtained by solving

$$W = Q(\, S_0 - S_e \,) = -r_s \, V \qquad\qquad (\; .28)$$

where $-r_s$ is rate of BOD removal per unit volume of mixed liquor in the tank.

As was shown by Fujie et al. (1983a), under conditions of low BOD concentrations the following approximate relation among $-r_s$, S_e, and activated sludge concentration X can be used to correlate the operating data of the aeration tank.

$$-r_s = k \, S_e \, X \qquad\qquad (\; .29)$$

where k is specific rate of reaction.

4. For a given set of S_0, N_0, and X, values of S_e are obtained from the above and N_e is obtained from Eq. (.25). These are substituted into Eq. (.26) to

afford overall oxygen yield $\widehat{Y}_0$. Using this $\widehat{Y}_0$, the relation between Q_{O_2} and W is then correlated by Eq. (.24).

5. The power required for the aeration P_w is calculated from Eq. (.27), and finally, the value of power economy W/P_w is evaluated under varied operating conditions in the tank.

(iii) <u>Illustrative calculation</u> Illustrative calculations were attempted for the aeration tank of a municipal wastewater treatment plant. Observed values of ϕ for the aeration tank, by the method described in 3.1 (iii) were directly applied instead of taking the procedure 1 of 3.3 (ii). The air blower, equi-pped for a tank of volume 8700 m^3, had the efficiency shown in Fig. .5. Other conditions and values of constants used in calculations are listed in Table .3.

Calculated values of power economies W/P_w are plotted against rates of BOD removal per unit volume W/V in the tank in Fig. .6. This figure also shows the relations between W/P_w and W/V for the case when a constant efficiency air blow-er efficiency is assumed. If a number of blowers are equipped to cover the range of oxygen demand in the tank, and an on-off control mechanism can keep al-most a constant dissolved oxygen concentration, the decrease in blower efficien-cy will be minimized.

As seen in Fig. .6, the reduction of BOD loading W/V in the aeration tank naturally improves the water quality of the effluent. That is, it brings about a low effluent BOD, but causes a considerable decrease in power economy. As the BOD loading decreases, the required air feed rate into the tank is reduced, and

TABLE .3

Conditions and values of constants used for illustrative calculations shown in Fig. .6

Depth of diffuser, h (m)	5
Diffuser performance, ϕ (m)	Observed values (1.2 - 1.6)*
Overall efficiency of air blower,	Fig. .5 and 0.545 (constant)
Dissolved oxygen concentration, C (mg/ℓ)	2
Oxygen yield for BOD removal, Y_0	0.84
Oxygen yield for nitrification, Y_N	4.3
Endogenous respiration rate constant, b (1/hr)	5×10^{-3}
First-order reaction rate constant, k (1/hr)	0.83×10^{-2}
Constant in Eq. (.25), k_N	1.47
Constant in Eq. (.25), β (mg/ℓ)	19.7
Influent BOD concentration, S_0 (mg/ℓ)	150
Influent ammonium nitrogen, N_0 (mg/ℓ)	20
Temperature, T (°C)	20

* Refer to Kubota et al., 1979.

the value of air diffuser performance Φ decreases considerablly. That is the oxygen utilization in the tank η increases, because the spiral liquid velocity $U_{\ell s}$ decreases (see Eq. (.13)). On the other hand, as lowering the effluent BOD S_e, an abrupt increase in $\tilde{Y}_0$ and a large oxygen consumption result in, owing to nitrification which occurs only at S_e 20 mg/ℓ (see Eqs. (.24) and (.25)).

The increase of activated sludge concentration X, at the same hydraulic retention time $\bar{\theta}$ in tha tank, reduces the effluent BOD, but also adds a considerable amount of oxygen consumption because of the endogenous respiration of the sludge (see Eq. (.24)). It should be noted that operating conditions, which improve the effluent concentration, always cause a decrease in power economy.

In Fig. .6, the effect of blower efficiency on the power economy is also shown. In most municipal wastewater treatment plants, actual BOD loadings in aeration tanks are usually less than the designed values. This improves the effluent quality but will lower the power economy. The relation between W/P_W and W/V calculated from statistical data of municipal wastewater treatment plants, as reported by Japan Sewage Works Association (1982), is plotted in Fig. .7. Here, power consumption for the aeration is assumed to be 60% of the total power consumption of the plant reported. In the figure, the range of calculated

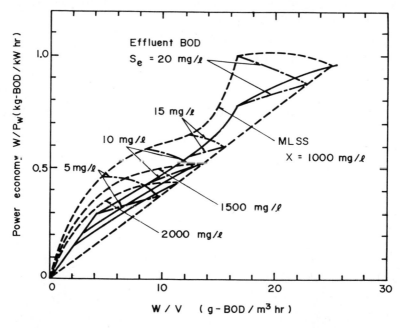

Fig. .6 Relations between power economy W/P_W and rate of BOD removal per unit volume W/V in a typical activated sludge aeration tank. Operating conditions and constants used for calculations are shown in Table .3. Lower values of W/P_W for specified S_e and X are in the case with a varing air blower efficiency (see Fig. .5), and higher values are in the case with a constant air blower efficiency.

results shown in Fig. .6 is superimposed.

It is apparent that the illustratively calculated range covers the practical operating data rather well.

(iv) <u>Power economy characteristics</u> Characteristics of the power economy in the conventional activated sludge process are summarized as follows:

1. Conditions lowering BOD in the effluent of the aeration tank considerably reduce the power economy.

2. The reduction of the power economy is brought about by the increase in oxygen consumption owing to nitrification and the endogenous respiration of activated sludge, and by the decrease of air blower efficiency.

4.   MODIFIED ACTIVATED SLUDGE PROCESS

4.1   Air-lift type deep shaft

An air lift type deep shaft is an aerator consisting of a long vertical shaft, reaching a depth of 100 to 300 m beneath the ground, and was developed by ICI Limited. As shown schematically in Fig. .8, liquid circulation is maintained by the gas lift action of injected air, and carries air bubbles deep into

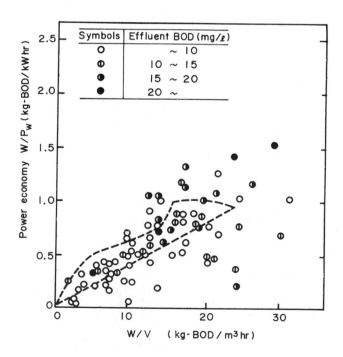

Fig. .7 Comparison of calculated results shown in Fig. .6 with operating data of municipal activated sludge wastewater treatment plants, reported by Japan Sewage Works Association (1982).

the bottom of the shaft. Effective oxygen transfer from the gas phase to the liquid phase is achieved, under elevated pressure at that point.

(i) <u>Mathematical modeling</u> In the air-lift type deep shaft, the static head difference caused by the difference in gas hold-up distributions between the downcomer and the riser in the shaft, is the driving force of liquid circulation. Under steady-state liquid circulation, the pressure drop, caused by the hydraulic friction of the flow circulation along the flow path, is balanced with the static head difference. Therefore, the condition of air injection into the shaft, necessary to maintain liquid circulation velocity, can be obtained by solving this momentum balance equation.

To obtain information of oxygen consumption, when the deep shaft is applied to the biological wastewater treatment as a modified activated sludge process, the mass balance equation with respect to oxygen, which is transfered from the injected air to liquid phase, must be solved simultaneously with the momentum balance equation. These equations were derived by Kubota et al. (1978), assuming plug flows for both the gas and liquid phases in the downcomer and the riser.

Theoretical and empirical relations among gas and liquid velocities, gas-liguid slip velocity and gas hold-up, and information of oxygen transfer through gas-liquid interface were also collected to solve the equations. For details of

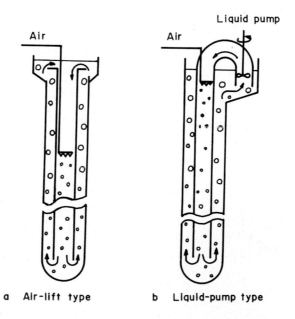

a   Air-lift type   b   Liquid-pump type

Fig. .8 Schemes of deep shaft aerators applied to modified activated sludge process.

the method of modeling and procedure of calculations, consult Kubota et al.'s
(1978) original report. Outlines of results of illustrative calculations are
reviewed in the following.

   (ii) <u>Operating limitation</u> Calculations were carried out for two types of
the shafts, the concentric type and the divider type, shown in Fig. .9. The
relation between the superficial liquid velocity U_ℓ and the air velocity required to maintain steady liquid circulation in the shaft $U_{go}{}^*$, for a divider type
shaft, is shown in Fig. .10. Here, $U_{go}{}^*$ denotes the value of the superficial
gas yelocity converted under atmospheric pressure. In Fig. .10, two different
values of U_ℓ are given for the same values of $U_{go}{}^*$ and h. Around a U_ℓ value,
U_ℓ increases with an increase of $U_{go}{}^*$ and decreases with a decrease of $U_{go}{}^*$,
while around another U_ℓ value, the opposite is observed. The region of no solution widely covers the left half of the diagram, in which U_ℓ increases with a
decrease of $U_{go}{}^*$. It should be mentioned that in order to maintain a constant
liquid circulation velocity for a fixed location of air injection, a minimum
value of $U_{go}{}^*$ exists. Beyond this minimum, stable liquid circulation cannot be
obtained, and operation near this minimum will thus be impractical. In conclusion, the region of practical stable liquid circulation must be restricted to
the right half of the diagram.

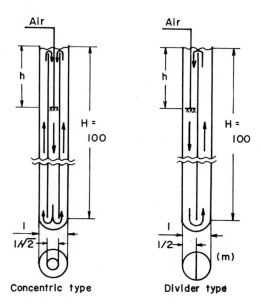

Fig. .9 Configuration and dimensions (m) of air-lift type deep shafts for
illustrative calculations

708

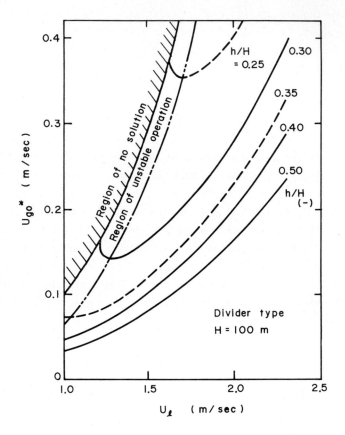

Fig. .10 Relations among superficial air velocity u_{go}*, liquid velocity U_ℓ, and depth of air injection h in an air-lift type deep shaft (Kubota et al.,1978).

(iii) <u>Power economy characteristics</u> It was assumed that the dissolved oxygen concentration in the liquid throughout the shaft was negligible. The maximum available value of oxygen utilization for a given value of air feed rate G_0 was thus obtained.

Calculated values of the maximum available oxygen transfer rate per unit power required $G_0\eta/P_w$ plotted against oxygen demand in the shaft $G_0\eta/V$ under various operating conditions for a concentric type shaft, are illustratively shown in Fig. .11, where P_w is power required and V is shaft volume. If a conversion factor relating the BOD removal to oxygen consumption, which is usually around unity under conditions of high oxygen demand in mixed liquor, is multiplied to $G_0\eta/P_w$ and $G_0\eta/V$, the values of the power economy W/P_w defined in Eq. (.1) and the BOD removal rate in the shaft W/V, respectively, can be obtained.

The solid curve ① in Fig. .11 shows the value of $G_0\eta/P_w$ for the case where

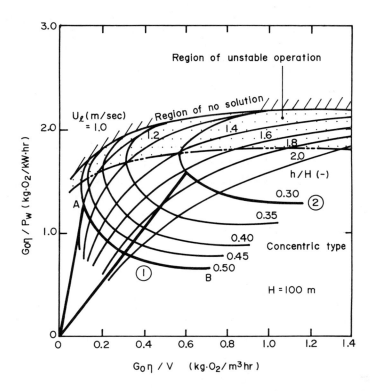

Fig. .11 Relations between maximum available oxygen transfer rate per unit power required $G_O\eta/P_W$ and oxygen demand in the liquid $G_O\eta/V$ in a concentric airlift type deep shaft (Kubota et al., 1978)

the depth of air injection is 50 m in a shaft 100 m in depth. The operation for a liquid circulation rate of U_ℓ = 1.0 m/sec and a dissolved oxygen concentration in liquid of C = 0 mg/ℓ, corresponds to a point A. Even when the oxygen demand in the liquid decreases from the value shown by point A, however, the supplied air velocity cannot be reduced, in order to maintain stable liquid circulation, set here as U_ℓ= 1.0 m/sec. Thus, a large part of the air supply power is unnecessarily spent to increase the dissolved oxygen concentration in the liquid C, while the power consumption P_W is kept almost constant. Therefore, the value of $G_O\eta/P_W$ decreases linearly along the $\overline{AO}$ line. On the contrary, if the oxygen demand in the liquid is greater than at point A, the air supply rate must be increased in order to supply dissolved oxygen to the shaft. In this case, the oxygen utilization decreases according to the increased U_{go}^*, and consequently the value of $G_O\eta/P_W$ decreases along the curve $\overline{AB}$.

In the case of high oxygen demand, the location of air injection h must be decreased to obtain a high value of $G_O\eta/P_W$. This is illustratively shown by the curve ②. In order to improve the power economy, the selection of the air injection location is an important design factor. With this in mind, it can also be concluded that a high power economy will be given only where a high oxygen demand exists in the liquid.

For this type of deep shaft, it must be pointed out, that the maximum value of power economy is considerably affected by the oxygen demand in liquid. Thus, for the rational shaft design the selection of the air sparger location is very important.

It should also be noted that in the calculations shown in Fig. .11, the dissolved oxygen concentration in the liquid C was zero. Therefore, in practical operations where C > 0, a considerable decrease in the value of $G_o \eta / P_w$ must occur.

4.2 Liquid-pump type deep shaft

A liquid-pump type deep shaft is schematically shown in Fig. .8. The concentric type is applied to divide the shaft section. The inner tube is used as the downcomer, while the annulus is used as the riser. The liquid pump is used at the top of the shaft to pump liquid in the riser part up to the top of the downcomer tube. Since air bubbles are sparged near the top of the downcomer, the air flow path in the liquid-pump type shaft is considerably longer than that in the air-lift type shaft.

(i) <u>Mathematical modeling</u> A similar mathermatical model for operation of the liquid-pump type deep shaft with that for the air-lift type deep shaft was derived by Hosono et al. (1978). The momentum and oxygen balance equations were set and solved simultaneously (for details see Hosono et al., 1978).

(ii) <u>Power requirement</u> Illustrative calculations were carried out on the shaft with a configuration and dimensions as shown in Fig. .12.

The relation among the required power P_w, the superficial liquid velocity U , and superficial air velocity at the shaft inlet U_{go}^* is shown in Fig. .13. The total required power P_w is the sum of the power to the liquid pump P_w and to the air blower P_{wg}. As seen in the

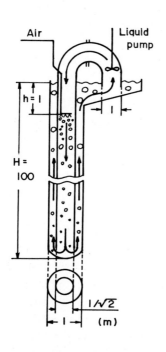

Fig. .12 Configuration and dimensions of a liquid-pump type deep shaft for illustrative calculations

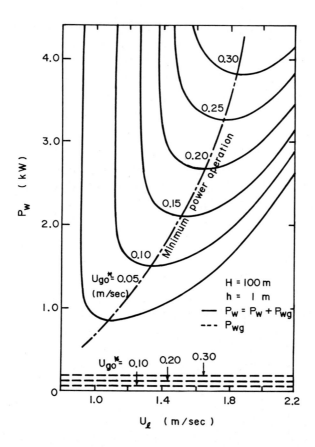

Fig. .13 Total power requirement under various values of U_{go}^* and U_ℓ for a liquid-pump type deep shaft. $P_{w\ell}$ and P_{wg} are requisite powers for the liquid-pump and the blower, respectively (Hosono et al., 1978)

figure, P_{wg} is less than 5% of the total power P_w. From Fig. .13, it is evident that there exist combinations of U_ℓ and U_{go}^* which give the minimum power P_w. The figure also shows that there is a critical value of U_ℓ which requires infinite P_w for a constant U_{go}^*. This means that one can not arbitrarily select both values of U_{go}^* and U_ℓ. There is an unacceptable operating condition, which does not yield stable liquid circulation in the shaft. Fortunately, as seen in the figure, such conditions are considerably apart from the conditions of the minimum power operation of the liquid-pump type shaft.

(iii) <u>Power economy characteristics</u> Calculated values of maximum available oxygen transfer rate per unit power required $G_0\eta/P_w$ are plotted against the oxygen demand $G_0\eta/V$ with various values of U_ℓ and U_{go}^* in Fig. .14. There exist two conditions to give a value of $G_0\eta/P_w$, because two stable operating conditions are available corresponding to two different values of U_ℓ for a constant U_{go}^*, as can be seen in Fig. .13.

Therefore, the relations between $G_0\eta/P_W$ and $G_0\eta/V$ are given for these two conditions of $U_\ell \geq U_{\ell c}$ and $U_\ell < U_{\ell c}$ in Fig. .14. $U_{\ell c}$ is the superficial liquid velocity which gives the maximum value of $G_0\eta/P_W$ for a constant value of U_{go}^*. If the value of $G_0\eta/V$ is given, the selection of the condition which gives the maximum value of $G_0\eta/P_W$, corresponding to the optimal power economy condition, can be made easily from this figure.

In Fig. .14 the maximum values of $G_0\eta/P_W$ for the air-lift shaft with a similar configuration are also shown. Compared with the air-lift shaft, the maximum values of $G_0\eta/P_W$ in the liquid-pump type shaft are considerably lower. It should be noted, however, that in the air-lift type shaft, operations which are close

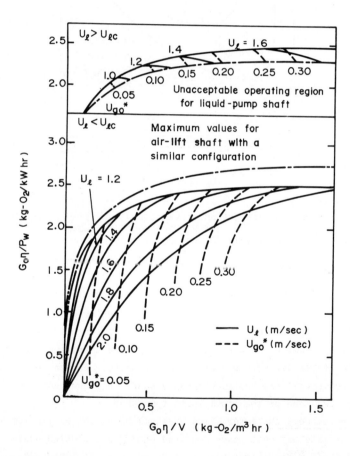

Fig. .14 Relations between maximum available oxygen transfer rate per unit power required $G_0\eta/P_W$ and oxygen damand $G_0\eta/V$ for a liquid-pump type deep shaft (Hosono et al., 1978)

to the maximum $G_o \eta / P_W$ must be avoided, because such operations are on the boundary of unstable operations, as seen in Fig. .11. On the other hand, in the liquid-pump shaft, conditions giving the optimal power economy are considerably apart from the unacceptable operating region. The following must also be mentioned. In the air-lift type shaft, since it is difficult to alter the location of air injection, it is not easy in practice to follow the optimum condition corresponding to the vairous values of oxygen demand. In the liquid-pump type, on the other hand, the operation is much more flexible, and one can easily select the optimum condition to obtain the maximum power economy under various conditions of oxygen demand in liquid.

5 TRICKLING FILTER

5.1 Performance of organic removal

(i) Mathematical modelling To predict the fraction of BOD removal of wastewater in the trickling filter, Hosono et al. (1980) proposed a mathematical model under the following assumptions.

1. The amount of microbial matter on the filter media is constant, and a steady state organic removal rate is maintained.

2. The wastewater passes through the packed filter media in a plug flow manner.

3. Sufficient oxygen is available in the free spaces in the filter bed, so that oxygen deficiency affecting the biochemical reaction rate does not occur.

4. The temperature is constant throughout the bed.

5. First-order kinetics can apply to the BOD removal rate.

To express the rate of BOD removal within the attached microbial film, per unit surface area of the packed media, R_S is assumed to be expressed by

$$R_S = -kS \qquad (.30)$$

where S is BOD concentration in the liquid on the microbial film surface and k is the rate constant.

6. Only a fraction of the specific surface area, expressed by α, is effective for the biological reaction in the filter. The value of α is a function of the operating conditions, such as specific surface area of filter media a_e, filter depth h, volumetric liquid flow rate per unit cross-sectional area of filter bed q, and influent BOD S_0.

The differential mass balance for the removable of BOD in the filter is thus

$$q (dS/dh) = \alpha R_S a_e \qquad (.31)$$

(ii) Empirical rate constants Substituting Eq. (.28) into Eq. (.29) and integrating under boundary condition of

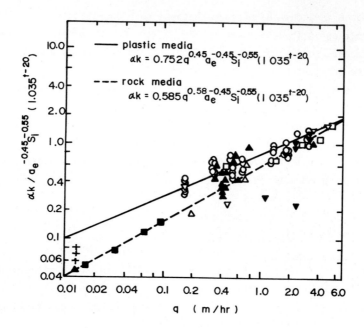

Fig. .15 Plots of previously reported data on standard trickling filters according to Eq. (.35) (Hosono et al., 1980)

$$h = 0 \quad ; \quad S = S_i \qquad (.32)$$

yields

$$S / S_i = 1 - x = \exp (-k \alpha a_e h / q) \qquad (.33)$$

where S_i is inlet value of BOD at the top of the filter and x is a fraction of the BOD removal.

The effective fraction of filter media surface α will be affected greatly by operating conditions. Since the actual value of k cannot be estimated separately, the effective reaction rate constant $k\alpha$ is correlated as a function of operational variables q, a_e, h, S_i and temperature t. That is,

$$k\alpha = f (q, a_e, h, S_i, t) \qquad (.34)$$

After surveying the available data previously reported on trickling filters, the values of $k\alpha$ were determined from Eq. (.33) under various operating conditions, and were correlated with the following empirical formula (Hosono et al., 1980).

$$k\alpha = k_0 q^n a_e^{-0.45} S_i^{-0.55} (1.035^{t-20}) \qquad (.35)$$

where constants k_0 and exponents n are:

plastic media : $K_0 = 0.752$, $n = 0.45$
rock media : $K_0 = 0.585$, $n = 0.58$

The dimensions of each variables are (m/h) for q, (1/m) for a_e, (m) for h, and (mg/ℓ) for S_i.

To verify the adaptability of Eq. (.33) with Eq. (.35) for previously reported data on standard trickling filters without liquid recycle, plots of

$$k\alpha / a_e^{-0.45} S_i^{-0.55} (1.035^{t-20})$$

against q are shown in Fig. .15.

(iii) <u>Fraction of BOD removal</u> Eqs. (.33) and (.35) can be directly applied to predict the fraction of BOD removal x of standard trickling filters without liquid recycling, where S_i is equal to the influent BOD concentration S_0.

In high-rate trickling filters, recycling is used to enhance BOD removal under high BOD loading. In trickling filters with liquid recycle, the actual liquid flow rate is given by

$$Q = Q_0 (1 + \gamma) \tag{ .36}$$

The inlet value of BOD at the top of the filter S_i is given by

$$S_i = \frac{Q_0 S_0 + \gamma Q_0 S_e}{(1 + \gamma) Q_0} = \frac{S_0 + \gamma S_e}{1 + \gamma} \tag{ .37}$$

where S_e is the effluent value of S. The net value of the fraction of BOD removed x, based on the influent waste, is defined as

$$x = (S_0 - S_e)/ S_0 \tag{ .38}$$

Therefore, the following relation can be derived :

$$\frac{S_e}{S_i} = \frac{(1 + \gamma)(1 - x)}{1 + \gamma - \gamma x} = \exp (-k \alpha a_e h / q) \tag{ .39}$$

5.2 <u>Power economy</u>

(i) <u>Power requirement</u> In trickling filters, the power required is usually governed by the power needed to pump wastewater up to the top of the filter. This should be added to that required for recirculation of liquid in the high-rate filter. The power P_w is given by

$$P_w = \Delta H \rho Q / \varepsilon \tag{ .40}$$

where Q is the liquid flow rate to be pumped up, ε is the pump efficiency and ρ

716

is the liquid density. The net head ΔH is given by

$$\Delta H = H + \ell + \Delta H_1 + \Delta H_2 \tag{.41}$$

Here H is filter depth, ℓ is distance from the sprinkler to the top of the fil-

TABLE .4

Conditions and constants used for calculation in trickling filter process

Specific surface area of filter media, a_e (m^2/m^3)	110
Equivalent length of pipe and bends, ($L + L_e$)/D	200
Distance between irrigation nozzle and top of filter media, ℓ_1 (m)	0.3
Distance between bottom of filter and water surface in reservoir tank, ℓ_2 (m)	1.2
Superficial liquid velocity in pipe, U_ℓ (m/sec)	1.5
Reaction rate constant at 20°C, k_0	0.380
Friction factor in pipes, f	0.005
Overall efficiency of pump, ϵ	0.6

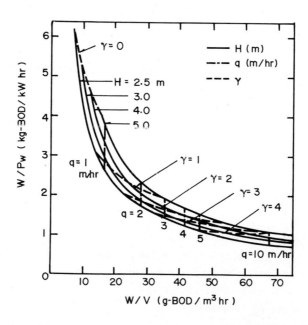

Fig. .16 Relation between power economy W/P_W and BOD removal rate per unit volume of filter media W/V on a trickling filter under steady state loading conditions with n = 1, a_e = 110 m^2/m^3, S_0 = 150 mg/ℓ and S_e = 20 mg/ℓ (Fujie et al., 1982)

ter, ΔH_1 is velocity head in lift pipe and ΔH_2 is friction head in pipe. For ΔH_1 and ΔH_2,

$$\Delta H_1 = U_\ell^2 / 2g \qquad\qquad (.42)$$

$$\Delta H_2 = 4f \left(U_\ell^2 / 2g \right) \frac{h + \ell}{d} + \kappa \qquad\qquad (.43)$$

where f is friction factor, g is acceleration of gravity, U_ℓ is the superficial liquid velocity in the lift pipe, κ is equivalent length corresponding to head loss at pipe bends, and d is diameter of the lift pipe.

(ii) <u>Steady state operation</u> Illustrative calculations of the power economy of the trickling filter with plastic media under varied operating conditions were carried out by Fujie et al. (1982). For domestic wastewater with inlet BOD of 150 mg/ℓ and required outlet BOD of 20 mg/ℓ at constant temperature of 20°C, effects of stage number n, filter depth H, and of recycle ratio γ were investigated. Other conditions and the values of constants used for their calculations are shown in Table .4. The value of k_0 in Eq. (.33) was taken as 0.380, which was determined from operating data of practical operating plants.

For a single stage filter, that is n = 1, calculated relations between the

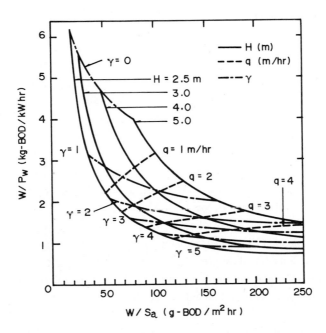

Fig. .17 Relations between power economy W/P_w and BOD removal rate per unit cross-sectional area of filter W/S_a on a trickling filter under steady state loading conditions with n = 1, a_e = 110 m^2/m^3, S_0 = 150 mg/ℓ and S_e = 20 mg/ℓ (Fujie et al., 1982)

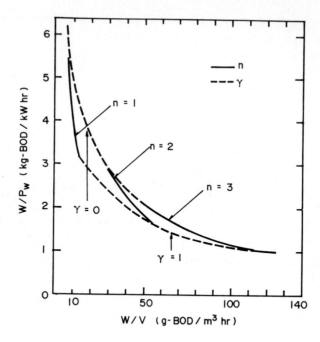

Fig. .18 Effect of staging of a trickling filter on power economy W/P_w under steady state loading conditions with $a_e = 110$ m^2/m^3, $H = 2.5$ m, $S_0 = 150$ mg/ℓ and $S_e = 20$ mg/ℓ

power economy and the BOD removal rate per unit volume of filter media are shown in Fig. .16. In Fig. .17, the power economy is plotted against the BOD removal rate per unit cross-sectional area of filter media. In comparison with other processes, the BOD removal rate in the trickling filter is sometimes discussed on the basis of its rate per unit floor area, rather than per unit volume of filter media. Figs. .16 and .17 show that increases in both the filter depth H and the recycle ratio of effluent γ enhance both the BOD removal rate per unit volume and unit cross-sectional area of the filter media, but bring about a reduction in the power economy.

The effect of staging of filter bed on both the power economy and BOD removal rate per unit volume of filter media, under steady state loading condition with filter depth H = 2.5 m, is shown in Fig. .18. The staging brings about an increase in BOD removal rate per unit volume of filter, giving the same effluent BOD concentration S_e, but brings about a considerable reduction in power economy.

The relations between the value of effluent BOD S_e and BOD loading W/S_a, for a trickling filter with bed depth of 4 m, is shown in Fig. .19. If the liquid flow rates per unit cross-sectional area of filter bed q are taken as parameters, the relations of Fig. .19 is obtained regardless of influent BOD. It is noted that the power economy varies widely depending on the values of q at a constant value of S_e. In the case where the trickling filter is used as a pre-treatment facility of an industrial wastewater with high BOD, predicted power economy values are large even under high BOD loading conditions.

(iii) <u>Effect of diurnal fluctuation of hydraulic loading</u> In the trickling filter, it is possible to maintain a constant irrigation rate in the filter bed by varing the recycle ratio γ, depending on fluctuating influent hydraulic loading.

Under such operating conditions, the relations between power economy W/P_w and BOD removal rate per unit cross-sectional area of the bed W/S_a, calculated by Fujie et al. (1982), is shown in Fig. .20. Values of power economy ranging from 0.5 to 1.0 kg-BOD/kW hr are considerably low, in comparison with those obtained under steady state loading conditions (see Fig. .17), and are almost the same as those of the activated sludge process. Therefore, one advantage of the trickling filter, which is its rather high power economy, is all but lost.

To avoid the reduction of power economy, the application of the balancing tank to equalize the diurnal fluctuation of hydraulic loading is effective. It

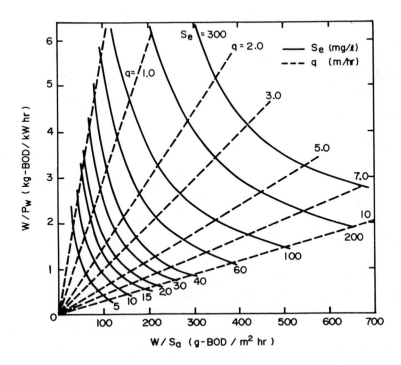

Fig. .19 Effect of effluent BOD concentration on power economy W/P_w of a trickling filter under steady state loading conditions with n = 1, a_e = 110 m^2/m^3, H = 4 m, and n = 1 (Fujie et al., 1982)

720

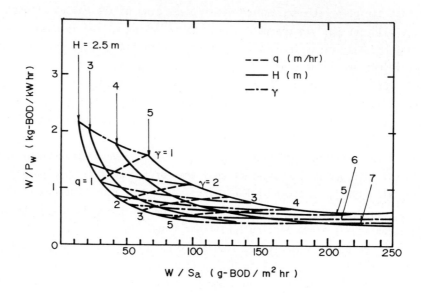

Fig. .20 Values of power economy W/P_W of a trickling filter when irrigation rate on filter bed q is maintained constant by recycling effluent, corresponding to a diurnal fluctuation of inflow rate of wastewater. Recycle ratios γ are averaged values, n = 1, a_e = 110 m^2/m^3 and weighted average of effluent BOD S_e = 20 mg/ℓ (Fujie et al., 1982)

is noted, however, that the volume of the balancing tank is sometimes larger than the volume of the filter bed itself, as illustratively shown by Fujie et al. (1982).

(iv) Rock media filter In trickling filters with rock media, which have restricted bed hights of usually less than 2 m, the standard operation and the high rate operation with low recycling ratio are applied. The BOD removal rate per unit floor area of the plants is naturally very low, but these restricted applications assure high values of power economy under conditions yielding reasonably good quality of the effluent (see Fujie et al., 1984).

(v) Power economy characteristics Power economy characteristics of the trickling filter are summarized as follows:

1. Increases in recycle ratio of effluent, filter bed depth, and number of stage enhance the BOD removal rate per unit volume and per unit floor area of the bed, but considerably reduce the power economy.

2. Characteristically high values of power economy for the trickling filter process can be obtained only at sufficiently low BOD loading conditions. It should be noted that the trickling filter, which has no or a little recycle of effluent, and thus is applied to low BOD loading conditions, gives considerably high values of power economy.

6   ROTATING BIOLOGICAL CONTACTOR

6.1   Performance of BOD removal

(i) <u>Mathematical modeling</u>   To remove BOD effectively in the rotating biological contactor (RBC), a multistage operation where several troughs with rotating
discs are connected in series, shown in Fig. .21, is usually applied. A mathematical model, corresponding to the process shown in Fig. .21, was derived to
predict the fraction of BOD removal by Hoshino and Kubota (1977a), under the
following assumptions:

1. Liquid in the troughs under the discs is completely mixed.

2. The amount of attached microorganisms on the discs is kept constant, because the amount increased is balanced with that sloughed off from the discs.

3. BOD removal by suspended microorganisms in the troughs can be neglected
compared with that by attached microorganisms on the discs.

The mass balance of BOD in i-th stage trough, under steady state condition,
gives

$$QS_{i-1} - QS_i - (-R_{si}) A_i = 0 \tag{ .44}$$

where Q is flow rate of wastewater, S is BOD concentration, $-R_{si}$ is rate of BOD
removal per unit surface area of discs, A is surface area of discs and subscript i denotes i-th stage.

It was shown by Fujie et al. (1983c) that in the RBC the rate of BOD removal
by attached microorganisms per unit surface area of discs can be approximated
by a first-order reaction, with respect to BOD concentration of wastewater in
the trough. That is,

$$-R_{Si} = \phi k S_i \tag{ .45}$$

where k is the first-order reaction rate constant and ϕ is the rotation factor
showing the effect of rotational speed of discs on the rate of BOD removal discussed later. By substituting Eq. (.45) into Eq. (.44) and solving, the effluent BOD concentration of the RBC is given by

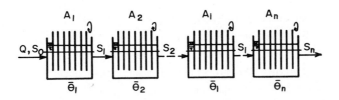

Fig. .21 Schematic diagram of staged rotating biological contactor (RBC) units

722

$$S_n = \frac{S_0}{(1 + \phi k A_1/Q)(1 + \phi k A_2/Q) \cdots (1 + \phi k A_n/Q)} \qquad (.46)$$

In practical operations, since the surface area of the discs in each stage is usually the same, Eq. (.46) can be simplified as:

$$S_n = S_0/\{ 1 + \phi k (A/nQ) \}^n \qquad (.47)$$

where A is total surface area of discs. Thus the fraction of BOD removal, $x = 1 - S_n / S_0$, is given by

$$x = 1 - 1/\{ 1 + \phi k (A/nQ) \}^n \qquad (.48)$$

This indicates that an increase in the stage number of troughs in the RBC brings about an increase in the fraction of BOD removed under the same hydraulic loading per unit surface area of discs, Q/A.

(ii) <u>Adaptability of the model to experimental data</u> Although the value of the first order reaction rate constant k depends upon the type of wastewater, apparent values of ϕk, experimentally obtained from domestic RBC wastewater treatment plants, are in the range of 7.1×10^{-3} to 26.4×10^{-3} m/h (Hoshino and Kubota, 1977a). The effect of temperature t on the reaction rate constant is shown by

$$k = k_0 (1.040)^{t-20} \qquad t \geq 12°C$$
$$k = 0.731 k_0 (1.152)^{t-12} \qquad t < 12°C \qquad (.49)$$

where k_0 is the reaction rate constant at t = 20°C.

Fig. .22 shows the comparison of the model prediction to the experimentally observed data in commercial scale RBC units, under steady state operations. As can be seen from Fig. .22, the relationship between the fraction of BOD removal and the hydraulic loading per unit surface area, estimated by Eq. (.48) fits the experimentally observed data well.

(iii) <u>Effect of rotational speed of discs</u> To evaluate the effect of rotational speed of the discs on the BOD removal rate in RBC processes, a more sophisticated mathematical model was used by Odai et al. (1981). The amount of organic substrate and dissolved oxygen, consumed by biochemical reaction within the microbial film attached on the disc surface, are supplied by diffusion through the microbial film. The differential mass balance of organic substrate and dissolved oxygen within the microbial film was taken under rotating disc conditions. By substituting appropriate rates of reaction within the microbial film into the mass balance equations and solving numerically, periodic changes in concentration profiles of organic substrate and dissolved oxygen,

within the microbial film, were obtained. The time-average rate of organic substrate consumption in the microbial film $-R_S$ was evaluated as a function of
rotational speed N of the disc.

The rotation factor ϕ was defined as

$$\phi = (-R_S)/(-R_S)_\infty \qquad\qquad (.50)$$

where $(-R_S)_\infty$ is the removal rate of BOD per unit surface area of disc at infinite rotational speed. When the rotational speed of disc N becomes infinite,

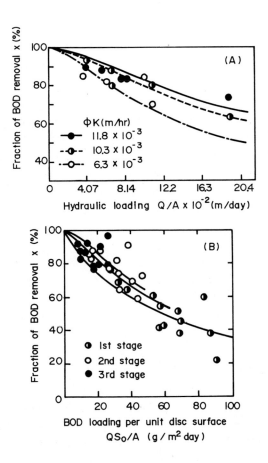

Fig. .22 Illustrative comparison between experimental data and calculated
values from Eq. (.48) on RBC units (Fujie et al., 1983c). (A) Experimental
data are obtained for domestic wastewater and for predicted curves, values of
k are selected as follows: $\phi k = 11.8 \times 10^{-3}$, 10.3×10^{-3} and 6.3×10^{-3} m/hr.
(B) Experimental data obtained for wastewater from a wool-spinning mill with
experimental conditions: D = 3 m, n = 3, N = 13 rpm, average effluent BOD S_0 =
160 mg/ℓ, t = 25-30 °C and for calculated curves, $\phi k = 15 \times 10^{-3}$ m/hr.

724

ϕ approaches unity and when N is zero, ϕ must be zero. The relation between ϕ and N, evaluated by the theoretical approach using the mathematical model mentioned above, interprets the experimental data fairly well for bio-oxidation of glucose with RBC. The semi-empirical formula of rotation factor ϕ was thus given as

$$\phi = N^{0.61}/(\ N^{0.61} + \psi\ ) \tag{.51}$$

where, $\psi = 1.6$ (25°C), $\psi = 0.6$ (20°C) and $\psi = 0.47$ (15°C). For details refer to Odai et al. (1981).

6.2 Power economy of rotating biological contactor

(i) Power requirement of disc rotation By assuming the analogy between the power requirement for disc rotation in RBC and that for stirrer in a stirred tank, Fujie et al. (1983c) suggested the following relations for the requisite power per unit surface area of discs P_W/A, in the RBC. If the rotational speed of the discs N is low, that is, the laminar flow of liquid in the trough is prevailing,

$$P_W/A = \lambda_1 (\ ND\ )^2 \tag{.52}$$

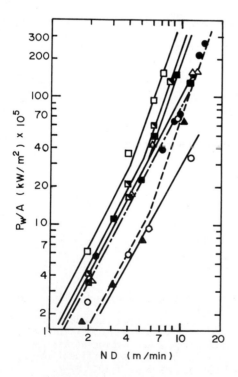

key	D (m)	$\lambda_1 \times 10^3$ $(\frac{kW^2}{min^4})$	$\lambda_2 \times 10^3$ $(\frac{kW^3}{min^5})$	Surface of disc*
	0.5	0.33	–	S
	0.5	1.10	–	P
	0.6	1.73	0.45	P
	0.5	1.10	0.20	P
	0.5	1.30	0.26	P
	0.3	0.86	–	P
	1.75	0.39	0.052	P

* S and P indicate smooth and processed, respectively.

Fig. .23 Plots of requisite power per unit surface area P_W/A against product of rotational speed N and diameter D of discs on RBD units (Fujie et al., 1983c)

and if N is high, i.e., the turbulent flow of liquid is prevailing,

$$P_w/A = \lambda_2 (ND)^3 \qquad\qquad (\ .53)$$

where D is disc diameter, and λ_1 and λ_2 are empirical constants which depend on the distance between discs and the processing conditions on the disc surface.

Only limited observed data are available to verify the adaptability of Eqs. (.52) and (.53). The correlation between P_w/A and ND obtained in previous reports is shown in Fig. .23. The transition from the region where P_w/A is proportional to $(ND)^2$, to the region where P_w/A is proportional to $(ND)^3$, can be clearly seen in the figure. It should be noticed that the processing condition on the disc surface affects the transient point from the laminar flow to the turbulent flow region, and also affects the values of λ_1 and λ_2.

(ii) <u>Steady state operation</u> In the estimation of power economy of the RBC on treatment of domestic wastewater under steady-state operational condition, the specifications of the RBC, the operational conditions, and empirical constants used are listed and shown in Table .5.

The power economy W/P_w, calculated for a single stage RBC, was plotted against the removal rate of BOD per unit surface area of discs as shown in Fig. .24. It shows that as the disc diameter decreases, the removal rate of BOD

TABLE .5
Conditions and values of constants used for illustrative calculations

Specifications of RBC and operating conditions:					
Disc diameter, D (m)	1	2	3	4	5
Shaft length, L (m)			2.5D		
Space between discs, d (m)			0.03		
Surface area of discs per unit floor area of RBC, A/S_a (m^2/m^2)	42.6	86.0	129	172	215
Peripheral speed of discs, v_r (m/min)			10 - 20		
Floor area of RBC, S_a (m^2)		$= (L + 0.15D)\times(D + 0.15D)$			

Water quality:		
Operation	Influent BOD (mg/ℓ)	Effluent BOD (mg/ℓ)
Steady loading	$S_0 = 150$	$S_n = 20$
Unsteady loading (see Fujie et al., 1983c),	$S_0 = 150$	$S_n = 20$ or $S_{n_{max}} = 20$

Constants:	
Reaction rate constant, k_0 (m/hr)	13.1×10^{-2}
Friction factors for RBC,	
λ_1 ($kW/min^2 m^4$)	0.86×10^{-5} for ND 10 m/min
λ_2 ($kW/min^3 m^5$)	0.052×10^{-5} for ND 10 m/min

per unit surface area of disc increases at a constant peripheral speed of disc. It is obvious, however, that a reduction in the disc diameter brings about a reduction in the removal rate of BOD per unit floor area of RBC. A reduction in the disc size for the same peripheral speed of disc brings about an increase in the power economy. It should be noted that the peripheral speed of disc V_r is an important factor in maintaining a suitable thickness of microbial film on the disc surface. In practice, V_r is recommended to be within the range of 10 to 20 m/min independent of the disc diameter (Antonie, 1976). Fig. .24 shows that an increase in peripheral speed V_r brings about a remarkable reduction in the power economy. Thus, the minimum peripheral speed V_r, able to maintain a stable condition of the microbial film, should be selected to bring about an increase in power economy.

The effect of staging of the RBC on the power economy was calculated and the results are shown in Fig. .25, where W/P_W was plotted against the removal rate

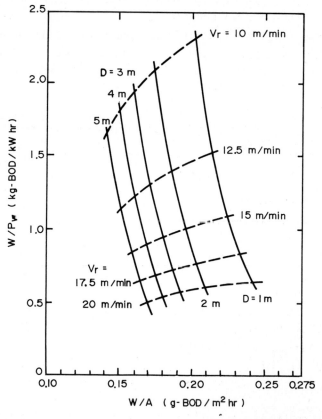

Fig. .24 Relations between BOD removal rate per unit surface area of disc W/A and power economy W/P_W for a single stage RBC unit, under steady state operating conditions with S_0 = 150 mg/ℓ, S_n = 20 mg/ℓ (Fujie et al., 1983c)

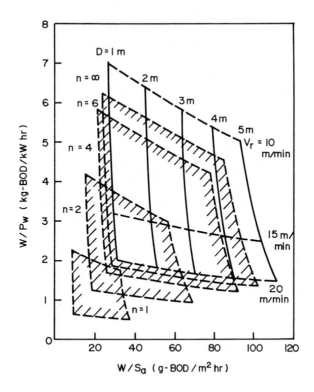

Fig. .25 Effect of staging on the BOD removal rate per unit floor area of a RBC unit and power economy W/P_W, under steady state operating conditions with S_0 = 150 mg/ℓ and S_n = 20 mg/ℓ. For the stage number n = 1 to 6, ranges of calculated values are shown as D = 1 to 5 m and V_r = 10 to 20 m/min (Fujie et al., 1983c)

of BOD per unit floor area of RBC, W/S_a. The liquid volume in the trough of the RBC was assumed to be the same for each stage. It can easily be predicted from Eq. (.48) that the staging of the RBC results in an increase in the hydraulic loading per unit surface area of disc, giving the same fraction of BOD removal. Calculated results presented in Fig. .25 show clearly that the staging brings about an increase in the hydraulic loading per unit disc surface under the same fraction of BOD removal; thus increasing the BOD loading per unit floor area of the RBC unit and increasing the power economy. While the effect of staging is rather obvious when the number of stages is less than 4, the increase of stages becomes much less effective after the number of stages is equal to or greater than 5. Hence, for practical use, the number of stages should be in the range of 3 to 5.

When the BOD loading increases for an RBC in operation, effluent BOD concentration might be maintained constant by increasing the rotational speed of disc. This brings about, however, a reduction in the power economy. As mentioned

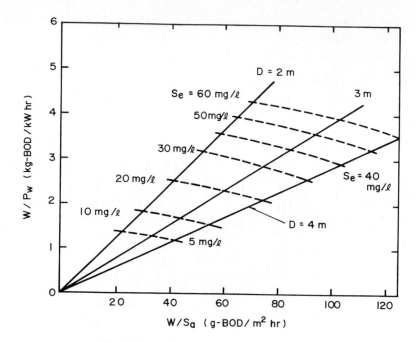

Fig. .26 Effect of effluent BOD concentration S_e on power economy W/P_w in a RBC with different disc diameter D. Operating conditions assumed are n = 4, V_r = 15 m/min and influent BOD S_0 = 150 mg/ℓ (Fujie et al., 1983c)

above, the peripheral speed of the disc should be kept within a range, in order to maintain the microbial film conditions constant. Besides, as seen in Fig. .25 a change in the peripheral speed of the disc results in only a slight change in the BOD removal rate per unit surface area in the RBC.

The relations between effluent BOD S_n and BOD loading W/S_a, in 4 staged RBC units with different disc diameters and a constant V_r = 15 m/min, is illustratively shown in Fig. .26. To obtain low values of S_n, a reduction in power economy can not be avoided. Therefore, RBC units with unnecessary excess capacity will give low power economy, although they might assure high quality of the effluent.

(iii)   Effect of diurnal fluctuation of BOD loading   The effect of diurnal fluctuation of BOD loading on the fraction of BOD removal can be evaluated by the modification of the mathematical model mentioned in .6.1. (i). The time course of effluent BOD concentration for the outlet is given by solving the following equations simultaneously from i = 1 to n, for a given change of inflow loading.

$$\frac{dS_i}{d\theta} = \frac{n \left(Q/\overline{Q} \right)}{\overline{\theta}_i} \left(S_{i-1} - S_i \right) - \frac{\phi k S_i}{\overline{\theta}_n} \left(A_i/\overline{Q} \right) \tag{.52}$$

where $\bar{Q}$ is the daily average inflow rate of wastewater and $\bar{\theta}_i$ is the average hydraulic retention time, given by $V_i/\bar{Q}$ in the i-th trough.

The relationship between power economy W/P_w and BOD removal rate per unit floor area of RBC unit W/S_a, calculated under a given condition of diurnal influent fluctuations (see Fujie et al., 1983c), is shown in Fig. .27. Conditions and constants used are shown in TABLE .5, except for the influent fluctuations. The number of stages for the RBC was selected as n = 4, and the ratio of trough volume to disc surface area V/A was varied from 0 to 0.01 m^3/m in order to ascertain the effect of hydraulic retention time in the trough on the power economy.

Fig. .27 shows that the power economy, under operational conditions of diurnal fluctuations of BOD loading, is less than a half of that under steady-state operational conditions, when the maximum effluent BOD concentration Sn_{max} was set as 20 mg/ℓ. It can also be observed that the effect of hydraulic retention time in the trough, which is proportional to the value of V/A, does not have so

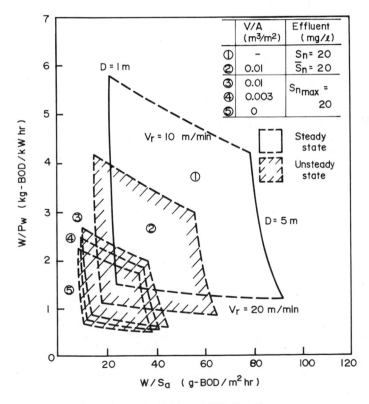

Fig. .27 Effect of diurnal fluctuation of BOD loading on power economy in a RBC unit. Operating conditions assumed for 2 to 5 are D = 1 - 5 m and V_r = 10 - 20 m/min and for other conditions see TABLE .5 (Fujie et al., 1983c)

730

much an effect on the power economy. In practical operations, the values of V/A are set at 0.005 to 0.01 m^3/m.

The relations between the power economy and the removal rate of BOD per unit floor area of RBC, estimated under conditions of daily average effluent concentration S_n = 20 mg/ℓ, are shown in Fig. .27. The decrease in power economy in this case, as compared with that of steady-state operational condition, is naturally minimized.

(iv) <u>Power economy in practical operations</u> Recently the RBC has been widely applied, especially in small scale wastewater treatment plants, because of its easy maintenance. In the design stage of the treatment facilities, the need of the correct prediction of BOD loading is essentially important to obtain a high value of power economy. The overestimation of BOD loading gives a low value of effluent BOD but results in high power consumption per unit amount of BOD removed. This is because the rotational speed of discs can not usually be varied, depending on the BOD loading of the plant. From results of a survey of practical operating conditions of RBC (Fujie et al., 1981), observed values of power economy are plotted against ratios of actual BOD loadings to designed BOD loadings in Fig. .28. This clearly shows that a poorly designed RBC operation gives considerably low values of power economy, as predicted from illustrative

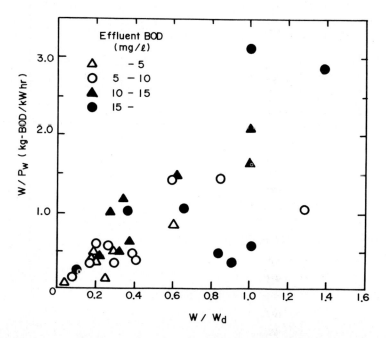

Fig. .28 Relations between power economy W/P_w and ratio of amount of BOD removed .to designed value W/W_d in commercial RBC units

calculations in Fig. .26.

In small scale wastewater treatment plants, the diurnal fluctuation of BOD loading especially of inflow rate of wastewater may be unavoidable. To compensate the diurnal fluctuation of influent flow rate and maintain stable effluent BOD, balancing tanks prior to the RBC units are usually applied in actual operations. However, the requisite volume of the balancing tank is usually very large and the investment cost becomes comparable to that of the RBC unit. In addition, the balancing tank is usually aerated to keep aerobic conditions of wastewater. The requisite power consumption of aeration in the balancing tank sometimes exceeds the power consumption of the disc rotation in the RBC unit (Fujie et al., 1981).

(v) Power economy characteristics The prominent features of the RBC processes, in view of the power economy characteristics, are summarized as follows:

1. A reduction in the disc diameter brings about an increase in power economy, but results in a considerable reduction in the removal rate of BOD per unit floor area of the RBC. This renders the RBC unsuitable for use in large scale wastewater treatment operations, but it is accceptable for small scale treatment operations where the floor area of the RBC is not an important factor in the plant.

2. A change in the disc's rotational speed does not result in a large change in its treating capacity, but affect the power economy to a greater extent. Thus to obtain a high power economy, the RBC should be operated with a peripheral speed close to its minimum.

3. The diurnal influent fluctuation of wastewater brings about a considerable reduction in the RBC power economy, especially if the effluent BOD concentration must be strictly set at a certain maximum value.

4. Overdesign of the RBC unit results in a reduction of power economy, while it gives a naturally high effluent quality.

5. The adaptation of a flow balancing tank prior to the RBC unit can considerably reduce the required floor area of the RBC unit, but the power consumption necessary to keep aerobic conditions in the balancing tank is sometimes very large in practical RBC plants.

7.   SUMMARY AND CONCLUSION

Mathematical models which were proposed to predict the oxygen consumption rate or the BOD removal rate in typical biological wastewater treatment processes : conventional activated sludge process, modified activated sludge process, trickling filter process, and rotating biological contactor (RBC) process were reviewed. Using calculated values of BOD removal rates under usual operating conditions, the power economy, which is defined as the amount of BOD removal per unit power consumption, was evaluated as a function of operating conditions of each process and discussed.

In order to characterize the processes, the relations between power economy and the effluent BOD are shown in Fig. .29. The relation for each process is depicted by a region, corresponding to the designed and operating conditions of the process, which are listed in TABLE .6. For the modified activated sludge process, i.e. deep shafts, to the calculated values of power economy, for the BOD removal rates under practical operating conditions, were assumed to be 60% of the maximum available BOD removal rates, as shown in Figs. .11 and .13. These were obtained under the condition of negligible dissolved oxygen concentration in liquid. Values of the power economy for the trickling filter and RBC processes are relatively high when compared to the activated sludge and

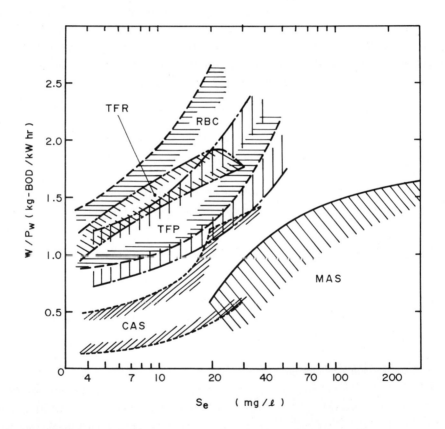

Fig. .29 Comparison of relations between power economy and effluent BOD for various biological wastewater treatment processes, where CAS is conventional activated sludge process, MAS is modified activated sludge process, TFP is trickling filter with plastic media, TFR is trickling filter with rock media and RBC is rotating biological contactor

modified activated sludge processes. It should be noted, however, that under the condition of diurnal fluctuation in BOD loading, a considerable decrease in power economy results from the regulation of the maximum value of the effluent BOD, as metioned in sections 5.2 and 6.2. In addition, these processes require considerable larger floor areas in the plants than those of the activated sludge process, in order to obtain the same value of effluent BOD.

In conclusion, the comparison of processes, as shown in Fig. .29, may not

TABLE .6

Typical operating conditions of the processes shown in Fig. .29

Conventional activated sludge (CAS):		
Diffuser performance, Φ (m)		1 - 2.5
Diffuser depth, h (m)		3 - 5
Overall efficiency of blower,		≤ 0.55
Influent BOD concentration, S_0 (mg/ℓ)		150 - 200
Influent ammonium nitrogen concentration, N_0 (mg/ℓ)		20 - 30
Mixed liquor suspended solid, X (mg/ℓ)		1000 - 2000
Constants used for calculations:		see TABLE .3
Modified activated sludge (MAS)*:		
Shaft length, H (m)		100 - 200
Superficial liquid velocity, U_ℓ (m/sec)		1 - 2
Mixed liquor suspended solid, X (mg/ℓ)		2000 - 4000
Trickling filter process:	TFP**	TFR**
Hight of filter media, H (m)	2.5 - 4.0	1.8
Number of stages, n		1
Specific surface area of filter media, a_e (1/m)	80 - 120	80 - 130
Hydraulic loading rate, q (m/hr)	1 - 4	0.6 - 1 H***
Constants used for calculations:		see Table .4
Rotating biological contactor (RBC):		
Disc diameter, D (m)		2 - 4
Number of stages, n		4
Peripheral speed of disc, V_r (m/min)		10 - 20
Influent BOD concentration, S_0 (mg/ℓ)		100 - 150
Constants used for calculations:		see TABLE .5

*   For details of calculations see Kubota et al.(1978) and Hosono et al.(1979)
**   TFP and TFR denote trickling filters with plastic media and with rock media, respectively.
*** H denotes high rate operations

be unconditional, but the information of power economy characteristic will be useful in selecting the correct process, according to the demand for energy conservation under various designated conditions.

NOMENCLATURE

A Total surface area of RBC discs, m^2

a_e Specific surface area of trickling filter media, $1/m$

b Rate constant of endogenous respiration of activated sludge, $1/hr$

C Dissolved oxygen concentration in aeration tank, mg/ℓ

C_o Value of C saturated with pure oxygen gas, mg/ℓ

D Diameter of RBC disc, m

d_{BO} Bubble diameter at diffuser depth in aeration tank, m

E_z Longitudinal dispersion coefficient in aeration tank, m^2/hr

G_o Mass. flow rate of supplied oxygen, kg/hr

G_v Volumetric air feed rate, m^3/hr

H Liquid depth of aeration tank and deep shaft, and height of trickling filter bed, m

ΔH Net head of liquid pump, m

h Liquid depth of air diffuser, m

K_{La} Volumetric oxygen transfer coefficient, $1/hr$

k First-order reaction rate constant, m/hr (for surface base rate), $1/hr$ (for volume base rate)

L Length of aeration tank, m

ℓ Length, m

N Rotational speed of RBC disc, $1/min$

n Number of stages

P Pressure, atm

P_w Requisite power, kW

Q Flow rate of wastewater, m^3/hr

Q_{O_2} Oxygen consumption rate in aeration tank, kg/m^3 hr

q Hydraulic loading per unit cross-sectional area of trickling filter bed, m/hr

R_s Removal rate of BOD per unit surface area of microbial film, g/m^2 hr

S BOD concentration, kg/m^3 or mg/ℓ

S_0, S_e Influent and effluent BOD concentrations, respectively, kg/m^3 or mg/ℓ

S_a Floor area of wastewater treatment unit, m^2

U_g Superficial air velocity, air flow rate per unit cross-sectional area of aeration unit, m/h

$U_{\ell s}$ Spiral flow velocity at water surface in aeration tank, m/hr or cm/sec

V   Volume, m^3

V_r Peripheral speed of RBC disc, m/min

W   Removal rate of BOD in wastewater treatment unit, kg/hr

X   Mixed liquor suspended solid concentration (MLSS), g/m^3 or mg/ℓ

x $= 1 - S_e/S_0$, fraction of BOD removal

Y_0 Overall value of oxygen yield, mass of oxygen consumed per mass of BOD removed

z   Mole fraction of oxygen in air bubble

Greek symbols

α Active fraction of filter media surface in trickling filter

γ Recycle ratio of effluent

ε Overall efficiency of air blower or liquid pump

η Oxygen utilization, fraction of oxygen consumed

Φ Diffuser performance, m

ϕ Rotation factor

REFERENCES

Antonie, R.L., 1976. Fixed Biological Surfaces Wastewater Treatment. CRC press.

Fujie, K., Ishihara, N. and Kubota H., 1980. Mass transfer coefficient of dispersed fine bubbles in electrolytic solutions. Journal of Fermentation Technology, 58: 477-484.

Fujie, K., Li, Q.J. and Kubota, H., 1981. A Survey of operating conditions of rotating biological contactors. Environmental Conservation Engineering (Japanese), 10: 1002-1008.

Fujie, K., Kubota, H. and Yushina, Y., 1982. Power economy of wastewater treatment on trickling filter -- plastic filter media --. Japan Journal of Water Pollution Research (Japanese), 5: 319-327.

Fujie, K., Tsubone, T., Kubota, H. and Shibuya, S., 1983a. Estimation of power economy on activated sludge wastewater treatment. Journal of Japan Sewage Works Association (Japanese), 20: No.227, 25-32.

Fujie, K., Sekizawa, T. and Kubota, H., 1983b. Liquid mixing in activated sludge aeration tank. Journal of Fermentation Technology, 81: 295-304.

Fujie, K., Bravo, H.E. and Kubota, H., 1983c. Operational design and power economy of rotating biological contactor. Water Research, 17: 1153-1162.

Fujie, K., Moriya, H. and Kubota, H., 1984. Effluent BOD and power economy on trickling filter with rock media. Japan Journal of Water Pollution Research (Japanese), 7: 310-318.

Hoshino, S. and Kubota, H., 1977a. Operation of rotating disc wastewater treatment. Water Purification and Liquid Wastes Treatment (Japanese), 18: 29-36.

Hoshino, S., Kubota, H., Kasakura, T. and Koyama, T., 1977b. Spiral liquid circulation rate in activated sludge aeration tank. Journal of Japan Sewage Works Association (Japanese), 14: No.160, 22-30.

Hosono, Y., Fujie, K. and Kubota, H., 1979. Operational characteristic evaluation of liquid-pump type deep shaft aerator. Journal of Chemical Engineering of Japan, 12: 136-142.

Hosono, Y., Kubota, H. and Miyaji, Y., 1980. Characteristic evaluation of trickling filter process. Water Research, 14: 581-590.

Kubota, H. and Miyaji, Y., 1977. Nitrification in activated sludge process. Journal of Japan Sewage Works Association (Japanese), 14: No.159, 16-20.

736

Kubota, H., Hosono, Y. and Fujie, K., 1978. Characteristic evaluations of ICI air-lift type deep shaft aerator. Journal of Chemical Engineering of Japan, 11: 319-325.

Kubota, H., Hoshino, S. and Kasakura, T., 1979. Oxygen transfer in activated sludge aeration tank -- proposition of diffuser performance --. Journal of Japan Sewage Works Association (Japanese), 16: No.184, 12-18.

Kubota, H., Fujie, K. and Kasakura, T., 1981. Automatic monitoring of exhaust gas analysis in activated sludge wastewater treatment process. Water Science and Technology, 13: 159-164.

Odai, S., Fujie, K. and Kubota, H., 1981. Effect of rotation speed on reaction rate on a rotating biological contactor. Journal of Fermentation Technology, 59: 227-234.

Sekizawa, T. and Kubota, H., 1979. Dissolved oxygen transfer with spiral liquid circulation in aeration tank of activated sludge. Journal of Fermentation Technology (Japanese), 57: 158-163.

Sekizawa, T., Fujie, K., Kubota, H., Kasakura, T. and Mizuno, A., 1984. Air diffuser performance in activated sludge aeration tanks. Journal of Water Pollution Control Federation, (in printing).

MATHEMATICAL MODELS FOR COST-EFFECTIVE BIOLOGICAL WASTEWATER TREATMENT

DANIEL TYTECA
Department of Public and International Affairs
Université Catholique de Louvain
B-1348 Louvain-la-Neuve (Belgium)

1.   INTRODUCTION

The importance of costs incurred for the conventional or advanced treatment of wastewater fully justifies that a particular attention be given to any means of reducing the impact of these costs, while properly meeting the water quality requirements. Mathematical models for cost-effective wastewater treatment appear in the literature as early as 1962 (Lynn et al., 1962) and since then various studies have been devoted to this problem. Some of these indicate that the potential savings through use of mathematical programming techniques can be as high as 25 % with respect to conventional methods (see Dick et al., 1976, 1978, or Tyteca and Smeers, 1981), as a consequence of the principle that the combination of independently designed elements does not lead to the same result as designing the combination as a whole. The subject of this chapter will be to explain more precisely how mathematical models can be proposed and developed as an aid to cost effective wastewater treatment, especially in the scope of biological treatment.

Cost savings in wastewater treatment can be achieved mainly in two ways : (1) by optimizing the design of the treatment plant, (2) by optimizing the operation of the plant. Ideally, the design phase should account for the expected operation characteristics of the plant during its lifespan, whenever the latter are predictable : this can yield additional savings as recent studies indicate (Herbay et al., 1983). Also, it is meaningful to optimize the operation of a previously designed plant. This in turn can be effected in two manners : (1) by looking at the seasonal or monthly variations of the plant influent and the characteristics of the receiving water body, (2) by concentrating on the daily or hourly variations and attempting to cope with these through automatic control. In the first case a mathematical model based on cost minimization will indicate how to readjust the operational variables, while in the second case cost savings will result from actions whose main purpose is the minimization of shock loads, plant failures or energy consumption. Only the first case will be considered in this chapter. It is apparent from the above comments that we will

738

concentrate on steady-state operation of wastewater treatment plants. Short-term and instantaneous variations will not be considered in this chapter as they are the subject of other parts of the book.

The chapter will be organized as follows (Fig. 1). In section 2 the main concern will be to provide mathematical relationships describing the operation of a treatment plant, with special emphasis on biological treatment. Mathematical relationships can be subdivided into four categories indicated in Fig. 1. Section 3 will be devoted to cost studies and cost optimization of treatment plants, which will require introducing cost functions for each of the process units composing a plant, as well as mathematical optimization techniques. In section 4 we will give a few examples illustrating the whole scheme.

2. MATHEMATICAL MODELS OF BIOLOGICAL WASTEWATER TREATMENT PLANTS

2.1. General framework and basic assumptions

Various types of biological treatment have been proposed and applied in practice. In this study we will retain only four of them, among the most wide-

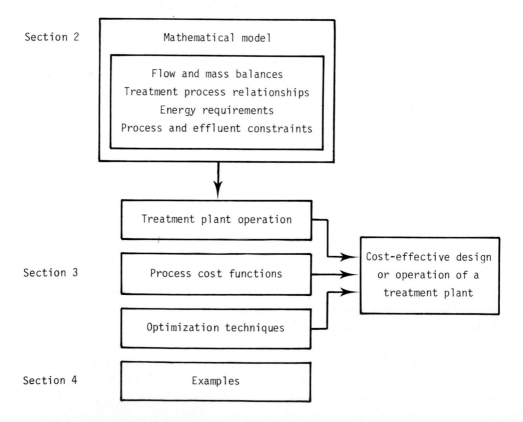

Fig. 1. Organization of the chapter.

spread, namely, the activated sludge system, the trickling filter, the rotating biological contactor, and the anaerobic digester. These are thought to be representative of most variants of biological treatment, except perhaps the aerated lagoons, which are thought to be less amenable to cost optimization studies due to the absence or lesser possibility of flow recycles and feedbacks. In addition, we will restrict the application of anaerobic digestion to sludge treatment, thus leaving water treatment to the three aforementioned aerobic processes. This may seem too restrictive but as will be seen the proposed framework allows for the easy replacement of a given process by any other one with different characteristics but with the same flow arrangement.

The chapter will be based on the treatment plant scheme represented in Fig. 2, where the "aerobic biological process" stands for any one of the three processes mentioned above : activated sludge, trickling filter, rotating biological contactor. In Fig. 2, various possible flow arrangements have been considered but not all will be retained in a given solution. Fig. 2 is centered on the process units where cost-effective studies can be really meaningful, through various possible trade-offs. Processes located upflow (screening, grit removal, ...) or downflow (tertiary treatment, incineration, ...), for which there are no significant possibilities of trade-offs or interactions with the processes of Fig. 2, and more especially with biological processes, have not been included in this study. In subsequent optimization examples, these upflow or downflow processes are assumed to be either absent or to have fixed performance characteristics.

In each numbered point of Fig. 2, the following components of wastewater, sludge or water will be considered :
- the flow rate, denoted by Q, in m^3/h
- the biodegradable pollutant load, subdivided into a dissolved part, L_b (in mg/ℓ of chemical oxygen demand, COD, or five-days biological oxygen demand, BOD_5) and a suspended part M_{bs} (in mg/ℓ of suspended solids, SS)
- the nonbiodegradable suspended load, M_s (in mg/ℓ SS)
- the active biomass, denoted by S or S* (in mg/ℓ SS) in the activated sludge aerator and in the anaerobic digester, respectively.

The assumptions underlying the subdivision of the waste into the components L_b, M_{bs}, M_s, S and S* have been previously examined (Tyteca and Nyns, 1979, Tyteca et al., 1977). In the following we will frequently use aggregates of these components, recalling the two main functions of a biological treatment plant : biological waste removal and physico-chemical solids removal. These aggregates are defined as

$$L^* = L_b + \omega M_{bs} \qquad (1)$$

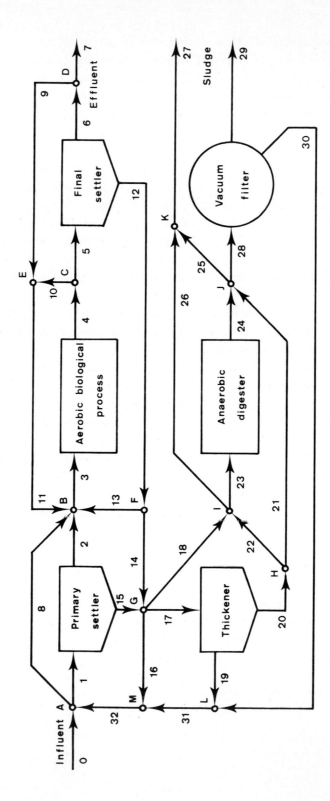

Fig. 2. Basic treatment plant scheme for cost-effective analyses.

$$M^* = M_{bs} + M_s + S + S^* \tag{2}$$

The first of these relationships gives the total biodegradable load (in mg/ℓ COD) and the ω factor gives the COD equivalent of biodegradable suspended solids (ω = 1.42 mg COD/mg SS). Eq. 2 yields the total suspended solids concentration (mg/ℓ SS).

A set of assumptions will be made before developing a mathematical model for the plant depicted in Fig. 2. First, it is assumed that clarification in the final settler and the thickener, as well as filtration in the filter, can be achieved almost perfectly, or equivalently, that the solids concentration at points 6 (and therefore 7, 9), 19, 30 (and therefore 31) of Fig. 2 are negligible :

$$M_6^* = M_7^* = M_9^* = M_{19}^* = M_{30}^* = M_{31}^* \cong 0 \tag{3}$$

For the thickener and the filter, these assumptions are in accordance with the facts that (1) models allowing for the calculation of M_{19}^* and M_{30}^* do not currently exist, and (2) the occurence of small concentrations of suspended solids in the thickener overflow and in the filtrate will have a negligible influence on design. For the final settler, suspended solids in the effluent are actually a serious problem in the plant daily operation and control, as analyzed e.g. by Roper and Grady (1974). However, as no satisfactory model presently exists for describing the evolution of M_6^*, it is quite reasonable to design the settler or operating it on a long-term basis, while imposing a constraint expressing that the overflow suspended solids concentration be as low as possible. The structure of the model to be presented would easily allow for the introduction or replacement of any equation describing these phenomena (clarification, filtration) as soon as such equations become available.

In addition, the influent data (point o in Fig. 2) will be assumed to be known through specification of either Q_0, L_0^* and M_0^*, or Q_0, L_{b_0}, M_{bs_0} and M_{s_0}. In the first case, when dealing with a standard type of wastewater (e.g., urban domestic wastewater), the components L_0^* and M_0^* can be easily redistributed into L_{b_0}, M_{bs_0} and M_{s_0} through the following approximate relationships :

$$M_{bs_0} = 60 \ldots 70 \% \text{ of } M_0^* \tag{4}$$

$$M_{s_0} = M_0^* - M_{bs_0} \tag{5}$$

$$L_{b_0} = L_0^* - \omega M_{bs_0} \tag{6}$$

In other situations (e.g., industrial wastewaters), the specification of M_{bs_0}, M_{s_0} and L_{b_0} will be required. The model can thus be conveniently applied to industrial wastewaters whose characteristics are not too different from an urban wastewater and/or whose decomposition into the terms M_{bs_0}, M_{s_0} and L_{b_0} is meaningful. In specific cases, where other types of pollutants (e.g. phenols, heavy metals, ...) become important, a different type of processes, models and relationships would have to be adopted. Finally, it is reasonable to admit that in most situations, there are no biological solids in the plant influent :

$$S_0 = S_0^* = 0 \tag{7}$$

2.2. Flow and mass balance relationships

A first type of relationships governing the operation of the plant depicted in Fig. 2 includes the flow and mass balance equations. These simply express that in each point of the plant where mixing and/or separation of flows occur, the volumetric and massic flow rates entering the point must equal those leaving it. Such equations of course do not hold in process units where material transformation occurs through biological activity, that is, in the two biological process units of Fig.2. Flow and mass balance relationships will thus be specified for each lettered point of Fig. 2, as well as for the physical treatment units (settlers, thickener, filter). Their general formulation is as follows :

$$\sum_{\to i} Q = \sum_{i \to} Q \tag{8}$$

$$\sum_{\to i} M_{bs} Q = \sum_{i \to} M_{bs} Q \tag{9}$$

$$\sum_{\to i} M_s Q = \sum_{i \to} M_s Q \tag{10}$$

$$\sum_{\to i} L_b Q = \sum_{i \to} L_b Q \tag{11}$$

$$\sum_{\to i} S Q = \sum_{i \to} S Q \tag{12}$$

where the symbols "$\to i$" and "$i \to$" stand for the flows entering and leaving a given point i, respectively. As an example, the flow balance relationships for points A (plant influent) and B will be, respectively

$$Q_0 + Q_{32} = Q_1 + Q_8 \tag{13}$$

$$Q_2 + Q_8 + Q_{11} + Q_{13} = Q_3 \qquad (14)$$

To give another example, as $M_{31}^* \cong 0$, the biodegradable suspended solids balance around point M will be

$$M_{bs16} \, Q_{16} = M_{bs32} \, Q_{32} \qquad (15)$$

In this context, the suspended solids concentrations in all flows leaving a lettered mixing point are the same, and equal to the concentration in the entering flow whenever it is unique. Thus, for example,

$$M_{S1} = M_{S8} \qquad (16)$$

$$S_{16} = S_{17} = S_{18} \qquad (17)$$

$$M_{bs12} = M_{bs13} = M_{bs14} \qquad (18)$$

This is also the case for the dissolved waste (L_b), which in addition is conserved in physical treatment units :

$$L_{b1} = L_{b2} = L_{b15} \qquad (19)$$

By proceeding in this way, it is easy to check the values of all flows and suspended and dissolved solids concentrations at each point of Fig. 2, as indicated in Table 1. Thus at this point we need 9 values for L_b, 15 values for M_{bs}, M_S and S, 4 values for S^*, and 30 values for Q. These are related by the aforementioned flow and mass balance equations. In addition, these values will be governed by the treatment process relationships, to be examined in the next section.

2.3. Treatment process relationships

In this section the models describing the processes of Fig. 2 will be briefly reviewed, with a special emphasis on the biological treatment processes. Thus we will first tackle the four biological processes retained in the present chapter. A fifth subsection will be devoted to the physico-chemical treatment units.

2.3.1. Activated sludge

When adopting an activated sludge system for the "aerobic biological process" of Fig. 2, the meaningful flows are flows 2 to 7 and 12 to 14. The flows 9 to 11 are usually not considered in such a system; therefore, $Q_9 = Q_{10} = Q_{11}$

TABLE 1

Values of flows and concentrations at each point of the plant in Fig. 2.

Point	0	1	2	3	4	5	6	7	8	9	10	11	12	13	14	15	16	17
L_b	L_{b_0}	L_{b_1}	L_{b_1}	L_{b_3}	L_{b_4}	L_{b_4}	L_{b_4}	L_{b_4}	L_{b_1}	L_{b_4}	L_{b_4}	L_{b_4}	L_{b_4}	L_{b_4}	L_{b_4}	L_{b_1}	$L_{b_{16}}$	$L_{b_{16}}$
M_{bs}	M_{bs_0}	M_{bs_1}	M_{bs_2}	M_{bs_3}	M_{bs_4}	M_{bs_4}	0	0	M_{bs_1}	0	M_{bs_4}	$M_{bs_{11}}$	$M_{bs_{12}}$	$M_{bs_{12}}$	$M_{bs_{12}}$	$M_{bs_{15}}$	$M_{bs_{16}}$	$M_{bs_{16}}$
M_s	M_{s_0}	M_{s_1}	M_{s_2}	M_{s_3}	M_{s_4}	M_{s_4}	0	0	M_{s_1}	0	M_{s_4}	$M_{s_{11}}$	$M_{s_{12}}$	$M_{s_{12}}$	$M_{s_{12}}$	$M_{s_{15}}$	$M_{s_{16}}$	$M_{s_{16}}$
S	0	S_1	S_2	S_3	S_4	S_4	0	0	S_1	0	S_4	S_{11}	S_{12}	S_{12}	S_{12}	S_{15}	S_{16}	S_{16}
S^*	0	0	0	0	0	0	0	0	0	0	0	0	0	0	0	0	0	0
Q	Q_0	Q_1	Q_2	Q_3	Q_3	Q_5	Q_6	Q_7	Q_8	Q_9	Q_{10}	Q_{11}	Q_{12}	Q_{13}	Q_{14}	Q_{15}	Q_{16}	Q_{17}

Point	18	19	20	21	22	23	24	25	26	27	28	29	30	31	32
L_b	$L_{b_{16}}$	$L_{b_{16}}$	$L_{b_{16}}$	$L_{b_{16}}$	$L_{b_{18}}$	$L_{b_{16}}$	$L_{b_{24}}$	$L_{b_{25}}$	$L_{b_{16}}$	$L_{b_{27}}$	$L_{b_{25}}$	$L_{b_{25}}$	$L_{b_{25}}$	$L_{b_{31}}$	$L_{b_{32}}$
M_{bs}	$M_{bs_{16}}$	0	$M_{bs_{20}}$	$M_{bs_{20}}$	$M_{bs_{20}}$	$M_{bs_{23}}$	$M_{bs_{24}}$	$M_{bs_{25}}$	$M_{bs_{23}}$	$M_{bs_{27}}$	$M_{bs_{25}}$	$M_{bs_{29}}$	0	0	$M_{bs_{32}}$
M_s	$M_{s_{16}}$	0	$M_{s_{20}}$	$M_{s_{20}}$	$M_{s_{20}}$	$M_{s_{23}}$	$M_{s_{24}}$	$M_{s_{25}}$	$M_{s_{23}}$	$M_{s_{27}}$	$M_{s_{25}}$	$M_{s_{29}}$	0	0	$M_{s_{32}}$
S	S_{16}	0	S_{20}	S_{20}	S_{20}	S_{23}	S_{24}	S_{25}	S_{23}	S_{27}	S_{25}	S_{29}	0	0	S_{32}
S^*	0	0	0	0	0	0	S^*_{24}	S^*_{25}	0	S^*_{27}	S^*_{25}	S^*_{29}	0	0	0
Q	Q_{18}	Q_{19}	Q_{20}	Q_{21}	Q_{22}	Q_{23}	Q_{23}	Q_{25}	Q_{26}	Q_{27}	Q_{28}	Q_{29}	Q_{30}	Q_{31}	Q_{32}

= 0. Before defining the model for the activated sludge system, a set of definitions can be made in order to write the system equations in a convenient and condensed way. The hydraulic retention time is defined as the total time (in h) that wastewater is maintained in the aerator :

$$\theta = \frac{V}{Q_2} \qquad (20)$$

in which V = the aerator volume (in m^3), and Q_2 = the inflow rate to the aerator (m^3/h). The sludge or mean cell retention time, or sludge age, is the total time (days) spent by the active biomass in the activated sludge system (aerator + settler) :

$$\theta_s = \frac{S}{\Delta S/\Delta t} = \frac{S_4 \ (Ah + V)}{Q_{14} \ S_{12} - Q_2 \ S_2} \qquad (21)$$

in which A and h = the final settler area (m^2) and height (m), respectively. This expression differs from expressions usually found in the literature by the fact that biomass is partially recycled into the primary settler (see Fig. 2), which contributes to increase the retention time, as reflected by the $- Q_2 S_2$ term in the denominator. The most significant part of the sludge age corresponds, in fact, to retention in the aerator, as confirmed in a study by Stall and Sherrard (1978) :

$$\theta_{sa} = \frac{S_4 \ V}{Q_{14} \ S_{12} - Q_2 \ S_2} = \alpha \ \theta_s \qquad (22)$$

$$\alpha = V \ / \ (V + Ah) \qquad (23)$$

Let us define next the thickening factor in the final settler as

$$\gamma = M_{12}^{\star} \ / \ M_4^{\star} \qquad (24)$$

and the activated sludge recycle and wasting ratios, respectively, as

$$\rho = Q_{13} \ / \ Q_2 \qquad (25)$$

$$\pi = Q_{14} \ / \ Q_2 \qquad (26)$$

Finally, the following dimensionless variables will be useful in relating the hydraulic retention time to the sludge age :

$$\phi = S_2 \ / \ S_4 \qquad (27)$$

$$\eta = 1 + \rho - \rho \gamma \qquad (28)$$

$$\eta' = \eta - \phi \qquad (29)$$

Indeed, one can easily check, through the above definition and a mass balance around the activated sludge system (see Tyteca, 1981a), that

$$\alpha \; \theta_s = \frac{\theta}{\eta'} \qquad (30)$$

The first part of the activated sludge model described hereafter was developed by the author and coworkers (Tyteca et al., 1977, Tyteca, 1981a). It includes equations allowing one to distinguish the dissolved and suspended parts in the biodegradable load which remains after biological treatment in the aerator. This part of the model has not been demonstrated experimentally; however, it is not incompatible with Daigger and Grady's observation (1977) that the remaining degradable load involves not only undegraded influent load, but also byproducts of the reactions underlying the biomass metabolic activity. In most activated sludge models, this problem is overlooked by neglecting the suspended part of the remaining biodegradable load.

A fundamental assumption for this part of the model is that the dissolved and suspended parts of the biodegradable load are equally available to the biomass and are degraded at the same rate. Though adsorption and a primary phase of extracellular degradation mainly affect the suspended solids (see e.g. Eckenfelder, 1966, and Jacquart et al., 1973), the metabolization phase, which is important in determining the retention time and the required oxygen amount, equally affects the dissolved and suspended parts.

The specific decay rate of bacteria (measured per hour or per day) is defined as the quantity of bacteria which lose their viability per unit time and per unit biomass :

$$b = - \frac{\left(\frac{\Delta S}{\Delta t}\right)_m}{S} \qquad (31)$$

in which m is set for "mortality". The total amount of dead biomass during one pass through the aerator is

$$(S)_m = b \; \frac{V \; S_4}{Q_3} \qquad (32)$$

By using the above definitions, and observing that $Q_3 = (1 + \rho) \; Q_2$,

$$(S)_m = b \, \frac{\theta \, S_4}{1 + \rho} \tag{33}$$

The degradable part of this quantity is then

$$(S)_{mdeg} = x \, b \, \frac{\theta \, S_4}{1 + \rho} \tag{34}$$

in which x = the biodegradable part of materials resulting from microbial lysis. Its value is near 0.765 (Goodman and Englande, 1974). The concentration of biodegradable load resulting from mortality, expressed in milligrams of COD per liter, is

$$(L_4^*)_m = \omega \, x \, b \, \frac{\theta \, S_4}{1 + \rho} \tag{35}$$

in which ω has been defined previously (see Eq. 1). If one defines κ as the part of the total degradable substrate which results from microbial lysis,

$$\kappa = \frac{(L_4^*)_m}{L_4^*} = \omega \, x \, b \, \frac{\theta \, S_4}{(1 + \rho) \, L_4^*} \tag{36}$$

A κ value equal to 1 means that the whole remaining degradable substrate results from mortality, and therefore no longer contains any trace of influent substrate, L_3^*. Defining λ and λ' as the dissolved parts of the total degradable substrate at points 4 and 3 of the plant in Fig. 2, respectively, one has

$$\lambda = \frac{L_{b_4}}{L_4^*} \tag{37}$$

$$\lambda' = \frac{L_{b_3}}{L_3^*} \tag{38}$$

Further defining λ'' as the dissolved part of degradable substrate resulting from mortality :

$$\lambda'' = \frac{(L_{b_4})_m}{(L_4^*)_m} \tag{39}$$

the following equality can be written :

$$\lambda = (1 - \kappa) \, \lambda' + \kappa \, \lambda'' \tag{40}$$

which simply expresses that, without mortality, $\kappa = 0$, the dissolved part of the degradable load in the aerator is the same as in the inflow (point 3 in

Fig. 2; $\lambda = \lambda'$), whereas, if mortality occurs, that part is progressively influenced by the distribution of degradable materials resulting from mortality
into dissolved and suspended parts. At the limit, when all of the degradable
substrate comes from dead materials, the distribution within dissolved and suspended parts is conditioned by the parameter λ'' ($\kappa \to 1 : \lambda \to \lambda''$).

The biodegradable load removal and the microbial growth in the aerator can
be described by Monod relationships. When assuming complete mixing steady-state
conditions, one has (see Tyteca et al., 1977) :

$$L_4^* = \frac{(1 + b\,\alpha\,\theta_s)\,K_M}{(\mu_{max} - b)\,\alpha\,\theta_s - 1} \tag{41}$$

$$S_4 = \frac{Y\,\alpha\,\theta_s}{\theta\,(1 + b\,\alpha\,\theta_s)}\,[L_2^* - L_4^*\,(\lambda + \eta - \lambda\,\eta)] \tag{42}$$

in which K_M, b, μ_{max} and Y = kinetic and thermodynamic constants; λ = the dissolved part of the biodegradable load L_4^* remaining after degradation; the other
symbols have been defined previously. The model in Eqs. 41 and 42 is similar to
other models generally found in the literature (see e.g. Lawrence and McCarty,
1970), with a slight modification to take into account the accumulation of degradable SS resulting from final settling and recycle, which is expressed
through the factor $(\lambda + \eta - \lambda\,\eta)$ in Eq. 42. Due to microbial decay, the nondegradable load increases in the aerator :

$$M_{S4} = \frac{M_{S2} + (1 - x)\,b\,\theta\,S_4}{\eta} \tag{43}$$

in which x has been defined previously (see Eq. 34).

The above model for activated sludge includes a set of constants or parameters. In previous studies, the following values have been adopted (Tyteca,
1981) : b = 0.07/day; K_M = 100 mg/ℓ; μ_{max} = 3.84/day; Y = 0.5; λ'' = 0.5, corresponding to a "standard" municipal wastewater composition.

2.3.2. Trickling filters

As opposed to activated sludge where Monod relationships were used in the
derivation of the model, there exists no satisfactory fundamental model for
trickling filters. Therefore most models of trickling filter systems are empirical. Such models have been reviewed and tested for cost optimization purposes
by Gotaas and Galler (1973, 1975) who were among the most active in this research field. In such models the main variables are the biodegradable load concentrations (i.e., L^* as defined above), while the suspended solids or active
biomass are usually not taken into account. Among the factors affecting most
significantly the filter's performance, are the recirculation ratio, the depth

and the radius of the filter. Galler and Gotaas themselves (1964, 1966) proposed an empirical model incorporating these factors. In the case of trickling filters, recirculation can be effected through either flow 10 or flow 12 in Fig. 2. Therefore if flow 10 is adopted, Eq. 25 has to be replaced by

$$\rho = Q_{11} / Q_2 \qquad (44)$$

when deriving the filter model. Moreover, it is generally implicitly assumed in trickling filter models that $M_{bs_4} \cong 0$, and therefore

$$L_{12}^* \cong L_{10}^* = L_4^* \qquad (45)$$

With these restrictions Galler and Gotaas' model for trickling filters can be written

$$L_4^* = \frac{K \ (L_3^*)^{1.19} \ (1 + \rho)^{0.28} \ (Q_3/A)^{0.13}}{(0.3048 + D)^{0.67} \ (T)^{0.15}} \qquad (46)$$

where K stands for a constant including unit conversions (= 0.319 in metric units), A is the filter area (in m^2), Q_3 is the flow rate through the filter (in m^3/h) and T is the temperature (in °C). The K value as well as the values of the exponents in Eq. 46 depend upon the type of wastewater and the authors provide different sets of values for other wastewater characteristics (Gotaas and Galler, 1975). The values indicated above are for domestic sewage.

As mentioned above, suspended solids are not explicitly taken into account in the trickling filter equations. As we need an estimation of suspended solids concentration for designing the final settler, and further, the thickener and the sludge treatment units, we can make a rough assumption by considering in a simplified way that the uptake of biodegradable suspended solids (for which no specific model exists) is compensated by the loss of dead biomass detached from the filter. Additionally, we can consider that the nondegradable suspended solids concentration flows unchanged through the filter. Therefore, the overall suspended solids concentration remains unchanged, and its value at point 4 of Fig. 2 (flow from trickling filter) can simply be obtained from a mass balance around the filter. First, if recirculation is effected through flow 10,

$$M_4^* \ Q_3 = M_2^* \ Q_2 + M_{11}^* Q_{11} \qquad (47)$$

If one observes that $Q_3 = (1 + \rho) \ Q_2$ and $Q_{11} = \rho \ Q_2$ (see Eq. 44), and $M_{11}^* = M_4^*$ (since $Q_9 = 0$ in Fig. 2), this yields

$$M_4^* \ (1 + \rho) = M_2^* + M_4^* \ \rho \qquad (48)$$

and thus simply

$$M_4^* = M_2^* \tag{49}$$

Second, if recirculation is effected through flow 12,

$$M_4^* \ Q_3 = M_2^* \ Q_2 + M_{12}^* \ Q_{13} \tag{50}$$

Here, Eqs. 24 and 25 are to be considered, yielding

$$M_4^* \ (1 + \rho) = M_2^* + \gamma \ M_4^* \ \rho \tag{51}$$

and finally

$$M_4^* = \frac{M_2^*}{\eta} \tag{52}$$

(see Eq. 28).

2.3.3. Rotating biological contactors

In the case of rotating biological contactors (RBC), various flow configurations can be adopted. In this chapter we will consider that recirculation is effected through flow 9, which will result in $Q_{10} = Q_{13} = 0$, and $Q_{11} = Q_9$, while $Q_{14} = Q_{12}$. The value to be selected for the recycle ratio is therefore given by Eq. 44. As in the case of trickling filters, no model is generally used for describing the evolution of suspended solids through RBC units. Therefore the same assumptions can be made (see section 2.3.2.). A mass balance around the RBC will result in the following, when observing that $M_9^* = 0$ (perfect clarification : see Table 1) :

$$M_4^* \ Q_3 = M_2^* \ Q_2 \tag{53}$$

or, noting that $Q_3 = (1 + \rho) \ Q_2$,

$$M_4^* = \frac{M_2^*}{1 + \rho} \tag{54}$$

The performance of RBC units will be described mainly through the removal of biodegradable load (L^*, as in the case of trickling filters). Various equations have been proposed for calculating the rotating biological contactor performance. In this chapter, the model of Kornegay and Andrews (Kornegay, 1975, Kornegay and Andrews, 1968) was adopted for its simplicity, flexibility and yet fairly good accuracy (see e.g. Edeline and Vandevenne, 1979). Substrate degradation in one RBC stage can be described by the following equation :

$$L_{in}^* - L_{out}^* = \frac{N\,A\,K}{Q}\ \frac{L_{out}^*}{K_f + L_{out}^*} \tag{55}$$

where L_{in}^* and L_{out}^* = biodegradable load concentration before and after treatment in the RBC stage, respectively, in mg/ℓ BOD_5, N is the number of disks in the stage, A is the lateral surface of one disk, in m^2, K is a constant including the biomass growth rate, the growth yield, the unit mass of biological film and the active film thickness, in g/m^2.day for instance, Q is the flow rate through the stage, in m^3/h, and K_f is a saturation constant, in mg/ℓ BOD_5. In subsequent examples, the values of N and K_f have been taken as 40 and 25 mg/ℓ, respectively, while K has been varied from stage to stage since the components of this "constant" are changing with changing substrate concentrations. In the case of the plant depicted in Fig. 2, taking a four stage RBC process as an example, the equations of the rotating biological contactor will be :

$$L_3^* - L_{31}^* = \frac{40\ A\ K_1}{(1+\rho)Q_2}\ \frac{L_{31}^*}{25 + L_{31}^*} \tag{56}$$

$$L_{31}^* - L_{32}^* = \frac{40\ A\ K_2}{(1+\rho)Q_2}\ \frac{L_{32}^*}{25 + L_{32}^*} \tag{57}$$

$$L_{32}^* - L_{33}^* = \frac{40\ A\ K_3}{(1+\rho)Q_2}\ \frac{L_{33}^*}{25 + L_{33}^*} \tag{58}$$

$$L_{33}^* - L_4^* = \frac{40\ A\ K_4}{(1+\rho)Q_2}\ \frac{L_4^*}{25 + L_4^*} \tag{59}$$

where L_{31}^*, L_{32}^*, L_{33}^* = the values of L^* after stage 1, after stage 2 and after stage 3, respectively. It has been assumed that each stage has the same number of disks N and the same disk area A, while the expression substituted for Q in Eq. 55 results from the fact that $Q_3 = (1 + \rho)\ Q_2$ (see Eq. 44).

A study of existing literature (Edeline and Vandevenne, 1979) shows that the "constant" K varies much as a function of substrate type, degradation stage and various other factors. It can be seen, however, to decrease from the first stage to the last (Clark et al., 1978). This constant has to be measured in each particular situation and with each given waste composition. In this research, the following values were taken only as an example :

$K_1 = 160$ g/m^2.day , $K_2 = 80$ g/m^2.day

$K_3 = 40$ g/m^2.day , $K_4 = 20$ g/m^2.day

$K_5 = 10$ g/m^2.day

2.3.4. Anaerobic digester

The anaerobic digester can be modeled through relationships similar to those used in the activated sludge process, based on Monod kinetics. An additional assumption in this case is that b_d (the specific decay rate for digestion) $\cong 0$. Therefore, the nondegradable load remains unchanged during digestion :

$$M_{S_{24}} = M_{S_{23}} \qquad (60)$$

Monod kinetics imply that the global biodegradable load (dissolved and suspended), and the concentration of biomass effecting digestion are respectively given by (see e.g. Tyteca et al., 1977) :

$$L_{24}^* = \frac{K_{Md}}{\mu_{maxd} \frac{V}{Q_{23}} - 1} \qquad (61)$$

$$S_{24}^* = Y_d (L_{23}^* - L_{24}^*) \qquad (62)$$

where V = the digester volume, in m^3. The components of degradable load after digestion, as for the aerator, are assumed to be proportional to their concentration before digestion :

$$L_{b_{24}} = L_{b_{23}} \frac{L_{24}^*}{L_{23}^*} \qquad (63)$$

$$M_{bs_{24}} = M_{bs_{23}} \frac{L_{24}^*}{L_{23}^*} \qquad (64)$$

The activated sludge biomass, S, involving degradable and nondegradable parts, will have the following concentration after digestion :

$$S_{24} = S_{23} [(1 - x) + x \frac{L_{24}^*}{L_{23}^*}] \qquad (65)$$

where x has the same meaning as in the activated sludge system (see section 2.3.1., Eq. 34). The following values have been taken for the constants : $Y_d = 0.044$, $\mu_{maxd} = 0.29/day$, $K_{Md} = 2200mg/\ell$.

In order to quantify the economic value of methane recovery in the digester, one more equation is needed, giving the total amount of methane produced in anaerobic digestion. This equation has been taken from Dick et al. (1978) :

$$CH_4 = r Q_i [s (BOD_i - BOD_e) - 1.42 S_d] \qquad (66)$$

in which CH_4 is the methane production rate, in Nm^3/day, r is a specific production factor, equal to 0.00840 $Nm^3/day/(mg/\ell \ BOD_u)/(m^3/h)$, Q_i is the flow rate through digester, in m^3/h, s is a factor (= 1.5) for converting BOD_5 to ultimate BOD (BOD_u), BOD_i and BOD_e are the influent and effluent five-days BOD, respectively, and S_d is the concentration of biomass effecting digestion. Assuming that

$$BOD_u \cong COD_b = L^* \tag{67}$$

that is, the biodegradable COD equals the ultimate BOD, and accounting for the fact that

$$S_d = Y_d (L_i^* - L_e^*) = Y_d (L_{23}^* - L_{24}^*) = S_{24}^* \tag{68}$$

in which Y_d = digester sludge yield factor, $\cong 0.044$ mgVSS/mgCOD, the following is obtained :

$$CH_4 = 0.00840 \, Q_{23} \, [(L_{23}^* - L_{24}^*) - 0.0625 (L_{23}^* - L_{24}^*)]$$

where CH_4 is expressed in m^3/day, or

$$CH_4 = 2.876 \, Q_{23} \, (L_{23}^* - L_{24}^*) \tag{69}$$

if CH_4 is expressed in m^3/year.

2.3.5. Physico-chemical treatment units

Models for the nonbiological treatment units of Fig. 2 will be briefly reviewed next. The primary settler performance is usually measured by the suspended solids removal. If we define ε as

$$\varepsilon = M_2^* / M_1^* \tag{70}$$

the suspended solids removal will be given by $1 - \varepsilon$ (in %). Most models for the primary settler are empirical and relate ε to the inflow rate, the settler area and the influent solids concentration. As an example, Voshel and Sak's model (1968) has been used in previous research by the author :

$$\varepsilon = 1 - a \frac{(M_1^*)^n}{(Q_1/A)^m} \tag{71}$$

where A is the settler area (in m^2), and a, m, n are empirical constants which have been measured for a standard domestic wastewater as a = 0.1395, n = 0.27, m = 0.22 (Voshel and Sak, 1968).

The final settler and the thickener (when using a gravity thickener) can be described through a semi-fundamental, semi-empirical model based on the limiting flux theory. The model also uses an empirical relationship :

$$v = v_0 \exp (-k_e M^*) \tag{72}$$

giving the settling velocity, v, in meters per second, as a function of the SS concentration, M^*. The "constants" v_0 and k_e have to be estimated in each particular wastewater situation, and belong to the category of parameters which are obviously influenced by biological variables (mainly the sludge age, $\alpha\ \theta_s$; see Bisogni and Lawrence, 1971). Unfortunately, the present state-of-the-art does not permit satisfactory modeling of these influences. If one assumes that v_0 and k_e have constant values, then the limiting flux theory implies that the settler area be at least equal to the following quantity (see Tyteca et al., 1977) :

$$A \geqslant \frac{M_4^* \ Q_5}{k_e \ v_0 \ M_L^{*2} \ \exp \ (-k_e M_L^*)} \tag{73}$$

in which M_L^* = the limit concentration, given by

$$M_L^* = \frac{M_{12}^* + \sqrt{M_{12}^{*2} - (4/k_e) \ M_{12}^*}}{2} \tag{74}$$

For perfect clarification to be achieved, the overflow velocity has to be lower than the initial settling velocity of inflowing solids, which gives another constraint on the area :

$$A \geqslant \frac{Q_6}{v_0 \ \exp \ (-k_e M_4^*)} \tag{75}$$

These relationships refer to the final settler as regards the variables subscripts. For the thickener, it suffices to replace in the equations M_4^* by M_{17}^*, M_{12}^* by M_{20}^*, Q_5 by Q_{17} and Q_6 by Q_{19}. It is usually accepted that thickening is by far the limiting step in the thickener, so that Eq. 75 does not need to be used for that unit. Values for the parameters v_0 and k_e have been taken in previous research (Tyteca, 1981) as 6.096 m/h and 6 10^{-4} ℓ/mg for the final settler, and 6.096 m/h and 2 10^{-4} ℓ/mg for the thickener, respectively.

The vacuum filter can be described by the Kozeny-Carman model which after expansion leads to (see Tyteca et al., 1977) :

$$L^2 = \frac{c^2 \ Q_{30}^2}{A^2} = \frac{2 \ P^{1-s} \ c \ y}{\mu \ r_0 \ \theta_t} \tag{76}$$

in which L = the applied filter load (in $g/m^2.h$), c = the concentration of filtered solids with respect to the filtrate volume (in mg/ℓ), A = the filter area (in m^2), P = the applied vacuum (in kg/cm^2), μ = the viscosity (centipoises), θ_t = the total filtration cycle time, y = the part of that time during which a given point is immersed, r_0 and s = empirical constants reflecting the cake specific resistance. As an example, the following values of the constants and parameters were considered : P = 7.35 psi = 5.17 g/m^2, μ = 1 centipoise, θ_t = 3 min, y = 0.7, s = 0.85, r_0 = 1.3 10^7 sec^2/g.

2.4. Energy requirements

In order to operate properly, the plant in Fig. 2 requires energy to be supplied for various purposes. In this study we will explicitly consider only the most significant energy needs, that is, aeration in the activated sludge basins and sludge or water pumping. Other energy needs, which mainly concern rotating devices in settlers, trickling filters or rotating biological contactors, directly depend on the size of the process unit. Therefore energy costs for these units are usually incorporated in the cost function (see section 3.2.2.), where the cost is expressed as a function of the process size.

Let us first consider activated sludge aeration requirements. Microbial growth in the aerator can be effected only in the presence of oxygen. The following equation can be used to relate the oxygen needs to the capacity of aeration equipment :

$$K_L a = C_1 \frac{S_4 L_4^*}{K_M + L_4^*} + C_2 S_4 \tag{77}$$

in which $K_L a$ = the oxygen transfer coefficient through liquid film (per hour), which reflects the capacity of the aeration devices, and C_1 and C_2 = parameters accounting for the oxygen concentration to be maintained, and the aforementioned kinetic and thermodynamic constants (Tyteca and Nyns, 1979, Tyteca et al., 1977). The air flow rate to be supplied in the aerator can be given, when adopting air diffusion devices, by (see Tyteca, 1979a, or Tyteca et al., 1977) :

$$G_s = C_3 (K_L a)^p V \tag{78}$$

in which C_3 and p = parameters involving various factors, such as the aerator depth and width. The installed power for aeration has to be proportional to G_s :

$$HP = C_4 G_s \tag{79}$$

When operating the activated sludge units at the design capacity, the energy required will be proportional to the quantity given by Eq. 79. If the units must be operated at a different steady-state, the value of G_s can be reevaluated from Eq. 77 and 78 and then substituted into Eq. 79 to yield the new energy requirements, provided the value of G_s does not exceed the design capacity. The values used in previous research for the parameters and constants appearing in these equations are as follows : C_1 = 0.015227 ℓ/mg.h, C_2 = 0.67957 10^{-3}

ℓ/mg.h, C_3 = 0.00365, C_4 = 1.7614, and p = 1.087.

The lift and recycle pumps can be designed on the basis of installed power. When neglecting head losses (which are not significant on the low distances prevailing in a treatment plant), the following equation reflects the installed power, W, as well as the energy requirements during a given steady-state :

$$W = \frac{\Phi\ Q\ Z}{\zeta} \tag{80}$$

in which Φ is the specific weight of water or sludge (kg/m^3), and may include a constant for unit conversion, Q is the flow to be recycled or lifted (m^3/h) over a height Z (m), while ζ is the pumping efficiency. In previous research the following values were adopted : Φ = 0.003704 (for converting into HP units) and ζ = 0.6.

2.5.   Process and effluent constraints

A set of constraints will complete the model of the treatment plant illustrated in Fig. 2. These include constraints on each process unit, originating from various biological, engineering, ... standpoints, and constraints on the required effluent waste concentrations, in order to satisfy given effluent quality standards. The latter constraints can be expressed simply, since our basic assumptions include perfect clarification of the final settler (Eq. 3). Hence, a constraint must be placed only on the value of the dissolved biodegradable load concentration :

$$L_{b_6} = L_{b_4} \leqslant \overline{L_b} \tag{81}$$

where $\overline{L_b}$ is a quality standard in terms of maximum allowable concentration of COD or BOD$_5$ (mg/ℓ).

No additional constraint must affect the activated sludge system previously modeled (section 2.3.1.). On the other hand, a limit can be placed on the maximum allowable hydraulic load on rotating biological contactors or trickling filters. Only the first situation will be considered here; the second is quite analogous. After Antonie (1979), the hydraulic load on RBC units should not exceed 4.5 gpd/sqft or 0.00764 m/h; therefore,

$$\frac{Q_3}{N\ n\ A} = \frac{Q_3}{40\ n\ A} \leqslant 0.00764 \tag{82}$$

where n is the number of stages, and Q_3 and A are expressed in m^3/h and m^2, respectively. A last constraint expresses that the diameter D of biodisks lies generally between 2.5 and 3.5 meters (Schroeder, 1977) :

$$2.5 \leqslant D \leqslant 3.5 \tag{83}$$

and since the lateral surface of one disk is given by (taking into account both sides of the disk) :

$$A = \frac{\pi}{2} (D^2 - d^2) \tag{84}$$

where d is the unsubmerged diameter of the disk, the following constraint is obtained on A :

$$\frac{\pi}{2} (6.25 - d^2) \leqslant A \leqslant \frac{\pi}{2} (12.25 - d^2) \tag{85}$$

where d appears as a parameter.

For the anaerobic digester, a constraint can be set for imposing a limit $\mathcal{L}_d$ on the weight load (in $g/m^3.h$) on the digester :

$$\frac{L_{23}^* Q_{23}}{V} \leqslant \mathcal{L}_d \tag{86}$$

For example, a value of 180 g $COD/m^3.h$ can be adopted for $\mathcal{L}_d$ (see for instance Berthouex and Polkowski, 1970).

The final settler and the thickener are constrained by the thickening ability of sludge. The sludge volume index (SVI) gives a measurement of the capacity of a given sludge to be thickened under standard conditions. The SVI yields a constraint on the maximum possible concentration in the settler bottom :

$$M_{12}^* \leqslant \frac{1.2 \ 10^6}{SVI} \tag{87}$$

in which the factor 1.2 accounts for the fact that the retention times in final settling are longer than the 30 min of the SVI test. The definition of M_L^* in Eq. 74 also implies that

$$M_{12}^* \geqslant \frac{4}{k_e} \tag{88}$$

The filter load is generally limited by a physical constraint of the type :

$$L = \frac{c \ Q_{30}}{A} \leqslant \mathcal{L}_f \tag{89}$$

One last constraint requires that the concentration of the filtered cake be higher than a given value, or that the cake be sufficiently dewatered in order to be properly conveyed to other sites (incineration, land-filling, dumping, ...) :

758

$$M^*_{29} = \frac{c\ Q_{30}}{Q_{29}} \geqslant (SS)_{min} \tag{90}$$

This constraint can alternatively be placed on flow 27 in Fig. 2, in situations where the filter is by-passed :

$$M^*_{27} \geqslant (SS)_{min} \tag{91}$$

3. COST STUDIES

Various trade-offs are possible within the plant depicted in Fig. 2, for which a steady-state model was developed and briefly described in previous sections. Such trade-offs, when correctly effected, can yield substancial cost savings. In order to study cost-effective configurations, two more mathematical tools must be developed (see Fig. 1). These are the cost functions, describing the cost of a process unit as a function of design or operating parameters (section 3.1.), and mathematical programming techniques, for searching the most cost-effective (from a design and/or operating standpoint) lay-out(s) of a given treatment plant (section 3.2.2.). In order to properly derive process cost functions, design and/or operating parameters and variables must first be identified for each of the processes composing a plant.

3.1. Process cost functions

The general form of cost functions used in treatment plants is

$$C = \sum_{i=1}^{n} a_i\ X_i^{b_i} \tag{92}$$

in which C = the total capital or operation and maintenance cost of a given treatment unit (in \$ or \$ per year), X_i = the design and/or operating parameters of that unit which most significantly influence the cost, a_i and b_i = estimated parameters, and n = number of terms in the sum. Table 2 gives the list of parameters used in calibrating Eq. 92, for each process unit appearing in Fig. 2 and modeled in section 2.

A crucial point lies in adequate data collection for determining the estimated parameters, the a_i's and b_i's. One methodology consists in collecting project cost estimates generated by a given equipment manufacturer. Such a method is valid at the local level and provided that the equipments will effectively be installed by the same manufacturer. Extrapolations to other regions or countries, other firms and other periods are rather daring. However, local data may be lacking; in such situations, cost escalation methods can be exploited, as the one proposed by Blecker et al. (1974). This method also provides for time escalations (whenever there is no other choice than exploiting old

TABLE 2

Design and operating parameters used to quantify process unit cost functions

Process unit	Parameters	
	Capital cost	Operation and maintenance cost
Activated sludge	V, G_s	V, G_s, HP
Trickling filter	A, D, Q_3	A, D, Q_3
Rotating biological contactor	A, N, number of stages, Q_3	A, N, number of stages, Q_3
Anaerobic digester	V	V, CH_4
Settlers and thickener	A	A
Vacuum filter	A	L (load)
Pumps	Q	Q, W

data, or conversely, when estimates for future periods are required).

Another methodology, much more widely used, is to collect data for various equipments previously installed, eventually at different periods, at a regional or national level, incorporating all manufacturers of the same type of equipments. The heterogeneity of data collected in such a manner is of course important and allows only for the generation of preliminary estimates. Such data cannot be considered to be specifically applicable to any particular situation. Moreover, even then one is not relieved from readjusting cost data with respect to time : price indexes are then best suited for this purpose (as will be discussed below).

As the scope of this chapter is rather general, the second methodology has been used. After collecting a set of cost data, the calibration of cost functions such as Eq. 92 involves the estimation of parameters (the a_i's and the b_i's) through regression analyses such as the least-squares method. A number of studies were published on the estimation of cost functions such as Eq. 92. One of the most authoritative is that of Patterson and Banker (1971). Although it was published in 1971, it is still the most attractive for our purpose, since more recent studies (Environmental Protection Agency, 1978 a and b, Metcalf and Eddy, 1975) either give cost equations as functions of global plant design parameters, instead of those shown in Table 2, or are based on too many scattered samples, leading in some cases to diseconomies of scale (the unitary cost is an increasing function of size), which is contrary to the usual observations.

For these reasons, use was made of the cost information provided by Patterson and Banker (1971) by deriving cost functions from the various cost diagrams given in that study. These cost functions, given in Table 3, are analogous to those used by Middleton and Lawrence (1975), derived from the same study. Since

TABLE 3

Investment and operating costs

k	Process unit	Investment costs, IC_k ($)	Fixed operating costs, FOC_k ($/year)	Variable operating costs, VOC_k ($/year)
1	Activated sludge	$461 \, \ell \, V^{0.71} + 6530 \, \ell \, G_s^{0.71}$	$53.9 \, w_{ma} \, G_s^{0.6}$	$121 \, w_{op} \, G_s^{0.55} + 65 \, P_c \, HP$
2	Rotating biological contactor	$7.51 \, \ell \, Q_3^{0.102} \, (nNA)^{0.898}$	$0.348 \, Q_3^{0.195} \, (nNA)^{0.805}$	
3	Digester	$2320 \, \ell \, V^{0.59}$	$2.31 \, w_{ma} \, V^{0.72} + 14.4 \, \ell \, V^{0.66}$	$4.14 \, w_{op} \, V^{0.71} - 0.0777 \, CH_4$
4	Settlers and thickener	$824 \, \ell \, A^{0.77}$	$9.23 \, w_{ma} \, A^{0.6} + 8.62 \, \ell \, A^{0.76}$	$17.1 \, w_{op} \, A^{0.6}$
5	Vacuum filter	$29200 \, \ell \, A^{0.71}$	$0.182 \, w_{ma} (cQ_{30})^{0.67} + 1.71 \, \ell \, (cQ_{30})^{0.71}$	$1.29 \, w_{op} (cQ_{30})^{0.65} + 0.481 \, \ell \, (cQ_{30})^{0.86}$
6	Sludge pumps	$9870 \, \ell \, Q^{0.53}$	$112 \, w_{ma} \, Q^{0.43} + 214 \, \ell \, Q^{0.64}$	$257 \, w_{op} \, Q^{0.41} + 65 \, P_c \, W$
7	Water pumps	$1710 \, \ell \, Q^{0.53}$	$0.0951 \, w_{ma} \, Q + 6.11 \, \ell \, Q^{0.8}$	$0.133 \, w_{op} \, Q + 65 \, P_c \, W$

the purpose of this chapter is twofold, i.e., studying the cost-effective design and the cost-effective long term operation, the operating costs have been
split into two categories indicated in Table 3, namely, the fixed operating
costs and the variable operating costs. Only the latter can be controlled in
current operation of a previously designed treatment plant. The fixed operating
costs comprise normal maintenance and repair costs, including material and supply, and are usually a function of the size of the treatment units. Once the
plant has been designed, these costs are a fixed part of the annual operation
and maintenance costs. On the other hand, the variable operating costs originate from expenses which directly depend on the rate at which the units are operated. They include wages for operating labor, and more importantly energy and
chemical costs.

While most of the cost information in Table 3 was derived from Patterson and
Banker, the costs indicated for the rotating biological contactor were taken
from Antonie (1979), whose data do not allow for a splitting between fixed and
variable operating costs. On the other hand, the variable operating costs for
the digester include a negative term indicating the monetary value of methane
recovered from sludge digestion. The value of 0.0777 $/m^3 was derived from Dick
et al. (1978).

The cost equations in Table 3 include a set of economical parameters, namely,
a cost index ι, affecting investment and material and supply costs, wage parameters w_{ma} and w_{op} for maintenance and operation, respectively, and the cost of
energy P_c. The values of these parameters were taken as follows in previous research (Tyteca, 1979, 1981). For the cost functions quantified after Patterson
and Banker (1971), thus excluding the RBC costs, use was made of the 1971 and
1977 values of the construction cost from the Engineering News Record, giving :

$$\iota = \frac{230}{140.5} = 1.637 \tag{93}$$

for estimating costs on a 1977 basis. The parameters w_{op} and w_{ma} indicate the
hourly wages for operation and maintenance, respectively, and were taken on the
same 1977 basis as

$$w_{op} = 11.4 \ \$/h \tag{94}$$

$$w_{ma} = 10.2 \ \$/h \tag{95}$$

The energy cost was taken, again only for illustrative purposes, as

$$P_c = 6.00 \ \text{cents/kWh} \tag{96}$$

This value is, of course, a question of debate. In his analysis of pumping systems, Deb (1978) uses a value of 3.00 ¢/kWh but mentions that the energy cost could rapidly increase to 10.0 ¢/kWh.

3.2. Cost optimization

3.2.1. Literature review

A number of studies have been devoted to the cost-effective design of waste-water treatment plants. Much less was made in the scope of cost-effective operation. Two important syntheses have been published on the optimal design. The first one, by Mishra et al. (1975), provides a complete survey of studies published before 1974. The other paper (Tyteca et al., 1977) includes a critical analysis of the principal investigations performed before 1976. An updated short synthesis is proposed hereafter (Table 4), including most works on the same subject up to 1982.

In Table 4, the "degree of difficulty" synthetically reflects the complexity of the mathematical relationships used in the various models. A value of 1 is adopted whenever the evolution of only one waste load component is described. A typical example is given by the empirical relationships describing the efficiency of a treatment unit as a function of its size, in terms of the percent BOD remaining after treatment. The value 2 symbolizes relationships including two components; a classical example is the Monod model, giving the concentrations of waste load and biomass resulting from biodegradation. A degree of 3 indicates that three components are taken into account, for example the organic load, the biomass and the nondegradable material load, or the nitrate concentration, etc. A value of 4 represents models which are still more complex, by including e.g. the dissolved and suspended biodegradable load, or incorporating the effect of factors such as temperature, heavy metals concentration, etc.

The order adopted in Table 4 reflects the increasing size of the problems to be solved : starting from studies on very simple systems (including only one treatment unit), Table 4 proceeds to more and more complex works, which in the end are devoted to a whole treatment plant. By doing so, one easily identifies a clear distinction between two research trends, as described in both of the aforementioned review papers. The first trend, which we will call the "Biologist - Mathematician" trend, is characterized by an attempt to describe the operation of the plant by using accurate mathematical models. The optimization phase is performed by simplistic techniques, which in some cases are even replaced by simple enumeration. The second trend, the "Economist - Operations Researcher" trend, is opposite : the main effort is on developing and demonstrating the use of powerful and/or sophisticated optimization techniques, while the operation of the plant is modeled through oversimplified relationships. Only in the last few years the gap between these two trends was progressively attenuated, by putting more and more the accent on the necessary adjustment between an

TABLE 4

Studies on cost-effective wastewater treatment

Author(s), year	System to be optimized	Degree of difficulty of the model	Optimization method	Goals and main conclusions
A. "Economist - Operations Researcher" trend				
Lynn et al. (1962)	2 types of primary settlers (differing by their overflow rates) + aerator + trickling filter	1	Problem equivalent to a transshipment problem, solved by a network algorithm	Detailed development of a general method for optimizing the combination of treatment processes with given BOD removal rate. Tested on a network with 4 treatment units.
Evenson et al. (1969a, 1969b)	Selection of the optimal treatment chain from various possible combinations of primary settler, chemical and biological treatment processes + vacuum filter or digestion	1	Dynamic programming	Elaboration of a dynamic programming algorithm for selecting the optimal treatment sequence. In the example of a cannery, the optimal sequence was found to be "some system of ponds". In case of very high influent waste load, it is preferable to treat in-site rather than in municipal plants.
Shih & DeFilippi (1970), Shih & Krishnan (1969, 1973)	Units to be selected from primary settlers, trickling filter, activated sludge, aerated lagoon, filtration, active carbon adsorption, thickening, vacuum filter, centrifugation, incineration, etc	1	Dynamic programming	Development of a method for computing the optimal sequence (like above). Examples for the cases of municipal and paper mill wastewaters, for illustrating the algorithm rather than providing an optimal design tool.
Schulte & Loehr (1971)	Selection of an optimal treatment chain from a given set (like above)	1	Dynamic programming	Illustrated by a duck farm wastewater. The most cost-effective sequence includes a primary settler, an aerated lagoon and a tertiary treatment (phosphate reduction). The flow rate and the phosphorus standards affect the optimal system choice and total cost.

TABLE 4 (continued)

Reference			Method	
Ecker & McNamara (1971)	Same as Shih & Krishnan (1969)	1	Geometric programming	Demonstrates the use of geometric programming in the optimal design of treatment plants (which is facilitated by the simplified form of equations). Same results as Shih & Krishnan (1969).
Marsden et al. (1972)	Illustrated by activated sludge + trickling filter + final settler, each with given size; only the arrangement is optimized	2	Non-linear programming algorithm from Graves, Pingry and Whinston	Application of the production theory. Gives the bases for an analytical approach of treatment cost functions. In the example reported, the optimal sequence is trickling filter + final settler.
Sterling (1976)	Selection of an optimal sequence with primary settler + biological treatment (activated sludge and/or trickling filters) + final settler	1	Dynamic programming	The main goal is to demonstrate the application of dynamic programming to the optimization of treatment plants, under given technological constraints.
Adams & Panagiotakopoulos (1977)	Basic network similar to Shih and Krishnan's (1969)	1	Network algorithm inspired from dynamic programming	Illustrated by Shih and Krishnan's example (1969). Compares the results obtained by the methods of Shih and Krishnan, Ecker and McNamara (1971) and the authors' algorithm. The advantages of the network method include flexibility, interactivity and higher efficiency, in addition to the similarity between the network and the treatment plant scheme.
Takama & Loucks (1981a, 1981b)	A first subsystem with "a number of treatment units" followed by a second subsystem with "control units to attain the maximization of system reliability"	1	Algorithm for multi-objective, multilevel optimization. Includes two interactive techniques: (1) generation of noninferior solutions, (2) search for a best compromise among them	The main purpose is to develop a solution method for multiobjective problems characterized by a hierarchical structure of decision-making and a large number of alternatives. The method is illustrated by a small example of wastewater treatment in the chemical industry.

TABLE 4 (continued)

B. "Biologist - Mathematician" trend

McBeath & Eliassen (1966)	Aerator (activated sludge) + final settler	2	Simulation	Sensitivity analyses. Variables most significantly affecting total cost : efficiency, suspended solids concentration in the aerator, influent flow rate and waste load.
Erickson & Fan (1968), Erickson et al. (1968); Erickson et al. (1968)	Aerators in series + final settler (no settling model)	2	Golden section search; modified simplex pattern search; discrete maximum principle	Studies of optimal flow mode. Step feed and plug flow are more advantageous than complete mixing in one basin.
Chen et al. (1970)	idem + account of uncertainty	3	Golden section search	Sensitivity analyses. The settler efficiency, the recycle rate and the specific growth rate most deeply affect the optimal design.
Lee et al. (1971)	Tower type activated sludge system	2	Golden section search; simplex pattern search	Study of the effect of partial sedimentation and back-flow on optimal design.
Naito et al. (1969), Fan et al. (1970, 1972)	Aerator + final settler; diffusion model (non-perfect flow)	3	Pattern search technique	Plug flow more advantageous than complete mixing. Effect of mixing increases under higher organic loads.
Erickson et al. (1977)	Aerator and final settler for clarification and thickening	2	Iterative procedure of the type trial-and-error	Effect of thickening on optimal design of the system. Consequences on the design in case of poor physical and/or biological behaviour.
Heydweiler et al. (1977a)	Tower-type activated sludge system + possibility of step feed	2	Trial-and-error + linearization by Taylor series	Advantage of increasing the number of stages in a tower-type system.
Heydweiler et al. (1977b)	idem + diffusion model (partial sedimentation)	3	idem	Sedimentation in the tower allows reducing the total volume and the settler area. Ignoring this sedimentation leads to a conservative design.

TABLE 4 (continued)

Lauria et al. (1977)	Aerator + final settler	2	Setting partial derivatives equal to 0 + Newton-Raphson procedure	Importance of thickening; importance of the effect of kinetic parameters on sludge settling. Advantage of adopting a simple optimization technique.
Keinath et al. (1977)	Aerator + final settler	2	Graphical representation in 2 dimensions	Extensive study of the limiting flux theory and its implications on design and operation.
Lee et al. (1976)	Final settler with two or three stages	2	Random search technique	Extensive study of the effect of various factors (resuspension, density currents, etc) on optimal design; advantage of a multistage settling system.
Galler & Gotaas (1964, 1966), Gotaas & Galler (1973, 1975)	Trickling filters	2	Linear programming	Comparison of various trickling filter models; study of the effect of various factors (depth, recirculation, filter radius, etc) on efficiency.
Fan et al. (1973)	Two stage anaerobic digestion	4	Simplex search technique	Two stage system much more advantageous than one stage system; increasing recycle rates increases efficiency and methane production.
Tikhe (1976)	Chlorination system	2	Setting derivatives equal to 0 + Newton-Raphson procedure	Optimal chlorine amount is a function of the volume of water to be treated; treating unsettled effluents is more expensive than settled effluents.
Mishra et al. (1973)	Primary settler + activated sludge system - diffusion model	3	Simulation	On the basis of construction costs alone, the primary settler is not advantageous. Addition of flocculant decreases total cost.
Fan et al. (1974), Kuo et al. (1974)	Primary settler + activated sludge system including the "masking effect" of suspended solids	3	Golden section search + pattern search technique + dynamic programming (2 steps)	The optimal size of primary settler increases with the masking effect and with the dissolved waste load. Addition of flocculant considerably affects the optimal design.

TABLE 4 (continued)

Reference	System		Method	Comments
Grady (1977)	Primary settler + activated sludge system	3	Dynamic programming	Interest of dynamic programming as an especially didactic tool, allowing for the easy comparison of the optimum with neighbouring solutions.
Hughes (1978)	Activated sludge system (with final settler) + two-stage sludge digestion + sludge drying beds	2	The Dembo algorithm for generalized geometric programming	Example of applying geometric programming to a reasonably realistic system described by a simple model. Various sensitivity analyses. Optimal design is most sensitive to changes in the costs of aeration tank, settling tank, and sludge disposal facilities.
Gemmell & Smith (1982)	Preliminary treatment (screens, detritors, comminutors), primary settler, activated sludge system, sludge treatment (thickening, anaerobic digestion)	2	Interactive computer program incorporating conventional design methods	The goal is twofold : to provide an interactive method allowing the user to test various design alternatives, and to remain at a reasonably simple level to keep control on the method. Will be further developed to incorporate more units.
Berthouex & Polkowski (1970)	Primary settler + activated sludge + anaerobic digester + propagation of variance	3	Enumerative pattern search technique	At the optimum, independence of the sludge treatment system with respect to the water treatment system. Accounting for uncertainty implies 3 to 8 % increase of optimal cost.
Tarrer et al. (1976)	Primary settler + activated sludge + chlorination + sludge treatment considered as a whole + uncertainty	4	Conjugate gradient (Powell; 2 degrees of freedom) + golden section search	The optimal value of sludge age results from settling constraints. Prime importance of a clarification model for the final settler. Accounting for uncertainty implies little change in total cost ($\leq 8\%$) but the optimal design is significantly affected.
Fan et al. (1973a), Mishra et al. (1973a, 1973b, 1974), Fan et al. (1975)	Trickling filter + activated sludge system + flow distribution	3	Structural parameter method + modified simplex pattern search	Technique for the optimal expansion of an existing treatment plant. The most economical system excludes the trickling filter (except for some particular values of parameters). The system with trickling filter, however, is recommended since it is more

TABLE 4 (continued)

				stable, better absorbs shocks and improves settling.
Smith (1968, 1969)	Primary settler + activated sludge (including nitrification) + thickener + anaerobic digester + vacuum filter + incinerator	4	Simulation	First attempt of complete modelling of a whole treatment plant. The sludge treatment system is the most expensive part of the plant.
Parkin & Dague (1972)	Primary settler + activated sludge + digester (aerobic or anaerobic)	3	Enumeration	The least-cost solution includes the least efficient primary settler; aerobic and anaerobic digesters lead to similar costs.
Middleton & Lawrence (1973, 1974, 1975, 1976); Craig et al. (1978)	Primary settler (last three studies) + activated sludge + thickener + digester + vacuum filter	3	Graphical representation in 2 dimensions. Box-complex algorithm (the last study)	Selection of an optimal "region" (rather than an optimal point) is recommended and facilitated by graphical representation. Cost is relatively unsensitive to sludge age; hence high sludge ages are recommended for safety reasons. At the optimum, independence of the sludge treatment system with respect to the liquid treatment system.
Stenstrom (1975), Stenstrom & Andrews (1980)	Primary settler + activated sludge (step feed; nitrification) + chlorination + thickener + anaerobic digester	4	Dynamic simulation	Study of the possibilities of optimal control of a whole treatment plant. SCOUR (specific oxygen utilization rate) is recommended as a control parameter rather than the sludge age. Existence of significant interactions between the two treatment chains (liquid, sludge). Adequate selection of mean cell retention time can result in significant operational cost savings, and can be achived in most existing treatment systems without or with few design changes.

TABLE 4 (continued)

Reference		Method	Process units	Comments
Dick & Simmons (1976), Dick et al. (1976, 1978)	4	Sequential method based on computation of the first and second derivatives; use of penalties	Primary settler + activated sludge + thickener + anaerobic digester + chemical conditionning + vacuum filter + sludge transportation + discharge (includes nitrogen, phosphorus and heavy metal compounds)	Study of optimal integration of sludge treatment units in a treatment plant and in regional projects for sludge management. Primary settling appears little economical. Study still in progress; attempts to include the effect of various interactions between the process units.
CIRIA (1973, 1975), Bowden et al. (1976)	4	Pattern search technique	Primary settler + activated sludge or trickling filter + tertiary treatment + thickener + anaerobic digester + chemical conditionning + press filter or drying beds	The goal is to elaborate and promote a computer code for the optimal design of whole treatment plants, leaving to the user the possibility of specifying the selection and combination of units as well as the value of various parameters (among which are cost parameters). Work still in progress.
Voelkel (1978)	4	Golden complex search (simplex search improved by the author)	Primary settler + activated sludge + gravity and flotation thickeners + anaerobic digester + chemical conditionning + vacuum filter + sludge discharge	Study on the sensitivity of optimal design and cost to various parameters and variables. The most significant parameters are found in settling and dewatering properties. The variables inducing the highest sensitivity are the influent characteristics, the oxygen concentration at saturation and the temperature.
Rossmann (1980)	> 4 see comments	Implicit enumeration coupled with a penalty method	A set of 12 possible treatment units from raw wastewater pumping to chlorination and sludge disposal (15000 possible system configurations). The components are selected from a list of candidate process units with fixed design characteristics.	Allows for the selection of processes whose combination was not fixed in advance but with fixed design characteristics. A strategy for refining the design levels is proposed by solving the problem in two phases.

efficient optimization technique and a sufficiently realistic mathematical description of the plant.

A few more remarks can be made about recent studies mentioned in Table 4 :

- The study by Voelkel (1978) represents one of the most advanced attempts to apply a relatively simple optimization technique (golden simplex search) to a plant described by a realistic and rather complex model. This attempt apparently led to a saturation of the method, since a number of erratic results were observed during sensitivity analyses, which were the main object of that work. In some cases the method was even unable to yield any feasible solution. This may correspond to the impossibility to obtain a global optimum, an unsufficient normalization of the model, or more probably to the unadaptedness of the optimization technique.

- Another significant fact is provided by the study by Rossmann (1980). In this case, the description of the plant has been made as realistic as possible, by incorporating e.g. as many as 18 components in the waste stream vector. As a consequence other parts of the method have been condemned, such as the possibility of selecting the process design characteristics, which are given predetermined values at the outset.

- In contrast with these attempts towards hyper-realistic description of the plant, it is interesting to note that in 1982 there is a trend to come back to simple models and simple methods, where the accent is on allowing the user to keep control on the method, as well as providing an efficient interactive way to design a plant (Gemmell and Smith, 1982).

The research conducted by the author and co-workers has been conceived as an attempt to reconcile the two aforementioned trends ("Biologist - Mathematician" and "Economist - Operations Researcher"). This was performed through use of a reasonably realistic mathematical model (of the type described in section 2., leading to a "degree of difficulty" of 4 in the scale used in Table 4), together with an efficient and powerful optimization technique, especially well adapted to the model structure. Additionally, the method has been made as interactive as possible by adopting a structure which considerably eases the data manipulation. The following sections will be devoted to a brief presentation of this method (sections 3.2.2. - 3.2.4.) and examples of application (section 4.).

3.2.2. Optimization method

As can be seen in Table 4, practically the whole range of optimization techniques used in the field of Operations Research have been tested for solving the problem of optimally designing a wastewater treatment plant. Most of these techniques are appropriate whenever the model describing the plant is not too complex and the selected plant structure corresponds to some remarkable characteristics. Thus, for example,

- Linear programming techniques will be useful if the model includes only

linear relationships, which in some cases can be obtained through approximation of nonlinear equations.
- Dynamic programming is of interest when there are not too many cycles and
  feedbacks in the hydraulic pattern.
- Network algorithms are useful when only the flow distribution within the
  process units has to be optimized, the size of each unit being predetermined.
- Heuristics, exploration or implicit enumeration techniques can provide
  very efficient methods, but as has been discussed they may lead to "saturation" and divert from the global optimum, especially when the size of the
  model becomes important.

The highly nonlinear model presented in section 2., corresponding to the plant
scheme in Fig. 2, which includes various cycles and return flows, and in which
furthermore the design of each unit has to be specified, precludes the use of
the three first classes of methods. The nonlinear structure of the model and
the hope to obtain a global optimum in an efficient way naturally leads to take
nonlinear programming methods into consideration. There exist various efficient
nonlinear programming techniques which were coded and are widely available on
most computers. One of the most powerful is the generalized reduced gradient
technique (GRG). In this research we have used the GRG algorithm implemented by
Abadie (1978).

Additionally, the model has been transcribed as a geometric program in order
to take advantage of the facilities provided by this formulation, for function
and gradient evaluation as well as for data manipulation, which in turn allows
for a very easy interactive use. Geometric programming was successfully applied
to a broad class of relatively simple engineering problems, including chemical,
civil, mechanical, ... engineering (see Rijckaert, 1973). It was used in treatment plant design in only two instances : first by Ecker and McNamara (1974),
whose model for the treatment units was rather simplified, and more recently by
Hughes (1978), who conducted an interesting analysis of the activated sludge
system. In all those studies, the problems tested have in common a relatively
moderate size (about 10 to 15 variables). In order to maintain the chapter within reasonable limits, only a brief presentation of geometric programming and
the way it was exploited will be proposed hereafter; the interested reader is
refered to Tyteca and Smeers (1981) or Smeers and Tyteca (1984) for further
details. An important part of the text below comes from Tyteca and Smeers (1981).

The main tools of geometric programming (see e.g. Duffin et al., 1967) are
mathematical expressions called posynomials, which can be defined as weighted
sums of products of variables raised to given powers. The general form of a
posynomial in the variables x_j is

$$f(x) = \sum_{i=1}^{m} C_i \prod_{j=1}^{n} (x_j)^{a_{ij}} \tag{97}$$

in which all the C_i coefficients are positive, while the exponents, a_{ij}, can take any real value. A typical geometric program can be written as :

$$\text{minimize} \quad g_0(x) = \sum_{i=1}^{m} C_{io} \prod_{j=1}^{n} (x_j)^{a_{ijo}} \tag{98}$$

$$\text{such that} \quad g_k(x) = \sum_{i=1}^{m_k} C_{ik} \prod_{j=1}^{n} (x_j)^{a_{ijk}} \leqslant 1 \qquad k = 1, 2, \ldots, p \tag{99}$$

$$x_j \geqslant 0 \qquad j = 1, 2, \ldots, n \tag{100}$$

The class of geometric programming can be extended by allowing, in Eq. 99, the C_{ik} and the right-hand side to be nonpositive. The obtained expressions are then called signomials and the resulting class of mathematical programs is denoted as signomial programming. Signomial programming provides a very general class of nonlinear programs, in the sense that all mathematical programs involving only the four algebraic operations +, -, x and : can be readily transformed into signomial programs. From a mathematical point of view, signomial programs are essentially different from posynomial programs, in the sense that posynomial programs can be shown to be convex up to a monotonic transformation of the variables. As a consequence, every local optimal solution of a posynomial program is global. On the other hand, combinatorial procedures must be called upon in order to guarantee a global optimal solution of a signomial program.

Various procedures have been presented in the literature for approximating signomial programs by posynomial ones. These proposals will not be examined here; only some of the tests that have been performed in this work will be briefly mentioned. Two versions of the problem have been constructed. In the first case, an approximating "convexified" problem was constructed on the basis of a proposal by Charnes and Cooper (1966). The approach followed is to first reduce all negative-signed terms to 1-x expressions (in which x may be a single variable or a product of two or more variables). Exploiting this method was found (Smeers and Tyteca, 1984; Tyteca, 1979a) to be the most convenient as well as the most efficient way of isolating the "irregularities" or nonconvexities caused by negative signs, which can divert the problem from a global optimum. These 1-x expressions can then be readily approximated by posynomials. It was found (Smeers and Tyteca, 1984; Tyteca, 1979a) that the best approximations are provided by single-term posynomials; i.e., positive monomials : $1 - x \cong a \, x^b$. The approximations, however, were found to be satisfactory only on a small interval. The second version of the model was the natural signomial problem. From various numerical experiments made with both versions of the model, it was not found advantageous to go first through an approximate model for finding a local optimum. The direct optimization of the original problem (in signomial form) always proved to be superior, from the point of view of both numerical efficiency and accuracy of the solution found. A full examination of this problem would lead us too far from the object of the chapter. Different tests were made to assess the globality of the optimum by changing the initial point in the optimization procedure. The optimal point found was identical in all cases, and thus, one can suspect that global optimality was indeed reached by the procedure. This conclusion is reinforced by the smoothness of the variations in the optimal solution and objective function during sensitivity analyses (see section 4.4.). On the other hand, the exponential functions encountered in the model (Eqs. 73 and 75) were replaced in all situations by posynomial expressions, which provide quite satisfactory approximations (less than 0.1 % error on large intervals).

As mentioned before, one can take advantage of the geometric programming formulation of the model, even when using a general purpose code such as GRG. Nonlinear programming codes usually require user-provided routines to compute the various functions appearing in the objective and in the constraints, as well as their derivatives. These functions can be of a very general nature, and thus the corresponding routines are specific to the model. They may also be difficult to code because derivatives usually exhibit complex analytical expressions. This is especially damaging if the model is intended for design purposes; indeed different configurations of the process units lead to different models. Similarly, modeling of wastewater treatment operations has not reached such a state that existing models can be taken as the final word in the field. In particular, slight changes in the process equations may require a complete recoding of the routines for computing the function and gradient values. This drawback is bypassed by first reformulating the problem as a geometric program. Indeed the functions appearing in those programs exhibit a systematic structure (Eqs. 98-100) that considerably eases the data manipulation and makes the user provided routines of a nonlinear programming code quite easy to write; moreover once coded, these routines can be used for all models that will be constructed on the same scheme. In this way, the optimization technique is not rigidly bound to the model structure.

The main difficulty in transcribing a given equation into geometric form (Eq. 99) is in replacing it by an inequality; thereafter it suffices to rearrange the terms on the left-hand side, so as to leave only the expression "$\leqslant 1$" on the right-hand side. No theoretical rule is available for choosing the sense of the inequality, accounting for the fact that, in the optimal solution, the inequality has to hold with equality. For this reason, one has to use empirical rules, which have the principle to proceed against the spontaneous evolution of the variables. Let us take Eq. 14 as an example. If we consider a rotating biological contactor, in which case $Q_{13} = 0$ (see section 2.3.3.), then, accounting for Eq. 44,

$$Q_2 + Q_8 + \rho\, Q_2 = Q_3 \qquad\qquad (101)$$

which we write as

$$Q_2\, Q_3^{-1} + Q_8\, Q_3^{-1} + \rho\, Q_2\, Q_3^{-1} = 1 \qquad\qquad (102)$$

and finally

$$Q_2\, Q_3^{-1} + Q_8\, Q_3^{-1} + \rho\, Q_2\, Q_3^{-1} \leqslant 1 \qquad\qquad (103)$$

which has exactly the form of Eq. 99. It is easy to show that Eq. 103 will hold with equality in any optimal solution to the problem. Indeed, Q_3 appears only with a positive exponent, a positive coefficient, and therefore with a positive derivative, in the cost function to be minimized (see Table 3). Thus, any solution in which Eq. 103 would hold with strict inequality could be improved by decreasing Q_3.

By generalizing the preceding example, it can be submitted that if a given variable x appearing in the objective function can be isolated in the right-hand side of an equation, one will write :

$\dots \leqslant x$ if most of the exponents of this variable in the objective function are positive,

$\dots \geqslant x$ if most of the exponents of this variable in the objective function are negative.

The validity of this principle for the particular problem on hand must obviously be verified by checking that these inequality constraints really hold as equalities in the optimal solution. When this is not the case, an adequate solution can often be found after a few trial-and error steps by changing the inequality senses.

In the same way as the preceding, the whole model can be rewritten as a geometric program. This transcription is rather systematic and straightforward for most of the equations and constraints. The interested reader is refered to Tyteca (1979a) or Tyteca and Smeers (1981) for additional examples.

3.2.3. Cost-effective design of treatment plants

A model for the cost-effective design of a wastewater treatment plant can be formulated as follows :

- Minimize the objective function, given as a weighted sum of capital and operation and maintenance (O & M) costs :

$$TC = \sum_{k=1}^{N} IC_k (X_k) + \Gamma \sum_{k=1}^{N} [FOC_k(X_k) + VOC_k(X_k)] \qquad (104)$$

in which X_k = the design and/or operating parameters of unit k; TC = the total discounted cost to be minimized; IC_k = the capital or investment cost for unit k; FOC_k and VOC_k = the fixed and the variable operating costs of unit k, respectively (IC_k, FOC_k and VOC_k are given by expressions appearing in Table 3); N = the total number of cost units to be considered; Γ = a discount factor, defined as a function of the discount rate i and the number n of years in the planned horizon :

$$\Gamma = \sum_{j=1}^{n} \frac{1}{(1+i)^j} = \frac{1 - (1+i)^{-n}}{i} \qquad (105)$$

Situations in which the O & M costs are not constant over the planned horizon, and more especially when these vary during the year, can be dealt with by replacing the preceding discount factor by a continuous discount function. This would not modify the form of the objective function given in Eq. 104.
- Subject to the following set of constraints : Flow and mass balance relationships (section 2.2.), Treatment process relationships (section 2.3.), Energy requirements (section 2.4.), Process and effluent constraints (section 2.5.).

In this version of the problem a steady-state is assumed for the whole treatment plant; the O & M costs (i.e., FOC_k and VOC_k appearing in Eq. 104) are assumed to be constant over the plant lifespan. Therefore a critical point lies in the correct determination of the inflow parameters and the effluent quality level to be maintained. Indeed designing the plant based on averaged influent components would be unsufficient during high flow and/or waste load conditions, while incorrect specification of the effluent quality level may lead to over- or underdesign. There exist various procedures for determining the influent and effluent conditions on which the plant should be designed, taking into account sufficient controlability for coping with influent variablity. Among these procedures are the safety factors, the design based on the use of equalization basins, the explicit consideration of uncertainty and the use of probabilities, the simulation of the influent characteristics with time series analysis, or long term analyses specifying when additional capacities should be installed. It would lead us too far to examine each of these procedures; however they are compatible and could be tested with the model formulation as given in previous sections, provided adequate data are available. In this chapter, only the first procedure, the safety factors, will be exploited for illustrative purposes in section 4.

3.2.4. Cost-effective operation of treatment plants

There exist a number of studies devoted to dynamic simulation of wastewater treatment plants submitted to various types of shock loads or influent variations, as reported for example in a book edited by Andrews et al. (1974). In such studies, however, the emphasis is on the automation or the control of day-by-day variations to avoid plant failures. The cost-effective operation is seldom the main preoccupation in these cases, though it must be recognized that considerable cost savings may result from adequate control of the plant. Dynamic modelling is not the speciality of the author and will therefore no longer be discussed in this chapter; the interested reader is refered to other parts of the book where this subject is dealt with in detail.

In this section, however, we will briefly mention how the model previously developed can be exploited for studying the cost-effective operation of a wastewater treatment plant. If we consider a given plant with predetermined design characteristics, it can be operated at various possible steady-states which can be significantly different from the steady-state considered in the design phase, as a consequence of variations in the influent characteristics, other specifications for the effluent quality level, or modifications in the operating scheme (e.g., changes in the recycle or sludge wasting rates). Therefore one can be interested in cost-effective operation of the plant through an adequate readjustment of the operational variables in response to various possible changes with respect to the initial design conditions (Smeers and Tyteca, 1985).

In order to study the cost-effective operation, we will consider some of the design variables of the model as fixed, and allow the other variables to be recomputed through the same optimization scheme as previously. Once the optimal size of the treatment units has been computed for a given design capacity, the following design variables are fixed : the areas of the primary and final settlers, the thickener and the vacuum filter, the volumes of the biological treatment units, the capacity of the air blowers in the aerator (when adopting activated sludge), and the capacity of the recycle and lift pumps. When attempting subsequently to optimize the variable operating costs, these variables are simply replaced by their design values in each constraint of the model where they appear. Some of these variables, however, reflect maximum capacity values, namely, the capacities of the blower and the pumps. Therefore, these variables can be readjusted in operating the plant, to any level below their maximum design values.

The most significant change in the model affects the objective function,

that is, the total cost to be minimized. For our purpose, the total costs of
the plant have been split into three categories : the capital costs, the fixed
operating costs, and the variable operating costs (Table 3). The distinction
between these cost categories and the exepenses included in each of them has
been explained in section 3.1.

To summarize, the proposed model for studying the cost-effective operation
of the plant can be formulated as follows :

- Minimize the objective function, given as the sum of variable operating
  costs (Table 3) over a given period (one year, one season, one month, one
  week, ...) during which a given steady-state (with constant influent and
  effluent characteristics) has to be maintained :

$$TC = \sum_{k=1}^{N} VOC_k (X_k) \tag{106}$$

  in which X_k = the operating variables of process unit k; VOC_k = the variable operating cost of unit k over the considered period; N = the total
  number of units to be considered.

- Subject to the same constraint set as previously (section 3.2.3.), in
  which the design variables (areas and volumes) are replaced by their adopted design value.

An example of use of this formulation will be given in section 4.3.

4.   EXAMPLES

4.1.   Optimal design of an activated sludge plant

   4.1.1. Introduction

This section briefly reports on application of the model presented above to
the optimal design of a wastewater treatment plant including six treatment
units, among which an activated sludge process (Fig. 3). The results indicated
below come from various studies by the author and coworkers (Tyteca, 1979a,
1979b, Smeers and Tyteca, 1984) but mainly form the paper by Tyteca and Smeers
(1981).

When adopting the scheme of Fig. 3, the reformulation of the model as a geometric program leads to a problem with 33 variables and 35 constraints in the
form of Eq. 99. The model includes 138 monomials of the form

$$C_{ik} \prod_{j=1}^{n} (x_j)^{a_{ijk}}$$

34 of which appear in the objective function (Eq. 98) and 22 others have a negative sign (see the discussion of section 3.2.2.). The model in final geometric

778

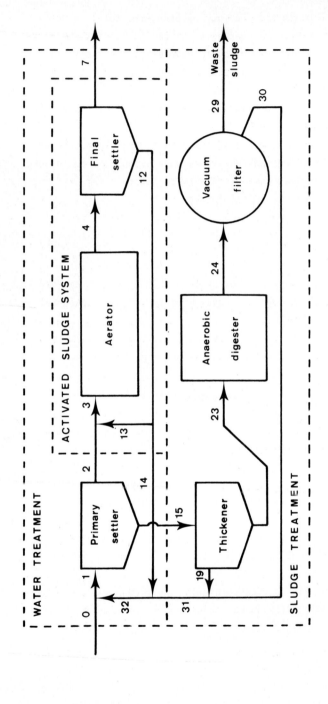

Fig. 3. Outline of the activated sludge treatment plant (Tyteca and Smeers, 1981).

form is given in full in Tyteca (1979a) and Smeers and Tyteca (1984). Twenty-six of the 35 constraints originate from equalities, and will therefore have to hold with equality at the optimal solution.

4.1.2. Sample situation

The aforementioned method has been applied to the case of an urban community generating essentially domestic wastewaters, with a small fraction originating from schools, hotels, stores, and a slaughterhouse. The wastewater collecting system is combined (rainwaters are collected in the same sewers as wastewaters), and the total volume amounts to approximately 30,000 equivalent-inhabitants. Other data are given in the following : (1) Daily flow - 5,400 m^3/day; (2) Dry weather diurnal average flow - 300 m^3/h; (3) Dry weather peak flow - 900 m^3/h; (4) Storm flow - 1,800 m^3/h; (5) Polluting load - L^* (COD) - 1,000 mg/ℓ and M^* (SS) - 500 mg/ℓ; and (6) effluent standards - L^* (COD) $\leqslant$ 30 mg/ℓ, and M^*(SS) $\leqslant$ 30 mg/ℓ. From Eqs. 4 to 6 (section 2.1.), one can easily derive the following values for the influent waste load components : L_{b_0} = 500 mg/ℓ, M_{bs_0} = 300 mg/ℓ, M_{s_0} = 200 mg/ℓ.

As a benchmark for the model, use was made of a treatment plant design as proposed by a specialized engineering firm. The plant was intended to provide complete treatment in dry-weather peak-flow conditions. The design proposed by the firm did not specify the value of all variables considered in the model; only the usual design variables were quantified. As an example, the primary settler was given an area of 360 m^2 and the activated sludge tank, a volume of 2,400 m^3. This solution had been obtained by empirical design methods and did not involve any modelling approach. In order to make it comparable to the solution generated by the optimization algorithm, the data provided by the firm were completed by using the equations of the model (Tyteca, 1979a). These computations were effected in three different cases :

1. Dry-weather peak flow (900 m^3/h), full load (L_b = 500 mg/ℓ, M_{bs} = 300 mg/ℓ, M_s = 200 mg/ℓ).
2. Dry-weather peak flow (900 m^3/h), 2/3 of the load (L_b = 333 mg/ℓ, M_{bs} = 200 mg/ℓ, M_s = 133 mg/ℓ).
3. Intermediate flow (600 m^3/h), full load (L_b = 500 mg/ℓ, M_{bs} = 300 mg/ℓ, M_s = 200 mg/ℓ).

The reason why the two last cases were tested is that the first case, being based on peak flow conditions at full load, could clearly lead to overdesign. As discussed above (section 3.2.3.), there is no theoretical means of determining the exact influent conditions on which the plant should be designed. The three cases proposed above will be tested for comparing three ways of adopting safety factors. One can note that the second and third cases give the same hourly pollutant load (i.e., L_b x Q = 333 x 900 = 600 x 500 = 300,000g/h). The three initial solutions obtained in the three cases appear in Table 5.

TABLE 5

Comparison of initial and optimal solutions obtained in cases 1, 2 and 3 (sample situation) (Tyteca and Smeers, 1981)

Variable and meaning	Case 1 (Q = 900, L_b = 500)			Case 2 (Q = 900, L_b = 333)		Case 3 (Q = 600, L_b = 500)	
	Initial solution	Optimal solution	Optimal solution ($\theta \leqslant 2.67$)	Initial solution	Optimal solution	Initial solution	Optimal solution
A_{ps}, primary settler area, m²	360	101	252	360	14.9	360	49.8
A_{fs}, final settler area, m²	776	281	790	465	267	343	190
A_{th}, thickener area, m²	591	243	239	52.8	164	261	162
A_{vf}, vacuum filter area, m²	71.7	74.7	74.9	71.7	49.1	48.2	49.8
K_La, oxygen transfer coefficient, h^{-1}	11.3	4.45	11.5	7.76	4.00	7.38	4.54
M_1^*, suspended solids before primary settler, mg/ℓ	905	1070	931	640	1050	825	1120
M_4^*, suspended solids in aerator, mg/ℓ	2770	1070	2770	1910	987	1780	1100
M_{24}^*, suspended solids in digester, mg/ℓ	19800	17500	17000	19800	18600	17500	17600
Q_1, flow rate to primary settler, m³/h	1000	1010	1000	1000	1010	680	675
V_d, digester volume, m³	6800	2920	3000	6800	1790	2590	1930
Q_{30}, filtrate flow rate, m³/h	14.0	15.6	15.9	14.0	9.90	10.0	10.3
$1 - \epsilon$, primary settler efficiency, %	70.0	55.4	65.2	63.8	36.1	74.3	52.4
θ, hydraulic retention time, h	2.67	7.20	2.67	2.67	5.79	4.00	7.14
ρ, recycle ratio, %	58.6	6.23	55.2	31.5	3.11	22.4	5.85
π, wasting ratio, %	6.70	8.08	7.24	5.00	9.39	5.00	8.78
TC, total cost, 10³ $	6797	5777	6048	6035	4474	4802	4299
Computer time, min : sec	-	2:35	4:02	-	4:22	-	2:04

4.1.3. Results

The model was run on the three cases, with the solutions indicated in Table 5 as initial solutions. The optimal solutions appear in the same table. Costs are computed from the aforementioned cost functions in initial as well as optimal solutions. Therefore, direct comparison between both is possible, allowing for an appreciation of the cost reduction effected by the optimal design algorithm with respect to the conventional design methods. The cost reduction in the three cases amounts to 15.0 %, 25.9 % and 10.5 %, respectively. This is quite typical of optimization methods as applied to engineering problems. The same range of cost reduction was observed by Dick et al. (1976, 1978) in their studies on optimal integration of sludge process units. This appears as a significant reduction with respect to conventional design. In absolute numbers, the economy is quite appreciable, since, for a medium size treatment plant (30,000 equivalent-inhabitants), this reduction may reach $ 1,000,000, or even $ 1,500,000. The computer time (Table 5), ranging from 2 min to 4 min, can be considered as negligible compared to such amounts.

The cost reduction was accomplished through various changes in design. The following significant trends can be observed with respect to the initial solution : (1) the final settler area A_{fs} and the aerator suspended solids M_4^* decrease; (2) the primary settler area A_{ps} drastically decreases, accompanied by a reduction of its efficiency, $1 - \varepsilon$; (3) variables reflecting energy expenses, such as the oxygen transfer coefficient $K_L a$, and the recycle ratio ρ, decrease; (4) the dramatic decrease of ρ is compensated by an increase of the wasting ratio π, which indirectly restores recycling through the transmission of wasted solids to the plant influent (see Fig. 3) and the reduced efficiency of the primary settler; and (5) the hydraulic retention time θ and thus the aerator volume V_a increase. These design changes were also such that process units on which two different constraints were imposed (i.e., final settler : clarification and thickening; digester : organic content reduction and load limitation; vacuum filter : cake moisture reduction and minimum required solids concentration) were optimally designed in such a way that both of these constraints are exactly satisfied (verified with equality) (Tyteca, 1979a, Smeers and Tyteca, 1984).

The proposed optimal solutions may be such that some practical or operational constraints, not accounted for in the primary version of the model, are not satisfied. The optimal values obtained for the retention time θ, e.g., are on the order of 6 h to 7 h, with the influent flow rate maintained to its peak or intermediate value. This implies that, under mean flow conditions, the hydraulic retention time returns to values that may exceed 20 h. Therefore, the model was also run, in the first case, with θ constrained to be less than 2.67 h, which corresponds to the initial design peak value (Table 5). Many variables then go back to values nearby the initial solution (A_{fs}, K_La, M_u^*, ε,ρ), while the cost reduction remains significant (11.0 % instead of 15.0 %). Computations not shown in Table 5 indicate that when θ is constrained to be less than 4 h in peak conditions (or 12 h in mean flow conditions), the cost reduction is 13.9 %, which is close to the 15 % of the first solution (Tyteca, 1979a).

As a consequence, it should be emphasized that the proposed optimization technique is not a rigid tool; any imperfection appearing in the "optimal" solution can be easily corrected by manipulating the constraints of the model, without substancially affecting the cost reduction.

4.2. Optimal design of a rotating biological contactor treatment plant

4.2.1. Introduction

A model has been calibrated for a plant including rotating biological contactor (RBC) units, corresponding to the hydraulic pattern shown in Fig. 4. Two examples have been tested, the first with four RBC stages, the second with five RBC stages. Recycling is effected from the effluent (see Fig. 4) and not from the thickened sludge as was the case for activated sludge. The model includes 13 + n equality constraints (n being the number of stages), 7 inequality constraints, and 19 + n variables. The model as well as the results indicated below are described in more detail in de Morais and Tyteca (1983).

4.2.2. Sample situation

The model was run with the following influent data, corresponding to domestic wastewater with standard composition :
- flow rate : Q_0 = 20 m³/h, 50 m³/h, 100 m³/h and 200 m³/h, corresponding to about 1000, 2500, 5000 and 10000 inhabitant-equivalents, respectively;
- biodegradable load concentration : L_0^* = 300 mg/ℓ BOD₅;
- suspended solids concentration : M_0^* = 5UU mg/ℓ.

The effluent load concentration was imposed in each case to be less than 30 mg/ℓ BOD₅.

Constraints 82 and 85 can provide a lower bound on the number n of stages in each of the above influent situations. First, it can be seen in Fig. 4 that the flow rate through the RBC units is about equal to that in the plant influent, since the lateral flows leaving or entering the principal line (from point 0 to point 7) are generally low and compensate each other, as will be confirmed in the optimization results. Therefore, taking the highest possible value for A in Eq. 85 : A = (π/2) 12.25 = 19.24, will yield with Eq. 82 a lower bound for n in each of the above influent situations :

Q_0 = 20 $\Rightarrow$ n ⩾ 3.40
Q_0 = 50 $\Rightarrow$ n ⩾ 8.50
Q_0 = 100 $\Rightarrow$ n ⩾ 17.0

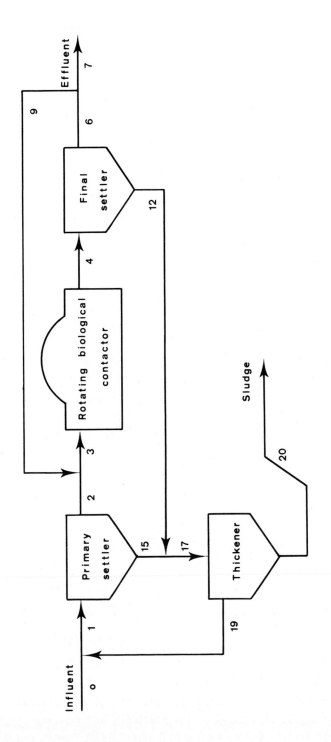

Fig. 4. Outline of the rotating biological contactor treatment plant (de Morais and Tyteca, 1983).

$Q_0 = 200 \Rightarrow n \geqslant 34.0$

Consequently, in the first case, one line of 4 stages was adopted; in the second case, two parallel lines of 5 stages; in the third case, both 4 parallel lines of 5 stages and 5 parallel lines of 4 stages were tested; and in the last case, both 9 and 10 lines of 4 stages were tested.

4.2.3. Results

Table 6 shows the complete results obtained with the sample data given above. A few remarks can be made about these results :

1) Both the primary settler area and the recycle ratio show a nought value in every situation : thus neither primary settling nor recycle into RBC appear economical and/or necessary for treatment in the situations analyzed. This is relevant since both systems without recycle (Antonie, 1979) and without primary settling (Rushbrook & Wilke, 1980) are encountered in practice. The absence of primary settler of course implies that the suspended solids concentration flows unchanged ($\varepsilon = 1$ in Eq. 70) and that the overflow is identical to the inflow. Thus in Fig. 4 both flows 15 and 9 can be dropped, while flow 12 (sludge wasting) is identical to flow 17 (thickener inflow), and flow 6 is identical to flow 7 (effluent).

2) In each situation, constraint 82 on the hydraulic load on RBC units is binding, and therefore the ratio of inflow rate to total RBC area is always equal to 0.00764 m/h. As the RBC cost itself is a function of the total disk area (Table 3), this explains that problems with different numbers of lines and stages but with same total RBC area lead to identical optimal cost levels (for $Q_0 = 100$ and 200 m³/h). It can be seen in Table 6 that in most cases the effluent quality constraint (30 mg/ℓ BOD₅) is largely satisfied : thus it is mainly the hydraulic load constraint that affects RBC design.

3) Except for small variations, problems with identical number of stages in a line show the same BOD₅ degradation sequence within the RBC stages (see values of L_3^*, L_{31}^* to L_{34}^* and L_4^* in Table 6). This is due to the fact that flow rate is proportional to total RBC area, so that the ratio of RBC area to flow rate is constant (see remark 2); hence the BOD₅ degradation within one stage, predicted by Eq. 55, will be the same provided that the constant K keeps the same value. Problems with 5 stages in a line show smaller overall degradation of BOD₅ than those with 4 stages in a line (see Table 6). This can be explained by the choice of the K values and the fact that when 5 stages are adopted, the flow pattern tends to reduce degradation in each stage, while the K value in the last stage is half of that in a 4 stage system.

5) The results are of course influenced by constant and parameter values, as can be expected from analogous studies on the influence of parameter uncertainty on treatment plant design (Berthouex and Polkowski, 1970, Tyteca and Nyns,

TABLE 6

Results of the optimization runs for the sample situation (de Morais and Tyteca, 1983)

Variable and meaning	Value at optimum					
Q_0 = influent flow rate [1]	20	50	100		200	
n. of lines/n. of stages	1/4	2/5	5/4	4/5	9/4	10/4
A_{ps} = primary settler area [2]	0	0	0	0	0	0
A_{fs} = final settler area [2]	4.34	10.9	21.7	21.7	43.4	43.3
A_{th} = thickener area [2]	6.51	16.3	32.5	32.5	65.1	65.1
A_b = RBC disk lateral area [2]	16.8	16.8	17.0	17.0	18.9	17.0
Total RBC area [2]	2690	6720	13600	13600	27200	27200
Q_1 = flow rate at point 1 [1]	20.6	51.4	104	104	208	208
L_1^* = BOD_5 at point 1 [3]	291	293	290	289	289	289
L_3^* = BOD_5 at point 3 [3]	291	293	289	289	289	289
L_{31}^* = BOD_5 after RBC stage 1 [3]	113	142	112	142	112	112
L_{32}^* = BOD_5 after RBC stage 2 [3]	44.2	76.4	43.2	76.2	43.1	43.1
L_{33}^* = BOD_5 after RBC stage 3 [3]	20.0	48.3	19.5	47.7	19.4	19.4
L_{34}^* = BOD_5 after RBC stage 4 [3]	-	35.8	-	35.1	-	-
L_4^* = BOD_5 at point 4 (effluent) [3]	11.5	30.0	11.1	29.2	11.1	11.1
L_{17}^* = BOD_5 at point 17 [3]	11.5	30.0	11.1	29.2	11.1	11.1
M_1^* = SS at point 1 [3]	486	486	478	478	478	478
M_4^* = SS at point 4 [3]	484	484	479	479	479	479
M_{17}^* = SS at point 17 [3]	12400	12400	9850	9850	9850	9850
γ = thickening factor in final settler	26.2	26.2	20.9	20.9	20.9	20.9
ε = % solids remaining after primary settling	100	100	100	100	100	100
ρ = recycle ratio into RBC units, %	0	0	0	0	0	0
π = wasting ratio from final settler, %	3.82	3.82	4.79	4.79	4.79	4.79
TC = total cost, 10^3	112.40	200.60	333.06	333.06	541.11	540.92
Computer time, seconds	64.9	100.6	35.5	19.6	44.9	17.0

[1] in m^3/h; [2] in m^2; [3] in mg/ℓ

1979). This is especially significant for the absence of primary settler and recycle flow in the optimal solution (see remark 1) : these results could have been quite different with another choice of parameter values. Therefore, it must be stressed that each design study should be preceeded by an analysis of the wastewater characteristics under concern, in order to properly determine the values to be given to the model constants and parameters, and more especially in the present case, for correct calibration of rotating biological contactor equations.

4.3. Optimal operation of an activated sludge plant
4.3.1. Introduction

The model has been tested in the perspective of section 3.2.4., in order to study the optimal operation of a previously designed wastewater treatment plant. These tests were performed for the activated sludge plant depicted in Fig. 3. As six variables are now given fixed values (i.e., A_{ps}, A_{fs}, A_{th}, A_{vf}, V_a and V_d) and the same constraints are used for the model as in the optimal design studies (section 4.1.), there are in this case 33 - 6 = 27 variables and 35 constraints. Two series of tests were made : in the first one the influent wastewater conditions were varied; in the second one, the effluent quality level was given various possible values different from the design value. The sample situation was that of section 4.1., case 3 (Q_0 = 600 m^3/h, L_{b_0} = 500 mg/ℓ, M_{bs_0} = 300 mg/ℓ, M_{s_0} = 200 mg/ℓ, L_{b_4} = 30 mg/ℓ).

4.3.2. Varying influent conditions

The six design variables were given the following values, obtained after the design optimization study and rounded to commercial values : A_{ps} = 120 m^2, A_{fs} = 200 m^2, A_{th} = 160 m^2, A_{vf} = 50 m^2, V_d = 2000 m^3, V_a = 4500 m^3. Table 7 shows the results obtained when varying Q_0, L_{b_0}, M_{bs_0} and M_{s_0} as indicated. The data included in Table 7 correspond to important operational variables and indicate how these variables should be readjusted in each of the inflow conditions considered. The cost figures correspond to one year operation but can be adapted to any shorter relevant period (season, month, week, ...) by dividing by 4, 12, or 52, ..., respectively.

A main result from the tests reported in Table 7 is that a plant designed for a given purpose can only operate at regimes below or equal to the design regime: attempts to optimize it at higher regimes (Table 7 : Q_0 x 1.2, or Q_0 x 1 with L_0^* x 2 and M_0^* x 2) resulted in unfeasible solutions. As a consequence, if a plant is to operate at a given regime during any period of its lifespan, it should be designed for that regime at the outset. What is important too is that in case of constant waste load with inflow doubling (Q_0 x 2, L_0^*/2, M_0^*/2 in Table 7), the operation of the plant is also impossible. Therefore only situations where the waste load (in g/h, i.e. Q_0 x L_0^*, Q_0 x M_0^*) and the flow load (in m^3/h, i.e. Q_0) are not increased with respect to the design conditions can be accommodated for in the plant. The waste concentration can be increased, provided that the flow rate is decreased in the same proportion (Table 7 : Q_0/2, L_0^* x 2, M_0^* x 2).

In general the operational variables are readjusted as could be expected un-

TABLE 7

Readjustment of operational variables and variable operating costs under various inflow situations

Variable	Basic situation $Q_0 \times 1$	Flow variations $Q_0 \times 0.5$	Flow variations $Q_0 \times 0.8$	Flow variations $Q_0 \times 1.2$	Constant waste load $Q_0 \times 2$, $L_0^*/2$, $M_0^*/2$	Constant waste load $Q_0/2$, $L_0 \times 2$, $M_0 \times 2$	Waste load variations $Q_0 \times 1$ $L_0^*/2$, $M_0^*/2$	Waste load variations $Q_0 \times 1$ $L_0 \times 2$, $M_0 \times 2$
$K_L a$, oxygen transfer coefficient, h^{-1}	4.4727	2.1071	3.5065	unfeasible	unfeasible	4.4490	2.2500	unfeasible
G_s, air flow rate in aerator, Nm^3/sec	83.690	36.928	64.237			83.208	39.658	
HP, power for aeration, kW	109.94	48.510	84.383			109.30	52.097	
Q_1, flow rate before primary settler, m^3/h	658.91	326.91	519.69			598.73	633.27	
θ, hydraulic retention time in aerator, h	7.0044	14.125	8.8849			7.7290	7.1969	
ρ, recycle ratio, %	7.2772	0.7720	3.5794			6.1905	2.7510	
Q_{13}, recycle flow rate, m^3/h	46.753	2.4595	18.129			36.043	17.201	
Q_{15}, primary settler underflow rate, m^3/h	16.453	8.3297	13.216			16.507	7.9982	
cQ_{30}, filter sludge production, g/h	193290	90387	149670			193350	91002	
Variable operating costs, 10^3 \$/year	161.98	86.729	131.34			160.52	94.176	

der the different inflow conditions represented. It appears that the recycle ratio (ρ) is especially sensitive to these conditions : for example, almost no recycle is required when the flow rate is half of the design value (Q_0 x 0.5), which compensates the considerable increase in the hydraulic retention time (θ). Situations with the same hourly waste load (i.e., basic situation and $Q_0/2$, L_0^* x 2, M_0^* x 2 on the one hand, Q_0 x 0.5 and Q_0 x 1, $L_0^*/2$, $M_0^*/2$ on the other hand) show very similar operational conditions and costs, except for the hydraulic retention time and the recycle flow rate in the second case.

4.3.3. Varying effluent quality level

In the same way as Table 7, Table 8 indicates results obtained when varying the effluent quality level L_{b_4} with respect to its design value, i.e. 30 mg/ℓ. The same type of conclusions can be drawn as in the previous tests : only quality levels below or very slightly above (27.5 mg/ℓ) the design level can be considered during operation (unless the influent conditions become less stringent). Most variables show a little sensitivity to the effluent quality specification, except once again the recycle ratio (ρ) and therefore the recycle flow rate (Q_{13}). There is also little cost variation for the range of conditions tested.

TABLE 8

Readjustment of operational variables and variable operating costs with various effluent quality levels (See Table 7 for the meaning of the variables)

L_{b_4}† Variable	25	27.5	30	35	40	45	50
$K_L a$		4.5266	4.4727	4.3784	4.2946	4.2197	4.1507
G_s		84.787	83.690	81.774	80.074	78.557	77.162
HP		111.38	109.94	107.43	105.19	103.20	101.36
Q_1		660.28	658.91	657.52	655.21	653.93	652.87
θ	unfeasible	6.9895	7.0044	7.0193	7.0444	7.0581	7.0695
ρ		8.5126	7.2772	5.5922	4.2337	3.3201	2.6240
Q_{13}		54.806	46.753	35.851	27.045	21.167	16.703
Q_{15}		16.461	16.453	16.431	16.406	16.374	16.335
cQ_{30}		193540	193290	192780	192270	191720	191100
Cost		163.04	161.98	160.16	158.50	157.04	155.70

† in mg/ℓ.

The same type of experiment has been repeated systematically for various design and operating effluent quality levels. The results of these studies have been published elsewhere (Smeers and Tyteca, 1985). They constitute the basis for cost functions developed in the scope of a research on regional water quality management with time varying control (Herbay et al., 1983). Another important result from these studies is that the variable operating costs are a significant part of the total discounted cost, since they amount to about 35 % of the total discounted cost, and 67 % of the total operating costs (Smeers and Tyteca, 1985).

4.4. Sensitivity analyses

The model described in sections 2. and 3. can be the basis for various sensitivity analyses, which is greatly facilitated by its interactive structure. Sensitivity analyses are an essential part of modelling studies, since they can help determining where the main efforts should be made to correctly assess the value of various influent or process parameters. The sensitivity of design and operational variables and costs to the influent characteristics has already been commented for a sample situation in sections 4.1. and 4.3., as well as the sensitivity of operational variables and costs to the effluent quality level in section 4.3. In this section a few more examples of sensitivity analyses will be given for a plant including an activated sludge system (Fig. 3). Most of the material of this section comes from a presentation by the author (Tyteca, 1981b).

4.4.1. Effect of the effluent quality level on the design variables and costs

Sensitivity of the plant design to the effluent quality level L_7^* is certainly one of the main aspects to be investigated, since this parameter binds together the design of the plant and the quality standards imposed on the effluent before its discharge to the receiving water body.

Table 9 gives results obtained in four different situations where the influent characteristics were maintained constant. The optimal cost can be considered as little sensitive to L_7^* as far as the required effluent quality is not too high; this sensitivity increases more than proportionally when L_7^* decreases from 15 to 10 and 5 mg/ℓ.

The effect of varying L_7^* on the optimal design of the plant depends on the variables considered. Four classes have been identified, on the basis of their sensitivity to L_7^* (Tyteca, 1979a) : (1) highly sensitive variables (they do more than double when L_7^* goes from 30 to 5 mg/ℓ) : primary settler area A_{ps}, hydraulic retention time in aerator θ, recycle ratio ρ, sludge age θ_s; (2) moderately sensitive variables (their increase is comprised between 50 and 100 % when L_7^* varies from 30 to 5 mg/ℓ) : final settler area A_{fs}, activated sludge biomass S_4, suspended solids concentration in the aerator M_4^*; (3) slightly sensitive variables (increase or decrease less than 50 % when L_7^* goes from 30 to 5 mg/ℓ) : primary settler efficiency ϵ, wasting ratio π, suspended solids before and after primary settling, respectively M_1^* and M_{15}^*, suspended solids after final settling M_{12}^* ; (4) unsensitive variables : thickener area A_{th}, digester volume V_d, filter

TABLE 9

Effect of the effluent quality level on optimal cost (in 10^3 \$)

	Case 1	Case 2	Case 3	Case 4
Q_0 (m^3/h)	600	600	100	5000
L_0^* (mg/ℓ)	1000	1000	500	1000
M_0^* (mg/ℓ)	500	500	300	300
$L_7^* = 40$	4233	4310	803	20645
$L_7^* = 30$	4299	4375	819	21025
$L_7^* = 20$	4390	4463	840	21510
$L_7^* = 15$	4453	4524	855	21824
$L_7^* = 10$	4543	4608	879	22435
$L_7^* = 5$	4848	4903	930	24434

area A_{vf}, organic and suspended load after digestion, respectively L_{24}^* and M_{24}^*, filtrate flow rate Q_{30}, suspended solids after thickening M_{20}^*, sludge production of the plant cQ_{30}.

It can be noted that the first two categories include key variables of the liquid treatment system, while the third category includes less important variables of the same system, and the fourth category comprises all variables of the sludge treatment system and only these. Therefore it can be concluded that the sludge treatment units and the phenomena occuring in them are practically unaffected by the water quality standards to be achieved after treatment. This and other comments of the previous section may suggest that the sludge treatment system be optimized independently of the liquid system, as other authors also indicate (Berthouex and Polkowski, 1970, Middleton and Lawrence, 1974, 1976, Tarrer et al., 1976).

Sensitivity analyses as described in the above lines have also been the basis for the computation of cost functions. These are discussed in detail in Tyteca (1981b).

4.4.2. Effect of the discount rate on the optimal design

The selected value of the discount rate in all aforementioned tests is 8.6 %. It is interesting to seek the sensitivity of the optimal design and cost to this parameter, in order to check whether or not a very refined value of the discount rate has to be specified before the optimization tests can be conducted. Two other values were tested, i.e., 6 % and 11.2 %, for a situation in which $Q_0 = 600$ m^3/h, $L_0^* = 1000$ mg/ℓ COD, $M_0^* = 500$ mg/ℓ suspended solids, $L_7^* = 30$ mg/ℓ COD. Some of the most meaningful results appear in Table 10.

It can be concluded from Table 10 that the optimal cost is insensitive to the discount rate : indeed, the last line indicates what would be the cost of

TABLE 10

Effect of the discount rate on the optimal solution

Optimal values	$i = 6\%$	$i = 8.6\%$	$i = 11.2\%$
A_{ps} = primary settler area, m^2	57.6	49.8	43.0
A_{fs} = final settler area, m^2	180	190	199
A_{th} = thickener area, m^2	158	162	165
A_{vf} = vacuum filter area, m^2	50.0	49.8	49.6
V_a = aerator volume, m^3	5080	4690	4410
V_d = digester volume, m^3	1940	1930	1910
HP = installed power for air blowers in aerator , kW	114	116	118
ε = % solids after primary settling	46.2	47.6	49.1
θ = hydraulic retention time, h	7.77	7.14	6.68
ρ = recycle ratio, %	4.67	5.85	6.87
π = wasting ratio, %	8.29	8.78	9.25
Total discounted cost, 10^3 \$	4804	4299	3925
Total discounted cost of the central solution (computed for $i = 8.6\%$), 10^3 \$	4804	4299	3926

the optimal ("central") solution obtained for i = 8.6 %, in the cases of both
i = 6 % and i = 11.2 %. One can see that this cost is then practically identical to the cost of the optimal solution obtained for i = 6 % and i = 11.2 %,
respectively.

Nevertheless, some variables are significantly readjusted when proceeding
from one optimal solution to the others : the most sensitive are the primary
settler area A_{ps}, for reasons discussed in the next section, and the recycle
ratio ρ, probably because of the modification of the impact of energy costs
when the discount rate varies. Other variables are slightly sensitive, i.e.,
those of the water treatment system. Finally, the sludge treatment system variables (A_{th}, A_{vf}, V_d) are practically unsensitive to the discount rate : this
rejoins previous observations on sludge treatment units.

4.4.3. Role of the primary settler

Most of the previously mentioned tests have shown an extreme sensitivity of
the primary settler area A_{ps} to changing conditions and parameters. As can be
seen for example in Table 5, this sensitivity is such that A_{ps} tends to very
low values when the incoming waste load is low. This fact was confirmed in other
experiments not shown herein (Tyteca, 1979a). In these cases where the influent
waste concentration is low, it can be concluded that the primary settler is un-

economical, and must be dropped from the treatment plant scheme, thus leaving the organics and solids removal to subsequent treatment phases (activated sludge, final settler).

The high sensitivity of the primary settler area is certainly due to the particular form of Eq. 71, describing the primary settler efficiency as a function of the influent suspended solids concentration M_1^* and the overflow rate (Q_1/A_{ps}). In this highly nonlinear equation, the value of parameter m is taken as 0.22, which indicates that, inversely, the area A_{ps} varies as about the fifth power of the required efficiency $(1 - \varepsilon)$! Eq. 71, which was taken from Voshel and Sak (1968), was based on a given type of wastewaters and operational conditions. Therefore it may be that this equation is not the most appropriate for a general optimization purpose. Other primary settling models exist in the literature (Tyteca et al., 1977), but all are empirical as the one used in this chapter, and there is no way of discriminating between them. A general model for primary settling is therefore certainly a good subject for further experimental work and research. Additionally, it can be recommended to give sufficient attention in correct calibration of any model in each wastewater situation encountered.

4.4.4. Effect of methane recovery in anaerobic digester

The economic impact of methane production in the anaerobic digester was until now seldom investigated (Dick et al., 1978, Pfeffer and Quindry, 1978). More and more, however, the advantages of exploiting this free energy source are recognized, and any analysis of the economics of treatment plants should incorporate this aspect.

The equation giving the variable operating costs of the anaerobic digester in Table 3 includes a negative term corresponding to the total potential benefit available from methane recovery (see section 3.1.). This term does not account for the fact that a part of methane can be directly used in the treatment plant, and therefore be of no direct economic value, and for such factors as gas losses, operational imperfections, ... Accordingly, a second expression has been tested, in which the monetary value of methane is taken as half the value adopted in section 3.1., that is, 0.0389 $/m³.

The model was run with the following input parameters : $Q_0 = 600$ m³/h, $L_0^* = 1000$ mg/ℓ COD, $M_0^* = 500$ mg/ℓ suspended solids, $L_7^* = 30$ mg/ℓ COD, and this in three situations : benefit from methane recovery = 0 ("central" situation); value of methane = 0.0389 $/m³; and value of methane = 0.0777 $/m³. The results appear in Table 11 and give rise to the following observations. First, the cost effect is significant, since the benefit amounts to 8 % in the second situation, and to 16.2 % in the third situation. However, this effect appears solely due to incorporating the methane benefit in the plant accounts : the change in design has only negligible influence on the cost, as can be seen when recomputing

TABLE 11

Effect of methane recovery on the optimal design of the plant

Optimal value	Value of CH₄ = 0	Value of CH₄ = 0.0389$/m³	Value of CH₄ = 0.0777$/m³
A_{ps} = primary settler area, m²	49.8	89.2	137
A_{fs} = final settler area, m²	190	189	188
A_{th} = thickener area, m²	161	177	192
A_{vf} = vacuum filter area, m²	49.8	49.3	48.9
V_a = aerator volume, m³	4690	4540	4410
V_d = digester volume, m³	1930	1960	1990
HP = installed power for air blowers in aerator, kW	116	112	109
ε = % solids after primary settling	47.6	41.7	36.9
θ = hydraulic retention time, h	7.14	7.00	6.87
ρ = recycle ratio, %	5.88	7.15	8.01
π = wasting ratio, %	8.77	7.53	6.67
Total discounted cost, 10³ $	4299	3954	3601
Total discounted cost of the central solution (value of CH₄ = 0), 10³ $	4299	3958	3618
CH₄ production, m³/day	2550	2620	2670

the cost of the "central" solution in both the second and third situations (see Table 11). The benefit then amounts to only 0.1 % and 0.5 %, respectively. This is in accordance with the very slight increase in methane production as indicated in Table 11.

Nevertheless, some variables show a significant change, more especially those involved in process units located upstreams of the digester. That is, while the digester itself is only slightly affected, the thickener (A_{th}), the primary settler (A_{ps}, ε), the recycle and wasting ratios (ρ and π) are substancially modified. The filter (A_{vf}), the aerator (V_a) and the final settler (A_{fs}) are little affected. Once again, the primary settler area is by far the most sensitive variable in this analysis of the benefit gained from methane recovery. This can be interpreted here from the central role played by the primary settler in the distribution of solids within the plant : as the primary settler area and efficiency are increased, more solids (especially the organic solids) are conveyed to the sludge treatment units, where they can be degraded, thus leading to more methane production. As a consequence, the activated sludge is somewhat relieved, and as less substrate is provided to the aerobic community of the aerator, the recycle ratio (ρ) has to increase, in order to allow for proper operation of the system.

5.   SUMMARY AND CONCLUSIONS

The cost-effective design of a wastewater treatment plant, and to a lesser extent its cost-effective operation, have been the subject of many diversified studies since 1962. Two main components of these studies are the mathematical model describing the operation of the various process units of the plant, and the optimization procedure allowing for the derivation of cost-effective design and/or operation of the plant. Depending upon the specialization of the authors, the emphasis has been placed, in most cases, on only one of these components, without developing satisfactorily if at all the other aspect (see Table 4 in section 3.2.1.).

In this chapter the mathematical modelling of the plant is first dealt with (section 2.), with a special emphasis on biological treatment processes. Since the object of the chapter was to study the cost-effective design and operation of a whole treatment plant, models for the other treatment units, as well as relationships binding together the plant components, have also been briefly reviewed. In a further step (section 3.), cost functions describing the cost of each process unit as a function of design or operational parameters were joined to the model, and techniques for studying the least-cost design and operation briefly discussed. The research conducted in this scope by the author and others has more especially been presented. In view of the aforementioned gap between "Biologists - Mathematicians" and "Economists - Operations Researchers", the effort has been placed in this research on fully developing both components of any cost-effective study, i.e., the mathematical model and the optimization technique, and the need for a perfect adjustment of the second component to the first one has been emphasized. This has resulted in a very flexible and efficient technique for studying the cost-effective design and operation of a treatment plant. The interactivity of the method, due to the easy data manipulation of the geometric programming transcription, has been presented as a main advantage.

As illustrations of this fact, the last section (section 4.) was devoted to a few results obtained from the model developed. The cost-effective design has been studied first, for a plant including an activated sludge process (section 4.1.) and for a plant with a rotating biological contactor (section 4.2.). In both situations the influent characteristics have been varied and for the rotating biological contactor two configurations have been tested (with 4 and 5 stages). These examples illustrate the recurrent use of the model, as a useful tool for providing the practicing engineer with a set of possible design solutions which have a cost close to the optimum. Thus it should be emphasized again that an optimization technique such as presented does not serve to yield a definitive design, but rather constitutes an efficient aid to preliminary cost-effective design. The sensitivity analyses reported in the last section (section 4.4.) also substantiate this fact : in this case the preoccupation is to identify which factors and parameters most significantly influence the design of the plant, and as a consequence where the main efforts should be made to improve the estimation of the parameters and the calibration of the equations. Finally section 4.3. showed how the same model can be exploited in studying the cost-effective operation of a previously designed treatment plant.

At several places in the chapter it was emphasized that the model response strongly depends on the data collected for its calibration. As a meaningful example with regard to the subject of this chapter, the discussion of section 3.1. stressed the importance of collecting adequate cost data : the use of the cost functions given herein can only yield preliminary indications as to the solution to be adopted. For more accurate results, in any particular situation, one should collect cost data effectively reflecting the actual situation with respect to the available equipment manufacturers, and the time and the country under consideration.

6. ACKNOWLEDGEMENTS

Most of the work reported in this chapter was the subject of the author's doctoral thesis. Sincere thanks are due to the directors of the research, Profs. E.J. Nyns and Y. Smeers from the Université Catholique de Louvain. Additional thanks to Y. Smeers for its idea of exploiting the model for cost-effective operation purposes (sections 3.2.4. and 4.3.). Parts of the work benefited from collaboration with other personalities, among whom Profs. F.X. de Donnea, J. Hermia and E. Loute (Université Catholique de Louvain) and Mr. L.A.J. de Morais (Pollution Control Division, Ministry of Public Works, Lisbon, Portugal). Special thanks to E. Loute for his contribution to the implementation of the optimization techniques. The research was supported by grants F.D.S. nrs 146 and 1/3 from the University of Louvain.

SYMBOLS

A	- area, m^2 (see indices below)
C	- cost, $ or 10^3 $
C_1, C_2, C_3, C_4	- composite parameters and constants for aeration devices (Eqs. 77-79)
D	- trickling filter depth, m, or diameter of rotating biological contactor, m
FOC_k	- fixed operating costs of unit k, $ or 10^3 $
G_s	- air flow rate in aerator, Nm^3/sec
HP	- installed power for aeration, kW

IC_k	- investment cost of unit k, $ or 10^3 $
K	- composite constant for trickling filter (Eq. 46) or rotating biological contactor (Eq. 55)
K_f	- saturation constant in rotating biological contactors, mg/ℓ BOD_5
$K_L a$	- oxygen transfer coefficient through liquid film, h^{-1}
K_M	- Michaelis-Menten saturation coefficient, mg/ℓ COD
L	- load applied to vacuum filter, g/m³.h
L_b	- dissolved biodegradable waste load, mg/ℓ COD (see indices in Fig. 2 and Table 1)
$\overline{L_b}$	- effluent standard on L_b, mg/ℓ COD
L^*	- total biodegradable waste load, mg/ℓ COD (see indices in Fig. 2 and Table 1)
ℓ_d	- upper limit on the load applied to digester, g COD/m³.h
ℓ_f	- upper limit on the load applied to vacuum filter, g SS/m³.h
M_{bs}, M_s, M^*	- biodegradable, nonbiodegradable and total suspended solids, respectively, mg/ℓ SS (see indices in Fig. 2 and Table 1)
M_L^*	- limiting suspended solids concentration in final settler and thickener, mg/ℓ SS
N	- total number of units in cost equations (Eq. 104 or 106), or number of disks in one stage of rotating biological contactor (Eq. 55)
P	- vacuum applied to filter, kgf/cm²
P_c	- cost of energy, cents/kWh
Q	- flow rate, m³/h (see indices in Fig. 2 and Table 1)
S, S^*	- active biomass in aerator and digester, respectively, mg/ℓ (see indices in Fig. 2 and Table 1)
SS	- suspended solids
SVI	- sludge volume index, mℓ/g
T	- temperature, °C
TC	- total cost, $ or 10^3 $
V	- volume, m³ (see indices below)
VOC_k	- variable operating costs of unit k, $ or 10^3 $
W	- installed power for pumping, kW
X_i	- design or operating parameter of a process unit (in cost equations : see Eq. 92)
Y	- biomass growth yield, g biomass/g substrate removed
Z	- height difference, m
a	- empirical constant of primary settler efficiency (Eq. 71)
b	- microbial decay coefficient, day^{-1}
a_i, b_i	- estimated parameters of cost functions (Eq. 92)
c	- concentration of filtered solids with respect to filtrate volume, mg/ℓ
d	- unsubmerged diameter of disk in rotating biological contactor, m
h	- height of settler, m

i	- discount rate
k_e	- empirical constant of sludge settling velocity, ℓ/mg (Eq. 72)
m, n	- empirical constants of primary settler (Eq. 71)
n	- number of stages in rotating biological contactors, or number of years in cost equations
p	- empirical constant of aeration device capacity (Eq. 78)
r_0, s	- empirical constants of cake specific resistance (Eq. 76)
v, v_0	- sludge settling velocity and empirical constant, m/h
w_{ma}, w_{op}	- wage parameters for maintenance and operation, respectively, \$/h
x	- biodegradable part of materials resulting from microbial lysis
y	- time during which a point of vacuum filter is immersed, percentage
Γ	- discount factor (Eq. 105)
ϕ	- water or sludge specific weight, kg/m^3
α	- aerator volume in total activated sludge system volume, percentage
γ	- thickening factor in final settler
ε	- complement of primary settler efficiency, $1 - \varepsilon$
ζ	- pump efficiency
η, η'	- composite variables (Eq. 28-29), expressing the ratio between retention times (Eq. 30)
θ, θ_s	- hydraulic and sludge retention times, h and day, respectively
θ_{sa}	- sludge retention time in aerator, days
θ_t	- total filtration cycle time, minutes
ι	- parameter for price indexation (Table 3)
κ	- part of total degradable substrate resulting from microbial lysis (activated sludge)
λ, λ'	- dissolved part of biodegradable load in aerator and aerator inflow, respectively
λ''	- dissolved part of biodegradable substrate resulting from mortality
μ	- viscocity of filtered sludge, centipoises
μ_{max}	- maximum specific growth rate, h^{-1}
π	- wasting ratio
ρ	- recycle ratio
ϕ	- auxiliary variable (Eq. 27)
ω	- COD equivalent of biomass and biodegradable SS, mg COD/mg SS

Indices (see also Fig. 2 and Table 1)

a	- aerator		m	- mortality
b	- rotating biological contactor		ps	- primary settler
d	- digester		th	- thickener
fs	- final settler		vf	- vacuum filter

REFERENCES

Abadie, J., 1978. The GRG method for nonlinear programming. In: H.J. Greenberg (Editor), Design and Implementation of Optimization Software. Sijthoff & Noordhoff, Alphen-aan-den-Rijn (Netherlands), pp. 335-362.

Adams, B.J. and Panagiotakopoulos, D., 1977. Network approach to optimal wastewater treatment system design. J. Water Pollut. Control Fed. 49: 623-632.

Andrews, J.F., Briggs, R. and Jenkins, S.H., 1974. Instrumentation, Control and Automation for Wastewater Treatment Systems. Progress in Water Technology, Vol. 6. Pergamon Press, Oxford, 570 pp.

Antonie, R.L., 1979. Applying the rotating biological contactor. Water & Sewage Works, ref. number: R-69-R-75.

Berthouex, P.M. and Polkowski, L.B., 1970. Optimum waste treatment plant design under uncertainty. J. Water Pollut. Control Fed. 42: 1589-1613.

Bisogni, J.J. and Lawrence, A.W., 1971. Relationships between biological solids retention time and settling characteristics of activated sludge. Water Res. 5: 735-763.

Blecker, H.G., Epstein, H.S. and Nichols, T.M., 1974. How to estimate and escalate costs of wastewater equipment. Chem. Eng., Deskbook Issue: 115-121.

Bowden, K., Gale, R.S. and Wright, D.E., 1976. Evaluation of the CIRIA prototype model for the design of sewage-treatment works. Water Pollut. Control 75: 192-205.

Charnes, A. and Cooper, W.W., 1966. A convex approximant method for nonconvex extensions of geometric programming. Proc. Natl. Acad. Sci. 56: 1361-1364.

Chen, M.S.K., Erickson, L.E. and Fan, L.T., 1970. Consideration of sensitivity and parameter uncertainty in optimal process design. Ind. Eng. Chem. Process Design Dev. 9: 514-521.

CIRIA (Construction Industry Research and Information Association), 1973. Cost-effective sewage treatment - the creation of an optimising model. Report 46, Vol. 1: General introduction. London, 34 pp.

CIRIA (Construction Industry Research and Information Association), 1975. Cost-effective sewage treatment - the creation of an optimising model. Report 46, Vol. 2: Performance and cost calculation (revised). 2nd ed., London, 50 pp. + tables.

Clark, J.H., Moseng, E.M. and Asano, T., 1978. Performance of a rotating biological contactor under varying wastewater flow. J. Water Pollut. Control Fed. 50: 896-911.

Craig, E.W., Meredith, D.D. and Middleton, A.C., 1978. Algorithm for optimal activated sludge design. J. Environ. Eng. Div. Am. Soc. Civ. Eng. 104: 1101-1117.

Daigger, G.T. and Grady, C.P.L., 1977. A model for the bio-oxidation process based on product formation concepts. Water Res. 11: 1049-1057.

Deb, A.K., 1978. Optimization in design of pumping systems. J. Environ. Eng. Div. Am. Soc. Civ. Eng. 104: 127-136.

de Morais, L.A.J. and Tyteca, D., 1983. Conception optimale de stations d'épuration d'eaux résiduaires par biodisques. La Tribune du Cebedeau 36: 519-526.

Dick, R.I. and Simmons, D.L., 1976. Optimal integration of processes for sludge management. In Proceedings of the Third National Conference on Sludge Management, Disposal and Utilization, Miami Beach, Florida, December 14-16. pp. 20-27.

Dick, R.I., Simmons, D.L., Ball, R.O. and Perlin, K., 1976. Process selection for optimal management of regional wastewater residuals. Dept. of Civil Engineering, Univ. of Delaware, Newark, Delaware, NSF-RA-760570. 10 + 259 pp.

Dick, R.I., Simmons, D.L., Ball, R.O., Perlin, K. and O'Hara, M.W., 1978. Process selection for optimal management of regional wastewater residuals. Dept. of Civil Engineering, Univ. of Delaware, Newark, Delaware. Final Project Report, 16 + 484 pp.

Duffin, R.J., Peterson, E.L. and Zener, C., 1967. Geometric Programming - Theory and Applications. Wiley & Sons, New York, 11 + 278 pp.

Eckenfelder, W.W., 1966. Industrial Water Pollution Control. McGraw-Hill, New York, 7 + 275 pp.

Ecker, J.G. and McNamara, J.R., 1971. Geometric programming and the preliminary design of industrial wastewater treatment plants. Water Resour. Res. 7: 18-22.

Edeline, F. and Vandevenne, L., 1979. Cinétique de l'épuration dans les biodisques. La Tribune du Cebedeau 32: 3-22.

Environmental Protection Agency, 1978a. Construction costs for municipal wastewater treatment plants : 1973-1977. EPA report 430/9-77-013, MCD-37. Washington, D.C., 213 pp.

Environmental Protection Agency, 1978b. Analysis of operations and maintenance costs for municipal wastewater treatment systems. EPA report 430/9-77-015, MCD-39. Washington, D.C., 312 pp.

Erickson, L.E., Chen, G.K.C. and Fan, L.T., 1968. Modeling and optimization of biological waste treatment systems. Chem. Eng. Prog. Symp. Ser. 64: 97-110.

Erickson, L.E. and Fan, L.T., 1968. Optimization of the hydraulic regime of activated sludge systems. J. Water Pollut. Control Fed. 40: 345-362.

Erickson, L.E., Heydweiller, J.C. and Fan, L.T., 1977. Consideration of suspended solids transport in activated sludge systems synthesis - I. Simultaneous use of clarification and sludge thickening models. AIChE Symp. Ser. 73(167), Water 1976: II. Biological Wastewater Treatment, p. 105.

Erickson, L.E., Ho, Y.S. and Fan, L.T., 1968. Modeling and optimization of step aeration waste treatment systems. J. Water Pollut. Control Fed. 40: 717-732.

Evenson, D.E., Orlob, G.T. and Monser, J.R., 1969. Preliminary design of waste treatment systems. Ind. Water Eng. 6: 16-21.

Evenson, D.E., Orlob, G.T. and Monser, J.R., 1969. Preliminary selection of waste treatment systems. J. Water Pollut. Control Fed. 41: 1845-1858.

Fan, L.T., Chen, G.K.C., Erickson, L.E. and Naito, M., 1970. Effects of axial dispersion on the optimal design of the activated sludge process. Water Res. 4: 271-284.

Fan, L.T., Erickson, L.E., Baltes, J.C. and Shah, P.S., 1973. Analysis and optimization of two-stage digestion. J. Water Pollut. Control Fed. 45: 591-610.

Fan, L.T., Kuo, M.T. and Erickson, L.E., 1974. Effect of suspended wastes on system design. J. Environ. Eng. Div. Am. Soc. Civ. Eng. 100: 1231-1247.

Fan, L.T., Mishra, P.N., Chen, G.K.C. and Erickson, L.E., 1972. Application of systems analysis techniques in biological waste treatment. Proc. IV IFS: Ferment. Technol. Today: 555-562.

Fan, L.T., Mishra, P.N. and Erickson, L.E., 1975. Optimal synthesis in the design of water pollution control systems. AIChE Symp. Ser. 71(145), Water 1974: II. Municipal Waste Treatment: 125-127.

Fan, L.T., Mishra, P.N., Shastry, J.S., Osakada, K. and Erickson, L.E., 1973. Synthesis of bioengineering systems. AIChE Symp. Ser. 69(132), Food Preservation: 123-133.

Galler, W.S. and Gotaas, H.B., 1964. Analysis of biological filter variables. J. Sanit. Eng. Div. Am. Soc. Civ. Eng. 90: 59-79.

Galeer, W.S. and Gotaas, H.B., 1966. Optimization analysis for biological filter design. J. Sanit. Eng. Div. Am. Soc. Civ. Eng. 92: 163-182.

Gemmell, J.S. and Smith, P.G., 1982. Interactive computer design for wastewater treatment plants. Public Health Eng. 10: 175-177.

Goodman, B.L. and Englande, A.J., 1974. A unified model of the activated sludge process. J. Water Pollut. Control Fed. 46: 312- 332.

Gotaas, H.B. and Galler, W.S., 1973. Design optimization for biological filter models. J. Environ. Eng. Div. Am. Soc. Civ. Eng. 99: 831-850.

Gotaas, H.B. and Galler, W.S., 1975. Biological filter design optimization. In: A. Charnes and W.R. Lynn (Editors), Mathematical Analysis of Decision Problems in Ecology. Springer-Verlag, Berlin-Heidelberg, pp. 204-247.

Grady, C.P.L., 1977. Simplified optimization of activated sludge process. J. Environ. Eng. Div. Am. Soc. Civ. Eng. 103: 413-429.

Herbay, J.P., Smeers, Y. and Tyteca, D., 1983. Water quality management with time varying river flow and discharger control. Water Resour. Res. 19: 1481-1487.

Heydweiller, J.C., Erickson, L.E. and Fan, L.T., 1977. Consideration of suspended solids transport in activated sludge systems synthesis - II. Tower-type activated sludge systems. AIChE Symp. Ser. 73(167), Water 1976: II. Biological Wastewater Treatment, p. 112.

Heydweiller, J.C., Erickson, L.E. and Fan, L.T., 1977. Consideration of suspended solids transport in activated sludge systems synthesis - III. Effect of organism sedimentation. AIChE Symp. Ser. 73(167), Water 1976: II. Biological Wastewater Treatment, p. 122.

Hughes, M.M., 1978. The least cost design of the activated sludge process with geometric programming. Rept. submitted for the degree of Master of Engineering, Dept. of Environmental Sciences and Engineering, Univ. of North Carolina at Chapel Hill, 8 + 116 pp.

Jacquart, J.C., Lefort, D. and Rovel, J.M., 1973. An attempt to take account of biological storage in the mathematical analysis of activated sludge behaviour. In: S.H. Jenkins (Editor), Advances in Water Pollution Research. Pergamon Press, Oxford, pp. 367-375.

Keinath, T.M., Ryckman, M.D., Dana, C.H. and Hofer, D.A., 1977. Activated sludge - unified system design and operation. J. Environ. Eng. Div. Am. Soc. Civ. Eng. 103: 829-849.

Kornegay, B.H., 1975. Modeling and simulation of fixed film biological reactors for carbonaceous waste treatment. In: T.M. Keinath and M. Wanielista (Editors), Mathematical Modeling for Water Pollution Control Processes. Ann Arbor Science, Ann Arbor, Mich., Chap. 8, pp. 271-318.

Kornegay, B.H. and Andrews, J.F., 1968. Kinetics of fixed-film biological reactors. J. Water Pollut. Control Fed. 40: R460-R472.

Kuo, M.T., Fan, L.T. and Erickson, L.E., 1974. Effects of suspended solids on primary clarifier size optimization. J. Water Pollut. Control Fed. 46: 2521-2535.

Lauria, D.T., Uunk, J.B. and Schaefer, J.K., 1977. Activated sludge process design. J. Environ. Eng. Div. Am. Soc. Civ. Eng. 103: 625-645.

Lawrence, A.W. and McCarty, P.L., 1970. Unified basis for biological treatment design and operation. J. Sanit. Eng. Div. Am. Soc. Civ. Eng. 96: 757-778.

Lee, C.R., Fan, L.T. and Takamatsu, T., 1976. Optimization of multistage secondary clarifier. J. Water Pollut. Control Fed. 48: 2578-2589.

Lee, S.S., Erickson, L.E. and Fan, L.T., 1971. Modeling and optimization of a tower-type activated sludge system. Biotechnol. Bioeng. Symp. n° 2: 141-173.

Lynn, W.R., Logan, J.A. and Charnes, A., 1962. Systems analysis for planning wastewater treatment plants. J. Water Pollut. Control Fed. 34: 565-581.

Marsden, J.R., Pingry, D.E. and Whinston, A., 1972. Production function theory and the optimal design of waste treatment facilities. Applied Economics 4: 279-290.

McDeath, B.C. and Eliassen, R., 1966. Sensitivity analysis of activated sludge economics. J. Sanit. Eng. Div. Am. Soc. Civ. Eng. 92: 147-167.

Metcalf and Eddy, Inc., 1975. Water pollution abatment technology : capabilities and costs - Volume 3. Rept. to National Commission on Water Quality on Assessment of Technologies and Costs for Publicly Owned Treatment Works, PB-250-690-03. National Technical Information Service, Springfield, Virginia, 2261, 650 pp.

Middleton, A.C. and Lawrence, A.W., 1973. Cost optimization of activated sludge wastewater treatment systems. EPM Tech. Report n. 73-1, Dept. of Environmental Engineering, Cornell University, Ithaca, N.Y., 68 pp.

Middleton, A.C. and Lawrence, A.W., 1974. Cost optimization of activated sludge systems. Biotechnol. Bioeng. 16: 807-826.

Middleton, A.C. and Lawrence, A.W., 1975. Least cost design of activated sludge wastewater treatment systems. EPM Tech. Report n. 75-1, Dept. of Environmental Engineering, Cornell University, Ithaca, N.Y., 70 pp.

Middleton, A.C. and Lawrence, A.W., 1976. Least cost design of activated sludge systems. J. Water Pollut. Control Fed. 48: 889-905.

Mishra, P.N., Fan, L.T. and Erickson, L.E., 1973. Biological wastewater treatment system design. Part I. Optimal synthesis. Canadian J. Chem. Eng. 51: 694-701.

Mishra, P.N., Fan, L.T. and Erickson, L.E., 1973. Biological wastewater treatment system design. Part II. Effect of parameter variations on optimal process system structure and design. Canadian J. Chem. Eng. 51: 702-708.

Mishra, P.N., Fan, L.T. and Erickson, L.E., 1974. Optimal capacity expansion of a secondary treatment system. J. Water Pollut. Control Fed. 46: 2704-2718.

Mishra, P.N., Fan, L.T. and Erickson, L.E., 1975. Application of mathematical optimization techniques in computer aided design of wastewater treatment systems. AIChE Symp. Ser. 71(145), Water 1974: II. Municipal Waste Treatment: 136-153.

Mishra, P.N., Fan, L.T., Erickson, L.E. and Kuo, M.C.T., 1973. Simulation studies of the activated sludge system including a primary clarifier. Indian Chem. Eng. 15(3): 18-29.

Naito, M., Takamatsu, T. and Fan, L.T., 1969. Optimization of the activated sludge process - Optimum volume ratio of aeration and sedimentation vessels. Water Res. 3: 433-443.

Parkin, G.F. and Dague, R.R., 1972. Optimal design of wastewater treatment systems by enumeration. J. Sanit. Eng. Div. Am. Soc. Civ. Eng. 98: 833-851.

Patterson, W.L. and Banker, R.F., 1971. Estimating costs and manpower requirements for conventional wastewater treatment facilities. Black & Veatch Consulting Engineers, U.S. Environmental Protection Agency, Water Pollution Control Research Series, 17090 DAN 10/71, 14 + 253 pp.

Pfeffer, J.T. and Quindry, G.E., 1978. Biological conversion of biomass to methane - Beef lot manure studies. Report n. UILU-ENG-78-2011, Dept. of Civil Engineering, University of Illinois, Urbana, Illinois, 268 pp.

Rijckaert, M.J., 1973. Engineering applications of geometric programming. In: Int. Summer School on The Impact of Optimization Theory on Technological Design, Leuven (Belgium), July 16-August 6, 1971, pp. 196-220.

Roper, R.E. and Grady, C.P.L., 1974. Activated sludge hydraulic control techniques evaluation by computer simulation. J. Water Pollut. Control Fed. 46: 2565-2578.

Rossman, L.A., 1980. Synthesis of waste treatment systems by implicit enumeration. J. Water Pollut. Control Fed. 52: 148-160.

Rushbrook, E.L. and Wilke, D.A., 1980. Energy conservation and alternative energy sources in wastewater treatment. J. Water Pollut. Control Fed. 52: 2477-2483.

Schroeder, E.D., 1977. Water and Wastewater Treatment. McGraw-Hill, New York, 11 + 370 pp.

Schulte, D.D. and Loehr, R.C., 1971. Analysis of duck farm waste treatment systems. In: Proc. Int. Symp. on Livestock Wastes, American Soc. of Agri. Eng., St. Joseph, Michigan, pp. 73-80.

Shih, C.S. and DeFilippi, J.A., 1970. System optimization of waste treatment plant process design. J. Sanit. Eng. Div. Am. Soc. Civ. Eng. 96: 409-421.

Shih, C.S. and Krishnan, P., 1969. Dynamic optimization for industrial waste treatment design. J. Water Pollut. Control Fed. 41: 1787-1798.

Shih, C.S. and Krishnan, P., 1973. System optimization for pulp and paper industrial wastewater treatment design. Water Res. 7: 1805-1820.

Smeers, Y. and Tyteca, D., 1984. A geometric programming model for the optimal design of wastewater treatment plants. Operations Research 32: 314-342.

Smeers, Y. and Tyteca, D., 1985. Variable operating costs of wastewater treatment plants. To appear in Transactions of the Institute of Measurement and Control.

Smith, R., 1968. Preliminary design and simulation of conventional wastewater renovation systems using the digital computer. Report n. WP-20-9, U.S. Dept. of the Interior, Federal Water Pollution Control Administration, Cincinnati, Ohio, 92 pp.

Smith, R., 1969. Preliminary design of wastewater treatment systems. J. Sanit. Eng. Div. Am. Soc. Civ. Eng. 95: 117-145.

Stall, T.R. and Sherrard, J.H., 1978. Evaluation of control parameters for the activated sludge process. J. Water Pollut. Control Fed. 50: 450-457.

Stenstrom, M.K., 1975. A dynamic model and computer compatible control strategies for wastewater treatment plants. PhD thesis, Environmental Systems Engineering, Clemson University, 13 + 310 pp.

Stenstrom, M.K. and Andrews, J.F., 1980. Cost interactions in activated sludge systems. J. Environ. Eng. Div. Am. Soc. Civ. Eng. 106: 787-796.

Sterling, R.A., 1976. Computerized algorithms simplify meeting federal effluent standards. Water Sewage Works 123(10): 68-72.

Takama, N. and Loucks, D.P., 1981a. A multilevel model and algorithm for some multiobjective problems. Water Resour. Bull. 17: 448-453.

Takama, N. and Loucks, D.P., 1981b. Multi-level optimization for multiobjective problems. Appl. Math. Modelling 5: 173-178.

Tarrer, A.R., Grady, C.P.L., Lim, H.C. and Koppel, L.B., 1976. Optimal activated sludge design under uncertainty. J. Environ. Eng. Div. Am. Soc. Civ. Eng. 102: 657-673.

Tikhe, M.L., 1976. Optimal design of chlorination systems. J. Environ. Eng. Div. Am. Soc. Civ. Eng. 102: 1019-1028.

Tyteca, D., 1979a. Modélisation et optimisation économique des stations d'épuration d'eaux résiduaires. Doctoral thesis, Institut des Sciences Naturelles Appliquées, Université Catholique de Louvain, Louvain-la-Neuve (Belgium), 11 + 224 + 3 pp.

Tyteca, D., 1979b. Modélisation et optimisation économique d'une station d'épuration d'eaux résiduaires urbaines. Etudes et Expansion 279: 101-112.

Tyteca, D., 1981a. Nonlinear programming model of wastewater treatment plant. J. Environ. Eng. Div. Am. Soc. Civ. Eng. 107: 747-766.

Tyteca, D., 1981b. Sensitivity analysis of the optimal design of a municipal wastewater treatment plant. In: D. Dubois (Editor), Progress in Ecological Engineering and Management by Mathematical Modelling. Editions Cebedoc, Liège (Belgium), pp. 743-766.

Tyteca, D. and Nyns, E.J., 1979. Design and operational charts for complete mixing activated sludge systems. Water Res. 13: 929-948.

Tyteca, D. and Smeers, Y., 1981. Nonlinear programming design of wastewater treatment plant. J. Environ. Eng. Div. Am. Soc. Civ. Eng. 107: 767-779.

Tyteca, D., Smeers, Y. and Nyns, E.J., 1977. Mathematical modeling and economic optimization of wastewater treatment plants. CRC Crit. Rev. Environ. Control 8: 1-89.

Voelkel, K.G., 1978. Sensitivity analyses of wastewater treatment plant independent state variables and technological parameters. PhD thesis, Civil and Environmental Engineering, University of Wisconsin-Madison, 13 + 282 pp.

Voshel, D. and Sak, J.G., 1968. Effect of primary effluent suspended solids and BOD on activated sludge process. J. Water Pollut. Control Fed. 40: R203-R212.